Handbook of Experimental Pharmacology

Volume 103

Physiology and Pharmacology of the Blood-Brain Barrier

Contributors

N.J. Abbott, D. Barnes, D.J. Begley, A.L. Betz
M.W.B. Bradbury, M.W. Brightman, D.J. Brooks, H.F. Cserr
P.A. Fraser, A. Gjedde, G.W. Goldstein, J. Greenwood
N.H. Greig, M.K. Gumerlock, J.J. Laterra
J.-M. Lefauconnier, P.J. Luthert, D.K. Male, E.A. Neuwelt
C.S. Patlak, O.E. Pratt, S.I. Rapoport, P.J. Robinson
N.R. Saunders, G.P. Schielke, Q.R. Smith

Editor

Michael W.B. Bradbury

Springer-Verlag
Berlin Heidelberg New York London Paris
Tokyo Hong Kong Barcelona Budapest

Professor MICHAEL BRADBURY, M.A., D.M. (Oxon)
Physiology Group
Biomedical Sciences Division
King's College London, University of London
Strand, London WC2R 2LS
Great Britain

With 95 Figures and 40 Tables

ISBN-13: 978-3-642-76896-5 e-ISBN-13: 978-3-642-76894-1
DOI: 10.1007/978-3-642-76894-1

Library of Congress Cataloging-in-Publication Data. Physiology and pharmacology of the blood-brain barrier/contributors, N.J. Abbott ... [et al.]; editor, Michael W.B. Bradbury. p. cm. – (Handbook of experimental pharmacology; v. 103) Includes bibliographical references and index. ISBN 3-540-54492-5. – ISBN 0-387-54492-5 1. Blood-Brain Barrier. I. Abbott, N.J. (N. Joan) II. Bradbury, M.W.B. (Michael William Blackburn) III. Series. [DNLM: 1. Blood-Brain Barrier – drug effects. 2. Blood-Brain Barrier – physiology. W1 HA51L v. 103 / WL 200 P578] QP905.H3 vol. 103 [QP375.5] 615'.1 s – dc20 [615'.7] DNLM/DLC for Library of Congress 92-2227 CIP

Softcover reprint of the hardcover 1st edition 1992

27/3130-5 4 3 2 1 0 – Printed on acid-free paper

List of Contributors

ABBOTT, N.J., Biomedical Sciences Division, Physiology Group, King's College London, University of London, Strand, London WC2R 2LS, Great Britain

BARNES, D., National Hospital for Nervous Diseases, Queen Square, London WC1N 3BG, Great Britain

BEGLEY, D.J., Biomedical Sciences Division, Physiology Group, King's College London, University of London, Strand, London WC2R 2LS, Great Britain

BETZ, A.L., Department of Pediatrics, Surgery and Neurology, Crosby Neurosurgery Research Laboratories, University of Michigan Medical Center, D3227 Medical Professional Building, Box 0532, 1500 E. Medical Center Drive, Ann Arbor, MI 48109-0718, USA

BRADBURY, M.W.B., Physiology Group, Biomedical Sciences Division, King's College London, University of London, Strand, London WC2R 2LS, Great Britain

BRIGHTMAN, M.W., National Institute of Neurological Communicative Diseases and Stroke, National Institutes of Health, Laboratory of Neurobiology, Building 36, Room 2A29, Bethesda, MD 20892, USA

BROOKS, D.J., MRC Cyclotron Unit, Hammersmith Hospital, DuCane Road, London W12 OHS, Great Britain

CSERR, H.F., Section of Physiology, Brown University, Box G-B318, Providence, RI 02912, USA

FRASER, P.A., Physiology Group, Biomedical Science Division, King's College London, University of London, Campden Hill Road, London W8 7AH, Great Britain

GJEDDE, A., Positron Imaging Laboratories, Montreal Neurological Institute and Hospital, Webster 2B-218, 3801 University Street, Montreal, Quebec H3A 2B4, Canada

GOLDSTEIN, G.W., Department of Pediatrics and Neurology, Kennedy Research Institute and Johns Hopkins Medical Institutions, 707 North Broadway, Baltimore, MD 21205, USA

GREENWOOD, J., Department of Clinical Science, Institute of Ophthalmology, Cayton Street, London EC1V 9AT, Great Britain

GREIG, N.H., Unit of Drug Development and Delivery, Laboratory of Neurosciences, Building 10, Room 6C103, National Institute of Health, Bethesda, MD 20892, USA

GUMERLOCK, M.K., Neurological Surgery 4SP200, University of Oklahoma Health Sciences Center, P.O. Box 26901, 920 S.L. Youth Blvd., Oklahoma City, OK 73190, USA

LATERRA, J.J., Department of Neurology, Kennedy Krieger Institute and Johns Hopkins Institutions, 707 North Broadway, Baltimore, MD 21205, USA

LEFAUCONNIER, J.-M., Unité de Neurotoxicologie, INSERM U.26, Hôpital Fernand Widal, 200, rue du Faubourg Saint-Denis, F-75475 Paris Cedex 10

LUTHERT, P.J., Department of Neuropathology, Institute of Psychiatry, De Crespigny Park, Denmark Hill, London SE5 8AF, Great Britain

MALE, D.K., Department of Neuropathology, Institute of Psychiatry, De Crespigny Park, Denmark Hill, London SE5 8AF, Great Britain

NEUWELT, E.A., Department of Neurology, Division of Neurosurgery, The Oregon Health Sciences University, 3181 SW Sam Jackson Pike Road, L603, Portland, OR 97201-3098, USA

PATLAK, C.S., Department of Neurosurgery, SUNY at Stony Brook, Stony Brook, NY 11794-8122, USA

PRATT, O.E., Department of Clinical Biochemistry, King's College London, School of Medicine and Dentistry, Bessemer Road, London SE5 9PJ, Great Britain

RAPOPORT, S.I., Laboratory of Neurosciences, National Institute on Aging, National Instiutes of Health, Building 10, Room 6C103, Bethesda, MD 20892, USA

ROBINSON, P.J., Human and Environmental Safety Division, Miami Valley Laboratories, The Procter & Gamble Co., P.O. Box 398707, Cincinnati, OH 45239-8707, USA

SAUNDERS, N.R., Clinical Neurological Sciences Group (Developmental), The University of Southampton, Room LF73B, Level F, South Block,

Southampton General Hospital, Tremona Road, Southampton SO9 4XY, Great Britain
Present Address: Department of Physiology, University of Tasmania, Sandy Bray, G.P.O. Box 252C, Hobalt, Tasmania 7001, Australia

SCHIELKE, G.P., Department of Surgery, University of Michigan Medical Center and Department of Pharmacology, Parke-Davis Pharmaceutical Research Division, 2800 Plymouth Road, Ann Arbor, MI 48106, USA

SMITH, Q.R., Laboratory of Neurosciences, National Institute on Aging, National Institutes of Health, Building 10, Room 6C-0103, Bethesda, MD 20892, USA

Preface*

Movement of substances into and out of the brain generally involves transport across the tight endothelium of the cerebral microvessels (the blood–brain barrier). The ultrastructure and general properties of the blood–brain barrier were delineated by the end of the 1960s, though the history of the idea is much older. Notable amongst early observations of a restriction on transport between blood and brain were those of LEWANDOWSKY (1900) and of GOLDMANN (1909, 1913). The concept was further developed by KROGH (1946) and by HUGH DAVSON in his seminal book on *The Physiology of the Ocular and Cerebrospinal Fluids* (1956).

The topic was reviewed in two monographs, each by a single author: *Blood-Brain Barrier in Physiology and Medicine* (RAPOPORT 1976) and *The Concept of a Blood-Brain Barrier* (BRADBURY 1979). Since then the field has expanded explosively and it would be very difficult for one author to do justice to the developments.

This Handbook aims both to describe and discuss the new approaches and to bring the older aspects of the subject up to date. It contains, on the one hand, information on the brain endothelium in vitro and in culture and, on the other, details of its investigation by the new imaging techniques in human subjects and patients. There are chapters on peptide transport, on trace metal transport, on the immunology of the barrier and on the manipulation of drugs to enhance their penetration into brain.

I am most grateful to the authors for the high quality of their contributions, and in most cases for their prompt submission of manuscripts. My thanks are due to Springer-Verlag, especially to Mrs. Doris Walker, for most efficient and supportive interaction. Mrs. Carol Matthews gave secretarial and administrative help of quite exceptional quality. Mr. Bob Roberts kindly produced photographic prints for several of the chapters.

London M.W.B. BRADBURY

---

* We heard with deep regret of the death of OLIVER PRATT after a period of illness and after the manuscripts had gone to press. He made a not inconsiderable contribution to the field of the blood-brain barrier and to other aspects of neurochemistry. His enthusiasm influenced many, and his Chapter 7 in this Handbook provides a fitting tribute.

Contents

CHAPTER 5

Transport of Amino Acids

CHAPTER 7

The Movement of Vitamins Into the Brain

CHAPTER 8

Electrolyte Transport

CHAPTER 13

Clinical Assessment of the Blood-Brain Barrier: Positron Emission Tomography

CHAPTER 14

Ontogenetic Development of Brain Barrier Mechanisms

CHAPTER 15

Comparative Physiology of the Blood-Brain Barrier

CHAPTER 18

Opening of the Barrier in Cerebral Pathology
P.J. Luthert. With 2 Figures ... 439

CHAPTER 19

Experimental Manipulation of the Blood-Brain and Blood-Retinal Barriers
J. Greenwood. With 4 Figures ... 459

CHAPTER 1

Ultrastructure of Brain Endothelium

M.W. BRIGHTMAN

A. Endothelium as a Blood-Brain Barrier

In vertebrate brains, either fixed in aldehydes directly or rapidly frozen first and then substituted with organic solvent before fixation at low temperatures, tracer molecules that had been infused intravascularly are prevented from reaching the interstitial fluid (IF) of the central nervous system (CNS). This barrier to the passage of hydrophilic solutes is due to the inability of the molecules to pass through zonular or circumferential junctions between adjacent endothelial cells. The second basis for the barrier is the inability of the pits or caveolae of the endothelial cells to transfer the molecules across the endothelium (REESE and KARNOVSKY 1967; BRIGHTMAN and REESE 1969). Unlike the mammal, some of the pial vessels in the anuran brain are capillaries. Consisting as they do of one cell layer, the endothelium, these pial vessels of the frog can be rapidly frozen. As the depth of rapid freezing that is free of ice-crystal artefact is only about 15–30 μm, the tunica media of arterioles or venules would be preserved but not the underlying endothelium. For this reason, the frog was selected to assess the possible artefacts that could be introduced by primary fixation in aldehydes. Another constraint of the rapid-freeze method is that tracers such as horseradish peroxidase (HRP), the demonstration of which depends on its enzymatic activity, cannot be used at such low temperatures. Instead, a molecule such as ferritin (M_r 950 000) is infused into the heart of normal frogs and those in which the blood-cerebrospinal fluid (CSF) barrier is opened by the topical application to the pial surface of 3 M urea (NAGY et al. 1988).

Abbreviations

APC	Antigen presenting cell
BBB	Blood-brain barrier
CNS	Central nervous system
CSF	Cerebrospinal fluid
CVO	Circumventricular organ
EAE	Experimental allergic encephalomyelitis
IF	Interstital fluid
HRP	Horseradish peroxidase
MHC	Major histocompatibility complex
PDGF	Platelet-derived growth factor
PMN	Polymorphonuclear leucocyte
R	Electrical resistance
SCG	Superior cervical ganglion

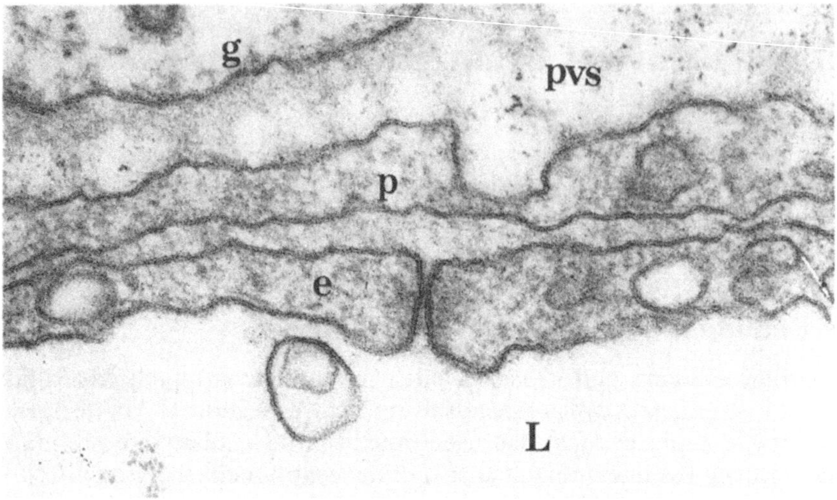

Fig. 1. An open, short, straight junction between two cerebral endothelial cells in a cerebral vessel of a nurse shark. There is direct communication between the blood space or lumen (*L*) of the capillary and the periendothelial space bordered by the endothelium (*e*) and a pericyte (*p*). Above the pericyte is a perivascular space (*pvs*) mured by pericyte and an astrogial cell (*g*), containing a coated vesicular profile that is probably a pit. ×100 000

B. Astroglia as a Blood-Brain Barrier

As the endothelium is to the mammalian blood-brain barrier (BBB), so is the perivascular astroglia to the elasmobranch blood-brain interface. Whereas the endothelium of the mammalian brain is responsible for creating a sharp gradient between HRP circulating in the plasma and the IF of the CNS, it is the perivascular astrocyte in elasmobranchs that appears to establish the gradient. Exogenous HRP, infused as a bolus into the blood of the nurse shark, Gingliostoma (BRIGHTMAN et al. 1971), or of skates, Raja (BUNDGAARD and CSERR 1981), passes unimpeded through open clefts between contiguous endothelial cells. In the shark, the narrowest portion of the patent clefts are about 3 nm wide (Fig. 1) and there are no belts of tight junctions to occlude the clefts. In several species of skates, punctate junctions do connect adjacent endothelial cells, but a few are open and about 15–20 nm wide. The endothelium of capillaries in the skate brain is exceptional in being fenestrated (BUNDGAARD and CSERR 1981a), a feature characteristic of the permeable vessels supplying circumventricular organs of the brain. In both shark and skate, blood – borne HRP rapidly enters a perivascular connective space but is stopped from further progress into the interstitial clefts of the cerebral parenchyma, at the clefts between adjacent astroglial foot processes, by tight junctions in tandem with gap junctions (Figs. 2, 3) (BRIGHTMAN et al. 1971). Tight junctions have been depicted in the skate endothelium, but the only evidence of their being zonular or cir-

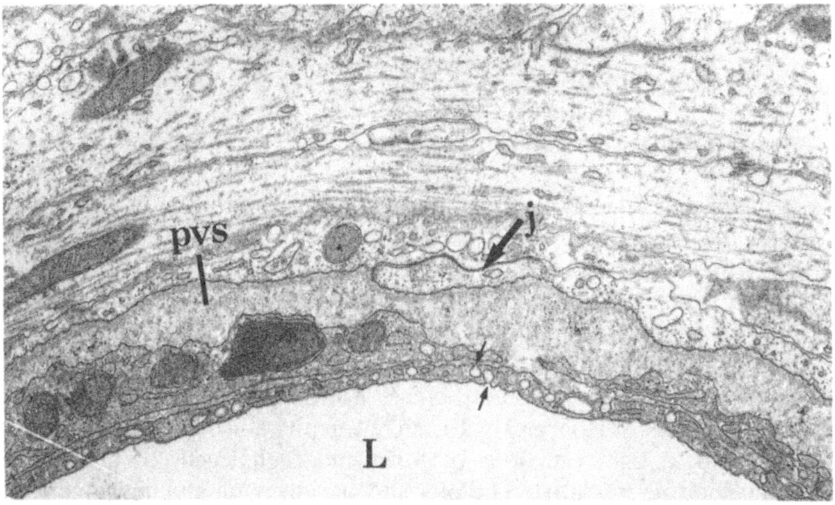

Fig. 2. Cerebral endothelium of a nurse shark with typically numerous pits or caveolae (*arrows*). Luminal pits communicate with capillary lumen (*L*), while abluminal pits communicate with collagen containing perivascular space (*pvs*). The space is also bordered by two astroglial cells joined by a gap junction (*j*) which is in tandem with tight junctions, not discernible at this magnification. ×20 000

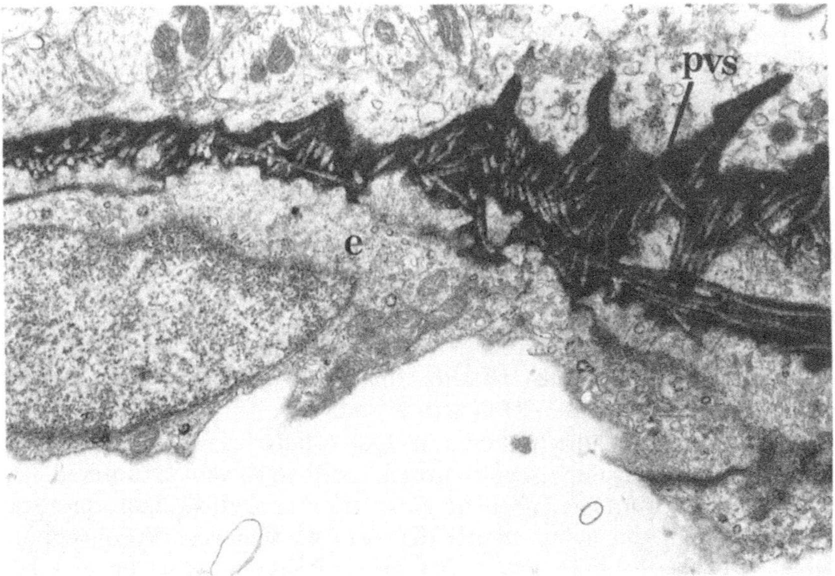

Fig. 3. Circulating horseradish peroxidase (HRP) has rapidly crossed the endothelial cell (*e*) of a cerebral vessel in the nurse shark to fill the perivascular space (*pvs*), in which collagen fibrils are outlined in negative contrast by the HRP. The gradient between the HRP in the perivascular space and the interstitial clefts in the upper portion of the figure is extremely sharp. The HRP stops abruptly at the glial surface facing the pvs. Only a few vesicular profiles in the endothelial cell have become labelled with HRP (*arrows*). ×20 000

cumferential is their exclusion of HRP from the periglial clefts (Bundgaard and Cserr 1981a).

Notwithstanding this patent paracellular route for direct exchange between cerebral blood and IF, the elasmobranch endothelium is replete with numerous pits and vesicular profiles and the logical inference was drawn that they too could be involved in the passive exchange of protein between blood and IF (Hashimoto 1972). However, even an extensive system of pits and vesicles may not necessarily connote that they are involved in transcellular passage. The cerebral endothelium of a primitive vertebrate, the cyclostome hagfish (Myxine glutinosa), is also riddled with plasmalemmal pits, apparent vesicles and tubules, yet these structures do not comprise a transcellular route. As in higher vertebrates with far fewer pits and vesicles, circulating HRP is halted at tight junctions between endothelial cells of the hagfish (Bundgaard et al. 1979a,b). The pits and the deep tubular invaginations, having an outer diameter of approximately 40–100 nm, readily accommodate HRP, but do not transfer it from blood to perivascular IF. Nor were the smaller molecules, radiolabelled polyethylene glycols of 900 and 4000 Da and microperoxidase of about 1900 Da, transported (Bundgaard and Cserr 1981b; Cserr and Bundgaard 1984). However, microperoxidase binds to proteins (Feder 1971) and, when injected into the blood stream, may reflect the mobility of its complex with serum albumin rather than of the smaller molecule, free microperoxidase.

The vessels of the CNS in amphibia also stand as barriers between blood and IF. In the tailed amphibian Necturus maculosus, the blood supply to the CNS consists of simple capillary loops. Unlike the brain capillaries of most other vertebrates, the *Necturus* capillary loops are surrounded by a collagen-containing connective tissue space. This arrangement served as a model to test the hypothesis that the cerebral endothelium was permeable to hydrophilic tracer molecules but, as the extracellular compartment was so small, not enough tracer molecules could accumulate to be detected. Contrary to this hypothesis, intravascularly injected HRP was prevented from reaching the perivascular spaces by the endothelium even though the spaces could be readily filled with HRP that had been injected via the cerebral ventricles (Bodenheimer and Brightman 1968).

The pial vessels of the frog brain include capillaries which, being accessible surface vessels, can be rapidly frozen and then solvent substituted. This alternative way of processing tissue preserves their fine structure in a state closer to that of the living animal than can be achieved by conventional fixation. The vesicles in rapidly frozen endothelial cells are fewer than those in conventionally fixed endothelium (McGuire and Twietmeyer 1983; Wagner and Andrews 1985). In rapidly frozen pial vessels of the frog, there is a system of pits and vesicles of variable size but no continuous channel could be traced across the cells (Latker et al. 1986). However, unless long series of continuous ultrathin plastic sections are cut, it cannot be certain whether or not a few such channels might have been captured by the rapid freezing technique. Accordingly, a possible transcapillary route for large molecules was reassessed with this method in frog brains, the pial vessels of which had been topically treated with hyperosomtic 3 *M* urea, known to open the endothelial junctions in mammals (BRIGHTMAN et al. 1973) and to increase the number of vesicular profiles. In the frog, such treatment resulted in the passage of blood–borne ferritin through, it appeared, open junctions. There were fewer pits in the rapidly frozen than in the conventionally aldehyde-fixed endothelium. Some of these pits were very large, being about 0.08–0.32 μm wide, but short series of thin plastic sections did not reveal any continuity across the endothelium (NAGY et al. 1988). Despite such evidence, the role of pits and vesicles in the passive non-specific transfer of large hydrophilic solutes across the cerebral endothelium is still in contention, as we shall see.

A comparable glial barrier has been defined morphologically and physiologically in insects (TREHERNE and PICHON 1972; LANE and TREHERNE 1972; LANE 1981). Insects, the neurons of which have the most evolutionarily advanced physiological behaviour of all invertebrates, do not have blood vessels within their ganglia. They do have a sheath of perineural, glial cells that envelopes their nervous system. The relatively small tracer, ionic lanthanum, is impeded by junctional complexes in its passage through the interstitial clefts of the sheath and it is the occluding tight junctions of these complexes that comprise the barrier (TREHERNE and PICHON 1972; LANE and TREHERNE 1972). Thus, in certain arthropods, some adjacent glial cells of the perineurium are fastened together by apparently zonular tight junctions that are responsible for establishing permeability gradients between haemolymph and the IF of ganglia and connectives (LANE 1981; ABBOTT et al. 1986; ABBOTT et al. 1977). The neuroglial cells of other invertebrates are without such junctions and do not form a barrier. However, certain molluscs, like the insect members of the arthropods, also have a glial–IF barrier without, apparently, the intervention of the usual tight junction. In cephalopod molluscs, such as cuttlefish, squid and octopus, the cerebral ganglia are invaded by microvessels, an arrangement more akin to higher forms. Their endothelium can be crossed by HRP but further progress of the glycoprotein is halted by perivascular glial cells. The glial cells of the sheath apparently form a continuous "seamless cuff" around the blood space (ABBOTT et al. 1986). However, the blood-CNS barrier appears to be a reticulum of extracellular fibrils rather than tight junctions. The barrier then, may be in the form of zonular intercellular junctions, as in the case of some arthropods, or the barrier may be the extracellular matrix, as in some molluscs. The perceptive conclusion has been drawn that these barriers not only function to protect neurons from the vegaries of solute concentration in blood, but may also bring about the subtle, local control or "fine tuning" of the microenvironment around specific neuronal groups. Such discrete isolation would be required by the more complex nervous system of these higher invertebrates (ABBOTT et al. 1977).

C. Blood-Brain Access

I. Vascularization of Grafts to the Brain

The tenet, established by the significant experiments of STEWART and WILEY (1981), that the graft determines the type of vasculature supplying it has withstood a number of tests. When embryonic avian brain that has not yet been vascularized is transplanted to the membranes of the chorioallantois, the permeable vessels of the membrane become impermeable when they enter the brain graft. Conversely, the normally impermeable vessels of a brain fragment grafted to peripheral tissue become permeable where the brain vessels enter this tissue (STEWART and WILEY 1981).

1. Grafts of Peripheral Tissues

The permeable vessels of the superior cervical ganglion (SCG) retain their permeability when the ganglion is translocated to the IV cerebral ventricle, a manoeuvre that causes very little damage to the ependymal or pial-glial surface. Grafts to the IV ventricle are most successful when placed upon or within the choroid plexus, the vessels of which are fenestrated and permeable. The SCG grafts apparently induce the formation of fenestrae and these new vessels, being permeable, enable the graft to circumvent the BBB (ROSENSTEIN and BRIGHTMAN 1983). Intravascular HRP leaves the vessels of the graft to enter its extracellular clefts, which are confluent with the CSF and the extracellular clefts of the adjacent brain. The HRP is thereby able to move from the graft into the interstitial compartment of the brain. As the interstitial clefts of the medulla underlying the graft are chiefly disposed longitudinally, between longitudinal fibre tracts, the spread of HRP is along the long axis of the brain stem. This longitudinal spread is reminiscent of the path taken by dextrans injected directly into the brain parenchyma (CSERR 1980).

2. Similarity Between Ventricle and Anterior Chamber of the Eye

The IV ventricle, as a site for tissue transplantation, is comparable to the anterior chamber of the eye. At both sites, collateral sprouts of axons can ensure survival of the graft. Thus, the size of a surviving muscle graft in the IV ventricle depends upon its becoming innervated. Capillaries, typical of those in undisturbed skeletal muscle – those with an endothelium containing numerous pits and junctions permeable to protein – survive among striated muscle cells that have become innervated by myelinated axons that form myoneural junctions with them (WAKAI et al. 1986). The remaining non-innervated muscle fibres undergo fatty degeneration. The unknown source of the myelinated fibres may either by pial or choroidal sprouts from the host's own intact SCG which sends axons to brain surfaces or from a cranial nerve, such as the facial.

HRP, visualized by the sensitive method utilizing tetramethyl benzidine, becomes widely disseminated within the brain once it enters the CSF or the interstitial compartment in the normal brain (WAKAI et al. 1986; BALIN et al. 1986) or in the brain bearing a graft of peripheral tissue (WAKAI et al. 1986). As illustrated in Fig. 4, HRP escapes from the graft as well as from the circumventricular organs (CVO) such as the median eminence and becomes widely disseminated with time. The HRP also spreads widely from the CVO in normal, non-graft bearing, brains (WAKAI et al. 1986; BALIN et al. 1986).

The certainty that HRP has penetrated the IF compartment is the labelling of the ubiquitous pericyte, a phagocytic, macrophage-like cell situated perivascularly throughout the CNS (BROADWELL and BRIGHTMAN 1976; WAKAI et al. 1986). While the uptake by pericytes of HRP was not surprising after its intraventricular or intrathecal administration, such widespread labelling after its intravenous infusion was unexpected. The explanation of how the pericytes came to be labelled has been provided by significant observations on rodents and primates (BALIN et al. 1986). Intravenously administered HRP leaves cerebral blood by leaking from normal vessels within CVOs and the meninges, to enter the cerebral IF from which it is endocytosed by pericytes (BALIN et al. 1986).

The sensitive method of detecting the HRP utilized tetramethyl benzidine and the actual fraction of intravenously administered HRP that penetrated the IF clefts was probably only a small fraction of the injected amount. Whether the penetration of a more physiological but larger molecule such as gamma-globulin (~160 kDa) matches that of HRP (~40 kDa) is doubtful. Exogenous gamma-globulin escapes from an isogenic muscle graft on the surface of the medulla oblongata but penetrates the medulla's IF compartment for only a short distance (WAKAI et al. 1986).

The HRP is propelled throughout the perivascular spaces of the brain by arterial pulsations (RENNELS et al. 1985). When the pulse pressure of the brachiocephalic artery of the cat is abolished without cutting off the flow of blood through it, the perivascular penetration of intrathecally infused HRP is halted (RENNELS et al. 1985). When the pulse is not obstructed, HRP rapidly moves along the wall of penetrating arterioles, thence along the perivascular spaces of the capillary bed and back into the subarachnoid space via the perivascular space of venules. However, this rapid flow may not disseminate solutes throughout the cerebral parenchyma. Even with the use of the sensitive substrate tetramethyl benzidine, the HRP appears to be restricted to the perivascular clefts. None of the glycoprotein, according to the published micrographs, is depicted in the interstital clefts around neurons and glial cells. Rather than being a means of dispersing the exogenous glycoprotein throughout the interstitial compartment, the arterial pulse appears to be a means of rapidly clearing such solutes by propelling them along the perivascular component of the interstitial space (BRIGHTMAN 1989).

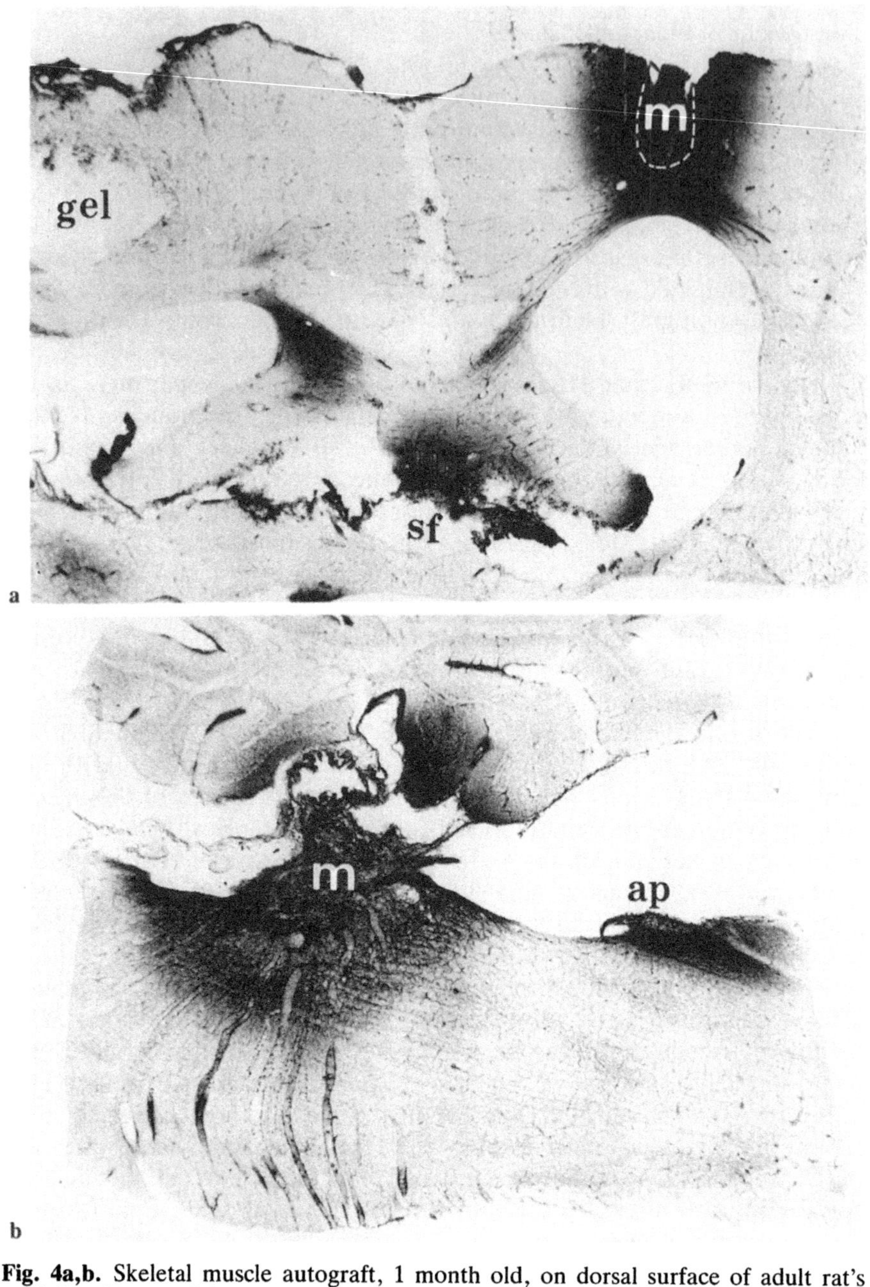

Fig. 4a,b. Skeletal muscle autograft, 1 month old, on dorsal surface of adult rat's medulla. Horseradish peroxidase (HRP) permitted to circulate for 60 min after intravenous infusion. Sections incubated with tetramethyl benzidine. **a** HRP has penetrated brain interstitium for a distance of about 0.5 mm from the edge of the muscle graft (*m*). HRP has also left the vessels of a CVO, the subfornical organ (*sf*), which lacks a BBB., The barrier has been reestablished in the vessels surrounding a piece of gel foam (*gel*) that had been inserted 1 month previously. ×18. **b** HRP has infiltrated medullary tissue adjacent to autogenic muscle graft (*m*) for a distance of 2.5 mm horizontally and 0.75 mm vertically. HRP has also flooded the interstitium of another CVO, the area postrema (*ap*), the fenestrated vessels of which are permeable to HRP. ×18

3. CNS Versus Peripheral Tissue Grafts

While there is agreement that the vessels supplying grafts of peripheral tissue retain their permeability (WAKAI et al. 1986), there is some question as to whether the vessels supplying CNS grafts to the brain retain their relative impermeability. When fetal cerebral cortex is transplanted to the cerebral cortex of mature host rats, the grafts' vessels remain permeable to both endogenous serum albumin and to exogenous HRP. Such grafts, therefore, are not physiologically integrated into the host brain because of a permanent dysfunction of the barrier. Circulating substances would have access to neurons and glia and could disrupt their normal activity (ROSENSTEIN 1988). A further consequence of a permanently disrupted barrier would be the initiation of an immunological cascade that could result in rejection of the graft (BROADWELL 1988).

There is further indication that the barrier is retained by the vessels of a CNS graft. In transplants of fetal mouse forebrain to the III ventricle of adult mice, the grafts' endothelium has a fine structure like that of normal barrier vessels and does not exude circulating HRP (BROADWELL et al. 1989).

II. Immunological Aspects of the Endothelium

The cerebral endothelium functions as an antigen presenting cell (APC), and as a cell that cooperates in the adhesion of immunologically competent cells and in their direct passage across certain vessels of the brain. The very morphological basis for the view that the CNS is an immunologically privileged site (BARKER and BILLINGHAM 1977) is the BBB itself (BILLIAU 1981). The role of the barrier has been summarized by WEKERLE et al. (1986). Because of the BBB and the absence from the CNS of a lymphatic drainage, circulating pathogens, immunologically competent cells and humoral immune solutes cannot reach the neural parenchyma. One consequence is that there is often only a partial rejection of foreign tissue grafts. However, an autoimmune response, experimental allergic encephalitis (EAE), can be elicited, both humorally and cellularly, within the CNS. The main aspect of this response considered here will be the role of the endothelium in the presentation of antigen and in the entry of activated lymphocytes into the brain substance.

In order for lymphocytes to recognize a specific antigen, it must be processed by a competent cell. The processing includes the presentation of the antigen in association with a major histocompatability complex (MHC) antigen situated on the plasma membrane of the competent APC. The expression of class II MHC (associated with Ia antigens) within the CNS is slight or undetectable (WONG et al. 1985), yet activated T lymphocytes can cross the BBB to confront astrocytes which can act as APCs (WEKERLE et al. 1986). When T lymphocytes are selected in vitro for their specificity to the antigen, myelin basic protein, they can be isolated and cloned. The intravenous infusion of these antigen activated lymphocytes triggers EAE.

In order for this autoimmune response to develop, the antigen must be presented to the activated T cells and these cells must be able to leave the blood to enter the brain substance. The cerebral endothelium is the first cell encountered by the circulating T cells and itself appears to play an active, but only partially understood role as an APC. The endothelial's presentation of auto-antigens, such as myelin basic protein, is incomplete. In highly purified cultures of rat brain endothelial cells, incubation with interferon induces the formation of Ia antigens within the cytoplasm and upon the surface of some endothelial cells. The cells recognize MHC-restricted antigens but cannot fully activate T lymphocytes. Thus, when the endothelial cells were challenged with lymphocytes from a T cell line specific for myelin basic protein, the endothelial cells were unable to induce proliferation of the T cells and were also unable to suppress T cell activation by other APC, such as macrophages (RISAU et al. 1990).

A less specific, yet essential role of the CNS endothelial cell is its association with circulating leucocytes before and during their entry into the neural substance. The same barrier endothelium that restricts the passage of large and small hydrophilic solutes from the blood is penetrated by leucocytes. When alpha-bungarotoxin was applied to the pial surface of the medulla oblongata of anaesthetized cats, the resulting acute inflammation was manifested by extravasation of polymorphonuclear leucocytes (PMN). The extravasation was exclusively from postcapillary venules (FAUSTMANN and DERMIETZEL 1985). The first event was margination of the PMN along the luminal surface of the endothelium, followed by adhesion between the two cell types.

Subsequent migration across the endothelium was exclusively transcellular rather than intercellular through junctions. Endothelial pores, 1–2 μm wide, were then formed and through which the PMN squeezed. The blood cells became highly deformed as they passed through the pores that, in turn, became enlarged by the migrating PMN. As the authors state, it is highly likely that the creation of the pores compromised the BBB. A second, less frequent mode of extravasation was the envelopment of PMN by endothelial cell processes and the concurrent formation of discontinuities in the abluminal portion of the cell membrane. The PMN were likewise able to elicit the formation of gaps in the endothelial basal lamina, through which they passed into the adventital space. The leucocytes may have been able to penetrate the basal lamina by elaborating matrix-degrading enzymes, such as endoglycosidase (NAPARSTEK et al. 1984; SAVION et al. 1984). It would appear that neither vesicles nor junctions are involved in the trans-endothelial passage of leucocytes into the perivascular interstitium of the CNS.

III. Circumventricular Organs

The CVO comprise seven small regions in which the capillaries are fenestrated and permeable to serum proteins (WEINDL 1973). Circulating HRP escapes from these vessels and is widely disseminated throughout the IF of the brain, as discussed above. However, HRP is prevented from reaching the CSF by a layer of epithelial cells that overlie the CVO's permeable vessels. The epithelial cells are tethered to each other by tight junctions that permit ionic lanthanum to pass (BOULDIN and KRIGMAN 1975) but that impede the passive extracellular flow of HRP between the cells BRIGHTMAN and REESE 1969; BALIN and Broadwell 1988). Macromolecules can cross the epithelial cells of one CVO, the choroid plexus, when infused into the cerebral ventricles; the solutes are endocytosed by the cells and transferred to the extracellular clefts between them. The only portion of the luminal surface of the cells available for endocytosis is the very small fraction of the plasmalemma that lies between the numerous microvilli and cilia (BRIGHTMAN 1977). Consequently, when HRP is perfused ventriculocisternally, only a small portion is incorporated by the epithelium and most of that portion is deposited within lysosomes for eventual enzymatic digestion. The incorporation and transfer of another macromolecule, ferritin, is similarly limited but can be appreciably augmented if the ferritin is made cationic. While some circulating macromolecules can cross the permeable vessels of the CVO to reach the overlying epithelium, they also encounter cells or their processes residing within the CVO's perivascular space.

The permeability of the microvessels within the CVOs to the passage of such substances from blood to brain has been put on a quantitative footing by GROSS and associates. The rate of blood-to-tissue flux of the small, neutral amino acid alpha-amino-isobutyric acid has been demonstrated, by quantitative autoradiography, to be 100–400 times faster in, e.g., the subfornical organ than across the barrier vessels of either gray or white matter of the brain (GROSS et al. 1986). As in other CVO, the vessels of the subfornical organ are fenestrated and contain about seven times more vesicular profiles than do barrier vessels (GROSS et al. 1986).

The resolution of this quantitative method is sufficiently high to differentiate differences in focal metabolic activity, in vivo, between small, adjacent regions of the brain. Thus, within the subfornical organ, examined morphometrically, several subdivisions with topographically distinct blood vessels may also be functionally distinctive. The capillaries of the rostral region are typical "barrier" microvessels. Neurons in this region are supplied by impermeable vessels and, accordingly, would be suited for modulating osmoregulation (SHAVER et al. 1990b). The more caudal portion is more complex morphologically. Here, the capillaries are permeable, the caudal most being fenestrated, with generous perivascular spaces. Here, too, are axons, their terminals and neurons that are immunoreactive for angiotensin (LIND et al. 1986). Perivascular axonal terminals would also be in a position to receive and transport, retrogradely, blood-borne substances. The implied neurohumoral activity of the caudal regions would be concordant with their high metabolic activity (SHAVERS et al. 1990a).

1. Fenestrae

The transendothelial passage of large solutes across fenestrated endothelium may or may not be through its tenuous, 5–6 nm thick diaphragms bridging the round holes or fenestrae. The composition of this very thin diaphragm is not known except that it is a site where anionic sites are aggregated, at least in the capillaries of the choroid plexus (DERMIETZEL et al. 1983). The pathway across the endothelial cell here may be canalicular. Morphometric comparisons of microvessels in choroid plexus and another CVO, the area postrema, suggest that clusters of vesicles within the endothelium may momentarily fuse with the abluminal or luminal plasma membrane so as to form a transient channel (COOMBER and STEWART 1988). As in fenestrated endothelia of other organs, fusion of vesicles with either one or the other pole of the cell, but not with both simultaneously, have been depicted.

2. Pits and Vesicles

The focal, minute, Ω-shaped invaginations of the luminal and abluminal cell membrane of endothelial cells in many organs, form narrow-necked caveloae or pits, about 50–100 nm in diameter, that have long been implicated in the transfer of solutes between blood and IF. According to this view, a pit forms a vesicle which moves across the cell to the opposite side of the cell, where the vesicle fuses with the cell membrane to release the vesicular contents into the IF (PALADE 1961). The vesicles came to be regarded as the equivalent of the large pores through which large solutes, i.e., those greater than 10000 M_r, were transferred across the endothelium (RENKIN 1964). This vesicular traffic has been brought into question. Three-dimensional reconstructions, derived from uninterrupted series of extremely thin plastic sections of fixed tissue that were examined by electron microscopy, reveal that what appear as independent cytosolic vesicles are actually pits or racemose clusters of communicating pits (FROKJAER-JENSEN 1983; BUNDGAARD et al. 1979b). According to this arrangement, the vesicular profiles seen in a single thin plastic section are members of a fixed immobile system of intercommunicating caveolae.

Good evidence has reinforced yet another concept, that of vesicle fusion and fission, to account for the transport of macromolecules across the endothelium of permeable vessels. In this scheme, pits and vesicles are not static but fuse momentarily with one vesicle or an intercommunicating cluster of them, then break off as separate vesicles. Fluid and its contents are thereby transferred from one vesicle to another or to a cluster until the contents are exocytosed by fusion of the final vesicle with the opposite plasma membrane. Thus, in the microvessels of the frog mesentery, a steady state is reached for circulating exogenous ferritin; at this time, the ferritin is most concentrated within luminal pits. With time, cytoplasmic vesicles become labelled with the ferritin and, ultimately, the abluminal pits on the opposite part of the cell membrane (CLOUGH and MICHEL 1981). If, according to the first concept (PALADE 1961) outlined above, a pit formed a vesicle that ferried its content across the cell, luminal pits, vesicles and abluminal pits should all have contained about the same amount of ferritin. That the ferritin content was unequal, suggested that the protein was transcytosed by a series of transient fusions and fissions (CLOUGH and MICHEL 1981).

Endocytosis in endothelial cells may, as in other cell types, be nonselective bulk fluid uptake or it may be selective. In nonselective uptake, solutes and fluid are internalized passively during formation of vesicles from membrane invaginations. The pits or caveolae are translocated when the cell membrane-bound vesicles move to other organelles in the cytosol. The extensive work of BROADWELL and colleagues (1987) has delineated clearly the endocytotic pathways in cerebral endothelium. In this endothelium, fluid phase endocytosis of circulating HRP is vectorial, being from blood into cell, rather than from IF to cell. Other molecules required by neurons and glia include transferrin and insulin. These ligands bind specifically to the cell membrane at sites where their receptors are situated. It is assumed that such molecules must be transferred across the endothelial cell into the IF and thence to neuronal and glial surfaces (BRIGHTMAN 1989; BROADWELL 1989).

3. Tubules

In the endothelium of pathological brain tissue, the tubules that develop have been regarded as transcellular channels (LOSSINSKY et al. 1979). It is plausible that, as with vesicles, a series of momentary fusions of a luminal pit with a tubule at one moment can bring plasma solutes into the lumen of the tubule. The same progression of fusions and fissions would eventually bring the tubule's content to the perivascular clefts (COOMBER and STEWART 1988). The fusion of pits and, perhaps, vesicles in the brain endothelium of the chameleon, Anolis carolinesis, has been seen in thick plastic sections of fixed tissue examined with the high voltage electron microscope (SHIVERS and HARRIS 1984). Nonetheless, the actual proof of such translocation of solutes might be obtained without benefit of tracers: what has to be demonstrated, in the way of proof, is an unequivocal simultaneous continuity of the membrane bounding a tubule with the cell membrane at luminal and abluminal faces of the cell membrane.

In all of these schemes, entailing the vesicular or vesiculo – tubular traffic of solutes across the cell, a persistent question is how a vesicle is either "marked" for fusion with other structures involved in transcellular passage or for shunting to the lysosomal compartment.

D. Ontogeny

The coincidental development of tight junctions and the blood – CSF barrier to proteins has been taken as a causal relationship. Such a relationship, between the ontogenetic appearance of endothelial and choroidal epithelial tight junctions and gradients of plasma proteins between (1) blood, (2) CSF and (3) IF compartments, has not been proven. The endothelial cells and the choroidal epithelial cells in the brain of fetal sheep and human specimens are tethered to their neighbours by complex tight junctions which, in freeze–fracture replicas, appear as interconnecting loops of strands. It is this sysem of anastomotic ridges that restricts the intercellular flow of solutes between the three compartments (Møllgård and Saunders 1975). However, the discontinuities in the component strands of the junctions are comparable to those in bovine cerebral endothelium in vitro and therefore suggest that such junctions may not be impervious to molecules (Tao-Cheng et al. 1987). Even continuous, long and complex tight junctions may be permeable to solutes. In the epithelium of the toad urinary bladder, the complex pattern of interwoven strands that make up the tight junctions are unchanged after the junctions are rendered permeable to ions by exposure of the epithelium to hyperosomotic lysine (Martinez-Palomo and Erlij 1975).

The very high content of proteins in the CSF of the fetal human and sheep brains coincides with the presence of a complex "barrier" type of tight junctions that would appear to be impervious to such molecules (Møllgård and Saunders 1975). Rather than being able to move between barrier cells, plasma proteins may reach CSF and IF by moving across the cells. An intimation of transcytotic movement has been the penetration of circulating alcian blue (M_r 1390) into a cytoplasmic tubular system of the cerebral endothelium and choroidal epithelium and thence into the CSF. The transcellular flux of solutes from one pole of a cell to an opposite pole can only be inferred, at present, from sections of fixed tissue. The participation of vesicles and tubules in a transcellular route during passive uptake is doubtful. Even in an uninterrupted series of very thin plastic sections, such confluent aggregates of vesicles may be traced as openings at *either* one pole *or* another of a cell, but not as continuous channels across the entire cell (Frokjaer-Jensen 1983). Moreover, in brain endothelium, cytoplasmic tubules with or without HRP label do not communicate with the plasma membrane (Balin et al. 1987). However, in a perturbed endothelium containing many pits and racemose clusters, the fusion of only one cytosolic vesicle with a tubule may be sufficient to create a transient transcellular channel which could momentarily bring two fluid compartments into communication (Coomber and Stewart 1988; Brightman 1989).

An alternative to either an intercellular or transcellular route that could account for the high protein of CSF would be a secretory path. Rather than being derived from plasma, the proteins in the CSF of fetal brains may be secreted by brain cells, including choroidal epithelium (MØLLGÅRD and SAUNDERS 1975; MØLLGÅRD et al. 1987). The choroidal epithelium is capable of secreting protein into the CSF. This epithelium in rats contains mRNA that encodes for the synthesis of transthyretin ("prealbumin") in vitro (GUDEMAN et al. 1987). Apocrine secretion of epithelial fragments or "aposomes" from choroidal epithelium are expelled into the CSF where they are unable to synthesize transthyretin but are able to produce four other proteins isolated by gel electrophoresis (GUDEMAN et al. 1989).

E. Astrocytic Modulation of the Barrier

Both in vivo and in vitro evidence suggests that the perivascular astrocyte modulates the structure and function of the cerebral endothelium and, therefore, the development and maintenance of the barrier. The impetus for the experiments that led to this awareness was the important observation that the avascular, embryonic, avian brain, when invaded by permeable vessels of the chorio-allantois, induced the capillaries to become impermeable to solutes (STEWART and WILEY 1981). Something in the brain fragment altered the properties of the once permeable vessels. The component of the tissue responsible for this profound alteration appears to be the astrocyte. When cell cultures of type 1 astrocytes are implanted upon the iris of the eye, normally supplied by vessels permeable to HRP or Evans blue – albumin, the vessels become impermeable (JANZER and RAFF, 1987).

This astrocytic effect on brain endothelium has also shown by in vitro experiments that identify the endothelial alteration. The tight junctions between endothelial cells, as revealed in freeze-fracture replicas, consist of interwoven strands or ridges within the cell membrane. Bovine brain endothelial cells in solo culture are joined by tight junctions that consist of patches of short, discontinuous junctional strands interspersed with gap junctions (Fig. 5a,c), an arrangement uncharacteristic of zonular tight junctions (TAO-CHENG et al. 1987). When the endothelial cells were co-cultured with rat-derived astrocytes, their junctional strands became longer, more continuous and interconnected with other strands (Fig. 5b,d). The number of gap junctions, which are characteristically discontinuous and not part of a barrier tight junction, was greatly reduced in the co-cultures. Apparently, the tight junctions were "normalized" into the barrier type by the astrocytes (TAO-CHENG et al. 1987). In comparable experiments, gap junctions persisted among the tight junctions between endothelial cells co-cultured with astrocytes (ARTHUR et al. 1987). When cells other than astrocytes were co-cultured with brain endothelial cells, their tight junctions were not enhanced. Contrary to the tenet that the recipient tissue milieu determines the nature of the endothelial junction, tight junctions did not develop in the endothelium from other organs, e.g. umbilical vein, aorta and pulmonary artery, when their endothelial cells were co-cultured with astrocytes

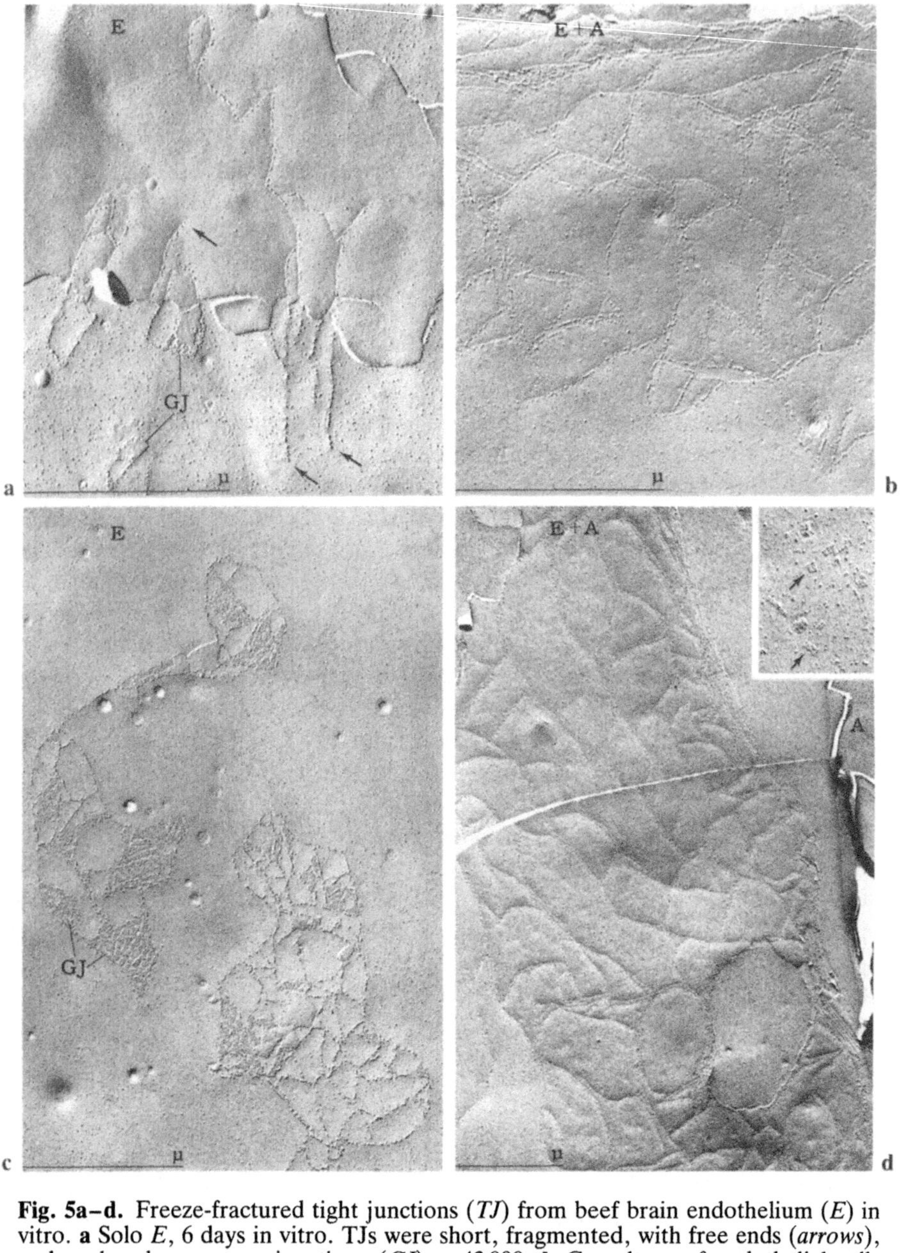

Fig. 5a–d. Freeze-fractured tight junctions (*TJ*) from beef brain endothelium (*E*) in vitro. **a** Solo *E*, 6 days in vitro. TJs were short, fragmented, with free ends (*arrows*), and enclosed many gap junctions (*GJ*). ×43 000. **b** Co-culture of endothelial cells and astroglia (*E* + *A*), 6 days. TJs were longer, broader, with far fewer gap junctions. ×43 000. **c** Solo *E*, 17 days in vitro. TJs were still in separate patches and were associated with many gap junctions (*GJ*). ×34 000. **d** *E* + *A* coculture, 14 days. An astrocyte, distinguished from other cell types by its orthogonal arrays of particles (*arrows in inset*, ×62 500), was next to this enhanced endothelial TJ. ×22 000. (From Tao-Cheng et al. 1987)

(Tao-Cheng et al. 1987). Although the tight junctions in co-cultures of brain endothelial cells and astrocytes approximate the barrier type, the electrical resistance (R) across these combined layers is extremely low. The $50\,\Omega\,cm^2$ that we have obtained is only about 1/38th of that expected from a barrier endothelium (Crone and Olesen 1982). A far higher R, more in keeping with a tight cell layer restrictive to passive ion flux, has been achieved recently.

The significant achievement of growing monolayers of brain endothelial cells in vitro with a relatively high R lends further support to the inference that astroglia augments the impermeability of brain endothelium. This R has been attained by two different methods, both of which require the presence of astrocytes or their conditioned medium.

One technique of obtaining endothelial layers of much higher R is based on the cloning of endothelium derived from capillaries, in distinction to larger microvessels, and plating them on a collagen-coated filter with an astrocyte layer on the opposite side of the filter (Meresse et al. 1989). The average R across the two cell layers was $661\,\Omega\,cm^2$ (Dehouck et al. 1990). The second method entails the continued presence of medium conditioned by astrocytes, the stimulation of adenylyl cyclase and the immunological removal of pericytes (Rubin et al. 1991). In such endothelial monolayers, grown on an appropriate collagen substrates, the R also had an average value of about $600\,\Omega\,cm^2$. Both systems provide a ready means of screening substances for their ability to perturb the endothelial barrier. Such systems should also enable the antigen presenting properties of endothelium to be evaluated without the participation of other cells and to define the conditions by which monocytes may cross the endothelial layer.

Despite such reinforcement of the notion that astrocytes may act to form and maintain the endothelial barrier, there is room for doubt. First, as discussed above, the cerebral endothelium of some elasmobranchs are ensheathed by astrocytes, but their endothelial cells are joined not by tight junctions but by open ones permeable to HRP. Either these particular astrocytes ensure that the endothelial junctions remain open or the junctional configuration is an inherent one that is not determined by the astrocytic investment. Secondly, astrocytes of the hypothalamic median eminence and astrocyte-like cells, the pituicytes, in the neural lobe of the pituitary gland are neightbours to an endothelium that is not tight but, instead, fenestrated and permeable to peroxidase. Thirdly, dissociated, cloned, brain endothelial cells in vitro can establish a continuous tight layer with an R ranging from $157–783\,\Omega\,cm^2$ without benefit of astroglial cells or their conditioned medium (Rutten et al. 1987). The precise way in which perivascular astrocytes might modulate the structural barrier of brain endothelium is not completely understood.

Another recent development in appreciating the possible relationship between brain endothelium and its astrocytic sheath is the remarkable achievement of bringing the two cell types that had been dissociated back together in vitro. By coating the plastic dish with an appropriate extracellular matrix, dissociated endothelial cells from brain microvessels are able to reform not only into capillary-like tubes but anastomotic ones as well (Minakawa et al. 1991). When dissociated astrocytes were added to such cultures, many of them appeared to preferentially settle upon the walls of the newly formed endothelial tubes (Minakawa et al. 1991).

A mechanism for this selective association is suggested by the possible ability of the brain endothelium to do what other endothelia can do. Bovine aortic endothelial cells secrete, from their basal or abluminal surface, an agent akin to platelet-derived growth factor (PDGF). In vitro, PDGF is chemotactic for adventitial cells that constitute the remainder of the vessel wall: smooth muscle cells and fibroblasts (Zerwes and Risau 1987). PDGF also acts as a chemo-attractant for astrocytes in vitro (Bressler et al. 1985). It is, therefore, likely that the capillary tubes, reassembled from dissociated endothelium, secrete a PDGF-like substance which attracts astrocytes to the site of the factor's release: the abluminal face of the endothelial tubes. It is also conceivable that the PDGF-like material attracts pericytes to the endothelium. A further indication of whether the PDGF substance effects these cell associations is to see whether the associations can be prevented by adding PDGF antibody to the medium either before or shortly after seeding the astrocytes onto the culture containing the capillary tubes.

F. Constituents of Barrier Cell Membranes

A monoclonal antibody specific for barrier endothelium in the central and peripheral nervous systems of rats immunostains the luminal surface of endothelial cells at blood-brain and blood-nerve interfaces but does not stain the fenestrated endothelium of CVOs (Sternberger and Sternberger 1987). This protein triplet (23.5, 25, 30 kDa) appears to overlap another one (46 kDa) that has been localized to the lateral junctional area of brain endothelium (Pardridge et al. 1986). A third highly glycosylated protein (45–52 kDa), the "HT7" antigen, has been associated with barrier endothelium and barrier epithelial cells of kidney tubules, retinal pigment layer and choroid plexus, and is likewise absent from CVO capillaries (Risau et al. 1986).

As a member of the immunoglobulin superfamily of cell surface receptors, the HT7 antigen is regarded as a recognition receptor that could conceivably be involved in the binding and transcytosis of selective ligands (Seulberger et al. 1990). Alternatively, the HT7 antigen may have to do with cell adhesion; it has some homology with members of cell adhesion molecules that belong to the immunoglobulin superfamily and it is concentrated in the regions of cell-cell contact (Seulberger et al. 1990). A more convincing localization awaits labelling the HT7 antibody with colloidal gold and the ultrastructural depiction of where the complex is situated.

The characterization of antigens intrinsic to barrier endothelium and barrier epithelia is contemporaneous with the advent of methods for obtaining such endothelium with known polarity and relatively high electrical resistance. This exciting coincidence holds promise for the identification of factors that may regulate the formation, organization and maintenance of the normal barrier endothelium, for screening those ligands selected for transcytosis, for interactions with lymphocytes and other cells in a defined immunological setting and for the precise changes in the behaviour of a perturbed endothelium.

References

Abbott NJ, Pichon J, Lane NJ (1977) Primitive forms of potassium homeostasis: observations on crustacean central nervous system with implications for bertebrage brain. Exp Eye Res [Suppl] 25:259–271

Abbott NJ, Lane NJ, Bundgaard M (1986) The blood-brain interface in invertebrates. Ann NY Acad Sci 481:20–42

Arthur FE, Shivers RR, Bowman PD (1987) Astrocyte-mediated induction of tight junctions in brain capillary endothelium: an efficient in vitro model. Dev Brain Res 36:155–159

Balin BJ, Broadwell RD (1988) Transcytosis of protein through the mammalian cerebral epithelium and endothelium. I. Choroid plexus and the blood-cerebrospinal fluid barrier. J Neurocytol 17:809–826

Balin BJ, Broadwell RD, Salcman M, El-Kalliny M (1986) Avenues of entry of peripherally administered protein to the CNS in mouse, rat, and squirrel monkey. J Comp Neurol 251:260–280

Balin BJ, Broadwell RD, Salcman M (1987) Tubular profiles do not form transendothelial channels through the blood-brain barrier. J Neurocytol 16:721–735

Barker CF, Billingham RE (1977) Immunologically privileged sites. In: Kunkel HG, Dixon FJ (eds) Advances in immunology. Academic, New York, pp 1–54

Billiau A (1981) Interferon therapy – pharmacokinetic and pharmacological aspects. Arch Virol 67:121–133

Bodenheimer TS, Brightman MW (1968) A blood-brain barrier to peroxidase in capillaries surrounded by perivascular spaces. Am J Anat 122:249–267

Bouldin TW, Krigman MR (1975) Differential permeability of cerebral capillary and choroid plexus to lanthanum ion. Brain Res 99:444–448

Bressler J, Grotendorst GR, Leviov C, Hjelmaland LM (1985) Chemotaxis of rat brain astrocytes to platelet derived growth factor. Brain Res 344:249–254

Brightman MW (1977) Morphology of blood-brain barrier interfaces. Exp Eye Res 25:1–25

Brightman MW (1989) The anatomic basis of the blood-brain barrier. In: Neuwelt EA (ed) Implications of the blood-brain barrier and its manipulation, vol 1. Plenum, New York, pp 53–83

Brightman MW, Reese TS (1969) Junctions between intimately apposed cell membranes in the vertebrate brain. J Cell Biol 40:648–677

Brightman MW, Reese TS, Olsson Y, Klatzo I (1971) Morphological aspects of the blood-brain barrier to peroxidase in elasmobranchs. Prog Neuropathol 1:146–161

Brightman MW, Hori M, Rapoport SI, Reese TS, Westergaard E (1973) Osmotic opening of tight junctions in cerebral endothelium. J Comp Neurol 152:317–326

Broadwell RD (1988) Addressing the absence of a blood-brain barrier within transplanted brain tissue. Science 241:473–474

Broadwell RD (1989) Transcytosis of macromolecules through the blood-brain barrier: a cell biological perspective and critical appraisal. Acta Neuropathol 79:117–128

Broadwell RD, Brightman, MW (1976) Entry of peroxidase into neurons of the central and peripheral nervous systems from extracerebral and cerebral blood. J Comp Neurol 166:257–284

Broadwell RD, Balin BJ, Salcman M (1987) Polarity of the blood-brain barrier to the endocytosis of the exogenous protein. Wiss Z Karl-Marx-Univleipe 36: 170–174

Broadwell RD, Charlton HM, Ganong WF, Salcman M, Sofroniew M (1989) Allografts of CNS tissue possess a blood-brain barrier. I. Grafts of medial preoptic area in hypogonadal mice. Exp Neurol 105:135–151

Bundgaard M, Cserr HF (1981a) A glial blood-brain barrier in elasmobranchs. Brain Res 226:61–73

Bundgaard M, Cserr H (1981b) Impermeability of hagfish cerebral capillaries to radiolabelled polyethylene glycols and to microperoxidase. Brain Res 206:71–81

Bundgaard M, Cserr H, Murray M (1979a) Impermeability of hagfish cerebral capillaries to horseradish peroxidase. Cell Tissue Res 198:65–77

Bundgaard M, Frokjaer-Jensen J, Crone C (1979b) Endothelial plasmalemmal vesicles as elements in a system of branching invaginations from the cell surface. Proc Natl Acad Sci USA 76:6439–6442

Clough G, Michel CC (1981) The role of vesicles in the transport of ferritin through frog endothelium. J Physiol (Lond) 315:127–142

Clough G, Michel CC (1988) Quantitative comparisons of hydraulic permeability and endothelial intercellular cleft dimensions in single frog capillaries. J Physiol (Lond) 405:563–576

Coomber BL, Stewart PA (1988) Three-dimensional reconstruction of vesicles in endothelium of blood-brain barrier versus highly permeable microvessels. Anat Rec 215:256–261

Crone C, Olesen SP (1982) Electrical resistance of brain microvascular endothelium. Brain Res 241:49–55

Cserr HF, Bundgaard M (1984) Blood-brain interfaces in vertebrates: a comparative approach. Am J Physiol 246:R277–R288

Cserr HR (1980) Convection of brain interstitial fluid. In: Kovach AGB, Hamar J, Szabb L (eds) Cardiovascular physiology microcirculation and capillary exchange. Proceedings of the 28th congress of physiological sciences, Budapest. Pergamon, New York, pp 337–341

Dehouck MP, Meresse S, Delorme P, Fruchart J-C, Cecchelli R (1990) An easier, reproducible, and mass-production method to study the blood-brain barrier in vitro. J Neurochem 54:1798–1801

Dermietzel R, Thurauf N, Kalweit P (1983) Surface charges associated with brain capillaries. II. In vivo studies on the role of molecular charge in endothelial permeability. J Ultrastruct Res 84:111–119

Faustmann PM, Dermietzel R (1985) Extravasation of polymorphonuclear leukocytes from the cerebral microvaculature. Cell Tissue Res 242:399–407

Feder N (1971) An ultrastructural tracer of low molecular weight. J Cell Biol 51:339–343

Frokjaer-Jensen J (1983) The plasmalemmal vesicular system in capillary endothelium. Prog Appl Microcirc 1:17–34

Gross PM, Sposito NM, Pettersen SE, Fenstermacher JD (1986) Differences in function and structure of the capillary endothelium in gray matter, white matter and a circumventricular organ of rat brain. Blood Vessels 23:261–270

Gudeman DM, Nelson SR, Merisko EM (1987) Protein secretion by choroid plexus: isolated apical fragments synthesize proteins in vitro. Tissue Cell 19:101–109

Gudeman DM, Brightman MW, Merisko EM, Merrill CR (1989) Release from live choroid plexus of apical fragments and electrophoretic characterization of their synthetic products. J Neurosci Res 24:184–191

Hashimoto PH (1972) Intracellular channels as a route for protein passage in the capillary endothelium of the shark brain. Am J Anat 134:41–58
Janzer RC, Raff MC (1987) Astrocytes induce blood-brain properties in endothelial cells. Nature 325:253–257
Lane JC (1981) Invertebrate neuroglia-junctional structure and development. J Exp Biol 95:7–33
Lane NJ, Treherne JE (1972) Studies on perineurial junctional complexes and the sites of uptake of microperoxidase and lanthanum in the cockroach central nervous system. Tissue Cell 4:427–436
Latker CH, Lynch KI, Rapoport SI (1986) The morphology of pial vessels of the frog preserved by rapid freezing and freeze substitution. Brain Res 375:186–192
Lind RW, Swanson LW, Ganton D (1986) Angiotensin II immunoreactivity in the neural afferents and efferents of the subfornical organ of the rat. Brain Res 321:209–215
Lossinsky AS, Garcia GH, Iwanowski L, Lightfoot WE (1979) New ultrastructural evidence for a protein transport system in endothelial cells of gerbil brains. Acta Neuropathol (Berl) 47:105–110
Martinez-Palomo A, Erlij D (1975) Structure of tight junctions in epithelia with different permeability. Proc Natl Acad Sci USA 72:4487–4491
McGuire PG, Twietmeyer TA (1983) Morphology of rapidly frozen endothelial cells. Circ Res 53:424–429
Meresse S, Dehouck MP, Delorme PM, Bensaid JP, Tauber C, Delbart C, Fruchart JC, Cecchelli R (1989) Bovine brain endothelial cells express tight junctions and monoamine oxidase activity in long-term culture. J Neurochem 53:1363–1371
Minakawa T, Bready J, Berliner J, Fisher M, Cancilla P (1991) In vitro interaction of astrocytes and pericytes with capillary – like tubular structures of brain microvessel endothelium. In: Abbott J, Lieberman EM, Raff M (eds) Glial-neuronal interaction. NY Acad Sci (in press)
Møllgård K, Saunders NR (1975) Complex tight junctions of epithelial and of endothelial cells in early foetal brain. J Neurocytol 4:453–468
Møllgård K, Balslev Y, Lauritzen B, Saunders NR (1987) Cell junctions and membrane specializations in the ventricular zone (germinal matrix) of the developing sheep brain: a CSF-brain barrier. J Neurocytol 16:433–444
Nagy Z, Pettigrew KD, Meiselman S, Brightman MW (1988) Cerebral vessels cyrofixed after hyperosmosis or cold injury in normothermic and hypothemic frogs. Brain Res 440:315–327
Naparstek Y, Cohen IR, Fuks Z, Vlodavsky I (1984) Activated T lymphocytes produce a matrix-degrading heparan sulphate endoglycosidase. Nature 310:241–244
Palade GE (1961) Blood capillaries of the heart and other organs. Circulation 24:368–384
Pardridge WM, Yang J, Eisenbers J, Mietus LJ (1986) Antibodies to blood-brain barrier bind selectively to brain capillary endothelial lateral membranes and to a 46K protein. J Cereb Blood Flow Metab 6:203–211
Reese TS, Karnovsky MJ (1967) Fine structural localization of a blood-brain barrier to exogenous peroxidase. J Cell Biol 34:207–217
Renkin EM (1964) Transport of large molecules across capillary walls. Physiologist 7:13–28
Rennels ML, Gregory TF, Blaumanis OR, Fujimoto K, Grady PA (1985) Evidence for a paravascular fluid circulation in the mammalian central nervous system, provided by the rapid distribution of tracer protein throughout the brain from the subarachnoid space. Brain Res 326:47–63
Risau W, Hallmann R, Albrecht U, Henke-Fahle S (1986) Brain induces the expression of an early cell surface marker for blood-brain barrier-specific endothelium. EMBO J 5:3179–3183
Risau W, Engelhardt B, Wekerle H (1990) Immune function of the blood-brain barrier: incomplete presentation of protein (auto-) antigens by rat brain microvascular endothelium in vitro. J Cell Biol 110:1757–1766

Rosenstein JM (1988) Addressing the absence of a blood-brain barrier within transplanted brain tissue. A response. Science 241:473–474

Rosenstein JM, Brightman MW (1983) Circumventing the blood-brain barrier with autonomic ganglion transplants. Science 221:879–881

Rubin LL, Barbu K, Bard C, Cannon DE, Hall H, Horner M, Janatpour C, Liaw C, Manning K, Morales J, Porter S, Tanner L, Tomaselli K, Yednock T (1991) Differentiation of brain endothelial cells in cell culture. In: Abbott J, Lieberman EM, Raff, M (eds) Glial-neuronal interaction. NY Acad Sci (In press)

Rutten MJ, Hoover RL, Karnovsky MJ (1987) Electrical resistance and macromolecular permeability of brain endothelial monolayer cultures. Brain Res 425:301–310

Savion N, Vlodavsky I, Fuks A (1984) Interaction of T lymphocytes and macrophages with cultured vascular endothelial cells: attachment, invasion, and subsequent degradation of the subendothelial extracellular matrix. J Cell Physiol 118:169–178

Seulberger H, Lottspeich F, Risau W (1990) The inducible blood-brain specific molecule HT7 is a novel immunoglobulin – like cell surface glycoprotein. EMBO J 9:2151–2158

Shaver SW, Kadekaro M, Gross PM (1990a) Differential rates of glusose metabolism across subregions of the subfornical organ in Brattleboro rats. Regul Pept 27:37–49

Shaver SW, Sposito NM, Gross PM (1990b) Quantitiative fine structure of capillaries in subregions of the rat subfornical organ. J Comp Neurol 294:145–152

Shivers RR, Harris RJ (1984) Opening of the blood-brain barrier in Anolids carolinensis. A high voltage electron microscope protein tracer study. Neuropathol Appl Neurobiol 10:343–356

Sternberger NW, Sternberger LA (1987) Blood-brain barrier protein recognized by monoclonal antibody. Proc Natl Acad Sci USA 84:8169–8173

Stewart PA, Wiley MJ (1981) Developing nervous tissue induces formation of blood-brain characteristics in invading endothelial cells: a study using quail-chick transplantation chimeras. Dev Biol 84:183–192

Tao-Cheng J-H, Nagy Z, Brightman MW (1987) Tight junctions of brain endothelium in vitro are enhanced by astroglia. J Neurosci 7:3293–3299

Treherne JE, Pichon Y (1972) The insect blood-brain barrier. Adv Insect Physiol 9:257–313

Wagner RC, Andrews S (1985) Ultrastructure of the vesicular system in rapidly frozen endothelium of the rete mirabile. J Ultrastruct Res 90:172–182

Wakai S, Meiselman SE, Brighman MW (1986) Focal circumvention of blood-brain barrier with grafts of muscle, skin and autonomic ganglia. Brain Res 386:209–222

Weindl A (1973) Neuroendocrine aspects of circumventricular organs. In: Ganong WF, Martini L (eds) Frontiers in neuroendocrinology. Oxford University Press, New York, pp 3–32

Wekerle H, Linington C, Lassmann H, Meyermann R (1986) Cellular immune reactivity within the CNS. Trends Neurosci 9:271–277

Wong GHW, Bartlett PF, Clark-Lewis I, McKimm-Breschkin JL, Schrader JW (1985) Interferon-γ induces the expression of H-2 and Ia antigens on brain cells. J Neuroimmunol 7:255–278

Zerwes HG, Risau W (1987) Polarized secretion of a platelet-derived growth factor – like chemotactic factor by endothelial cells in vitro. J Cell Biol 105:2037–2041

CHAPTER 2

Methods of Study

Q.R. Smith

A. Introduction

Man's knowledge concerning the blood-brain barrier has exploded during the past 25 years due in part to the development of a vast array of methods to examine barrier characteristics and transport properties both in vivo and in vitro. Barrier permeability and uptake have been studied in vivo using a diverse series of peripheral injection, single pass extraction, and brain perfusion techniques, which have established many of the basic principles of barrier transport physiology. Barrier structure and function have been probed at the morphologic level with electron microscopy and immunohistochemistry. Quantitative autoradiography and morphometry have established that brain regions differ in solute transfer due in part to differences in brain capillary density and possibly to differences in capillary transport function. Questions have arisen concerning the extent to which all brain capillaries are perfused, and whether the perfused fraction is regulated to meet the needs of cerebral metabolism. In the past few years, techniques in cellular and molecular biology have been developed to culture brain endothelia, to identify and isolate specific transport carriers, and to study factors that control expression of barrier characteristics in vitro. Such studies may, in the future, identify the genes that define the blood-brain barrier, thereby allowing physicians at some future date to selectively "turn on" or "turn off" barrier functions to aid clinical therapy.

This chapter will review and analyze the basic methods for examining blood-brain barrier structure and function. Only limited coverage will be made of the more classic techniques for studying the blood-brain barrier, including the intravenous (i.v.) injection, indicator dilution and brain uptake index techniques, as these methods have been recently summarized in detail

---

Abbreviations

BUI	Brain uptake index	PET	Positron emission tomography
FITC	Fluorescein isothiocyanate	p.o.	By mouth
i.m.	Intramuscular	s.c.	Subcutaneous
i.v.	Intravenous	SPECT	Single photon emission computerized tomography
i.p.	Intraperitoneal		
MRI	Magnetic resonance imaging	UV/Vis	Ultraviolet/visible (light)

elsewhere (SMITH 1985, 1989). Instead, more emphasis will be placed on newer approaches to examine barrier function, such as brain microdialysis, autoradiography, positron emission tomography (PET) and magnetic resonance imaging (MRI), brain perfusion, single capillary permeability, endothelial culture, and cellular and molecular biology, as these new methods will likely be in the forefront of major advances in the future. Further, though the primary focus of the review will be on the brain capillary endothelium, some mention will also be made of other barrier sites, such as the choroid plexus epithelium, and the blood-nerve and blood-retinal barriers, as these tissues also have critical roles in regulating solute uptake into the nervous system and are often ignored.

B. Morphologic Techniques

I. Organic Dye Tracers

The inability of organic dyes, such as Evans blue, trypan blue, and fluorescein, to stain brain following peripheral injection forms the basis of the oldest method for studying blood-brain barrier permeability (GOLDMANN 1913). Such dyes are often charged and bind tightly to plasma proteins (WOLMAN et al. 1981). As a consequence, they do not readily cross cell membranes. Following i.v. injection, they circulate primarily in blood and pass minimally into brain, except in "non-barrier" regions. Staining in brain is observed only when barrier permeability is altered causing significant protein or fluid extravasation, or when complications arise, such as changes in protein binding, pH, or transcellular transport.

The dye method, though qualitative, has been employed extensively to demonstrate enhanced barrier permeability in brain tumors and in brain tissue following stroke, hypertension, hyperosmotic infusion, trauma, infection, and disease (ITO et al. 1976; RAPOPORT 1976; RAPOPORT et al. 1980; WOLMAN et al. 1981; GREIG et al. 1983; KUROIWA et al. 1985). The method can be made semiquantitative either by visually ranking the extent of brain staining (RAPOPORT 1976) or by extracting and measuring the dye by ultraviolet/visible light (UV/Vis) or fluorescence spectrometry (SARIA and LUNDBERG 1983). Results with these latter methods have been shown to correlate well with more exacting techniques (RAPOPORT et al. 1980). Organic dyes can also be covalently attached to proteins or dextrans (e.g., fluorescein isothiocyanate [FITC]-dextrans) to produce more defined markers of barrier permeability (HULISTROM et al. 1983; MAYHAN and HEISTAD 1985; MAYHAN et al. 1989).

II. Electron Microscopy

Electron microscopy has also been used extensively to study barrier permeability and ultrastructure in vivo. In most such studies, an electron-dense tracer is injected into an animal, followed by a brief uptake period to allow distribution. The animal is then killed, and the brain is perfused, fixed, sectioned, and processed for microscopy. Blood-brain barrier permeability and integrity are evaluated from the relative presence or absence of tracer in the brain interstitial fluid space.

A number of tracers are available for electron microscopy which differ in molecular size and weight. The most commonly used are: ferritin (400 000 daltons [Da], diameter ≥100 Å), horseradish peroxidase (40 000 Da, 50–60 Å), cytochrome c (13 000 Da, 25–30 Å), microperoxidase (1900 Da, ~15 Å), and colloidal and ionic lanthanum (139 Da, 1.15 Å) (REESE and KARNOVSKY 1967; BRIGHTMAN and REESE 1969; FEDER 1971; MILHORAT et al. 1975; VAN DEURS and AMTORP 1978; BUNDGAARD 1982; DOROVINI-ZIS et al. 1983). All such compounds are markedly hydrophilic and cross the barrier only slowly by passive diffusion. As was mentioned for the organic dye tracers, the electron microscopic method can be made semiquantitative using optical techniques (GREIG et al. 1983).

Electron microscopy has proved extremely valuable for the study of barrier integrity at the cellular level and in the evaluation of the relative contributions of transport pathways (e.g., tight junctions, vesicles). Recent studies have shown that vesicular profiles that are often observed in systemic and brain capillaries may not represent true vesicles, but are instead caveolae or membrane invaginations (BUNDGAARD 1986). The rarity of true free vesicles supports the hypothesis that vesicular transport contributes minimally to brain uptake of small, hydrophilic solutes.

III. Quantitative Morphometry

Quantitative morphometry has been used to obtain insights on vascular density, capillary surface area, capillary radius, tight junctional area, mitochondrial density, and relative density of capillary fenestra (BAR 1980; BELL and BALL 1981; COOMBER and STEWART 1985; STEWART et al. 1987; GROSS et al. 1987). Such studies allow the integration of knowledge concerning the transport properties of the barrier with data concerning barrier ultrastructure, capillary geometry, and brain blood flow. The approach has been particularly valuable in explaining regional differences in barrier function with regard to capillary growth and surface area (KLEIN et al. 1986; FENSTERMACHER et al. 1988).

C. In Vivo Transport Methods

Blood-brain barrier permeability and transport can be measured quantitatively using any number of in vivo physiological methods. Such methods are all based essentially on the same principles. A solute or drug is delivered into the vascular system and allowed to circulate for a brief period of time. A certain fraction of the compound is taken up into brain, and that fraction is quantitated either by (i) the relative removal or "extraction" of the compound from blood during a single pass through the brain or by (ii) direct measurement of brain concentration. The amount taken up is compared to that which was delivered to the central nervous system to calculate a blood-to-brain transfer coefficient. For compounds that are taken up slowly relative to cerebral blood flow, this transfer coefficient (K_{in}) will be equivalent to the product of blood-brain barrier permeability (P) and surface area (A) (i.e., the PA product). The PA can be converted to a permeability coefficient if the barrier surface area available for exchange is known. However, because barrier surface area is difficult to accurately measure (see below), most studies simply express results as blood-brain barrier PA values. As CRONE (1984a) pointed out, the PA gives an estimate of the relative "resistivity" or "transport capacity" of the barrier and is equivalent to influx (J_{in}) normalized to unit concentration (C_{cap}) [$PA = J_{in}/C_{cap}$].

Other terms that are often used to describe brain uptake are "clearance," which is equivalent to K_{in}, and "extraction," which is employed in studies using single-pass uptake methods. Extraction is the fraction of compound taken up into brain during a single pass through the cerebral circulation.

The following is a brief review of the primary in vivo methods that are currently used to examine blood-brain barrier transport.

I. Intravenous Administration/Compartmental Analysis

The intravenous administration/compartmental analysis technique is the backbone of most blood-brain barrier studies because of its simplicity, versatility, and sensitivity. With this procedure, a solute is delivered into the vascular system, either by direct i.v. injection or by peripheral administration (subcutaneous, per os, intramuscular, or intraperitoneal; s.c., p.o., i.m., or i.p.), and then allowed to circulate for a given period of time. Blood samples are collected at frequent intervals in order to document the time course of plasma concentration. At the end of the uptake period, the animal is killed and brain solute concentration is determined by direct assay. If uptake is unidirectional (i.e., if all compound that goes into brain does not come out), then a blood-brain barrier K_{in} or PA product is calculated as the simple ratio of the quantity of solute in brain (Q_{br}) divided by the plasma concentration integral ($\int C_p dt$) (OHNO et al. 1978):

$$K_{in} = Q_{br} \Big/ \int C_p dt \tag{1}$$

If uptake is not unidirectional (i.e., if some efflux has occurred), then PA must be calculated using compartmental analysis. K_{in} can be converted to PA using the Crone-Renkin model of capillary transfer as

$$PA = -v_f F \ln(1 - K_{in}/v_f F) \tag{2}$$

where F represents cerebral blood flow and v_f the effective volume fraction of blood which contributes to solute uptake. For solutes that do not readily cross red cells, $v_f F$ equals cerebral plasma flow.

The exact shape of the plasma concentration curve is not critical. Plasma concentration may fall, rise, or remain constant depending on the method of delivery. The only important thing is that the plasma concentration curve be carefully measured for accurate determination of plasma integral or computer modelling. In this regard, a constant plasma concentration can simplify calculations, as with constant C_p, $\int C_p dt = C_p T$. Further, a constant plasma concentration can also simplify problems with vascular binding and distribution. However, the difficulties with maintaining a constant plasma concentration are not trivial, and thus most studies use a simple bolus i.v. injection.

In addition to carefully documenting the plasma time course, the investigator must correct the measured quantity of tracer in brain for that fraction which is intravascular (i.e., which is trapped in residual blood in brain). The simplest way to do this is to wash the intravascular tracer out of the brain by perfusion with tracer-free fluid (Carlsson and Johansson 1978; Harik and McGunigal 1984; Bradbury and Deane 1986). However, such washouts are not always possible and are rarely complete, thus leaving behind some tracer in the vascular compartment. As a result, most studies either calculate the vascular correction from the measured "blood" volume or examine uptake over several time points so that the vascular component drops out in the calculations.

With the "blood" volume correction technique, the residual blood in brain is quantitated, either simultaneously or separately, using a blood marker (^{125}I-albumin, ^{3}H-inulin, ^{3}H-dextran, ^{14}C-sucrose or ^{51}Cr-erythrocytes) and is expressed as a "volume" per gram brain tissue (ml or µl/g). The brain "blood" volume (V_b) is then multiplied by the arterial blood concentration (C_b) of solute and the product subtracted from the total measured brain quantity (Q_{tot}) to obtain Q_{br} – the quantity that has actually crossed into the nervous system:

$$Q_{br} = Q_{tot} - V_b C_b \qquad (3)$$

Alternatively, for solutes that do not penetrate appreciably into blood cells, residual vascular solute can be expressed in terms of a brain "plasma" volume and "plasma" concentration.

Care must be employed when using this correction to ensure that the "blood" or "plasma" volume of the vascular marker matches that of the test compound. Small solutes, such as sucrose, mannitol, and sodium, appear to distribute in a greater brain "plasma" space than large polysaccharides and proteins, for a reason that has not been adequately determined (Smith et al. 1988). This difference in spaces has lead to some confusion (Preston et al. 1983; Preston and Haas 1986; Lucchesi and Gosselin 1990), though the observation is well established in the literature (see rapidly equilibrating space; Bradbury 1979), and merits additional investigation. Further, the subtraction technique does not correct for solute specific binding to brain endothelial cells or brain endothelial accumulation. Some proteins, ions, and metals have been observed to concentrate in brain endothelial cells, so that the fraction of solute that crosses the barrier is lower than actually measured. To correct for this, Toews et al. (1978) and Triguero et al. (1990) have introduced a "capillary depletion" technique which utilizes centrifugation to separate brain into "capillary" and "brain parenchymal" fractions. By determination of the quantity of tracer in each, a first estimate can be made of solute that which has crossed the blood-brain barrier. However, it should be noted that this separation technique may be subject to artifacts due to tracer redistribution during sample isolation, centrifugation, or filtration. Autoradiography, X-ray microanalysis, and local mass spectrometry provide alternative ways to evaluate the intracerebral distribution of tracer at the cellular level (Thomas et al. 1973; Duffy and Pardridge 1987).

Vascular correction can also be performed by examining solute uptake at two or more time points and then analyzing the data by linear regression. Because total brain tracer is the sum of intravascular (Q_v) and brain parenchymal (Q_{br}) components, uptake at early time points (when transfer is unidirectional) can be given by the following equations:

$$Q_{tot} = Q_{br} + Q_v \tag{4}$$

$$Q_{tot} = K_{in} \int C_p dt + Q_v \tag{5}$$

Dividing by plasma concentration gives (Gjedde 1981; Blasberg et al. 1983)

$$Q_{tot}/C_p = K_{in}\left[\left(\int C_p dt\right) \Big/ C_p\right] + Q_v/C_p \tag{6}$$

This last equation is in the form of a straight line with slope K_{in} and intercept Q_v/C_p. If uptake is examined over several times and the results expressed as Q_{tot}/C_p and $[(\int C_p dt)/C_p]$ ratios, then K_{in} can be calculated by linear regression without any explicit correction for residual intravascular compound. Good examples of this technique are found in Gjedde (1981), Blasberg et al. (1983), Fuglsang et al. (1986), Smith et al. (1988), Rechthand et al. (1988), Ennis et al. (1990) and Pullen et al. (1990). It is an extension of earlier methods (Sarna et al. 1977; Bradbury 1979) for the case where plasma concentration is not constant. The advantages of this regression method are that it requires no vascular marker tracer and, in theory, provides optimal correction for residual blood. The disadvantages are that the method requires two or more animals for a single PA or K_{in} determination (unless uptake is measured with an in vivo scanning technique) and that individual-to-individual variability in PA, V_b, or F can invalidate the method if variation is too large.

Most studies that use the intravenous administration method work on the initial, unidirectional phase of tracer uptake, before brain-to-blood backflux becomes significant. To verify that uptake is unidirectional, most investigators examine transfer at several time points and plot the results (Q_{tot}/C_p vs $(\int C_p dt)/C_p$) to determine the linear range. Backflux is marked by a significant fall-off from linearity at later time points (SARNA et al. 1977; BLASBERG et al. 1983). An alternative method which is valid when the investigator knows the brain distribution volume of tracer is to compare the average plasma concentration ($C_p(\text{avg}) = \int C_p dt/T$) to the terminal brain concentration corrected for the brain volume of distribution ($C_{br} = Q_{br}/V_{br}$). If the terminal brain concentration (Q_{br}/V_{br}) is less than 20% of the average plasma concentration, then it is likely that backflux can be ignored (<10%).

When tracer uptake is examined over a wide time frame, the results can be analyzed with any number of compartmental models to provide additional data on brain efflux, intracerebral distribution, and CSF exchange (DAVSON and WELCH 1971; OHNO et al. 1978; RAPOPORT et al. 1982).

A semiquantitative ^{125}I-albumin method has been used extensively to evaluate barrier integrity in various experimental and disease states (CARLSSON and JOHANSSON 1978; HEISTAD and MARCUS 1979; HARIK and McGUNIGAL 1984). With this method, ^{125}I-albumin is injected i.v. and allowed to circulate for a short time. The animal is then killed, the brain is washed free of intravascular tracer, and the brain and blood tracer contents are determined. Relative permeability is expressed as (%) = 100[brain cpm/g]/[blood cpm/ml]. The limitations of this method are (1) that it makes no correction for possible differences in ^{125}I-albumin exposure (i.e., the $\int C_p dt$ is not determined), (2) that it relies on washout to remove residual intravascular tracer, which may or may not be effective under differing experimental conditions, and (3) that the washout may also remove parenchymal (extravascular) brain tracer if the barrier is not intact, leading to underestimation of permeability.

In summary, the primary advantages of the intravenous administration method are its versatility, simplicity, and sensitivity. K_{in} values as low as 10^{-7} ml s^{-1} g^{-1} can be measured, which represents an estimated single-pass brain extraction of only 0.0006%. The method measures transport under the most physiological conditions (i.e., no intrarterial injections and no required anesthesia), and can readily be adapted for autoradiography (see below). Potential problems are that the method often requires numerous animals, and allows only limited control of plasma concentration for studies of saturable transport, protein binding, and metabolism. Further, the method is subject to errors due to tracer metabolism in tissues other than the brain when chromatography of tracer in both plasma and brain is not performed.

II. Brain Perfusion

Brain perfusion presents an alternative methodological approach which complements the standard i.v. administration procedure. Brain perfusion offers the advantages of total control of perfusate composition and flow rate for studies of saturable brain uptake, protein binding, and metabolism. The method is similar to the intravenous administration method in that it offers high sensitivity and relative simplicity, but with much greater control of experimental conditions.

The brain perfusion approach is not new: the method has been applied on and off in neurochemical research for over 30 years. Significant progress was made in the 1970s by Gilboe, Betz, Drewes and colleagues using the isolated, perfused dog brain (for review, see GILBOE 1982). However, prior to 1982, brain perfusion methods used primarily isolated brains, which required long surgical preparation times (2–4h just to set up a perfusion), and measured transport by relatively insensitive, single-pass unidirectional extraction methods (ANDJUS et al. 1967; ZIVIN and SNARR 1972; BETZ et al. 1973, 1975; WOODS and YOUDIM 1978). The major advance with the more recent brain perfusion methods is that transport is measured by continuous uptake, much like that employed in the i.v. administration procedure. As a result, these new methods have far greater sensitivity and thus can measure brain transport for more-slowly penetrating compounds. TAKASATO et al. (1984) with a brain perfusion technique measured the *PA* to ^{14}C-sucrose, which is on the order of $5 \times 10^{-6}\,ml\,s^{-1}\,g^{-1}$ (brain extraction =0.03%). FISHMAN et al. (1987), a few years later, examined the transport of ^{125}I-transferrin ($PA < 1 \times 10^{-6}\,ml\,s^{-1}\,g^{-1}$) using a similar procedure.

Since the first report of a brain perfusion procedure of this new class (TAKASATO et al. 1982), several papers have come out describing variations of the method for different species and under differing experimental conditions (HERVONEN and STEINWALL 1984; GREENWOOD et al. 1985, 1989a,b; ZLOKOVIC et al. 1986; FISHMAN et al. 1987; MOORHOUSE et al. 1988; IVES and GARDINER 1990; DEANE and BRADBURY 1990; TRIGUERO et al. 1990). At present, methods have been developed for rabbits, rats, mice, and guinea pigs. Perfusion has been carried out with 1-week-old rats (NAGASHIMA et al. 1987) and with the "energy-depleted brain" (GREENWOOD et al. 1985).

In all of these methods the basic procedure is to infuse fluid into the carotid artery or systemic circulation so as to completely take over blood flow to the brain. The surgery is not necessarily difficult. In fact, in some papers investigators simply insert a catheter or needle directly into the common carotid artery (HERVONEN and STEINWALL 1984; NAGASHIMA et al. 1987; DEANE and BRADBURY 1990). An alternative approach is to insert the cannula into the ascending aorta via the left ventricle (GREENWOOD et al. 1985). With the original Takasato technique, the heart was left pumping, but more recently it has been found valuable to stop the heart to avoid mixing (NAGASHIMA et al. 1987; SMITH et al. 1990, 1991). After a finite time, the perfusate is changed to a fluid which contains the solute of interest, and then blood-brain barrier transport is measured from the accumulation of solute in brain during perfusion for a short period (15 s–30 min). The theory and calculations for transport are essentially the same as those given for the intravenous administration procedure, except perfusate plasma concentration ($C_{\text{perfusate}}$) is substituted for arterial plasma concentration in Eq. 1. Hence, K_{in} is calculated as

$$K_{\text{in}} = Q_{\text{br}}/[C_{\text{perfusate}}T] \tag{7}$$

where T represents net uptake time. Like the intravenous method, one has to correct for residual intravascular tracer and verify the linear, unidirectional portion of the uptake curve. K_{in} can be converted to PA using Eq. 2.

The barrier remains remarkably intact during perfusion even when non-blood infusion fluids are used (TAKASATO et al. 1984; GREENWOOD et al. 1985). Measured blood-brain barrier PA values for many nonelectrolytes correspond well to those obtained in vivo (TAKASATO et al. 1984; GREENWOOD et al. 1985; TRIGUERO et al. 1990; PARDRIDGE et al. 1990). A number of perfusion fluids have been tested, including whole blood, artificial blood, plasma, saline and isotonic sucrose solution. All have been found to provide reasonable transport rates for short periods. Both constant rate and peristaltic pumps have been used, thus far, with comparable results.

Although the brain perfusion approach has been employed primarily to study barrier transport systems for nutrients, peptides, metals and proteins, the control provided by the method is also valuable in studies of brain metabolism, drug action, protein binding, and modulation of barrier function. One important resource of the method that is just beginning to be tapped is the control the method provides of cerebral blood flow. Preliminary studies by SMITH et al. (1990) showed that flow could be varied over a 30-fold range in the anesthetized rat at physiological pressures. Subsequent evaluations have found that variation is really possible over a far greater range (>300-fold) with maximal flow rates for saline of ~0.4–0.5 $\text{ml s}^{-1}\text{g}^{-1}$ (24–30 $\text{ml min}^{-1}\text{g}^{-1}$). Over the bulk of the range, PA values to nonelectrolytes such as ethylene glycol and thiourea are essentially constant (Q. Smith, unpublished observations). The high flow rates enable determination of PA values that are far greater than previously allowed (~1–2 $\text{ml s}^{-1}\text{g}^{-1}$)(SMITH et al. 1991). [Note: Normal cerebral blood flow in the normal awake rat is on the order of $2–3 \times 10^{-2} \text{ml s}^{-1}\text{g}^{-1}$ (1–2 $\text{ml min}^{-1}\text{g}^{-1}$), which limits accurate PA determinations to $\sim 6–9 \times 10^{-2} \text{ml s}^{-1}\text{g}^{-1}$. The PA to the lipophilic molecule antipyrine is $\sim 1.4 \times 10^{-2} \text{ml s}^{-1}\text{g}^{-1}$ (TAKASATO et al. 1984), whereas the PA to the cerebral blood flow marker iodoantipyrine is $\sim 5 \times 10^{-2} \text{ml s}^{-1}\text{g}^{-1}$ (SAWADA et al. 1989).] It is expected that this flow control capability will prove especially valuable in studies of drug uptake into brain and the influence of plasma protein binding (LEVITAN et al. 1984; SMITH et al. 1990, 1991). The brain perfusion method has also been used to study precursor relations in brain protein synthesis and lipid incorporation (HARGREAVES-WALL et al. 1990; WASHIZAKI et al. 1991).

III. Indicator Dilution

The indicator dilution method is an intra-arterial injection, venous-sampling method that is designed to measure extraction from a single pass of a compound through the cerebral circulation. The method was developed by CRONE (1963) and has been extensively used to measure solute transport in various tissues.

With the indicator dilution technique, a buffered solution containing the test compound and an impermeant reference tracer is injected into the carotid artery as a bolus. Commonly used reference tracers are ^{125}I-albumin, ^{24}Na, ^{36}Cl, and ^{57}Co-DPTA, none of which measurably crosses the blood brain barrier in a single pass. Immediately following injection, serial blood samples are collected from the superior sagittal sinus or the internal juglar vein to measure tracer concentrations in venous effluent from brain. Then, for each sample, an apparent extraction (E) is calculated from the difference in the arterial (C_A) and venous (C_V) concentrations as

$$E = (C_A - C_V)/C_A \tag{8}$$

C_A, however, is not directly measured but is obtained indirectly from the venous concentration of impermeant reference tracer. In most instances, E, is calculated as

$$E = 1 - (C'_{test}/C'_{ref}) \tag{9}$$

where $C' = (C_V/C_{injectate})$. Equation 9 corrects for the fact that injectate concentrations of test and reference tracers may differ. E can be converted to K_{in} or PA with the following equations (CRONE 1963) when uptake is unidirectional:

$$K_{in} = v_f FE \tag{10}$$

$$PA = -v_f F \ln(1 - E) \tag{11}$$

Recent reports have applied the indicator dilution technique to study brain capillary heterogeneity (HERTZ and PAULSON 1980; SAWADA et al. 1989) and to evaluate asymmetry in blood-brain barrier transport (KNUDSEN et al. 1990a,b). Such studies are possible because the outflow curve provides much detailed information on the time course of uptake and efflux. The technique has the benefit of allowing several sequential measurements of transport in the same animal, and has been applied to man (LASSEN et al. 1971; KNUDSEN et al. 1990a,b). However, the indicator dilution technique, like other single pass extraction methods, has limited sensitivity and can only be used to accurately measure extractions of ~5% to ~85%. This corresponds to an estimated PA range of $\sim 1 \times 10^{-3}$ to $6 \times 10^{-2}\,\mathrm{ml\,s^{-1}\,g^{-1}}$. The technique is limited at low extractions by the difficulty of accurately determining small differences (1%–5%) in venous effluent concentrations. Further, problems can arise due to intravascular separation of test and reference tracers as a result of intralaminar or Taylor diffusion or of erythrocyte carriage (LASSEN et al. 1971; HERTZ and PAULSON 1980). Failure to correct for tracer separation can lead to significant errors in estimated extraction for slowly penetrating compounds. Extracerebral contamination of sinus blood can also be a problem and may require surgical modification (SAWADA et al. 1989) or use of a second reference tracer (HERTZ and BOLWIG 1976). Finally, the technique does not allow regional evaluation of transport and provides only a single value for the entire cerebral hemisphere.

IV. Brain Uptake Index

The brain uptake index technique, like the indicator dilution technique, is an intracarotid injection, single-pass extraction method. It was introduced by OLDENDORF in 1970 as a fast, simple method to measure blood-brain barrier transport in small animals. The method differs from the indicator dilution technique in that extraction is determined not from the tracer concentration in venous blood, but from the net quantity of tracer taken up into brain. Its primary advantage is that because the method is so simple it can be used to gain considerable information on blood-brain transfer in a short period of time. In addition, the method requires only small quantities of tracer and little or no expensive equipment (aside from a liquid scintillation or gamma counter). As a result, the brain uptake index technique has been used extensively in the blood-brain barrier field to study transport under normal physiologic and pathologic conditions. In fact, in Oldendorf's laboratory alone, the method has been used on over 10000 animals to study the transport of at least 200 different compounds (OLDENDORF 1981).

With the standard brain uptake index procedure, a 200 µl bolus of buffered saline containing containing a test tracer and a permeant reference tracer is injected rapidly (<1 s) into the carotid artery of an anesthetized rat. The bolus fills the carotid artery and is carried through the brain by cerebral blood flow. After a single capillary pass (5–15 s), the rat is decapitated, and the brain is removed and analyzed for tracer contents. The brain uptake index (BUI, %) is calculated as the ratio of test and reference tracer concentrations in brain divided by that in injectate:

$$\mathrm{BUI} = 100[C_{\mathrm{test}}/C_{\mathrm{ref}}]_{\mathrm{br}}/[C_{\mathrm{test}}/C_{\mathrm{ref}}]_{\mathrm{inj}} \tag{12}$$

where "br" and "inj" stand for "brain" and "injectate," respectively. The BUI, thus, is a relative measure of transport in that uptake is expressed as a percentage of that for a known, permeant control (usually ^{3}H-water or ^{14}C-butanol).

The BUI can be corrected for residual intravascular tracer by subtracting the BUI of an impermeant tracer, such as sucrose, inulin, or dextran. Normal values for impermeant markers are in the range of 1%–3% (OLDENDORF 1970, 1981). In some studies, an impermeant vascular tracer, such as ^{113m}In-EDTA, is included in the bolus, along with the test and permeant reference tracers, to simultaneously measure the vascular correction.

One great strength of the BUI technique is that, because the bolus transiently fill the cerebral blood vessels, it allows ready variation of capillary solute concentration for studies of saturable influx, competition, and inhibition. Injectate solute concentrations can be varied over a great range (~0–100 m*M*), and in most cases the resultant capillary concentration follows the same level. Because the injectate can be manipulated easily, the method is also valuable for studies of the effects of protein binding on brain uptake and of the effects of pH on saturable and passive transfer. In addition, the technique can be used on conscious animals to avoid complications of the anesthetized state (BRAUN et al. 1985).

One problem with the BUI technique is that it is often difficult to accurately convert the BUI to a blood-brain barrier PA or K_{in} value. This arises because there is uncertainty in the precise extraction value of various reference tracers, including ^{3}H-water and ^{14}C-butanol. Extractions for these reference tracers are less than 100% and vary with flow and metabolic condition (BRAUN et al. 1985; PARDRIDGE and FIERER 1985). Furthermore, backflux of test tracer during the single pass can be a difficulty unless the entire time course of transport is determined. In 1985, PARDRIDGE and FIERER identified an iodoamphetamine compound which has an apparent net brain extraction of ~100% and, therefore, may prove useful as a reference tracer in future studies. And lastly, the BUI technique has problems due to mixing or efflux of endogenous compound from brain as the bolus passes through the cerebral circulation (SMITH et al. 1984; PARDRIDGE et al. 1985; MOMMA et al. 1987). This "mixing" can lead to underestimation of influx rates for solutes that are taken up into brain by saturable transport.

Regardless of these limitations, the BUI is very valuable for a quick, semiquantitative indication of uptake for compounds measured under similar conditions. However, because of possible changes in flow, reference tracer extraction, or mixing, care must be exercised when trying to interpret BUIs measured under differing conditions, especially in various disease states.

V. Other Techniques

There are a number of additional in vivo methods which, though less frequently used, still merit examination. For example, RAICHLE et al. (1974) developed a single-injection external registration method for measurement of single pass extraction of gamma- or positron-emitting isotopes. The method has similarities to the BUI method in that the tracer is injected intra-arterially, and then the brain concentration is monitored. As indicated by the title, the method uses external detection to measure the time course of brain tracer activity. From the time course, it is possible to calculate an apparent extraction and estimate a *PA*. The method requires no internal reference tracer and is limited primarily to large animals for which external radiotracer detection is possible. Like the indicator dilution technique, the calculated *PA* represents the average for whole brain.

Another method, developed by Mayhan and Heistad, uses intravital fluorescence microscopy to examine barrier permeability to fluorescent-labeled polysaccharides and proteins (MAYHAN and HEISTAD 1985; MAYHAN et al. 1989). With this technique, extravasation of i.v. injected fluorescent compound is quantitated directly either from the number of fluorescent or "leakage" sites observed in brain tissue with fluorescence microscopy, or from the clearance of fluorescent product from pial vessels. For the latter, the pial surface is superfused with saline, and the fluid is collected and analyzed by fluorescence spectrometry. With this method, MAYHAN et al. (1989) showed that changes in barrier permeability to dextrans following acute hypertensive opening depended on molecular size and charge.

Efflux from brain, though less regularly examined, can be studied with any of three methods. PATLAK and FENSTERMACHER (1975) developed a ventriculocisternal perfusion method which introduces a compound into the nervous system via the ventricular system and then analyzes the compound's distribution and efflux by tissue sampling. The method examines the tissue concentration gradient of solute after 15 min–6 h of continuous ventriculo-cisternal perfusion. Tissue concentrations (C_x) are expressed relative to that at the ventricular surface (C_{vs}) and then the data are fit to the equation

$$C_x/C_{vs} = S\exp[-(k_o/D_t)^{1/2}x] \qquad (13)$$

In Eq. 13, S is the apparent brain distribution space of the compound, k_o is the brain-to-blood transfer rate constant, D_t is the effective diffusion co-efficient of the tracer in the tissue, and x is the distance from the ventricular surface. k_o is related to *PA* as

$$k_o = (v_f F/V_{br})[1 - \exp(-PA/v_f F)] \qquad (14)$$

where V_{br} is the brain distribution volume of the tracer and the volume from which efflux occurs.

BRADBURY et al. (1975) developed a similar method to examine efflux except that the method of Bradbury and colleagues depends not on ventri-culocisternal perfusion but on BUI pulse labeling to load the compound into the central nervous system. The quantity of tracer in brain at various times after injection is determined and fit to the equation

$$Q_{br}(t) = Q_{br}(0)[\exp(-k_o t)] \qquad (15)$$

where $Q_{br}(0)$ is the quantity of tracer in brain at the start of efflux ($T = 0$) and k_o is the brain-to-blood efflux rate constant. As with the ventriculocisternal perfusion technique, k_o can be converted to PA using Eq. 14.

These last two methods – like any efflux method – have many assumptions and limitations that restrict their general application. For example, the results can be complicated by intracerebral metabolism, binding, or active cellular accumulation. Further, it is often difficult to know the exact volume from which efflux occurs. Information on efflux can also be obtained by examining the time course of tracer activity following loading by i.v. administration or brain perfusion. The results, then, are analyzed to obtain brain distribution volumes and efflux coefficients by compartmental analysis (DUNCAN et al. 1991). Analysis of outflow curves from indicator dilution studies also provides an alternative way to get information on efflux from the central nervous system (KNUDSEN et al. 1990a,b).

D. New Dimensions

Several new methods have been developed during the past 10–15 years that merit special mention. Some complement or extend existing in vivo methods (e.g., autoradiography, positron emission tomography, microdialysis), whereas others are totally independent (e.g., single capillary studies) and allow investigators to address completely new sets of questions.

I. Evaluation of Barrier Transport and Permeability in Humans

The recent development of positron emission tomography (PET) and magnetic resonance imaging (MRI) now allow in vivo quantitation of regional blood-brain barrier transport in humans. Though not perfect at present, the methods offer the hope that ready evaluation will be available in the future for experimental and clinical studies.

1. Positron Emission Tomography

Several reports have been published during the past decade on the use of PET to examine blood-brain barrier permeability and transport in humans. Most have used [^{68}Ga]EDTA or ^{82}Rb as markers of passive barrier permeability (HAWKINS et al. 1984; BROOKS et al. 1984; JARDEN et al. 1985; SCHLAGETER et al. 1987). Transfer rates for these markers are low and similar to those measured in experimental animals. Both tracers have shown marked elevation in blood-brain barrier transfer rates in human brain tumors (HAWKINS et al. 1984; JARDEN et al. 1987).

Nutrient transport at the blood-brain barrier has been studied in PET using ^{11}C- or ^{18}F-labeled monosaccharides and amino acids (BROOKS et al. 1986; LEENDERS et al. 1986; O'TUAMA et al. 1988; HAWKINS et al. 1989; HERHOLZ et al. 1989). Reports demonstrate saturable uptake of both sets of compounds and sensitivity to competitive inhibition. Because radiotracers of naturally occurring solutes, such as D-[^{11}C]glucose and L-[^{11}C]methionine, are broken down intracerebrally or incorporated into metabolic products, several "nonmetabolizable" analogues have been developed and tested for transport activity. KOEPPE et al. (1990) proposed the use of [^{11}C]aminocyclohexane carboxylic acid as a marker of cerebrovascular neutral amino acid transport activity. For barrier hexose transport, 3-*O*-methyl-D-[^{11}C]glucose has been used (BROOKS et al. 1986; HERHOLZ et al. 1989).

Positron emission tomography is based on the principle that when a positron combines with an electron to disintegrate, two 0.5 MeV photons are given off in opposite directions. These photons are detected with an external array of scintillators, and the results analyzed through coincident circuits to produce a three-dimensional picture of brain tracer activity. The property of dual emission combined with coincidence detection affords PET a far greater level of resolution and accuracy than other external counting methods (e.g., single photon emission computerized tomography, SPECT). The resolution of PET scanners has improved considerably in recent years and is now less than 1 cm. Other commonly used positron-emitting isotopes are ^{13}N and ^{15}O. Tracer delivery and data analysis in PET are based primarily on the intravenous administration technique.

2. Magnetic Resonance Imaging

MRI may also provide an alternative tool to obtain quantitative information on blood-brain barrier transport and permeability in humans. Recently, LARSSON et al. (1990) examined the use of MRI and gadolinium-DPTA to determine barrier permeability in patients with multiple sclerosis and brain tumors. Though a nonlinear relation between gadolinium-DPTA concentration and signal was obtained, LARSSON et al. (1990) were able, through the use of a model, to calculate *PA* values for multiple sclerosis plaques and brain tumors. The results suggested that this technique may be suitable for evaluation of barrier integrity in a number of disorders, though additional work is required to validate the model and improve the level of quantitation.

II. Autoradiography

Disease-induced changes in barrier permeability are often focal in nature and do not occur uniformly throughout brain. As a result, methods that measure transfer into whole brain, or use only gross regional dissection, can often miss important changes in barrier permeability and certainly distort the local magnitude of alteration in barrier transfer.

Difficulties in examining regional barrier transport were overcome by Sokoloff and colleagues in the late 1970s with the development of quantitative autoradiography. With autoradiography, it is now possible to examine local changes in transport at the regional, subregional, or cellular level, and to better understand the accompanying alterations in brain function.

Most standard autoradiographic protocols utilize the i.v. administration technique with either ^3H or ^{14}C radiotracer. At the end of the experiment, the brain is promptly removed from the skull and frozen in an organic solvent (Freon, isopentane) cooled to −20° to −60°C. The frozen brain is then cut into 20 µm sections with a cryostat, and the sections are exposed to autoradiographic film. Methylmethacrylate standards of known radioactivity are included along with the sections to allow determination of a standard curve. After sufficient exposure, the film is removed, developed and assayed for optical density to calculate local brain tracer concentrations. Procedures have also been developed for dual-label, ^3H and ^{14}C autoradiography (GJEDDE and DIEMER 1985).

The autoradiographic approach is especially valuable for analyzing regional differences in transport and metabolism, and for evaluating localized changes in barrier transport with disease. Using autoradiography, Blasberg and colleagues have demonstrated marked heterogeneity in brain tumor permeability, both within tumors and between tumors, in rats using various brain tumor models (for review, see GROOTHUIS et al. 1984).

III. Brain Vasculature/Perfused Capillaries

Regional differences in blood-to-brain transfer are known to occur in normal brain and are thought to be due, at least in part, to local differences in capillary density. For example, cortical grey matter is reported to have 2–3-fold greater capillary density than white matter, and observed differences in blood-brain barrier *PA* products between the two regions are also 2–3-fold (BRADBURY 1979; OHNO et al. 1978).

Regional variation has recently been confirmed on a broader level in studies using quantitative morphometry and autoradiography. Such studies have demonstrated direct correlations between regional brain capillary surface area and in vivo blood-brain barrier transport for glucose and amino acids (HAWKINS et al. 1982; GJEDDE and DIEMER 1985; FENSTERMACHER et al. 1988). Preliminary data on brain uptake of [^{14}C]sucrose also conform to this pattern (Q. Smith, unpublished observations).

Although all studies agree that that there is regional variation in brain capillary density, there is considerable controversy concerning the extent of perfusion of local brain capillary beds. Some studies suggest that the only a fraction of brain capillaries are normally perfused in vivo (WEISS 1988) and that the fraction can vary with metabolic state and anesthesia (BRAUN et al. 1985; BUCHWEITZ and WEISS 1986; FRANCOIS-DAINVILLE et al. 1986; SHOCKLEY and LAMANNA 1988; FENSTERMACHER et al. 1988). Others suggest that essentially all capillaries are perfused, and that changes in brain blood flow can be completely explained by changes in the linear velocity of blood flow within the capillary bed (GOBEL et al. 1989). This controversy has important implications for the regulation of blood flow and nutrient delivery to brain.

Several techniques have been employed to determine the extent of brain vasculature perfusion. Some studies examine the extent of capillary filling by dyes at short times following intravenous dye injection (WEISS et al. 1982; FRANCOIS-DAINVILLE et al. 1986; GOBEL et al. 1989). Others measure differences in local *PA* products under conditions that are expected to either close down (pentobarbital anesthesia) or recruit (hypercapnia, hypoxia, metabolic activation) brain capillaries (GJEDDE and RASMUSSEN 1980; HERTZ and PAULSON 1982; BRAUN et al. 1985; SAIJA et al. 1989). One group of investigators has measured in vivo brain blood volume using radiotracers, and then compared the value to that measured morphometrically from tissue slices (FENSTERMACHER et al. 1988). A fourth has evaluated changes in brain capillary mean transit time to radiotracers using external detection (SHOCKLEY and LAMANNA 1988). And finally, some investigators have examined the extent of perfusion by direct vascular observation (RACKL et al. 1981). This last approach has been pursued recently using confocal laser microscopy.

IV. Other Barrier Sites

Although the cerebral capillaries represent the largest surface area for exchange between blood and neural tissue (CRONE 1984a), and therefore constitute the major site of transport for many solutes, some compounds appear to gain access to brain from other interfaces, including the choroid plexus epithelium, arachnoid membrane, and nonbarrier regions. In addition, solute uptake into nerve and retina can follow multiple pathways and is more complex than simple capillary transfer. As interest is expanding in routes solutes use to enter the nervous system and as there is increased interest in the function of the blood-nerve and blood-retinal barriers, a brief review will be given of recent advances in quantitation in these areas.

1. Choroid Plexus Epithelium

The choroid plexuses are a major site of entry for inorganic ions such as sodium, chloride, and calcium into the nervous system (BAKAY 1960; SMITH and RAPOPORT 1986; MURPHY et al. 1988). Transfer across the plexuses can be estimated from direct cerebrospinal fluid sampling and from ventriculocisternal perfusion. More precise information is obtained using isolated, in situ preparations where the secreted choroid plexus fluid is collected directly in a chamber without exchange with nearby brain tissue (for review, see JOHANSON 1988).

Uptake into choroid plexus epithelial cells can also be examined by direct tissue sampling (JOHANSON 1988). In this regard, PRESTON et al. (1989) have demonstrated saturable transfer of neutral amino acids into choroid plexus epithelial cells using an isolated perfused sheep choroid plexus system.

2. Blood-Nerve Barrier

Solute transfer into nerve can proceed via two sites: the first is the endoneurial capillaries and the second is the perineurial sheath which surrounds and isolates the endoneurial space. Recently, RECHTHAND and colleagues (1987, 1988) developed a compartmental model to simultaneously examine solute transfer into nerve endoneurium taking into account the individual contributions of the two pathways. Nerve uptake of [^{14}C]sucrose was found to proceed primarily via the nerve capillaries even though the surface area of the perineurium exceeds that of the endoneurial capillaries by 2–3-fold. The contribution of nerve capillaries was explained by a greater permeability – 5–10-fold that of perineurium or brain capillaries. The combined *PA* of the blood-nerve barrier system to [^{14}C]sucrose was ~1.3–1.5 × 10^{-5} ml s^{-1} g^{-1} (RECHTHAND et al. 1988).

Transfer across perineurium can be examined separately using either an in situ incubation procedure (RECHTHAND et al. 1988) or an isolated frog perineurial sheath preparation (WEERASURIYA et al. 1980). An in situ perfusion procedure was developed to examine saturable transfer of ions and nutrients at the blood-nerve barrier (RECHTHAND et al. 1985). Facilitated uptake systems were demonstrated in nerve for glucose and large neutral amino acids (RECHTHAND et al. 1985; WADHWANI et al. 1990).

A critical feature of all these studies is that the epineurial/perineurial sheath must be carefully removed from the nerve prior to endoneurial tracer content determination. Because of rapid blood-to-epineurium equilibration and a large epineurial extracellular space, the epineurial/perineurial content of hydrophilic tracer can exceed that in endoneurium by 5–50-fold. To avoid loss of fluid or tracer during dissection, it is best to freeze the whole nerve and then peel off the epi-perineurium.

3. Blood-Retinal Barrier

The retina, like the nerve, has multiple sites for solute uptake and exchange. Uptake can occur via the retinal capillary endothelial cells and via the retinal pigment epithelium (BRADBURY 1979). Further, solute can diffuse from the vitreous humor system.

To assess the permeability of the blood-retinal barrier, LIGHTMAN et al. (1987) developed an i.v. injection technique using radiolabeled sucrose or mannitol. A comparable system was presented in 1986 by ENNIS and BETZ. At the end of the uptake period, the eye is rapidly removed and the retina dissected from the choroid and sclera. Tracer concentrations in retina and plasma are then determined by scintillation counting. In the study by LIGHTMAN et al. (1987), the K_{in} for sucrose at the blood-retinal barrier (~4 × 10^{-6} ml s^{-1} g^{-1}) was similar to that at the blood-brain barrier and differed significantly from zero. A critical feature in all studies of blood-retinal barrier permeability is that effort must be made to avoid retinal contamination from tracer in aqueous or vitreous humor.

V. In Vivo Microdialysis

In vivo microdialysis, though not used extensively thus far, does hold significant promise for studies of blood-brain barrier transfer (BENVENISTE 1989). With microdialysis it is possible to continuously monitor the brain interstitial fluid concentration of a solute following i.v. administration. Further, the method allows direct application of solutes and drugs within the central nervous system. Such an approach may prove useful in resolving questions concerning relations between plasma, brain extracellular, and brain intracellular fluid. Most blood-brain barrier studies up to this point have treated brain as a simple single compartment system. In vivo microdialysis may allow a more realistic description of the system. Preliminary studies have already examined the use of microdialysis to evaluate blood-brain transfer (HUTSON et al. 1985; DURING et al. 1989). It is expected that further studies will employ the method in the future.

VI. Single Capillary Studies

Although most blood-brain barrier research to date has concentrated on bulk tissue methods, recent reports suggest that single capillary studies are also possible. CRONE and OLESEN in 1982 used a single capillary approach to measure the electrical resistance of the brain capillary endothelium of the frog in vivo. With this method, two microelectrodes were inserted into a capillary in situ. Then, current was injected into the capillary via one microelectrode and the induced intravascular potential change was measured using the second microelectrode at various distances along the capillary. The electrical resistance was then calculated according to the theory for "leaky cables" using the equations

$$V(x) = V(0)\exp(-x/\partial) \tag{16}$$

$$R_m = r_i \partial^2 2\pi a \tag{17}$$

where $V(x)$ is the intracapillary potential at distance x from the current electrode, $V(0)$ is the potential at the tip of the current electrode, a is the capillary radius, and R_m is the membrane resistance (CRONE and OLESEN 1982). r_i is a constant for the capillary and is determined by the resistivity of blood and the cross sectional area (CRONE and OLESEN 1982). ∂ is a "length constant" which describes how rapidly the potential decays with distance from the current source. With this approach, the average measured resistance was $\sim 1900\,\Omega\,cm^2$, which is comparable to that of some of the tightest epithelial tissues. Subsequent studies have found similar values for brain capillary resistance in mammals (BUTT et al. 1990) and have demonstrated that cerebrovascular resistance is modulated by hormones and neurotransmitters (OLESEN 1985).

CRONE in a later study (1984b) used a similar technique to examine diffusion potentials across brain capillaries in response to ionic gradients.

The results suggested that the ions cross brain capillaries predominantly through paracellular pores that show little ionic selectivity

More recently, FRASER and DALLAS (1990) determined the filtration coefficient and osmotic reflection coefficient of single brain microvessels in frogs. In the future it may be possible to more thoroughly analyze ion transport at brain capillaries using electrophysiologic techniques. Further, endothelial cell ion channels and pumps may be isolated and examined using patch clamp technology.

E. Molecular/Cellular Biology

I. Isolated Microvessels/Endothelial Monolayers

Isolated brain microvessels and cultured brain endothelial monolayers provide a new arena in which to study blood-brain barrier physiology and function. With these preparations, endothelial transporters and enzymes can be assayed directly without complications from other tissue elements. In addition, conditions can be varied in vitro over a far greater range than would ever be tolerated in vivo.

Numerous methods have been published for the isolation and purification of brain capillaries, including capillaries from human brain (for review, see WILLIAMS et al. 1980; GOLDSTEIN et al. 1984; TSUJI et al. 1987). Most methods rely on an initial homogenization step, followed by a centrifugation or filtration, for purification. Purity is never 100%, and preparations usually contain small quantities of erythrocytes, pericytes, smooth muscle cells, and/or astrocytic processes (WILLIAMS et al. 1980; WHITE et al. 1981). Cellular viability, as marked by exclusion of vital dyes, is often variable and low.

Regardless of these difficulties, microvessel preparations are markedly enriched with respect to brain endothelial cells and thus are well suited for the characterization of barrier enzymes, proteins, and receptors. The preparations can be used for transport studies, but there has been wide variation in reported uptake kinetics among papers, especially for amino acids (HJELLE et al. 1978; CARDELLI-CANGIANO et al. 1981; CHOI and PARDRIDGE 1986; CANGIANO et al. 1988; HARGREAVES and PARDRIDGE 1988). This variation may be explained partly by differences in cell viability and passive uptake. In most in vitro studies of amino acid transport, the "nonsaturable" component of uptake is extremely large.

In recent years, there has been considerable interest in development of cultured brain endothelial monolayers as models of the blood-brain barrier. Most of the preparations studied thus far have relatively low electrical resistances and exhibit high permeabilities, as compared to the in vivo microvascular barrier (RUTTEN et al. 1987; VANBREE et al. 1988; SHAH et al. 1989; PARDRIDGE et al. 1990). DEHOUCK et al. (1990), using an endothelial cell/astrocyte coculture system, obtained a monolayer electrical resistance of ~660 $\Omega\,cm^2$, which is the highest reported value for an in vitro endothelial barrier system. The importance of the astrocytes was shown by the fact that the electrical resistance was significantly lower when the brain endothelial cells were cultured alone. The results suggest that, at present, cultured brain endothelial monolayers do not represent a complete model of the blood-brain barrier, due to dedifferentiation and loss of critical trophic factors in vitro. Consistent with this is the observation that facilitated transport systems for glucose and amino acids, though present in brain endothelial cultures, operate at rates far reduced from those of the in vivo barrier system (PARDRIDGE et al. 1990).

II. Identification of Barrier Transporters and Enzymes

Techniques for the isolation and purification of brain endothelial cells have stimulated research in the identification and characterization of specific blood-brain barrier transport proteins. Using cytochalasin B and antibodies against the human erythrocyte glucose transporter. DICK et al. (1984) identified a 53-kDa protein in brain capillaries that appeared to share many of the properties of the blood-brain barrier glucose transporter, including affinity for 2-deoxy-D-glucose, D-glucose, 3-*O*-methyl-D-glucose, D-mannose, and D-galactose, but not L-glucose. IC_{50} values for inhibition of cytochalasin B binding by hexoses directly correlated with reported K_m values for hexose transport across the blood-brain barrier (DICK et al. 1984). Subsequent isolation, cloning and sequence analysis of the porcine blood-brain barrier glucose transporter revealed 97% sequence homology with the human erythrocyte glucose carrier (WEILLER-GUTTLER et al. 1989). Future studies may utilize similar methods to definitively identify other blood-brain barrier transport proteins. Receptors for possible blood-brain barrier transcytosis of insulin and transferrin have been identified in isolated brain capillary endothelial cells (PARDRIDGE 1986).

III. Expression of Barrier Characteristics

Two of the biggest questions in the blood-brain barrier field are: 1) what are the genes and factors that control expression of barrier characteristics and 2) can these processes be modulated in vivo. Transplantation experiments have shown that blood-brain barrier characteristics are induced in endothelial cells of different organs that invade brain tissue grafts (STEWART and WILEY 1981). Brain cells likely release specific trophic factors that have an important role in the induction and maintenance of barrier characteristics in the cerebral endothelium. Insight into the specific cells that control these processes can be gained from consideration of the fact that astrocytic end feet cover >80% of the abluminal surface of brain endothelial cells. Consistent with this, blood-brain barrier characteristics have been observed in capillaries of invading blood vessels in pure astrocytic transplants (JANZER and RAFF 1987). Work over the next two decades may concentrate on identifying specific blood-brain barrier trophic factors and the genes they control. In this regard, cultured brain endothelial monolayers may prove extremely useful for identifying factors.

A recent study by HARIK et al. (1990) suggests that expression of occluding junctions and barrier glucose transport protein may be tightly linked. All barrier cells of the peripheral and central nervous systems that have occluding tight junctions were also found to express high glucose transporter activity. If this is so, then other proteins may be linked as well, such as the neutral amino acid carrier which is also found at the blood-nerve barrier (WADHWANI et al. 1990). In this regard there may be a barrier capillary gene complex which is expressed as a unit throughout the body where ever nervous tissue is found. Knowledge of the specific factors that control this complex may allow, at some date in the future, temporary modulation of barrier properties in the clinic in order to aid therapeutic outcome.

F. Summary and Perspectives

The past two decades have seen striking advances in technology for evaluation of blood-brain barrier structure and function. Quantitative methods now exist for the regional evaluation of blood-brain barrier permeability and transport in man under normal and pathologic conditions. Such methods may prove useful in the future for the diagnosis and treatment of human brain diseases. In addition, a myriad of in vivo and in vitro methods have been developed for the examination of various barrier properties by experimental scientists. Through the application of these methods, it may be possible one day to induce expression of the blood-brain barrier in vitro through addition of various peptides and proteins to endothelial cell cultures. Further, it will be of interest to continue to explore differences among the various barrier systems, as such differences likely underlie specific functional roles. Fundamentally, the past 20 years have seen the transformation in perception of the barrier from a simple exclusionary interface, to an active complex that has a critical role in the maintenance and regulation of brain function.

References

Andjus RK, Suhara K, Sloviter HA (1967) An isolated perfused rat brain preparation, its spontaneous and stimulated activity. J Appl Physiol 22:1033–1039

Bakay L (1960) Studies in sodium exchange. Experiments with plasma, cerebrospinal fluid, and normal, injured, and embryonic brain tissue. Neurology 10:564–571

Bar T (1980) The vascular system of the cerebral cortex. In: Brodal A, Hild W, Van Limborgh J, Ortmann R, Schiebler TH, Tondury G, Wolff E (eds) Advances in

anatomy, embryology and cell biology, vol 59. Springer, Berlin Heidelberg New York, pp 1–61
Bell MA, Ball MJ (1981) Morphometric comparison of hippocampal microvasculature in ageing and demented people: diameters and densities. Acta Neuropathol (Berl) 53:299–318
Benveniste H (1989) Brain microdialysis. J Neurochem 52:1667–1679
Betz AL, Gilboe DD, Yudilevich DL, Drewes LR (1973) Kinetics of unidirectional glucose transport into the isolated dog brain. Am J Physiol 225:586–592
Betz AL, Gilboe DD, Drewes LR (1975) Kinetics of unidirectional transport into brain; effects of isoleucine, valine and anoxia. Am J Physiol 228:895–902
Blasberg RG, Fenstermacher JD, Patlak CS (1983) Transport of alpha-aminoisobutyric acid across brain capillary and cellular membranes. J Cereb Blood Flow Metab 3:8–32
Bradbury MWB (1979) The concept of a blood-brain barrier. Wiley, Chichester
Bradbury MWB, Deane R (1986) Rate of uptake of lead-203 into brain and other soft tissues of the rat at constant radiotracer levels in plasma. Ann N Y Acad Sci 481:142–160
Bradbury MWB, Patlak CS, Oldendorf WH (1975) Analysis of brain uptake and loss of radiotracers after intracarotid injection. Am J Physiol 229:1110–1115
Braun LD, Miller LP, Pardridge WM, Oldendorf WH (1985) Kinetics of regional blood-brain barrier glucose transport and cerebral blood flow determined with the carotid injection technique in conscious rats. J Neurochem 44:911–915
Brightman MW, Reese TS (1969) Junctions between intimately apposed cell membranes in the vertebrate brain. J Cell Biol 40:648–676
Brooks DJ, Beaney RP, Lammertsma AA, Leenders KL, Horlock PL, Kensett MJ, Marshall J, Thomas DG, Jones T (1984) Quantitative measurement of blood-brain barrier permeability using rubidium-82 and positron emission tomography. J Cereb Blood Flow Metab 4:535–545
Brooks DJ, Beaney RP, Lammertsma AA (1986) Glucose transport across the blood-brain barrier in normal human subjects and patients with cerebral tumours studied using [11C]-3-O-methyl-D-glucose and positron emission tomography. J Cereb Blood Flow Metab 6:230–239
Buchweitz E, Weiss HR (1986) Alterations in perfused capillary morphometry in awake vs anesthetized brain. Brain Res 377:105–111
Bundgaard M (1982) Ultrastructure of frog cerebral and pial microvessels and their impermeability to lanthanum ions. Brain Res 241:57–65
Bundgaard M (1986) Pathways across the vertebrate blood-brain barrier: morphological viewpoints. Ann N Y Acad Sci 481:7–19
Butt AM, Jones HC, Abbott MJ (1990) Electrical resistance across the blood-brain barrier in anaesthetized rats: a developmental study. J Physiol (Lond) 429:47–62
Cangiano C, Cardelli-Cangiano P, Cascino A, Ceci F, Fiori A, Mulieri M, Muscaritoli M, Barberini C, Strom R, Fanelli FR (1988) Uptake of amino acids by brain microvessels isolated from rats with experimental chronic renal failure. J Neurochem 51:1675–1681
Cardelli-Cangiano P, Cangiano C, James JH, Jeppson B, Brenner W, Fischer JE (1981) Uptake of amino acids by brain microvessels isolated from rats after portacaval anastomosis. J Neurochem 36:627–632
Carlsson C, Johansson BB (1978) Blood-brain barrier dysfunction after amphetamine administration in rats. Acta Neuropathol (Berl) 41:125–129
Choi TB, Pardridge WM (1986) Phenylalanine transport at the human blood-brain barrier. Studies with isolated human brain capillaries. J Biol Chem 261:6536–6541
Coomber BL, Stewart PA (1985) Morphometric analysis of CNS microvascular endothelium. Microvasc Res 30:99–115
Crone C (1963) The permeability of capillaries in various organs as determined by use of the "Indicator Diffusion" method. Acta Physiol Scand 58:292–305

Crone C (1984a) The function of capillaries. In: Baker PF (ed) Recent advances in physiology, vol 10. Churchill Livingstone, Edinburgh, pp 125–162
Crone C (1984b) Lack of selectivity to small ions in paracellular pathways in cerebral and muscle capillaries of the frog. J Physiol (Lond) 353:317–337
Crone C, Olesen SP (1982) Electrical resistance of brain microvascular endothelium. Brain Res 241:49–55
Davson H, Welch K (1971) The permeation of several materials into the fluids of the rabbit's brain. J Physiol (Lond) 218:337–351
Deane R, Bradbury MWB (1990) Transport of lead-203 at the blood-brain barrier during short cerebrovascular perfusion with saline in the rat. J Neurochem 54:905–914
Dehouck MP, Meresse S, Delorme P, Fruchart JC, Cecchelli R (1990) An easier, reproducible, and mass-production method to study the blood-brain barrier in vitro. J Neurochem 54:1798–1801
Dick APK, Harik SI, Klip A, Walker DM (1984) Identification and characterization of the glucose transporter of the blood-brain barrier by cytochalasin B binding and immunological reactivity. Proc Natl Acad Sci USA 81:7233–7237
Dorovini-Zis K, Sato M, Goping G, Rapoport S, Brightman M (1983) Ionic lanthanum passage across cerebral endothelium exposed to hypertonic arabinose. Acta Neuropathol (Berl) 60:49–60
Duffy KR, Pardridge WM (1987) Blood-brain barrier transcytosis of insulin in developing rabbits. Brain Res 420:32–38
Duncan MW, Villacreses NE, Pearson PG, Wyatt L, Rapoport SI, Kopin IJ, Markey SP, Smith QR (1991) 2-Amino-3-(methylamino)-propanoic acid (BMAA) pharmacokinetics and blood-brain barrier permeability in the rat. J Pharmacol Exp Ther 258:27–35
During MJ, Heyes MP, Freese A, Markey SP, Martin JB, Roth RH (1989) Quinolinic acid concentrations in striatal extracellular fluid reach potentially neurotoxic levels following systemic L-tryptophan loading. Brain Res 476:384–387
Ennis SR, Betz AL (1986) Sucrose permeability of the blood-retinal and blood-brain barriers. Invest Ophthalmol Vis Sci 27:1095–1102
Ennis SR, Keep RF, Schielke GP, Betz AL (1990) Decrease in perfusion of cerebral capillaries during incomplete ischemia and reperfusion. J Cereb Blood Flow Metab 10:213–220
Feder N (1971) Microperoxidase: an ultrastructural tracer of low molecular weight. J Cell Biol 51:339–343
Fenstermacher J, Goss P, Sposito N, Acuff V, Pettersen S, Gruber K (1988) Structural and functional variations in capillary systems within the brain. Ann N Y Acad Sci:21–30
Fishman JB, Rubin JB, Handrahan JV, Connor JR, Fine RE (1987) Receptor-mediated transfer of transferrin across the blood-brain barrier. J Neurosci Res 18:299–304
Francois-Dainville E, Buchweitz E, Weiss H (1986) Effect of hypoxia on percent of arteriolar and capillary beds perfused in the rat brain. J Appl Physiol 60:280–288
Fraser PA, Dallas AD (1990) Measurement of filtration coefficient in single cerebral microvessels of the frog. J Physiol (Lond) 423:343–361
Fuglsang A, Lomholt M, Gjedde A (1986) Blood-brain transfer of glucose and glucose analogs in newborn rats. J Neurochem 46:1417–1428
Gilboe DD (1982) Perfusion of the isolated brain. In: Lajtha A (ed) Handbook of neurochemistry, vol 2. Plenum, New York, pp 301–330
Gjedde A (1981) High- and low-affinity transport of D-glucose from blood to brain. J Neurochem 36:1463–1471
Gjedde A, Diemer NH (1985) Double-tracer study of the fine regional blood-brain glucose transfer in the rat by computer-assisted autoradiography. J Cereb Blood Flow Metab 5:282–289

Gjedde A, Rasmussen M (1980) Pentobarbital anesthesia reduces blood-brain glucose transfer in the rat. J Neurochem 35:1382–1387
Gobel U, Klein B, Schrock H, Kuschinsky W (1989) Lack of capillary recruitment in the brains of awake rats during hypercapnia. J Cereb Blood Flow Metab 9:491–499
Goldmann EE (1913) Vitalfarbung am Zentralnervesystem. Abhandl Preuss Akad Wiss Math K1 1:1–13
Goldstein GW, Betz AL, Bowman PD (1984) Use of isolated brain capillaries and cultured endothelial cells to study the blood-brain barrier. Fed Proc 43:191–195
Greenwood J, Luthert PJ, Pratt OE, Lantos PL (1985) Maintenance of the integrity of the blood-brain barrier in the rat during an in situ saline-based perfusion. Neurosci Lett 56:223–227
Greenwood J, Hazell AS, Pratt OE (1989a) The transport of leucine and aminoacyclopen tanecarboxylate across the intact, energy-depleted rat blood-brain barrier. J Cereb Blood Flow Metab 9:226–233
Greenwood J, Hazell AS, Luthert PH (1989b) The effect of low pH saline perfusate upon the integrity of the energy-depleted rat blood-brain barrier. J Cereb Blood Flow Metab 9:234–242
Greig NH, Jones HB, Cavanagh JB (1983) Blood-brain barrier integrity and host responses in experimental metastatic brain tumours. Clin Exp Metastasis 1:229–246
Groothuis DR, Molnar P, Blasberg RG (1984) Regional blood flow and blood-to-tissue transport in five brain tumor models. Prog Exp Tumor Res 27:132–153
Gross PM, Blasberg RG, Fenstermacher JD, Patlak CS (1987) The microcirculation of rat circumventricular organs and pituitary gland. Brain Res Bull 18:73–85
Hargreaves KM, Pardridge WM (1988) Neutral amino acid transport at the human blood-brain barrier. J Biol Chem 263:19392–19397
Hargreaves-Wall KM, Buciak JL, Pardridge WM (1990) Measurement of free intracellular and transfer RNA amino acid specific activity and protein synthesis in rat brain in vivo. J Cereb Blood Flow Metab 10:162–169
Harik SI, McGunigal T (1984) The protective influence of the locus ceruleus on the blood-brain barrier. Ann Neurol 15:568–574
Harik SI, Kalaria RN, Andersson L, Lundahl P, Perry G (1990) Immunocytochemical localization of the erythroid glucose transporter: abundance in tissues with barrier functions. J Neurosci 10:3862–3872
Hawkins RA, Mans AM, Biebuyck JF (1982) Amino acid supply to individual cerebral structure in awake and anesthetized rats. Am J Physiol 242:E1–E11
Hawkins RA, Phelps ME, Huang SC, Wapenski JA, Grim PD, Parker RG, Juillard G, Greenberg P (1984) A kinteic evaluation of blood-brain barrier permeability in human brain tumors with [^{68}Ga]EDTA and positron emission tomography. J Cereb Blood Flow Metab 4:507–515
Hawkins RA, Huang SC, Barrio JR, Keen RE, Feng D, Mazziotta JC, Phelps ME (1989) Estimation of local protein synthesis rates with L-[1-^{11}C]leucine and PET: methods, model and results in animals and humans. J Cereb Blood Flow Metab 9:446–460
Heistad DD, Marcus ML (1979) Effect of sympathetic stimulation on permeability of the blood-brain barrier to albumin during acute hypertension in cats. Circ Res 45:331–341
Herholz K, Weinhard K, Pietrzyk, Pawlik G, Heiss WD (1989) Measurement of blood-brain hexose transport with dynamic PET: comparison of [^{18}F]2-fluoro-2-deoxyglucose and [^{11}C]O-methylglucose. J Cereb Blood Flow Metab 9:104–110
Hertz MM, Bolwig TG (1976) Blood-brain barrier studies in the rat: an indicator dilution technique with tracer sodium as an internal standard for estimation of extracerebral contamination. Brain Res 107:333–343
Hertz MM, Paulson OB (1980) Heterogeneity of cerebral capillary flow in man and its consequences for estimation of blood-brain barrier permeability. J Clin Invest 65:1145–1151

Hertz MM, Paulson OB (1982) Transfer across human blood-brain barrier: evidence for capillary recruitment and for paradox glucose permeability increases in hypocapnia. Microvasc Res 24:364–376

Hervonen H, Steinwall O (1984) Endothelial surface sulfhydryl-groups in blood-brain barrier transport of nutrients. Acta Physiol Scand 121:343–351

Hjelle JT, Baird-Lambert J, Cardinale G, Spector S, Udenfriend S (1978) Isolated microvessels: the blood-brain barrier in vitro. Proc Natl Acad Sci USA 75:4544–4548

Hultstrom D, Malmgren L, Gilstring D, Olsson Y (1983) FITC-dextrans as tracers for macromolecular movements in the nervous system. Acta Neuropathol (Berl) 59:53–63

Hutson PH, Sarna GS, Kantamaneni BD, Curzon G (1985) Monitoring the effect of a tryptophan load on brain indole metabolism in freely moving rats by simultaneous cerebrospinal fluid sampling and brain dialysis. J Neurochem 44:1266–1273

Ito U, Go KG, Walker JT, Spatz M, Klatzo I (1976) Experimental cerebral ischemia in mongolian gerbils III. Behaviour of the blood-brain barrier. Acta Neuropathol (Berl) 34:1–6

Ives NK, Gardiner RM (1990) Blood-brain barrier permeability to bilirubin in the rat studied using intracarotid bolus injection and in situ brain perfusion. Pediatr Res 27:436–441

Janzer RC, Raff MC (1987) Astrocytes induce blood-brain barrier properties in endothélial cells. Nature 325:253–257

Jarden JO, Dhawan V, Poltorak A, Posner JB, Rottenberg DA (1985) Positron emission tomographic measurement of blood-to-brain and blood-to-tumor transport of ^{82}Rb: the effect of dexamethasone and whole-brain radiation therapy. Ann Neurol 18:636–646

Jarden JO, Dhawan V, Moeller JR, Strother SC, Rottenberg DA (1989) The time course of steroid action on blood-to-brain and blood-to-tumor transport of ^{82}Rb: a positron emission tomographic study. Ann Neurol 25:239–245

Johanson CE (1988) The choroid plexus-arachnoid membrane-cerebrospinal fluid system. In: Boulton AA, Baker GB, Walz W (eds) Neuromethods; the neuronal microenvironment. Humana, Clifton, pp 33–104

Klein B, Kuschinsky W, Schrock H, Vetterlein F (1986) Interdependency of local capillary density, blood flow and metabolism in rat brains. Am J Physiol 251:H1333–H1340

Knudsen GM, Pettigrew K, Patlak CS, Hertz MM, Paulson OB (1990a) Kinetic analysis of blood-brain barrier transport in man: evaluation of tracer backflux and capillary heterogeneity. Microvasc Res 39:28–49

Knudsen GM, Pettigrew KD, Patlak CS, Hertz MM, Paulson OB (1990b) Asymmetrical transport of amino acids across the blood-brain barrier in humans. J Cereb Blood Flow Metab 10:698–706

Koeppe RA, Mangner T, Betz AL, Shulkin BL, Allen R, Kollros P, Kulh DE, Agranoff BW (1990) Use of ^{11}C-aminocyclohexanecarboxylate for the measurement of amino acid uptake and distribution volume in human brain. J Cereb Blood Flow Metab 10:727–739

Kuroiwa T, Ting P, Martinez H, Klatzo I (1985) The biphasic opening of the blood-brain barrier to proteins following temporary middle cerebral artery occlusion. Acta Neuropathol (Berl) 68:122–129

Larsson HBW, Stubgaard M, Frederiksen JL, Jensen M, Henriksen O, Paulson OB (1990) Quantiation of blood-brain barrier defect by magnetic resonance imaging and gadolinium-DPTA in patients with multiple sclerosis and brain tumors. Magn Reson Med 16:117–131

Lassen NA, Trap-Jensen J, Alexander SC, Olesen J, Paulson OB (1971) Blood-brain barrier studies in man using the double-indicator method. Am J Physiol 220:1627–1633

Leenders KL, Poewe WH, Palmer AJ, Brenton DP, Frackowiak RSJ (1986) Inhibition of L-[^{18}F]fluorodopa uptake into human brain by amino acids demonstrated by positron emission tomography. Ann Neurol 20:258–262

Levitan H, Ziylan Z, Smith QR, Takasato Y, Rapoport SI (1984) Brain uptake of a food dye, erythrosin B, prevented by plasma protein binding. Brain Res 322:131–134

Lightman SL, Palestine AG, Rapoport SI, Rechthand E (1987) Quantitative assessment of the permeability of the rat blood-retinal barrier to small water-soluble non-electrolytes. J Physiol (Lond) 389:483–490

Lucchesi KJ, Gosselin RE (1990) Mechanism of L-glucose, raffinose and inulin transport across the intact blood-brain barrier. Am J Physiol 258:H695–H705

Mayhan WG, Heistad DD (1985) Permeability of the blood-brain barrier to various sized molecules. Am J Physiol 248:H712–H718

Mayhan WG, Faraci FM, Siems JL, Heistad DD (1989) Role of molecular charge in disruption of the blood-brain barrier during acute hypertension. Circ Res 64:658–664

Milhorat TH, Davis DA, Hammock MK (1975) Experimental intracerebral movement of electron microscopic tracers of various molecular sizes. J Neurosurg 42:315–329

Momma S, Aoyagi M, Rapoport SI, Smith QR (1987) Phenylalanine transport across the blood-brain barrier as studied using the in situ brain perfusion technique. J Neurochem 48:1291–1300

Moorhouse SR, Carden S, Drewitt PN, Eley BP, Hargreaves RJ, Pelling D (1988) The effect of chronic low level lead exposure on blood-brain barrier function in the developing rat. Biochem Pharmacol 37:4539–4547

Murphy VA, Smith QR, Rapoport SI (1988) Regulation of brain and cerebrospinal fluid calcium by brain barrier membranes following vitamin D-rlated chronic hypo-and hypercalcemia in rats. J Neurochem 51:1777–1782

Nagashima T, Lefauconnier JM, Smith QR (1987) Developmental changes in neutral amino acid transport across the blood-brain barrier. J Cereb Blood Flow Metab 7:S501

Oldendorf WH (1970) Measurement of brain uptake of radiolabeled substances using a tritiated water internal standard. Brain Res 24:372–376

Oldendorf WH (1981) Clearance of radiolabeled substances by brain after arterial injection using a diffusible internal standard. In: Marks N, Rodnight R (eds) Research methods in neurochemistry, vol 5. Plenum, New York, pp 91–112

Olesen SP (1985) A calcium-dependent reversible permeability increase in microvessels in frog brain, induced by serotonin. J Physiol (Lond) 361:103–113

Ohno K, Pettigrew KD, Rapoport SI (1978) Lower limits of cerebrovascular permeability to nonelectrolytes in the conscious rat. Am J Physiol 235:H299–H307

O'Tuama LA, Guilarte TR, Douglass KH, Wagner HN, Wong DF, Dannals RF, Ravert HT, Wilson AA, LaFrance ND, Bice AN, Links JM (1988) Assesment of [^{11}C]-L-methionine transport into the human brain. J Cereb Blood Flow Metab 8:241–345

Pardridge WM (1986) Mechanisms of neuropeptide interaction with the blood-brain barrier. Ann N Y Acad Sci 481:231–249

Pardridge WM, Fierer G (1985) Blood-brain barrier transport of butanol and water relative to N-isopropyl-p-iodoamphetamine as the internal reference. J Cereb Blood Flow Metab 5:275–281

Pardridge WM, Landaw EM, Miller LP, Braun LD, Oldendorf WH (1985) Carotid artery injection technique: bounds for bolus mixing by plasma and by brain. J Cereb Blood Flow Metab 5:576–583

Pardridge WM, Triguero D, Yang J, Cancilla PA (1990) Comparisons of in vitro and in vivo models of drug transcytosis through the blood-brain barrier. J Pharmacol Exp Ther 253:884–891

Patlak CS, Fenstermacher JD (1975) Measurements of of dog blood-brain transfer constant by ventriculocisternal perfusion. Am J Physiol 229:877–884

Preston E, Haas N (1986) Defining the lower limits of blood-brain barrier permeability: factors affecting the magnitude and interpretation of permeability-area products. J Neurosci Res 16:709–719

Preston E, Allen M, Haas N (1983) A modified method for measurement of radiotracer permeation across the rat blood-brain barrier: the problem of correcting brain uptake for intravascular tracer. J Neurosci Methods 9:45–55

Preston JE, Segal MB, Walley GJ, Zlokovic BV (1989) Neutral amino acid uptake by the isolated perfused sheep choroid plexus. J Physiol (Lond) 408:31–43

Pullen RGL, Candy JM, Morris CM, Taylor G, Keith AB, Edwardson JA (1990) Gallium-67 as a potential marker for aluminium transport in rat brain: Implications for Alzheimer's disease. J Neurochem 55:251–259

Rackl A, Pawlok G, Bing RJ (1981) Cerebral capillary topography and red cell flow in vivo. In: Cervos-Navarro J, Fritschka E (eds) Cerebral microcirculation and metabolism. Raven, New York, pp 17–21

Raichle ME, Eichling JO, Grubb RL (1974) Brain permeability of water. Arch Neurol 30:319–321

Rapoport SI (1976) Opening of the blood-brain barrier by acute hypertension. Exp Neurol 52:467–479

Rapoport SI, Fredericks WR, Ohno K, Pettigrew KD (1980) Quantitative aspects of reversible osmotic opening of the blood-brain barrier. Am J Physiol 238:R421–R431

Rapoport SI, Fitzhugh R, Pettigrew KD, Sundaram U, Ohno K (1982) Drug entry into and distribution within brain and cerebrospinal fluid: [^{14}C]urea pharmacokinetics. Am J Physiol 242:R339–R348

Rechthand E, Smith QR, Rapoport SI (1985) Facilitated transport of glucose from blood into peripheral nerve. J Neurochem 45:957–964

Rechthand E, Smith QR, Rapoport SI (1987) Transfer of nonelectrolytes from blood into peripheral nerve endoneurium. Am J Physiol 252:H1175–H1182

Rechthand E, Smith QR, Rapoport SI (1988) A compartmental analysis of solute transfer and exchange across blood-nerve barrier. Am J Physiol 255:R317–R325

Reese TS, Karnovsky MJ (1967) Fine structural localization of a blood-brain barrier to exogenous peroxidase. J Cell Biol 34:207–217

Rutten MJ, Hoover RL, Karnovsky MJ (1987) Electrical resistance and macromolecular permeability of brain endothelial monolayer culture. Brain Res 425:301–310

Saija A, Princi P, DiPasquale R, Costa G (1989) Modifications of the permeability of the blood-brain barrier and local cerebral metabolism in pentobarbital- and ketamine-anesthetized rats. Neuropharmacology 28:997–1002

Saria A, Lundberg JM (1983) Evans blue fluorescence: quantitative and morphological evaluation of vascular permeability in animal tissues. J Neurosci Methods 8:41–49

Sarna GS, Bradbury MWB, Cavanagh J (1977) Permeability of the blood-brain barrier after portocaval anastomosis in the rat. Brain Res 138:550–554

Sawada Y, Patlak CS, Blasberg RG (1989) Kinetics of cerebrovascular transport based on indicator diffusion technique. Am J Physiol 256:H794–H812

Schlageter NL, Carson RE, Rapoport SI (1987) Examination of blood-brain barrier permeability in dementia of the Alzheimer type with [^{68}Ga]EDTA and positron emission tomography. J Cereb Blood Flow Metab 7:1–8

Shah MV, Audus KL, Borchardt RT (1989) The application of bovine brain microvessel endothelial-cell monolayers grown onto polycarbonate membranes in vitro to estimate the potential permeability of solutes throught the blood-brain barrier. Pharmacol Res 6:624–627

Shockley RP, LaManna JC (1988) Determination of rat cerebral cortical blood volume changes by capillary mean transit time analysis during hypoxia, hypercapnia and hyperventilation. Brain Res 454:170–178

Smith QR (1985) Methods to determine blood-brain barrier permeability and transport. In: Boulton AA, Baker GB (eds) Neuromethods, vol 1; general neurochemical techniques. Humana, Clifton, pp 389–418
Smith QR (1989) Quantitation of blood-brain barrier permeability. In: Neuwelt EA (ed) Implications of the blood-brain barrier and its manipulation, vol 1. Plenum, New York, pp 85–118
Smith QR, Rapoport SI (1986) Cerebrovascular permeability coefficients to sodium, potassium and chloride. J Neurochem 46:1732–1742
Smith QR, Takasato Y, Smith QR (1984) Kinetic analysis of L-leucine transport across the blood-brain barrier. Brain Res 311:167–170
Smith QR, Ziylan YZ, Robinson PJ, Rapoport SI (1988) Kinetics and distribution volumes for tracers of different sizes in the brain plasma space. Brain Res 462:1–9
Smith QR, Fukui Shinsuke, Robinson P, Rapoport SI (1990) Influence of cerebral blood flow of tryptophan uptake into brain. In: Lubec G, Rosenthal GA (eds) Amino acids: chemistry, biology and medicine. ESCOM Science Publishers BU, Leiden, pp 364–369
Smith QR, Nagura H, Washizaki Y, DeGeorge J, Roibinson P, Rapoport SI (1991) Kinetics of fatty acid dissociation from albumin and transport into brain. Soc Neurosci Abstr 17:239
Stewart PA, Wiley MJ (1981) Developing nervous tissue induces formation of blood-brain barrier characteristics in invading endothelial cells: a study using quail-chick transplantation chimeras. Dev Biol 84:183–192
Stewart PA, Magliocco M, Hayakawa K, Farell CL, Del Maestro RF, Girvin J, Kaufmann JCE, Vinters HV, Gilbert J (1987) A quantitative analysis of blood-brain barrier ultrastructure in the aging human. Microvasc Res 33:270–282
Takasato Y, Rapoport SI, Smith QR (1982) A new method to determine cerebrovascular permeability in the anesthetized rat. Soc Neurosci Abstr 8:850
Takasato Y, Rapoport SI, Smith QR (1984) An in situ brain perfusion technique to study cerebrovascular transport in the rat. Am J Physiol 247:H484–H493
Thomas JA, Dallenbach FD, Thomas M (1973) The distribution of radioactive lead (210Pb) in the cerebellum of developing rats. J Pathol 109:45–50
Toews AD, Kolber A, Hayward J, Krigman MR, Morell P (1978) Experimental lead encephalopathy in the suckling rat: concentration of lead in cellular fractions enriched in brain capillaries. Brain Res 147:131–138
Triguero D, Buciak J, Pardridge WM (1990) Capillary depletion method for quantification of blood-brain barrier transport of circulating peptides and plasma proteins. J Neurochem 54:1882–1888
Tsuji T, Mimori Y, Nakamura S, Kameyama M (1987) A micromethod for the isolation of large and small microvessels from frozen autopsied human brain. J Neurochem 49:1796–1800
Van Bree JBMM, DeBoer AG, Danhof M, Ginsel LA, Breimer DD (1988) Characterization of an "in vitro" blood-brain barrier: effects of molecular size and lipophilicity on cerebrovascular endothelial transport rates of drugs. J Pharmacol Exp Ther 247:1233–1239
Van Deurs B, Amtorp O (1978) Blood-brain barrier in rats to the heme peptide microperoxidase. Neuroscience 3:737–748
Wadhwani KC, Smith QR, Rapoport SI (1990) Facilitated transport of L-phenylalanine across the blood-nerve barrier of rat peripheral nerve. Am J Physiol 258:R1436–R1444
Washizaki Y, Purdon D, DeGeorge J, Rapoport SI, Smith QR (1991) Fatty acid uptake and esterification by the in situ perfused rat brain. Soc Neurosci Abstr 17:864
Weerasuriya A, Rapoport SI, Taylor RE (1980) Ionic permeabilities of the frog perineurium. Brain Res 191:405–415
Weiler-Guttler H, Zinke H, Mockel B, Frey A, Gassen HG (1989) cDNA cloning and sequence analysis of the glucose transporter from porcine blood-brain barrier. Biol Chem Hoppe-Seyler 370:467–473

Weiss HR (1988) Measurement of cerebral capillary perfusion with a fluorescent label. Microvasc Res 36:172–180
Weiss HR, Buchweitz E, Murtha TJ, Auletta M (1982) Quantitative regional determination of morphometric indices of the total and perfused capillary network in the rat brain. Circ Res 51:494–503
White FP, Dutton GR, Norenberg MD (1981) Microvessels isolated from rat brain: localization of astrocyte processes by immunohistochemical techniques. J Neurochem 36:328–332
Williams SK, Gillis JF, Matthews MA, Wagner RC, Bitensky MW (1980) Isolation and characterization of brain endothelial cells: morphology and enzyme activity. J Neurochem 35:374–381
Wolman M, Klatzo I, Chui E, Wilmes F, Nishimoto K, Fujiwara K, Spatz M (1981) Evaluation of the dye-protein tracers in pathophysiology of the blood-brain barrier. Acta Neuropathol (Berl) 54:55–61
Woods HF, Youdim MBH (1978) The isolated perfused rat brain preparation – a critical assessment. Essays Neurochem 3:49–69
Zivin JA, Snarr JF (1972) A stable preparation for rat brain perfusion: effect of flow rate on glucose uptake. J Appl Physiol 32:658–663
Zloković BV, Begley DJ, Djuricic BM, Mitrovic DM (1986) Measurement of solute transport across the blood-brain barrier in the perfused guinea pig brain: method and application to N-methylaminoisobutyric acid. J Neurochem 46:1441–1451

CHAPTER 3

Diffusional and Osmotic Permeability to Water

P.A. FRASER

A. Introduction

The ionic composition of cerebral interstitial fluid is closely controlled to maintain a suitable microenvironment for neuronal functioning, and the blood-brain barrier contributes to this close homeostasis by the prevention of simple diffusion and convection of small hydrophillic substances, ions and neurotransmitters across capillary walls. It is now well established that the cerebral endothelium is the cellular element responsible for the blood-brain barrier in mammals (reviewed by BRADBURY 1979; CSERR and BUNDGAARD 1984), and freeze-fracture replicas of cerebral microvascular endothelial cells examined in the electron microscope show that there are tight junction complexes at the cell margins that are similar to those of tight epithelia (BRIGHTMAN and REESE 1969; VAN DEURS 1981). So although the probable route for water across the blood-brain barrier is through, rather than between, the endothelial cells, it has been suggested that there is a para-cellular as well as a transcellular route (CRONE 1984), and that this para-cellular route would be shared by small ions. It is also possible for the

---

Abbreviations

D	Diffusion coefficient	$\mathrm{cm\,s^{-2}}$
R	Gas constant	$\mathrm{erg\,K^{-1}\,mol^{-1}}$
T	Temperature	K
$C(C_v, C_a)$	Concentration	$\mathrm{mol\,l^{-1}}$
J_w, J_s, J_v	Flux (water, solute, volume)	$\mathrm{(mol,\ cm^3)\,s^{-1}}$
μ	Chemical potential	$\mathrm{mol\,l^{-1}}$
A	Area	$\mathrm{cm^2}$
Δx	Path length	cm
P_D, P_F	Permeability (diffusional, filtration)	$\mathrm{cm\,s^{-1}}$
V_w	Partial molar volume (water)	$\mathrm{cm^3\,mol^{-1}}$
p	Hydrostatic pressure	$\mathrm{dyn\,cm^{-2}}$
L_p	Hydraulic conductivity	$\mathrm{dyn\,cm^{-1}\,s^{-1}}$
F	Blood flow	$\mathrm{cm^3\,s^{-1}}$
PS	Permeability surface area product	$\mathrm{cm^3\,s^{-1}}$
E	Extraction $= (1 - C_v/C_A)$	
σ	Reflection coefficient to Ringer solution solutes	
π	Osmotic pressure of Ringer solution solutes	$\mathrm{dyn\,cm^{-2}}$
n	Quantity of solute	mol

endothelial cells to have a relatively high water permeability and yet restrict ions. The purpose of this article is to review a number of the experiments carried out in recent years that contribute to current understanding on the passage of water across this extremely tight endothelium.

A gradient in chemical potential of water across a membrane will result in a flow. The origin of the chemical potential gradient may be due to the concentration of isotopically labelled water, hydrostatic pressure, or to differences in total solute concentration. The flow may be either diffusional, when the solvent molecules cross by dissolving into the membrane, or bulk flow (osmotic, or quasi-laminar or more simply non-diffusional; MAURO 1965) when the molecules pass through channels which allow passage of more than one abreast. In some instances it is possible to measure the diffusive permeability independently from the bulk flow permeability, and differences in these measures have been used to detect the presence of water-filled membrane channels, most notably in mammalian erythrocytes which have been contrasted to those of birds. The diffusive and osmotic permeabilities have been estimated in the blood-brain barrier, but there are difficulties with these attempts which arise from extending the analysis derived from single artificial membranes on rigid supports between two well-mixed compartments, to two membranes in series and an effectively stagnant interstitial space.

B. The Nature of Permeability

The relationship between hydrostatic pressure and diffusion in causing flow across membranes with fluid-filled channels was extensively analysed by MAURO (1957, 1965). Essentially he showed that diffusion could account for only a small proportion of the total volume flux, but this depended on the type of membrane used. His argument may be outlined in the following way. The general form of the diffusion equation is:

$$\frac{\mathrm{d}n}{dt} = \frac{DA}{RT} C \frac{\mathrm{d}\mu}{\mathrm{d}x} \tag{1}$$

where $\mathrm{d}\mu/\mathrm{d}x$ is the gradient of chemical potential. If we consider the movement of tracer water in the absence of any pressure gradient as the indicator of diffusive permeability, we may consider the chemical potential alone, so for dilute solutions of labelled water the change in chemical potential for a fractional change in concentration will be:

$$\mathrm{d}\mu = RT \frac{\mathrm{d}C}{C} \tag{2}$$

This may be substituted in Eq. 1, so long as the concentration gradient across the membrane is uniform, and this gives the familiar equation:

$$\frac{dn}{dt} = J_w = D_w \frac{A}{\Delta x} \Delta C \tag{3}$$

which leads to the definition of diffusive permeability:

$$P_D = D_w \frac{A}{\Delta x} \tag{4}$$

If we now consider pressure driven flows, such that the flow is entirely diffusive (i.e. *not* through pores or channels) we may state:

$$C\overline{V_w} = 1 \tag{5}$$

and

$$d\mu = \overline{V_w} dp \tag{6}$$

since water only is being considered. If it is assumed that the pressure gradient across the membrane is constant, substitution in Eq. 1 gives:

$$\frac{dn}{dt} = J_w = D \frac{A}{\Delta x} \frac{\Delta p}{RT} \tag{7}$$

but

$$L_p = \frac{J_v}{\Delta p} = \frac{J_w \overline{V_w}}{\Delta p} = \frac{P_D \overline{V_w}}{RT} \tag{8}$$

hence:

$$L_p \frac{RT}{\overline{V_w}} = P_F \tag{9}$$

where P_F is known as the filtration permeability. Thus it is possible to obtain values for L_p from experiments in which a volume flux is obtained for a hydrostatic or osmotic pressure gradient and to compare them directly with values from diffusion experiments which measure P_D. Robbins and Mauro (1960) showed convincingly that for artificial collodion membranes the P_F values were far in excess of those for P_D, thereby demonstrating the presence of water channels.

C. Water Permeability of the Blood-Brain Barrier

The permeability of the blood-brain barrier has been estimated in a variety of ways to give measures of P_D and P_F, and Table 1 summarizes these findings and compares them with those from erythrocytes and cultured pulmonary artery endothelial cells. The erythrocyte values are particularly illuminating as they show the effects of water channels. Human red cells have a protein (localized in band 3 of sodium dodecyl sulphate gels; Ojcius

Table 1. Diffusive and osmotic permeability coefficients for cerebral endothelium and single membranes

Tissue	P_D $\times 10^5$ cm s^{-1}	P_F $\times 10^5$ cm s^{-1}	Reference
		BLOOD-BRAIN BARRIER	
Rabbit		117[a]	FENSTERMACHER and JOHNSON (1967)
Rat	9–25[a]		BOLWIG and LASSEN (1974)
Rhesus monkey	19		EICHLING et al. (1974)
Dog	4		PATLAK and FENSTERMACHER (1975)
Man	24	104	PAULSON et al. (1977)
Frog		276	FRASER and DALLAS (1990)
		MEMBRANES	
Edothelial cells	304	317	GARRICK et al. (1988)
Black lipid film	106	114	CASS and FINKELSTEIN (1967)
Human erythrocytes	530	1730	FETTIPLACE and HAYDON (1980) (review)
Human erythrocytes + CMBSA[b]	180	200	
Chicken erythrocytes	135	210	

[a] Recalculated for capillary surface area 100 $cm^2\,g^{-1}$.
[b] *p*-chloromericuribenzosulphonic acid.

and SOLOMON 1988) which apparently allows water to cross them in a non-diffusional manner. This protein may be blocked by *p*-chloromericuri-benzosulphonic acid, when the human red cells have P_F and P_D values similar to the chicken cells. At first inspection the values for the blood-brain barrier are consistent with the idea that there are water-specific channels, with the ratio P_F/P_D ranging from 4 to 70, but the experiments should be examined critically before the idea can be accepted.

I. Diffusive Permeability

It had been widely assumed that the uptake of labelled water into the brain was entirely flow-limited (e.g., YUDILEVICH and DE ROSE 1967; OLDENDORF 1970), as in the heart (YIPINTSOI and BASINGTHWAIGHTE 1970) and skeletal muscle (JOHNSON et al. 1952). But when blood flow was above 30 ml 100 g^{-1} min^{-1} there was some evidence of restriction to the free diffusion of water (EICHLING et al. 1973), and the measurement of this restriction became the basis for estimates of P_D. The theoretical basis upon which these measurements rest was developed by RENKIN (1959) and CRONE (1963). The ideas that underlie this technique are that the microvasculature of an organ may be regarded as being composed of exchange vessels organized in parallel, so that they may be represented as being one idealized capillary, and that the rate of loss of a substance across the capillary wall can, by using the Fick principle, be estimated from the rate of blood flow and the arterio-venous concentration difference.

$$J_s = F(C_a - C_v) = PS\,\bar{C} \tag{10}$$

Since the rate of loss of the substance is proportional to its concentration, its concentration profile along the axis of the capillary will be exponential, and hence the above expression will reduce to

$$C_v = C_a e^{-\frac{PS}{F}} \tag{11}$$

from which, by taking logarithms and rearranging:

$$PS = -F\ln\left(\frac{C_v}{C_a}\right) = -F\ln(1 - E) \tag{12}$$

The usual method for carrying out these permeability measurements is to administer a mixture of test and reference substances as an arterial bolus injection and to measure their relative venous concentrations in time as the bolus emerges from the organ. The tissue blood flow could either be measured directly in perfused preparations, or from the washout of a highly diffusible radioactive gas such as xenon. This method involves a number of assumptions, some of which are crucial to any successful analysis. One is that all the exchange vessels can be properly modelled as one capillary. This can be overcome by using a tracer that does not leave the vasculature as a reference, so that its concentration at the venous end of the microcirculation will give a good estimate of the arterial concentration of the test substance. The tissue must also be homogeneously perfused or else the interstitium may be rapidly loaded with test solute from fast flowing capillaries, and then either unload it to slower flowing ones, or shunt it directly into venous vessels ahead of that retained in the circulation. A third condition, which is difficult to overcome or to verify, is that the tissue concentration of the test substance is negligible at the time the measurements are made, otherwise the extraction will be a net value rather than an estimate of the unidirectional flux. Usually it is assumed that these conditions are fulfilled if the ratios of test to reference tracer are taken at early times of the outflow and the ratio permeability–surface area product/blood flow (PS/F) < 0.2 (FENSTERMACHER 1984), but it is much more satisfactory if it can be shown that the measurement of *PS* is independent of flow, i.e., that the loss of test substance is entirely diffusion limited (ALVAREZ and YUDILEVICH 1969).

Three studies have been carried out using this approach in order to estimate the diffusional permeability of water, in the rat by BOLWIG and LASSEN (1974), and in man (PAULSON et al. 1977). They used essentially similar methods for estimating the arterial concentration of water by using a non-diffusible tracer (sucrose and Cl^- respectively) from the venous concentrations. The rates of tissue perfusion were estimated from previously published data in the rat study and from ^{133}Xe clearance in the human experiments. The disadvantage with using a soft gamma emitter is that the blood flow was measured from cortical regions only, whereas the venous samples would have been derived from the whole brain. However, this is unlikely to have had much affect on the values obtained since, as the authors themselves appreciated, that although there was some diffusional restriction, there was also considerable flow limitation to water uptake by the tissue, which necessarily led to P_D being underestimated.

EICHLING et al. (1974) used a different approach, but seem to be mistaken in their interpretation of their data, and for this reason their analysis should be considered in some detail. They adopted an interesting technique: using the positron emitting isotope of oxygen ^{15}O to label water, an intracarotid bolus injection was given and the high energy gamma emission from positron-electron extinctions was measured from the whole brain by external monitoring. They found that the washout of the injected bolus was best described by two exponential curves, a rapid phase which had the same time course as labelled erythrocytes and a slower phase. They argued that the rapid phase was due to water that remained in the circulation and had not had time to diffuse into the tissue. They found that the relationship between blood flow and extraction was a simple linear one, of the form $E = a - bF$, where a and b are constants. The ratio between the extrapolated initial values of the two exponentials was taken to be the extraction, and this is probably an error since no account had been made for any inhomogeneity of the microvasculature. It is interesting to note that xenon washouts from the cerebral circulation also do not follow simple mono-exponential curves, but there has been no suggestion that the removal of this gas is limited by diffusion barriers (BRODERSEN et al. 1973). It has now become apparent that the cerebral microvasculature is heterogeneous: some regions with particularly high capillary tortuosity being bypassed by more direct pathways (PAWLIK et al. 1981; CHANG et al. 1981). More recently KNUDSEN et al. (1990) have shown that this circulatory arrangement has important implications for blood-tissue transfer of glucose. However, EICHLING et al. (1974) made a much more serious mistake when they extended Eq. 12 to generate a linear plot of $\ln(1 - E)$ against $1/F$ with $-PS$ as gradient without establishing whether the relationship was applicable: in other words, whether their measure of extraction was a true estimate of unidirectional flux. In fact, examination of Eq. 11 shows that when $PS/F \ll 1$, then $E \approx PS/F$. Thus extraction must be proportional to the reciprocal of flow and not, as their own data indicated, flow itself.

PATLAK and FENSTERMACHER (1975) developed an intriguing technique which they used to give an estimate of P_D. Essentially, they perfused the ventricles with solutions containing radioactive tracers for up to 4 h, after which the brain was rapidly removed and the caudate nucleus removed and the tissue concentration of the tracers measured orthogonally with the ventricular surface. This steady-state tissue concentration gradient was used to give an estimate of the rate at which tracer was removed to the blood and thence to calculate a transfer function from which a measure of *PS* was derived. As can be seen from Table 1, the P_D value is about one-fifth of that obtained from other techniques. It is not easy to give a good theoretical reason for this large discrepancy, particularly since the other techniques were likely to underestimate P_D. These results may be explained by a heterogeneous microvascular blood flow, or perhaps a lower tissue distribution volume than was expected. It is more likely that the 3 min delay between the cessation of blood flow and removal and freezing of the tissue would allow sufficient smearing of the original diffusion profile to give an erroneously low value and, furthermore, the process of freezing the tissue may have resulted in some movement of water and its tracer.

In conclusion there have been a number of attempts to quantify P_D for cerebral endothelium in vivo, but these have all given values considerably lower than would be expected from the measurement obtained from cells in culture (see Table 1), and this is arguably due not to any real difference in the nature of the cells but to the limitations of the techniques used in the measurements in the whole brain.

II. Osmotic Permeability

There have not been many attempts at measuring P_F, mainly because FENSTERMACHER and JOHNSON (1967) produced an appealing experimental design to attack this question. The first measure in the literature was that made by COULTER (1958) who used a volumeter sealed to the skull of the experimental animal, and measured the rate of fluid flux into the brain in response to changes in applied hydrostatic pressure. He assumed that the skull was a completely sealed box, and ignored other routes for fluid movement, such as CSF in the spinal cord, or even arteries and veins entering and leaving the brain. It is probably for this reason that his values for P_F were higher than that of even mesenteric microvessels. FENSTERMACHER and JOHNSON (1967) used osmotic pressure loads of between 15 and 65 $mmol\,l^{-1}$ in the blood to produce the volume changes recorded in a similarly designed volumeter. They found that when using polar lipid-insoluble substances that were not transported into the brain (sucrose and raffinose), the rate of fluid movement from the subdural space to the volumeter was consistent with a very high reflection coefficient, and thus argued that it represented the volume flux across the cerebral endothelium. This may be largely correct, but the method assumes that the brain tissue behaves as a perfect osmometer, and, as CSERR et al. (1987) have pointed out, the rate of cerebral cell volume regulation is important, and occurs at a similar rate to the gross fluid movements measured in the volumeter. Similarly, the assumption that CSF volume remained constant is probably wrong, since there will be some direct water movement across the choroid plexus and the arachnoid. It was also disappointing that they failed to measure the blood osmolarity in the experiments, and merely relied on calculations based on the concentration changes of the experimental test substances. Thus, as cells adjust to the effects of a hyperosmotic environment they take up Na^+ and K^+, so reducing the plasma osmolarity (Cserr et al. 1987). This was overlooked and so the trans-endothelial osmotic gradient was overestimated. These oversights probably had effects in opposite directions and the overall result may not have been seriously different because of them, but some doubt must be attached to the value of P_F obtained.

Paulson et al. (1977) estimated P_F in man by using a variant of the osmotic transient technique. They gave a short bolus intracarotid injection of hyperosmotic mannitol (1100 mOsm kg^{-1}) mixed with $^{36}Cl^-$ and measured the transendothelial water flux from their relative dilutions in jugular venous blood from the activities of circulating radiolabelled albumin and $^{36}Cl^-$. This was combined with the measured blood flow (xenon clearance) and venous osmolarity and hence they were able to estimate the volume flux for a known osmotic pressure gradient. This approach answers a number of criticisms levelled at Fenstermacher and Johnson (1967), but suffers from the fact that blood flow was measured in the cortex alone, while the dilutions occurred over the whole brain. Additionally, the problem with a single pass osmotic transient is that the bulk of the material effect would have been preferentially distributed to regions of high blood flow, and would so lead to some underestimate in the value obtained.

Heterogeneity is of course the great difficulty that can never be entirely overcome in whole organ experiments, and it is for this reason that the author, inspired by the late Christian Crone's experiments on measuring the electrical resistance of frog pial microvessels (Crone and Olesen 1982), embarked on measurements of filtration coefficient in the same preparation (Fraser and Dallas 1990). Single capillaries were viewed in the microscope and were cannulated and perfused with a frog Ringer solution which contained the small polar fluorescent dye carboxy-fluorescein. The surface of these vessels, which are essentially naked (Bundgaard 1982), was

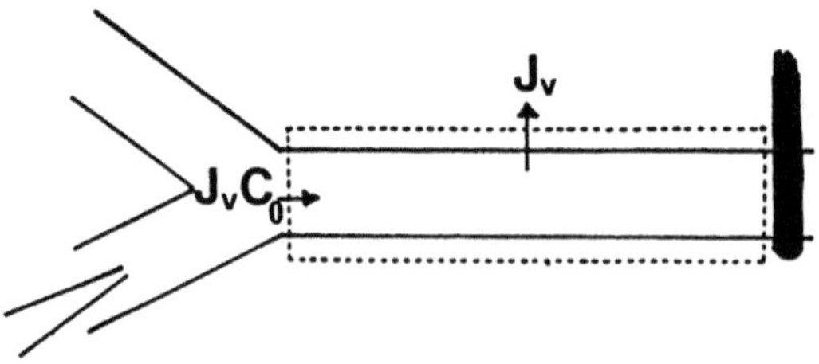

Fig. 1. A perfusion pipette and occluding needle during an experiment on a frog pial microvessel. Fluid is extracted across the wall at rate J_v and is replaced by dye at rate J_vC_o. The dye concentration was measured within the *dotted* area

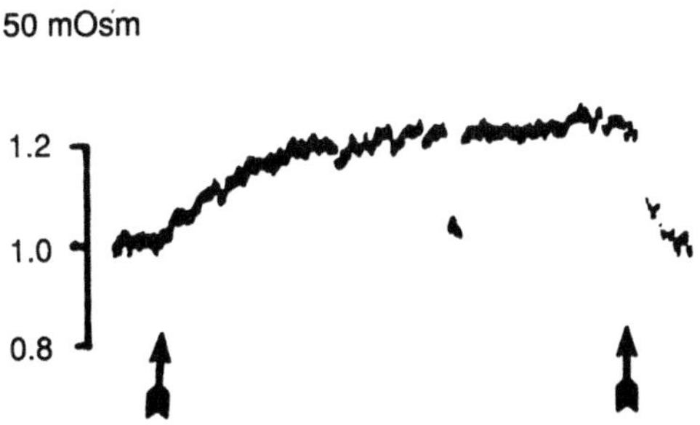

Fig. 2. Relative carboxyfluorescein concentration during an occlusion experiment. A $50\,\mathrm{mmol\,l^{-1}}$ sucrose solution was added to the normal Ringer solution, and the occluding needle blocked the vessel between the *two arrows*. Concentration rose as fresh dye containing fluid was drawn into the segment to replace that extracted by the external hyperosmolarity

superfused with a similar solution, but not containing dye, while the vessels were occluded by a fine glass needle. If the dye concentration remained constant the vessel was deemed to be tight and the superfusing solution was replaced by one made $50\,\mathrm{mosmol\,l^{-1}}$ hyperosmotic with sucrose. On re-occlusion the dye concentration in the trapped segment rose as fresh dye-containing solution replaced the water drawn across the vessel wall by the osmotic gradient (Figs. 1, 2). The concentration change followed the mono-exponential curve

$$C_t = C_\infty - (C_\infty - 1)\,e^{-kt} \tag{13}$$

where

$$K = \frac{2}{r} L_p\, \sigma \pi \tag{14}$$

and C_∞ was predicted from the ratio of the original osmotic pressures in the two solutions. The predicted value of C_∞ was used to check the experimentally observed value, and there was good agreement if the osmotic load was $50\,\mathrm{mosmol\,l^{-1}}$ or less. In these cases it was felt safe to assume that $\sigma = 1$, and hence the filtration coefficient and P_F were calculated from the rate constant of the exponential. The axial concentration of the dye in the occluded capillary segment was measured at different times during the experiment and was found to conform to a mathematical model that solved the convective diffusion equations that were expected to describe those conditions (Fig. 3).

Crone (1984) found that the bi-ionic potentials for Na^+ and K^+ were similar to those in free solution and argued that this demonstrated the

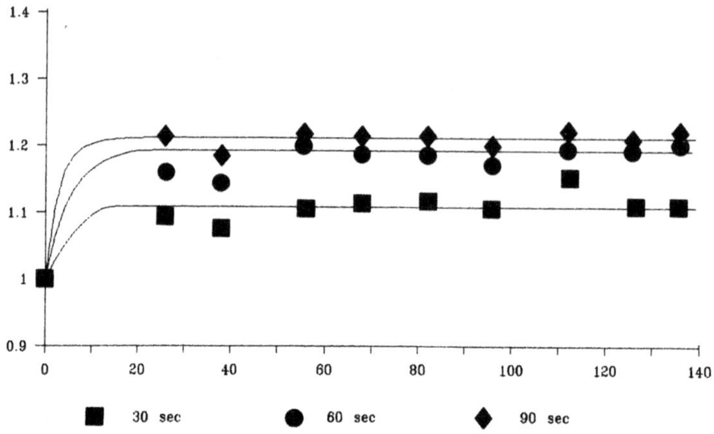

Fig. 3. Distribution of dye concentration at 30, 60 and 90 s after the beginning of an occlusion along the length of the vessel. The *horizontal scale* is in micrometers and the *vertical* is normalized arbitrary units of fluorescence. The *continuous lines* are those predicted from a model of these experiments

presence of a paracellular pathway. Such a pathway is not inconsistent with the occlusion studies if, at the maximum limit, it carries no more water than the transcellular route. However, Crone's results must be open to question since he may well have been using leaky microvessels. When he cannulated the microvessels for the ionic potential measurements he did not test them for the characteristic tightness of the barrier, and Fraser and Dallas (1990) found that only 9 out of 95 cannulated vessels were tight.

The main drawbacks to the single vessel occlusion technique are that it was limited to the brain's surface and that it proved to be very difficult to find tight vessels, presumably owing to trauma. This highlights the contrasting advantages of single vessel and whole organ experiments. In the single vessel experiments it is possible to perform well designed studies with clearly defined boundary conditions, but errors due to sampling give rise to uncertainty. On the other hand, in the whole organ experiments any heterogeneity vitiates the experimental design and creates the uncertainty of how well the values obtained truly represent a mean value for the organ as a whole.

D. Conclusion

The question whether there are specific water channels in the cerebral endothelium has not been answered definitively. Paulson et al. (1977) concluded that the difference in their P_D and P_F values did not represent a real difference, and this author concurs with that view, particularly since there is no difference in the values for cultured endothelial cells (Garrick et al. 1988). Patlak and Paulson (1981) argued that the slight differences in the values in brain endothelium may be due not to specific water channels, or to unstirred layer effects on diffusive permeability, but to unstirred layer effects on filtration permeability, if the P_F of the two sides of the endothelium is different. This could be tested by measuring P_F in the same vessels and changing the sign of the osmotic challenge. Indeed, the P_F values for the rabbit and man, where the water flow was from brain to blood, are lower than those found in the frog where the water flow was in the opposite direction. However, it is just as likely that the differences in technique, let alone that between species, are responsible for this result.

References

Alvarez AO, Yudilevich DL (1969) Heart capillary permeability to lipid-insoluble molecules. J Physiol (Lond) 202:45–58

Bolwig TG, Lassen NA (1974) The diffusion permeability to water of the rat blood-brain barrier. Acta Physiol Scand 93:415–422

Bradbury MWB (1979) The concept of the blood-brain barrier. Wiley, New York

Brightman MW, Reese TS (1969) Junctions between intimately apposed cell membranes in the vertebrate brain. J Cell Biol 40:648–677

Broderson P, Sejrsen P, Lassen NA (1973) Diffusion bypass of xenon in brain circulation. Circulation 32:363–369

Bundgaard M (1982) Ultrastructure of frog cerebral and pial microvessels and their impermeability to lanthanum ions. Brain Res 241:57–65

Cass A, Finkelstein A (1967) Water permeability of thin lipid membranes. J Gen Physiol 50:1765–1784

Chang B-L, Yamakawa T, Nuccio J, Pace R, Bing RJ (1981) Microcirculation of left atrial muscle, cerebral cortex and mesentery of the cat. Circ Res 50:240–249

Coulter NA (1958) Filtration coefficient of the capillaries of the brain. Am J Physiol 195:459–464

Crone C (1963) The permeability of capillaries in various organs as determined by use of the 'indicator dilution' method. Acta Physiol Scand 58:292–305

Crone C (1984) Lack of selectivity to small ions in cerebral and muscle capillaries of the frog. J Physiol (Lond) 353:317–337

Crone C, Olesen S-P (1982) Electrical resistance of brain microvascular endothelium. Brain Res 241:49–55

Cserr HF, Bundgaard M (1984) Blood-brain interfaces in vertebrates: a comparative approach. Am J Physiol 246:R277–R288

Cserr HF, DePasquale M, Patlak CS (1987) Regulation of brain water and electrolytes during acute hyperosmolality in rats. Am J Physiol 253:F530–F537

Eichling JO, Reichle ME, Grubb RL, Ter-Pogossian MM (1974) Evidence of the limitations of water as a free diffusible tracer in brain of the rhesus monkey. Circ Res 35:358–364

Fenstermacher JD (1984) Volume regulation of the central nervous system. In: Staub NC, Taylor AE (eds) Edema. Raven, New York, pp 383–404

Fenstermacher JD, Johnson JA (1967) Filtration and reflection coefficients of the rabbit blood-brain barrier. Am J Physiol 211:341–346

Fettiplace R, Haydon DH (1980) Water permeability of lipid membranes. Physiol Rev 60:510–550

Fraser PA, Dallas AD (1990) Measurement of filtration coefficient in single cerebral microvessels in the frog. J Physiol (Lond) 423:343–361

Garrick RA, DiRisio DJ, Giannuzzi R, Cua WO, Ryan US, Chinard FP (1988) The osmotic permeability of isolated calf pulmonary artery endothelial cells. Biochim Biophys Acta 939:343–348

Johnson JA, Cavert HM, Lifson N (1952) Kinetics concerned with distribution of isotopic water in isolated perfused dog heart and skeletal muscle. Am J Physiol 171:687–693
Knudsen GM, Pettigrew KD, Paulson OB, Hertz MM, Patlak CS (1990) Kinetic analysis of blood-brain barrier transport in man: quantitative evaluation in the presence of tracer backflux and capillary heterogeneity. Microvasc Res 39:28–49
Mauro A (1957) Nature of solvent transfer in osmosis. Science 126:252–253
Mauro A (1965) Osmotic flow in a rigid porous membrane. Science 149:867–869
Ojcius DM, Solomon AK (1988) Sites of p-chloromericuribenzosulfonic acid inhibition of red cell urea and water transport. Biochim Biophys Acta 942:73–82
Oldendorf WH (1970) Measurement of brain uptake of radiolabeled substances using a tritiated water internal standard. Brain Res 24:372–376
Patlak CS, Fenstermacher JD (1975) Measurements of dog blood-brain transfer constants by ventriculocisternal perfusion. Am J Physiol 229:877–884
Patlak CS, Paulson OB (1981) The role of unstirred layers for water exchange across the blood-brain barrier. Microvasc Res 21:117–127
Paulson OB, Hertz MM, Bolwig TG, Lassen NA (1977) Filtration and diffusion of water across the blood-brain barrier in man. Microvasc Res 13:113–124
Pawlik G, Rackl A, Bing RJ (1981) Quantitative capillary topography and blood flow in the cerebral cortex of cats: an in vivo microscopic study. Brain Res 208:35–58
Renkin E (1959) Transport of potassium-42 from blood to tissue in isolated mammalian skeletal muscles. Am J Physiol 197:1205–1210
Robbins E, Mauro A (1960) Experimental study of the independence of diffusion and hydrodynamic permeability coefficients of colloidin membranes. J Gen Physiol 43:523–532
Van Deurs B (1980) Structural aspects of blood-brain barriers, with special reference to the permeability of the cerebral endothelium and choroidal endothelium. Int Rev Cytol 65:117–191
Yipintsoi T, Basingthwaighte JB (1970) Circulatory transport of iodoantipyrine and water in the isolated dog heart. Circ Res 27:461–477
Yudilevich DL, De Rose N (1967) Blood-brain transfer of glucose and other molecules measured by rapid indicator dilution. Am J Physiol 220:841–846

CHAPTER 4

Blood-Brain Glucose Transfer

A. GJEDDE

A. Brief History

Under normal circumstances, a mixture of α- and β-D-glucopyranose (D-glucose) is the only fuel of brain energy metabolism (PARDRIDGE 1983). The breakdown of D-glucose is regulated by complex mechanisms that influence the activities of phosphofructokinase and hexokinase. The metabolism depends indirectly on the glucose concentration of the brain intracellular and interstitial fluids. Since the tissue glucose concentration, in turn, depends directly on the glucose phosphorylation rate, the blood-brain transfer assumes a pivotal role in the supply of glucose to the brain.

The interface between the vascular and extravascular spaces in brain is the *barrière hèmato-encéphalique*, originally so named by STERN and GAUTIER (1921) because of its uncertain location. In a much quoted passage, AUGUST KROGH (1946) suggested that the endothelial lining of the brain vasculature may function as an extended cell membrane. The suggestion raised the question of whether the paracellular route past the cerebral capillary endothelium, i.e., the transport between the endothelial cells, is restricted to such an extent that it unmasks the properties of endothelial cell membranes in general, or whether the brain capillary endothelial membranes themselves have unusual properties that render them particularly restrictive to some classes of compounds, and specifically permeable to others.

The concept of facilitated diffusion evolved in parallel with the study of the transport of electrolytes and polar nonelectrolytes across cellular membranes. The nature of the specific glucose-transporting properties of all membranes remained a mystery until the discovery of insulin heralded the study of facilitated diffusion. By 1940, many workers speculated that insulin acted to facilitate the transfer of glucose across muscle cell membranes by interaction with a "specific" property of the membrane (LUNDSGAARD 1939; LEVINE et al. 1949). In 1948, LEFÈVRE showed that glucose efflux from human erythrocytes was a function of the intracellular glucose concentration in the manner predicted by the MICHAELIS-MENTEN equation (MICHAELIS and MENTEN 1913). Although it was apparent that human erythrocytes must possess these "specific" properties in abundance, the relation to insulin, if any, remained obscure.

Table 1. Mammalian glucose transporter genes

Isoform	Insulin effect	Tissue or membrane	K_t (mM)
glut-1	–	$HepG_2$/erythrocyte/BBB	1
glut-2	–	Hepatocytes, β-cells, basolateral membranes	20
glut-3	–	Fetal skeletal muscle cells/brain tissue	<1
glut-4	+	Muscle cells/adipocytes	5
glut-5		Brush-border membranes	1–2
glut-6		Nonexpressed pseudogene	

In an abstract, CRONE (1960) reported that the blood-brain transfer of D-glucose appeared to be much faster than predicted from the properties of the physically similar fructose. He also noted that the rate of transfer appeared to change in inverse proportion to the D-glucose concentration, as expected of substances subject to facilitated diffusion (CRONE 1965). In the years between 1960 and 1980, the transfer was shown to obey the Michaelis-Menten criteria of saturability, competitive inhibition, and stereospecificity. Since 1980, both insulin-dependent and non-insulin-dependent glucose transporters have been isolated, identified, sequenced, and cloned. Although a mechanism of insulin action has been proposed, it is uncertain how the insulin-insensitive glucose transport is regulated, if at all.

B. Brain Endothelial Glucose Transporter

I. Molecular Biology

Glucose transporters are grouped on the basis of their sodium-dependence and hence on the energy requirements of the transport. The sodium-dependent glucose transfer through epithelial brush-border membranes is driven by hydrolysis of adenosine triphosphate (ATP). The inward transport concentrates glucose sufficiently to allow this nutrient to cross the basolateral membrane by passive, facilitated diffusion. Of the putative family of passive, facilitative glucose transporter genes, the six members listed in Table 1 have been identified (KAYANO et al. 1988, 1990). GERHART et al. (1989) and PARDRIDGE et al. (1990b) showed that the glucose transporter of the cerebral capillary endothelium is of the glut-1 isoform. Unlike these investigators, BAGLEY et al. (1989) found evidence of glut-l in neuropil, as well as in endothelium, but the contribution may be quantitatively insignificant.

KASAHARA and HINKLE (1976, 1977) first purified the transporter in human erythrocyte membrane, using reconstitution of D-glucose transport in erythrocyte ghosts as a criterion. Cytochalasin B inhibits transport of hexoses across cell membranes (KLETZIEN et al. 1972) by specific binding to

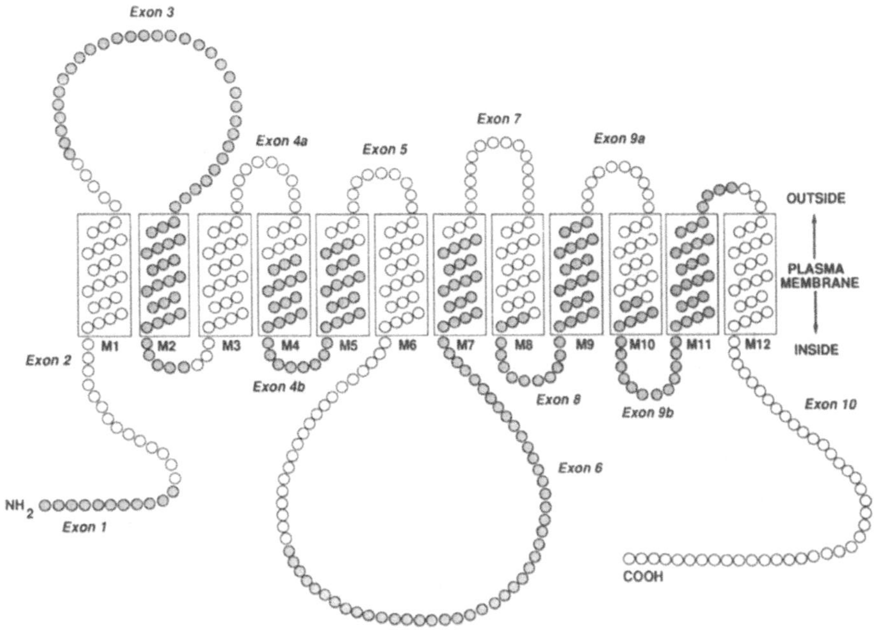

Fig. 1. Blood-brain glucose transporter. Model for exon-intron organization of ancestral facilitative glucose-transporter gene. *Circles* indicate amino acids encoded by each of 12 exons in ancestral gene. Putative membrane-spanning α-helices are numbered *M1–M12*. (From BELL et al. 1990)

a protein component of the membrane with an affinity constant of 100 n*M* (LIN and SPUDICH 1974). When cytochalasin B was found to be a potent non-competitive inhibitor of glucose transport in human erythrocytes by TAVERNA and LANGDON (1973), photoaffinity labeling with this mold metabolite, and purification, identified a 54000 $g\,mol^{-1}$ integral component of human erythrocyte membranes (BALDWIN et al. 1982; CARTER-SU et al. 1982) and isolated rat, porcine, and ovine brain microvessels (BALDWIN et al. 1984; DICK et al. 1984) with the properties of a glucose transporter. DICK et al. (1984) also found that the microvessel density of glucose transporters was 20-fold higher than that of the particulate fraction of cerebral cortex.

Using an antibody raised against the human erythrocyte transporter, MUECKLER et al. (1985) cloned a transporter cDNA from the human hepatoma HepG2 line and proposed the model shown in Fig. 1. BIRNBAUM et al. (1986) isolated a cDNA encoding a rat brain glucose transporter with an amino acid sequence 98% identical to that of the HepG2 protein (MATTAEI et al. 1987b; FUKUMOTO et al. 1988). Subsequently, the gene was shown to be expressed also in peripheral nerve and choroid plexus epithelium and to reside on chromosome 1 with 35000 bases, 10 exons and 9 introns

(FROEHNER et al. 1988; SHOWS et al. 1987; BALDWIN and LIENHARD 1989; GERHART et al. 1989).

The glut-1 transporter is expressed in most tissues (PILCH 1990), particularly in cultured cells and during fetal life (ASANO et al. 1988; FUKUMOTO et al. 1988; WERNER et al. 1989). Postnatally, glut-1 expression declines in most tissues other than placenta, kidney, and brain (BELL et al. 1990), although oncogenic transformation causes reexpression of the gene (THORENS et al. 1988). It is interesting that the glut-1 protein belongs to the glucose-regulated family of stress-inducible proteins (WERTHEIMER et al. 1991). In brain, the glut-1 transporter is confined to the capillary endothelium. In other endothelia, it is only expressed in the testes in which the capillary endothelium is similarly restrictive as in brain (HARIK et al. 1990).

If, in brain, the glut-1 transporter is expressed only at the blood-brain barrier (BOADO and PARDRIDGE 1990; PARDRIDGE et al. 1990b), it is possible that glut-3 may be the gene expressed in brain tissue itself. Glut-3 encodes for a glucose-transporter in fetal muscle and fat cells, and both glut-1 and glut-3 encode glucose transporters with comparatively low transport MICHAELIS constants ($K_t \sim 1\,\mathrm{m}M$).

It is possible that an as yet unidentified second glucose transporter with a higher K_t is present in brain endothelial cells but there is no evidence that the gene of this transporter is glut-2 which encodes the glucose transporter of hepatocytes, the basolateral membrane of small intestine epithelial cells, the distal tubule cells of the kidney, and the β-cells of the islets of Langerhans. This gene resides on chromosome 3 in humans and is 52% homologous with glut-1 (THORENS et al. 1988; FUKUMOTO et al. 1989). Glut-2 expresses a transporter with a K_t of $20\,\mathrm{m}M$ which is difficult to reveal in vivo in the presence of higher-affinity transporters, also because cytochalasin B binds to glut-2 with comparatively low affinity (AXELROD and PILCH 1983).

Glut-4, the gene of the insulin-sensitive glucose transporter of muscle and fat cells, has been localized to chromosome 17 in humans and appears to be 65% homologous with glut-1 (JAMES et al. 1988, 1989; KAYANO et al. 1988; BIRNBAUM 1989; CHARRON et al. 1989). In view of the discussion of an effect of insulin on blood-brain glucose transfer, it is interesting that there is evidence against the presence of the insulin-sensitive glucose transporter in endothelial membranes (SLOT et al. 1990).

II. Molecular Kinetics

The molecular mechanism underlying the facilitated diffusion of glucose has been the subject of much debate. The discovery of the transporter's molecular structure did not provide evidence to reject the mobile carrier model originally proposed for the glucose transfer by WIDDAS (1952). In fact, there is considerable evidence against the alternative single-file pore that the molecular structure appears to support (CUNNINGHAM et al. 1986, 1989; CARRUTHERS 1990). The mobile carrier theory has taken the form

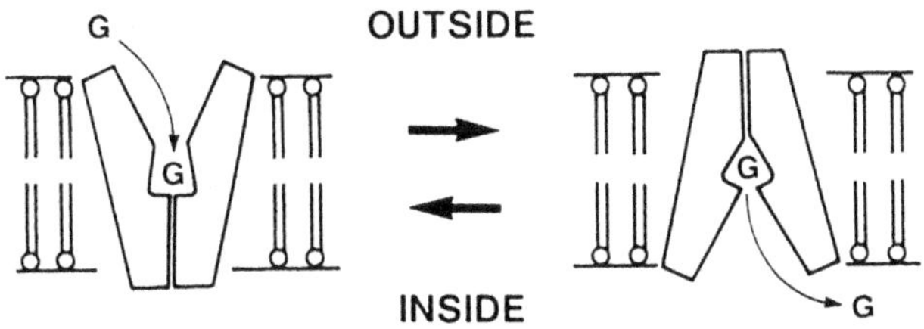

Fig. 2. Blood-brain glucose transporter. Schematic representation of the "alternating conformation" model of the glucose transporter (the inner and outer glucose-binding site may be composed of completely different domains.) (From OKA et al. 1990)

either of an "alternating conformation" model (Fig. 2), in which the carrier can be reached from only one side of the membrane at a time (WIDDAS 1952; OKA et al. 1989), or of a "fixed-sites" model, in which the carrier can be approached and occupied from both sides of the membrane simultaneously (BAKER and WIDDAS 1973; HELGERSON and CARRUTHERS 1989; NAFTALIN 1988). The latter model departs from the mobile carrier concept by allowing several transport sites to be exposed to both sides of the membrane simultaneously. CUNNINGHAM et al. (1989) tested the alternating-conformer model by simulation of two membranes in series, separated by a minute, well-mixed endothelial space. By using the nomenclature proposed by STEIN (1986), and kinetic constants estimated by LOWE and WALMSLEY (1986), the simulation confirmed the experimental estimates of an apparent MICHAELIS constant (K_t) of 8 mM for the transendothelial transport when the half-saturation concentration of the carrier, i.e., the affinity measured in the absence of glucose on the "trans" side of the membrane (zero-trans condition) was 7 mM. Again, this value is considerably higher than expected for the glut-1 transporter.

Studies of the binding of cytochalasin B to isolated brain microvessels have yielded densities of glucose transporters between 1 and 10 pmol mg^{-1} endothelial cells (DICK et al. 1984; BALDWIN et al. 1985; DICK and HARIK 1986; MATTHAEI et al. 1986; KASANICKI et al. 1987; HARIK et al. 1988; KALARIA et al. 1988; MORIN et al. 1988; PARDRIDGE et al. 1990b; MOORADIAN and MORIN 1991), 1 mg being the approximate quantity of endothelial cells in 1 g of brain tissue (GJEDDE 1983). CUNNINGHAM et al. (1989) calculated a transporter quantity of 40 pmol necessary to generate the known maximum transendothelial flux of 2–4 μmol g^{-1} min^{-1}. The significance of this discrepancy is not clear. It may indicate that cytochalasin binding is incomplete (MORIN et al. 1988).

III. Structural Requirements of Glucose Transport

OLDENDORF (1971), BETZ et al. (1975), and CREMER et al. (1983) studied a series of hexoses and other competitors for blood-brain glucose transport to

Table 2. Inhibitory constants of glucose transport inhibitors relative to Michaelis constant of glucose

Carbon	Inhibitor	BBB[a]	Other membranes[b]
C1	1-Deoxyglucose		10
	1-Methylglucose		High
C2	2-Deoxyglucose	0.7	
	2-Fluoro-2-deoxyglucose	0.6	
	2,2-Dichloro-2-deoxyglucose		1
	2-O-Methylglucose		15
C3	3-Fluoro-3-deoxyglucose		1
	3-O-Methylglucose	1.1	1
	Mannose	1.7	
	3-Deoxyglucose		10
C4	4-O-Propylglucose		1
	Galactose	2.7	13
C6	6-Fluoro-6-deoxyglucose		0.2
	6-Chloro-6-deoxyglucose	0.3	
	6-Chloro-6-deoxymannose	0.4	
	6-Tosyl-6-deoxygalactose	0.7	
	6-Chloro-6-deoxygalactose	0.9	
	6-Deoxyglucose		1
	Xylose	Infinity	Infinity
	Fucose	Infinity	Infinity

[a] From PARDRIDGE and OLDENDORF (1975); BETZ et al. (1975); CREMER and CUNNINGHAM (1979); FUGLSANG et al. (1986); CRANE et al. (1983); PARDRIDGE (1983); GJEDDE (1984a).
[b] From CARRUTHERS (1990).

determine the degree of its specificity. In addition, deoxy- and fluorodeoxyglucose have been particularly well studied because of their use in measurements of brain glucose metabolism (SOKOLOFF et al. 1977; PHELPS et al. 1979). As shown in Table 2, the relative affinities for the substrates compare well with those observed in erythrocytes and other organs. The affinities indicate that the hydroxyl groups of carbon atoms 3 and 4 are particularly important for identification of the molecule by the transporter. There is evidence that the orientation is preserved during transport such that the molecule enters the transporter from the extracellular space with C-6 facing forward and from the intracellular space with C-1 facing forward (BALDWIN and LIENHARD 1981).

C. Theory of Blood-Brain Glucose Transfer

The study of glucose transport across the blood-brain barrier has been complicated by several kinetic problems that include:

1. The flow-dependent relationship between clearance and permeability,
2. the time-dependent relationship between "first pass" and net extraction, i.e., the problem of bidirectional flux, and

3. the concentration-dependent relationship between apparent permeability and maximal transport.

These issues have never been dealt with in their entirety in a single publication, although important elements have been discussed by LUND-ANDERSEN (1979), CUNNINGHAM and CREMER (1981), GJEDDE (1982, 1983), LASSEN and GJEDDE (1983), PARDRIDGE (1983), and CUNNINGHAM et al. (1986, 1989).

I. Apparent Permeability and Flux

The quantity of glucose transported from blood to brain is a function of an apparent endothelial permeability (P, in units of $\mathrm{cm\,s^{-1}}$ or $\mathrm{nm\,s^{-1}}$), the area of endothelial surface available for transport (S, in $\mathrm{cm^2\,g^{-1}}$ or $\mathrm{ml^{-1}}$), and the concentrations of glucose on the two sides of the interface. The apparent permeability is measurable only as a clearance which is not directly proportional to the permeability–surface area product because tracer must escape from the capillary during transit.

A second assumption is made that the net rate of tracer transfer across the endothelium equals the difference between the unidirectional rates of transfer. From the definition of transport by passive diffusion, it follows that

$$\Delta J(t) = P_1 S \frac{\bar{C}_c(t)}{\alpha} - P_2 S \frac{M_e(t)}{V_d} \tag{1}$$

in which $\Delta J(t)$ is the net flux across the membrane, $\bar{C}_c(t)$ is the weighted average concentration of the tracer in capillary blood or plasma, P_1 the apparent blood-brain permeability, P_2 the permeability to tracer returning from the brain to blood, V_d the solvent (water) volume of distribution in brain, α the solubility of the tracer in blood or plasma relative to water, and $M_e(t)$ the quantity (mass) of exchangeable tracer glucose in brain.

Specific Activity. The average capillary tracer concentration $\bar{C}_c(t)$ varies both with the arterial tracer concentration $C_a(t)$ and the concentration of exchangeable tracer glucose in the brain interstitial fluid. This exchange causes the specific activity to fall from the arterial to the venous end of the capillary. CRONE (1963), JOHNSON and WILSON (1966), CHRISTENSEN et al. (1982), and GJEDDE (1980, 1982) derived solutions for the tracer concentration in the venous effluent, $C_v(t)$. When, for simplicity, $P = P_1 S/\alpha$, and $p = P_2 S/V_d$, the solution is

$$C_v(t) = C_a(t)\, e^{-P/F} + M_e(t) \frac{p}{P}\left(1 - e^{-P/F}\right) \tag{2}$$

where F is the blood flow. The equation indicates that the specific activity of tracer glucose will decline from the arterial to the venous end of the capillary and that this decline will be time-dependent (Fig, 3; LASSEN and

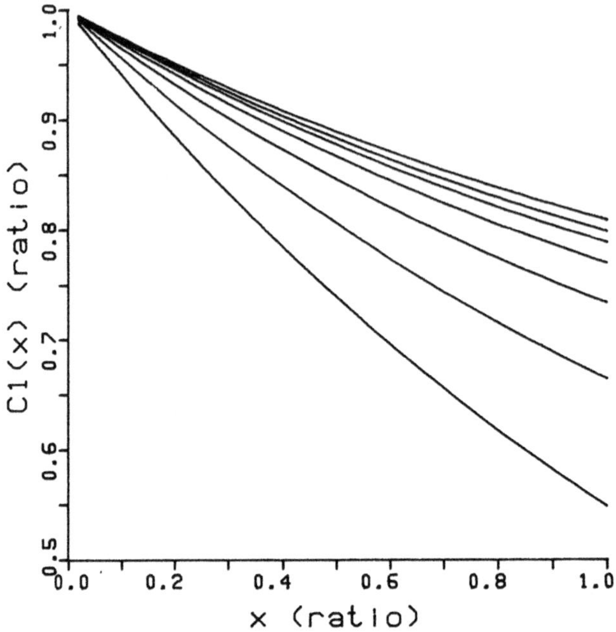

Fig. 3. Effect of capillary passage on specific activity of glucose tracer. Concentration profile in capillary of tracer glucose for T_{max} = 400 and glucose consumption = 70 µmol 100 g^{-1} min^{-1}, C_a = 9 mM, K_t = 7 mM, and F = 40 ml 100 g^{-1} min^{-1} The *lines* (from the *bottom*) represent the times "0", 1, 2, 3, 4, 5, and 45 min. (From Lassen and Gjedde 1983)

Gjedde 1983). After weighed integration of Eq. 2 and insertion in Eq. 1, the net exchange is given by

$$\Delta J(t) = P\left(\frac{1 - e^{-P/F}}{P/F}\right) C_a(T) - p\left(\frac{1 - e^{-P/F}}{P/F}\right) M_e(t) \tag{3}$$

where the term $1 - e^{-P/F}$ is the extraction fraction, $E(o)$, for unidirectional transfer of tracer. Equation 3 yields the definitions

$$K_1 \equiv P\left[\frac{1 - e^{-P/F}}{P/F}\right] = F(1 - e^{-P/F}) = E(o)F \tag{4}$$

and

$$k_2 \equiv p\left[\frac{1 - e^{-P/F}}{P/F}\right] = \frac{F}{V_e}(1 - e^{-P/F}) = \frac{E(o)F}{V_e} \tag{5}$$

where K_1 is the initial ("unidirectional") clearance of tracer glucose, and k_2 the fractional clearance of the brain interstitial fluid. From these definitions, it follows that $K_1/k_2 = P/p$, defined as the partition volume V_e.

Together, Eq. 2 and Eq. 3 indicate that the total gain of radioactivity in the brain equals the net transfer of tracer glucose across the capillary endothelium at all times. The gain of tracer per unit time is given by

$$\frac{dM(t)}{dt} = K_1 C_a(t) - k_2 M_e(t) \tag{6}$$

where $M(t)$ is the content of tracer glucose or tracer-derived molecules in brain. Equation 6 can be solved directly only when either (1) all extravascular tracer glucose is exchangeable for the duration of the study, i.e., $M(t) = M_e(t) + V_0 C_a(t)$, or (2) no extravascular tracer is exchangeable, i.e., $M_e(t) \simeq 0$. In the former case,

$$\frac{dM(t)}{dt} = (K_1 + k_2 V_0)\, C_a(t) - k_2 M(t) \tag{7}$$

where V_0 is a vascular volume of distribution, rather than the vascular volume itself. In the latter case,

$$\frac{dM(t)}{dt} = K_1 C_a(t) \tag{8}$$

Note that by definition, $M(t)$ is never zero.

Initial Versus Steady-State Influx. According to Eq. 1, the tracer flux is unidirectional when

$$\Delta J(t) \cong \frac{P_1 S}{\alpha} \bar{C}_c(t) \tag{9}$$

It is convenient to refer to this flux as the "influx", $\vec{J}(t)$. Initially when $M_e(t) = 0$, it follows from Eq. 2 that:

$$\bar{C}_c(t) = \left[\frac{1 - e^{-P/F}}{P/F}\right] C_a(t) \tag{10}$$

Hence, initially,

$$\vec{J}(t) = K_1 C_a(t) \tag{11}$$

At steady-state, when $M_e(t) \neq 0$, $\bar{C}_c(t) = C_a(t) - \dfrac{\Delta J}{2F}$

where ΔJ is the net transfer of native glucose across the endothelium. Hence, at steady-state,

$$\vec{J} = PC_a - \Delta J \frac{P}{2F} \tag{12}$$

where $\vec{J}$ is the influx and C_a the arterial concentration of native glucose. Equation 12 reduces to $\vec{J} \cong PC_a$ for $\Delta J \ll FC_a$. Steady-state rates of influx must be calculated from values of P obtained under unidirectional circumstances (Eqs. 7 or 8) and calculated by means of Eq. 12. In the past, experimental conditions were often far from unidirectional, and erroneous steady-state fluxes were calculated with Eq. 11.

The unidirectionality of transfer can be evaluated by solving Eq. 2 for the extraction fraction at time t, as defined by GJEDDE (1982):

$$E(t) \equiv \frac{C_a(t) - C_v(t)}{C_a(t)} = (1 - e^{-P/F})\left(1 - \frac{M_e(t)}{V_e C_a(t)}\right) \tag{13}$$

according to which $E(t) = E(o)$ for $M_e(t) = 0$. Hence, "first pass" extraction means negligible tissue content, allowing Eq. 6 to be solved for $M_e(t) \cong 0$ (Eq. 8; CRONE 1963).

The fraction of tracer extracted in (rather that at) the time T is

$$E(T) \equiv 1 - \left[\int_0^T C_v(t)\mathrm{d}t \Big/ \int_0^T C_a(t)\mathrm{d}t\right] = M_e(T) \Big/ \left[F\int_0^T C_a(t)\mathrm{d}t\right] \tag{14}$$

For low values of T, by integration of Eq. 2, and substitution (GJEDDE 1982),

$$\frac{E(T)}{E(o)} = 1 - E(T)\frac{F\int_0^T M_e(t)\mathrm{d}t}{V_e M_e(\mathrm{T})} \cong 1 - E(T)\frac{FT}{2V_e} \tag{15}$$

which describes the relative backflux in the period after the injection. If the measured extraction is 40% in 20 s, if flow is 0.5 ml min^{-1}, and if the partition volume is 0.5 ml, backflux is 7% in 20 s. In 3 min, it would be 60%. For longer periods, however, the equation underestimates the backflux.

II. Facilitated Diffusion

1. Michaelis-Menten Equation

Facilitation of diffusion must involve ligand recognition as the initial step. The criteria of facilitation are therefore the same as for pure receptor-ligand interactions and include saturability, stereospecificity, and competitive inhibition, as well as evidence of the consequent counter-transport.

Transport rates in vivo usually reveal a lower affinity between the hypothetical receptor and the substrate than described for pure receptor-ligand interactions (GJEDDE and BODSCH 1987). The reason for the low affinity is plain from MICHAELIS' and MENTEN's (1913) original steady-state solution of the enzyme kinetic equation, observed by LEFÈVRE (1961) also to describe the transport of tracer glucose into red blood cells:

$$\vec{J} = \frac{T_{max} C_a}{K_t + C_a} \tag{16}$$

where $\vec{J}$ is the unidirectional flux, K_t the MICHAELIS or half-saturation concentration, T_{max} the maximal transport rate, and C_a the external steady-state substrate concentration. The equation is the equilibrium solution of a differential equation describing the change per unit time of the quantity of substrate bound to the transporter. The change of bound substrate is the difference between substrate associating with, and dissociating from, the receptor. The speed of association reflects the likelihood of a ligand molecule joining the receptor from the vicinity of a site (k_a), multiplied by the number of molecules and the number of receptor sites. The speed of dissociation reflects the combined likelihood of dissociation (k_d) by return to the substrate solution (k_r) and transport through the membrane (k_t), multiplied by the number of receptor-ligand complexes. For "true" receptors, the k_d reflects only the rate of return to the solution and hence is much lower. For transporters, the k_d reflects both processes. The maximal transport rate is $k_t B_{max}$ where B_{max} is the total number of sites. The actual transport rate is $k_t B$ where B is the number of receptor-ligand complexes. If $\vec{J}$ is substituted for $k_t B$, T_{max} for $k_t B_{max}$, and K_t for $(k_r + k_t)$, Eq. 16 is obtained.

2. Michaelis-Menten Constants

Bachelard et al. (1973), Pardridge and Oldendorf (1975), and Gjedde (1980) defined the apparent permeability of the endothelium to substrates of facilitated diffusion on the basis of Eq. 16:

$$P = \frac{P_1 S}{\alpha} \equiv \frac{T_{max}}{K_t + C_a} \tag{17}$$

From this equation, it follows that simultaneous measurements of P and $\vec{J}$ are linearly related. Thus $\vec{J}$ is a linear function of P:

$$\vec{J} = -K_t P + T_{max} \tag{18}$$

in which T_{max} is the ordinate intercept and $-K_t$ the slope. Equation 18 is identical to the Eadie-Hofstee (Eadie 1942; Hofstee 1952) or Scatchard (1949) plots of enzyme reaction velocity. An example of the use of Eqs. 17 and 18 to evaluate T_{max} and K_t is given in Fig. 4 (Crone 1985).

3. Partition Volume

Assuming symmetrical transport, the fractional rate of transfer from the brain tissue to the circulation is given by

$$p = \frac{P_2 S}{V_d} \equiv \frac{\alpha T_{max}/V_d}{K_t + C_e} \tag{19}$$

in which K_t remains the half-saturation concentration of the tracer relative to whole-blood or plasma (hence α) and C_e is the concentration of native glucose in tissue water. Division yields the P/p ratio for facilitated diffusion of a single substrate, defining the tracer partition volume:

$$V_e = \frac{K_1}{k_2} = \frac{P}{p} = \frac{V_d}{\alpha}\left[\frac{K_t + C_e}{K_t + C_a}\right] \tag{20}$$

Equation 20 predicts that the partition volume must be the same for native glucose and all glucose tracers with identical values of T_{max}. For glucose for which $\alpha \cong 1$, the equation reduces to

$$V_e = \frac{V_d K_t + M_e}{K_t + C_a} \tag{21}$$

where M_e is the tissue content of native glucose. This equation was used by GJEDDE and DIEMER (1983) to calculate the glucose content of brain tissue on the basis of measurements of V_e and C_a when K_t and V_d were known.

III. Multiple Membranes

The estimates of K_t and T_{max} according to Eq. 17 or 18 are empirical measurements with an uncertain relation to the properties of the individual membranes interposed between the circulation and the cytosol. Glucose and glucose tracers must pass both membranes of the capillary endothelium as

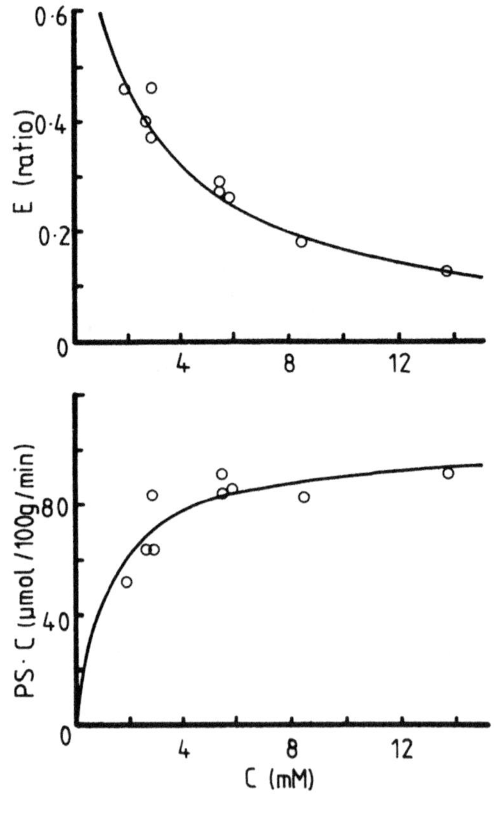

Fig. 4. Blood-brain glucose transport in dog. The *upper part* of the figure shows the initial unidirectional extraction of labeled D-glucose in the dog's brain during a single passage. The *abscissa* shows the concentrations of glucose in the blood. The *lower part* shows the calculated unidirectional transport rate of various glucose concentrations. (Modified from CRONE 1965)

well as the membranes of glial and neuronal cells before they can react with metabolic enzymes. At least four resistances may be important (i.e., both membranes of the endothelium, the cell membrane, and the hexokinase reaction), and probably one or more unstirred layers. The half-time of endothelial equilibration is of the order of 0.1 s (GJEDDE and CHRISTENSEN 1984); the half-time of extracellular fluid equilibration 0.5–2 min (LUND-ANDERSEN and KJELDSEN 1977), depending on the relative resistances of cell membrane transfer and phosphorylation by hexokinase; and the half-time of total brain water equilibration, in absence of CO_2 loss, close to 10 min (GJEDDE 1982; BRØNDSTED and GJEDDE 1988). It is a question whether the time of equilibration is defined entirely by the size of the water space, or whether the glial and neuronal cell membranes impose an additional delay, relative to the rate of diffusion in water. A problem is also whether the cell membranes impose a delay on the equilibration relative to the rate of metabolism which would tend to establish a steady-state concentration difference between the extra- and intravascular space.

1. Endothelial Membranes

The simple Michaelis-Menten kinetic analysis lumps the two membranes of the cerebral capillary endothelium. While the combination assumes that the endothelial transfer of glucose is so rapid, and the volume of the intra-endothelial space so small, that the tracer in the endothelium achieves almost instant equilibration with the tracer in the capillary, the kinetic constants nonetheless depend on the individual constants of the two membranes.

The consequences of using a single Michaelis-Menten expression for the quantitation of transendothelial flux were examined by PAPPENHEIMER and SETCHELL (1973) and GJEDDE and CHRISTENSEN (1984). The latter authors concluded that steady-state between the blood and the endothelial cells was reached in 0.1 s. They formulated a model of two endothelial cell membranes with Michaelis-Menten properties and showed that the constants of the transendothelial transport are simple functions of the properties of the individual membranes when transport across the endothelium is rapid ($P > 10\,\mathrm{nm\,s^{-1}}$):

$$T_{\max} = \frac{J^a_{\max} J^b_{\max}}{J^a_{\max} + J^b_{\max} + \dfrac{\Delta J(K^b_t - K^a_t)}{K^a_t}} \tag{22}$$

and

$$K_t = \frac{J^a_{\max} K^b_t + J^b_{\max} K^a_t + \Delta J(K^b_t - K^a_t)}{J^a_{\max} + J^b_{\max} + \dfrac{\Delta J(K^b_t - K^a_t)}{K^a_t}} \tag{23}$$

where a and b refer to the individual membranes, and $J_{\max}$ and K_t refer to their Michaelis constants. In the same analysis, the Michaelis-Menten constants determined in tracer experiments described the facilitated diffusion in the steady-state only when the two membranes had identical kinetic properties (Gjedde and Christensen 1984). By examination of published relations between blood and brain glucose, the authors discovered no discrepancy between tracer and steady-state measurements and hence concluded that the two membranes appeared to have identical glucose transport properties. A more detailed discussion of the molecular implications for the two-membrane endothelium has been presented by Cunningham et al. (1989). For symmetrical transport, i.e., $K_t^b = K_t^a$ and $J_{max}^b = J_{max}^a$, Eqs. 22 and 23 can be used to establish both the ratio between T_{max} and ΔJ and the value of K_t by nonlinear regression analysis. The $T_{max}/\Delta J$ ratio is 4 in awake rats (Gjedde and Christensen 1984).

2. Glial and Neuronal Cell Membranes

Almost all models of blood-brain glucose transfer have assumed that the transfer across cell membranes is so rapid that glucose concentrations are equal in extra- and intracellular fluid.

That this is not necessarily so was borne out by Raichle et al. (1975) who used coincident detection of positron emission to determine net and unidirectional blood-brain glucose transfer in the anaesthetized monkey. The uptake was followed every second for several minutes and yielded a glucose distribution space of no more than 15% of the brain water space. In a study using the double-indicator diffusion method (see Table 3), Knudsen et al. (1990) also concluded that the initial glucose distribution in brain roughly equals the volume of the extracellular space. Using the model assuming non-mixing of tracer in the extravascular compartment (unstirred compartment), the latter authors attempted to determine the distribution of tracer glucose in the endothelium but the size of the space yielded by their three-compartment model was too large to be consistent with endothelial distribution. Using a similar model, but now assuming complete mixing in the extravascular compartment, in order to determine the magnitude of initial distribution in the extracellular space, the authors obtained a value of approximately 15% of the water space.

In contrast to the high time-resolution studies that revealed a possible resistance at the cell membrane, low time-resolution studies have suggested that cell membrane transport is not rate limiting in the steady state. The studies of Lund-Andersen and Kjeldsen (1976, 1977) concluded that glucose tracer equilibration between the extra- and intracellular compartments is immeasurably fast in vitro, rendering diffusion in the extracellular space rate-limiting for transport into the intracellular space. The minimum value for the coefficient of membrane transfer was at least an order of magnitude greater than the phosphorylation coefficient of 2-deoxyglucose.

The question of the relationship between cell membrane transport of glucose and the cellular metabolic rate for this nutrient has not been answered by steady-state methods. Measurements of brain glucose concentrations in rodents indicate that the space in which native glucose is dissolved in the steady-state must be much greater than the extracellular space if a concentration gradient is to exist between plasma and tissue glucose. The steady-state distribution of 3-*O*-methylglucose, a nonmetabolizable hexose, averages 60% of total brain water in both rodents and humans, and has been shown to approach 100% of the brain water volume in hyperglycemic states (GJEDDE and DIEMER 1983; BROOKS et al. 1986b), indicating saturation of the luminal glucose transporter and distribution of 3-*O*-methylglucose in the entire water volume in the brain. Also, the study of DIEMER et al. (1985) speaks in favor of the distribution of glucose in the entire water phase of brain tissue because it yielded a glial and neuronal cell membrane permeability to glucose close to that observed for the blood-brain barrier per unit surface area, indicating a much greater total permeability because of the much greater cellular surface area (5000-fold).

D. Evidence of Blood-Brain Glucose Transfer

It has been extraordinarily difficult to obtain consistently accurate and precise measurements of blood-brain glucose transport in any species. In many studies, the change of glucose transport under abnormal circumstances turned out to be less than the variation observed for normal values between different studies. For this reason, some disagreement still remains about normal and abnormal values and in rodents and humans.

I. Methods

The final common path for all measurements of glucose transport is the determination of K_1, the unidirectional clearance, either by direct measurement, or as the product of an extraction fraction and blood or plasma flow. In most newer studies, the PS product and hence the flux (P_1SC_a/α), have been calculated from K_1 according to Eq. 18.

1. Operational Equations

Clearance. Direct measurements of clearance can be made by sampling arterial perfusate and brain tissue simultaneously. The equation underlying the calculation of K_1 is one of several solutions to Eq. 7:

$$M(T) = (K_1 + k_2V_0)e^{-k_2T}\int_0^T C_a(t)e^{k_2t}\mathrm{d}t \tag{24}$$

Equation 24 can be solved by nonlinear regression analysis of at least three separate observations. However, statistically, multilinear regression analysis may be preferable to nonlinear regression. By integration of Eq. 7, BLOMQVIST (1984) derived a sum of integrals:

$$M(T) = K_1\left(1 + \frac{V_0}{V_e}\right)\int_0^T C_a(t)\mathrm{d}t - \left(\frac{K_1}{V_e}\right)\int_0^T M(t)\mathrm{d}t \tag{25}$$

where $C_a(t)$ is the arterial concentration, V_e the partition volume, and V_0 the initial volume of distribution. If the magnitude of one or more of the parameters is independently known, the number of required observations can be reduced proportionately. In experiments with a single observation of $M(T)$, Eq. 24 can be solved directly for K_1 when k_2 is known:

$$K_1 = \frac{M(T)}{\int_0^T C_a(t)e^{k_2(T-t)}\mathrm{d}t} - V_0k_2 \tag{26}$$

For short periods of circulation, when k_2 is unknown, this equation reduces to the solution of Eq. 8 first introduced by SCHAEFER et al. (1976) and BLASBERG et al. (1978):

$$\lim_{T\to 0} K_1 \cong \frac{M(T)}{\left(1 + \frac{V_0}{V_e}\right)\int_0^T C_a(t)\mathrm{d}t} = \frac{M(T) - V_0C_a(T)}{\int_0^T C_a(t)\mathrm{d}t} \tag{27}$$

Equation 25 is soluble by multilinear regression only when a sufficient number of measurements are available. If each measurement is a separate experiment, Eq. 25 can be normalized to obtain comparable measures of virtual volumes of distribution in each experiment:

$$V(T) = \frac{M(T)}{C_a(T)} = K_1\left[\left(1 + \frac{V_0}{V_e}\right)\frac{\int_0^T C_a(t)\mathrm{d}t}{C_a(T)} - \left(\frac{1}{V_e}\right)\frac{\int_0^T M(t)\,\mathrm{d}t}{C_a(T)}\right] \tag{28}$$

When V_e is sufficiently large, the equation shrinks to the normalized solution of Eq. 14:

$$V(T) = K_1\,\Theta\,(T) + V_0 \tag{29}$$

where $\Theta(T)$ is the normalized integral $\int_0^T C_a(t)\mathrm{d}t/C_a(T)$ which has unit of time and equals time when the tracer concentration in plasma is constant. $V(T)$ is a linear function of $\Theta(T)$ with a slope of K_1 when the $\int_0^T M(t)\mathrm{d}t/V_e$ ratio remains negligible compared to $\int_0^T C_a(t)\mathrm{d}t$. Equation 29 underlies the "slope-intercept" or "Patlak" plot (GJEDDE 1981b, 1982; PATLAK et al. 1983). Continued linearity of this plot signifies absent backflux. Examples are shown in Fig. 5 for a series of glucose analogs entering neonatal rat brain (FUGLSANG et al. 1986).

Extraction Fractions. Extraction fractions can be measured directly only by sampling arterial and venous perfusate of the brain simultaneously, using Eq. 13 to calculate $E(t)$, or Eq. 14 to calculate $E(T)$. Extraction fractions can also be calculated indirectly as the ratio between clearance and perfusion rate.

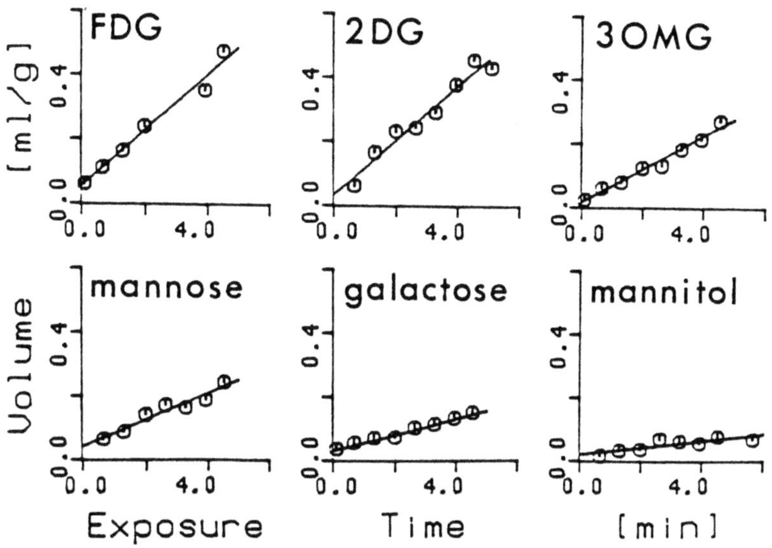

Fig. 5. Blood-brain glucose transport in rat. Apparent distribution volume in brain of five glucose analogs and mannitol (a plasma marker) as function of time. The five hexoses include fluorodeoxyglucose (*FDG*), deoxyglucose (*2DG*), and 3-*O*-methylglucose (*3OMG*). *Abscissae*, blood integrals, normalized against concentration, weighted exposure time (min); *ordinates*, brain/blood radioactivity ratios. Each *point* is one observation. Each *graph* represents one litter. (From FUGLSANG et al. 1986)

2. Experimental Procedures

The methods applied to living tissue can roughly be divided in two groups, including in the first group all methods requiring in or ex vivo sampling of tissue, and in the second group the methods that require sampling of cerebral venous blood. In the former group, the methods can be further subdivided in two groups, one including the methods in which the brain is perfused by native blood, the other including the methods that employ an artificial perfusate (Table 3). In Table 3, the abbreviations correspond to the descriptions below.

IND. The indicator diffusion method (CRONE 1963; LASSEN et al. 1971) requires that a mixture of test and reference substance be sampled in cerebral venous blood after intracarotid injection. The method yields an estimate of $E(t)$. The reference substance must remain in the vasculature to provide the measure of $C_a(t)$ in Eq. 13. The time resolution of the method is excellent but the blood-brain transfer values have an uncertain relation to individual brain regions or even the brain as a whole. The intracarotid injection also posed ethical problems in normal volunteers.

Table 3. Methods of studying blood-brain glucose transfer

Tissue sampling		Venous sampling
Artificial perfusate	Native perfusion	
Intracarotid bolus (BUI)	Intravenous bolus (INT)	Intracarotid bolus (IND) Intravenous bolus (IVB)
Intracarotid perfusion (ICP)	Intravenous infusion (INF)	

IVB. For the reasons indicated above, as well as to take capillary heterogeneity and backflux into account, the indicator diffusion method was recently modified for intravenous, rather than arterial, bolus injection in rats (SAWADA et al. 1989) and humans (KNUDSEN et al. 1990).

BUI. In 1970, OLDENDORF published a modified method of intracarotid bolus injection of a mixture of a test tracer and a reference tracer, usually labeled water. As such, it represents the inverse of the indicator diffusion method. According to Eq. 14, the ratio between the integrated extraction fraction [E(T)] of two tracers equals the ratio between the quantities of the tracers measured in brain tissue. If the first-pass extraction of the reference tracer is unity, the ratio becomes a measure of the extraction of the test tracer, assuming negligible backflux of both tracers. To fulfill the latter assumption, brain contents were measured 15 s (later 5 s) after the injection. Because the first-pass extraction of labeled water may be less than unity, Oldendorf referred to the ratio as the brain uptake index (BUI). Many subsequent users of the method attempted to calculate true integrated extraction fractions by correcting the BUI for an assumed first-pass extraction of water of less than unity, choosing fractions between 0.43 (GJEDDE and RASMUSSEN 1980a) and 0.85 (OLDENDORF and BRAUN 1976).

The particular attraction of Oldendorf's method was the claim that the cerebral vasculature is cleared of native blood by the intracarotid injection. This feature allowed Oldendorf and his associates to complete an impressive series of determinations of Michaelis constants for the blood-brain transfer of almost all known substrates of facilitated diffusion (OLDENDORF 1971; PARDRIDGE 1983). Only recently has the complete separation of native blood and injectate been disputed (PARDRIDGE et al. 1985) and the method is in any case unsuitable for blood flow determination, necessary for calculation of *PS* products.

INT. Several authors used SAPIRSTEIN's (1958) indicator fractionation principle to devise an intravenous bolus injection version of Oldendorf's method in which extraction fractions could be measured by comparing brain uptake of two different tracers and clearances could be measured by comparing the brain uptake with the integrated arterial concentration. This method, now commonly known as the integral method, appeared in different forms, one designed for blood flow measurements (SCHAEFER et al. 1976; VAN UITERT and LEVY 1978) but with no provision for backflux of highly diffusible tracers, and another designed to yield blood-brain transfer rates (BACHELARD et al. 1973; DANIEL et al. 1978; GJEDDE et al. 1980). The intravenous bolus injection method obtains the tracer delivery by mechanical integration of the arterial concentration and subsequent calculation of the kinetic circulation time Θ on the basis of the integral and the final concentration (GJEDDE 1981b, 1982; PATLAK et al. 1983). The advantage of this method is the feasibility of measuring blood flow simultaneously. In the later experiments, tracer glucose was administered as a rapid intravenous bolus and the experiment terminated after 10 or 20 s to avoid backflux (GJEDDE 1980).

INF. In the earlier experiments, the tracer often circulated for a minute or more. The glucose tracer was infused continuously in a manner calculated to yield a constant arterial level of tracer glucose (PATLAK et al. 1983; DANIEL et al. 1978) according to Eq. 29 in which Θ is then real time.

MET. In a variation of the integral method based on the glucose metabolism method of SOKOLOFF et al. (1977), CUNNINGHAM and CREMER (1981) determined glucose clearance (K_1) and metabolism simultaneously by neurochemical separation of glucose tracer and metabolites in brain. In a later treatment, the authors extended the method to include calculation of the *PS* product from the clearance (HARGREAVES et al. 1986).

ICP. The method of isolating and independently perfusing the vasculature of the brain via the intracarotid artery provided a combination of the intracarotid and intravenous administration methods, by means of which the rate of perfusion and composition of the perfusate can both be controlled and glucose transfer determined, either by bolus administration of the tracer in the perfusate and venous rather than tissue sampling (BETZ et al. 1973), or by continuous infusion of the tracer and tissue sampling at the end of the experiment (TAKASATO et al. 1984; ZLOKOVIC et al. 1986). Despite the controlled infusion, the method may require that the investigators measure perfusate flow simultaneously. The advantage is the control of perfusate composition.

3. The Blood-Brain Barrier In Vitro

Microvessels have been separated from other components of brain tissue and studied in vitro (BETZ et al. 1979; JOÓ 1985). Glucose and glucose analog 'uptake' has been measured with labeled tracers, and the number of glucose binding sites has been estimated with cytochalasin B (BETZ et al. 1983a). As discussed above, isolated microvessels with a surface–volume ratio at least 1000-fold higher in vitro than in vivo are not suited for transport studies that last more than a few seconds. For this reason, the results of the uptake studies (e.g., KOLBER et al. 1979) throw little light on

Table 4. Blood-brain glucose transfer studies in different species

Species	Reference	Year(s)
Mouse	GROWDON et al.	1971
	HORTON	1973
Gerbil	BETZ and IANNOTTI; BETZ et al.	1983; 1983b
Rat	BIDDER	1968
	BUSCHIAZZO et al.	1970
	OLDENDORF	1971
	BACHELARD et al.	1973
	BRENDER et al.; PARDRIDGE and OLDENDORF	1975
	DANIEL et al.	1975b, 1978, 1980
	CREMER et al.	1976, 1979, 1981, 1983
	SOKOLOFF et al.	1977
	NEMOTO et al.	1978
	CUNNINGHAM and SARNA; CREMER and CUNNINGHAM; POLLAY and STEVENS; SARNA et al.	1979
	GJEDDE	1980, 1981b, 1982, 1984a
	GJEDDE and RASMUSSEN	1980a,b
	GJEDDE et al.	1980, 1981
	T.G. CHRISTENSEN et al.; CUNNINGHAM and CREMER; GJEDDE and CRONE	1981
	CORNFORD et al.; MCCALL et al.	1982
	PARDRIDGE et al.	1982, 1990a
	GJEDDE and DIEMER	1983, 1985b
	HAWKINS et al.	1984
	BRAUN et al.; CRANE et al.; GJEDDE and LAURITZEN	1985
	LAMANNA and HARIK	1985, 1986
	FUGLSANG et al.; CUNNINGHAM et al.; HARGREAVES et al.; MANS et al.	1986
	PLANAS and CUNNINGHAM; NAMBA et al.	1987
	DUCKROW; HARIK and LAMANNA	1988
	PELLIGRINO et al.	1990a,b
Cat	CUTLER and SIPE	1971
Rabbit	AGNEW and CRONE	1967
	PAPPENHEIMER and SETCHELL	1973
	BERSON et al.	1975
	BRAUN et al.	1980
	CORNFORD et al.	1982
	CORNFORD and CORNFORD	1986
Dog	CRONE	1965
	YUDELEVICH and ROSE	1971
	BETZ et al.	1973, 1974, 1975
	BETZ and GILBOE	1974
	DREWES et al.	1977
	KINTNER et al.	1980
	O. CHRISTENSEN et al.	1982
Sheep	PAPPENHEIMER and SETCHELL	1973
Monkey	RAICHLE et al.	1975
Man	HERTZ et al.	1981
	HERTZ and PAULSON	1982
	BLOMQVIST et al.	1985, 1991
	GJEDDE et al.	1985
	BROOKS et al.	1986a,b
	FEINENDEGEN et al.	1986
	GUTNIAK et al.; KNUDSEN et al.	1990

the blood-brain transfer of hexoses (BRADBURY 1985). In the future, cultured endothelial cells from the brain may provide an alternative to in vivo studies of transendothelial flux.

II. Normal Values in Awake Subjects

Table 4 lists most of the studies of blood-brain glucose transfer in a number of species. Originally, the observations with the tissue sampling method were made after a minute or more, without correction for backflux (BACHELARD et al. 1973; DANIEL et al. 1978). In addition, glucose transfer rates were calculated from the clearance rather than the *PS* product, and the subjects were invariably anaesthetized. For these reasons, only values obtained in awake subjects after about 1980 can reasonably be argued to represent normal values, as listed in Tables 5–7.

1. Permeability and Flux in Normoglycemia

The first demonstration of facilitated diffusion of glucose across the blood-brain barrier was provided by CRONE (1960) who used the indicator diffusion method. The result is illustrated in Fig. 4, redrawn from CRONE (1965) by CRONE (1985). The graph shows a Michaelis-Menten-type relation between the unidirectional transport of glucose from the circulation to brain tissue and the plasma glucose concentration at which the transport occurs. The flux of glucose was calculated from the fractions of extracted glucose shown in the upper graph, indicating a maximal glucose transport rate close to 90 μmol $100\,g^{-1}\,min^{-1}$ and a Michaelis half-saturation constant close to 2 m*M*.

Average Gray Matter Values. From Table 5, the following conclusions can be drawn: In the awake adult rat, the gray matter blood-brain glucose transfer is close to 200 μmol $100\,g^{-1}\,min^{-1}$ at a glucose concentration of about 8 m*M*, indicating a *PS* product of approximately 25 ml $100\,g^{-1}\,min^{-1}$, depending on the region of the brain. The glucose influx is therefore approximately twice as high as the net utilization of 100 μmol $100\,g^{-1}\,min^{-1}$ (SOKOLOFF et al. 1977). In resting humans, the influx, measured with glucose, 3-*O*-methylglucose, or fluorodeoxyglucose, is close to 50 μmol $100\,g^{-1}\,min^{-1}$ at 4 m*M*, or approximately twice the net consumption as well, corresponding to a *PS* product of 12.5 ml $100\,g^{-1}\,min^{-1}$ and a value of K_1 of 10 ml $100\,g^{-1}\,min^{-1}$, as listed in Table 6. Measurements of blood-brain fluorodeoxyglucose transfer are particularly abundant because of the use of fluorodeoxyglucose in positron tomography.

Regional Variation. In vivo, the regional variation of blood-brain glucose transport generally follows the distribution of blood flow and glucose consumption (RAICHLE et al. 1975; CREMER et al. 1983; HAWKINS et al. 1983; GJEDDE and DIEMER 1985b; PLANAS and CUNNINGHAM 1987; GJEDDE et al.

Table 5. Blood-brain glucose transport in awake subjects

Reference	Species	C_a (mM)	PS (ml 100 g^{-1} min^{-1})	PSC_a (µmol 100 g^{-1} min^{-1})	K_1 (ml 100 g^{-1} min^{-1})	K_1C_a (µmol 100 g^{-1} min^{-1})
SOKOLOFF et al. (1977)	Rat	9			19	168
NEMOTO et al. (1978)	Rat	5			10	48
GJEDDE and RASMUSSEN (1980a)	Rat	9			18	160
GJEDDE and RASMUSSEN (1980b)	Rat	8			19	160
CREMER et al. (1981)[a]	Rat	10	17	170		
BRAUN et al. (1985)[a]	Rat	[10]	11	108		
CRANE et al. (1985)[a]	Rat	8	14	127		
LAMANNA and HARIK (1985)	Rat	10	17	170		
HARGREAVES et al. (1986)	Rat	10	18	180		
LAMANNA and HARIK (1986)	Rat	11	17	186		
PLANAS and CUNNINGHAM (1987)	Rat	10	16	158		
PELLIGRINO et al. (1990a)	Rat	9			15	136
PELLIGRINO et al. (1990b)	Rat	9	16	144		
HERTZ et al. (1981)	Man	5			9	45
HERTZ and PAULSON (1982)	Man	5	7	36		
BLOMQVIST et al. (1985)	Man	5			6	30
BROOKS et al. (1986a)	Man	5			6	30
BROOKS et al. (1986b)	Man	4			5	22
FEINENDEGEN et al. (1986)	Man	4			16	67
GUTNIAK et al. (1990)	Man	6			6	39

[a] After subtraction of nonsaturable component.

Table 6. Michaelis constants of blood-brain glucose transfer in awake subjects

Reference	Species	T_{max} (µmol 100 g^{-1} min^{-1})	K_t (mM)
Nemoto et al. (1978)	Rat	190	12
Gjedde and Rasmussen (1980b)	Rat	410	9
Braun et al. (1985)[a]	Rat	160	8
Crane et al. (1985)[a]	Rat	190	8
LaManna and Harik (1985)	Rat	250	5
LaManna and Harik (1986)	Rat	250	5
Hargreaves et al. (1986)	Rat	410	8
Brooks et al. (1986a)	Man	40	4
Feinendegen et al. (1986)	Man	200	4
Ribeiro et al. (1991)	Man	80	2
Blomqvist et al. (1991)	Man	60	4

[a] After subtraction of nonsaturable component.

Table 7. Blood-brain clearance of fluorodeoxyglucose in man

Reference	K_1 (ml 100 g^{-1} min^{-1})
Phelps et al. (1979)	10
Friedland et al. (1983)[a]	13
Hawkins et al. (1983)[a]	
Young adults	8
Older subjects	10
Kato et al. (1984)	15
Gjedde et al. (1985)	10
Reivich et al. (1985)	
k_3-model	11
k_4-model	10
Evans et al. (1986)	8
Hawkins et al. (1986)	4
Lammertsma et al. (1987)	
k_3-model	6
k_4-model	7
Redies et al. (1989)	8
Eastman et al. (1990)	10
Kuwabara et al. (1990)	9
Shapiro et al. (1990)	14
Jagust et al. (1991)	16
Hasselbalch et al. (1991)	9
Kuwabara and Gjedde (1991)	9

[a] No corrections for radioactivity within vessels.

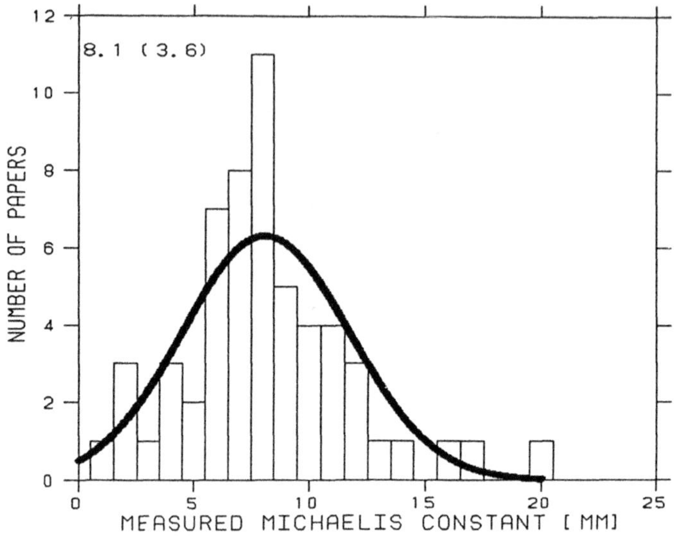

Fig. 6. Blood-brain glucose transport in rat. Frequency of measured K_t values for blood-brain glucose transfer in rat, using a variety of techniques and anaesthetic states. *Abscissa*, K_t estimate; *ordinate*, number of publications

1990). In vitro the distribution of glucose transporters frequently has been shown to be different from that of the in vivo transport and metabolism mechanisms (DICK and HARIK 1986; TUCKER and CUNNINGHAM 1988; BAGLEY et al. 1989), perhaps indicating regional differences of glucose transporter reserves. Somewhat inappropriately, this phenomenon has been termed the flow-metabolism couple (ROY and SHERRINGTON 1890). More correctly, the couple refers to the preservation of the relationship between the three variables during functional changes. They will be considered below.

2. Michaelis-Menten Constants

The maximal transport capacity of the cerebral capillary endothelium has no strong interspecies variation, although it tends to be inversely proportional to the size of the subject in (Table 6). In rats, the estimates of T_{max} range from 40 μmol $100\,g^{-1}\,min^{-1}$, observed by PARDRIDGE et al. (1982) during heavy pentobarbital anesthesia, to 600 μmol $100\,g^{-1}\,min^{-1}$, observed by GJEDDE and DIEMER (1985b) in fine structures of the brain, with a median between 200 and 400 μmol $100\,g^{-1}\,min^{-1}$, depending on the duration of the experiment, the crudeness of tissue sampling (manual versus autoradiography), the absence or presence of anesthesia, the stage of postnatal development, and the method of calculation (i.e., permeability–surface area product or clearance). The estimates of the half-saturation concentration, K_t, vary from 1 to 12 mM, with a definite median at 8 mM in rats (Fig. 6),

Table 8. Blood-brain glucose transfer in pentobarbital-anesthetized rats

Authors	Flux (concentration) [μmol 100 g^{-1} min^{-1} (m*M*)]		T_{max} (K_t) [μmol 100 g^{-1} min^{-1} (m*M*)]	
	Awake	Pentobarbital	Awake	Pentobarbital
GJEDDE and RASMUSSEN (1980b)	160 (8)	114 (9)	410 (9)	240 (5)
CRANE et al. (1985)				
fed	127 (8)	49 (8)	190 (8)	100 (10)
2-day fast	167 (6)	50 (7)	180 (7)	70 (8)
HARGREAVES et al. (1986)	174 (6)	95 (6)	—	—
LAMANNA and HARIK (1986)	186 (11)	144 (12)	250 (5)	210 (4)

and 4 m*M* in humans (Table 6). Thus, it is reasonable to conclude that the maximal transport is close to 400 μmol 100 g^{-1} min^{-1} in the awake restrained rat and 100 μmol 100 g^{-1} min^{-1} in the resting awake human. The maximally possible apparent permeability-surface area product, equal to the T_{max}/K_t ratio, is 50 ml 100 g^{-1} min^{-1} in awake restrained rats and 25 ml 100 g^{-1} min^{-1} in awake, resting humans. Estimates of the endothelial surface area average 50 cm^2 g^{-1} in humans (HUNZIKER et al. 1979) and 100 cm^2 g^{-1} in rats (GJEDDE and DIEMER 1985b). These values yield identical apparent maximal permeabilities in the two species of approximately 800 nm s^{-1} (Table 8). Part of the variation of T_{max} may be due to a formal flaw in the logic underlying T_{max} and K_t measurements. By lowering glucose concentrations with insulin or raising them with glucose, it cannot be assumed that the conditions of blood-brain transfer remain the same; blood flow or metabolic changes may alter the kinetic properties of the cerebral capillary endothelium and hence invalidate the use of the Michaelis-Menten equation.

In studies in which transfer rates in both directions, as well as the glucose concentrations on both sides of the endothelium, are known, it is possible to calculate apparent values for T_{max} and K_t for a single physiological state. These calculations assume symmetrical transport and have been attempted only in a few cases (RIBEIRO et al. 1991; BLOMQVIST et al. 1991). For reasons that are not understood, these measurements tend to be lower than those obtained at different arterial plasma concentrations.

3. Non-Saturable Glucose Transfer

In many studies of glucose transfer, the data appeared to support the presence of a nonsaturable diffusion component of glucose transport, ranging in magnitude from 0.5 to 5 ml 100 g^{-1} min^{-1}. It is interesting that in vitro studies of glucose transport by the glut-1 transporter indicate an intrinsic Michaelis half-saturation constant of 1 m*M* (CARRUTHERS 1990). As discussed above and shown in Fig. 6, most in vivo studies of blood-brain glucose transfer in rat reveal an apparent Michaelis constant close to 8 mM. It is tempting to speculate that two transporters may be present, one expressed by glut-1, having a K_t value close to 1 mM, the second expressed by a gene with a value of higher K_t.

GJEDDE (1981a,b) showed that blood-brain glucose transport data can be analyzed for two transport processes rather than one. According to this analysis, one transporter has a K_t of 1 mM and a T_{max} of 160 µmol 100 g^{-1} min^{-1} in anesthetized rats, the second transporter a T_{max}/K_t ratio of 5 ml 100 g^{-1}min^{-1}. Assuming a K_t of 20 mM for the second transporter, the T_{max} would be 100 µmol 100 g^{-1}min^{-1}. When analyzed for a single transporter, the two transporters yield an average K_t close to 8 mM at commonly employed glucose concentrations. The second transporter may be responsible for the "non-saturable" glucose transport believed to be specific because L-glucose is subject to no comparable transport. According to GJEDDE (1981a,b), the low affinity transporter is responsible for the majority of the net transport at high concentrations of glucose in plasma and hence would be the one expected to be sensitive to consistently elevated glucose concentrations.

4. Metabolism-Flux Ratio and the Lumped Constant

The observation that glucose influx normally exceeds glucose consumption by a factor of two is reflected not only in the brain glucose content but also in certain more specialized indicators, such as the metabolism index $[k_3/(k_2 + k_3)]$, and the "lumped constant" (SOKOLOFF et al. 1977). The lumped constant is the net rate of deoxyglucose transfer, relative to the net glucose transfer. It depends exclusively on the relation between glucose demand and supply and can therefore be calculated from the ratio between unidirectional and net glucose transfer. In turn, when the lumped constant is known, the ratio between glucose demand and supply, i.e., the metabolism-flux couple, can be determined. The observation that lumped constants are similar in a wide variety of species and conditions confirms the uniform relationship between unidirectional and net glucose transfer (GJEDDE 1987).

At a net flux of half the unidirectional flux, the brain interstitial glucose concentration is one third of the K_t, or 2.7 mM in rats and 1.3 mM in humans (GJEDDE and CHRISTENSEN 1984). This is generally the range observed in awake rats and calculated in humans. A number of studies of the relationship between brain and plasma glucose as a function of the plasma glucose concentration was analyzed by GJEDDE and CHRISTENSEN (1984) who concluded that the relationship supported symmetrical blood-brain glucose transfer with a $T_{max}/\Delta J$ ratio of 3–5, and a K_t value of 7 mM. The question is whether the glucose concentration of the brain tissue changes under normal circumstances, i.e., of how finely tuned the regulation of glucose supply is under changing physiological conditions, and of how sensitive it is to pathological influences.

III. Acute Changes of Glucose Transport

The factors that influence glucose transfer acutely are included in Eqs. 12 and 20. They are blood flow (F), which may increase the average capillary glucose concentration, albeit almost negligibly; whole-blood solubility (α), which may increase the apparent permeability by lowering the distribution in red blood cells; the endothelial surface available for facilitated diffusion (S), which may increase by recruitment of reserve capillaries; and the apparent permeability (P), which may be increased by recruitment of transporters and by changes of transporter affinity, for example by changes of ATP concentration (CARRUTHERS 1990). According to this theory, ATP regulates the affinity of the glucose transporter, causing the affinity to decline when ATP concentrations are low (JACQUEZ 1984; CARRUTHERS 1986; HELGERSON et al. 1989).

1. Activation and Deactivation

Anesthesia. The interest in the effect of anaesthetic agents on blood-brain barrier characteristics has been two-fold, focused on one hand on the consquences of the ability of these agents to depress brain function in general, and on the other on the consequences of specific effects of the agents on membrane properties, if any. These factors are not easily separated in vivo. Most investigators of the blood-brain transfer of glucose in animals naturally used anesthetics. Only in some studies did the animal, usually a rat, remain awake during the application of a noninvasive procedure, but these studies were often designed for other purposes. As a result, the number of studies in which the differences between the awake and the anesthetized states were specifically recorded are comparatively few.

Before 1980, BETZ et al. (1973) in dogs, HORTON (1973) in mice, and NEMOTO et al. (1978) in rats, observed no consistent changes of the blood-brain glucose transfer in pentobarbital or halothane anesthesia. The significance of these findings is difficult to assess because opposite changes of blood flow and metabolism were expected with anesthetic agents like halothane, and because the methods ignored the effect of blood flow changes on the measured unidirectional clearance (i.e., as distinct from changes of the apparent PS-product or Michaelis constants of the transport). The absent change of glucose transfer was consistent with measurements of brain glucose content which generally went up as, e.g., in Fig. 7, adapted from MAYMAN et al. (1964). The applications of Eqs. 22 and 23 to the findings of these authors revealed that the ratio between T_{max} (the maximal transport) and ΔJ (the net transport) increased from 2.5 to 5 in phenobarbital anaesthesia, suggesting a decline of metabolism in the absence of a decline of transport.

After 1980, several studies, listed in Table 9, showed that pentobarbital, which consistently reduces both brain metabolism and blood flow, causes blood-brain glucose transfer to decline as well (GJEDDE and RASMUSSEN

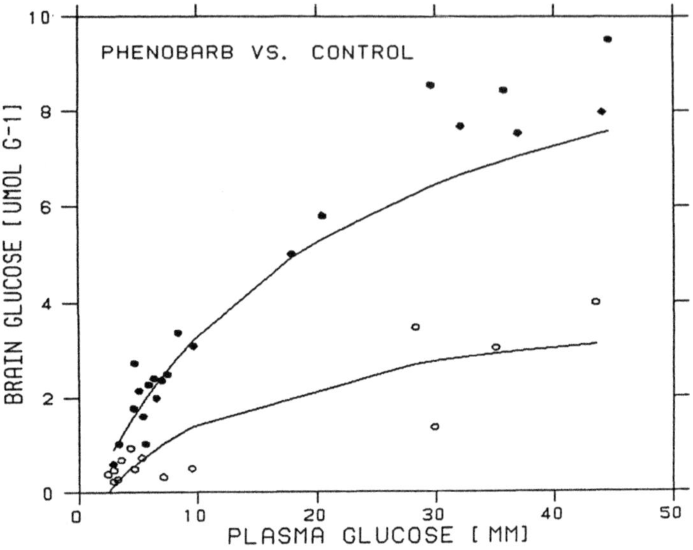

Fig. 7. Blood-brain glucose transfer during anaesthesia. The relation of brain glucose to plasma glucose in hyperglycemia. Phenobarbitone (250 mg/kg i.p.) was given 1 h before decapitation. At the same time 25 mmol kg^{-1} glucose was injected i.p. into mice to obtain plasma glucose concentrations above 10 m*M*. The *points* represent experimentally determined values. For simplicity, only the particular control values which were determined concurrently with the phenobarbital values are shown. The *curves* represent values calculated for a reversible, carrier-mediated glucose transport system as described by GJEDDE and CHRISTENSEN (1984). (From MAYMAN et al. 1964)

1980b; CRANE et al. 1985; HARGREAVES et al. 1986; LAMANNA and HARIK 1986). The decline varied from 25% to 75% of the control value observed in awake rats, depending on the depth of anesthesia and the condition of the awake rats. The estimates of T_{max} declined approximately 50% and the estimates of K_t did not change significantly. In contrast to the earlier reports, these studies suggested that blood-brain glucose transfer declines in proportion to the decline of blood flow and metabolism observed during pentobarbital anesthesia. Also, SOKOLOFF et al. (1977) found a small but insignificant decline of the lumped constant in pentobarbital anesthesia, indicating parallel declines of transport and metabolism. Together, the early and the late studies show that both metabolism and flux decline in anesthesia but the metabolism perhaps more so than the flux. However, earlier measurements of increased brain glucose contents in anesthesia may have been influenced by postmortem breakdown of glucose, likely to affect rapidly metabolizing brains more than slowly metabolizing brains (GJEDDE 1984b).

Activation. There are few studies of glucose transport changes during normal physiological activation of the brain and the metabolism-flux couple has

Table 9. Fundamental blood-brain glucose transport characteristics in rat and man

Species	T_{max} (μmol $100\,g^{-1}$ min^{-1})	K_t (mM)	$P_{max}S$ (ml $100\,g^{-1}$ min^{-1})	S (cm^2/g)	P_{max} (nm/s)	C_a (mM)	$PS\,C_a$ (μmol $100\,g^{-1}$ min^{-1})
Rat	400	8	50	100	800	8	200
Man	100	4	25	50	800	4	50

rarely been tested by comparing two states of different functional activation of the same region or regions of the brain. The reason is that physiological excitation of neurons is difficult to study in rats, and humans; in humans because physiological excitation may not be maintained without attenuation for more than a few minutes, in rats because a condition comparable to that of the resting human is likely to be replaced with a state of excitation due to physical restraint. The sedated rat may actually be a more correct parallel to the relaxed, resting human. Also, one result of excitation may be increased blood glucose. Systemic excitation may cause blood glucose to rise and hence elevate blood-brain glucose transfer without changing the transport characteristics. Despite these limitations, in the only study of its kind, GJEDDE et al. (1991) observed a 15% increase of both glucose influx and metabolism during somatosensory stimulation in humans, apparently due to increased capillary diffusion capacity (recruitment) rather than increased blood flow.

In a few studies, attempts were made to enhance neuronal activation further by less physiological means. CREMER et al. (1981, 1983) compared glucose transport and consumption in brain of rats administered a pyrethroid insecticide. The insecticide caused the transport and metabolism of glucose to increase by 50%–100% in most regions suggesting an increase of the capillary diffusion capacity for glucose, as shown in Fig. 8. Calculated on the basis of a constant K_t, the T_{max} increased somewhat less. In rats first subjected to an overnight fast, during which the plasma glucose concentration declined by 50%, the glucose clearance but not the net flux increased. In this group, the brain glucose content declined, accompanied by a marked increase of T_{max} from 337 μmol $100\,g^{-1}\,min^{-1}$ in the cerebellum of control rats to more than 700 μmol $100\,g^{-1}\,min^{-1}$ in fasted rats in which a strong tremor was evident (CREMER et al. 1983). The decline of brain glucose content signifies an increase of the lumped constant and hence an insufficient increase of glucose transport. The studies suggest that some capillaries may be recruited to allow increased glucose transport in conditions in which glucose consumption is increased but strong evidence for an actual increase of the total number of perfused capillaries is lacking. Less heterogeneity of capillary perfusion velocities may be sufficient to account for these results.

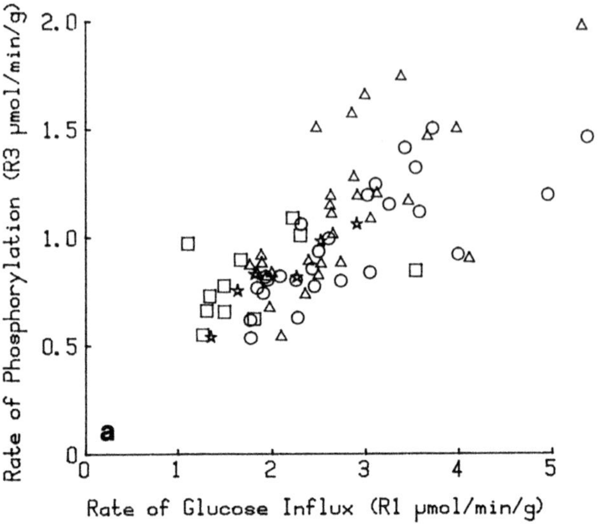

a

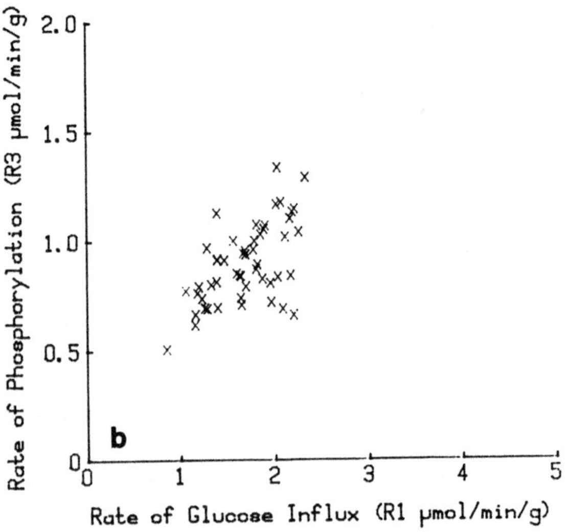

b

Fig. 8a,b. Blood-brain glucose transfer during activation. The relationship between rate of glucose influx and rate of glucose phosphorylation during excitation of rat brain with insecticide. *Ordinate*, rate of glucose phosphorylation (μmol g^{-1} min^{-1}). **a** Rats given pyrethroid insecticide. **b** Control rats. Note preserved correlation at increased values of influx and metabolism. *Abscissa*, rate of glucose influx (μmol g^{-1} min^{-1}). (From CREMER et al. 1981)

Seizures. Generalized seizures in young and adult mammals are generally believed to be accompanied by substantial increases of oxygen and glucose metabolism and blood flow but there are few reliable reports that directly or indirectly reveal the relationship between unidirectional and net blood-brain glucose transfer in this condition of massive departure from steady-state. A few studies indicate that the lumped constant rises significantly in certain types of seizures, suggesting isolated increases of metabolism in these conditions. GJEDDE and DIEMER (1985a) used 3-*O*-methylglucose to calculate a 50% increase of rat hippocampal lumped constant in kainic-acid-induced seizures, while FUJIKAWA et al. (1989) used measurements of brain glucose content to calculate lumped constant increases of more than 100% in newborn marmosets given bicuculline. Also, CREMER et al. (1988) noted an increase of the distribution of 2-deoxyglucose relative to glucose in rats given kainic acid, a change that signified an increase of the lumped constant. It is evident that glucose transport fails to keep pace with glucose metabolism during increases of the latter that exceed 20%–50% of normal levels. These findings alone suggest that the reserve of capillaries available for recruitment, if any, cannot be substantial.

2. Carbon Dioxide and Spreading Depression

Although the regional variation of glucose transport from blood to brain appears to be correlated to differences of blood flow and metabolism, primary blood flow changes do not necessarily lead to changes of blood-brain glucose transfer.

Hypercapnia. Carbon dioxide is a strong stimulant of blood flow change. Since it is unlikely that carbon dioxide is the agent responsible for the changes of blood flow observed during neuronal excitation, it is not a priori certain that carbon-dioxide-induced changes of blood flow must change blood-brain glucose transfer. In fact, in rabbits, in 1975, BERSON et al. measured a decline of glucose extraction during massive hypercapnia (100 mmHg). Low extraction was also observed in two more recent studies of glucose transport during hypercapnia (HERTZ and PAULSON 1982; GJEDDE and LAURITZEN 1985). The low extraction fractions show that carbon dioxide-induced blood flow changes are not accompanied by parallel changes of glucose transport capacity, i.e., of the number of capillary glucose transporters. Originally, the calculation of blood-brain glucose flux on the basis of simultaneous measurements of blood flow suggested increased glucose transport (HERTZ and PAULSON 1982) but later reanalysis of the same data (KNUDSEN et al. 1991) revealed no increase of blood-brain glucose flux. As shown in Table 10, the same conclusion was drawn by GJEDDE and LAURITZEN (1985) for rats.

Hypocapnia. In hypocapnia, leading to low blood flow, it is the expectation that the first-pass glucose extraction must go up, as also observed by HERTZ

Table 10. Effect of carbon dioxide on blood-brain glucose transfer in rat

Group	P_aCO_2 (mm/tg)	C_a (m*M*)	*PS* (ml 100 g^{-1} min^{-1})	*E* (%)	(*n*)
Hypocapnia	25 ± 1.0	11.8 ± 0.6	17.0 ± 3.2	44 ± 6	(5)
Normocapnia	38 ± 1.5	12.2 ± 1.6	23.4 ± 3.0	44 ± 5	(7)
Hypercapnia	67 ± 2.0	12.7 ± 0.2	20.8 ± 3.4	12 ± 3	(4)

From GJEDDE and LAURITZEN (1985).

and PAULSON (1982). In rat, GJEDDE and LAURITZEN (1985) observed no significant change of glucose transport, while in man the transport appeared to go up. The explanation for the latter change is not clear.

Spreading Depression. GJEDDE et al. (1981) measured the changes of glucose transfer during Leao's spreading cortical depression. The condition is characterized by massively elevated potassium concentrations and transiently elevated blood flow. No change of glucose transfer, measured as the *PS* product, occurred during the wave of increased blood flow, confirming the conclusion drawn above that the blood flow changes per se cause no change of blood-brain glucose transfer.

3. Anoxia and Ischaemia

The question of blood-brain glucose transfer modulation in anoxic or ischemic states has been particularly difficult to answer because the marked changes of blood flow, metabolism, and glucose concentrations on both sides of the blood-brain barrier may affect blood-brain glucose transfer without changing the fundamental characteristics of the transport.

Anoxia. Anoxia causes cerebral blood flow to increase. In the absence of other changes, the increased blood flow lowers the extraction of glucose. A reduction of the first-pass extraction indicates a reduced flux-flow couple and hence insufficient recruitment. AGNEW and CRONE (1967) and BERSON et al. (1975) observed the decline of glucose extraction in anoxic rabbits. To circumvent the problem of increased blood flow, BETZ et al. (1974) and KINTNER et al. (1980) mechanically perfused the head of dogs without oxygen, adding radioactive glucose to the perfusate to measure blood-brain glucose transfer. The transfer was calculated as the product of the measured extraction fraction and the rate of outflow through a venous catheter, relative to the weight of the brain. Calculated on this basis, the unidirectional glucose influx appeared to decline by 50% within 15 min of the onset of anoxic perfusion. The reason for the decline is not clear but may be associated with loss of ATP from brain tissue and endothelial cells. It is also possible that the blood flow rate was underestimated by the venous outflow method; unless venous blood is collected in toto, outflow is in reality a pressure measurement, perhaps reflecting the low resistance during anoxia. During reoxygenation, glucose transport slowly returned to normal.

Ischaemia. During complete ischaemia, it is impossible to measure glucose transport because of the lack of perfusion. Studies have therefore focused on the changes of glucose transport after ischemia. In the postischemic phase, Betz et al. (1983b), in gerbils, and Gjedde et al. (1990), in living humans after an episode of stroke, reported significantly reduced glucose transfer, related to a reduction of the capillary diffusion capacity, as expressed by a calculated density of "standardized" capillaries. Although the latter result suggested that the change is not specifically related to glucose transport but to general changes of the microvasculature, part of the decline may also be attributed to low transport affinity according to Gjedde and Siemkowicz (1978), who reported an increase of K_t from 6 to 12 mM, again perhaps as a result of ATP loss from endothelial cells.

The separate changes of blood flow and glucose transport observed after ischemia, and during spreading depression and hypo- or hypercapnia, indicate that blood flow and glucose transport are regulated by different mechanisms in the vascular bed, possibly in different sectors of the vasculature. Blood flow regulation may be an arteriolar function, glucose transport regulation a capillary function.

IV. Chronic Changes

1. Pre- and Postnatal Development

A continuous change of blood-brain glucose transfer occurs as a function of the age of the subject, i.e., during pre- and postnatal development. Glut-1 seems to be ubiquitous in fetal tissues but declines after birth. This change appears to reflect absolute and relative adjustments of the numbers of glucose transporters. Thus, erythrocyte glut-1 transporters are lost after birth in many species (Widdas 1955; Goodwin 1956; Jacquez 1984). The signal for the continued expression of the glut-1 transporter in brain capillary endothelium has not been identified but it is possible that the increasing tightness of the blood-brain barrier in postnatal life, and the consequent decline of glucose concentration in brain extracellular space relative to plasma, maintains the expression of the glut-1 transporters, perhaps by influence of a humoral agent released by glial foot processes (Maxwell et al. 1989).

Several studies show an increase of blood-brain glucose transfer in parallel with the postnatal increase of glucose consumption, with little change of the lumped constant, indicating constant relations between unidirectional and net glucose transfer (Fuglsang et al. 1986; Dyve and Gjedde 1991). Cremer et al. (1976) noted an increase of Oldendorf's Brain

Uptake Index of first-pass extraction from 19% to 32% between 18 days and 9 weeks of age in rats. DANIEL et al. (1978) noted at least a doubling of the blood-brain cleareance between 2 and 9 weeks of age in rats and, interestingly, a decline toward neonatal values in senescent (64-week-old) rats. CREMER et al. (1979) described a 40% increase of both blood-brain glucose transfer and T_{max} between the ages of 2.5 weeks and adulthood in rats. CORNFORD and CORNFORD (1986) noted a several-fold increase of blood-brain glucose transfer between 1-day-old and 28-day-old rabbits, the result of an almost 20-fold increase of the T_{max} of the glucose transporter, in contrast to two earlier studies in which the BUI of glucose changed very little during postnatal development of rabbits (BRAUN et al. 1980; CORNFORD et al. 1982). CREMER et al. (1979) and CORNFORD and CORNFORD (1986) observed very little increase of K_t with age, in contrast to FUGLSANG et al. (1986) who calculated a much higher value of K_t in newborn rats than in adult rats. Both CREMER et al. (1979) and CORNFORD and CORNFORD (1986) determined the so-called nonsaturable component of glucose transport separately. If it is correct that the nonsaturable component is in reality a less-high-affinity transporter, it is possible that the glut-1 transporter may be not only maintained but specifically induced during development, in response to the declining ratio between glucose in brain and blood. This speculation predicts that the glut-1 transport system is particularly sensitive to changes of the extracellular glucose concentration.

2. Dementia

There is evidence of a decline of glucose transport in cerebral cortex in Alzheimer's disease in parallel with the decline of functional activity and metabolism, related both to the cortical atrophy and to the decline of functional activity of the surviving tissue (KALARIA and HARIK 1989; JAGUST et al. 1991). It is unlikely that this phenomenon is related to any direct mechanism of regulation of glucose transport. It is more likely that the decline represents an acceleration of the normal decline of cerebral metabolism in senescence.

3. Hypo- and Hyperglycaemia

The Michaelis-Menten nature of the blood-brain glucose transfer dictates that glucose clearance (K_1) and apparent permeability (PS) must increase when blood glucose declines, or decline when blood glucose increases. With continued rediscovery of this aspect of the Michaelis-Menten equation, the real question is whether additional, specific changes occur when the state of abnormal glucose concentrations persists. Glucopenia caused an upregulation of glucose transport in cultured fibroblasts (ULLREY et al. 1975; ULLREY and KALCKAR 1981) and low extracellular glucose caused glucose transporter numbers to increase in chicken embryo fibroblasts (PESSIN et al. 1982; KLETZIEN and PERDUE 1985). The concept of glucose-starvation enhancement of sugar transport in animal cell cultures (GSE; for review, see GERMINARIO et al. 1982) is by now so entrenched that the question is not whether it oocurs but whether it is a feature of all glucose transporters and whether it occurs in vivo. In the rat L6 cell line of skeletal muscle, basal glucose transport is subserved by glut-1 type transporters (WALKER et al. 1989). By elevating glucose in the medium from 5 to 15 m*M*, glucose transport was reduced 30% by depletion of glut-1 transporter protein and cytochalasin B binding sites (KLIP and PAQUET 1990). Thus, in vitro, the glut-1 transporter appears to be down-regulated (KLIP et al. 1991). However, in vivo, the situation may be different. In mildly diabetic rats, glut-4 transporters in muscle cells but not glut-1 were diminished when measured by cytochalasin B binding (RAMLAL et al. 1989) but mRNA levels were not reduced, suggesting a translational block (KLIP and PAQUET 1990). The significance of this difference between in vivo and in vitro conditions in not clear.

Hyperglycaemia. Previously, there were no confirmed inborn or idiopathic abnormalities of blood-brain glucose transfer in vivo. Both WYKE (1959) and HARRIS and PROUT (1970) published case reports suggesting the existence of an idiopathic or acquired state of relative cerebral hypoglycemia, i.e., a condition in which the blood glucose level is normal, yet blood-brain glucose transfer appears to be abnormal. Similar observations were made by DEFRONZO et al. (1977, 1980) and LILAVIVATHANE et al. (1979) in diabetics in which the blood sugar level was normalized rapidly. Recently, De Vivo et al. (1991) reported the existence of two children with persistent CSF glycopenia, seizures and delayed development that seemed to be associated with a genetically deficient glut-1 transporter. This report would seem to confirm the existence of Wyke's syndrome.

The finding of a pre- and postnatal development of glucose transport supported the prediction that elevated glucose in brain may inhibit the expression of the glut-1 transporter. Observations of this kind led GJEDDE and CRONE (1981) to examine the blood-brain glucose transfer in rats with maintained hyperglycemia to test whether prolonged exposure of the blood-brain barrier to elevated glucose will inhibit the transport. The results, obtained with the intravenous bolus injection method and tissue sampling, indicated a down-regulation of the blood-brain glucose transfer in the hyperglycemic state. The finding can be criticized because only two (glucose clearance and blood flow) of the three variables required to calculate blood-brain glucose transfer (i.e., glucose clearance, blood flow, and blood volume) were simultaneously measured.

The result was initially confirmed by other studies, both directly with another method (MCCALL et al. 1982, using intracarotid bolus injection and tissue sampling) and indirectly by determination of the number of cytochalasin B binding sites in isolated cerebral microvessels in vitro (MATTHAEI et al. 1986). At the same time, BROOKS et al. (1986b) found no

Table 11. Blood-brain glucose transfer in chronic hyperglycaemia

Reference	C_a (mM)	Duration (weeks)					
		0	1	2	3	4–5	6–8
		PS products (ml $100\,g^{-1}\,min^{-1}$)					
GJEDDE and CRONE (1981)	10	21			17		
PARDRIDGE et al. (1990a)	10	28	18				
PELLIGRINO et al. (1990b)[a]	8–9	15					37
GJEDDE and CRONE (1981)	30	10			7		
DUCKROW (1988)	27–29	11	11		11		
HARIK and LAMANNA (1988)	27–32	7		9		8	

[a] K_1 rather than PS.

difference of unidirectional blood-brain glucose transfer between normal humans and diabetics, and DUCKROW (1988) and HARIK and LAMANNA (1988), in separate reports, found no change of glucose transfer in chronically diabetic rats. Likewise, HARIK et al. (1988) reported an increase, rather than a decrease, of cytochalasin B binding sites in isolated brain microvessels, and in erythrocytes from diabetics (HARIK et al. 1991a). However, HARIK and LAMANNA (1988) also observed a significantly reduced L-glucose space in the chronically hyperglycemic rats.

To address the issue, using the carotid perfusion method first introduced by TAKASATO et al. (1984), PARDRIDGE et al. (1990a) controlled most of the factors that affect glucose transport. In rats, 1 week after the elevation of a plasma glucose from 5.8 to 24.6 mM, at a perfusate glucose concentration of 10 mM, the PS product of glucose transport had declined from 28 to 16 ml $hg^{-1}\,min^{-1}$. The rate of tissue perfusion also declined in these rats, despite identical inflow rates in the two groups. The same group of investigators also found an increase of mRNA, confirming a translational block in this condition (CHOI et al. 1989). Using the BUI method, also MOORADIAN and MORIN (1991) noted a significant decline of 3-O-methylglucose uptake from 42% to 33%, 8 days after induction of hyperglycaemia. Acutely hyperglycaemic animals were not affected.

Using the method designed to yield glucose metabolic rate and first introduced by CUNNINGHAM and CREMER (1981), PELLIGRINO et al. (1990b) observed an increase of the unidirectional glucose clearance (K_1) in rats hyperglycaemic at 33 mM for 6–8 weeks with acutely normalized glucose levels; from an average of 16 ml $100\,g^{-1}\,min^{-1}$ (8.5 mM glucose) in normal rats to 35 ml $100\,g^{-1}\,min^{-1}$ (7.9 mM glucose) in the chronically hyperglycaemic rats. Glucose concentration in brain and glucose consumption were also elevated. Although neither glucose PS product, glucose influx, nor blood or plasma flow were measured, the glucose concentration and metabolism measurements are difficult to reconcile with a reduction of glucose transport. The duration of hyperglycaemia may play a role; the original

Table 12. Blood-brain glucose transfer in chronic hyperglycaemia

Reference	C_a (mM)	Duration (weeks)	
		0	1
		BU Indices (%)	
McCall et al. (1982)	25	40	27
Pardridge et al. (1990, unpubl.)	24	26	20
Mooradian and Morin (1991)	25	42	33

observation by Gjedde and Crone (1981) was made 3 weeks after the induction of hyperglycaemia and predicted a modest decline of the glucose *PS* product from 21 to 17 ml 100 g^{-1} min^{-1} at a plasma glucose concentration of 10 mM. The findings have been summarized in Tables 11 and 12.

Insulin. Originally, a fourth factor was not considered in these studies, i.e., the insulin level in chronic hyperglycaemia, but for apparently good reasons: Crone (1965), Buschiazzo et al. (1970), Betz et al. (1973), Daniel et al. (1975b) and Betz et al. (1979) had concluded that insulin had no measurable effect on unidirectional blood-brain glucose transfer. Although some of these studies can be criticized on methodological grounds (Bradbury 1985), the conclusion was later supported by the finding that the glucose transporter of the cerebral capillary endothelium is of the insulin-insensitive kind, and that the insulin-sensitive glucose transporter is not expressed in endothelial cells (Slot et al. 1990).

In 1981, Hertz et al. (1981) measured unidirectional blood-brain glucose transfer in humans with the intracarotid bolus injection method and arteriovenous sampling. Insulin at a 100-fold elevated concentration increased the unidirectional blood-brain glucose transfer by 50%. The study appeared to confirm that blood-brain glucose transfer may be low in the absence of massively elevated insulin levels. The finding received support from in vitro measurements of glucose transfer into, and cytochalasin B binding to, isolated cerebral microvessels (Djuričić et al. 1983; Matthaei et al. 1986). Both studies have been criticized, the former because glucose transfer into isolated microvessles cannot be unidirectional (Bradbury 1985), the latter by Harik et al. (1988), who observed a 30% increase of cytochalasin B binding in hyperglycaemia and suggested that sonication of tissue in the experiments of Matthaei et al. in part abolished specific cytochalasin B binding.

In 1987, Namba et al. reported no effect of insulin on the unidirectional blood-brain transfer of 3-*O*-methylglucose, a nonmetabolizable glucose analog. Since the intracarotid bolus injection method in humans does not distinguish between glucose transport in different brain regions, it is possible

Table 13. Effect of insulin on blood-brain glucose transfer

Reference	Species	C_a (mM)	Tracer (ml 100^{-1} min^{-1})	K_1	Plasma insulin (pM)
HERTZ et al. (1981)	Man	5	glc	9	39
		5		13	4200
		5		13	4950
		5		11	1620
		5		12	660
NAMBA et al. (1987)	Rat	7	3OMG	33	42
		7		17	687
SHAPIRO et al. (1990)	Man	5	FDG	14	41
		5		14	455
		3		15	421
HASSELBALCH et al. (1991)	Man	5	FDG	9	44
		5		10	490

that the effect on glucose transport was caused by uptake into regions with a specific sensitivity to glucose (REITH et al. 1987). When the problem was reinvestigated with a regional method (positron emission tomography, PET) at a more physiological insulin level, no change of the K_1 of fluorodeoxyglucose was seen (HASSELBALCH et al. 1991). These findings have been summarized in Table 13. Concluding from 10 years of study of unidirectional blood-brain glucose transfer in prolonged hyperglycaemia and insulin-deficient states, it appears unlikely that insulin has a significant effect at physiological levels but likely that prolonged hyperglycemia may reduce blood-brain glucose transfer under some circumstances.

Hypoglycaemia. It is of interest to observe the change of blood-brain glucose transfer under the opposite circumstance of prolonged hypoglycemia and potentially glucose-deficient states like inanition. Contrary to the original report by DANIEL et al. (1971), inanition with ketoacidosis was shown to stimulate blood-brain transfer of ketone bodies (GJEDDE and CRONE 1975; MOORE et al. 1976) and glucose (T.G. CHRISTENSEN et al. 1981) while hypoglycaemia stimulated glucose transfer (MCCALL et al. 1986; PELLIGRINO et al. 1990b). If confirmed, the studies of PELLIGRINO et al. (1990a,b) indicate somewhat implausibly that both hypo- and hyperglycaemia stimulate unidirectional blood-brain glucose transfer. The stimulation of glucose transfer during inanition was not confirmed by CRANE et al. (1985) in rats, or REDIES et al. (1989) in obese humans undergoing therapeutic weight loss. In conclusion, it appears that sustained hypoglycaemia may stimulate blood-brain glucose transfer, whereas inanition, with or without ketoacidosis, probably has little effect. For the reasons discussed above, it is not likely that insulin is responsible for the stimulation of unidirectional blood brain glucose transfer during sustained insulin-induced hypoglycaemia.

Acknowledgments. This review is dedicated to the memory of Christian Crone. The author wishes to thank Anna Di Pancrazio for her untiring help with the preparation of the typescript. Jill E. Cremer, Vincent Cunningham, Sami Harik, Lester Drewes, Gitte Moos Knudsen, Joseph LaManna, Olaf Paulson, Hanna Pappius, and William M. Pardridge kindly offered suggestions and advice.

References

Agnew WF, Crone C (1967) Permeability of brain capillaries to hexoses and pentoses in the rabbit. Acta Physiol Scand 70:168–175

Asano T, Shibasaki Y, Kasuga M, Kanazawa Y, Takaku F, Akanuma Y, Oka Y (1988) Cloning of a rabbit brain glucose transporter cDNA and alteration of glucose transporter mRNA during tissue development. Biochem Biophys Res Commun 154:1204–1211

Axelrod JD, Pilch PF (1983) Unique cytochalasin B binding characteristics of the hepatic glucose carrier. Biochemistry 22:2222–2227

Bachelard HS, Daniel PM, Love ER, Pratt OE (1973) The transport of glucose into the brain of the rat in vivo. Proc R Soc Lond B 183:990–993

Bagley PR, Tucker SP, Nolan C, Lindsay JG, Davies A, Baldwin SA, Cremer JE, Cunningham VJ (1989) Anatomical mapping of glucose transporter protein and pyruvate dehydrogenase in rat brain: an immunogold study. Brain Res 499:214–224

Baker GF, Widdas WF (1973) The asymmetry of the facilitated transfer system for hexose simple kinetics of a two component model. J Physiol 231 (Lond):143–165

Baldwin SA, Lienhard GE (1981) Glucose transport across plasma membranes: facilitated diffusion systems. Trends Biochem Sci 6:208–211

Baldwin SA, Lienhard GE (1989) Purification and reconstitution of glucose transporter from human erythrocytes. Methods Enzymol 174:39–50

Baldwin SA, Baldwin JM, Lienhard GE (1982) Monosaccaride transporter of the human erythrocyte. Characterization of an improved preparation. Biochemistry 21:3836–3842

Baldwin SA, Brewster F, Cairns MT, Gardiner RM, Ruggier R (1984) Identification of a D-glucose sensitive cytochalasin B binding component of isolated ovine cerebral microvessels. J Physiol (Lond) 357:75P

Baldwin SA, Cairns MT, Gardiner RM, Ruggier R (1985) A D-glucose sensitive cytochalasin B binding component of cerebral microvessels. J Neurochem 45:650–652

Barnett JEG, Holman RA, Chalkey RA, Munday KA (1975) Evidence for two asymmetric conformational states in the human erythrocyte sugar transport system. Biochem J 145:417–429

Bell GI, Kayano T, Buse JB, Burant CF, Takeda J, Lin D, Fukumoto H, Seino S (1990) Molecular biology of mammalian glucose transporters. Diabetes Care 13:198–208

Berson FG, Spatz M, Klatzo I (1975) Effects of oxygen saturation and PCO_2 on brain uptake of glucose analogues in rabbits. Stroke 6:691–696

Betz AL, Gilboe DD (1974) Kinetics of cerebral glucose transport in vivo: inhibition by 3-*O*-methylglucose. Brain Res 65:368–372

Betz AL, Goldstein GW (1981) Developmental changes in metabolism and transport properties of capillaries isolated from rat brain. J Physiol (Lond) 312:365–376

Betz AL, Iannotti F (1983) Simultaneous determination of regional cerebral blood flow and blood-brain glucose transport kinetics in the gerbil. J Cereb Blood Flow Metab 3:193–199

Betz AL, Gilboe DD, Yudilevich DK, Drewes LR (1973) Kinetics of unidirectional glucose transport into the isolated dog brain. Am J Physiol 225:586–592

Betz AL, Gilboe DD, Drewes LR (1974) Effects of anoxia on net uptake and unidirectional transport of glucose into the isolated dog brain. Brain Res 65:307–316

Betz AL, Drewes LR, Gilboe DD (1975) Inhibition of glucose transport into brain by phlorizin, phloretin and glucose analogues. Biochim Biophys Acta 406:505–515
Betz AL, Csejtey J, Goldstein GW (1979) Hexose transport and phosphorylation by capillaries isolated from rat brain. Am J Physiol 236:C96–C102
Betz AL, Bowman PD, Goldstein GW (1983a) Hexose transport in microvascular endothelial cells cultured from bovine retina. Exp Eye Res 36:269–277
Betz AL, Iannotti F, Hoff JT (1983b) Ischemia reduces blood-to-brain glucose transport in the gerbil. J Cereb Blood Flow Metab 3:200–206
Bidder TG (1968) Hexose translocation across the blood-brain interface: configurational aspects. J Neurochem 15:867–874
Birnbaum MJ (1989) Identification of a novel gene encoding an insulin-responsive glucose transporter protein. Cell 57:305–315
Birnbaum MJ, Haspel HC, Rosen OM (1986) Cloning and characterization of a cDNA encoding the rat brain glucose-transporter protein. Proc Natl Acad Sci USA 83:5784–5788
Blasberg RG, Patlak CS, Jehle JW, Fenstermacher JD (1978) An autoradiographic technique to measure the permeability of normal and abnormal brain capillaries. Neurology 28:363
Blomqvist G (1984) On the construction of functional maps in positron emission tomography. J Cereb Blood Flow Metab 4:629–632
Blomqvist G, Bergström K, Bergström M, Ehrin E, Eriksson L, Garmelius B, Lindberg B, Lilja A, Litton J-E, Lundmark L, Lundqvist H, Malmborg P, Moström U, Nilsson L, Stone-Elander S, Widén L (1985) Models for ^{11}C-glucose. In: Greitz T, Ingvar DH, Widén L (eds) The metabolism of the human brain studied with positron emission tomography. Raven, New York, p 185
Blomqvist G, Gjedde A, Gutniak M, Grill V, Widén L, Stone-Elander S, Hellstrand E (1991) Facilitated transport of glucose from blood to brain in man and the effect of moderate hypoglycemia on regional cerebral glucose utilization. Eur J Nucl Med 18:834–837
Boado RJ, Pardridge WM (1990) The brain-type glucose transporter mRNA is specifically expressed at the blood-brain barrier. Biochem Biophys Res Commun 166:174–179
Bohr C (1909) Über die spezifische Tätigkeit der Lungen bei der respiratorischen Gasaufnahme und ihr Verhalten zu der durch die Alveolarwand stattfindenden Gasdiffusion. Skand Arch Physiol 22:221–280
Bradbury MWB (1985) The blood-brain barrier in vitro. Neurochem Int 7:27–28
Braun LD, Cornford EM, Oldendorf WH (1980) Newborn rabbit blood-brain barrier is selectively permeable and differs substantially from the adult. J Neurochem 34:147–152
Braun LD, Miller LP, Pardridge WM, Oldendorf WH (1985) Kinetics of regional blood-brain barrier glucose transport and cerebral blood flow determined with the carotid injection technique in conscious rats. J Neurochem 44:911–915
Brender J, Andersen PE, Rafaelsen OJ (1975) Blood-brain transfer of D-glucose, L-leucine, and L-tryptophan in the rat. Acta Physiol Scand 93:490–499
Brooks DJ, Beaney RP, Lammertsma AA, Herold S, Turton DR, Luthra SK, Frackowiak RSJ, Thomas DGT, Marshall J, Jones T (1986a) Glucose transport across the blood-brain barrier in normal human subjects and patients with cerebral tumours studied using $[^{11}C]$3-*O*-methyl-D-glucose and positron emission tomography. J Cereb Blood Flow Metab 6:230–239
Brooks DJ, Gibbs JSR, Sharp P, Herold S, Turton DR, Luthra SK, Kohner EM, Bloom SR, Jones T (1986b) Regional cerebral glucose transport in insulin-dependent diabetic patients studied using $[^{11}C]$3-*O*-methyl-D-glucose and positron emission tomography. J Cereb Blood Flow Metab 6:240–244
Brøndsted HE, Gjedde A (1988) Measuring brain glucose phosphorylation with labeled glucose. Am J Physiol 254 (Endocrinol Metab 17):E443–E448

Bryan RM, Hawkins RA, Mans AM, Davis DW, Page RB (1983) Cerebral glucose utilization in awake unstressed rats. Am J Physiol 244:C270–C275
Buschiazzo PM, Terrell EB, Regen DM (1970) Sugar transport across the blood-brain barrier. Am J Physiol 219:1505–1513
Carruthers A (1986) ATP regulation of the human red cell sugar transporter. J Biol Chem 261:11028–11037
Carruthers A (1990) Facilitated diffusion of glucose. Physiol Rev 70:1135–1176
Carter-Su C, Pessin JE, Mora R, Gitomer W, Czech MP (1982) Photoaffinity labeling of the human erythrocyte D-glucose transporter. J Biol Chem 257:5419–5425
Charron M, Brosius FC, Alper SL, Lodish HF (1989) A glucose transport protein expressed predominantly in insulin-responsive tissues. Proc Natl Acad Sci USA 86:2535–2539
Choi TB, Boado RJ, Pardridge WM (1989) Blood-brain barrier glucose transporter mRNA is increased in experimental diabetes mellitus. Biochem Biophys Res Commun 164:375–380
Christensen O, Andersen HL, Betz AL, Gilboe DD (1982) Transport of glucose across the blood-brain barrier: reevaluation of the accelerative exchange diffusion. Acta Physiol Scand 115:233–238
Christensen TG, Diemer NH, Laursen H, Gjedde A (1981) Starvation accelerates blood-brain glucose transfer. Acta Physiol Scand 112:221–223
Cornford EM, Cornford ME (1986) Nutrient transport and the blood-brain barrier in developing animals. Fed Proc 45:2065–2072
Cornford EM, Braun LD, Oldendorf WH (1982) Developmental modulations of blood-brain barrier permeability as an indicator of changing nutritional requirements in the brain. Pediatr Res 16:324–328
Crane PD, Pardridge WM, Braun LD, Oldendorf WH (1983) Kinetics of transport and phosphorylation of 2-fluoro-2-deoxy-D-glucose in rat brain. J Neurochem 40:160–167
Crane PD, Pardridge WM, Braun LD, Oldendorf WH (1985) Two-day starvation does not alter the kinetics of blood-brain barrier transport and phosphorylation of glucose in rat brain. J Cereb Blood Flow Metab 5:40–46
Cremer JE, Cunningham VJ (1979) Effects of some chlorinated sugar derivatives on the hexose transport system of the blood-brain barrier. Biochem J 180:677–679
Cremer JE, Braun LD, Oldendorf WH (1976) Changes during development in transport processes of the blood-brain barrier. Biochim Biophys Acta 448:633–637
Cremer JE, Cunningham VJ, Pardridge WM, Braun LD, Oldendorf WH (1979) Kinetics of blood-brain transport of pyruvate, lactate and glucose in suckling, weanling and adult rats. J Neurochem 33:439–446
Cremer JE, Ray DE, Sarna GS, Cunningham VJ (1981) A study of the kinetic behaviour of glucose based on simultaneous estimates of influx and phosphorylation in brain regions of rats in different physiological states. Brain Res 221:331–342
Cremer JE, Cunningham VJ, Seville MP (1983) Relationship between extraction and metabolism of glucose, blood flow, and tissue blood volume in regions of rat brain. J Cereb Blood Flow Metab 3:291–302
Cremer JE, Seville MP, Cunningham VJ (1988) Tracer deoxyglucose kinetics in brain regions of rats given kainic acid. J Cereb Blood Flow Metab 8:244–253
Crone C (1960) The diffusion of some organic non-electrolytes from blood to brain tissue. Acta Physiol Scand 50 [Suppl 175]:33–34
Crone C (1963) The permeability of capillaries in various organs as determined by use of the "indicator diffusion" method. Acta Physiol Scand 58:292–305
Crone C (1965) Facilitated transfer of glucose from blood into brain tissue. J Physiol (Lond) 181:103–113

Crone C (1975) General properties of the blood-brain barrier with special emphasis on glucose. In: Cserr HF, Fenstermacher JD, Fencl V (eds) Fluid environment of the brain. Academic, New York, p 33
Crone C (1985) The blood-brain barrier: a modified tight epithelium. In: Suckling AJ, Rumsby MG, Bradbury MWB (eds) The blood-brain barrier in health and disease. Ellis Horwood, Chichester, p 17
Cunningham VJ, Cremer JE (1981) A method for the simultaneous estimation of regional rates of glucose influx and phosphorylation in rat brain using radiolabelled 2-deoxyglucose. Brain Res 221:319–330
Cunningham VJ, Sarna GS (1979) Estimation of the kinetic parameters of unidirectional transport across the blood-brain barrier. J Neurochem 33:433–437
Cunningham VJ, Hargreaves RJ, Pelling D, Moorhouse SR (1986) Regional blood-brain glucose transfer in the rat: a novel double-membrane kinetic analysis. J Cereb Blood Flow Metab 6:305–314
Cunningham VJ, Cremer JE, Hargreaves RJ (1989) Relationships between neuronal activity, energy metabolism and cerebral circulation. In: Buattaini F, Govoni S, Maghoni MS, Trabucchi M (eds) Regulatory mechanisms of neuron to vessel communication in the brain. Springer, Berlin Heidelberg New York, p 325 (NATO ASI Ser H, vol 33)
Cutler RWP, Sipe JC (1971) Mediated transport of glucose between blood and brain in the cat. Am J Physiol 220(5):1182–1186
Daniel PM, Moorehouse SR, Love ER, Pratt OE (1971) Factors influencing utilization of ketone-bodies by brain in normal rats. Lancet ii:637–638
Daniel PM, Donaldson J, Pratt OE (1975a) A method for injecting substances into the circulation to reach rapidly and to maintain a steady level. Med Biol Eng 13:214–227
Daniel PM, Love ER, Pratt OE (1975b) Insulin and the way the brain handles glucose. J Neurochem 25:471–476
Daniel PM, Love ER, Pratt OE (1978) The effect of age upon the influx of glucose into the brain. J Physiol (Lond) 274:141–148
Daniel PM, Love ER, Pratt OE (1980) Inhibition of glucose transport into the brains of suckling rats by raising the level of galactose in the blood. J Physiol (Lond) 305:44–45
DeFronzo RA, Andres R, Bledsoe TA, Boden G, Faloona GA, Tobin JD (1977) A test of the hypothesis that the rate of fall in glucose concentration triggers counterregulatory hormonal responses in man. Diabetes 26:445–452
DeFronzo RA, Hendler R, Christensen N (1980) Stimulation of counterregulatory hormonal responses in diabetic man by a fall in glucose concentration. Diabetes 29:125–131
DeVivo DC, Trifiletti RR, Jacobson RI, Ronen GM, Behmand RA, Harik SI (1991) Defective glucose transport across the blood-brain barrier as a cause of persistent hypoglycorrhachia, seizures, and developmental delay. N Engl J Med 325:703–709
Dick APK, Harik SI (1986) Distribution of the glucose transporter in the mammalian brain. J Neurochem 46:1406–1411
Dick APK, Harik SI, Klip A, Walker DM (1984) Identification and characterization of the glucose transporter of the glucose transporter of the blood-brain barrier by cytochalasin B binding and immunological reactivity. Proc Natl Acad Sci USA 81:7233–7237
Diemer NH, Benveniste H, Gjedde A (1985) In vivo cell membrane permeability to deoxyglucose in rat brain. Acta Neurol Scand 72:87
Djuričić DM, Kostić VS, Mršulja BB (1983) Insulin increases entrance of 2-deoxy-D-[^{3}H]glucose in isolated rate brain microvessels. Brain Res 275:186–188
Drewes LR, Horton RW, Betz AL, Gilboe DD (1977) Cytochalasin B inhibition of brain glucose transport and the influence of blood components on inhibitor concentration. Biochim Biophys Acta 471:477–486

Duckrow RB (1988) Glucose transfer into rat brain during acute and chronic hyperglycemia. Metab Brain Dis 3:201–209
Dyve S, Gjedde A (1991) Glucose metabolism of fetal rat brain in utero measured with labeled deoxyglucose. Acta Neurol Scand 83:14–19
Eadie GS (1942) The inhibition of cholinesterase by physostigmine and prostigmine. J Biol Chem 146:85–93
Eadie GS (1952) On the evaluation of the constants V_m and K_M in enzyme reactions. Science 116:688
Eastman RC, Carson RE, Gordon ME, Berg GW, Lillioja S, Larson SM, Roth J (1990) Brain glucose metabolism in noninsulin-dependent diabetes mellitus: a study in Pima Indians using positron emission tomography during hyperinsulinemia with euglycemic glucose clamp. J Clin Endocrinol Metab 71:1602–1610
Evans AC, Diksic M, Yamamoto YL, Kato A, DAgher A, Redies C, Hakim A (1986) Effect of vascular activity in the determination of rate constants for the uptake of ^{18}F-labeled 2-fluoro-2-deoxy-D-glucose: error analysis and normal values in older subjects. J Cereb Blood Flow Metab 6:724–738
Feinendegen LE, Herzog H, Wieler H, Patton DD, Schmid A (1986) Glucose transport and utilization in the human brain: model using carbon-11 methylglucose and positron emission tomography. J Nucl Med 27:1864–1877
Friedland RP, Budinger TF, Yano Y, Huesman RH, Knittel B, Derenzo SE, Koss B, Ober BA (1983) Regional cerebral metabolic alterations in Alzheimer-type dementia: kinetic studies with 18-fluorodeoxyglucose. J Cereb Blood Flow Metab 3[Suppl 1]:S510–S511
Froehner SC, Davis A, Baldwin SA, Leinhard GE (1988) The blood-nerve barrier is rich in glucose transporter. J Neurocytol 17:173–178
Fuglsang A, Lomholt M, Gjedde A (1986) Blood-brain transfer of glucose and glucose analogs in newborn rats. J Neurochem 46:1417–1428
Fujikawa DG, Dwyer BE, Lake RG, Wasterlain CG (1989) Local cerebral glucose utilization during status epilepticus in newborn primates. Am J Physiol 256 (Cell Physiol 25):C1160–C1167
Fukumoto H, Seino S, Imura H, Seino Y, Bell GI (1988) Characterization and expression of human HepG2/erythrocyte glucose-transporter gene. Diabetes 37:657–661
Fukumoto H, Kayano T, Buse JB, Edwards Y, Pilch PF, Bell GI, Seino S (1989) Cloning and characterization of the major insulin-responsive glucose transporter expressed in human skeletal muscle and other insulin-responsive tissues. J Biol Chem 264:7776–7779
Furler SM, Jenkins AB, Stolien LH, Kraegen EW (1991) In vivo location of the rate limiting step of hexose uptake in muscle and brain tissue of the rat. Am J Physiol 261:E337–E347
Gerhart DZ, LeVasseur RJ, Broderuis MA, Drewes LR (1989) Glucose transporter localization in brain using light and electron immunocytochemistry. J Neurosci Res 22:464–472
Germinario RJ, Rockman H, Oliveira M, Manuel S, Taylor M (1982) Regulation of sugar transport in cultured diploid human skin fibroblasts. J Cell Physiol 112:367–372
Gjedde A (1980) Rapid steady-state analysis of blood-brain glucose transfer in rat. Acta Physiol Scand 108:331–339
Gjedde A (1981a) Regulation and adaptation of substrate transport to the brain. Adv Physiol Sci 7:307–315
Gjedde A (1981b) High- and low-affinity transport of D-glucose from blood to brain. J Neurochem 36:1463–1471
Gjedde A (1982) Calculation of glucose phosphorylation from brain uptake of glucose analogs in vivo. A re-examination. Brain Res Rev 4:237–274
Gjedde A (1983) Modulation of substrate transport to the brain. Acta Neurol Scand 67:3–25

Gjedde A (1984a) Blood-brain transfer of galactose in experimental galactosemia, with special reference to the competitive interaction between galactose and glucose. J Neurochem 43:1654–1662
Gjedde A (1984b) On the measurement of glucose in brain. Neurochem Res 9:1665–1669
Gjedde A (1987) Does deoxyglucose uptake in the brain reflect energy metabolism? Biochem Pharmacol 36:1853–1861
Gjedde A, Bodsch W (1987) Facilitated diffusion across the blood-brain barrier: interactions between receptors and transporters. Karl Marx U. Math Natur Wiss R 36:67–71
Gjedde A, Christensen O (1984) Estimates of Michaelis-Menten constants for the two membranes of the brain endothelium. J Cereb Blood Flow Metab 4:241–249
Gjedde A, Crone C (1975) Induction processes in blood-brain transfer of ketone bodies during starvation. Am J Physiol 229:1165–1169
Gjedde A, Crone C (1981) Blood-brain glucose transfer: repression in chronic hyperglycemia. Science 214:456–457
Gjedde A, Diemer NH (1983) Autroadiographic determination of regional brain glucose content. J Cereb Blood Flow Metab 3:303–310
Gjedde A, Diemer NH (1985a) Relationship between unidirectional and net uptake of glucose and glucose analog into brain: on the variability of transfer and lumped constants. In: Greitz T et al. (eds) The metabolism of the human brain studied with positron emission tomography. Raven, New York, p 207
Gjedde A, Diemer NH (1985b) Double-tracer study of the fine regional blood-brain glucose transfer in the rat by computer-assisted autoradiography. J Cereb Blood Flow Metab 5:282–289
Gjedde A, Lauritzen M (1985) The CO_2-reactivity of blood-brain glucose transport in rat, 12th International Symposium on Cerebral Blood Flow and Metabolism. Studentlitteratur, Lund
Gjedde A, Rasmussen M (1980a) Blood-brain glucose transport in the conscious rat: comparison of the intravenous and intracarotid methods. J Neurochem 35:1375–1381
Gjedde A, Rasmussen M (1980b) Pentobarbital anesthesia reduces blood-brain glucose transfer in the rat. J Neurochem 35:1382–1387
Gjedde A, Siemkowicz E (1978) Post-ischemic coma in rat: effect of glucose and insulin treatment on cerebral metabolic recovery. Trans Am Neurol Assoc 103:45–47
Gjedde A, Hansen AJ, Siemkowicz E (1980) Rapid simultaneous determination of regional blood flow and blood-brain glucose transfer in brain of rat. Acta Physiol Scand 108:321–330
Gjedde A, Hansen AJ, Quistorff B (1981) Blood-brain glucose transfer in spreading depression. J Neurochem 37:807–812
Gjedde A, Wienhard K, Heiss W-D, Kloster G, Diemer NH, Herholz K, Pawlik G (1985) Comparative regional analysis of 2-fluorodeoxyglucose and methylglucose uptake in brain of four stroke patients. With special reference to the regional estimation of the lumped constant. J Cereb Blood Flow Metab 5:163–178
Gjedde A, Kuwabara H, Hakim AM (1990) Reduction of functional capillary density in human brain after stroke. J Cereb Blood Flow Metab 3:317–326
Gjedde A, Kuwabara H, Ohta S, Brust P, Meyer E (1991) Density of perfused capillaries in living human brain during functional activation. Prog Brain Res 19:209–215
Goodwin RFW (1956) Distribution of sugar between red cells and plasma: variations associated with age and species. J Physiol 134 (Lond): 88–101
Growdon WA, Bratton TS, Houston MC, Tarpley HL, Regen DM (1971) Brain glucose metabolism in the intact mouse. Am J Physiol 221:1738–1745
Gutniak M, Blomqvist G, Stone-Elander S, Widén L, Hamberger B, Grill V (1990) Cerebral blood flow and substrate utilization in insulin-treated diabetic subjects. Am J Physiol 258:E805–E812

Hargreaves RJ, Planas AM, Cremer JE, Cunningham VJ (1986) Studies on the relationship between cerebral glucose transport and phosphorylation using 2-deoxyglucose. J Cereb Blood Flow Metab 6:708–716
Harik SI (1988) Glucose transporter of the blood-brain barrier and brain in chronic hyperglycemia. J Neurochem 51:1930–1934
Harik SI, LaManna JC (1988) Vascular perfusion and blood-brain glucose transport in acute and chronic hyperglycaemia. J Neurochem 51:1924–1929
Harik SI, Gravina SA, Kalaria RN (1988) Glucose transporter of the blood-brain barrier and brain in chronic hyperglycaemia. J Neurochem 51:1930–1934
Harik SI, Kalaria RN, Andersson L, Lundahl P, Perry G (1990) Immunocytochemical localization of the erythroid glucose transporter: abundance in tissues with barrier functions. J Neurosci 10:3862–3872
Harik SI, Behmand RA, Arafah BM (1991a) Chronic hyperglycaemia increases the density of glucose transporters in human erythrocyte membranes. J Clin Endocrinol Metab 72:814–818
Harik SI, Behmand RE, LaManna JC (1991b) Chronic hypobaric hypoxia increases the density of cerebral capillaries and their glucose transporter protein, J Cereb Blood Flow Metab 11 [Suppl 1]:S496
Harris M, Prout BJ (1970) Relative hypoglycaemia. Lancet ii:317
Hasselbalch S, Knudsen GM, Jakobsen J, Holm S, Høgh P, Paulson O (1991) The effect of insulin on brain FDG transport and metabolism in man studied by PET. J Cereb Blood Flow Metab 11 [Suppl 2]:S465
Hawkins RA, Mans AM, Davis DW, Hibbard LS, Lu DM (1983) Glucose availability to individual cerebral structures is correlated to glucose metabolism. J Neurochem 40:1013–1018
Hawkins RA, Phelps ME, Huang SC (1984) Effects of temporal sampling, glucose metabolic rates, and disruptions of the blood-brain barrier on the FDG model with and without a vascular compartment: studies in human brain tumor with PET. J Cereb Blood Flow Metab 6:170–183
Hawkins RA, Phelps ME, Huang SC (1986) Effects of temporal sampling, glucose metabolic rates, and disruptions of the blood-brain barrier on the FDG model with and without a vascular compartment: studies in human brain tumor with PET. J Cereb Blood Flow Metab 6:170–183
Helgerson AL, Carruthers A (1989) Analysis of protein-mediated 3-O-methylglucose transport in rat erythrocytes: rejection of the alternating conformation carrier model for sugar transport. Biochemistry 28:4580–4594
Helgerson AL, Hebert DN, Naderi S, Carruthers A (1989) Characterization of two independent modes of action of ATP on human erythrocyte sugar transport. Biochemistry 28:6410–6417
Hertz MM, Paulson OB (1982) Transfer across the human blood-brain barrier: evidence for capillary recruitment and for a paradox glucose permeability increase in hypocapnia. Microvasc Res 24:364–376
Hertz MM, Paulson OB, Barry DI, Christensen JS, Svendsen PA (1981) Insulin increases glucose transfer across the blood-brain barrier in man. J Clin Invest 67:597–604
Hofstee BHJ (1952) On the evaluation of the constants V_m and K_M in enzyme reactions. Science 116:329–331
Hofstee BHJ (1954–56) Graphical analysis of single enzyme systems. Enzymologia 17:273–278
Horton RW (1973) The kinetics of glucose influx into the mouse brain in vivo and the effects of pentobarbitone anaesthesia and hypothermia. 54th Meeting Transactions of the Biochemical Society 1:127–128
Hunziker O, Abdel'al S, Schulz U (1979) The aging human cerebral cortex: a stereological characterization of changes in the capillary net. J Gerontol 34:345–350
Jacquez JA (1984) Red blood cell as glucose carrier: significance for placental and cerebral glucose transfer. Am J Physiol 246:R289–R298

Jagust WJ, Seab JP, Huesman RH, Valk PE, Mathis CA, Reed BR, Coxson PG, Budinger TF (1991) Diminished glucose transport in Alzheimer's disease: dynamic PET studies. J Cereb Blood Flow Metab 11:323–330

James DE, Brown R, Navarro J, Pilch PF (1988) Insulin-regulatable tissues express a unique insulin-sensitive glucose transport protein. Nature 333:183–185

James DE, Strube M, Mueckler M (1989) Molecular cloning and characterization of an insulin-regulatable glucose transporter. Nature 338:83–87

Johnson JA, Wilson TA (1966) A model for capillary exchange. Am J Physiol 210:1299–1303

Joó F (1985) The blood-brain barrier in vitro: ten years of research on microvessels isolated from the brain. Neurochem Int 7:1–25

Kalaria RN, Harik SI (1989) Reduced glucose transporter at the blood-brain barrier and in cerebral cortex in Alzheimer's disease. J Neurochem 53:1083–1088

Kalaria RN, Gravina SA, Schmidley JW, Perry G, Harik SI (1988) The glucose transporter of the human brain and blood-brain barrier. Ann Neurol 24:757–764

Kasahara M, Hinkle PC (1976) Reconstitution of D-glucose transport catalyzed by a protein fraction from human erythrocytes in sonicated liposomes. Proc Natl Acad Sci USA 73:396–400

Kasahara M, Hinkle PC (1977) Reconstitution and purification of the D-glucose transporter from human erythrocytes. J Biol Chem 252:7384–7390

Kasanicki MA, Cairns MT, Davis A, Gardiner RM, Baldwin SA (1987) Identification and characterization of the glucose-transport of the bovine blood/brain barrier. Biochem J 247:101–108

Kato A, Diksic M, Yamamoto YL, Strother SC, Feindel W (1984) An improved approach for measurement of regional cerebral rate constants in the deoxyglucose method with positron emission tomography. J Cereb Blood Flow Metab 4:555–563

Kayano T, Fukumoto H, Eddy RL, Fan Y-S, Byers MG, Shows TB, Bell GI (1988) Evidence for a family of human glucose transporter-like proteins. J Biol Chem 263:15245–15248

Kayano T, Burant CF, Fukumoto G, Gould W, Fan Y-S, Eddy RL, Byers MG, Shows TB, Scine S, Bell GI (1990) Human facilitative glucose transporters. J Biol Chem 265:13276–13282

Kety SS (1951) The theory and application of the exchange of inert gas at the lungs and tissues. Pharmacol Rev 3:1–41

Kety SS (1960) Theory of blood-tissue exchange and its application to measurement of blood flow. In: Bruner HD (ed) Methods in medical research, vol 8. Year Book, New York, p 223

Kintner D, Costello DJ, Levin AB, Gilboe DD (1980) Brain metabolism after 30 minutes of hypoxic or anoxic perfusion or ischemia. Am J Physiol 239:E501–E509

Kletzien RF, Perdue J (1985) Induction of sugar transport in chick embryo fibroblasts by hexose starvation: evidence for transcriptional regulation of transport. J Biol Chem 250:593–600

Kletzien RF, Perdue J, Springer A (1972) Inhibition of sugar uptake in cultured cells. J Biol Chem 247:2964–2966

Klip A, Paquet MR (1990) Glucose transport and glucose transporters in muscle and their metabolic regulation. Diabetes Care 13:228–243

Klip A, Dimitrakoudis D, Ramlal T, Burdett E, Marette A, Bilan P, Mitsumoto Y, Kojvisto U-M, Sarabia V (1991) Regulation of glucose transport and transporters in muscle: acute and chronic effects of insulin and glucose. Symposium on advances in regulation of carbohydrate metabolism, Jerusalem

Knudsen GM, Pettigrew KD, Paulson OB, Hertz MM, Patlak CS (1990) Kinetic analysis of blood-brain barrier transport of D-glucose in man: quantitative evaluation in the presence of tracer backflux and capillary heterogeneity. Microvasc Res 39:28–49

Knudsen GM, Hertz M, Paulson OB (1991) Does capillary recruitment exist in the human brain? J Cereb Blood Flow Metab 11 [Suppl 2]:S442

Kolber AR, Bagnell CR, Krigman MR, Hayward J, Morell P (1979) Transport of sugars into microvessels isolated from rat brain: a model for the blood-brain barrier. J Neurochem 33:419–432

Krogh A (1946) The active and passive exchanges of inorganic ions through the surface of living cells and through membranes generally. Proc R Soc Lond Ser B 133:140–200

Kuwabara H, Gjedde A (1991) Measurements of glucose phosphorylation with FDG and PET are not reduced by dephosphorylation of FDG-6-phosphate. J Nucl Med 32:692–698

Kuwabara H, Evans AC, Gjedde A (1990) Michaelis-Menten constraints improved cerebral glucose metabolism and regional lumped constant measurements with [^{18}F]fluroodeoxyglucose. J Cereb Blood Flow Metab 10:180–189

LaManna JC, Harik SI (1985) Regional comparisons of brain glucose influx. Brain Res 326:299–305

LaManna JC, Harik SI (1986) Regional studies of blood-brain transport of glucose and leucine in awake and anesthetized rats. J Cereb Blood Flow Metab 6:717–723

Lammertsma AA, Brooks DJ, Frackowiak SJ, Beaney RP, Herold S, Heather JD, Palmer AJ, Jones T (1987) Measurement of glucose utilization with [^{18}F]2-fluoro-2-deoxy-D-glucose: a comparison of different analytical methods. J Cereb Blood Flow Metab 7:161–172

Langmuir I (1916) The constitution and fundamental properties of solids and liquids. J Am Chem Soc 37:2221–2295

Lassen NA, Gjedde A (1983) Kinetic analysis of the uptake of glucose and some of its analogs in the brain, using the single capillary model: comments on some points of controversy. In: Lambrecht RM, Rescigno A (eds)) Lecture notes in biomathematics. Springer, Berlin Heidelberg New York, pp 387–410

Lassen NA, Trap-Jensen AJ, Alexander SC, Olesen J, Paulson OB (1971) Blood-brain barrier studies in man using the double-indicator method. Am J Physiol 220:1627–1633

LeFèvre PG (1948) Evidence of active transfer of certain non-electrolytes across the human red cell membrane. J Gen Physiol 31:505–527

LeFèvre PG (1961) Sugar transport in the red blood cell: structure activity relationships in substrates and antagonists. Pharmacol Rev 16:39–70

Levine R, Goldstein M, Klein S, Huddlestun B (1949) The action of insulin on the distribution of galactose in eviscerated nephrectomized dogs. J Biol Chem 179:985–986

Lilavivathane U, Brodows RG, Woolf PD, Campbell RG (1979) Counterregulatory hormonal responses to rapid glucose lowering in diabetic man. Diabetes 28:873–877

Lin S, Spudich JA (1974) Biochemical studies on the mode of action of cytochalasin B. J Biol Chem 249:5778–5783

Lowe AG, Walmsley AR (1986) The kinetics of glucose transport in human red blood cells. Biochim Biophys Acta 857:146–154

Lund-Andersen H (1979) Transport of glucose from blood to brain. Physiol Rev 59:305–352

Lund-Andersen H, Kjeldsen CS (1976) Uptake of glucose analogues by rat brain cortex slices: a kinetic analysis based upon a model. J Neurochem 27:361–368

Lund-Andersen H, Kjeldsen CS (1977) Uptake of glucose analogues by rat brain cortex slices: membrane transport versus metabolism of 2-deoxy-D-glucose. J Neurochem 29:205–211

Lundsgaard E (1939) On the mode of action of insulin. Uppsala Läk Fören Förh 45:1–4

Mans AM, Davis DW, Biebuyck JF, Hawkins RA (1986) Failure of glucose and branched chain amino acids to normalize brain glucose use in portacaval shunted rats. J Neurochem 47:1434–1443
Matthaei S, Horuk R, Olefsy JM (1986) Blood-brain glucose transfer in diabetes mellitus. Decreased number of glucose transporters at blood-brain barrier. Diabetes 35:1181–1184
Matthaei S, Garvey WT, Horuk T, Hueckstadt TP, Olefsky JM (1987a) Human adipocyte glucose transport system: biochemical and functional heterogeneity of hexose carriers. J Clin Invest 79:703–709
Matthaei S, Olefsky JM, Horuk R (1987b) Biochemical characterization and subcellular distribution of the glucose transporter from rat brain microvessels. Biochim Biophys Acta 905:417–425
Maxwell K, Berliner JA, Cancilla PA (1989) Stimulation of glucose analogue uptake by cerebral microvessel endothelial cells by a product released by astrocytes. J Neuropathol Exp Neurol 48:69–80
Mayman CI, Gatfield PD, Breckenbridge BM (1964) The glucose content of brain in anaesthesia. J Neurochem 11:483–487
McCall AL, Millington WR, Wurtman RJ (1982) Metabolic fuel and amino acid transport into the brain in experimental diabetes mellitus. Proc Natl Acad Sci USA 79:5406–5410
McCall AL, Fixman LB, Fleming N, Tornheim K, Chick W, Ruderman NB (1986) Chronic hypoglycemia increases brain glucose transport. Am J Physiol 251:E442–E447
Michaelis L, Menten ML (1913) Die Kinetik der Invertinwirkung. Biochem Z 49:333–369
Mooradian AD, Morin AM (1991) Brain uptake of glucose in diabetes mellitus: the role of glucose transporters. Am J Med Sci 301:173–177
Moore TJ, Lione AP, Sugden MSM, Regen DM (1976) β-hydroxybutyrate transport in rat brain: developmental and dietary modulations. Am J Physiol 230:619–630
Morin AM, Dwyer BE, Fujikawa DG, Wasterlain CG (1988) Low [3H]cytochalasin B binding in the cerebral cortex of newborn rat. J Neurochem 51:206–211
Mueckler MC, Caruso C, Baldwin SA, Panico M, Blench I, Morris HR, Allard WJ, Lienhard GE, Lodish HF (1985) Sequence and structure of a human glucose transporter. Science 229:941–945
Naftalin RJ (1988) Pre-steady-state uptake of D-glucose by the human erythrocyte is inconsistent with a circulating carrier mechanism. Biochim Biophys Acta 946:431–438
Namba H, Lucignani G, Nehlig A, Patlak C, Pettigrew K, Kennedy C, Sokoloff L (1987) Effects of insulin on hexose transport across blood- brain barrier in normoglycemia. Am J Physiol 252:E299–E303
Nemoto EM, Stezoski SW, MacMurdo D (1978) Glucose transport across the rat blood-brain barrier during anesthesia. Anesthesiology 49:170–176
Oka Y, Asano T, Shibasaki Y, Kasuga M, Kanazawa Y, Takaku F (1989) Studies with antipeptide antibody suggest the presence of at least two types of glucose transporter in rat brain and adipocyte. J Biol Chem 263:13432–13439
Oka Y, Asano T, Shibasaki Y, Lin J-L, Tsukuda K, Katagiri H, Akanuma Y, Takaku F (1990) C-terminal truncated glucose transporter is locked into an inward-facing form without transport activity. Nature 345:550–553
Oldendorf WH (1970) Measurement of brain uptake of radiolabeled substances using a tritiated water internal standard. Brain Res 24:372–376
Oldendorf WH (1971) Brain uptake of labeled amino acids, amines, and hexoses after arterial injection. Am J Physiol 221:1629–1639
Oldendorf WH, Braun LD (1976) [^{3}H]tryptamine and ^{3}H-water as diffusible internal standards for measuring brain extraction of radiolabeled substances following carotid injection. Brain Res 113:219–224
Pappenheimer JR, Setchell (1973) Cerebral glucose transport and oxygen consumption in sheep and rabbits. J Physiol (Lond) 233:529–551

Pardridge WM (1983) Brain metabolism: a perspective from the blood-brain barrier. Physiol Rev 63:1481–1535
Pardridge WM, Oldendorf WH (1975) Kinetics of blood-brain barrier transport of hexoses. Biochim Biophys Acta 382:377–392
Pardridge WM, Oldendorf WH (1977) Transport of metabolic substrates through the blood-brain barrier. J Neurochem 28:5–12
Pardridge WM, Crane PD, Mietus LJ, Oldendorf WH (1982) Kinetics of regional blood-brain barrier transport and brain phosphorylation of glucose and 2-deoxyglucose in the barbiturate-anesthetized rat. J Neurochem 38:560–568
Pardridge WM, Landaw EM, Miller LP, Braun LD, Oldendorf WH (1985) Carotid artery injection technique: bounds for bolus mixing by plasma and by brain. J Cereb Blood Flow Metab 5:576–583
Pardridge WM, Boado RJ, Farrell C (1990a) Downregulation of blood-brain barrier glucose transporter in experimental diabetes. Diabetes 39:1040–1044
Pardridge WM, Boado RJ, Farrell C (1990b) Brain-type glucose transporter (GLUT-1) is selectively localized to the blood-brain barrier. J Biol Chem 265:18035–18040
Patlak CS, Blasberg RG, Fenstermacher JD (1983) Graphical evaluation of blood-to-brain transfer constants from multiple-time uptake data. J Cereb Blood Flow Metab 3:1–7
Pelligrino DA, Segil LJ, Albrecht RF (1990a) Brain glucose utilization and transport and cortical function in chronic vs. acute hypoglycemia. Am J Physiol 259:E729–E735
Pelligrino DA, Lipa MD, Albrecht RF (1990b) Regional blood-brain glucose transfer and glucose utilization in chronically hyperglycemic, diabetic rats following acute glycemic normalization. J Cereb Blood Flow Metab 10:775–780
Pessin JE, Tillotson LG, Yamada K, Gitomer W, Carter-Su C, Mora R, Isselbacher KJ, Czech M (1982) Identification of the stereospecific hexose transporter from starved and fed chicken embryo fibroblasts. Proc Natl Acad Sci USA 79:2286–2290
Phelps ME, Huan SC, Hoffman EJ, Selin C, Sokoloff L, Kuhl DE (1979) Tomographic measurement of local cerebral glucose metabolic rate in humans with [F^{18}] 2-fluoro-2-deoxy-D-glucose: validation of method. Ann Neurol 6:271–388
Pilch PF (1990) Glucose transporters: what's in a name? Endocrinology 126:3–5
Planas AM, Cunningham VJ (1987) Uncoupling of cerebral glucose supply and utilization after hexane-2,5-dione intoxication in the rat. J Neurochem 48:816–823
Pollay M, Stevens FA (1979) Starvation-induced changes in transport of ketone bodies across the blood-brain barrier. J Neurosci Res 5:163–172
Raichle ME, Larson KB, Phelps ME, Grubb RL Jr, Welch MJ, Ter-Pogossian MM (1975) In vivo measurement of brain glucose transport and metabolism employing glucose-^{11}C. Am J Physiol 228:1936–1948
Ramlal T, Rastogi S, Vranic M, Klip A (1989) Decrease in glucose transporter number in skeletal muscle of mildly diabetic (streptozotocin-treated) rats. Endocrinology 125:890–897
Redies C, Hoffer LJ, Beil C, Marliss EB, Evans AC, Lariviere F, Marrett S, Meyer E, Diksic M, Gjedde A, Hakim AM (1989) Generalized decrease in brain glucose metabolism during fasting in humans studied by PET. Am J Physiol 256:E805–E810
Reith J, Ermisch A, Diemer NH, Gjedde A (1987) Saturable retention of vasopressin in vivo, associated with inhibition of blood-brain transfer of large neutral amino acids. J Neurochem 49:1471–1480
Reivich M, Kuhl D, Wolf A, Greenberg J, Phelps M, Ido T, Casella V, Fowler J, Hoffman E, Alavi A, Som P, Sokoloff L (1979) The [18F]fluorodeoxyglucose method for the measurement of local cerebral glucose utilization in man. Circ Res 44:127–137

Reivich M, Alavi A, Wolf A, Fowler J, Russell J, Arnett C, MacGregor RR, Shiue CY, Atkins H, Anand A, Dann R, Greenberg JH (1985) Glucose metabolic rate kinetic model parameter determination in humans: the lumped constants and rate constants for [18F]fluorodeoxyglucose and [^{11}C]deoxyglucose. J Cereb Blood Flow Metab 5:179–192

Rescigno A, Beck JS, Goren HJ (1983) Determination of dependence of binding parameters on receptor occupancy. Bull Math Biol 44:477–489

Ribeiro L, Bercovic SF, Kuwabara H, Andermann F, Gjedde A (1991) Functional capillary density is reduced in brain of patients with MERFF (myoclonus epilepsy, ragged-red fiber syndrome). 21st Annual Meeting for Society for Neuroscience, 10–15 Nov 1991, New Orleans

Roy CS, Sherrington CS (1890) On the regulation of the blood supply of the brain. J Physiol 11:85–108

Sapirstein LA (1958) Regional blood flow by fractional distribution of indicators. Am J Physiol 193:161–165

Sarna GS, Bradbury MWB, Cremer JE, Lai JCK, Teal HM (1979) Brain metabolism and specific transport at the blood-brain barrier after portocaval anastomosis in the rat. Brain Res 160:69–83

Sawada Y, Patlak CS, Blasberg RG (1989) Kinetic analysis of cerebrovascular transport based on indicator diffusion technique. Am J Physiol 256:794–812

Scatchard G (1949) The attractions of proteins for small molecules and ions. Ann NY Acad Sci 51:660–672

Schaefer JA, Gjedde A, Plum F (1976) Regional cerebral blood blow using n-(14C)butanol. Neurology 26:394

Shapiro ET, Cooper M, Chen C-T, Given BDF, Polonsky KS (1990) Change in hexose distribution volume and fractional utilization of [18F]-2-deoxy-2-fluoro-D-glucose in brain during acute hypoglycemia in humans. Diabetes 39:175–180

Sheppard CW (1948) The theory of the study of transfers within a multi-compartmental system using isotopic tracers. J Appl Physiol 19:70–76

Shows TB, Eddy RL, Byers MG, Fukushima Y, Dehaven CR, Murray JC, Bell GI (1987) Polymorphic human glucose transport gene (GLUT) is on chromosome 1p31.3→p35. Diabetes 36:546–549

Slot JW, Moxley R, Geuze HJ, James DE, (1990) No evidence for expression of the insulin-regulatable glucose transporter in endothelial cells. Nature 346:369–371

Sokoloff L, Reivich M, Kennedy C, Des Rosiers MH, Patlak CS, Pettigrew KD, Sakurada O, Shinohara M (1977) The ^{14}C deoxyglucose method for the measurement of local cerebral glucose utilization: theory, procedure, and normal values in the conscious and anaesthetized albino rat. J Neurochem 28:897–916

Stein WD (1986) Transport and diffusion across cell membranes. Academic, New York

Stern L, Gautier R (1921) Recherches sur le liquide céphalo-rachidien. I. Les rapports entre le liquide céphalo-rachidien et la circulation sanguine. Arch Int Physiol 17:138–192

Takasato Y, Rapoport SI, Smith QR (1984) An in situ brain perfusion technique to study cerebrovascular transport in the rat. Am J Physiol 247:H484–H493

Taverna RD, Langdon RG (1973) Reversible association of cytochalasin B with the human erythrocyte membrane: inhibition of glucose transport and the stoichiometry of cytochalasin binding. Biochim Biophys Acta 323:207–219

Thorens B, Sarkar HK, Kaback HR, Lodish HF (1988) Cloning and functional expression in bacteria of a novel glucose transporter present in liver, intestine, kidney, and beta pancreatic islet cells. Cell 55:281–290

Tucker SP, Cunningham VJ (1988) Autoradiography of [^{3}H]cytochalasin B binding in rat brain. Brain Res 450:131–136

Ullrey DB, Kalckar HM (1981) The nature of regulation of hexose transport in cultured mammalian fibroblasts: Aerobic ‘repressive’ control by D-glucosamine. Arch Biochem Biophys 209:168–174

Ullrey DB, Gammon BMT, Kalckar HM (1975) Uptake patterns and transport enhancements in cultures of hamster cells deprived of carbohydrates. Arch Biochem Biophys 167:410–416

Van Uitert RL, Levy DE (1978) Regional brain blood flow in the conscious gerbil. Stroke 9:67

Walker PS, Ramial T, Donovan JA, Doering TP, Sandra A, Klip A, Pessin JE (1989) Insulin and glucose-dependent regulation of the glucose transport system in the rate L6 skeletal muscle cell line. J Biol Chem 264:6587–6595

Werner H, Adamo M, Lowe WL, Roberts CT, LeRoith D (1989) Developmental regulation of rat brain/HepG2 glucose transporter gene expression. Mol Endocrinol 3:273–279

Wertheimer E, Sasson S, Cerasi C, Ben-Neriah Y (1991) The ubiguitous glucose transporter Glut-1 belongs to the glucose-regulated protein family of stress inducible protein. Proc Natl Acad Sci USA 88:2525–2529

Widdas WF (1952) Inability of diffusion to account for placental glucose transfer in the sheep and consideration of the kinetics of a possible carrier transfer. J Physiol (Lond) 118:23–39

Widdas WF (1955) Hexose permeability of foetal erythrocytes. J Physiol (Lond) 127:318–327

Wyke BD (1959) Electroencephalographic studies in the syndrome of relative cerebral hypoglycemia. Electroencephalogr Clin Neurophysiol 11:602

Yudelevich DL, DeRose N (1971) Blood-brain barrier transfer of glucose and other molecules measured by rapid indicator dilution. Am J Physiol 220:841–846

Zloković BV, Begley DJ, Djuričić BM, Mitrovic DM (1986) Measurement of solute transport across the blood-brain barrier in the perfused guinea pig brain: method and application to N-methyl-α-aminoisobutyric acid. J Neurochem 46:1444–1451

CHAPTER 5

Transport of Amino Acids

J.-M. LEFAUCONNIER

Transport of amino acids is very important for brain function as some of them are the precursors of neurotransmitters (serotonin, catecholamines, histamine) and of small polypeptides, and they participate in protein synthesis which is very active in some areas of the brain. The physiology of the endothelial cells responsible of the blood-brain barrier (BBB) has been compared to that of an epithelium (BETZ and GOLDSTEIN 1978; BRADBURY 1984; FENSTERMACHER and RAPOPORT 1984; CRONE 1985). This means that the transport systems at the luminal membrane may be different to those at the abluminal membrane.

A. Amino Acid Transport Systems in Animal Cells

The amino acid transport systems in different types of tissue have been studied principally by CHRISTENSEN et al. (for review see COLLARINI and OXENDER 1987). Neutral amino acids are transported mainly by one of the three systems described below but they are generally also transported to a lesser extent by the other systems.

1. *The A system* (OXENDER and CHRISTENSEN 1963). This is a sodium-dependent system and serves mainly for cell uptake of amino acids with short, polar, or linear side chains, such as alanine, glycine, proline, and the non-metabolizable 2-aminoisobutyric acid (AIB) and its methylated derivative, 2-(methylamino)isobutyric acid (MeAIB). Uptake of amino

Abbreviations

AIB	α-amino-isobutyric acid	CBF	cerebral blood flow
AMPc	cyclic adenosine monophosphate	CSF	cerebrospinal fluid
		DOPA	dihydroxyphenylalanine
ASC	alanine, serine, cysteine	GABA	γ-aminobutyric acid
ATP	adenosine triphosphate	γ-GT	γ-glutamyltranspeptidase
BBB	blood–brain barrier	HTP	hydroxytryptophan
BCAA	branched chain amino acids	LNAA	large neutral amino acids
BCH	2-aminobicyclo-2,2,1-heptane-2-carboxylic acid	MeAIB	methyl-α-aminoisobutyric acid
BUI	brain uptake index	PCA	portacaval anastomosis

acids by system A is often subject to inhibition by its own intracellular substrates (*trans*-inhibition) and is greatly decreased at lowered extracellular pH. This system has been described in a wide variety of cells and is subject to significant regulation by amino acid availability and by hormones.

2. *The L system*. This is a sodium independent system and is reactive with the large neutral amino acids (LNAA): branched chain (BCAA) and aromatic amino acids, such as leucine, isoleucine, valine, phenylalanine, tyrosine, tryptophan, and the nonmetabolizable analog 2-aminobicyclo-2,2,1-heptane-2-carboxylic acid (BCH). It is mainly an exchange system. It can regulate its transport activity in response to the availability of system L amino acids (*trans*-stimulation). This system is present in a variety of cell types. Subsequently, other sodium-independent systems have been described: in hepatocytes WEISSBACH et al. (1982) distinguished a system L1 with a very high affinity for its substrates and a system L2, and system T for aromatic amino acids (ROSENBERG et al. 1980).

3. *The alanine, serine, cysteine (ASC) system* (CHRISTENSEN et al. 1967). This is also a sodium-dependent system and transports mainly alanine, serine, cysteine, and homologous amino acids with up to five carbons. In addition to its substrate preference, this system is distinguished from system A by its intolerance to N-methylated substrates, its higher stereospecificity, and its different pH sensitivity. Intracellular substrates of the ASC system increase its activity (*trans*-stimulation). It is the major sodium-dependent system in most cell types.

Cationic amino acids such as lysine and arginine (WHITE et al. 1982) can be transported by a sodium-independent system called system y^+. It is subject to *trans*-stimulation. Anionic amino acids such as glutamate and aspartate are transported by many different systems including the x^-_{AG} system which is sodium dependent and prefers aspartate and glutamate (CHRISTENSEN 1984, 1985). For taurine, a β-amino acid, a sodium-dependent system has been suggested. (CHRISTENSEN 1984)

B. Neutral and Cationic Amino Acids

In the transport of amino acids, it is necessary to know their influx (rate at which they enter the brain), their efflux (rate at which they leave the brain), and their net uptake which is the difference between influx and efflux.

I. Influx

1. Quantification of Influx

The influx is equal to the permeability–surface area product (*PS*) multiplied by the plasma concentration of the amino acids. However, in practice, it is transfer constants that are measured: either the extraction fraction (*E*) or the unidirectional transfer constant for influx (K_i). They must be measured over a short period of time to exclude the backflux phenomenon.

The extraction fraction E is determined in experiments in which the test material is injected intraarterially as a bolus. It is equal to the amount of test material which enters the brain parenchyma, measured per unit mass of brain tissue (A), divided by the amount of test material presented to the relevant capillary bed during the same time (B):

$$E = \frac{A}{B}. \tag{1}$$

The unidirectional transfer constant K_i is measured in multiple-pass experiments in which the test substance is administered intravenously. It is the total amount of test substance (A) which has passed into brain parenchyma divided by the arterial concentration–time integral. A is expressed per unit mass of brain tissue at the termination of experiment. C_p is the concentration of the test material in arterial plasma at time t. T is the time of the end of the experiment.

$$K_i = \frac{A}{\int_0^T C_p \, dt} \tag{2}$$

The units of K_i are volume mass^{-1} time^{-1}.

The two values thus determined are linked by the formula:

$$K_i = FEV_c, \tag{3}$$

where V_c is the fractional distributional volume of the test material in intracapillary fluid and F is the cerebral blood flow.

E is linked to the capillary surface area by the formula:

$$E = 1 - \exp(-PS/FV_c) \tag{4}$$

Influx has been studied by in vivo experiments using different methods, as described in Chap. 2, generally involving radioactive amino acids. Two different approaches, (intraarterial and intravenous injection) have been used.

With *intra-arterial injection*, the amino acids are injected via an artery and there is theoretically no interference by competitors. There is often a reference substance that remains intravascular or is diffusible. There are three techniques:

1. The *indicator diffusion technique* (Yudilevich et al. 1972), where the extraction of the test material is calculated from the arteriovenous difference. However, arterial concentration is not measured directly. Instead, a reference substance that remains within the intracerebral circulation is used as an indicator of the arterial concentration of the test material. This technique has also been used in humans (Knudsen et al. 1990) and results obtained for influx and also for backflux.

2. The *brain uptake index* (BUI) technique (OLDENDORF 1971), in which the ratio of the brain concentration of a radioactive amino acid to that of a radioactive diffusible reference substance is calculated (AA/R). The BUI is the ratio as calculated above divided by the ratio in the injected material and expressed as a percentage.
3. The *in situ brain perfusion technique* (TAKASATO et al. 1984), in which a catheter is inserted retrogradely in the external carotid artery and the hemisphere perfused with buffer or plasma after ligature of the common carotid artery. The radioactivity is measured in different brain areas and no reference substance is required as transfer depends only on the perfusion characteristics. There is a minimal amino acid contribution from systemic blood (MOMMA et al. 1987).

With *intravenous injection*, the solute is injected using one of three techniques: (1) a bolus injection (OHNO et al. 1978; BLASBERG et al. 1983); (2) a continuous infusion; or (3) a programmable infusion to maintain a constant plasma concentration (DANIEL et al. 1975a). The disadvantage of these methods is the obligatory presence of plasma amino acids which are competitors for the amino acids being studied; their advantage is that they allow measurements to be made in physiological or pathological conditions.

Autoradiography can be used with either method, but has mainly been used with the intravenous method (HAWKINS et al. 1982). The methods can also been used in humans thanks to positron emission tomography, which gives images like those of autoradiography (COMAR et al. 1976; HAWKINS et al. 1989).

These techniques have revealed some important characteristics of luminal transport. The transport of essential amino acids and tyrosine is saturable and stereospecific, while the transport of nonessential amino acids is very low (OLDENDORF 1971; BAÑOS et al. 1973). The main systems for the transport of amino acids at the BBB are the L system for neutral amino acids and the cationic amino acid system (RICHTER and WAINER 1971; WADE and KATZMAN 1975a). Some transport of anionic amino acids has also been described (OLDENDORF and SZABO 1976). Transport by the ASC system has also been shown by SERSHEN and LAJTHA (1979) and TOVAR et al. (1988), while studies of cysteine transport by WADE and BRADY (1981) concluded that if the ASC system existed it was not quantitatively important.

2. Summary of the Results Obtained

The L system has been extensively studied. It transports the amino acids phenylalanine, leucine, isoleucine, tyrosine, tryptophan, methionine, histidine, valine, and threonine and also some drugs which are synthetic amino acids: dihydroxyphenylalanine (DOPA; WADE and KATZMAN 1975a,b) and methyl-DOPA (MARKOVITZ and FERNSTROM 1977).

As in other cells, the transport rate is given by the Michaelis-Menten equation, and it has generally been necessary to add a small nonsaturable component:

$$V = \frac{V_{max}\,[AA]}{K_m + [AA]} + K_d\,[AA] \tag{5}$$

where V_{max} is the maximum rate of transport, [AA] is the liquid concentration of the amino acid, K_m is the Michaelis constant for transport, and K_d is the constant for transport which occurs by simple diffusion. Using the BUI technique, PARDRIDGE and OLDENDORF (1975) found K_m values for amino acids of the L system indicating a high affinity. More recently, the results obtained by SMITH et al. (1987) with the in situ brain perfusion technique indicate a still higher affinity. The difference between the two techniques is that in the BUI technique the amino acid is injected in a bolus of known composition and blood flow returns to normal immediately after the injection. In the in situ brain perfusion technique, the amino acid is injected in a buffered physiological saline, which can be perfectly controlled and gives a much higher flow than in the BUI technique. It is possible that in the BUI technique the injection modifies the blood flow and there is some mixing of the bolus with blood at the moment of injection. In addition, during measurement of transport, there may be some efflux of cold amino acids across the endothelium from the brain. These amino acids may interact with the radioactive amino acids. This hypothesis is supported by the results of MILLER et al. (1985), who used the BUI technique to measure the transport of amino acids in conscious animals. They found that the cerebral blood flow (CBF) was higher than in anaesthetized animals and that the K_m for the different amino acids was much lower. The variations of K_m among the amino acids with measurable affinity for the cerebrovascular transport system depend on differences in side-chain hydrophobicity (SMITH et al. 1987).

When the influx of one amino acid, for example phenylalanine, has been studied in plasma instead of buffer, its transport has been found to be greatly decreased (OLDENDORF 1971). This appeared to be due to competition from the amino acids transported by the same system.

The phenomenon of competition among the different amino acids of the L system for blood–brain transport appears to be very important. This has already been shown by FERNSTROM and WURTMAN (1972a), who proposed that the brain tryptophan concentration depends on the ratio of plasma tryptophan concentration to the sum of the plasma concentrations of competing amino acids (tyrosine, phenylalanine, leucine, isoleucine, valine): [tryptophan]/[LNAA].

Later studies (PARDRIDGE 1977) showed that for calculations concerning amino acids it was necessary to take into account the concentration of all

Table 1. K_m (mM) and apparent K_m ($K_{m,app}$) of various amino acids according to three techniques:

	BUI (anesthetized)[a]		BUI (conscious)[b]		In situ perfusion[c]	
	K_m	$K_{m,app}$	K_m	$K_{m,app}$	K_m	$K_{m,app}$
Phenylalanine	0.12	0.45	0.032	0.213	0.011	0.17
Tryptophan	0.19	0.71	0.052	0.356	0.015	0.33
Tyrosine	0.16	0.58	0.086	0.615	0.064	1.42
Leucine	0.15	0.53	0.087	0.614	0.029	0.50
Isoleucine	0.33	1.30	0.145	1.120	0.056	1.21
Methionine	0.19	0.77	0.083	0.642	0.040	0.86
Histidine	0.28	1.10	0.164	1.296	0.10	2.22
Valine	0.63	2.50	0.168	1.239	0.21	4.69
Threonine	0.73	3.00			0.22	4.86
Glutamine					0.88	19.90

[a] In the anaesthetized animal: from PARDRIDGE and OLDENDORF (1975) and PARDRIDGE (1977).
[b] In the conscious animal: from MILLER et al. (1985).
[c] With the in situ brain perfusion technique: from SMITH et al. (1987).

other amino acids [AA] and their respective K_m values, i.e., K_m, AA, to calculate the apparent K_m according to the formula:

$$K_{m,app} = K_m \left(1 + \sum \frac{[AA]}{K_{m,AA}}\right) \tag{6}$$

The rate of unidirectional influx of a circulating amino acid into brain is then:

$$V = \frac{V_{max}\,[AA]}{K_{m,app} + [AA]} + K_d\,[AA] \tag{7}$$

Although competition among amino acids for a common transport system has been demonstrated for numerous cell membranes, physiological changes in plasma amino acid levels will cause competitive effects only if the K_m of the amino acid transport approximates the plasma level (PARDRIDGE and OLDENDORF 1977).

Table 1 gives the K_m values obtained by the BUI technique in anaesthetized animals (PARDRIDGE and OLDENDORF 1975) and in conscious animals (MILLER et al. 1985) and the K_m obtained with the in situ brain perfusion technique (SMITH et al. 1987).

Tryptophan is the only amino acid which is appreciably bound to albumin. The problem is to know whether only the fraction which is free in vitro enters the brain or whether a fraction from the albumin-bound tryptophan can also enter. The interactions of ligand and protein in vitro can be very different to events in vivo (PARDRIDGE 1983). The passage of plasma proteins through a capillary bed represents a nonequilibrium condition under most circumstances as the albumin–tryptophan complex dissociates and reassociates several times during capillary transit (PARDRIDGE 1983). Thus there is competition between the apparent tryptophan-binding capacity of the endothelial cell carrier and the tryptophan-binding capacity of albumin. If f is the fraction of the exchangeable tryptophan in vivo, then:

$$f = \frac{K_D^a}{K_D^a + [A]_t} \qquad (8)$$

where K_D^a is the apparent dissociation constant of in vivo albumin binding of tryptophan in the brain capillary milieu and $[A]_t$ total albumin concentration.

SMITH et al. (1990) showed that the free fraction in vivo exceeded by two-fold that measured in vitro by equilibrium analysis and varied inversely according to the blood flow. PARDRIDGE and FIERER (1990) showed that the K varied with the anaesthetic and thus with the blood flow.

This has also been studied in humans (GILLMAN et al. 1981). The concentrations of free and total tryptophan and of LNAA were measured in plasma, lumbar and ventricular cerebrospinal fluid (CSF), and brain of psychiatric patients undergoing a stereotactic subcaudate tractotomy. During the intervention it was necessary to remove ventricular CSF and a small amount of brain cortex. The results were that plasma free tryptophan was a better predictor of human brain tryptophan concentration than total tryptophan. There was no correlation between brain tryptophan and the ratio tryptophan/LNAA. In these patients conditions are different from those in animal experiments and it is possible that there is a difference between humans and laboratory animals in, for example, the K_m of amino acid transport. Some other amino acids are bound to albumin, in particular cysteine and homocysteine (RASSIN 1990).

An interesting problem is whether there are regional differences in blood-brain transport of an amino acid. HAWKINS et al. (1982) showed that blood-brain phenylalanine transport varied with capillary density, which CRAIGIE had already measured in 1920. SMITH et al. (1985) observed that V_{max} was higher in cortical regions. They also considered that this could be due to the capillary density. LAMANNA and HARIK (1986), in a study on the blood-brain transport of leucine and glucose, observed that the *PS* of leucine varied in the different brain regions in the same way as that of glucose. The regional differences disappeared with anaesthesia, as did differences in blood flow, while differences depending on vascular volume remained. They suggested that regional differences in blood-brain transport were due to local brain perfusion and to capillary perfusion characteristics.

An important factor in brain utilization of amino acids is the relative importance of blood-brain transport of amino acids through the BBB and transport at the membrane of brain cells (neurons, astrocytes). PARDRIDGE and OLDENDORF (1975) measured the K_m and V_{max} for the different amino acids at the BBB and compared them with the values at the brain cell membrane. They found that the K_m and V_{max} values for blood-brain transport of amino acids were 1/3 and 1/10 of values at the brain cell membrane, respectively. This indicates the great importance of the BBB in regulating brain amino acid concentration. In addition, the amino acid transport systems at the brain cell membranes are different from those at the BBB. SERSHEN and LAJTHA (1979) have found ten overlapping transport classes in brain cells.

The characteristics of the cationic amino acid (lysine, arginine) transport system are very similar to those of the L system (BAÑOS et al. 1974; PARDRIDGE et OLDENDORF 1975). The K_m and V_{max} have been measured and inhibition of the transport of one cationic amino acid by the other has also been observed.

All the above data refer to studies in the rat, but other species have also been studied, including dogs (BETZ et al. 1975), rabbits (CORNFORD and CORNFORD 1986), sheep, (BRENTON and GARDINER 1988), and monkeys (LEATHWOOD and FERNSTROM 1990). The important point is the competition between amino acids for blood-brain transport, since it depends on the K_m of the amino acid for the carrier, which differs according to the species.

II. Efflux

The problem of amino acid efflux is more complex than that of influx, as it does not occur only in brain capillaries. After amino acids have crossed the brain endothelial cells, they first enter the brain extracellular space, then the brain cells. In the extracellular space they could be mixed with amino acids leaving the brain cells and then have a small diffusion to the CSF, and a rapid uptake by choroidal, meningeal and ependymal associated tissue, as noted by BLASBERG et al. (1983). Efflux can also occur with amino acids synthesized in brain cells. It can thus occur (1) in brain capillaries, (2) after diffusion to the CSF by passive transport through the arachnoid villi, and (3) after diffusion to the CSF, by carrier transport in the choroid plexus or the arachnoid membrane.

1. Efflux in the Capillaries

Efflux in the capillaries has been estimated with various methods: BETZ and GILBOE (1973) measured the *concentration of amino acids in arterial and venous blood* in isolated dog brains. There was a net efflux if the concentration of an amino acid was higher in venous than in arterial blood, but under these conditions the net efflux must be marked to be detected. This was the case for glutamine asparagine, glutamic acid, and taurine.

If the amino acids belong to the L system or to the cationic amino acid system, efflux can be estimated using the *BUI technique* by prolonging the time interval between carotid injection and decapitation. With cycloleucine, which is not metabolized in brain, the rate constant (K_0) was measured. This has been compared to the values obtained by Eq. 7 to see if the K_m for efflux was the same as for influx. For this, it was necessary to estimate amino acid concentrations in the interstitial space. Since they are not known, they were presumed to be similar to concentrations in the CSF, which are very low except for glutamine. Since these concentrations are very low, $K_{m,app}$ is very similar to the absolute K_m. The values calculated in this way were similar to the measured efflux. This provided evidence for transport symmetry for amino acids of the L system at the BBB (PARDRIDGE 1983).

Amino acids that are not transported from blood to brain in general have a high concentration in brain, because of local synthesis. They can be transported from brain to blood by a system different from the L system. This was shown by BETZ and GOLDSTEIN (1978) on *isolated brain microvessels*. This technique was developed from 1973 to 1975 by Joó (1973), BRENDEL et al. (1974), and GOLDSTEIN et al. (1975). BETZ and GOLDSTEIN studied the transport of L-leucine (L system) and MeAIB (A system) in these isolated microvessels. They observed that both amino acids were transported, but the uptake of MeAIB was sodium dependent and was inhibited by some amino acids, e.g., proline, which did not inhibit leucine uptake. Thus the transport of an A system amino acid was clearly demonstrated on these microvessels. BETZ and GOLDSTEIN hypothesized that in their preparation the incubation medium had access to both the luminal plasma membrane, which is normally in contact with blood, and the abluminal plasma membrane, which is normally in contact with the basal lamina and the brain extracellular fluid. These results were compared to those of numerous in vivo studies which have indicated that the A transport system is not present at the luminal membrane of the brain capillary. BETZ and GOLDSTEIN thus concluded that this transport system was located on the abluminal membrane of the brain capillary endothelial cell and that this A system polarity could explain the active efflux of small neutral amino acids from brain to blood, despite their limited influx. They thus proposed a BBB polarity, and compared it to epithelial tissue where tight junctions constitute a barrier for the lateral diffusion of proteins in the membrane. This method of measuring transport in isolated brain microvessels at the same time gives information on the transport at the luminal and abluminal membrane. However, as transport at the luminal membrane is often already known it can indirectly give information on transport into the endothelium at the abluminal membrane, which partly determines efflux from the brain.

HJELLE et al. (1978) first studied the transport of the L system amino acids into isolated microvessels and found uptake affinities which were similar to those found in in vivo experiments. Uptake of L-glutamine has also been studied in rat brain microvessels by CANGIANO et al. (1983a). Its transport was partly sodium dependent (system A) and partly sodium independent (system L). The role of the transport of this amino acid will be explained later.

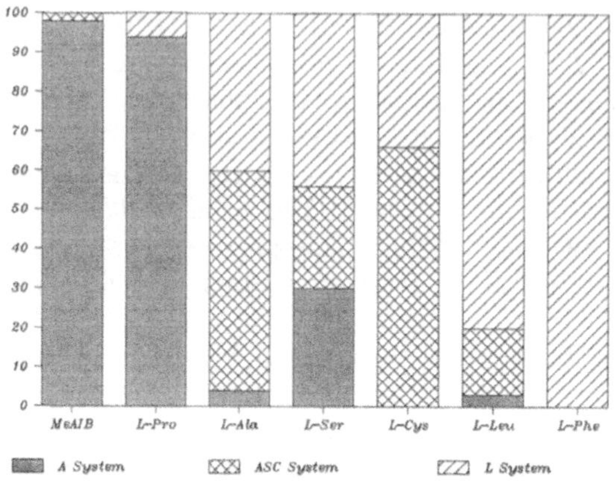

Fig. 1. Uptake of some amino acids expressed as a percentage of total activity by isolated brain microvessels. *A system*: sodium-dependent uptake inhibited by 10 m*M* MeAIB. *ASC system*: sodium-dependent uptake not inhibited by 10 m*M* MeAIB. *L system*: sodium-independent uptake inhibited by 10 m*M* BCH. The low nonsaturable component is not indicated

Transport of the amino acids studied was in general not known to occur by the L system. HWANG et al. (1980), studying the uptake of L-cystine, which cannot be measured in vivo, showed that cystine was taken up by a high-affinity sodium-dependent uptake system (K_m 3.6 μ*M*). Later (1983), they studied L-proline uptake and found it to be mediated mainly by the A system. The A system, however, is not the only sodium-dependent system in isolated microvessels. LEFAUCONNIER et al. (1985b) showed that some amino acids in isolated microvessels were transported by several systems (Fig. 1), and TAYARANI et al. (1987a) found that the ASC system, defined by sodium-dependent transport not inhibitable by MeAIB, was present. Taurine uptake was also measured by TAYARANI et al. (1989). It was sodium dependent and had a high affinity (28 μ*M*). This was lower than the plasma concentration of taurine. In addition, the V_{max} of transport was rather low. For the moment it is difficult to say whether this transport system is on the luminal membrane or on the abluminal or on both, but these results do not contradict results obtained in vivo: there is a saturable system (LEFAUCONNIER and TROUVÉ 1983) which probably has a K_m situated at a concentration lower than the plasma concentration of taurine (LEFAUCONNIER et al. 1978). Thus, it seems possible that this system represents transport partly at the luminal membrane and partly at the abluminal membrane, since an efflux has been measured by BETZ and GILBOE (1973).

Some experiments have been performed in isolated human brain microvessels (HARGREAVES and PARDRIDGE 1988). They can be isolated intact within 24 h after death and the K_m of uptake systems of several amino acids measured. These K_m were found to be very low when compared to plasma amino acid concentration. Leucine was transported equally by the L and ASC systems.

Neutral amino acid transport has also been studied in *cultures of endothelial cells*. These cultures give information on events during influx and efflux. Mouse cerebral endothelium cells in culture have been studied by CANCILLA and DEBAULT (1983). They showed that the presence of astrocytes was necessary to induce the activity of γ-glutamyltranspeptidase (γ-GT), an enzyme characteristic of brain endothelial cells. They found that the A and L systems of neutral amino acid transport were in a dynamic state and that they were influenced by the phase of cell growth and contact with astrocytes. A and L systems have also been found in rat (HUGHES and LANTOS 1989) and bovine cultures of endothelial cells (AUDUS and BORCHARD 1986) in the absence of astrocytes.

2. Passive Efflux Via the Arachnoid Villi

Passive efflux via the arachnoid villi occurs after diffusion of amino acids into the CSF.

3. Efflux Via the Choroid Plexus and Arachnoid Membrane

Amino acids are actively transported from CSF to blood through choroid and arachnoid cells. This was first predicted from the results of ventriculocisternal or ventriculolumbar perfusion (BITO et al. 1966; CUTLER and LORENZO 1968; SNODGRASS et al. 1969; CUTLER 1970; MURRAY and CUTLER 1970). In these studies, a radioactive amino acid was injected with artificial CSF into the lateral ventricle using a perfusion pump, and radioactivity was counted in the outflow from the cisterna magna. These experiments showed that various classes of amino acids were actively transported. The K_m could be measured and the effect of competitors and inhibitors tested. The problem was then to discover the site of this active uptake. Studies carried out on choroid plexus showed that all the amino acids tested had an uptake transport (LORENZO and CUTLER 1969; COBEN et al. 1971). For neutral amino acids system A was the most active, but there was also an ASC system and an L system. Anionic and cationic amino acids were also transported. This uptake was found to serve mainly in situ protein synthesis as after 1 h of perfusion 80% of leucine was bound and this binding did not occur after administration of a protein synthesis inhibitor (CARUTHERS and LORENZO 1974). This means that efflux via the choroid plexus is probably not very important for amino acids used in protein synthesis. γ-Aminobutyric acid (GABA) also had a very intense uptake (SNODGRASS and LORENZO 1973). It has also been shown that the arachnoid membrane of the bullfrog can transport some amino acids (WRIGHT 1974). The transport systems have also been studied in plasma membrane vesicles of the rabbit choroid plexus (ROSS and WRIGHT 1984) and the conclusion was that these systems were very similar to those described by COBEN et al. (1971) in the rabbit.

Amino acids are also transported from blood to CSF, but this transport has generally not been estimated because, as HAWKINS et al. wrote (1982), the areas that do not have a BBB are not described accurately by Eq. 2. However, PRESTON and SEGAL (1990) studied both the blood to CSF and the CSF to blood transport across the perfused choroid plexus of the sheep. The systems of transport existed in the two directions, with the exception of the A system, which did not exist on the blood-CSF barrier. For other amino acids the blood-CSF transport was higher than the CSF-blood transport at low concentrations of amino acids. The effect of raising the concentration of aspartate, glycine, phenylalanine, and serine had a variable effect on transport. When the concentration of aspartate and glycine was elevated to 0.1 m*M*, the flux which was from blood to CSF became from CSF to blood. This happened for phenylalanine and glycine only when the concentration attained 1 m*M*, which is not a physiological concentration.

III. Net Uptake

The net uptake is the difference between the influx and the efflux of amino acids. It has been measured in dogs by BETZ and GILBOE (1973), who found that only BCAA had an appreciable net uptake. They supposed that system L amino acids which entered the brain and were used for synthesis of neurotransmitters and proteins left the brain at the same rate they had entered. This could be explained by proteolysis which, in adult animals, can balance protein synthesis. Neurotransmitters are degraded but the amount of amino acids used for neurotransmitter synthesis is very low and cannot be detected by differences in concentration. The transport of glutamine can be used for the replacement of amino acids (BCAA and precursors of neurotransmitters). JAMES and FISHER (1981) proposed that because system L is an exchange system, to satisfy the condition of equal exchange, then the brain must produce some non-essential system L substrate to exchange for amino acids of which it is a net consumer (BCAA). System A activity could concentrate in the capillary cell a non-essential amino acid with a low affinity for the L system. Among the non-essential amino acids synthesized in brain, glutamine would seem the best candidate for exchange via system L. It is synthesized in astrocytes at an extremely rapid rate and its concentration is likely to be high in the vicinity of the abluminal capillary surface. In CSF its concentration is very high (0.5 m*M*) and it could be very easily concentrated in endothelial cells. This was demonstrated by CANGIANO et al. (1983a), who showed in addition that the presence of glutamine transported by the A system stimulated the transport of large amino acids of the L system. GORGIEWSKI-HRISOHO et al. (1986) observed that tryptophan uptake was stimulated by D- and L-glutamine and suggested that one could expect some difference in effect between the D- and L-forms if an exchange is mediated by the L system; in addition, D- and L-glutamine are γ-glutamyl donors for γ-GT which could be involved in the stimulation observed.

C. Anionic Amino Acids

Unlike the LNAA and the cationic amino acids, the concentration of anionic amino acids is very high in the brain. The techniques used to measure the influx and efflux of glutamate are the same as for the other amino acids but the results obtained have been completely different, i.e., the influx for anionic amino acids is very low (OLDENDORF and SZABO 1976). The experiments of BETZ and GILBOE (1973) showed that there was net efflux of glutamic acid. This led PARDRIDGE (1979) to say that comparison of glutamate influx and net release in the brain suggested that an active efflux system existed within the BBB. This efflux system was found by HUTCHINSON et al. (1985), who disclosed a very high affinity ($K_m = 1.9\,\mu M$) sodium-dependent system for glutamate in rat brain isolated microvessels. This system was strongly inhibited by ouabain, glutamate, and aspartate and slightly inhibited by glutamine and asparagine. It probably subserved the rapid unidirectional outward transport of glutamate from brain to blood. The energy seems to be coupled with the sodium gradient across the membrane of the capillary endothelial cells. It can be supposed that the very low in vivo uptake occurs at the luminal membrane of capillary endothelial cells and that the active system in the abluminal membrane prevents further endothelial cell-brain transport.

Anionic amino acids are considered excitotoxic, but this excitotoxicity does not seem related to blood-brain transport since excitotoxic amino acids are synthesized in brain and their blood-brain transport is very low. OLNEY (1969, 1986) reported that glutamic acid was toxic when administered subcutaneously to young animals, these animals developing endocrinological abnormalities when adult. This was due to the neurotoxicity of glutamic acid on cells of the arcuate nucleus which belong to the circumventricular organs. They are situated outside the BBB and it is not known why they are more accessible in young animals.

The problem of excitotoxicity of amino acids can nevertheless occur with synthetic amino acids injected intravenously, for example, kainic acid. The brain uptake of kainic acid is 20 times lower than that of glutamic acid measured by the intravenous route. Kainic acid does not seem to have a carrier as its uptake is not saturable and there is no uptake in isolated microvessels (LEFAUCONNIER et al. 1986). It can cause convulsions, probably because its very small influx into brain is due to the small diffusional component of blood-brain transport and because of its very high neurotoxicity.

D. Metabolism in the Endothelial Cells of Some Transported Amino Acids

Some transported amino acids can be extensively metabolized in the endothelial cells and this can modify the results of transport. The enzymes implicated in this metabolism are:

DOPA Decarboxylase. DOPA is the precursor of dopamine, and has been administered to parkinsonian patients in order to restore the depleted striatal dopamine compartment. It has been found that L-DOPA is transported into brain by the L system (OLDENDORF 1971), but that only a very low percentage of the injected radioactive amino acid entered mouse brain (WURTMAN et al. 1970). A histologic study of DOPA accumulation has shown that fluorescence developed in cerebral capillaries a few minutes after drug administration (BERTLER et al. 1963). WADE and KATZMAN (1975b) have suggested that L-DOPA could be transported to the endothelial cell, decarboxylated into dopamine, and then metabolized by monoamine oxidase. They have thus studied the penetration of DOPA into brain by using L-[2,3-^{3}H]-DOPA and L-[carboxy-^{14}C]-DOPA. If DOPA is decarboxylated, $^{14}CO_2$ from the radioactive group carboxyl will be removed by exchange with blood CO_2 and unlabelled dopamine will remain, whereas [^{3}H]DOPA radioactivity will remain. It was thus possible to calculate the percentage of DOPA decarboxylated, which ranged from 63% to 77% of the DOPA transported and showed no regional difference.

A study was also done in monkeys with [^{18}F]fluoro-DOPA, which behaves like L-DOPA and can be assayed using a γ-ray detector (GARNETT et al. 1980). This showed an influx of DOPA into the endothelium, a backflux from it, the formation of dopamine in brain, and the destruction of dopamine. These results explain the requirement for a decarboxylase inhibitor to be given in combination with DOPA when it is administered to humans.

4-Aminobutyrate-2-Oxoglutarate Aminotransferase (GABA Transaminase or GABA-T). This is the rate-limiting enzyme of GABA catabolism and histologic staining shows it to have a very high activity in cerebral blood vessel endothelium (VAN GELDER 1968).

γ-Glutamyl Transpeptidase (γ-GT). This enzyme catalyzes the transfer of the glutamyl moiety of glutathione to a variety of acceptor molecules such as amino acids. It has been found to have high activity in brain capillary endothelial cells and to be absent from endothelial cells of other organs. It may have a role in the transport of amino acids, in particular at the BBB (ORLOWSKI et al. 1974). However, such transport has not been directly demonstrated. In the cells of the intestinal brush border membrane, which also have high γ-GT activity, it has been shown that papain treatment of membrane vesicles results in the digestion of γ-GT, while transport of leucine is increased (BERTELOOT et al. 1980).

E. Regulation of Blood-Brain Amino Acid Transport

Amino acid transport is usually regulated to meet the nutritional requirements of a cell, but blood-brain amino acid transport is regulated to meet the nutritional requirements of an organ; thus it has to supply amino acids both for brain protein synthesis and for brain neurotransmitter synthesis as well as removing unwanted amino acids.

Brain amino acid concentration depends on the plasma concentration of amino acids and on the proportions and rates of transport of the amino acids between blood and brain (HARPER and TEWS 1988).

I. Availability of Amino Acids: Plasma Amino Acid Concentration

The plasma concentration of an amino acid tends to reflect protein or amino acid uptake during the period immediately after a meal but is quite stable in the postabsorptive state. The amino acids are regulated by systems that respond appropriately to correct deviations from the stable fasting concentrations. These concentrations are regulated by:

- Gastric emptying. This seems to be important as it is rapid for low protein intake and delayed with high protein intake
- Protein synthesis in the whole organism. Influx of amino acids into the liver stimulates synthesis and suppresses degradation of tissue proteins. Ingestion of a meal also stimulates release of insulin, which in turn stimulates uptake of amino acids and synthesis of proteins by muscle and heart
- Catabolism of amino acids. This is a major mechanism for control of plasma amino acid concentration when protein intake exceeds that necessary for protein synthesis.

II. Proportions and Rate of Transport of the Amino Acids Between Blood and Brain

The transport of amino acids from blood to brain is important for brain function as some amino acids of the L system are precursors of neurotransmitters, and amino acid competition for blood-brain transport seems to play an important role in the synthesis of neurotransmitters. Increase of brain tryptophan increases the synthesis of serotonin (FERNSTROM and WURTMAN 1971, 1972a). Tryptophan hydroxylase, the first enzyme of serotonin synthesis, has a rather high K_m which is about the value of the brain concentration of tryptophan. Tryptophan concentration is thus important, as the synthesis of serotonin is not subject to end-product inhibition. Brain tryptophan increase can occur in rats after a high-carbohydrate protein-free meal. This then leads to insulin secretion, which increases muscle uptake of most amino acids, specially BCAA. The plasma concentration of the competing amino acids then decreases and brain tryptophan increases. On the other hand, when food proteins increase, all the amino acids in the plasma increase, as does the concentration of tryptophan, but as its concentration is often lower than that of the other amino acids in proteins, its relative concentration decreases. Since there is competition for amino acid transport into brain, the brain concentration of tryptophan also decreases. The influence of the composition of meals containing various proportions of protein on brain amino acid composition has been studied by GLAESER et al. 1983. Similar results have been found by LEATHWOOD and FERNSTROM (1990) in cynomolgus monkeys, except that increases in serotonin were restricted to certain brain regions. Synthesis of catecholamines from tyrosine seemed to be different as brain dopamine and norepinephrine did not increase after tyrosine administration. However, ACWORTH et al. (1988), using in vivo cerebral microdialysis, showed that increasing the availability of tyrosine to some parts of the brain during neuronal firing increased the release of dopamine. This was true only for some groups of dopaminergic neurons. Histidine injected intraperitoneally (SCHWARTZ et al. 1972) has been shown to increase the level of histamine in the brain.

When rats consume widely different amounts of high-quality proteins, amino acid concentration does not change much. However, substantial deviations in the amino acid pattern of the diet can cause striking changes in the brain amino acid pool, which otherwise is relatively stable. If rats receive a low-protein diet and a mixture of indispensable amino acids without histidine, a 80% decrease in brain histidine is observed in 3–4 h (PENG et al. 1972). When protein intake is greatly increased or decreased, or when the pattern of amino acids is modified, food intake is depressed (HARPER and TEWS 1988).

III. Is Amino Acid Transport Modified by Hormones?

Hormones have been shown to modify amino acid transport in a great number of tissues (SHOTWELL et al. 1983). Hormones could act directly on the transport system, but such effects have been reported to occur mainly on the system A carrier or indirectly by modifications of the plasma amino acid concentrations.

The effect of insulin has been studied, perhaps because the increase of brain tryptophan with a carbohydrate-rich protein-free diet is believed to be due to insulin. This hormone lowers the plasma level of most LNAA (FERNSTROM and WURTMAN 1972b). As insulin is also the major antilipolytic hormone, it increases the transfer of free fatty acids from serum albumin to adipocytes. This increases the affinity of serum albumin for tryptophan. In rat brain, total plasma tryptophan is almost as accessible as free tryptophan (PARDRIDGE 1983), and brain tryptophan increases. Similar conclusions have also been suggested by DANIEL et al. (1981). The effect of insulin has also been studied on isolated microvessels (CANGIANO et al. 1983b). They showed that insulin increased the transport of tryptophan, but this increase was sodium dependent, like the transport of MeAIB, which was also increased by insulin; thus, the effect of insulin appeared to be on the A transport system. This differs from in vivo observations, which show that the effect of insulin seems to be mainly indirect.

The effect of thyroid hormones has also been investigated. Daniel et al. (1975b) reported that the blood-brain influx of leucine, valine, and lysine was lower in adult rats made hypothyroid at birth than in control rats. Hypothyroidism was also studied in immature (16-day-old) rats (Lefauconnier et al. 1985a). The BUI of small neutral amino acids was greatly enhanced but, as the CBF was decreased, transport was also measured by the intravenous method. The influx of alanine was found to be normal and that of phenylalanine decreased. In another study using adult rats (3 months old) which had been hypothyroid for 7 weeks, no alteration was noted in amino acid transport (Mooradian 1990). It seems possible that in young animals hypothyroidism modifies the maturation of the capillary bed, as has already been reported by Eayrs (1954), or of the neutral amino acid transport carrier. In adult animals the capillary bed is mature and cannot be modified further.

IV. Can Amino Acid Transport Be Modified by Neurotransmitters?

As was the case with hormones, it is also difficult to say whether amino acid transport is directly modified by neurotransmitter action or by an indirect action due to a modification of the transport conditions. For example, it has been reported by Knott and Curzon (1972) that 24-h food deprivation or 3-h immobilization increases serotonin metabolism due to an increase in brain tryptophan. This increase in brain tryptophan was explained by the authors as an increase in plasma free tryptophan which was probably due to an increased unesterified fatty acid concentration. Such fatty acids become bound to albumin and thus release tryptophan, which becomes free. This suggested to the authors that the increase in brain tryptophan could be mediated by increased sympathetic nervous activity since plasma fatty acid concentrations increase during norepinephrine infusion. Later studies by this group (Hutson et al. 1980) showed that an intravenous infusion of the β-adrenoreceptor agonist isoprenaline decreased plasma total tryptophan and tyrosine, increased plasma free tryptophan, and also increased brain tryptophan and tyrosine and thus increased the transport.

Eriksson and Carlsson (1988) also used isoprenaline and observed that its injection caused a decrease in plasma amino acids and an increase in brain amino acids. They concluded that the transport of LNAA into brain was regulated by a β-adrenergic mechanism. This was also suggested by the study of Edwards and Sorisio (1988).

It is interesting to compare these results with those already obtained for the regulation of BBB permeability by adrenergic agonists. There are β-adrenergic receptors on brain capillaries (Harik et al. 1980; Peroutka et al. 1980; Kobayashi et al. 1981; Culvenor and Jarrot 1981). These receptors usually mediate their action via cyclic adenosine monophosphate (AMPc) as second messenger. Adenylate cyclase exists in brain microvessels (Joó et al. 1975) and is stimulated by β-adrenergic receptors agonists (Baca and

PALMER 1978; HERBST et al. 1979; HUANG and DRUMMOND 1979; NATHANSON and GLASER 1979). With β-adrenergic agonists, an increase in BBB permeability to water has been observed; thus β-adrenergic agonists may cause an increase of BBB permeability to substances of low molecular weight such as water. But is this increase specific for water? It would be interesting to see whether stimulation by β-adrenergic agonists increases the transport of other substances from blood to brain. It is interesting that TAGLIAMONTE et al. (1971) investigated whether dibutyryl cyclic AMP affected serotonin synthesis and noted that the increased synthesis observed was not due to a modification of the activity of the enzyme but to an increase in its substrate, tryptophan. They also observed increased blood–brain transport of tryptophan.

Other studies have been carried out with the peptide arginine-vasopressin (BRUST 1986). It was found that its injection led to a decrease in leucine transport. REITH et al. (1987) showed that while the blood-brain transport of leucine was decreased in all brain regions, there was a saturable fixation of arginine-vasopressin on the receptors which had been already identified on microvessels in the hippocampus.

F. Physiological and Pathological Modifications

I. Blood-Brain Amino Acid Transport During Development

Protein synthesis is higher in developing brain than in adult brain (LAJTHA and DUNLOP 1981), while the capillary density is much lower (BÄR and WOLFF 1973). It thus seems probable that blood-brain transport in the postnatal period is different from that in adult life (for review see JOHANSON 1989). In addition, the blood–brain transport has been described as being like that in epithelial tissue and an asymmetry has been found in the transport systems at the luminal and abluminal membranes. Thus the study of blood-brain transport during development raises several problems.

1. Technical Problems

How can amino acid transport be studied in immature animals? It is not possible to use the BUI technique as such, because normal blood flow must follow the injection. In young rats this is impossible as the smallest needle completely occludes the carotid artery. The in situ brain perfusion technique and the intravenous technique also involve difficulties.

These problems have been solved in various ways. For the BUI technique: (a) immature and adult rabbits were used (CORNFORD et al. 1982); (b) animals were injected in the heart (SERSHEN and LAJTHA 1976); (c) the injection was made retrogradely in the right brachial artery, which gave a clear bolus in the carotid artery in the orthograde direction (LEFAUCONNIER and TROUVÉ 1983).

For the intravenous method, the first measurements were made in 6-day-old animals by a programmable perfusion (BAÑOS et al. 1978) or a bolus (LEFAUCONNIER et al. 1985c), and an arterial catheter could be placed for blood sampling. For the in situ brain perfusion technique, the perfusate was injected using a needle placed in the common carotid artery. The needle completely occluded the artery and thus prevented the arterial blood from contaminating the injectate (NAGASHIMA et al. 1987).

2. How Many Blood-Brain Transport Systems Have Been Found in Immature Animals?

In adult animals there are only three systems of transport at the luminal membrane for amino acids. Is this the same in immature animals, bearing in mind that all amino acids are needed for protein synthesis? Results obtained (SERSHEN and LAJTHA 1976; BAÑOS et al. 1978; CORNFORD et al. 1982) have shown that the amino acid transport systems have a higher capacity in immature animals than in adult animals. The L system and the cationic amino acid system are present, and two additional systems of transport have been observed.

An ASC transport system was demonstrated with the BUI technique (SERSHEN and LAJTHA 1979; LEFAUCONNIER and TROUVÉ 1983). This system promotes the transport of alanine, serine, cysteine, and threonine and is sodium dependent.

An A transport system was found with the in situ brain perfusion technique (NAGASHIMA et al. 1987).

These results have also confirmed a phenomenon already reported for other tissues (WISE 1975): developing tissue has a greater number of transport systems which are sodium dependent.

3. Coexistence of Higher Blood-Brain Transport with a Lower Density of Capillaries

A problem not solved by these experiments is the coexistence of higher blood-brain transport with a lower density of capillaries supplying blood for transport. In order to try to solve this problem, phenylalanine was injected intravenously (LEFAUCONNIER et al. 1991) into 6-day-old rats which were decapitated 20 s later. The resulting autoradiograms showed very intense radioactivity in the periphery of the brain and a much lower level in the centre. Since this did not seem specific to a blood-brain transport system, the experiment was repeated with sucrose (which has no carrier transport), phenylalanine, and alanine, and the animals were decapitated after 5, 20, and 90 s. At 5 s the greater part of the radioactivity was in the subarachnoid space, though there was some phenylalanine radioactivity in the brain (Fig. 2). The radioactivity in the subarachnoid space was not due to carrier transport as it was also present after injection of sucrose. At 90 s the subarachnoid radioactivity had decreased and brain phenylalanine and

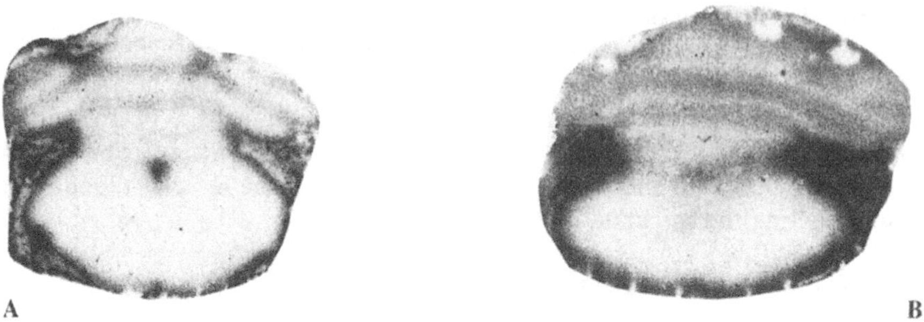

A B

Fig. 2A,B. Autoradiographic study of blood-brain transport of [^{14}C]sucrose (**A**) and L-[^{14}C]phenylalanine (**B**) in slices obtained in animals decapitated 5 s after injection

alanine radioactivity was higher than at 5 s (Fig. 3). However, this radioactivity was more intense at the periphery of the cortex than inside the brain. It thus seems probable that the high amino acid transport in immature animals could be partly due to leakage from vessels in the subarachnoid space, from where the amino acid would be taken up by the brain cells, which would show a higher level than those which obtain amino acids after transport through the BBB. The differences between immature and adult animals are greater when the amino acid is administred by infusion rather than as a bolus.

4. When Does the Asymmetry of Transport Between the Two Membranes of the Endothelial Cells Appear?

This has been studied by BETZ and GOLDSTEIN (1981). They observed that sodium-dependent transport of some amino acids through the abluminal membrane did not occur before 21 days of age. A morphological study by CORNFORD and CORNFORD (1986) showed that a high density of mitochondria was already visible in the capillary at day 1 and that a decrease in the thickness of the cytoplasm was evident at day 21. Thus, there may be a correspondence between morphology and function.

II. Effect of Drugs on Blood-Brain Transport of Amino Acids

Since the influence of anaesthetics is important to researchers, it has been studied. SAGE and DUFFY (1979) noted that the BUI of several amino acids was modified with pentobarbital anaesthesia. The BUI of LNAA was increased while the BUI of anionic and cationic amino acids was not modified. During anaesthesia the cerebral blood flow is decreased and, if the brain influx of amino acids is unchanged, the BUI must be increased. This was true only for the neutral amino acids. It seems possible that the permeability is modified for the other amino acids. HAWKINS et al. (1982)

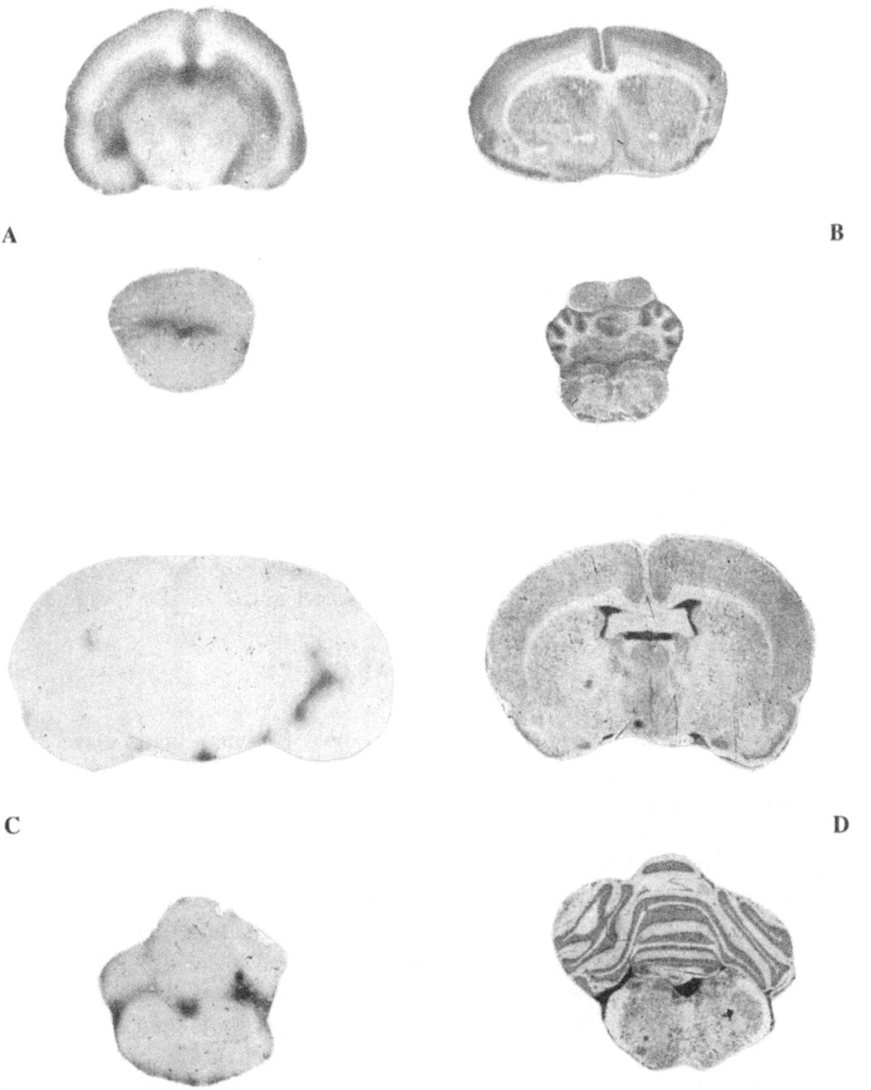

A B C D

Fig. 3A–D. Autoradiographic study of blood–brain transport of alanine (**A, C**) and phenylalanine (**B, D**) 90 s after injection in 6-day-old (**A, B**) and adult animals (**C, D**)

studied the effect of several anaesthetics on blood-brain transport of phenylalanine using the intravenous technique with autoradiography. The transport of this amino acid was modified, but this seemed to be due to differences in the plasma concentration of the amino acid and not to an effect on the L system.

The possibility of an action of lithium, an effective agent in the clinical treatment of manic-depressive psychosis, on tryptophan blood-brain transport has been explored, but no effect was found (YUWILER et al. 1979; EHRLICH et al. 1980).

III. Pathological Modifications of Amino Acid Transport

In view of what we now know concerning blood–brain transport of amino acids, modifications can be explained as follows:

- By an alternation of the concentration in plasma of one or several neutral amino acids, because transport, particularly of neurotransmitter precursors, depends on the ratio of amino acid plasma concentration to the sum of the concentrations of its LNAA competitors
- By a modification of the conditions of exchange, as the main system for neutral amino acid transport is the L system, which is mainly an exchange system
- By an alteration of the carriers at the plasma membrane
- By an alteration of the sodium gradient for an amino acid transported by a sodium-dependent system.

Currently, modifications can be mainly classified under the first and second possibilities.

1. Hepatic Encephalopathies

Marked alterations in blood-brain transport of amino acids have been observed in the hepatic encephalopathies. Much work has been done in this field, but only studies concerning transport of amino acids will be cited, and the possible significance in the pathogenesis of symptoms of hepatic encephalopathy will not be discussed (for reviews see BERNARDINI and FISCHER 1982; ROSSI FANELLI 1987; BUTTERWORTH et al. 1987).

In patients with cirrhosis there is a high concentration of tryptophan in the CSF but no change in total plasma tryptophan and only a very small elevation of free tryptophan (YOUNG et al. 1975). JAMES et al. (1978) studied the plasma and brain concentrations of amino acids in rats with a portacaval anastomosis (PCA), which is an animal model of cirrhosis, and evaluated blood–brain transport with the BUI technique. They found that the plasma concentration of BCAA was decreased while that of aromatic amino acids was increased, and that the brain concentration of BCAA was not modified while that of aromatic amino acids and methionine and histidine was increased. The BUI technique showed a marked increase for phenylalanine, tyrosine, tryptophan, and leucine and a decrease for arginine. Transport of inulin and tyramine was unchanged. The modifications observed were thus not due to an abnormal increase in the permeability of the BBB. Similar results were found by ZANCHIN et al. (1979). MANS et al. (1982) also found similar results, but using another technique (autoradiography after intravenous injection), and showed that the increase in transport of leucine was lower than that of phenylalanine and tryptophan.

In 1979, JAMES et al. reported that in humans with cirrhosis and in rats with PCA there was a marked increase in blood ammonia. This increased the ammonia influx into the brain, where it was rapidly detoxified in astrocytes by a reaction with glutamic acid yielding glutamine due to the action of glutamine synthetase. Glutamine is transported to endothelial cells by transport system A and is exchanged with LNAA by the L system (CANGIANO et al. 1983a). In 1984, CARDELLI-CANGIANO et al. studied directly the effect of ammonia on amino acid uptake by isolated brain microvessels. They showed that ammonium ion (0.25 m*M*) caused a stimulation of the uptake of LNAA, while the uptake of MeAIB, glutamic acid, and lysine was not modified. This stimulation occurred via an increase in the V_{max} of the saturable component. It was apparently mediated by the intracellular formation of glutamine, which was then exchanged for LNAA through the L transport system.

Analysis of correlations of plasma ratios of LNAA expressed as plasma competitor function, brain LNAA, and brain glutamine showed that there was a significant correlation between plasma competitor function and brain LNAA in control and PCA animals. An even better correlation was found between brain LNAA and brain glutamine (JEPPSON et al. 1985). The increase in LNAA transport was investigated to see whether it was due to glutamine or to ammonia by testing the effect of hyperammonia (which gives rise to brain glutamine) and by the injection of methionine sulfoximine, an inhibitor of glutamine synthesis, before infusion of ammonium salts. The administration of methionine sulfoximine prevented the increase in the blood-brain transport of leucine and phenylalanine (JONUNG et al. 1985). Thus glutamine seems to play an important role in the transport of amino acids. However, BUTTERWORTH et al. (1988) showed that glutamine synthetase is decreased by 15% in the cortex of shunted rats but not modified in the brain stem. Administration of ammonium acetate to shunted rats resulted in severe encephalopathy. Glutamine concentration in the brain stem of comatose rats was increased twofold, whereas it was unchanged in the cerebal cortex. Consequently, ammonia reached disproportionately high levels in cerebral cortex. These findings suggest a limitation in the capacity of the cerebral cortex to remove additional blood-borne ammonia by glutamine formation following PCA.

2. Uraemia

CANGIANO et al. (1988) speculated that the neurological symptoms observed in uraemia are similar to those occurring in hepatic encephalopathy and found support for this hypothesis in their experiments on isolated brain microvessels.

3. Diabetes

MANS et al. (1987) measured the blood–brain transport of phenylalanine and lysine in 4-week streptozotocin diabetic rats. The *PS* of phenylalanine was decreased and that of lysine was increased. These modifications seemed to be due mainly to the plasma concentrations of competing amino acids, which were considerably modified. The BCAA were generally increased, while aromatic amino acids were relatively unchanged and basic amino acids were decreased. Brain amino acids were also altered, with an increase in BCAA and a decrease in aromatic amino acids. These alterations seem to be due to modification of plasma amino acids and the strong competitive effect of BCAA and not to a modification in the carrier system (McCall et al. 1982; Crandall and Fernstrom 1983; Brosnan et al. 1984).

4. Aminoaciduria

Blood-brain amino acid transport can also be seriously altered in aminoaciduria, as large quantities of one amino acid are present in plasma. The blood-brain transport of this amino acid will predominate over transport of other amino acids. Protein synthesis will be decreased, especially in the young, and synthesis of neurotransmitter will be altered. This was first demonstrated by Oldendorf in 1971 for phenylketonuria. He measured brain uptake of selenomethionine, which can be studied by external γ-counting and can be used as a competitive amino acid analog. He found a reduction in the uptake of this amino acid in patients with phenylketonuria. Bustany et al. (1981) studied this disease by positron emission tomography (PET) using L-[^{11}C]methionine and also found only a very small uptake of methionine when the low-phenylalanine diet was stopped.

5. Intoxications

Amino acid transport perturbation has also been investigated in intoxications. For example, lead intoxication in young animals induces encephalopathy with lesions of the BBB and numerous hemorrhages (Pentschew and Garro 1966). These lesions do not occur in adult animals and it has been investigated whether a dose lower than that which causes encephalopathy administered at an age when encephalopathy no longer occurs could impair carrier function. Lorenzo and Gewirtz (1977) noted that in 20- and 30-day-old rabbits tryptophan extraction was lower in intoxicated than in control animals. This has also been studied in 19-day-old rats, where transport of alanine and phenylalanine was found to be unchanged (Lefauconnier et al. 1980). In 19- to 21-day-old rats, the *PS* product for lysine and histidine was greater in intoxicated animals than in controls in some regions of the cerebral hemisphere (Moorhouse et al. 1988). In adult rats a low level of lead exposure since birth did not modify tyrosine transport (Michaelson and Bradbury 1982). It was observed that, at least in rats, there was no impairment of the transfer of metabolic substrates. The increase in transport in some areas was perhaps due to a delay in maturation in susceptible regions.

The amino acid L transport system has been shown to allow mercury to enter brain when it is conjugated with L-cysteine (Aschner and Clarkson 1988). At a higher dosage, mercurials cause perturbations of amino acid transport at the BBB (Hervonen and Steinwall 1974). Pardridge (1976) showed that mercuric chloride ($10^{-4}\,M$) caused nearly complete inhibition of the transport of cycloleucine and tryptophan. The effect of mercurials was also tested on isolated brain microvessels (Tayarani et al. 1987b). Inhibition of the transport of alanine, phenylalanine, MeAIB, and glutamic acid by $10^{-4}\,M$ mercuric chloride or methylmercury was very marked. As amino acid transport at the abluminal membrane depends also on the sodium gradient, the activity of ATPase was measured. It was found to be inhibited by the same concentration of mercurials. It is thus difficult to say whether the carrier transport of an amino acid is inhibited or if it is the sodium gradient which is abnormal. However, it seems likely that the carrier transport of phenylalanine is inhibited since the L system is inhibited and is not sodium dependent and has previously been shown to be inhibited in in vivo experiments, as cycloleucine is also transported by the L system.

6. Depressive States

It has been shown that brain serotonin and, in some areas, brain dopamine are influenced by brain tryptophan and tyrosine concentrations. Because of the transport system at the BBB, the concentration of these amino acids in brain depends on their concentration in plasma or, more accurately (because of the competitiveness of these amino acids), on the ratio of the concentration of one amino acid to that of the other LNAA (Fernstrom and Wurtman 1972a). Pardridge (1977) has shown that the K_m of transport of each amino acid has to be taken into account. Another ratio has thus been calculated for the transport of these amino acids. As monoaminergic neurons are thought to be altered in the pathophysiology of depression, the plasma concentration of these amino acids and the [tryptophan]/[LNAA] ratio have been measured in depressed patients and found to be decreased (DeMyer et al. 1981; Møller et al. 1988; Maes et al. 1990). Whether the efficacy of an antidepressant is correlated with the ratios estimated before treatment has also been investigated. In his retrospective study, Møller (1988) found that there was a better correlation between the [tryptophan]/[LNAA] ratio than with the ratio taking into account the K_m of transport of the amino acids, but the K_m values used were based on those obtained from rats. Two years later, Møller et al. (1990) compared the clinical response in depressed people with either the plasma concentration of tryptophan alone or with the [tryptophan]/[LNAA] ratio. They noted that in the case of one treatment improvement was inversely correlated with the baseline tryptophan concentration and in the other was inversely correlated with the [tryptophan]/[LNAA] ratio.

GAILLARD and TISSOT (1979), instead of comparing the plasma amino acid pattern, evaluated the blood–brain transport of some amino acids in patients with depression or mania by measuring the arteriovenous difference (venous sampling was done from the jugular vein) in concentration of L-DOPA, L-5-hydroxytryptophan (L-5-HTP), tyrosine, and tryptophan during an infusion of L-DOPA or L-5-HTP. In depressive patients there was a small uptake of tryptophan, correlated with a large efflux of tyrosine. In manic patients there was an efflux of tryptophan and an uptake of tyrosine. Such modifications were not observed in peripheral tissues. They also found that the plasma/brain flux in these patients was highly correlated with the plasma/erythrocyte flux and they treated depressive patients according to the results of the latter (AZORIN et al. 1990). Treatment results were better than when the patients were treated according to clinical findings only. But what is surprising is that in transport into erythrocytes the problem of competitiveness does not exist, whereas competitiveness plays an important role in transport through the BBB. In the treatment of depressed patients competitiveness seems to play no role. This is perhaps related to the study of GILLMAN et al. (1986) in humans. They showed that brain tryptophan levels correlated much better with plasma free tryptophan than with the [tryptophan]/[LNAA] ratio.

References

Acworth IN, During MJ, Wurtman RJ (1988) Processes that couple amino acid availability to neurotransmitter synthesis and release. In: Huether (ed) Amino acid availability and brain function in health and disease. Springer, Berlin Heidelberg New York (NATO ASI Series, vol H20)

Aschner M, Clarkson TW (1988) Uptake of methylmercury in the rat brain: effects of amino acids. Brain Res 462:31–39

Audus KL, Borchardt RT (1986) Characteristics of the large neutral amino acid transport system of bovine brain microvessel endothelial cell monolayers. J Neurochem 47:484–488

Azorin JM, Bovier Ph, Widmer J, Jeanningros R, Tissot R (1990) L-Tyrosine and L-tryptophan membrane transport in erythrocytes and antidepressant drug choice. Biol Phychiatry 27:723–734

Baca GM, Palmer GC (1978) Presence of hormonally-sensitive adenylate cyclase receptors in capillary-enriched fractions from rat cerebral cortex. Blood Vessels 15:286–298

Baños G, Daniel PM, Moorhouse SR, Pratt OE (1973) The influx of amino acids into the brain of the rat in vivo: the essential compared with some non-essential amino acids. Proc R Soc Lond B 183:59–70

Baños G, Daniel PM, Pratt OE (1974) Saturation of a shared mechanism which transports L-arginine and L-lysine into the brain of the living rat. J Physiol (Lond) 236:29–41

Baños G, Daniel PM, Pratt OE (1978) The effect of age upon the entry of some amino acids into the brain, and their incorporation into cerebral protein. Dev Med Child Neurol 20:335–346

Bär Th, Wolff JR (1973) On the vascularization of the rat's cerebral cortex. Bibl Anat 11:515–519

Bernardini P, Fischer JE (1982) Amino acid imbalance and hepatic encephalopathy. Annu Rev Nutr 2:419–454
Berteloot A, Bennetts RW, Ramaswamy K (1980) Transport characteristics of papain-treated brush-border membrane vesicles. Non-involvement of γ-glutamyltransferase in leucine transport. Biochim Biophys Acta 601:592–604
Bertler A, Falck B, Rosengren E (1963) The direct demonstration of a barrier mechanism in the brain capillaries. Acta Pharmacol Toxicol 20:317–321
Betz AL, Gilboe DD (1973) Effect of pentobarbital on amino acid and urea flux in the isolated dog brain. Am J Physiol 224:580–587
Betz AL, Goldstein GW (1978) Polarity of the blood-brain barrier: neutral amino acid transport into isolated brain capillaries. Science 202:225–227
Betz AL, Goldstein GW (1981) Developmental changes in metabolism and transport properties of capillaries isolated from rat brain. J Physiol (Lond) 312:365–376
Betz AL, Gilboe DD, Drewes LR (1975) Kinetics of unidirectional leucine transport into brain: effects of isoleucine, valine, and anoxia. Am J Physiol 228:895–900
Bito L, Davson H, Levin E, Murray M, Snider N (1966) The concentrations of amino acids and other electrolytes in cerebrospinal fluid, in vivo dialysate of brain, and blood plasma in the dog. J Neurochem 13:1057–1067
Blasberg RG, Fenstermacher JD, Patlak CS (1983) Transport of α-aminoisobutyric acid across brain capillary and cellular membranes. J Cereb Blood Flow Metab 3:8–32
Bradbury MWB (1984) The structure and function of the blood-brain barrier. Fed Proc 43:186–190
Brendel K, Meezan E, Carlson EC (1974) Isolated brain microvessels: a purified, metabolically active preparation from bovine cerebral cortex. Science 185:953–955
Brenton DP, Cardiner RM (1988) Transport of L-phenylalanine and related amino acids at the ovine blood-brain barrier. J Physiol (Lond) 402:497–514
Brosnan JT, Forsey RGP, Brosnan ME (1984) Uptake of tyrosine and leucine in vivo by brain of diabetic and control rats. Am J Physiol 247:C450–C453
Brust P (1986) Changes in regional blood-brain transfer of L-leucine elicited by arginine-vasopressin. J Neurochem 46:534–541
Bustany P, Sargent T, Saudubray JM, Henry JF, Comar D (1981) Regional brain uptake and protein incorporation of ^{11}C-L-methionine studied in vivo with PET. J Cereb Blood Flow Metab 1 (Suppl 1):S17–S18
Butterworth RF, Giguère JF, Michaud J, Lavoie J, Pomier Layrargues G (1987) Ammonia: key factor in the pathogenesis of hepatic encephalopathy. Neurochem Pathol 6:1–12
Butterworth RF, Girard G, Giguère JF (1988) Regional differences in the capacity for ammonia removal by brain following portocaval anastomosis. J Neurochem 51:486–490
Cancilla PA, DeBault LE (1983) Neutral amino acid transport properties of cerebral endothelial cells in vitro. J Neuropathol Exp Neurol 42:191–199
Cangiano C, Cardelli-Cangiano P, James JH, Rossi-Fanelli F, Patrizi MA, Brackett KA, Strom R, Fischer JE (1983a) Brain microvessels take up large neutral amino acids in exchange for glutamine. J Biol Chem 258:8949–8954
Cangiano C, Cardelli-Cangiano P, Cascino A, Patrizi MA (1983b) On the stimulation by insulin of tryptophan transport across the blood-brain barrier. Biochem Intern 7:617–627
Cangiano C, Cardelli-Cangiano P, Cascino A, Ceci F, Fiori A, Mulieri M, Muscaritoli M, Barberini C, Strom R, Rossi Fanelli F (1988) Uptake of amino acids by brain microvessels isolated from rats with experimental chronic renal failure. J Neurochem 51:1675–1681
Cardelli-Cangiano P, Cangiano C, James JH, Ceci F, Fischer JE, Strom R (1984) Effect of ammonia on amino acid uptake by brain microvessels. J Biol Chem 259:5295–5300

Caruthers JS, Lorenzo AV (1974) In vitro studies on the uptake and incorporation of natural amino acids in rabbit choroid plexus. Brain Res 73:35–50
Christensen HN (1964) Relations in the transport of β-alanine and the α-amino acids in the Earlich cell. J Biol Chem 239:3584–3589.
Christensen HN (1984) Organic ion transport during seven decades. The amino acids. Biochim Biophys Acta 779:255–269
Christensen HN (1985) On the strategy of kinetic discrimination of amino acid transport systems. J Membrane Biol 84:97–103
Christensen HN, Liang M, Archer EG (1967) A distinct Na-requiring transport system for alanine, serine, cysteine and similar amino acids. J Biol Chem 242:5237–5246
Coben LA, Cotlier E, Beaty C, Becker B (1971) Transport of amino acids by rabbit choroid plexus in vitro. Brain Res 30:67–82
Collarini EJ, Oxender DL (1987) Mechanisms of transport of amino acids across membranes. Annu Rev Nutr 7:75–90
Comar D, Cartron JC, Maziere M, Marazano C (1976) Labeling and metabolism of methionine-methyl-^{11}C. Eur J Nucl Med 1:11–14
Cornford EM, Cornford ME (1986) Nutrient transport and the blood-brain barrier in developing animals. Fed Proc 45:2065–2072
Cornford EM, Braun LD, Oldendorf WH (1982) Developmental modulations of blood-brain barrier permeability as an indicator of changing nutritional requirements in the brain. Pediatr Res 16:324–328
Craigie EH (1920) On the relative vascularity of various parts of the central nervous system of the albino rat. J Comp Neurol 31:429–464
Crandall EA, Fernstrom JD (1983) Effect of experimental diabetes on the levels of aromatic and branched-chain amino acids in rat blood and brain. Diabetes 32:222–230
Crone C (1985) The blood-brain barrier: a modified tight epithelium. In: Bradbury, Rumsby, Suckling (eds) The blood-brain barrier in health and disease. Ellis Horwood, Chichester. Chapter 1:17–40
Culvenor AJ, Jarrott B (1981) Comparison of β-adrenoreceptors in bovine intracerebral microvessels and cerebral gray matter by [^{3}H]dihydroalprenolol binding. Neuroscience 6:1383–1388
Cutler RWP (1970) Transport of lysine from cerebrospinal fluid of the cat. J Neurochem 17:1017–1027
Cutler RWP, Lorenzo AV (1968) Transport of 1-aminocyclopentanecarboxylic acid from feline cerebrosphinal fluid. Science 161:1363
Daniel PM, Donaldson J, Pratt OE (1975a) A method for injecting substances into the circulation to reach rapidly and to maintain a steady level. Med Biol Engineering 1975:214–227
Daniel PM, Love ER, Pratt OE (1975b) Hypothyroidism and aminoacid entry into brain and muscle. Lancet 2:872
Daniel PM, Love ER, Moorhouse SR, Pratt OE (1981) The effect of insulin upon the influx of tryptophan into the brain of the rabbit. J Physiol (Lond) 312:551–562
DeMyer MK, Shea PA, Hendric HG, Yoshimura NN (1981) Plasma tryptophan and five other amino acids in depressed and normal subjects. Arch Gen Psychiatry 38:642–646
Eayrs JT (1954) The vascularity of the cerebral cortex in normal and cretinous rats. J Anat 88:164–173
Edwards DJ, Sorisio DA (1988) Effects of imipramine on tyrosine and tryptophan are mediated by β-adrenoceptor stimulation. Life Sci 42:853–862
Ehrlich BE, Diamond JM, Braun LD, Cornford EM, Oldendorf WH (1980) Effects of lithium on blood-brain barrier transport of the neurotransmitter precursors choline, tyrosine and tryptophan. Brain Res 193:604–607
Eriksson T, Carlsson A (1988) β-Adrenergic control of brain uptake of large neutral amino acids. Life Sci 42:1583–1589

Fenstermacher JD, Rapoport SI (1984) Blood-brain Barrier. In: Renkin E, Michel C (eds) Handbook of physiology, The Cardiovascular system, vol 4, Chapter 21. American Physiological Society, Bethesda, USA, pp 969–1000
Fernstrom JD, Wurtman RJ (1971) Brain serotonin content: increase following ingestion of a carbohydrate diet. Science 174:1023–1025
Fernstrom JD, Wurtman RJ (1972a) Brain serotonin content: physiological regulation by plasma neutral amino acids. Science 178:414–416
Fernstrom JD, Wurtman RJ (1972b) Elevation of plasma tryptophan by insulin in rat. Metabolism 21:337–342
Gaillard JM, Tissot T (1979) Blood-brain movement of tryptophan and tyrosine in manic depressive illness and schizophrenia. J Neural Transm [Suppl] 15:189–196
Garnett ES, Firnau G, Nahmias C, Sood S, Belbeck L (1980) Blood-brain barrier transport and cerebral utilization of dopa in living monkeys. Am J Physiol 238:R318–R327
Gillman PK, Barlett JR, Bridges PK, Hunt A, Patel AJ, Kantamaneni BD, Curzon G (1981) Indolic substances in plasma, cerebrospinal fluid, and frontal cortex of human subjects infused with saline or tryptophan. J Neurochem 37:410–417
Glaeser BS, Maher TJ, Wurtman RJ (1983) Changes in brain levels of acidic, basic, and neutral amino acids after consumption of single meals containing various proportions of protein. J Neurochem 41:1016–1021
Goldstein GW, Wolinski JS, Csejtey J, Diamond I (1975) Isolation of metabolically active capillaries from rat brain. J Neurochem 25:715–717
Gorgievski-Hrisoho M, Colombo JP, Bachman C (1986) Stimulation of tryptophan uptake into brain microvessels by D-glutamine. Brain Res 367:395–397
Hargreaves KM, Pardridge WM (1988) Neutral amino acid transport at the human blood-brain barrier. J Biol Chem 263:19392–19397
Harik SI, Sharma VK, Wetherbee JR, Warren RH, Banerjee SP (1980) Adrenergic receptors of cerebral microvessels. Eur J Pharmacol 61: 207–208
Harper AE, Tews JK (1988) Nutritional and metabolic control of brain amino acid concentrations. In: Huether (ed) Amino acid availability and brain function in health and disease. Springer, Berlin Heidelberg New York, pp 3–12 (NATO ASI Series, vol H20)
Hawkins RA, Mans AM, Biebuyck JF (1982) Amino acid supply to individual structures in awake and anesthetized rats. Am J Physiol 242:E1–E11
Hawkins RA, Huang S-C, Barrio JR, Keen RE, Feng D, Mazziota JC, Phelps ME (1989) Estimation of local cerebral protein synthesis rates with L-[1-^{11}C] leucine and PET: methods, model, and results in animals and humans. J Cereb Blood Flow Metab 9:446–460
Herbst TJ, Raichle ME, Ferrendelli JA (1979) β-Adrenergic regulation of adenosine-3′,5′-monophosphate concentration in brain microvessels. Science 204:330–332
Hervonen H, Steinwall O (1984) Endothelial surface sulfhydryl-group in blood-brain barrier transport of nutrients. Acta Physiol Scand 121:343–351
Hjelle JT, Baird-Lambert J, Cardinale G, Spector S, Udenfriend S (1978) Isolated microvessels: the blood-brain barrier in vitro. Proc Natl Acad Sci USA 75:4544–4548
Huang M, Drummond GI (1979) Adenylate cyclase in cerbral microvessels: action of guanine nucleotides, adenosine and other agonists. Mol Pharmacol 16:462–472
Hughes CCW, Lantos PL (1989) Uptake of leucine and alanine by cultured cerebral capillary endothelial cells. Brain Res 480:126–132
Hutchinson HT, Eisenberg HM, Haber B (1985) High-affinity transport of glutamate in rat brain microvessels. Exp Neurol 87:260–269
Hutson PH, Knott PJ, Curzon G (1980) Effect of isoprenaline infusion on the distribution of tryptophan, tyrosine and isoleucine between brain and other tissues. Biochem Pharmacol 29:509–516

Hwang SM, Weiss S, Segal S (1980) Uptake of L-[^{35}S]cystine by isolated rat brain capillaries. J Neurochem 35:417–424
Hwang SM, Miller M, Segal S (1983) Uptake of L-[^{14}C]proline by isolated rat brain capillaries. J Neurochem 40:317–323
James JH, Fischer JE (1981) Transport of neutral amino acids at the blood-brain barrier. Pharmacology 22:1–7
James JH, Escourrou J, Fischer JE (1978) Blood-brain neutral amino acid transport activity is increased after portocaval anastomosis. Science 200:1395–1397
James JH, Jeppsson B, Ziparo V, Fischer JE (1979) Hyperammonaemia, plasma amino acid imbalance, and blood-brain amino acid transport: a unified theory of portal-systemic encephalopathy. Lancet 2:772–775
Jeppsson B, James JH, Edwards LL, Fischer JE (1985) Relationship of brain glutamine and brain neutral amino acid concentrations after portacaval anastomosis in rats. Eur J Clin Invest 15:179–187
Johanson CE (1989) Ontogeny and phylogeny of the blood-brain barrier. In: Neuwelt EA (ed) Implications of the blood-brain barrier and its manipulation, vol 1. Plenum, New York, p 157
Jonung T, Rigotti P, James JH, Brackett K, Fischer JE (1985) Effect of hyperammonemia and methionine sulfoximine on the kinetic parameters of blood-brain transport of leucine and phenylalanine. J Neurochem 45:308–318
Joó F, Karnuschina I (1973) A procedure of the isolation of capillaries from rat brain. Cytobios 8:41–48
Joó F, Tóth I, Jancso G (1975) Brain adenylate cyclase: its common occurrence in the capillaries and astrocytes. Naturwissenschaften 62:397–398
Knott PJ, Curzon G (1972) Free tryptophan in plasma and brain tryptophan metabolism. Nature 239:452–453
Knudsen GM, Pettigrew KD, Patlak CS, Hertz MM, Paulson OB (1990) Asymmetrical transport of amino acids across the blood-brain barrier in humans. J Cereb Blood Flow Metab 10:698–706
Kobayashi H, Memo M, Spano PF, Trabucchi M (1981) Identification of β-adrenergic receptor binding sites in rat brain microvessels, using [^{125}I]iodohydroxybenzylpindolol. J Neurochem 36:1383–1388
Lajtha A, Dunlop D (1981) Turnover of protein in the nervous system. Life Sci 29:755–767
LaManna JC, Harik SI (1986) Regional studies of blood-brain barrier transport of glucose and leucine in awake and anesthetized rats. J Cereb Blood Flow Metab 6:717–723
Leathwood PD, Fernstrom JD (1990) Effect of an oral tryptophan/carbohydrate load on tryptophan, large neutral amino acid, and serotonin and 5-hydroxyindoleacetic acid levels in monkey brain. J Neural Transm (Gen Sect) 79:25–34
Lefauconnier JM, Trouvé R (1983) Development changes in the pattern of amino acid transport at the blood-brain barrier in rats. Develop Brain Res 6:175–182
Lefauconnier JM, Urban F, Mandel P (1978) Taurine transport into the brain in rat. Biochimie 4:381–387
Lefauconnier JM, Lavielle E, Terrien N, Bernard G, Fournier E (1980) Effect of various lead doses on some cerebral capillary functions in the suckling rat. Toxicol Appl Pharmacol 55:467–476
Lefauconnier JM, Lacombe P, Bernard G (1985a) Cerebral blood flow and blood-brain barrier influx of some neutral amino acids in control and hypothyroid 16-day-old rats. J Cereb Blood Flow Metab 5:318–326
Lefauconnier JM, Tayarani I, Roux F (1985b) Delineation of transport systems for some neutral amino acids in isolated rat brain microvessels. J Cereb Blood Flow Metab 5 (Suppl 1):S91–S92
Lefauconnier JM, Bernard G (1985c) Evolution of cerebral blood flow and blood-brain clearance of phenylalanine and alanine during rat development. J Cereb Blood Flow Metab 5 (Suppl 1):S93–S94

Lefauconnier JM, Tayarani I, Bernard G (1986) Blood-brain barrier permeability to excitatory amino acids. Adv Exp Med Biol 203:191–198

Lefauconnier JM, Bouchaud C, Bernard G (1991) Initial process of diffusion of small molecules from blood vessels to the meninges in the young rat. Neurosci Lett 121:9–11

Lorenzo AV, Cutler RWP (1969) Amino acid transport by choroid plexus in vitro. J Neurochem 16:577–585

Lorenzo AV, Gewirtz M (1977) Inhibition of [^{14}C]tryptophan transport into brain of lead exposed neonatal rabbits. Brain Res 132:386–392

Maes M, Jacobs MP, Suy E, Vandewoude M, Minner B, Raus J (1990) Effects of dexamethasone on the availability of L-tryptophan and on the insulin and FFA concentrations in unipolar depressed patients. Biol Psychiatry 27:854–862

Mans AM, Biebuyck JF, Shelly K, Hawkins RA (1982) Regional blood-brain barrier permeability to amino acids after portacaval anastomosis. J Neurochem 38:705–717

Mans AM, DeJoseph R, Davis DW, Hawkins RA (1987) Regional amino acid transport into brain during diabetes: effect of plasma amino acids. Am J Physiol 253:E575–E583

Markovitz DC, Fernstrom JD (1977) Diet and uptake of aldomet by the brain: competition with natural large neutral amino acids. Science 197:1014–1015

McCall AL, Millington WR, Wurtmann RJ (1982) Metabolic fuel and amino acid transport into the brain in experimental diabetes mellitus. Proc Natl Acad Sci USA 79:5406–5410

Michaelson A, Bradbury M (1982) Effect of early inorganic lead exposure on rat blood-brain barrier permeability to tyrosine or choline. Biochem Pharmacol 31:1881–1885

Miller LP, Pardridge WM, Braun LD, Oldendorf WH (1985) Kinetic constants for blood-brain barrier amino acid transport in conscious rats. J Neurochem 45:1427–1432

Møller SE (1988) Tryptophan and tyrosine ratios to neutral amino acids in depressed patient in regard to K_m: relation to efficacy of antidepressant treatments. In: Huether G (ed) Amino acid availability and brain function in health and disease. Springer, Berlin Heidelberg New York, pp 355–361 (NATO ASI series, vol H20)

Møller SE, Bech P, Berrum H, Bojholm S, Butler B, Folker H, Gram LF, Larsen JK, Lauritzen L, Loldrup D, Munk-Andersen E, Odum K, Rafaelsen OJ (1990) Plasma ratio tryptophan/neutral amino acids in relation to clinical response to paroxetine and clomipramine in patients with major depression. J Affect Disord 18, 59–66

Momma S, Aoyagi M, Rapoport SI, Smith QR (1987) Phenylalanine transport across the blood-brain barrier as studied with the in situ brain perfusion technique. J Neurochem 48, 1291–1300

Mooradian AD (1990) Metabolic fuel and amino acid transport into the brain in experimental hypothyroidism. Acta Endocrinol 122(2):156–162

Moorhouse SR, Carden S, Drewitt PN, Eley BP, Hargreaves RJ, Pelling D (1988) The effect of chronic low level lead exposure on blood-brain barrier function in the developing rat. Biochem Pharmacol 37:4539–4547

Murray JE, Cutler RWP (1970) Transport of glycine from the cerebrospinal fluid. Arch Neurol 23:23–31

Nagashima T, Lefauconnier JM, Smith QR (1987) Developmental changes in neutral amino acid transport across the blood-brain barrier. J Cereb Blood Flow Metab 7 (Suppl 1):S524

Nathanson JA, Glaser JH (1979) Identification of β-adrenergic-sensitive adenylate cyclase in intracranial blood vessels. Nature 278:567–569

Ohno K, Pettigrew KD, Rapoport SI (1978) Lower limits of cerebrovascular permeability to nonelectrolytes in the conscious rat. Am J Physiol 235:H299–H307

Oldendorf WH (1971) Brain uptake of radiolabeled amino acids, amines, and hexoses after arterial injection. Am J Physiol 221:1629–1639
Oldendorf WH, Szabo J (1976) Amino acid assignment to one of three blood-brain barrier amino acid carriers. Am J Physiol 230:94–98
Oldendorf WH, Sisson WB, Silverstein A (1971) Brain uptake of selenomethionine Se 75. Arch Neurol 24:524–528
Olney JW (1969) Brain lesions, obesity and other disturbances in mice treated with sodium glutamate. Science 164:719–721
Olney JW (1986) Excitotoxic amino acids. NIPS 1:19–23
Orlowski M, Sessa G, Green JP (1974) γ-Glutamyl transpeptidase in brain capillaries: possible site of a blood-brain barrier for amino acids. Science 184:66–68
Oxender DL, Christensen HN (1963) Distinct mediating systems for the transport of neutral amino acids by the Ehrlich cell. J Biol Chem 238:3686–3699
Pardridge WM (1976) Inorganic mercury: selective effects on blood-brain barrier transport systems. J Neurochem 27:333–335
Pardridge WM (1977) Kinetics of competitive inhibition of neutral amino acid transport across the blood-brain barrier. J Neurochem 28:103–108
Pardridge WM (1979) Regulation of amino acid availability to brain: selective control mechanism for glutamate. In: Filer LJ et al. (eds) Glutamic acid: advances in biochemistry and physiology. Raven, New York
Pardridge WM (1983) Brain metabolism: a perspective from the blood-brain barrier. Physiol Rev 63:1481–1535
Pardridge WM, Fierer G (1990) Transport of tryptophan into brain from the circulating albumin-bound pool in rats and in rabbits. J Neurochem 54:971–976
Pardridge WM, Oldendorf WH (1975) Kinetic analysis of blood-brain transport of amino acids. Biochim Biophys Acta 401:128–136
Pardridge WM, Oldendorf WH (1977) Transport of metabolic substrates through the blood-brain barrier. J Neurochem 28:5–12
Peng Y, Tews JK, Harper AE (1972) Amino acid imbalance, protein intake, and changes in rat brain and plasma amino acids. Am J Physiol 222:314–321
Pentschew A, Garro F (1966) Lead encephalo-myelopathy of the suckling rat and its implications on the prophyrinopathic nervous diseases. Acta Neuropathol 6:266–278
Peroutka SJ, Moskowitz MA, Reinhard JF, Snyder SH (1980) Neurotransmitter receptor binding in bovine cerebral microvessels. Science 208:610–612
Preston JE, Segal MB (1990) The steady-state amino acid fluxes across the perfused choroid plexus of the sheep. Brain Res 525:275–279
Rassin DK (1990) transport into brain of albumin-bound amino acids. J Neurochem 55:722
Reith J, Ermisch A, Diemer NH, Gjedde A (1987) Saturable retention of vasopressin by hippocampus vessels in vivo, associated with inhibition of blood-brain transfer of large neutral amino acids. J Neurochem 49:1471–1479
Richter JJ, Wainer A (1971) Evidence for separate systems for the transport of neutral and basic amino acids across the blood-brain barrier. J Neurochem 18:613–620
Rosenberg R, Young JD, Ellory JC (1980) L-Tryptophan transport in human red blood cells. Biochim Biophys Acta 598:375–384
Ross HJ, Wright EM (1984) Neutral amino acid transport by plasma membrane vesicles of the rabbit choroid plexus. Brain Res 295:155–160
Rossi-Fanelli F, Cascino A, Strom R, Cardelli-Cangiano P, Ceci F, Muscaritoli M, Cangiano C (1987) Amino acids and hepatic encephalopathy. Prog Neurobiol 28:277–301
Sage JJ, Duffy TE (1979) Pentobarbital anesthesia: influence on amino acid transport across the blood-brain barrier. J Neurochem 33:963–965
Schwartz JC, Lampart C, Rose C (1972) Histamine formation in rat brain in vivo: effects of histidine loads. J Neurochem 19:801–810

Sershen H, Lajtha A (1976) Capillary transport of amino acids in the developing brain. Exp Neurol 53:465–474
Sershen H, Lajtha A (1979) Inhibition pattern by analogs indicates the presence of ten or more transport systems for amino acids in brain cells. J Neurochem 32:719–726
Shotwell MA, Kilberg MS, Oxender DL (1983) The regulation of neutral amino acid transport in mammalian cells. Biochim Biophys Acta 737:267–284
Smith QR, Takasato Y, Sweeney DJ, Rapoport SI (1985) Regional cerebrovascular transport of leucine as measured by the in situ brain perfusion technique. J Cereb Blood Flow Metab 5:300–311
Smith QR, Momma S, Aoyagi M, Rapoport SI (1987) Kinetics of neutral amino acid transport across the blood-brain barrier. J Neurochem 49:1651–1658
Smith QR, Fukui S, Robinson P, Rapoport SI (1990) Influence of cerebral blood flow on tryptophan uptake into brain 364–369. In: Lubec G, Rosenthal GA (eds) Amino acids. ESCOM
Snodgrass SR, Lorenzo AV (1973) Transport of GABA from the perfused ventricular system of the cat. J Neurochem 20:761–769
Snodgrass SR, Cutler RWP, Song Kang E, Lorenzo AV (1969) Transport of neutral amino acids from feline cerebrospinal fluid. Am J Physiol 217:974–980
Tagliamonte A, Tagliamonte P, Forn J, Perez-Cruet J, Krishna G, Gessa GL (1971) Stimulation of brain serotonin synthesis by dibutyryl-cyclic AMP in rats. J Neurochem 18:1191–1196
Takasato Y, Rapoport SI, Smith QR (1984) An in situ brain perfusion technique to study cerebrovascular transport in rat. Am J Physiol 247:H484–H493
Tayarani I, Lefauconnier JM, Roux F, Bourre JM (1987a) Evidence for an alanine, serine and cysteine system of transport in isolated brain capillaries. J Cereb Blood Flow Metab 7:585–591
Tayarani I, Lefauconnier JM, Bourre JM (1987b) The effect of mercurials on amino acid transport and rubidium uptake by isolated rat brain microvessels. Neurotoxicol 8:543–552
Tayarani I, Cloez I, Lefauconnier JM, Bourre JM (1989) Sodium-dependent high-affinity uptake of taurine by isolated rat brain capillaries. Biochim Biophys Acta 985:168–172
Tovar A, Tews JK, Torres N, Harper AE (1988) Some characteristics of threonine transport across the blood-brain barrier of the rat. J Neurochem 51:1285–1293
Van Gelder NM (1968) A possible enzyme barrier for γ-aminobutyric acid in the central nervous system. Prog Brain Res 29:259–271
Wade LA, Katzman R (1975a) Synthetic amino acids and the nature of L-DOPA transport at the blood-brain barrier. J Neurochem 25:837–842
Wade LA, Katzman R (1975b) Rat brain regional uptake and decarboxylation of L-DOPA following carotid injection. Am J Physiol 228:352–359
Wade LA, Brady HM (1981) Cysteine and cystine transport at the blood-brain barrier. J Neurochem 37:730–734
Weissbach L, Handlogten ME, Christensen HN, Kilberg MS (1982) Evidence for two Na-independent neutral amino acid transport systems in primary cultures of rat hepatocytes. J Biol Chem 257:12006–12011
White MF, Gazzola GC, Christensen HN (1982) Cationic amino acid transport into cultured animal cells. J Biol Chem 257:4443–4449
Wise WC (1975) Maturation of membrane function: transport of amino acid by rat erythroid cells. J Cell Physiol 87:199–212
Wright EM (1974) Active transport of glycine across the frog arachnoid membrane. Brain Res 76:354–358
Wurtman RJ, Chou C, Rose C (1970) The fate of ^{14}C-dihydroxyphenylalanine (^{14}C-dopa) in the whole mouse. J Pharmacol Exp Ther 174:351–356
Young SN, Lal S, Sourkes TL, Feldmuller F, Aronoff A, Martin JB (1975) Relationships between tryptophan in serum and CSF, and 5-hydroxyindoleacetic acid in

CSF of man: effect of cirrhosis of liver and probenecid administration. J Neurol Neurosurg Psychiatr 38:322–330
Yudilevich DL, De Rose N, Sepúlveda FV (1972) Facilitated transport of amino acids through the blood-brain barrier of the dog studied in a single capillary circulation. Brain Res 44:569–578
Yuwiler A, Bennett BL, Brammer GL, Geller E (1979) Lithium treatment and tryptophan transport through the blood-brain barrier. Biochem Pharmacol 28: 2709–2712
Zanchin G, Rigotti P, Dussini N, Vassanelli P, Battistin L (1979) Cerebral amino acid levels and uptakes in rats after portocaval anastomosis. II Regional studies in vivo. J Neurosci Res 4:301–310

CHAPTER 6

Peptides and the Blood-Brain Barrier

D.J. BEGLEY

A. Introduction: Central Actions of Peripherally Administered Peptides

When nature, being oppressed,
commands the mind
To suffer with the body.
King Lear, Act II Scene IV

When Shakespeare put these words into Lear's mouth he clearly would never have heard of a peptide or for that matter of any other plasma solute that might affect central nervous activity. However, he clearly appreciated the ability of the humours to influence mood and behaviour. Interestingly the essential concept is still with us that blood-borne signals, in this case peptides, might alter central nervous activity.

This review will deal almost exclusively with peptides, that is peptide molecules and analogues with a molecular size less than that of insulin. This review is also selective, due to constraints on space, concentrating largely on recent significant developments and also concentrating exclusively on blood to brain transport and excluding comment on peptide transport out of the central nervous system (CNS). Mechanisms regulating peptide levels in cerebrospinal fluid (CSF) have recently been reviewed (BEGLEY and CHAIN 1992).

That peptides may influence central nervous activity is now extensively documented. A review of this literature is outside of the scope of this chapter but a useful and interesting compilation of peptides and their central nervous effects has been drawn together by ZADINA et al. (1986). This review lists 135 central nervous effects for some 80 peptides together with references to the original observations, culled from merely 5 years publication of a single journal, *Peptides*.

The specific mechanisms by which these peptides might exert central nervous effects are more controversial and can be classified into six general areas:

1. There may be a specific or non-specific uptake at the blood-tissue interfaces of the brain which constitute the blood-brain and blood-CSF barriers. This uptake might then be followed by transcellular transport,

either by a vesicular mechanism of transcytosis or by directional membrane transport at the luminal and abluminal membranes of the endothelial cells, both resulting in transfer into brain extracellular fluid. Alternatively, after uptake into the cerebral endothelial cells some or all peptides might be subject to enzymic hydrolysis.
2. There may be a relatively non-specific entry through the more permeable circumventricular organs with local nervous activity being affected, which could then be transmitted to other brain regions.

Abbreviations

A-V	Arterio-venous
ACTH	Adrenocorticotropin
AII	Angiotensin II
(r)ANP	Atrial natriuretic peptide (recombinant)
AVP	Arg^{8}-vasopressin
BCH	2-amino-bicyclo-2,2,1-heptane-2-carboxylic acid
B_{max}	Maximum binding (capacity)
BSA	Bovine sesum albumin
BUI	Brain uptake index
cAMP	Cyclic adenosine monophosphate
CSF	Cerebrospinal fluid
CNS	Central nervous system
DADLE	D-Ala^{2}-D-Leu^{5}-encephalin
DAGO	D-Ala^{2}-*N*-Me-Phe^{4}-Gly^{5}-OH-encephalin
DDAVP	Desamido-Cys^{1}-D-Arg^{8}-vasopressin (desmopressin)
DGAVP	Desglycineamide^{9}-arginine^{8}-vasopressin
DSIP	Delta-sleep-inducing peptide
E	Extraction
EDTA	Ethylenediaminetetra-acetic acid
GnRH	Gonadotropin-releasing hormone
HIV	Human immunodeficiency virus
HEPES	*N*-2-hydroxyethylpiperazine-*n'*-2-ethanesulphonic acid
HPLC	High performance liquid chromatography
HRP	Horseradish peroxidase
ID_{50}	Concentration of peptide required to inhibit receptor binding by 50%
IGF (I and II)	Insulin-like growth factors I and II
IgG	Immunoglobulin G
rIL-2	Interleukin 2 (recombinant)
K_D	Binding constant (affinity)
K_d	Non-saturable diffusion constant
K_{in}	Cerebrovascular permeability constant (cerebrovascular influx constant)
K_m	Michaelis-Menten constant
LDL	Low density lipoprotein
$\log P$	Logarithm of the partition coefficient
MeAIB	*N*-Methyl-α-aminoisobutyric acid
MIF-1	MSH-release inhibitory factor
MSH (α and β)	Melanocyte stimulating hormone
3-*O*-MG	3-*O*-methyl glucose
P	Cerebrovascular pemeability coefficient
PMA	Phorbol myrisate ester
PS	permeability – surface area (product)
PTS-1	Protein transport system-1
SDS	Sodium dodecyl sulphate
T	Experimental time at which brain permeability/uptake is determined
T_{50}	Plasma half-life
TRH	Thyrotropin-releasing hormone
U_{max}	Maximum uptake
V_1 receptor	Vasopressin 1 receptor
V_i	Initial volume of distribution
V_{max}	Maximum velocity

3. The peptides (first messengers), might interact with luminal receptors on brain capillary endothelia which then activate second intracellular messengers, resulting in the release of a third messenger into brain extracellular fluid.
4. Peptides could interact with luminal receptors on brain capillary endothelia which alter the permeability of the barrier to other blood solutes which enter the brain and produce the observed effects.
5. Peptides might act peripherally to alter the activity of afferent nerves.
6. Peptides might act via non-specific mechanisms which alter cerebral blood flow or result in altered states of arousal.

Certainly more is known concerning some of these possibilities than others, and this review will largely direct itself towards transport phenomena. A variety of techniques has been used to explore the permeability of the blood-brain interface and the following discussion is conveniently structured around the various experimental approaches used to investigate the problem.

B. Uptake of Peptides at the Blood-Brain Interfaces

I. Intracarotid Bolus Injection Studies (Brain-Uptake Index)

The brain-uptake index (BUI) method of OLDENDORF (1970) was a major step forward in quantifying the uptake of blood-borne solutes by the cerebral vasculature. Essentially, the method relies on the rapid injection of a mixture of tracer (test molecule) and reference molecule, in a single bolus, into the right carotid artery of an experimental animal, usually the rat, followed by decapitation and removal of the brain some 5–15 s after the injection. The method relies on the reference molecule having a relatively rapid penetration of the cerebral capillary endothelium compared to the test molecule. As applied to peptides the reference molecule is usually [^{14}C]butanol for a ^{3}H-labelled peptide and [^{3}H]H_2O for a ^{14}C-labelled peptide. The tracer remaining in the vascular space can be corrected for by including [^{113}In]ethylenediaminetetraacetic acid (EDTA) in the bolus (OLDENDORF and BRAUN 1976) or by subtracting the BUI for a non-penetrating molecule such as sucrose (ZLOKOVIĆ et al. 1983).

The BUI is usually expressed as a percentage uptake of the tracer molecule relative to the reference molecule. Alternatively, results can be expressed as an extraction fraction or percentage, which is the proportion of the injected tracer extracted during a single passage through the cerebral vasculature (Q_b), divided by the total amount available for extraction (Q_a), and has the advantage of being independent of the reference molecule (CLARK et al. 1981). The extraction (E) is given by the relationship $E = Q_b/Q_a$. However, in practice, with the OLDENDORF technique not all reference tracers used are flow limited and therefore the BUI is not equivalent to E (CLARK et al. 1981) and BUI $= E_{tracer}/E_{reference}$ (CEFALU and PARDRIDGE 1985; see Table 1).

Table 1. Extraction of radiolabelled peptides at the blood brain barrier of the rat determined by various intracarotid injection methods

Peptide	Residues	Extraction (E, %)	Reference
Carnosine	2	1.1	Cornford et al. (1978)
Gly-Phe	2	0.36[a]	Zloković et al. (1983)
Gly-Leu	2	0.51[a]	Zloković et al. (1983)
Glutathione	3	0.4–0.6	Cornford et al. (1978)
TRH	3	1.0	Cornford et al. (1978)
		0.02–0.1[b]	Zloković et al. (1985c)
Encephalin (leu)	5	2.4	Cornford et al. (1978)
		−0.56[b]	Zloković et al. (1985a)
DADLE	5	1.23	Zloković et al. (1985a)
Encephalin (met)	5	2.4–3.0	Cornford et al. (1978)
β-casomorphin-5	5	1.9	Ermisch et al. (1983)
DGAVP	8	1.2	Ermisch et al. (1985a,b)
CCK8	8	~0	Oldendorf (1981)
AVP	9	2.4	Ermisch et al. (1985a,b)
LVP	9	1.3	Ermisch et al. (1985a,b)
OT	9	1.3	Ermisch et al. (1985a,b)
GnRH	10	0.7	Verheugen et al. (1983)
		1.3	Ermisch et al. (1984)
Substance P	11	0.5	Landgraf et al. (1983)
Cyclosporin A	11	2.9	Cefalu and Pardridge (1985)
		0.94[b]	Begley et al. (1990)
Somatostatin	14	0.01	Oldendorf (1981)
β-endorphin	31	1.9	Ermisch et al. (1985b)
Insulin	51	−1.34[b]	Squires (1990)
Inert polar molecules	**MW**	**Apparent extraction (E, %) (residual activity in vasculature)**	**Reference**
D-mannitol	167	1.65[c]	Oldendorf (1971)
		1.4[b]	Zloković et al. (1985c)
Sucrose	342	1.7[a]	Zloković et al. (1985a)
Inulin	5000	1.66[c]	Oldendorf (1971)
		1.15	Verheugen et al. (1983)
		0.49[b]	Squires, (1990)
Dextran	70000	0.9	Cornford et al. (1978)

Based on a table by Ermisch et al. (1985b). An attempt has been made to make all of the values comparable by calculating tracer extraction fractions (E%) from BUI values where appropriate.

[a,b] BUI values determined in the author's laboratory and E calculated from BUI values using an extraction of 85% for [^{3}H]water (Cornford et al. 1978) and of 64% for [^{14}C]butanol (Pardridge and Fierer 1985). Hence, for ^{14}C-labelled tracers $^a E$ = BUI × 0.85 and for ^{3}H-labelled tracers $^b E$ = BUI × 0.64.

[c] Values from Oldendorf's laboratory are converted from BUI values, where necessary, using the percentage water extraction of Cornford et al. (1978). Other values are extraction values quoted in the literature.

Relative extractions obtained by carotid bolus injection methods for a large number of peptides investigated in a variety of studies are shown in Table 1. Generally, the extraction values are less than 3% and do not differ significantly from the extractions of inert polar molecules. The use of a carotid bolus injection for the estimation of peptide extraction has certain limitations, some of which are discussed by OLDENDORF (1981). As might be expected, hydrophilic molecules such as peptides will only penetrate the lipid membranes of the capillary endothelium slowly and do not show a significant extraction in the 5–15 s of a BUI study. Thus, although the technique is valuable for differentiating between rapidly penetrating lipid-soluble solutes and for amino acids which may have specific carrier-mediated uptake mechanisms the method is of limited usefulness when applied to peptides. Two studies have suggested that peptides might show significant brain uptake indices. GREENBERG et al. (1976), using a method faithfully reproducing that of Oldendorf, produced BUI values of 13.7% for ^{3}H-labelled melanocyte stimulating hormone (MSH) release inhibitory factor 1 ([^{3}H]MIF-1), 9.6% for [^{3}H]MSH-α and 13.0% for ^{14}C-labelled Arg^{8}-vasopressin ([^{14}C]AVP). Later studies have not produced comparable values for peptides. Also a study by KASTIN et al. (1976) suggested that the BUI for [^{3}H-Tyr]methionine encephalin was 15.0%. However, in this study the procedure of Oldendorf was not followed and the tracer encephalin and reference water were injected into different groups of rats.

The intracarotid bolus injection technique has, however, produced some very valuable insights into the interaction of peptide molecules with the blood-brain barrier. The study of ZLOKOVIĆ et al. (1983) investigated the possibility that small peptides might interact with amino acid carriers at the blood-brain interface. The dipeptides glycyl-L-leucine and glycyl-L-phenylalanine were investigated. Both leucine and phenylalanine are transported into brain principally by the L-transport system of CHRISTENSEN (1979). As the dipeptides investigated have the A-system amino acid glycine at the N-terminus and an L-system amino acid at the C-terminus it was of some interest to see if the peptides retained a reactivity with the L-system transporter. Because the A-system transporter is largely absent at the blood-brain barrier (CHRISTENSEN 1979), transport of these peptides by the L-system might prove a route of entry for A-system amino acids such as glycine into the brain. The ingenuity of the approach was to use millimolar concentrations of various competitor peptides combined in the bolus to compete with tracer concentrations of the L-system amino acids. This study shows quite clearly that that glycyl-L-phenylalanine, glycyl-L-leucine and glycine do not competitively inhibit the BUI of phenylalanine and, in addition, that glycyl-L-leucine, leucyl-L-glycine, and leucyl-L-leucyl-L-leucine have no effect on the BUI of leucine. BCH (2-amino-bicyclo-2,2,1)-heptane-2-carboxylic acid), the definitive substrate of the L-transporter, reduced the BUI of both phenylalanine and leucine by 88%, whilst MeAIB (*N*-methyl-α-aminoisobutyric acid), diagnostic of the A-transporter, had no effect. Both amino acids self-inhibited very effectively. These experiments show conclusively that the formation of a peptide by an L-system amino acid effectively abolishes reactivity with the L-system transporter. This is not the case in the intestine, where some peptides are transported as readily as their constituent amino acids and compete for amino acid transporters (MATTHEWS 1972).

Some naturally occurring peptides are rapidly hydrolysed by peptidases present in plasma and also by possible extrinsic peptidases on the luminal surface of the cerebral capillary endothelium. This enzymatic activity would result in the liberation of the labelled amino acid and other labelled and non-labelled peptide fragments and produce unreliably high BUI or extraction values. For instance, the apparent BUI for [^{3}H-Pro] thyrotropin releasing hormone (TRH) and [^{3}H-Tyr]leucine encephalin is of the order of 5.5% (ZLOKOVIĆ et al. 1985a–c; BEGLEY and ZLOKOVIĆ 1986) and can be reduced to approximately those of sucrose, mannitol and inulin by the addition of 2 mM bacitracin, an aminopeptidase inhibitor, to the injectate bolus. Thus, liberation of the rapidly transported amino acids proline and tyrosine, *en passage*, through the cerebral vasculature had produced falsely high BUI values.

In the case of TRH, free labelled proline is being generated, as demonstrated by a marked inhibition of the BUI value by 10 mM proline and 2 mM bacitracin, with 10 mM TRH producing only a minimal inhibition. These results suggest that the BUI of intact TRH is small and not significantly different from that of mannitol. In the case of leucine encephalin, self-inhibition is not very effective, whereas 2 mM bacitracin produces a huge reduction in the apparent BUI and 10 mM BCH and 5 mM tyrosine inhibit uptake very significantly, suggesting the liberation of free labelled tyrosine during passage through the cerebral vasculature. It is interesting to note that in the case of D-Ala^2-D-Leu^5-encephalin (DADLE), which is more resistant to enzymic hydrolysis (HAMBROOK et al. 1976), the BUI of the tracer alone is similar to that of leucine encephalin but 10 mM BCH and 2 mM bacitracin produce far smaller reductions in the BUI. This observation supports the concept of a greater stability for DADLE during a single pass of the brain vasculature. The residual BUI of DADLE in the presence of 2 mM bacitracin is higher than that of leucine encephalin, suggesting a higher permeability for the intact tracer of the capillary endothelium. This is in agreement with the greater lipid solubility of DADLE ($\log P - 1.05$) compared to leucine encephalin ($\log P - 1.64$), (BEGLEY and ZLOKOVIĆ 1986). However, it has been shown that leucine encephalin does in fact have a carrier-mediated uptake at the luminal surface of the capillary endothelium in brain perfusion experiments (ZLOKOVIĆ et al. 1987b, 1989a) (see Sect. B.III), whereas the reactivity of DADLE in this experimental situation has not been investigated. However, DADLE does show saturable carrier-mediated uptake at the choroid plexus (ZLOKOVIĆ et al. 1988b) (see Sect. B.IV).

The site of peptide hydrolysis during intracarotid bolus injection might be at the luminal surface of the capillary itself due to extrinsic peptidases and also to peptidases present in plasma. PARDRIDGE et al. (1985c) have shown that the fractional mixing of plasma with the 200-μl bolus injected during a typical BUI determination may be between 7% and 9%, thus ensuring adequate access of plasma peptidases to the tracer peptide. Some early studies which did not take the possibility of rapid hydrolysis of peptides into account have produced erroneously high values for brain uptake.

VERHEUGEN et al. (1983) found that the BUI for gonadotropin-releasing hormone (GnRH) rises from 1.1% when the bolus contains rat or human plasma and interpret this to indicate hydrolysis by plasma enzymes. ERMISCH et al. (1984) calculated an extraction of 1.3% for GnRH in rats using a carotid bolus injection technique where the tracer and reference were not coinjected. Using a similar technique, ERMISCH et al. (1985a) calculated extraction fractions for arginine vasopressin (AVP) lysine vasopressin, desglycineamide arginine vasopressin (DGAVP), and oxytocin (Table 1). They showed that the 1%–2.5% extraction values that they obtain are not saturable by including excess tracer in the bolus and that therefore no carrier is involved at the blood-brain barrier. Brain regions outside of the blood-brain barrier extracted up to 30 times more peptide than areas within the blood-brain barrier.

The BUI technique has also been used to study the brain uptake of the lipid-soluble peptide cyclosporin A (CEFALU and PARDRIDGE 1985; BEGLEY et al. 1990). The technique shows that the peptide has a remarkably small extraction (Table 1), in spite of its high lipid solubility ($\log P + 2.99$). The brain uptake of cyclosporin has been further investigated by BEGLEY et al. (1990) and is discussed in Sects. B.III and B.VI.

BANKS et al. (1984) in a study with delta-sleep-inducing peptide (DSIP) injected into the carotid artery, concluded that DSIP was taken up by a non-saturable mechanism, as large concentrations of DSIP, *N*-Tyr-DSIP and D-Ala^4-DSIP-NH_2 incorporated into the bolus were unable to inhibit uptake. Essentially similar results were obtained whether the injection was by the intracarotid or intravenous route.

FRANK et al. (1985) estimated a BUI for insulin with [^{3}H]water as a reference in neonatal rabbits. The BUI was 22.0 ± 1.1% in the newborn, 12.8 ± 0.6% at 3 weeks and 6.5 ± 0.1% at 11 weeks. These results suggest that the extraction of insulin is relatively large in the newborn and decreases as the animals age. This reduction in the BUI may be associated with a reduction in insulin receptor number and affinity as the animals age (see Sect. B.V).

SQUIRES (1990) has determined the BUI for insulin in the rat, with butanol as the reference. The brain extraction for insulin in a bolus of *N*-2-hydroxyethl piperazine-*n*′-2-ethanesulphonic acid (HEPES) buffered saline was not significantly different from the apparent extraction of inulin (Table 1).

II. Intravenous Bolus Injection Studies

An extension of the time-course over which brain-uptake studies can be made is possible using an intravenous bolus injection technique, which enables the brain uptake of more slowly penetrating solutes such as peptides to be quantified and differentiated. An animal is injected intravenously with a bolus of radioactively labelled tracer and arterial blood samples are taken at accurately timed intervals. At the end of the experimental period the last arterial sample is taken, a CSF sample may be obtained by cisternal puncture and the animal decapitated. Brain regions can then be removed and their radioactive content measured. This technique was first applied to brain uptake of non-electrolytes by OHNO et al. (1978). A cerebrovascular permeability constant (K_{in}) with the dimensions ml g^{-1} min^{-1} may be calculated and a correction may be made for tracer trapped in the vascular space by allowing for the intravascular inulin space of the various brain regions.

Alternatively, K_{in} may be estimated by a multiple time regression analysis method (BLASBERG et al. 1983; PATLAK et al. 1983) where the ratio of radioactivity of brain to plasma ($C_{br(T)}/C_{pl(T)}$) is plotted against an ordinate of the integral of the plasma radioactivity with time, divided by the plasma radioactivity at experimental time T. If the experiment is performed a number of times at varying experimental durations, the slope of the initial linear portion of the graph represents the unidirectional cerebrovascular permeability constant (K_{in}) in ml g^{-1} min^{-1}. Multiple time regression analysis, in addition to the unidirectional permeability constant (K_{in}) also yields an initial volume of distribution (V_i) for the test solute, which is given graphically by the ordinate intercept of the uptake plot.

Use of the intravenous bolus injection technique requires that the tracer is stable in plasma and is not rapidly metabolized or, in the case of a peptide, hydrolysed into fragments or individual amino acids. Thus, it is not suitable for many naturally occurring peptides but is well suited to peptide analogues, especially those designed to have extended half-lives in plasma. It has been used recently to investigate the interaction of somatostatin analogues with the blood-brain interface. The intravenous bolus injection technique also assumes that there is no significant backflux from brain to blood occurring during the experimental period. Certainly with some lipid-soluble peptide analogues such as cyclosporin some backflux might occur as the plasma concentration of the tracer falls. With lipid-soluble tracers therefore the K_{in} obtained is better termed 'an apparent permeability constant'.

A variety of tumours have been shown to have somatostatin receptors on their cell surfaces and various somatostatin analogues have been shown to be active in limiting growth in these tumours (SCHALLY et al. 1986). A number of tumours of the CNS have also been shown to possess somatostatin receptors (REUBI et al. 1987) and in their early stages may retain a blood-brain barrier. It is therefore of considerable interest to investigate the permeability of the blood-brain barrier to these somatostatin analogues.

Table 2. Cerebrovascular permeability constants ($K_{in(corr)}$), and permeability coefficients (P), for [^{14}C]Sandostatin (SMS 201 995) in various brain regions of the guinea pig determined by an intravenous bolus injection method

Brain region	n	$K_{in(corr)}$[a] ($\mu l\, g^{-1} min^{-2}$)	P[b] ($cm/s \times 10^{-7}$)
Whole brain	6	0.083 ± 0.021	0.138
Hippocampus	6	0.058 ± 0.017	0.097
Caudate nucleus	6	0.024 ± 0.005	0.040
Parietal cortex	6	0.050 ± 0.017	0.083
Hypothalamus	6	0.276 ± 0.055	0.460
Cerebrospinal fluid	6	0.362 ± 0.155	–
Choroid plexus	6	4.335 ± 0.388	–
Pituitary (anterior and posterior)	6	1.109 ± 0.161	–

From Begley, original data.
Values ± SEM.
[a] $K_{in(corr)}$) is calculated with a correction for the inulin space of the tissue, except in the case of CSF.
[b] The permeability coefficient P is calculated as $K_{in(corr)}/S$, assuming an endothelial surface area (S) of 100 cm^2/g brain tissue (Bradbury 1979).

An octapeptide analogue of somatostatin, Sandostatin, has found numerous clinical uses (Lamberts 1987) in the treatment of endocrine disturbances, especially those resulting from inappropriate hormone secretion by tumours. The half-life of Sandostatin in the human after subcutaneous administration is 113 min (Lamberts 1986, 1987), and this long half-life is conferred by the incorporation of D-isomers of amino acids into the molecule, thus reducing the effectiveness of plasma and tissue peptidases.

D-Phe-Cys-Phe-D-Trp-Lys-Thr-Cys-Thr-OH

Sandostatin (SMS 201-995; Octreotide)

Ala-Gly-Cys-Lys-Asn-Phe-Phe-Trp-Lys-Thr-Phe-Thr-Ser-Cys-OH

Somatostatin

The substitution of phenylalanine in position 4 of somatostatin gives Sandostatin a greater potency with respect to the inhibition of growth hormone release, it being some 45 times more active after intravenous administration than somatostatin in this respect, but the substitution reduces its activity in inhibiting insulin release.

Table 2 shows the uptake of Sandostatin by various brain regions and also by the choroid plexus and pituitary determined by the method of Ohno et al. (1978) 60 min after intravenous bolus injection. The table shows a permeability onstant (K_{in}) with a correction for inulin space in the brain tissues. The K_{in} for Sandostatin is very small, the K_{in} for whole brain (that is the remainder of the brain after dissection of the specific regions, minus structures caudal to the hemispheres) being 1/5 times less than that of

Table 3. Cerebrovascular permeability constants (K_{in}), initial volumes of distribution (V_i) and permeability coefficients (P)[a] for ^{125}I-octapeptide analogues of somatostatin for whole mouse brain[b], determined by an intravenous bolus injection method, using multiple time regression analysis[c]

Analogue	K_{in} ($\mu l\,g^{-1}\,min^{-1}$)	V_i (ml/100 g)	P (cm/s × 10^{-7})
RC161	0.232	1.64	0.39
RC160	0.092	1.37	0.15
RC121	0.040	1.36	0.07

Adapted from BANKS et al. (1990b).
Values ± SEM.
[a] The permeability coefficient P is calculated as K_{in}/S, assuming an endothelial surface area (S) of 100 cm^2/g of brain tissue (BRADBURY 1979).
[b] Brain minus pituitary and pineal.
[c] Multipletime regression analysis yields an initial volume of distribution value (V_i) and thus the K_{in} value does not have to be corrected for tissue inulin space.

sucrose (OHNO et al. 1978). The caudate nucleus shows a very small K_{in} for Sandostatin compared to other brain regions, and the K_{in} for the hypothalamus is significantly larger than other regions. Uptake into choroid plexus and pituitary structures is relatively large, perhaps suggesting that they are particularly rich in accessible somatostatin receptors. The sample taken as pituitary includes the adenohypophysis, the neurohypophysis and the median eminence (infundibulum), all of which are high-permeability structures.

BANKS et al. (1990b) have recently used multiple time regression analysis after intravenous bolus injection (BLASBERG et al. 1983; PATLAK et al. 1983) to investigate the permeability of the mouse blood-brain barrier to various octapeptide somatostatin analogues and to obtain a unidirectional permeability constant.

Ac-D-Phe-Cys-Tyr-D-Trp-Lys-Val-Cys-Thr-NH_2
RC 161

D-Phe-Cys-Tyr-D-Trp-Lys-Val-Cys-Trp-NH_2
RC 160

D-Phe-Cys-Tyr-D-Trp-Lys-Val-Cys-Thr-NH_2
RC 121

These analogues have been designed to have extended plasma half-lives and are resistant to plasma peptidases and thus have enhanced and extended somatostatin-like actions.

The results of the study by BANKS et al. are shown in Table 3. The data obtained are very comparable with the data for Sandostatin shown in Table 2, even though the methodological approach is somewhat different. The analogue RC 161 shows the greatest K_{in} value which is comparable to that of

Sandostatin using the technique of OHNO (see Table 2). The K_{in} for RC 121, which apparently enters least readily, is comparable to that of albumin estimated by the same technique (BANKS et al. 1990b). The K_{in} for RC 160 is some 2/5 smaller and RC 121 1/6 that for RC 161. The structure of RC 161 is identical to that of RC 121 except that the N-terminus is replaced with an acetyl group, thus presumably conferring greater lipid solubility to the molecule and possibly leading to greater brain uptake. RC 160 is otherwise identical to RC 121 except that the C-terminal threonine is replaced with tryptophan, which presumably confers the greater uptake of RC 160. Also, BANKS et al. (1990b) in their study could not saturate the uptake of RC 160 in a brain perfusion study (see Sect. B.III), which suggests that the limited brain uptake of these somatostatin analogues is not carrier mediated. Certainly the greater advantage, as far as brain uptake is concerned, is conferred by the addition of the acetyl group to the N-terminus in RC 161. It would be interesting to examine the effect of acetylating the N-terminus of RC 160 in relation to its brain uptake. All three RC analogues have a C-terminal amide group which will increase their lipid solubility relative to Sandostatin and native somatostatin.

It would appear in the case of the octapeptide analogues of somatostatin that changes in lipid solubility may produce some significant relative differences in the unidirectional permeability constant and that in the absence of specific carrier mechanisms lipid solubility is an important factor. In addition, binding to accessible somatostatin receptors will play a role in tissues such as the choroid plexus and pituitary.

The question of brain penetration for these somatostatin analogues is an important one as RC 160 has been shown to have powerful and prolonged analgesic effects after peripheral administration in the mouse (ESCHALIER et al. 1990) and a central or peripheral mode of action for this effect remains to be established. Also, analogues such as RC 161 appear to have a greater brain penetration than RC 160. In addition, the presence of somatostatin receptors on tumours both in nervous tissue and elsewhere raises the possibility of using these receptor agonists to localize the possibly treat these tumours (KRENNING et al. 1989). It is interesting to note that in the study of KRENNING et al. a meningioma could be localized with [^{125}I-Tyr^{3}]Sandostatin (SMS 201-995) but brain uptake was not seen in any patient.

Also important in this context is the observation of BANKS et al. (1990b) who showed by high performance liquid chromatography (HPLC) analysis that, 10 min after intravenous injection of RC 160, the majority of the circulating radioactivity represented intact peptide. The proportion of intact peptide fell to 5% in the blood and 37% in the brain 60 min after injection. They conclude that the somatostatin analogues studied can enter the brain in limited amounts after intravenous injection but that a significant proportion of the radioactive material found in the brain represents degradation products. Thus small differences in peptide structure may lead to differences in the rate and manner of degradation, thus giving rise to differing K_{in}

Table 4. Apparent[a] cerebrovascular permeability constants ($K_{in(corr)}$)[b] and permeability coefficients (P)[c] for [^{3}H]cyclosporin in various brain regions of the guinea pig determined by an intravenous bolus injection technique, at a single time point[d]

Brain region	*n*	$K_{in(corr)}$ ($\mu l\, g^{-1} min^{-1}$)	P (cm/s $\times 10^{-7}$)
Whole brain	9	3.52 ± 0.55	5.86
Hippocampus	9	2.53 ± 0.38	4.22
Caudate nucleus	9	3.03 ± 0.49	5.05
Parietal cortex	9	3.04 ± 0.47	5.05
Hypothalamus	9	4.07 ± 1.07	6.78
Cerebrospinal fluid	3	0.79 ± 0.07	–
Choroid plexus	3	345.25 ± 59.29	–
Pituitary (anterior and posterior)	5	299.84 ± 64.57	–

From BEGLEY et al. (1990).
Values ± SEM.
[a] As discussed in the text, with lipid-soluble tracers the K_{in} is best termed apparent as some tracer backflux may occur during the experimental period (60 min).
[b] $K_{in(corr)}$ is calculated with a correction for the inulin space of the tissue, except in the case of CSF.
[c] The permeability coefficient (P) is calculated as $K_{in(corr)}/S$, assuming an endothelial surface area (S) of 100 cm^2/g brain tissue (BRADBURY 1979).
[d] Single time point experiments (60 min) with an intravenous bolus injection method do not yield an initial volume of distribution (V_i).

values. Also, the manner of labelling the peptide may be significant, with ^{125}I leading to different results from ^{14}C or ^{3}H labelling.

However, it is gratifying to note that the two variants of the intravenous bolus injection technique with different isotopic labels on the molecules have in the case of somatostatin analogues produced essentially similar results. The initial volumes of distribution (V_i) given by these multiple time regression analyses are similar to those found in the study of BLASBERG et al. (1983) and in brain perfusion methods (see Sect. B.III).

BANKS et al. (1987) have employed the intravenous bolus injection technique with multiple time regression analysis to determine a cerebrovascular permeability constant for arginine vasopressin. They obtain a K_{in} value of 2.47 $\mu l\, g^{-1} min^{-1}$, which is significantly larger, some 2.6 times so, than the permeability constant produced by brain perfusion (see Table 5).

BEGLEY (1990) and BEGLEY et al. (1990) have investigated the brain uptake of the lipid-soluble immunosuppressant drug cyclosporin using the intravenous bolus injection technique of OHNO et al. (1978). Cyclosporin is a highly lipid soluble peptide with a $\log P$ of +2.99 and thus might be expected to cross the blood-brain barrier readily. Guinea pigs were given an intravenous bolus injection of [Abu-^{3}H]cyclosporin A and the K_{in} estimated for several brain regions. The results are shown in Table 4. Uptake into various brain regions was small, compared to some solutes, but significantly greater, for instance, than that of the somatostatin analogues, and some

8 times greater than the K_{in} for sucrose determined by the same method (OHNO et al. 1978). Uptake into the choroid plexus and the pituitary structures was significant, suggesting that areas outside of the blood-brain barrier readily accumulate this drug. It is interesting to note that although it was suggested earlier that backflux might occur with a lipid-soluble tracer such as cyclosporin, apparent permeability constants obtained with this tracer by the bolus injection method for brain regions with a barrier are larger than those obtained by brain perfusion techniques (see Sect. B.III and Tables 4, 5).

It is suggested that as cyclosporin is transported in plasma largely in association with red cells and lipoprotein (LEMAIRE and TILLEMENT 1982), its penetration of the brain is restricted by the presence of the tight endothelial junctions at the cerebral capillary endothelium and the ependymal surface of the choroid plexus which limit the movement of the protein/lipid carriers (BEGLEY 1990; BEGLEY et al. 1990). It is suggested that cyclosporin may enter the cerebral endothelial cells in association with low density lipoprotein (LDL) particles and is not re-exported at the abluminal surface of the cerebral capillary (BEGLEY et al. 1990). This suggestion is supported by the observation that cyclosporin is readily accumulated in vitro by cultured cerebral endothelial cells and by the choroid plexus in vivo (BEGLEY 1990; BEGLEY et al. 1990).

RAPOPORT et al. (1980) estimated the brain-uptake of several analogues of β-endorphin (Met-[D-Ala^2]enkephalinamide, β-[D-Ala^{62},^{14}C-Homoarg^{69}]-lipotropin61-69, α-[D-Ala^2, ^{14}C-Homoarg^9]endorphin, β-[D-Ala^2, ^{14}C-Homoarg]-endorphin) between 3 and 30 min after intravenous bolus injection. Calculated permeability coefficients (P) ranged from 14.0×10^{-7} to 39.0×10^{-7} cm/s, corresponding to K_{in} values of 8.4–23.4 μl g^{-1} min^{-1}, suggesting high permeabilities for these analogues which have not been paralleled in studies with other peptides. Although the substitution of D-isomers of amino acids does confer some resistance to hydrolysis in peptide molecules it remains possible that some significant hydrolysis might have occurred during these studies.

The intravenous bolus injection technique with multiple time uptake analysis has also been used to study an octapeptide analogue of peptide T ([^{125}I-Tyr]-D-[Ala^1]-peptide T-amide) and its transport into the brain of the mouse (BARRERA et al. 1987). Peptide T is an octapeptide derived from the human immunodeficiency virus (HIV) envelope glycoprotein 120. The K_{in} value ranged from 1.13 to 2.50 μl g^{-1} min^{-1} and the V_i from 1.99 to 2.27 ml/100 g measured over a 20-min exposure. The unidirectional permeability constant was reduced by 23% when the bolus contained 1.2 μM D-[Ala^1]-peptide T-amide labelled with non-radioactive iodine in the tyrosine residue. The addition of 100 μM iodo-L-tyrosine to the bolus was without effect. Pretreating the mice with 200 μl aluminium chloride intraperitoneally (2 mg elemental aluminium per mouse) reduced the K_{in} by 45%. A brain extract subjected to HPLC analysis showed that 63% of the recovered radioactivity eluted at the ^{125}I-peptide T analogue retention time, with 23% eluting at the position of ^{125}I-Tyr, indicating that much of the labelled analogue remained intact. It would appear that D-[Ala^1]peptide T-amide is extracted at the blood-brain barrier by a carrier-mediated mechanism.

BANKS et al. (1989) have also re-examined the brain uptake of ^{125}I-interleukin-1 alpha ($\log P - 1.63$) using an intravenous bolus injection technique with multiple time regression analysis. Unidirectional permeability constants were obtained ranging from: striatum, K_{in} 0.072 µl g^{-1} min^{-1}, V_i 0.24 ml/100 g; parietal cortex, K_{in} 0.072 µl g^{-1} min^{-1}, V_i 0.12 ml 100 g; hippocampus, K_{in} 0.084 µl g^{-1} min^{-1}; V_i 0.28 ml/100 g; and hypothalamus, K_{in} 0.13 µl g^{-1} min^{-1}; V_i 0.11 ml/100 g over a 100-min exposure period. These values are extremely small but significantly higher than those of albumin measured under the same conditions and similar to those for Sandostatin determined by the method of OHNO et al. (see Table 2). BANKS et al. (1989) feel that reversible binding to the luminal endothelial membranes is unlikely as this might be reflected in a raised value for V_i rather than the K_{in}. The results suggest that the hypothalamus has a higher permeability to interleukin-1 alpha than other brain structures.

III. Brain Perfusion Studies

Brain vascular perfusion studies offer a number of advantages over other methods for studying the interactions of peptides with the blood-brain interface. Firstly, the composition of the perfusion medium can be controlled precisely and held constant by excluding the animals own circulatory contribution to brain perfusion. Generally, the perfusion medium is dextran saline containing washed erythrocytes to improve the oxygen supply to the brain or dextran saline alone. This makes the method especially useful for peptides which have a short half-life in the intact animal as a result of the action of plasma and tissue enzymes not associated with the blood-brain barrier itself. Also, the effects of protein binding or other possible plasma carriers may be investigated by excluding them from or including them in the perfusate. In addition, peptides can be presented to the luminal surface of the of the blood-brain interface in physiological concentrations together with competitive and non-competitive inhibitors of transport. Perfusion can also be maintained for up to 30 min (ZLOKOVIĆ et al. 1986) or even longer (RICHERSON and GETTING 1990), enabling kinetic parameters for these slowly penetrating solutes to be established. Also of great practical importance, the absolute amounts of valuable labelled peptide required for a particular study is greatly reduced.

The majority of brain perfusion studies with peptides have been conducted using the method developed by BEGLEY and ZLOKOVIĆ in the guinea pig; the method is fully described in ZLOKOVIĆ et al. (1986) and BEGLEY et al. (1990). The arterial vascular supply to the anterior communicating artery of the circle of Willis in the guinea pig is unusual (BUGGE 1974) and makes a

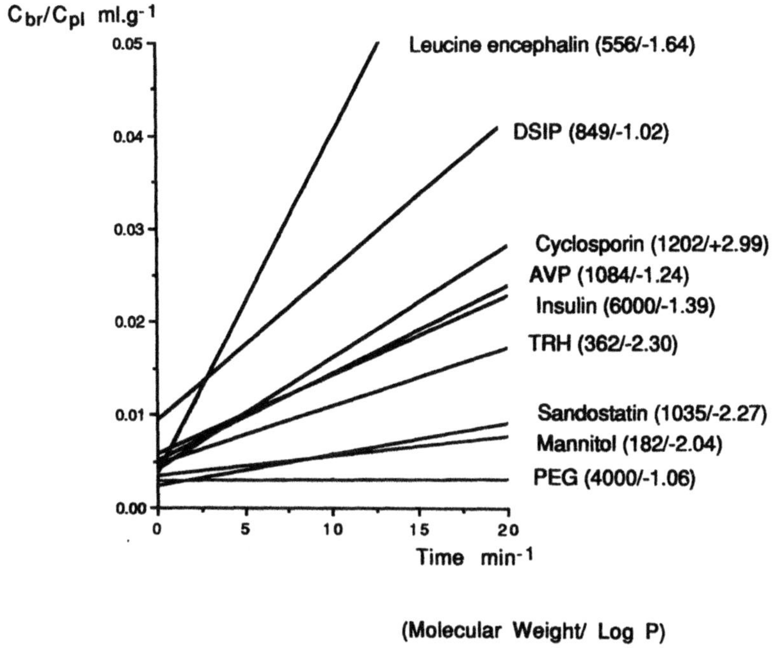

Fig. 1. Simple line plots illustrating the uptake of a number of tracers during brain perfusion in the guinea pig. The *slope* of the line gives the unidirectional cerebrovascular permeability constant, K_{in} ($\mu l\, g^{-1} min^{-1}$); and the *ordinate intercept* the initial volume of distribution, V_i (ml/100 g). There is no departure from linearity of the plots during the 20-min duration of the perfusions, illustrating a lack of backflux of tracer under the experimental conditions. Shown in *parentheses* are the molecular weights of the individual tracers and their log P value (lipid solubility), indicating that for these particular tracers there is no correlation of these values with the unidirectional cerebrovascular permeability constant. The same data are presented in numerical form in Table 5

stable vascular perfusion of the forebrain with a minimal contribution from the vertebral circulation easier to achieve.

The fact that the concentration of labelled peptide in the perfusion fluid is held constant during an experiment allows the unidirectional cerebrovascular permeability constant (K_{in}) and an initial volume of distribution (V_i) to be determined by a plot of C_{br}/C_{pl} against time (ZLOKOVIĆ et al. 1986; BEGLEY et al. 1990), where the slope of the line represents K_{in} and the ordinate intercept V_i and C_{br} and C_{pl} are the radioactive concentration of tracer in brain and plasma respectively.

Figure 1 and Table 5 illustrate the unidirectional permeability constant (K_{in}) and the initial volume of distribution (V_i) for a number of peptides in the hippocampus of the guinea pig brain, compared with the inert polar molecules mannitol and polyethylene glycol 4000, together with a calculated

Table 5. Unidirectional cerebrovascular permeability constants (K_{in}), initial volumes of distribution (V_i) and permeability coefficients (P)[a] for various tracers in guinea pig hippocampus determined by a brain vascular perfusion method

Tracer	n	K_{in} (μl g^{-1} min^{-1})	V_i (ml/100 g)	P (cm/s × 10^{-7})	Reference
[^{3}H]Leucine encephalin	11	3.63 ± 0.25	0.39 ± 0.06	6.05	(a)
^{125}I-DSIP	7	1.61 ± 0.26	0.95 ± 0.24	2.68	(b)
[^{3}H]Cyclosporin A	19	1.20 ± 0.10	0.42 ± 0.10	2.00	(c)
[^{3}H]AVP	8	0.94 ± 0.14	0.52 ± 0.17	1.57	(d)
^{125}I-Insulin	7	0.85 ± 0.22	0.59 ± 0.31	1.42	(e)
[^{3}H]TRH	8	0.63 ± 0.07	0.47 ± 0.05	1.05	(f)
[^{14}C]Sandostatin	7	0.34 ± 0.13	0.24 ± 0.18	0.57	(g)
[^{14}C]Mannitol	8	0.22 ± 0.02	0.34 ± 0.07	0.37	(h)
[^{3}H]PEG	8	0.004 ± 0.0005	0.29 ± 0.03	0.007	(i)

Values ± SEM.
[a] The permeability coefficient (P) is calculated as K_{in}/S, assuming an endothelial surface area (S) of 100 cm^2/g brain tissue (Bradbury 1979).
References:
(a) ZLOKOVIĆ et al. (1987, 1988d), (b) ZLOKOVIĆ et al. (1989b), (c) BEGLEY et al. (1990), (d) ZLOKOVIĆ et al. (1988d) (e) D.J. BEGLEY and A. VLADIĆ (original data) (f) ZLOKVIĆ et al. (1988a) (g) D.J. BEGLEY and A. VLADIĆ (original data), (h) ZLOKOVIĆ et al. (1986) (i) RAKIĆ et al. (1989)

permeability coefficient (P), assuming a brain vascular surface area of 100 cm^2/g brain (BRADBURY 1979).

The peptides investigated exhibit a variety of unidirectional permeability constants. The uptake of leucine encephalin, DSIP, AVP and insulin can be shown to be saturable (ZLOKOVIĆ et al. 1987, 1988d, 1989a,b; D.J. BEGLEY and A. VLADIĆ, original data). Saturability cannot be demonstrated for TRH (ZLOKOVIĆ et al. 1988a) and has not been investigated for cyclosporin and Sandostatin (BEGLEY et al. 1990; ZLOKOVIĆ et al. 1988b; D.J. BEGLEY and A. VLADIĆ, original data). The initial volumes of distribution are less than 1% (1 ml/100 g of brain).

The slope plots illustrated in Fig. 1 show in parentheses the molecular weight of the peptide and the log P value, demonstrating that, for this series of tracers, there is no relationship between molecular weight, lipid solubility and the unidirectional permeability constant.

The unidirectional permeability constant for leucine encephalin is remarkably consistent between brain regions, being between 3.62 and 3.63 μl g^{-1} min^{-1} in the hippocampus, caudate nucleus and parietal cortex (ZLOKOVIĆ et al. 1987, 1988d). The addition of 2 mM unlabelled leucine encephalin (0.13 mM in the final perfusate) resulted in a 74%–78% inhibition of leucine encephalin uptake at the luminal surface of the blood-brain interface. The N-terminal tetrapeptide of leucine encephalin (Tyr-Gly-Gly-Phe) produced only a 15%–29% inhibition of uptake in the brain regions studied. 5 mM L-tyrosine, 2 mM bacitracin (an amino peptidase inhibitor) or

5 mM Gly-Gly-Phe-Leu (the C-terminal tetrapeptide of leucine encephalin) produced no significant reduction of the K_{in} value (ZLOKOVIĆ et al. 1987). Inclusion of a second amino peptidase inhibitor, bestatin (0.5 mM), was also without effect (ZLOKOVIĆ et al. 1989a). Only self-inhibition and inhibition with des-leucine encephalin is effective, the intact molecule being three time more effective than a higher concentration of des-leucine encephalin. Also, Tyr-Gly-Gly and Tyr-Gly were without effect on the unidirectional uptake of leucine encephalin (ZLOKOVIĆ et al. 1989a).

These observations, together with the total lack of inhibition with L-tyrosine and inhibitors of aminopeptidase activity, would suggest that the leucine encephalin is transported an the luminal surface as the intact molecule and that there is no hydrolysis of the peptide at the blood-brain interface.

The carrier-mediated uptake mechanism for this peptide is therefore remarkably specific to leucine encephalin, and this specificity can be shown to be due to the presence of a transporter as opposed to a receptor. In the presence of the δ-opioid receptor antagonist ICI 174846 (allyl^2-Tyr-AIB-Phe-OH), and the υ-receptor agonist DAGO (Tyr-D-Ala-Gly-N-methyl-Phe-Gly-OH), the unidirectional uptake constant was unchanged (ZLOKOVIĆ et al. 1989a). Also, naloxone 0.3 mM was without effect (ZLOKOVIĆ et al. 1987). Thus the transporter for this uptake is clearly unrelated to the opiate receptor responsible for the CNS effects of morphine and related peptides.

ZLOKOVIC et al. (1989a) have published a kinetic analysis of leucine encephalin extraction at the blood-brain interface. This study demonstrates that the unidirectional uptake varies as a function of unlabelled peptide concentration in the perfusing medium. In the three brain regions analysed uptake is best modelled by a single saturable mechanism with Michaelis-Menten kinetics. Best fit analysis gives a maximum velocity (V_{max}) between 0.14 and 0.16 nmol min^{-1} g^{-1} and a half-saturation (K_m) of between 34 and 41 μM, well above the tracer levels of peptides in perfusate used to determine K_{in} values. These values compare closely to K_m values derived for neutral amino acids, e.g. L-leucine, obtained with a similar brain perfusion technique developed by SMITH and TAKASATO in the rat (Q.R. SMITH et al. 1984, 1987; TAKASATO et al. 1984, 1985), but the V_{max} value indicates a capacity of the system only one-third of that for neutral amino acids. The diffusional component (K_d) for leucine encephalin indicated by the ZLOKOVIĆ et al. (1989a) study is very small and only accounts of 1.8%–2.9% of the unidirectional permeability constant much less than the 24% non-saturable component remaining in the study of ZLOKOVIĆ et al. (1987) with the same highest saturating leucine encephalin concentration of 0.13 mM.

Delta-sleep-inducing peptide (DSIP) can also be shown to have a saturable uptake at the blood-brain interface (ZLOKOVIĆ et al. 1989b). Regional unidirectional permeability constants are 0.93, 1.33 and 1.61 μl g^{-1} min^{-1} in the parietal cortex, the caudate nucleus and the hippocampus, and in the presence of 7 μM DSIP the uptake of ^{125}I-DSIP (0.3 nM) was inhibited by 59%, 73% and 86%, respectively. The K_{in} value and the slope of the plot for the hippocampus in the presence of 7 μM DSIP were similar to that of mannitol. The amino acid L-tryptophan, which is the N-terminal residue of DSIP, at a concentration of 7 μM also inhibited K_{in} in the brain regions by 71%, 83% and 86%, respectively. It is also interesting to note that 1 μM L-tryptophan does not depress the K_{in} value further although this must be close to the usual plasma levels of this amino acid. Analysis by thin layer chromatography (ZLOKOVIĆ et al. 1989b) failed to show free labelled tryptophan or ^{125}I-iodide in the inflowing perfusate, the venous effluent or a forebrain extract at the end of the experiment, indicating that hydrolysis of DSIP is not taking place at the blood-brain interface. The authors conclude that DSIP is extracted at the blood-brain interface by a carrier-mediated transport mechanism which may be saturated at remarkably low levels of DSIP (7 μM). DSIP may, in addition, show a reactivity with the L-tryptophan transporter (L-system, CHRISTENSEN 1979), an affinity which is competitively inhibited at normal levels of plasma tryptophan. It is also interesting to note that the initial volume of distribution (V_i) is reduced to that of mannitol by 7 μM DSIP but is not affected by 7 μM L-tryptophan.

Perfusion studies in the guinea pig with cyclosporin (BEGLEY et al. 1990) show a measurable time-dependent uptake with a K_{in} value significantly greater than that of mannitol and polyethylene glycol (Table 5, Fig. 1). The K_{in} for the parietal cortex of 0.43 μl g^{-1} min^{-1} was smaller than that in both the hippocampus and caudate nucleus, both of which were 1.20 μl g^{-1} min^{-1}. The K_{in} values produced by the brain perfusion technique are of the same order of magnitude but some 0.14–0.45 times smaller than for the corresponding region estimated by an intravenous bolus injection technique (see Table 4).

The K_{in} value and kinetics for AVP have also been determined by the brain perfusion method (ZLOKOVIĆ et al. 1988d; HYMAN et al. 1990). Uptake into the parietal cortex, caudate nucleus and hippocampus were saturable, showing Michaelis-Menten type kinetics, with K_m values of 2.1 ± 0.3, 2.6 ± 0.5 and 2.7 ± 0.3 μM and V_{max} values of 5.5 ± 0.7, 5.6 ± 0.9 and 4.9 ± 0.5 fmol min^{-1} g^{-1}, respectively (HYMAN et al. 1990). The diffusional component K_d was not significantly different from zero in any of the brain regions, as for leucine encephalin studied in the same system (ZLOKOVIĆ et al. 1989a). The vasopressin uptake could only be reduced by competition with intact unlabelled AVP. Analogues and fragments of the peptide such as des-Gly-NH_2-Arg^8-VP, pressinoic acid and pGlu^4-Cyt^6-Arg^8-VP(Arg) were without effect, as were the amino acids L-tyrosine and L-phenylalanine and the aminopeptidase inhibitor bestatin at 0.5 mM concentration (HYMAN et al. 1990). These studies would suggest that the extraction of AVP at the blood-brain interface is dependent on a specific carrier-mediated transport mechanism which is remarkably specific to the peptide under study and that there is no hydrolysis of the peptide at the luminal surface of the blood-brain barrier. In this respect, the extractions of leucine encephalin, DSIP and AVP are similar. A possible cross-reactivity for transporters has been shown in the case of DSIP, where 0.2 m*M* AVP reduces the K_{in} for DSIP by between 26% and 54% depending on the brain region, whilst 8 μ*M* AVP is without significant effect (ZLOKOVIĆ et al. 1989b). No interaction between the other peptides in this respect has been investigated. However, the possibility remains that AVP in high concentration may influence regional blood flow and thus affect extraction by the cerebral vasculature (see Sect. D).

The technique of in situ brain perfusion in the guinea pig has recently been applied to insulin (Fig. 1, Table 5; D.J. BEGLEY and A. VLADIC, original data). In the hippocampus the K_{in} value was $0.85 \pm 0.22\,\mu l\,g^{-1}\,min^{-1}$ and the initial volume of distribution (V_i) 0.59 ± 0.31 ml/100 g. In a separate series of experiments it was shown that the unidirectional cerebrovascular permeability constant for ^{125}I-porcine insulin could be reduced by 72% with an excess (1000 μ*M*) of unlabelled porcine insulin from 0.91 to $0.25\,\mu l\,g^{-1}\,min^{-1}$. The K_{in} value for insulin was not affected by the addition of 4.0×10^{-4} – sodium perchlorate to the perfusate, suggesting that the uptake of intact insulin was being measured. Perfusion with low temperature perfusate (5°C) resulted in a marked reduction in the accumulation of insulin by the brain during an experiment, suggesting that vesicular internalisation of the insulin by the cerebral capillary endothelium is an important step in uptake (D.J. BEGLEY and A. VLADIĆ, original data).

The guinea pig brain perfusion technique has also been employed to investigate the interaction of TRH with the blood-brain interface (Fig. 1, Table 5; ZLOKOVIĆ et al. 1988a). The unidirectional cerebrovascular permeability constant for TRH ([L-proline-2,3,4,5-^{3}H(N)]-TRH) can be reduced by some 43%–48% by adding 2 m*M* bacitracin to the perfusate. K_{in} values for the parietal cortex, caudate nucleus and hippocampus in the presence of bacitracin were similar, being 0.63 ± 0.07, 0.62 ± 0.06 and $0.65 \pm 0.08\,\mu l\,g^{-1}\,min^{-1}$, respectively (ZLOKOVIĆ et al. 1988a). The further addition of 1 m*M* unlabelled TRH to the perfusate did not reduce the K_{in} for TRH to a greater extent than bacitracin alone. 10 m*M* prolinamide also had no effect on the kinetics of TRH entry. ZLOKOVIĆ et al. (1988a) interpret these results as demonstrating the absence of a specific saturable transport mechanism for TRH at the luminal surface of the blood-brain interface, a situation entirely analogous to that found in the perfused choroid plexus (ZLOKOVIĆ et al. 1985b,c; see Sect. B.IV). The K_{in} for TRH, however, is nearly three times greater than that for mannitol in spite of its greater molecular weight. This might possibly be related to their relative solubility in the lipid of natural membranes. TRH is distinct from the other peptide tracers investigated by the brain perfusion technique in exhibiting enzymic hydrolysis at the luminal surface of the blood-brain interface itself in the absence of plasma enzymes. Presumably this hydrolysis is the result of the activity of extrinsic membrane peptidases.

BARRERA et al. (1989) have used a brain perfusion technique to examine the brain uptake of ^{125}I-Tyr-MIF-1 at the blood-brain interface. This study used an erythrocyte-free perfusate, as these cells have been shown to sequester Tyr-MIF-1 (BANKS et al. 1988). In the BARRERA et al. (1989) study, rat brains were perfused for up to 6 min prior to decapitation and uptake into the right cerebral cortex assessed. The K_{in} obtained for Tyr-MIF-1 in the cerebral cortex was 4.44 $\mu l\,g^{-1}\,min^{-1}$ and the V_i 0.17 ml/100 g cortex. This value indicates quite substantial entry of the peptide into brain exceeding that of leucine encephalin (see Table 5). This uptake could not be inhibited by including non-radioactive ^{127}I-Tyr-MIF (25 nmol/ml) in the perfusate. Inclusion of 400 nmol/ml of leucine or methionine encephalin in the perfusate was also without effect. Tyr-MIF and methionine encephalin are transported by and compete for, a single transport mechanism, protein transport system-1 (PTS-1), out of the central nervous system (BANKS and KASTIN 1990). Pre-treatment of the animals with aluminium reduced the unidirectional permeability constant for Tyr-MIF-1 significantly. In a later study, BANKS et al. (1990a) examined the brain uptake of the stereoisomers ^{125}I-L-Tyr-MIF-1 and ^{131}I-D-Tyr-MIF-1 in the rat using the same methodology. They produced, for whole brain with pineal and pituitary removed, rather smaller unidirectional permeability constants in this study of K_{in} 2.62 $\mu l\,g^{-1}\,min^{-1}$, V_i 0.3 ml/100 g for L-Tyr-MIF-1 and K_{in} 2.42 $\mu l\,g^{-1}\,min^{-1}$, V_i 2.42 ml/100 g for D-Tyr-MIF-1. There was no significant difference in the permeability constants for the two isomers. The differences in permeability constant compared to the earlier study perhaps suggests that the cerebral cortex is more permeable than other brain regions to Tyr-MIF-1. These experiments with Tyr-MIF-1 indicate that brain uptake of this small peptide is significant, but it fails to show stereospecificity and saturation kinetics and is thus diffusion limited. This is a similar situation to that of TRH but Tyr-MIF-1 shows a greater penetrance, possibly related to its greater lipid solubility ($\log P$ + 0.06; BANKS and KASTIN 1985).

The octapeptide analogue of somatostatin, Sandostatin, has a K_{in} value for hippocampus determined by brain perfusion of $0.34 \pm 0.13\,\mu l\,g^{-1}\,min^{-1}$ with a V_i of 0.24 ± 0.18 ml/100 g brain (Table 5; D.J. BEGLEY and A. VLADIĆ, original data). This unidirectional permeability constant is not significantly different from that of mannitol determined with the same technique (ZLOKOVIĆ et al. 1986). This result is in full accordance with that produced by the intravenous bolus injection technique (see Sect. B.II), where the K_{in}, with or without correction for the insulin space, is about 0.24–0.56 times smaller than that produced by the perfusion technique (see Sect. B.II), and is smaller than that for sucrose determined by OHNO et al. (1978). Even though the intravenous bolus injection technique and the brain perfusion technique yield cerebrovascular permeability constants in the same units, and often of the same order of magnitude, direct comparison may not be justified.

BANKS et al. (1990b) used the brain perfusion technique developed in the rat by SMITH and TAKASATO (Q.R. SMITH et al. 1987; TAKASATO et al. 1984, 1985) to investigate the uptake of the octapeptide analogue of somatostatin RC 160. In the BANKS et al. (1990b) study, the authors obtained a K_{in} value of $20.3 \pm 4.6\,\mu l\,g^{-1}\,min^{-1}$ after 60 s perfusion, some 220 times larger than by the intravenous bolus technique used in the same investigation (see Table 3). Inclusion of 0.4 mM unlabelled RC 160 did not reduce this value, suggesting that penetration is not carrier mediated. When Tc-albumin was simultaneously infused with ^{125}I-RC 160, approximately 48% of the brain radioactivity would have derived from the vascular compartment, suggesting that for this short time course the measured K_{in} is about twice that of the true K_{in} (BANKS et al. 1990b). The authors estimate that 91.8% of the radioactivity entering the brain represented intact ^{125}I-RC 160. The larger unidirectional permeability constant estimated by BANKS et al. (1990b) with the brain perfusion method of TAKASATO et al. (1984) requires some comment as values are usually of the same order of magnitude as provided by the intravenous bolus method (TAKASATO et al. 1984; see also the previous discussion of Sandostatin). As BANKS et al. (1990b) suggest, a plasma factor which is absent from the perfusate may inhibit brain uptake in the intact animal. However, this would have to apply specifically to RC 160 as Sandostatin yields similar results with both an intravenous bolus and a brain perfusion technique, although the K_{in} determined by the intravenous bolus method is smaller (see above). Alternatively, it may be that the short time course of the SMITH/TAKASATO perfusion method does not offer a good discrimination for these slowly penetrating solutes. Reference to Fig. 1 shows that at 60 s of perfusion the C_{br}/C_{pl} ratio is little different from that at zero time and almost equivalent to the initial volume of distribution (V_i).

IV. Choroid Plexus Perfusion Studies

The more permeable blood-CSF barrier is partly produced by the ependymal epithelium of the choroid plexus; it as suggested by its low electrical resistance, $20\,\Omega/cm^2$ (ZEUTHEN and WRIGHT 1981; see Sect. B.VI), might prove a significant route of entry for some peptides into the CNS.

All studies of peptide transport at the choroid plexus-blood interface have been performed using the perfused choroid plexus of the sheep. This method was developed by POLLAY et al. (1972) and further extended by DEANE and SEGAL (1985) to study sugar transport at the choroid plexus. The method used to study peptide uptake is fully described in DEANE and SEGAL (1985). The choroid plexus is perfused through the carotid arteries of the isolated brain with all vessels from the circle of Willis excepting the anterior choroidal artery tied off or electrocoagulated. The effluent blood is collected at the great vein of Galen. The paired tracer indicator dilution method has been reviewed by YUDILEVICH and MANN (1982) and has been applied to peptides in the perfused choroid plexus by ZLOKOVIĆ et al. (1985b,c, 1988c) with [D-^{14}C]-mannitol as an extracellular reference. Alternatively, a steady-state analysis method described by DEANE and SEGAL (1985) involving infusion of a constant concentration of labelled peptide into the anterior choroidal artery is used to investigate DSIP uptake (ZLOKOVIĆ et al. 1988b).

A study with leucine encephalin and its analogue D-Ala^2-D-Leu^5-encephalin (DADLE) (ZLOKOVIĆ et al. 1988c) has demonstrated a cellular uptake at the choroid plexus blood/tissue interface for the two tracers. A single bolus of a mixture of labelled encephalin and [^{14}C]-D-mannitol are injected into the inflowing perfusate and samples of the venous outflow collected over 1 min, producing dilution profiles. The recovery of the [^{3}H-Tyr]leucine encephalin is reduced compared to that of [^{14}C]-mannitol, indicating a cellular retention of the encephalin. A percentage cellular uptake may then be calculated from the ratio of tracer molecule to reference molecule for the individual 1-min samples.

A maximal uptake (U_{max}) is calculated from the mean of the data points which form a plateau region of uptake. Alternatively, the total recovery of ^{3}H or ^{14}C radioactivity in the venous outflow can be summed and compared with the total radioactivity injected as a bolus in the perfusate to give the arterio-venous (A-V) loss for the tracer or reference. The A-V loss for mannitol in these experiments is 27%–32% (ZLOKOVIĆ et al. 1985c, 1988c). This loss of mannitol could be due to non-specific passage across the ependyma of the choroid plexus into the superfusate bathing the choroid plexus, to irreversible uptake into a compartment within the choroid plexus, or to leaks in the vascular perfusion circuit, as minor branches of the anterior choroidal artery which do not supply the choroid plexus cannot be occluded, leading to loss of part of the bolus injection (see later discussion).

Table 6 shows the maximal cellular uptake (U_{max}) relative to mannitol for leucine encephalin and its stable analogue DADLE using the paired tracer indicator dilution method. The study of ZLOKOVIĆ et al. (1988c) indicates a significant cellular uptake of both leucine encephalin and DADLE which can be significantly self-inhibited and also cross-inhibited, indicating that the two tracers are competing for the same carrier. BCH (2-amino-bicyclo-2,2,1-heptane-2-carboxylic acid) and tyrosine, which both compete with the tracer for the L-amino acid transport system, reduce U_{max} by 20%–36% with a greater degree of inhibition for leucine encephalin than for the stable analogue. The aminopeptidase inhibitor bacitracin also reduces the U_{max} value for leucine encephalin by 40% and for DADLE by 28%, suggesting that DADLE is less subject to hydrolysis in the choroid plexus. The results are interpreted by the authors as indicating that unidirectional uptake of encephalins at the blood-tissue interface consists of two components: a saturable component which represents uptake of intact peptide and a non-saturable component produced by enzymatic hydrolysis during the single pass liberating free [^{3}H]tyrosine. The authors suggest that a

Table 6. Blood-tissue extraction of encephalins during a single capillary passage through the choroid plexus of the sheep

Tracer	n	Maximal cellular uptake (U_{max} %)[a]	A-V loss (%)[b]	p[c]
[^{3}H]Leucine encephalin	6	54 ± 4	51 ± 5	<0.01
[^{3}H]-D-Ala^2-D-leu^5-encephalin	7	50 ± 5	45 ± 5	<0.01
[^{14}C]Mannitol	13		32 ± 5	

From ZLOKOVIĆ et al. (1988c).
Values ± SEM.
[a] Uptake is calculated as Uptake % = (1-[^{3}H]tracer/[^{14}C]mannitol) × 100 for each effluent sample and U_{max} is determined as the plateau region of uptake.
[b] The A-V loss is the difference in the outflowing and injected radioactivity for tracers and reference.
[c] p is the significance of the difference between the A-V loss of tracer compared to mannitol, unpaired Student's t.

greater proportion of intact DADLE than leucine encephalin extracted from the perfusate may potentially cross the blood-CSF barrier due to its greater stability. The A-V loss of leucine encephalin is 51 ± 5%, of DADLE 45 ± 5% and of mannitol 32 ± 55% in a single pass (ZLOKOVIĆ et al. 1988c).

Similar experiments have been performed with thyrotropin-releasing hormone ([^{3}H-prolinamide]TRH; ZLOKOVIĆ et al. 1985b,c). The U_{max} for TRH in the presence of bacitracin was 5.9 ± 1.6%, which probably represents the true cellular uptake of the peptide. The U_{max} for TRH alone was 38.1 ± 3.1%, indicating a significant hydrolysis of the peptide in a single pass which was prevented by the addition of bacitracin. The unidirectional uptake of the peptide could not be saturated with unlabelled TRH at 10 mM concentration, indicating an extraction produced by simple diffusion, and was markedly reduced 10 mM proline and 2 mM bacitracin in the perfusate, confirming hydrolysis of the TRH during the single pass. The A-V loss of TRH in the presence of bacitracin was 29.8 ± 3.1% compared to 26.9 ± 3.0% for mannitol (ZLOKOVIĆ et al. 1985c). This study provides evidence that TRH does not interact with a specific receptor or carrier at the blood-CSF barrier. The study in fact suggests a rapid hydrolysis in the tissue or by plasma enzymes in the perfusate. The true U_{max} is shown in the presence of bacitracin and is considerably lower than that for leucine encephalin or DADLE (Table 6).

Extraction of DSIP by the choroid plexus has also been studied in the perfused sheep choroid plexus preparation, but only using a steady-state analysis (ZLOKOVIĆ et al. 1988b). The extraction of DSIP during a steady-state perfusion is between 3.25 ± 0.32% and 5.62 ± 0.11% and increases with the perfused concentration of DSIP. The extraction shows Michaelis-Menten kinetics, with a K_m of 4.98 ± 0.44 nM and a V_{max} of 272 ± 10 fmol/min, indicating a high affinity, low capacity mechanism. It is important to note that the uptake is calculated from an A-V difference and does not necessarily represent transport across the choroid plexus into CSF. The study of Zloković et al. (1988b) also provides no evidence for degradation of DSIP in a single pass through the choroid plexus. This observation is in agreement with the brain perfusion studies of Zloković et al. (1989b) and studies with cultured endothelial cells (Raeissi and Audus 1989) (see Sects. B.III, B.V).

The perfused choroid plexus of the sheep is a valuable preparation which allows the transport properties of the plasma tissue interfaces at the cerebral capillary endothelium and the choroid plexus to be compared. One drawback of the technique is that transport across the ependyma of the plexus cannot be estimated accurately as transport of peptide into the choroid plexus superfusate cannot be quantitatively measured. Extraction is assessed in terms of the reference and tracer single pass dilution and/or A-V differences. As the A-V loss of mannitol in a single pass is 27%–32% (Zloković et al. 1985c, 1988c), this clearly cannot represent the flux of this solute across the blood-CSF barrier. If this were the case, mannitol would rapidly enter the CSF and reach equilibrium, which clearly does not happen. The greater part of the A-V loss of mannitol in these single pass studies must be largely due to a proportion of the perfusate containing the bolus not appearing in the venous drainage, for the reasons discussed earlier. The U_{max} value calculated from paired tracer indicator dilution techniques for peptides thus generally represents uptake into choroid plexus capillary endothelial cells, the stroma of the plexus and possibly the ependymal cells themselves, but provides no direct evidence for transport into CSF.

V. Isolated Cerebral Microvessels

Isolated brain microvessels may be prepared by a centrifugation method (Goldstein et al. 1975) which has been modified by Pardridge et al. (1985b). The technique has been applied to brain microvessels of a number of species and produces suspensions of microvessels of various lengths which are substantially free of glia and other brain cells and tissue fragments (Frank and Pardridge 1981; Pardridge et al. 1985b; Frank et al. 1986a,b). The microvessels retain a positive staining for factor VIII and gamma-glutamyl transpeptidase, which are specific markers for the brain capillary endothelium (Pardridge et al. 1985a). A variety of labelled peptides can then be incubated with the capillary fragments and binding, internalization and exocytosis studied. The results of various studies with isolated microvessels yielding affinities (K_D) and maximal binding (B_{max}) are summarized in Table 7.

Pardridge and Mietus (1981) used isolated cerebral microvessels to study the uptake of [^{3}H-Tyr]leucine encephalin and observed a time-dependent uptake of radioactivity. This uptake was totally abolished by the inclusion of 5 mM L-tyrosine in the medium. The authors interpretation of

Table 7. Binding affinities (K_D) and capacities (B_{max}) for peptides assessed in isolated cerebral microvessel preparations

Receptor	K_{D1} (nM)	B_{max1} (pmol/g protein)	K_{D2} (nM)	B_{max2} (pmol/g protein)	Source	Reference
Insulin	2.3 ± 0.3	0.18 ± 0.05	0.05 ± 0.02	1.4 ± 0.2	Bovine	(a)
Insulin	1.2 ± 0.5	1.7 ± 0.08	48.0	0.74	Human postmortem	(b)
IGFI	2.0	1.7			Bovine	(c)
IGFII	1.8	1.0			Bovine	(c)
Insulin	2.90 ± 0.74	0.32 ± 0.06	0.036 ± 0.028	1.59 ± 0.06	Newborn rabbit	(d)
Insulin	1.87 ± 0.27	0.17 ± 0.10	0.021 ± 0.017	1.07 ± 0.45	3-Week rabbit	(d)
Insulin	2.30 ± 0.91	0.14 ± 0.03	0.028 ± 0.011	0.43 ± 0.13	11-Week rabbit	(d)
Insulin	1.25 ± 0.39	0.103 ± 0.038	0.045 ± 0.006	2.25 ± 0.14	Rat	(e)
Insulin	2.30 ± 0.78	0.057 ± 0.014	0.021 ± 0.006	1.89 ± 0.36	Rat (streptozotocin diabetic)	(e)
AVP	2.7	0.4			Rat	(f)

Values ± SEM.
References:
(a) Frank and Pardridge (1981), (b) Pardridge et al. (1985b), (c) Frank et al. (1986b), (d) Frank et al. (1985), (e) Frank et al. (1986a), (f) Kretzschmar and Ermisch (1988).

this finding is that there is no high affinity binding or transport mechanism present in the microvessels and that aminopeptidase activity associated with the microvessels is hydrolysing the leucine encephalin to form free [^{3}H]-L-tyrosine, which is then taken up via the L-transport system of CHRISTENSEN (1979). This interpretation is undoubtably correct in this experimental situation, but at first sight is in contrast to the brain perfusion experiments of ZLOKOVIĆ et al. (1987, 1989a), where a high affinity transport mechanism for leucine encephalin is demonstrated and the addition of 5 m*M* L-tyrosine to the perfusate was without effect on leucine encephalin extraction. This contrast points to some essential differences between isolated microvessels and in situ perfusion studies. Firstly, the geometry of the isolated microvessels means that there is a large area of abluminal capillary surface exposed to the medium compared with the luminal surface area. If peptidases extrinsic to the capillary membranes are only to be found on the abluminal surface this would explain why degradation of leucine encephalin is seen in one study and not in the other. This considerable enzyme activity in isolated microvessels would mask any uptake mechanism for the encephalin. Also, as the capillary fragments are intact for substantial lengths it is possible that the incubation medium containing the peptide under study does not gain adequate access to the luminal surface and that peptide interaction at this surface therefore contributes little to the observed uptake. There is also a possibility that soluble aminopeptidases from the cytosol may leak into the incubation medium from broken microvessel ends.

Isolated bovine brain microvessels have also been used by FRANK and PARDRIDGE (1981) to demonstrate the presence of insulin receptors. These studies showed no breakdown of the insulin molecule by the microvessels, and a Scatchard analysis of ^{125}I-insulin binding suggests two binding sites (receptors), a high affinity, low capacity site and in addition a low affinity, high capacity site (Table 7). One half of maximal binding is achieved in 7 min and binding reaches a steady-state by 45 min. Approximately 25% of the binding was non-specific. TSH, growth hormone, prolactin and pro-insulin did not significantly displace labelled insulin from the microvessels.

A further study, using microvessels from human postmortem brain (PARDRIDGE et al. 1985b), also suggests two binding sites for insulin (Table 7). Non-specific binding in this study was very high and represented 75% of the total insulin bound at 37°C. Acid wash experiments indicated that insulin was internalized and that this could account for essentially all of the non-specific binding. Exocytosis was assessed by resuspending the microvessels in fresh buffer containing unlabelled insulin and sampling aliquots of the microvessels for up to 60 min to determine the rate at which radioactivity declined in the vessels. PARDRIDGE et al. (1985b) suggest that these findings are consistent with a model of endocytosis and exocytosis of insulin that might provide a mechanism for vesicular transport across the cerebral capillary endothelium.

A subsequent study with microvessels derived from fresh bovine brain (FRANK et al. 1986b) showed that microvessels preloaded with ^{125}I-insulin for 30 min, subjected to an acid wash, and then resuspended in buffer containing 1 µg/ml unlabelled insulin exported only ^{125}I-L-tyrosine into the medium plus a small labelled oligopeptide. This was confirmed by HPLC analysis of the medium. Thus it would appear that little or no intact insulin is exported by the vessels. In the same study, binding of ^{125}I-insulin-like growth factor I (IGF I) and ^{125}I-IGF II were investigated. The maximum specific binding was 48% for IGF I, 40% for IGF II and 15% for insulin. The 50% inhibitory dose (ID_{50}) was 2.9 nM (22 ng/ml) for IGF I, 3.3 nM (25 ng/ml) for IGF II and 1.2 nM (7 ng/ml) for insulin. Insulin could displace IGF I but not IGF II from the microvessels, suggesting a separate receptor for IGF II (FRANK et al. 1986b). A Scatchard analysis gave a straight line for both IGF molecules, K_D and B_{max} being as shown on Table 7 (FRANK et al. 1986b; PARDRIDGE 1986).

FRANK et al. (1986a) prepared cerebral microvessels from rats rendered diabetic with streptozotocin. Scatchard analysis of binding in these vessels showed that both high and low affinity insulin receptors decreased in absolute numbers in the diabetic animals (see Table 7). The affinity of the high affinity receptors decreased and that of the low affinity receptors increased, both by a factor of approximately 50%. Treatment of the diabetic rats with 2 units of protamine zinc insulin per day for 7 days increased insulin binding when the microvessels were prepared, indicating a recruitment or upregulation of receptors.

Cerebral microvessels from newborn rabbits have been isolated by FRANK et al. (1985). Again, two populations of insulin receptors were identified by Scatchard analysis. K_D values were highest in the newborn animals, declining in 3-week-old and 11-week-old animals. B_{max} also declined as the animals aged (Table 7). Thus both affinity and capacity declined with age for both classes of receptor. The total receptor number could also be shown to have declined. These changes correlate with a reduction in the BUI for insulin (FRANK et al. 1985) (ses Sect. B.I), perhaps reflecting a maturation of the blood-brain barrier.

KRETZSCHMAR and ERMISCH (1987) have used microvessels prepared by a similar procedure to investigate the binding of ^{125}I-AVP to vessels isolated from different regions of the rat brain. Binding was saturable in microvessels isolated from the hippocampus, with maximal inhibition being produced by a competitive concentration of 100 nmol/l of AVP. Non-specific binding was highest in the cortex, with specific binding in the hippocampus being reduced to the lower non-specific level of that in the striatum with maximum inhibition by unlabelled AVP. Scatchard analysis of binding in the hippocampal microvessels indicated a single binding site (Table 7). Hippocampal microvessels from dehydrated rats showed a higher specific binding than control rats, and hippocampal microvessels produced from Brattleboro rats showed not only a smaller maximal binding but also, perhaps surprisingly, a smaller K_D (affinity).

PARDRIDGE et al. (1985a) have investigated the degradation of somatostatin analogues by isolated bovine brain microvessels. The presumed aminopeptidase degraded [^{125}I-Tyr^{11}]somatostatin to free iodotyrosine with an apparent K_m of 76 μM and V_{max} of 7.4 nmol min/mg protein. [^{125}I-Tyr^{11}]Somatostatin was also converted to free iodotyrosine by the isolated microvessels, but at a much slower rate, indicating the involvement of more than one enzyme activity in somatostatin breakdown. [^{125}I-Tyr^{11}]Somatostatin was rapidly bound to the microvessels whilst [Tyr^8]bradykinin, cholecystokinin (CCK8) and leucine encephalin were bound less avidly. [^{125}I-Tyr^{11}]Somatostatin was bound non-competitively to the microvessels, reaching equilibrium in 5 s, and was not temperature sensitive, suggesting a lack of internalization. However, 20% of the analogue bound could not be displaced by an acid wash. In this case, hydrolysis of the somatostatin analogue was presumably still occurring during the binding studies as an aminopeptidase inhibitor was not added to the medium.

Isolated cerebral microvessels are a useful preparation for studying receptor ligand binding but the results become somewhat difficult to interpret in relation to possible transport across the endothelium. The location of the binding and internalisation is difficult to ascribe to the luminal or abluminal surface, and polarity and directionality of any possible transcytotic mechanism may be lost. Also, peptidases to which the peptides being studied do not normally come into contact, might be exposed or released. This release of enzymes would be the result of mechanical damage during preparation of the isolated microvessels as they must fracture somewhere and do not exclude trypan blue.

VI. Cultured Cerebral Endothelial Cells

A number of studies now describe the use of monolayers of cerebral endothelial cells, grown to confluence after 7–14 days in culture, to study transport processes in vitro. Brain capillary fragments are separated from dispersed brain, usually rat or bovine, by a centrifugation step and either plated onto collagen coated wells of tissue culture plates (HUGHES and LANTOS 1989; ABBOTT et al. 1990; BEGLEY et al. 1990) or grown on porous filters (AUDUS and BORCHARDT 1986a,b, 1987; DEHOUCK et al. 1989).

The cerebral capillary has, by virtue of its "tight junctions," an electrical resistance of approximately 2000 Ω/cm^2 in the frog (CRONE and OLESEN 1981) and 1462–1490 Ω/cm^2 in the rat (BUTT et al. 1990). Other capillaries such as those of mesentery and muscle have electrical resistances of 1–2 Ω/cm^2 and 20 Ω/cm^2, respectively (OLESEN and CRONE 1983). Cerebral endothelial cells grown to confluence and with visible "tight junctions" under the electron microscope have electrical resistances in the region of

$100\,\Omega/cm^2$ and would thus be classified as leaky epithelia (CRONE 1986a). A recent study by DEHOUCK et al. (1989) has shown that, if bovine brain capillary cells are cocultured with rat astrocytes, after 1 week a confluent layer of cells is produced with an electrical resistance of $661 \pm 48\,\Omega/cm^2$. This is a powerful argument for astrocytes having an inductive influence on the formation of the blood-brain barrier. The cultured bovine endothelial cells also have ten times the γ-glutamyl transferase activity when cultured with the rat astrocytes than when cultured alone. The bovine cells when cultured alone had a electrical resistance of $416 \pm 58\,\Omega/cm^2$. In the same study, DEHOUCK et al. (1989) demonstrated that coculturing the bovine endothelial cells with astrocytes also significantly reduced the permeabilty of the cell layer to [^{3}H]inulin.

In cultured endothelial cells uptake of the fluid phase marker lucifer yellow is similar to that of the blood-brain barrier in vivo (GUILLOT et al. 1990) and fluid-phase endocytosis and subsequent turnover of the endocytic compartment is some 20 times less than in macrophages and 6 times less than in fibroblasts. Uptake of lucifer yellow was proportional to concentration over a wide concentration range (GUILLOT et al. 1987, 1990). The time-course of lucifer yellow uptake was curvilinear with an initially rapid uptake that after 20 min slowed to a steady-state (GUILLOT et al. 1990). Reduced temperature, 1 m*M* KCN and 50 m*M* 2-deoxyglucose all inhibit the uptake of lucifer yellow (AUDUS 1990). Efflux studies by pulse-chase experiments have indicated that efflux is rapid with 80% of the accumulated lucifer yellow being lost in 2 min (GUILLOT et al. 1990). This efflux was not temperature sensitive.

Electron microscopy has shown that horseradish peroxidase (HRP) is localized in 65-nm vesicles 10 min after presentation to the cultured endothelial cells and 300- to 500-nm diameter vesicles after 60 min (GUILLOT et al. 1987).

The peptides angiotensin II (AII), saralasin (AII agonist), bradykinin and phorbol myrisate ester (PMA) all stimulate the uptake of lucifer yellow by endothelial cell monolayers, by between 120% and 185% (AUDUS 1990). Sarathrin (AII antagonist) and indomethecin alone are without effect but pretreatment of the cells with indomethacin abolishes the effect of AII and greatly reduces that of saralasin whilst having no effect on the stimulation of bradykinin and PMA (AUDUS 1990). It is thus suggested that prostaglandin is involved in the action of AII and saralasin (AUDUS 1990). In similar experiments aluminium and vincristine also both reduce lucifer yellow uptake by 30% (AUDUS et al. 1988).

It is reasonable to assume that larger peptides and proteins such as insulin and albumin will require a transport mechanism involving internalisation via endocytosis to traverse the cerebral capillary endothelium, whereas smaller peptides and solutes could utilize a carrier transferase similar to those present for amino acids. Endocytosis could be a non-selective fluid-phase endocytosis, similar to that for lucifer yellow, or a specific receptor-mediated endocytosis. ^{125}I-albumin flux across confluent endothelial cells grown on polycarbonate filters and placed in the Ussing-type chamber was linear up to 200 μM albumin (K.R. SMITH and BORCHARDT 1987). Some 20% of the albumin transferred from the luminal to the abluminal compartment was degraded as assessed by trichloroacetic acid precipitation of samples from the abluminal compartment. Specific binding of the labelled albumin was 0.2%/mg of total protein. The authors suggest that there is a non-specific endocytic (pinocytic) transport across the endothelial cell layer with some lysosomal breakdown of the albumin during transport.

A study using rat cerebral endothelial cells grown to confluence in collage-coated tissue culture wells has also investigated ^{125}I-albumin binding and endocytosis (RAMLAKHAN 1990). At 4°C specific binding could be demonstrated with both a time and concentration dependency and labelled bovine serum albumin (BSA) was displaced by unlabelled albumin over a range of 0.6–100 mg/ml. Internalization was assessed after an acid wash of the cells, reaching 1.5 ± 0.27 ng/mg DNA after 60 min at 37°C, which was significantly greater than that at 4°C (0.59 ± 0.19 ng/mg DNA). Internalization of labelled BSA was reduced by the metabolic poison NaCN (1 mM) and was competitively inhibited by unlabelled BSA (10–100 mg/ml). Internalization was also stimulated by 100 μg/ml dibutyryl cyclic adenosine monophosphate (cAMP).

RAEISSI and AUDUS (1989) have used endothelial cell monolayers grown on support membranes to study the transport of the nonapeptide DSIP across the monolayer from the luminal surface (unattached border) to the abluminal surface (attached border). The membrane holding the cells is mounted in an Ussing-type chamber, which allows sampling of the luminal and abluminal compartments (AUDUS et al. 1990). The transport of DSIP appears to be by simple diffusion and cannot be saturated. The uptake was linear with time and increasing concentration and was not inhibited by 500 μM tryptophan or 50 mM 2-deoxyglucose (RAEISSI and AUDUS 1989). The calculated permeability coefficient (P) was the same whether transfer was from luminal to abluminal compartment or vice versa. These findings are in marked contrast to those obtained in brain perfusion studies (ZLOKOVIĆ et al. 1989b), where extraction of DSIP from the perfusate was saturable and competitive with both unlabelled DSIP and tryptophan (see Sect. B.III). The permeability coefficient (P) for DSIP of 1.1 ± 0.2 cm min $\times 10^{-4}$ for the cells in vitro (RAEISSI and AUDUS 1989) is some 11–20 times greater than the permeability coefficient for the peptide determined in vivo by the brain perfusion method (ZLOKOVIĆ et al. 1989b). It may be that in these in vitro studies the saturable components are submerged in a large diffusional component. In the study of RAEISSI and AUDUS (1989), there was a very limited degradation of DSIP by the cultured cells and the peptide had a 10-h half-life in the system. Other studies by this group (BARANCZYK-KUZMA and AUDUS 1987; BARANCZYK-KUZMA et al. 1989) have both shown soluble and membrane-associated aminopeptidase activity and acid hydrolase activity associated with lysosomal and microsomal fractions in cultured endothelial cells. Presumably, these peptidases do not come into contact with DSIP in these transport studies. A lack of hydrolysis by acid hydrolases may indicate that DSIP is not internalised by the cultured cells. In this context it is interesting to note that the in vivo brain perfusion studies of ZLOKOVIĆ et al. (1989b) and studies in the perfused choroid plexus (ZLOKOVIĆ et al. 1988b) indicate no hydrolysis of the DSIP suggesting that this peptide is not a substrate for cell surface aminopeptidases.

The lipid-soluble immunosuppressant drug cyclosporin A is rapidly accumulated by endothelial cells grown in collagen-coated tissue culture wells (BEGLEY et al. 1990). The cells accumulate cyclosporin, producing a tissue/medium ratio of 20 after 25 min incubation. Both heparin (5 mg/ml) and monodansylcadaverine (1.2 m*M*) added to the incubation medium do not significantly reduce the accumulation of cyclosporin A by the cells. These observations suggest that cyclosporin uptake in cultured endothelial cells is not associated with binding and internalisation of low density lipoprotein (LDL) particles containing cyclosporin or endocytic mechanisms generally. This study, in combination with in vivo measurements of uptake, suggests that little transfer of the accumulated cyclosporin occurs into the cerebral compartment (BEGLEY et al. 1990) (see Sects. B.II, B.III).

The use of cultured cerebral endothelial cells is a very promising technique in the investigation of blood-brain barrier permeability in which it should, ideally, be possible to control all other variables, leaving the permeability of the barrier itself to investigate. However, there remain concerns regarding the "tightness" of the barrier in terms of its electrical resistance and its polarization regarding transport processes, and the expression of some specific transport mechanisms. These factors may suggest the possibility of a de-differentiation of thc cells, although they have been shown to retain alkaline phosphatase and γ-glutamyl transpeptidase (transferase) activity and the presence of factor VIII antigen, all of which are characteristic of the endothelium in vivo (AUDUS and BORCHARDT 1986). It is vitally important that such in vitro studies be conducted in parallel with in vivo perfusion studies so that the results of the two methods may be directly compared.

C. Evidence for Transport Into Brain Extracellular Fluid

Evidence that peptides are extracted from plasma or perfusion fluid at the luminal surface of the blood-brain interface is not a proof that these peptides may gain access to receptors on neurons and glia which are exposed to brain interstitial fluid and CSF. Access to brain extracellular fluid may be via three routes. It may be either across the cerebral capillary endothelium by a mechanism of transcellular transport, or across the choroid plexus, which might also require a transcellular step at the ependymal surface adjacent to the CSF where the ependyma are joined by tight junctions.

Alternatively, localized groups of neurones or glia might be influenced by peptides penetrating clearly defined fenestrated areas of the cerebral capillary endothelium termed the circumventricular organs. Technically, there also remains what is termed the "functional leak" where the ependyma joined by tight junctions covering the choroid plexus changes to the fully permeable ependyma of the ventricles. Thus, peptides which have gained access to the extracellular space of the plexuses from blood could then theoretically diffuse around the transitional regions. In the short term both the functional leak and the circumventricular organs probably contribute little to the peptide content of CSF. The two remaining interfaces which might then contribute significantly to the peptide content of brain extracellular fluid are the tight capillary endothelium of the brain, if transcellular transport is a reality, and the choroid plexus, which is the secretory organ producing the major portion of the CSF volume. Because of its vastly greater surface area, the capillary endothelium, if permeable, would obviously make the greatest contribution. In this section some of the literature directed at supporting translocation of intact peptide into brain extracellular fluid will be reviewed.

The demonstration that when the concentration of a peptide in plasma is raised the CSF concentration of that peptide also rises is not a proof per se that the peptide has traversed the blood-brain barrier as the peptide may have been released in parallel from local sources within the brain. Absolute proof requires an isotopic label, or the detection of a peptide species or an analogue not normally present in the experimental animal.

KASTIN et al. (1979) studied the possible transfer of DSIP acrosss the blood-brain barrier, which used an antibody for radioimmunoassay of the peptide which required eight out of the nine amino acids in the peptide for binding. When 200 µg DSIP was injected into the carotid artery of rats and the animals were decapitated 5 s later, the levels of the peptide in a brain extract could be shown by radioimmunoassay to have doubled. Decapitation 1 min after intracarotid injection produced a raised DSIP level which was not significantly different from that in controls and after 5 min levels had returned to basal, indicating that the elevation in DSIP levels was transient. Although it could be shown that the retention of DSIP was greater than that simply retained in the vascular space and that this DSIP was substantially intact, this study gives no clue as to the location of the retained DSIP and the peptide may be accumulated onto receptors on the luminal surface of the brain capillary endothelium or sequestered within the endothelium itself without penetrating further into brain.

Using an identical approach, KASTIN et al. (1982) examined the penetration of five peptides closely related to DSIP: desTrp^1-DSIP, *N*-Tyr-DSIP, D-Ala^3-DSIP, D-Ala^4-DSIP and D-Ala^4-DSIP-NH_2. Of these peptides, desTrp^1-DSIP, D-Ala^4-DSIP and, after correction for antibody cross-reactivity, D-Ala^4-DSIP-NH_2 showed enhanced brain uptake compared to DSIP and the other analogues. Removal of the first amino acid residue tryptophan or the substitution of D-alanine in position 4 appeared to enhance brain uptake. The authors could not easily relate the structural changes to the relative penetration. The C-terminal amidation of D-Ala^4-DSIP-NH_2 would confer a greater lipid solubility and substitution of D-alanine in position 4 may render the molecule less susceptible to enzymic degradation both by plasma and brain tissue enzymes. It is difficult to envisage how removing the N-terminal tryptophan enhances uptake other than rendering the molecule smaller. It is interesting to note that, in the study of ZLOKOVIĆ et al. (1989b) (see Sect. B.III) using a brain perfusion method to study DSIP uptake in the guinea pig, tryptophan inhibited the uptake of the peptide. Possibly desTrp^1-DSIP is not subject to this inhibition and thus shows a larger relative uptake than the intact molecule.

WALSH et al. (1987) injected ^{125}I-prolactin into the jugular vein of rats and after 60 min took a CSF sample by cisternal puncture. The animal was then rapidly perfused with Ringer lactate, followed by fixative. Radioactivity was measured in the choroid plexus, median eminence and cerebral cortex. The CSF sample was separated by polyacrylamide gel electrophoresis in parallel with a prolactin reference. Coinjection of a 550-fold excess of unlabelled prolactin or human growth hormone produced a significant reduction in the ^{125}I-prolactin binding in the choroid plexus, whilst excess bovine growth hormone and insulin had no effect. The competitors had no effect on uptake in the median eminence and the cerebral cortex did not accumulate radioactivity above background. Microdensiometric scanning of the polyacrylamide gels showed a significant reduction in ^{125}I-insulin transport into CSF in the presence of excess unlabelled prolactin (73% inhibition), whilst excess human growth hormone produced a similar (75%) inhibition. Bovine growth hormone and insulin had no effect on ^{125}I-prolactin transport into CSF. This study suggest a receptor-mediated transport of prolactin from plasma into CSF located at the choroid plexus. It is interesting to note that the cerebral cortex does not appear to accumulate prolactin, suggesting that the capillary endothelium in this tissue is impermeable to prolactin and that the tissue does not accumulate prolactin from CSF.

LORENZO et al. (1988) has similarly concluded that prolactin enters CSF via the choroid plexus. Four hours after intravenous injection of ^{125}I-prolactin into pregnant rabbits at 28–30 days of gestation, labelled prolactin could be detected in the CSF of both the pregnant doe and the rabbit fetuses. The blood/CSF ratio for radioactivity was 0.10 ± 0.01 for the mother and 0.05 ± 0.02 for the fetuses; $68.9 \pm 8.4\%$ of the radioactivity in the maternal CSF was trichloroacetic acid precipitable, compared with 38.2 $\pm$ 2.7% in plasma; the comparable figures for the fetus was $43.2 \pm 7.7\%$ and $12.7 \pm 0.9\%$, respectively, indicating some breakdown of the labelled prolactin.

WILSON et al. (1984) investigated the transport of MSH-α into CSF by a ventriculocisternal perfusion technique in the rat. Immunoreactive MSH-α was injected intravenously into animals and its rate of appearance in the outflowing perfusate determined by radioimmunoassay. Simultaneously, the passge of [^{14}C]inulin coinjected with the MSH-α was estimated. The rate constant for the entry of MSH-α into CSF was 0.00087 ± 0.00034 min^{-1} and for inulin 0.00055 ± 0.00028 min^{-1}. The two rate constants were not significantly different. It was concluded that entry of MSH-α was by aqueous diffusion along with other water-soluble macromolecules. Endogenous MSH-α, in the outflowing perfusate in control animals was high (0.01 nmol/l), about 10% of the circulating level in plasma, suggesting a source of MSH-α from elsewhere in the CNS. Also, the calculated rate constant for removal of MSH-α from CSF was 1.30 ± 0.23 min^{-1}, considerably in excess of the removal constant for inulin, which was 0.118 ± 0.041 min^{-1}. This difference would ensure that any rise in CSF MSH-α levels were transient. Interestingly, the coinjection of MSH-α and inulin appeared to accelerate the clearance of inulin from CSF.

A pioneering study by MARGOLIS and ALTSZULER (1967) suggests that insulin rises in CSF when it is infused intravenously into the dog. Plasma samples were taken and CSF samples from the cisterna magna prior to an infusion of 0.2–1.0 U $kg^{-1}h^{-1}$ insulin. Glucose was also infused at a rate which prevented hypoglycaemia or produced hyperglycaemia. Insulin was found by radioimmunoassay in the CSF of all normal dogs investigated at a concentration of 3 μU/ml, or 27% of the average plasma concentration, which was 11 μU/ml. When concentrations of plasma insulin were raised by intravenous infusion, the CSF levels rose between 2 and 70-fold at plasma insulin levels of 1000–2000 μU/ml. In the physiological range of 150–300 μU/ml the CSF insulin levels rose by up to 10-fold. Plasma glucose levels did not affect insulin levels in the CSF and the transfer of insulin into CSF appeared to be saturable. Entry of insulin into CSF was slow and took 3–4 h to stabilize when insulin was infused.

WOODS and PORTE (1977), also in the dog, examined the relationship between plasma and CSF insulin levels. Immunoreactive insulin was measured in plasma and CSF after the intravenous administration of 0.2 U/kg of porcine insulin either as a pulse injection or as a 1-h infusion. These procedures caused a large increase in plasma insulin and a relatively small increase in CSF insulin. After a pulse injection plasma insulin levels rose from 20 μU/ml to over 3000 μU/ml; CSF insulin levels rose from 6.5 μU/ml to 16.6 μU/ml with a lag of 15 min. The same amount of insulin infused over 1 h increased plasma insulin from 20 μU/ml to 80 μU/ml for the duration of the infusion. CSF insulin rose from approximately 5 to 10 μU/ml with a lag of 15–30 min. The infusion of glucose to elevate endogenous plasma insulin levels also produced a rise in CSF insulin levels, but to a much lesser extent than intravenous insulin infusion.

SCHWARTZ et al. (1990) have investigated the transport of insulin into the CSF of the dog after intravenous infusion. Insulin levels were measured by radioimmunoassay. When plasma levels of insulin were maintained in the hyperinsulinaemic range of 500 pM, whilst maintaining euglycaemia by glucose infusion, CSF insulin levels did not alter for the first hour, but by 90 min a statistically significant rise was observed. When plasma insulin levels were increased step-wise from 500 to 15 000 pM, CSF insulin rose in a dose-dependent fashion. After bolus intravenous injection the plasma half-life (T_{50}) of insulin in the CSF was 143 ± 7 min, which is similar to bulk CSF turnover in the dog. The rate constant for insulin clearance from the CSF was 0.005 min^{-1}, much slower than that found for inulin in the rat (WILSON et al. 1984). The T_{50} value for insulin in dog CSF compares very favourably with the T_{50} for insulin in rabbit CSF determined in by ventriculocisternal perfusion (BEGLEY and CHAIN 1992). Thus, insulin is probably cleared from CSF by bulk flow alone.

In the study of SCHWARTZ et al. (1990) pro-insulin was transported into CSF five times less rapidly than insulin, suggesting the presence of a specific transport mechanism. The lag period of 1 h before CSF insulin levels begin to rise is of some interest. This might be occasioned by the delay introduced by a mechanism of transcellular transport at either the blood-brain or blood-CSF barrier, with the pathway from the cerebral capillaries to CSF being the more tortuous route.

SQUIRES (1990) performed experiments where porcine insulin was intravenously infused into rabbits for 2 h and plasma and CSF insulin levels estimated by HPLC. The infusion approximately doubled the circulating insulin levels from 56.8 ± 18.0 to 129.9 ± 12.0 μg/ml. The CSF level of insulin after peripheral insulin infusion was 1.4 ± 0.4 μg/ml, which was not significantly different from control or sham-infused animals. It was concluded that insulin had not entered the CSF from plasma during these experiments. Interestingly, although rabbit and porcine insulin standards could be distinguished by HPLC with distinct retention times, after infusion of porcine insulin only a single peak was detectable in plasma at the retention time of rabbit insulin. Possibly either the experimental animal or the preparative treatment of plasma for HPLC analysis was modifying the infused porcine insulin. Rabbit and porcine insulin differ by a single amino acid substituted at the B^{30} position.

DUFFY and PARDRIDGE (1987) used 1- to 7-day-old neonatal rabbits to investigate the uptake of insulin at the blood-brain barrier. Neonatal rabbits were used as it was hoped that plasma degradation of the ^{125}I-insulin used would be minimized in these animals, and the brain extraction of insulin is known to be greater in such animals (FRANK et al. 1985). ^{125}I-Insulin was introduced at 250 μl/min into the carotid artery of the rabbits for 1.5- or 10-min periods and the mean brain uptake relative to [^{3}H]albumin calculated. At 1 min this was 99.3 ± 5.5%, at 5 min 110.1 ± 4.3% and at 10 min 143.6 ± 7.9%. This result indicates a progressive preferential uptake for the insulin. Saturability could be demonstrated by simultaneously infusing 100 μg/ml of unlabelled insulin. Thaw-mount autoradiography of the brains 10 min after infusion of the ^{125}I-insulin revealed silver grains in the pericapillary spaces as well as brain parenchyma. HPLC analysis of radioactivity extracted from brain tissue showed that the major peak had the same retention time as an insulin standard, indicating minor degradation of the insulin. The radioactive peak extracted from brain by an acid ethanol extraction could also be removed by precipitation with an anti-insulin serum.

BROADWELL et al. (1989) has adopted a similar approach in adult rats. Insulin conjugated to HRP was infused into the carotid artery. After 10 min of infusion HRP reaction product could be found in endothelial endocytic vesicles, endosomes and secondary lysosomes. No reaction product was present in the perivascular clefts or perivascular phagocytic cells. After 1 h of perfusion, however, perivascular clefts in discrete areas such as the striatum were HRP positive. Presumptive exocytic vesicles could be seen apparently emptying HRP-positive material into the perivascular clefts. It perhaps significant to note the apparent lag before reaction product appears in the perivascular space is similar to that seen in the insulin transport study of SCHWARTZ et al. (1990). In the study of BROADWELL et al. (1989), the HRP conjugates had been shown to retain the specific receptor binding characteristics of insulin, so this particular aspect of biological activity was retained in the complex. However, HRP is a foreign protein to the mammal and the possibility cannot be excluded that the conjugates might react atypically with specific transport mechanisms or that disassociation between protein and HRP may occur (BROADWELL 1989).

Van HOUTEN et al. (1979) have shown in the rat that, after intracardiac injection followed by autoradiographic localization, ^{125}I-insulin binds readily to circumventricular regions lacking tight junctions between endothelial cells. In addition, the median basal hypothalamus, medial paravagal regions and the choroid plexus showed marked concentrations of silver grains. In the presence of a 500-fold excess of insulin, binding to the circumventricular regions, the medial basal hypothalamus and the paravagal region was reduced by 56%–78%, indicating the presence of competitive binding sites. In the external contact zone of the median eminence and the arcuate nucleus, increasing concentrations of unlabelled insulin incorporated into the intracardiac injection produced a progressive decrease in the number of silver grains seen by autoradiography. Desalanine insulin was as effective as insulin in inhibiting binding, proinsulin was 20–50 times less effective and ovine prolactin and porcine adrenocorticotropin (ACTH) were without effect.

The blood-CSF barrier to [^{3}H]arginine vasopressin (AVP), ^{125}I-desmopressin (DDAVP) and ^{125}I-desglycinamide arginine vasopressin (DGAVP) in the dog has been investigated by ANG and JENKINS (1982). After intravenous injection or infusion the radioactivity in CSF was between 0.5% and 1.4% of the total plasma radioactivity. However, only intact DDAVP could be detected in the CSF, with 7% of the total CSF radioactivity representing unmetabolized peptide. The authors ascribed this finding to the fact that DDAVP had a half-life in plasma of 50 min whilst those for AVP and DGAVP were 13 min and 8 min, respectively. Also, DDAVP may be metabolized more slowly within the CNS compared to the other peptides and may be transported out of CSF less rapidly.

MENS et al. (1983) have similarly estimated the appearance of AVP and oxytocin in CSF by radioimmunoassay after subcutaneous or intravenous administration. AVP was detectable in CSF 2 min after subcutaneous injection of 0.5 μg and peak concentrations were reached at 5 min. Administration of 0.5 μg oxytocin subcutaneously or intravenously produced an elevation of CSF levels, reaching a peak between 10 and 30 min. The authors calculated that 0.001% of the peripherally administered dose of AVP and between 0.002 and 0.003% of the peripherally administered dose of oxytocin reached the CNS 5–10 min after injection. Both neuropeptides are cleared from the CSF, with a T_{50} of 26 min for AVP and 19 min for oxytocin.

A competitive antagonist of AII, ^{125}I-Sar^1-Ile^8-angiotensin II, fails to penetrate into CSF after infusion into rats via the brachial artery (HARDING et al. 1988). Furthermore, this study failed to show any transport of ^{125}I-Sar^1-Ile^8-angiotensin II after intracerebroventricular injection. Although radioactive label reached the peripheral circulation rapidly, no intact Sar^1-Ile^8-angiotensin II could be detected in plasma. An HPLC technique was used to identify any intact Sar^1-Ile^8-angiotensin II in CSF and plasma. Thus, Sar^1-Ile^8-angiotensin II does not appear to traverse the blood-CSF barrier in either direction. Nephrectomy was without effect on the result. As AII increases blood pressure and stimulates AVP secretion (REID 1984) after either intravenous or intracerebroventricular administration (FITZSIMONS 1980), it would appear that the peptide can reach its sites of action from either side of the blood-brain interface. These observations suggest very strongly that the sites of action and receptor binding might be located in the circumventricular organs. The total sum of the permeabilities of the various circumventricular organs, plus their anatomical relationships with the brain ventricles and blood, would probably not be sufficient to cause detectable transport in either direction. If this explanation is correct the choroid plexus must be tight to AII and can also not be a site of action for the observed effects. Alternatively, the receptor sites for CSF and blood AII might be entirely separate and function independently.

Cox et al. (1990) have shown that the AVP enhancement of the baroreflex is mediated through the area postrema. Destruction of the area postrema with kainic acid greatly reduces the reflex bradycardia produced by AVP infusion but not by phenylephrine. Also, the acute pressor response to AII and bradykinin is dependent upon the integrity of the area postrema (Cox et al. 1990).

The peptide cyclo-Leu-Gly, a diketopiperazine, is derived from the C-terminus of oxytocin, is resistant to enzymic degradation and is claimed to affect memory processes (HOFFMAN et al. 1977). After an intravenous bolus injection of cyclo-Leu-[^{14}C]Gly in the cat, plasma radioactivity declined rapidly within the first 60 min. Cyclo-Leu-Gly was detectable in CSF within 10 min of intravenous injection and its concentration increased up to 60 min. After this time radioactivity in both plasma and CSF declined. During infusion experiments the plasma radioactivity was held almost constant and a CSF/plasma ratio of 2.7 was achieved. Levels of radioactivity in urine generally followed the pattern in plasma. Experiments where cyclo-Leu-Gly was incubated with brain, kidney and liver homogenates showed that the peptide was stable under these conditions, whereas [^{14}C]AVP was rapidly broken down. Chromatography of ethanol extracts of plasma, CSF and urine revealed only single spots corresponding to cyclo-Leu-Gly. Cyclo-Leu-Gly is a highly lipid soluble molecule.

WALTER et al. (1979) have shown that cyclo-Leu-Gly, cyclo-Pro-Phe, and the naturally occurring peptide Pro-Leu-Gly-NH_2 (MIF) are effective in blocking physical dependence to opiates as judged by naloxone-induced morphine withdrawal in mice, suggesting significant penetration of the blood-brain barrier.

DE WILDT et al. (1982) also investigated the recovery of intact Pro-[^{3}H-Leu]-Gly-NH_2 (MIF) from the brain of the rat after intravenous injection. The structure of MIF is identical to the C-terminal tripeptide of oxytocin, and it has been isolated from the hypothalamus of the bovine brain. After intravenous injection of Pro-[^{3}H-Leu]-Gly-NH_2 in the rat, HPLC showed that only 0.008%–0.001% of the administered dose per gram of tissue could be recovered from the brain. At longer times intact peptide could not be demonstrated. As with the study of KASTIN et al. (1979) with DSIP, the location of the intact peptide is uncertain.

In order to overcome this difficulty TRIGUERO et al. (1990) have recently introduced a "capillary depletion method" to study the transfer of peptides and proteins into brain extracellular fluid, thus providing support for exocytosis at the abluminal surface of the cerebral capillary endothelium. Radiolabelled tracers were perfused into rat brain and in vivo extraction determined by the brain perfusion technique of TAKASATO et al. (1984). After perfusion, 10 s was allowed to enable the animal's blood to displace perfusate from the vascular lumen and the rat was then decapitated. The brain was then subjected to a homogenization process and a capillary pellet centrifuged out. Volumes of distribution could then be calculated for the homogenate, the capillary pellet and the supernatant after centrifugation. Radioactivity in the supernatant was held to represent transfer into the post-capillary compartment of brain parenchyma. This study showed that cationized albumin and cationized immunoglobulin G (IgG) were present in the supernatant in significant quantities whilst native albumin and acetylated LDL were not. Although perfusion showed LDL binding to the luminal surface of the cerebral endothelium, it did not appear to be transcytosed. Hence this tracer is remaining either bound to, or sequestered within, the capillary endothelium. Tracers did not appear to be liberated from the capillaries by the homogenization process and the centrifugation procedure as evidenced by a lack of LDL, native albumin and [^{14}C]sucrose in the supernatant after perfusion. This approach to blood-brain barrier transport may be helpful in quantifying transport through the capillary endothelium.

D. The Influence of Peptides on Blood-Brain Barrier Permeability to Specific Solutes

One way in which peptides might alter central nervous activity is by binding to receptors at the cerebral capillary lumen and thus altering the blood-brain barrier permeability in a specific or non-specific manner to other plasma solutes which might then secondarily bring about central nervous effects. Although ample evidence now exists that the luminal surface of the cerebral capillary endothelium is studded with peptide receptors and transporters, relatively little has been published with regard to the possible influence of peptides on blood-brain barrier permeability to specific molecules. CRONE (1986b) suggests, for instance, that bradykinin might increase cerebral endothelial cell permeability in a non-specific way by elevating intracellular free calcium levels, which may produce changes in the shape of cells by activating the actin-myosin system. AVP and related peptides might act on cerebral endothelium in a similar way by raising intracellular calcium via V_1 receptors, thus changing cell shape or altering the activity of the L-system amino acid transporter by calcium-activated protein kinases (ERMISCH et al. 1988).

One of the most obvious peptides that might act to change cerebral endothelial permeability is insulin. Several studies have investigated the possible influence of insulin on glucose uptake by the brain. The permeability of the blood-brain barrier to glucose is generally believed to be maximal and not to present a rate-limiting step to brain metabolism (LUND-ANDERSEN 1979), any increased demand for glucose being met by an increase in cerebral blood flow and capillary recruitment (CRONE and THOMPSON 1970). Using the paired tracer indicator dilution technique, BETZ et al. (1973) came to the conclusion that 800 μU/ml insulin had no effect on the uptake of glucose by the perfused dog's brain, under both single-pass and steady-state conditions. DANIEL et al. (1975) reported that insulin increased the net uptake of glucose by the brain of the rat but did not affect the unidirectional influx. However, some doubt has been cast on these results (HERTZ et al. 1981), as the measurements were made over 1 min when the animal was not in a steady-state, blood glucose first having been depressed by insulin injection and then raised by glucose injection, creating an increased gradient for glucose into the brain.

HERTZ et al. (1981) have investigated the effect of insulin on glucose transfer across the blood-brain barrier in man. The unidirectional extraction (E) for [^{14}C]-D-glucose was determined by the indicator dilution technique using $^{36}Cl^-$ as the intravascular reference. 0.4 U insulin/kg body weight was infused and blood glucose was maintained constant by glucose infusion. An increased extraction of glucose from 14% to 21% was found, together with a significantly increased unidirectional flux from 0.46 to 0.66 μmol g^{-1} min^{-1} during the insulin infusion. The infusion raised the plasma insulin level to 1500 μU/ml, far in excess of the physiological range. The net brain uptake of glucose, however, was unaltered during the experimental period of 45 min. The authors suggest that this is the result of an increased backflux, coincident with the increased influx, and the effect of the insulin might be a speeding up of the glucose carrier in both directions. HERTZ et al. (1981) also suggest that the effect of the insulin is at the level of the cerebral endothelium , as the 45 min duration of the study is too short for insulin penetration of the brain, and the increase in the *E* value reached significance at 30 min after the start of infusion, together with the unidirectional influx increasing within 20 min. However, it is interesting to note in this context that SCHWARTZ et al. (1990), in their study in the dog (see Sect. C), found a lag of 60 min before insulin levels rose in CSF after infusion. This observation raises the possibility that insulin might be acting within the blood-brain barrier, thus creating larger gradients for glucose into the brain.

A study with isolated cerebral microvessels of the rat (BETZ et al. 1979) showed no effect of insulin (0.01 U/ml in the medium) on 3-*O*-methyl glucose (3-*O*-MG) accumulation and uptake was 80% of control values. They concluded that in the isolated microvessels 3-*O*-MG uptake was entirely consistent with studies of blood-brain barrier sugar transport in vivo.

In a study in the rat, DE MONTIS et al. (1978) concluded that insulin 2 U/kg injected subcutaneously increased the uptake of large neutral amino acids into brain. Brain leucine levels in fact fell, as did brain isoleucine levels. The brain levels of valine and phenylalanine remained constant, whilst those of tyrosine and tryptophan rose. At the same time the plasma levels of all of these amino acids fell, with the exception of tryptophan, total tryptophan rising and free tryptophan fulling. The total plasma content of these amino acids fell from 593 to 511 nmol/ml and the brain content remained constant at 300 nmol/g. However, the experimental animals were in no sense in a metabolic steady-state, making any firm conclusion difficult to draw. The authors themselves state that the data do not allow a decision as to whether the stimulatory effect of insulin is a primary effect on the transport system or is secondary to an increase in brain amino acid utilization. Any conclusions concerning the effect of insulin on amino acid transport at the blood-brain barrier must remain speculative.

BRUST (1986) has examined the effect of AVP on L-leucine uptake at the blood-brain barrier. Leucine was injected into the right carotid artery of the rat in concentrations from trace to 6.4 m*M*. In a separate series of animals, AVP was coinjected together with the varied concentrations of leucine at the physiological concentration of 40 p*M*. Leucine uptake by the brain was calculated as an accumulation of the injected dose (%g). The data allowed

K_m, V_{max} and K_d to be calculated for leucine uptake in the presence and absence of AVP in nine brain regions. In the cerebral hemispheres, for example, the V_{max} decreased from 21.0 to 7.6 nmol min^{-1} g^{-1} and the K_m declined from 0.11 to 0.029 mM. The control values in the cerebral cortex were similar to those quoted by PARDRIDGE and MIETUS (1982). In all nine brain regions the V_{max} and K_m were reduced compared to control values. The non-saturable component (K_d) was increased in all brain regions but was not significantly greater than control values. Control values for V_{max} and K_m varied between the brain regions by about a factor of 100%. It is not clear whether these differences represent real differences in transport constants in differing brain regions or whether the bolus mixes to differing extents with blood from the contralateral carotid artery in different parts of the brain, producing varying concentrations of leucine. Cerebral blood flow in the regions, measured with iodo-[^{14}C]-antipyrine at various times after AVP injection, was not changed. The differences in the determined constants did not correlate with cerebral blood flow for the region and thus may represent real differences in transport properties for the region. The authors conclude that AVP interacts either directly or indirectly with the transporter for large neutral amino acids.

In a similar study, REITH et al. (1987) demonstrated a saturable retention of AVP by the capillary endothelium of the hippocampus which was associated with a reduction in leucine uptake by the structure. Animals were injected with an intravenous bolus of either tracer leucine, tracer AVP or tracer leucine containing 50 nmol unlabelled AVP. Brain uptake was determined after either 20 s or after periods ranging from 10 s to 2 min using a multiple time regression analysis. The coinjection of AVP produced a reduction in leucine transport in the hippocampus, with the *PS* product falling from 72.0 to 42.0 µl g^{-1} min^{-1} (42%), and smaller reductions of 17% and 19% in the olfactory bulb and colliculi respectively. In the olfactory bulb and the hippocampus, the endothelial permeability to AVP did not exceed that of mannitol. In the hippocampus and non-blood-brain barrier regions (pituitary and pineal) AVP was retained to a greater extent than mannitol and this uptake was inhibited by excess AVP. Studies with [^{14}C]iodoantipyrine showed no changes in blood flow. The authors take the view that the AVP is being bound by specific receptors at the luminal surface of the cerebral capillary endothelium and this is also the surface at which the inhibitory effect on leucine uptake is being exerted. However, as the AVP is injected peripherally into the femoral vein, the possibility remains that there may be breakdown of the peptide in plasma, in contact with other tissues and at the blood-brain barrier itself, to produce other possible active fragments or amino acids which exert the observed effect.

REVEST et al. (1991) have examined the effects of AVP on leucine uptake in cultured endothelial cells. AVP at 5 µM had no effect on the accumulation of leucine by confluent cultures. These experiments suggest either that the inhibitory effects of AVP noted by BRUST (1986) and REITH et al. (1987) are not exerted at the level of the capillary endothelial cell or that the cultured cells have dedifferentiated to the extent that they have lost this particular interaction.

BRUST and ZICHA (1988) have examined the regional uptake of L-leucine in Brattleboro rats which lack endogenous vasopressin. Homozygous and heterozygous Brattleboro rate were compared. Regional blood flow determined by [^{14}C]iodoantipyrine uptake was the same in both rats except in the olfactory bulb of the homozygous animals, which was 22% lower. The V_{max} values for leucine uptake, estimated by a brain uptake index (BUI) technique, in heterozygous animals were 28%–64% lower than in homozygous (Brattleboro) animals. The K_m for leucine uptake in heterozygous animals was 14%–58% lower than in homozygous animals These differences were most marked in the hippocampus and hypothalamus. The authors suggest that the presence of endogenous vasopressin in the heterozygous animals is suppressing the uptake of large neutral amino acids by the brain. Brattleboro rats have been shown to have vasopressin receptors on isolated hippocampal microvessels (KRETZSCHMAR and ERMISCH 1987).

There are various reports in the literature that peptides derived from proopiomelanocortin (ACTH and MSH) might influence the permeability of the blood-brain or blood-CSF barriers. RUDMAN and KUTNER (1978) investigated the effect of ACTH, MSH-α, MSH-β and "choroid plexus peptide IIF" injected intracisternally on the entry of ^{125}I-albumin, [^{14}C]mannitol, [^{14}C]sucrose, [^{14}C]insulin, [^{14}C]glucose, [^{14}C]-α-aminoisobutyric acid and [^{14}C]valine into CSF. Intracisternal injection of 1000 μg of the peptides increased the rate of appearance, within 15 min, of albumin, mannitol, insulin and sucrose in the CSF two to four time over that of controls. Minimal effective doses were 1 μg for MSH-α and MSH-β, 10 μg for ACTH and 100 μg for choroid plexus peptide IIF. The minimum effective dose was 100 times higher intravenously, suggesting that the effect was being exerted from the CSF side of the barrier. The melanotropic peptides did not alter the rate of appearance of glucose, α-aminoisobutyric acid or valine into CSF, nor did AVP, norepinephrine, histamine and serotonin. When injected intracisternally, 100 μg of each had no effect on the penetration of any of the solutes. In these experiments glucose and valine may have undergone metabolism. However the lack of acceleration of α-aminoisobutyric acid entry, which is non-metabolizable, suggests that the mechanism has some specificity.

GOLDMAN and MURPHY (1981) investigated the effect of intravenous administration of ACTH/MSH 4–9 (ORG-2766) on the regional uptake of the diffusion-limited antipyrine relative to the highly diffusible iodoantipyrine in the brains of conscious unrestrained rats, using a modified OLDENDORF (1970) technique. ACTH/MSH 4–9 exerts behavioural effects in rats, enhancing motor behaviour, visually evoked responses and memory. Within 10 min of intravenous injection the relative extraction of antipyrine was reduced significantly in the hypothalamus, parietal cortex and frontal cortex. The average of the regional *PS* values for antipyrine was $0.026\,cm^3\,s^{-1}\,g^{-1}$ ($1560\,\mu l\,g^{-1}\,min^{-1}$) and was reduced by 15.3% in the hypothalamus, 18.2% in the parietal cortex and 21.2% in the frontal cortex. Regional blood flow in the various brain areas, as estimated by iodoantipyrine extraction, was not affected. Assuming that the capillary surface area remains constant, the authors conclude that the peptide is producing a reduction in the permeability of these particular brain regions and suggest that these changes in permeability may underlie the behavioural effects of the peptide.

Similarly KASTIN and FABRE (1982) have shown that MSH-α administered intravenously to rats in doses of 200 µg/kg increased the penetration of ^{99m}Tc-pertechnetate into the brain after discarding the hypothalamic, pineal and pituitary regions. SANKAR et al. (1981) showed selective effects of MSH-α and MIF-1 on blood-brain barrier permeability in rats. Intravenously injected MSH-α (80, 160 µg/kg or 5 mg/kg) or MIF-1 (1 mg/kg or 10 mg/kg) did not alter the permeability of the barrier to ^{131}I-albumin. However, MSH-α selectively increased permeability to ^{99m}Tc-pertechnetate. Permeability to ^{99m}Tc-pertechnetate was increased by MSH-α but not by MIF-1 at doses known to produce behavioural and electroencephalographic responses in the animals.

ELLISON et al. (1987), in a study in cats, have suggested that human recombinant interleukin-2 (rIl-2) increases permeability of the blood brain barrier to HRP (M_r 40 000) and IgG (M_r 160 000). Eight animals were injected intravenously with 100 000 U rIL-2/kg and showed extravasation of HRP or IgG at 1 h (five animals) and 4 h (three animals) assessed by histological methods. However, great doubt must be cast on these findings as four out of six control animals injected with the injection vehicle containing 135 mg of mannitol and 54 µg of SDS (sodium dodecyl sulphate) also showed an increased cerebrovascular permeability to the markers.

IBARAGI et al. (1989) showed, using isolated rat cerebral microvessels, that α-rat atrial natriuretic peptide 1–28 [atrial natriuretic peptide (ANP)-(99–126); rANP)], influenced amiloride-sensitive Na^+ transport into the tissue. With ^{125}-I-rANP, binding to cortical microvessels had a K_D of 173 p*M* and a B_{max} of 159 fmol/mg protein. In the presence of $10^{-4}\,M$ amiloride these values were reduced to 33 p*M* and 88 fmol/g protein respectively. $10^{-7}\,M$ rANP reduced the uptake of $^{22}Na^+$ into microvessels in the presence of 1 m*M* ouabain, 1 m*M* furosemide and 2 m*M* LiCl in the buffer, suggesting that there was a specific effect of rANP on amiloride-sensitive Na^+ transport. They suggest that circulating ANP may regulate Na^+ transport across the blood-brain barrier in vivo. However, as the experiments are conducted with isolated microvessels, reservations must be held regarding whether the ANP is acting at the luminal or abluminal surface, or both, as discussed in Sect. B.V. The ANP receptors on the cortical microvessels may also be different to those in other tissues, as amiloride increases the affinity but decreases the number of receptors in the cerebral cortical microvessel preparation, whereas in adrenal zona glomerulosa cells there is an increase in affinity with no change in receptor number (IBARAGI et al. 1989).

E. General Conclusions

Clearly, peptides interact with the blood-brain interface in a number of distinct ways. Numerous receptors exist for peptides on cerebral capillary endothelial cells. Some of these receptors will be linked to intracellular second messengers which will alter endothelial cell activity and other will be transporters or transferases. Transporters associated with smaller peptides such as encephalin and vasopressin may function in a similar manner to the more familiar carriers associated with amino acid transport and function to translocate the peptide through the cell membrane. Other receptors which bind larger peptides such as insulin would appear to function as part of a mechanism of receptor-mediated endocytosis. The receptors/transporters associated with the blood-accessible ependymal surface of the choroid plexus and accessible to blood-borne solutes are poorly characterised and cannot be distinguished with current methods from binding and uptake to other cells of the choroid plexus.

Once a blood-borne solute is internalized by the cerebral capillary endothelium, there is very limited evidence to confirm that peptides are exported intact at the abluminal surface into brain extracellular fluid. Indeed, for some circulating peptides it may prove to be the case that internalisation is a dead end, and that once internalised the peptide is destroyed by the endothelial cells as part of a deactivation mechanism. In the case of TRH, extrinsic luminal cell-surface peptidases hydrolyse the peptide on contact. With peptides such as MSH-α and DSIP and insulin there is a good case for the transport of detectable amounts into the CSF, although the case for parallel release of peptide into CSF from brain-synthesised stores still have to be fully addressed. It may be that for many endogenous circulating peptides the blood-brain interface effectively neutralises the peptide, allowing an isoneurohormone/neuroregulatory milieu to exist within the CNS and, for instance, a given peptide to act as a hormone peripherally and a neurotransmitter centrally. The case for penetration is better proven with analogues, as in the case of certain vasopressin analogues, which can have no source other than their intravenous introduction into plasma.

Evidence with a number of peptide analogues clearly points to the fact that lipid solubility is an important factor in extraction at the blood-brain interface and penetration into the CSF, with the more lipid soluble molecules being favoured. However, this may have limitations as the highly lipid soluble peptide cyclosporin only shows a limited penetration into the CNS. It may be that substances may become too lipid soluble to be favoured for transport, especially if they require a carrier such as LDL or albumin to maintain any measurable concentration in an aqueous medium. Thus, if the carrier is not transported in quantity through the blood-brain barrier and there is no equivalent carrier present in brain extracellular fluid to accept the highly lipid soluble substance, it effectively becomes trapped in the lipid of the endothelium and progresses no further. Lipid solubility will obviously be the major consideration for peptides and analogues which do not exhibit receptor-mediated uptake.

However, the case for the penetration of endogenous peptides must not be dismissed too lightly as many are extremely potent substances and the question arises of how much is enough. It may well require the penetration of only minute amounts through a relatively permeable area such as the choroid plexus to generate direct central nervous effects. With peptides and their analogues, shown to have central effects after peripheral administration, the effect is often several orders of magnitude greater on a dose-response basis if the peptide is injected intracerebroventricularly. The demonstration of carrier-mediated extraction at the blood-brain interface is only a first step in possible transcellular transport and it requires the development of much more sensitive techniques to detect the solutes in brain extracellular fluid.

When it comes to the development of analogues of peptides with central nervous effects for possible therapeutic use, a number of lines suggest themselves. Firstly, an analogue should ideally be resistant to blood and tissue enzymes to extend its half-life in the circulation. The selective substitution of D-isomers of amino acids into the molecule has application here. Also, the cyclisation of a peptide effectively removes the reactivity of the N- and C-termini, greatly reducing their susceptibility to exopeptidases. If an existing transport mechanism for an endogenous peptide or transported substance is to be used, the analogue should still retain reactivity with that mechanism. If natural permeability to cell membranes is to be exploited increased lipid solubility may favour uptake, but probably only within certain limits. Acylation, amidation and cyclisation all improve lipid solubility. The ideal substance, if a dependence on membrane transporters is to be avoided, would have a moderate partition between lipid and water, such as is shown by ethanol. In a similar manner, if transporters are involved, the analogue should be designed so that its reactivity with mechanisms transporting peptides into the brain is maximised, and with transporters which clear or transport peptides out of the CSF, including soluble and membrane-bound peptidases, minimised. There is evidence that the mechanisms transporting peptides from CSF have different characteristics from those extracting peptides at the luminal surface of the blood-brain interface (Begley and Chain 1992). Finally, the analogue must still show reactivity, preferably enhanced, with receptors for an endogenous ligand within the CNS to exert an effect. It may well be that all of these requirements cannot be satisfied, certainly for a range of analogues, and it may be necessary to design pro-drugs which have the desired transport characteristics and are then converted to an active peptide or analogue within the CNS.

In the further future investigation of peptide transport at the blood-brain barrier, the influences of peptides on barrier activity, and the screening of peptide analogues and other drugs for potential use, three main areas recommend themselves for attention. Firstly and secondly, in vivo perfusion studies must be continued parallel with those in confluent endothelial cell cultures, especially where studies on peptide uptake appear to suggest that the cultured cells have lost saturable carrier-mediated uptake mechanisms. At present, perfusion techniques offer the only methods of investigating the interaction of the intact barrier in situ with various solutes free from the influence of plasma elements and metabolism elsewhere in the body. Thus, the characterisation of the barrier under these conditions must proceed together with the in vitro investigations to ensure that the blood-brain barrier is being faithfully recreated in the culture system. Thirdly, methods for the sensitive detection of peptides and analogues in brain extracellular fluid must be developed and applied to determine whether transcellular transport in the intact barrier is in fact occurring. Once the factors governing transport are fully understood, analogue design can move forward to produce effective drugs with central nervous action and, conversely but perhaps equally importantly, to ensure that desired drug action remains peripheral where blood-brain barrier penetration might produce neurotoxic or other unwanted side effects.

References

Abbott NJ, Hughes CCW, Greenwood J, Orsmond P, Ramlakhan N, Revest PA (1990) Primary culture of rat brain capillary endothelial cells for physiological studies. J Physiol (Lond) 423:2P

Ang VTY, Jenkins IS (1982) Blood-cerebrospinal fluid barrier to arginine-vasopressin, desmopressin and desglycinamide arginine-vasopressin in the dog. J Endocrinol 93:319–325

Audus KL (1990) Blood-brain barrier: mechanisms of peptide regulation and transport. J Controll Release 11:51–59

Audus KL, Borchardt RT (1986a) Characterization of an in vitro blood-brain barrier model system for studying drug transport and metabolism. Pharm Res 3:81–87

Audus KL, Borchardt RT (1986b) Characteristics of the large neutral amino acid transport system of bovine brain microvessel endothelial cell monolayers. J Neurochem 47:484–488

Audus KL, Borchardt RT (1987) The use of isolated epithelial and cultured endothelial cells to elucidate drug transport mechanisms. In: Rand MJ, Raper C (eds) International pharmacology symposium, Elsevier, Amsterdam, pp 615–618

Audus KL, Bartel RL, Hidalgo IJ, Borchardt RT (1990) The use of cultured epithelial and endothelial cells for drug transport and metabolism studies. Pharm Res 7:435–451

Audus KL, Shinogle JA, Holthaus SR, Guillott FL (1988) Aluminum effects on brain microvessel endothelial cell monolayer permeability. Int J Pharmacol 45:249–257

Banks WA, Kastin AJ (1985) Peptides and the blood-brain barrier: lipophilicity as a predictor of permeability. Brain Res Bull 15:287–292

Banks WA, Kastin AJ (1990) Peptide transport systems for opiates across the blood-brain barrier. Am J Physiol 259:E1–E10
Banks WA, Kastin AJ, Coy DH (1984) Evidence that [^{125}I-N-Tyr]-delta-sleep-inducing peptide crosses the blood-brain barrier by a non-competitive mechanism. Brain Res 301:201–207
Banks WA, Kastin AJ, Horvath A, Michals EA (1987) Carrier-mediated transport of vasopressin across the blood-brain barrier of the mouse. J Neurosci Res 18:326–332
Banks WA, Kastin AJ, Fasold MB (1988) Differential effects of aluminum on the blood-brain barrier transport of peptides, technecium and albumin. J Pharmacol Exp Ther 244:579–585
Banks WA, Kastin AJ, Durham DA (1989) Bidirectional transport of interleukin-1 alpha across the blood-brain barrier. Brain Res Bull 23:433–437
Banks WA, Kastin AJ, Michals EA, Barrera CM (1990a) Stereospecific transport of Tyr-MIF-1 across the blood-brain barrier by peptide transport sytem 1. Brain Res Bull 25:589–592
Banks WA, Schally AV, Barrera CM, Fasold MB, Durham DA, Csernus VJ, Groot K, Kastin AJ (1990b) Permeability of the murine blood-brain barrier to some octapoptide analogs of somatostatin. Proc Natl Acad Sci USA 87: 6762-6766
Baranczyk-Kuzma A, Audus KL (1987) Characteristics of aminopeptidase activity from bovine brain microvessel endothelium. J Cereb Blood Flow Metab 7: 801–805
Barancyk-Kuzma A, Raub TJ, Audus KL (1989) Demonstration of acid hydrolase activity in primary cultures of bovine brain microvessel endothelium. J Cereb Blood Flow Metab 9:280–289
Barrera CM Kastin AJ, Banks WA (1987) D-[Ala^1]-peptide T-amide is transported from blood to brain by a saturable system. Brain Res Bull 19:629–633
Barrera CM, Banks WA, Kastin AJ (1989) Passage of Tyr-MIF-1 from blood to brain. Brain Res Bull 23:439–442
Begley DJ (1990) Interactions of the lipid-soluble drug cyclosporin with the blood-brain barrier of the anaesthetized guinea-pig. J Physiol (Lond) 423:39P
Begley DJ, Chain DG (1992) Mechanisms regulating peptide levels in the cerebrospinal fluid. In: Segal MB (ed) The fluids and barriers of the eye and brain. Macmillan, Basingstoke, pp 82–105
Begley DJ, Zloković BV (1986) Neuropeptides and the blood-brain barrier. In: Suckling AJ, Rumsby MG, Bradbury MWB (eds) The blood-brain barrier in health and disease. Ellis Horwood, Chichester VCH, New York, pp 98–108
Begley DJ, Squires LK, Zloković BV, Mitrović DM, Hughes CCW, Revest PA, Greenwood J (1990) Permeability of the blood-brain barrier to the immunosuppressive cyclic peptide cyclosporin A. J Neurochem 55:1222–1230
Betz AL, Gilboe DD, Yudilevich DL, Drewes LR (1973) Kinetics of unidirectional glucose transport into the isolated dog brain. Am J Physiol 225:586–592
Betz AL, Csejtey J, Goldstein GW (1979) Hexose transport and phosphorylation by capillaries isolated from rat brain. Am J Physiol 236:C96–C102
Blasberg RG, Fenstermacher JD, Patlak CS (1983) Transport of α-aminoisobutyric acid across brain capillary and cellular membranes. J Cereb Blood Flow Metab 3:8–32
Bradbury MWB (1979) The concept of a blood-brain barrier. Wiley, Chichester, pp 137–140
Broadwell R (1989) Transcytosis of macromolecules through the blood-brain barrier: a cell biological perspective and critical appraisal. Acta Neuropathol (Berl) 79:17–128
Broadwell R, Wolf A, Tangoren M (1989) Transcytosis of blood-borne ferrotransferrin and insulin through the blood-brain barrier. Soc Neurosci Abstr 15:821

Brust P (1986) Changes in regional blood-brain transfer of L-leucine elicited by arginine-vasopressin. J Neurochem 46:534–541
Brust P, Zicha J (1988) Kinetics of regional blood-brain barrier transport of L-leucine in Brattleboro rats. Biomed Biochim Acta 47:1013–1021
Bugge J (1974) The cephalic arteries of hystricomorph rodents. Symp Zool Soc Lond 43:61–78
Butt AM, Jones HC, Abbott NJ (1990) Electrical resistance across the blood-brain barrier in anaesthetised rats: a developmental study. J Physiol (Lond) 429: 47–62
Cefalu WT, Pardridge WM (1985) Restrictive transport of a lipid-soluble peptide (cyclosporin) through the blood-brain barrier. J Neurochem 45:1954–1956
Christensen H (1979) Developments in amino acid transport illustrated for the blood-brain barrier. Biochem Pharmacol 28:1989–1992
Clark HB, Hartman BK Raichle ME, Preskorn SH, Larson KB (1981) Measurement of cerebral vascular extraction fractions in the rat using intracarotid injection techniques. Brain Res 208:311–323
Cornford EM, Braun LD, Crane PD, Oldendorf WH (1978) Blood-brain barrier restriction of peptides and the low uptake of enkephalins. Endocrinology 103: 1297–1303
Cox BF, Hay M, Bishop VS (1990) Neurons in area postrema mediate vasopressin-induced enhancement of the baroreflex. Am J Physiol 258:H1943–H1946
Crone C (1986a) The blood-brain barrier: a modified tight epithlium. In: Suckling AJ, Rumsby MG, Bradbury MWB (eds) The blood-brain barrier in health and disease. Ellis Horwood, Chichester VCH, New York, pp 17–40
Crone C (1986b) Modulation of solute permeability in microvascular endothelium. Fed Proc 45:77–83
Crone C, Olesen SP (1981) The electric resistance of brain capillary endothelium. J Physiol (Lond) 316:53P–54P
Crone C, Thompson AM (1970) Permeability of brain capillaries. In: Crone, C, Lassen NA (eds) Capillary permeability. Munksgaard, Copenhagen, pp 447–453
Daniel PM, Love ER, Pratt OE (1975) Insulin and the way the brain handles glucose. J Neurochem 25:471–476
Deane R, Segal MB (1985) The transport of sugars across the perfused choroid plexus of the sheep. J Physiol (Lond) 362:245–260
Dehouck M-P, Meresse PD, Fruchart J-C, Cecchelli R (1989) An easier, reproducible, and mass-production method to study the blood-brain barrier in vitro. J Neurochem 54:1798–1801
De Montis MG, Olianas MC, Haber B, Tagliamonte A (1978) Increase in large neutral amino acid transport into brain by insulin. J Neurochem 30:121–124
De Wildt D, Verhoef J, Witter A (1982) H-Pro-[^{2}H]Leu-Gly NH_2: uptake and metabolism in rat brain. J Neurochem 38:67–74
Duffy KR, Pardridge WM (1987) Blood-brain barrier transcytosis of insulin in developing rabbits. Brain Res 420:32–38
Ellison MD, Povlishock JT, Merchant RE (1987) Blood-brain barrier dysfunction in cats following recombinant interleukin-2 infusion. Cancer Res 47:5765–5770
Ermisch A, Rühle H-J, Neubert K, Hartrodt K, Landgraf R (1983) On the blood-brain barrier to peptides: [^{3}H]-beta-casomorphin-5 uptake by eighteen brain regions in vivo. J Neurochem 41:1229–1223
Ermisch A, Rühle H-J, Klauschenz E, Kretzschmar R (1984) On the blood-brain barrier to peptides: [^{3}H]-gonadotropin releasing hormone accumulation by eighteen regions of the rat brain and by anterior pituitary. Exp Clin Endocrinol 84:112–116
Ermisch A, Barth T, Rühle H-J, Skopkova J, Hibas P, Landgraf R (1985a) On the blood-brain barrier to peptides: accumulation of radioactive labelled vasopressin, desGlyNH_2-vasopressin and oxytocin by brain regions. Endocrinol Exp 19:29–37
Ermisch A, Rühle H-J, Landgraf R, Hess J (1985b) Blood-brain barrier and peptides. J Cereb Blood Flow Metab 5:350–358

Ermisch A, Landgraf R, Brust P, Kretzschmar R, Hess J (1988) Peptide receptors of the cerebral capillary endothelium and the transport of amino acids across the blood-brain barrier. In: Rakić LJ, Begley DJ, Davson H, Zloković BV (eds) Peptide and amino acid transport mechanisms in the central nervous system. Macmillan Stockton, Basingstoke, pp 41–53

Eschalier A, Ardid D, Aumaître O, Fialip J, Duchêne-Marullaz P (1990) Central long-lasting analgesic effect of RC 160, an analog of somatostatin. Eur J Pharmacol 183:1067

Fitzsimons JT (1980) Angiotensin stimulation of the central nervous system. Rev Phys Biochem Pharmacol 87:117–149

Frank HJL, Pardridge WM (1981) A direct in vitro demonstration of insulin binding to isolated brain microvessels. Diabetes 30:757–761

Frank HJL, Jankovic-Vokes T, Pardridge WM, Morris WL (1985) Enhanced binding to blood-brain barrier in vivo and to brain microvessels in vitro in newborn rabbits. Diabetes 43:728–733

Frank HJL, Pardridge WM, Jankovic-Vokes T, Vinters HV, Morris WL (1986a) Insulin binding to the blood-brain barrier in the streptozotocin diabetic rat. J Neurochem 47:405–411

Frank HJL, Pardridge WM, Morris, WL, Rosenfeld RG, Choi TB (1986b) Binding and internalization of insulin and insulin-like growth factors by isolated brain microvessels. Diabetes 35:654–661

Goldman H, Murphy S (1981) An analog of ACTH/MSH4-9, ORG-2766, reduces permeability of the blood-brain barrier. Pharmacol Biochem Behav 14:845–848

Goldstein GW, Wolinsky JS, Csejtey J, Diamond I (1975) Isolation of metabolically active capillaries from rat brain. J Neurochem 25:715–717

Greenberg R, Whalley CE, Jourdikian F, Mendelson IS, Walter R, Nikolics K, Coy DH, Schally AV, Kastin AJ (1976) Peptides readily penetrate the blood-brain barrier: uptake by synaptosomes is passive. Pharmacol Biochem Behav 5 [Suppl 1]:151–158

Guillot FL, Raub TJ, Audus KL (1987) Fluid phase endocytosis by bovine brain capillary endothelial cells in vitro. Cell Biol 105:312a

Guillot FL, Audus KL, Raub TJ (1990) Fluid-Phase endocytosis by primary cultures of bovine brain microveobsel endothelial cell monolayers. Microvasc Res 39:1–14

Hambrook JM, Morgan BA, Rance MJ, Smith CFC (1976) Mode of inactivation of the enkephalins by rat and human plasma and rat brain homogenates. Nature 262:782–783

Harding JW, Sulliva MJ, Hanesworth JM, Cushing LL, Wright JW (1988) Inability of [^{125}I]-Sar^{1}-Ile^{8}-Angiotensin II to move between the blood and cerebrospinal fluid compartments. J Neurochem 50:554–557

Hertz MM, Paulson OB, Barry DI, Christiansen JS, Svendsen PA (1981) Insulin increases glucose transfer across the blood-brain barrier in man. J Clin Invest 67:597–604

Hoffman PL, Walter R, Bulat M (1977) An enzymatically stable peptide with activity in the central nervous system: its penetration through the blood-CSF barrier. Brain Res 122:87–94

Hughes CCW, Lantos PL (1989) Uptake of leucine and alanine by cultured cerebral cpaillary endothelial cells. Brain Res 480:126–132

Hyman S, Lipovac MN, McComb JG, Tang G, Zloković BV (1990) Kinetic analysis of vasopressin-arginine uptake at the luminal side of the blood-brain barrier studied in an in situ perfused brain of the anaesthetised guinea-pig. J Physiol (Lond) 423:37P

Ibaragi M, Niwa M, Ozaki M (1989) Atrial natriuretic peptide modulates amiloride-sensitive Na^+ transport across the blood-brain barrier. J Neurochem 53: 1802–1806

Kastin AJ, Fabre LA (1982) Limitations to effect of α-MSH on permeability of blood-brain barrier to i.v. ^{99m}Tc-pertechnetate. Pharmacol Biochem Behav 17:1199–1201

Kastin AJ, Nissen C, Schally AV, Coy DH (1976) Blood-brain barrier, half-time disappearance, and brain distribution for labeled enkephalin and a potent analog. Brain Res Bull 1:583–589

Kastin AJ, Nissen C, Schally AV, Coy DH (1979) Additional evidence that small amounts of a peptide can cross the blood-brain barrier. Pharmacol Biochem Behav 11:717–719

Kastin AJ, Banks WA, Castellanos PF, Nissen C, Coy DH (1982) Differential penetration of DSIP peptides into rat brain. Pharmacol Biochem Behav 17: 1187–1191

Krenning EP, Breeman WAP, Kooij PPM, Lameris JS, Bakker WH, Koper JW, Ausema L, Reubi JC, Lamberts SWJ (1989) Localisation of endocrine-related tumours with radioiodinated analogue of somatostatin. Lancet i:242–243

Kretzschmar R, Ermisch A (1987) Arginine vasopressin binding to isolated cerebral microvessels. Wiss Z Karl Marx Univ Leipzig Math-Naturwiss R 36(1):78–80

Lamberts SWJ (1986) Non-pituitary action of somatostatin. A review on the therapeutic role of SMS 201–995 (Sandostatin). Acta Endocrinol 112 [Suppl 276]:41–55

Lamberts SWJ (1987) A guide to the clinical use of the somatostatin analogue SMS 201–995 (Sandostatin). Acta Endocrinol 116 [Suppl 286]:54–66

Landgraf R, Klauschenz E, Bienert M Ermisch A, Oehma P (1983) Some observations indicating a low brain uptake of [^{3}H]-Nle^{11}-substance P. Pharmazie 38:108–110

Lemaire M, Tillement JP (1982) Role of lipoproteins and erythrocytes in the in vitro binding and distribution of cyclosporin A in the blood. J Pharm Pharmacol 34:715–718

Lorenzo AV, Winston KR, Adler J (1988) The uptake by choroid plexus and passage of ^{125}I-Prolactin in preterm rabbits. In: Rakić LJ, Begley DJ, Davson H, Zlokovic BV (eds) Peptide and amino acid transport mechanisms in the central nervous system. Macmillan/Stockton, Basingstoke, pp 67–78

Lund-Andersen H (1979) Transport of glucose from blood to brain. Physiol Rev 59:305–352

Margolis RU, Altszuler N (1967) Insulin in the cerebrospinal fluid. Nature 215: 1375–1376

Matthews DM (1972) Rates of peptide uptake by small intestine. In: Elliot K, O'Connor A (eds) Peptide transport in bacteria and mammalian gut. Ciba Foundation Symposium. Elsevier, Amsterdam, pp 71–88

Mens WBJ, Witter A, van Wimersma-Greidanus T (1983) Penetration of neurohypophysial hormones from plasma into cerebrospinal fluid: half-times of disappearance of these neuropetides from CSF. Brain Res 262:143–149

Ohno K, Pettigrew KD, Rapoport SI (1978) Lower limits of cerebrovascular permeability to non-electrolytes in the conscious rat. Am J Physiol 235: H229–H307

Oldendorf WH (1970) Measurement of brain uptake of radiolabelled substances using a tritiated water internal standard. Brain Res 24:372–376

Oldendorf WM (1971) Brain uptake of radio-labeled amino acids, amines and hexoses after arterial injection. Am J Physiol 221:1629–1639

Oldendorf WM (1981) Blood-brain barrier permeability to peptides, pitfalls in measurement. Peptides 2 [Suppl 2]:109–111

Oldendorf WH, Braun LD (1976) [^{3}H]-Tryptamine and [^{3}H]-water as diffusible internal standards for measuring brain extraction of radio-labeled substances following carotid injection. Brain Res 113:219–224

Olesen SP, Crone C (1983) Electrical resistance of muscle capillary endothelium. Biophys J 42:31–41

Pardridge WM (1986) Mechanisms of neuropeptide ineraction with the blood-brain barrier. Ann NY Acad Sci 481:231–249

Pardridge WM, Mietus LJ (1981) Enkephalin and blood-brain barrier: studies of binding and degradation in isolated brain microvessels. Endocrinology 109:1138–1143

Pardridge WM, Mietus LJ (1982) Kinetics of neutral amino acid transport through the blood-brain barrier of the newborn rabbit. J Neurochem 38:955–962
Pardridge WM, Fierer G (1985) Blood-brain barrier transport of butanol and water relative to N-isopropyl-p-iodoamphetamine as the internal reference. J Cereb Blood Flow Metab 5:275–281
Pardridge WM, Eisenberg J, Yamada T (1985a) Rapid sequestration and degradation of somatostatin by isolated brain microvessels. J Neurochem 44:1178–1184
Pardridge WM, Eisenberg J, Yang J (1985b) Human blood-brain barrier insulin receptor. J Neurochem 44:1771–1778
Pardridge WM, Landaw EM, Miller LP, Braun LD, Oldendorf WM (1985c) Carotid artery injection technique: bounds for bolus mixing by plasma and by brain. J Cereb Blood Flow Metab 5:576–583
Patlak CS, Blasberg RG, Fenstermacher JD (1983) Graphical evaluation of blood-to-brain transfer constants from multiple-time uptake data. J Cereb Blood Flow Metab 3:1–7
Pollay M, Stevens A, Estrada E, Kaplan R (1972) Extracorporeal perfusion of choroid plexus. J Appl Physiol 32:612–617
Raeissi S, Audus KL (1989) In-vitro characterization of blood-brain barrier permeability to delta sleep-inducing peptide. J Pharm Pharmacol 41:848–852
Rakić Lj, Zloković BV, Davson H, Segal MB, Begley DJ, Lipovac MN, Mitrović DM (1989) Chronic amphetamine intoxication and the blood-brain barrier to inert polar molecules studied in the vascularly perfused guinea pig brain. J Neurol Sci 94:41–50
Ramlakhan N (1990) Albumin binding and endocytosis by cultured rat brain endothelium. J Physiol (Lond) 423:34P
Rapoport SI, Klee WA, Pettigrew KD, Ohno K (1980) Entry of opioid peptides into the central nervous system. Science 207:84–86
Reid IA (1984) Actions of angiotensin II on the brain: mechanisms and physiological role. Am J Physiol 206:F533–F543
Reith J, Ermisch A, Diemer NH, Gjedde A (1987) Saturable retention of vasopressin by hippocampus vessels in vivo, associated with inhibition of blod-brain transfer of large neutral amino acids. J Neurochem 49:1471–1479
Reubi JC, Lang W, Maurer R, Koper JW, Lamberts SWJ (1987) Distribution and biochemical characterisation of somatostatin receptors in tumours of the human central nervous system. Cancer Res 47:5758–5764
Revest PA, Greenwood J, Abbott NJ (1992) Amino acid transport by rat brain capillary endothelial cells in culture. J Neurochem (in press)
Richerson GB, Getting PA (1990) Preservation of integrative function in a perfused guinea pig brain. Brain Res 517:7–18
Rudman D, Kutner MH (1978) Melanotropic peptides increase permeability of plasma/cerebrospinal fluid barrier. Am J Physiol 234:E327–E332
Sankar R, Domer F, Kastin AJ (1981) Selective effects of α-MSH and MIF-1 on the blood-brain barrier. Peptides 2:345–347
Schally AV, Cai R-Z, Torres-Aleman I, Redding TW, Szoke B, Fu D, Hierowski MT, Konturek S (1986) Endocrine, gastrointestinal and antitumor activity of somatostatin analogues. In: Moody TW (ed) Neural and endocrine peptides and receptors. Plenum, New York, pp 73–88
Schwartz NW, Sipols A, Kahn SE, Latteman DF, Taborsky GJ, Bergman RN, Woods SC, Porte D (1990) Kinetics and specificity of insulin uptake from plasma into cerebrospinal fluid. Am J Physiol 259:E378–E383
Smith KR, Borchardt RT (1984) Permeability and mechanism of albumin transport across blood-brain capillary endothelial cells. Pharm Res 4:S–41
Smith QR, Takasato Y, Rapoport SI (1984) Kinetic analysis of L-leucine transport across the blood-brain barrier. Brain Res 311:167–170
Smith QR, Momma S, Aoyagi M, Rapoport SI (1987) Kinetics of neutral amino acid transport across the blood-brain barrier. J Neurochem 49:1651–1658

Squires LK (1990) A study of insulin in plasma and cerebrospinal fluid by high performance liquid chromatography. PhD Thesis, University of London
Takasato Y, Rapoport SI, Smith QR (1984) An in situ brain perfusion technique to study cerebrovascular transport in the rat. Am J Physiol 247:H484–H493
Takasato Y, Momma S, Smith QR (1985) Kinetic analysis of cerebrovascular isoleucine transport from saline and plasma. J Neurochem 45:1013–1020
Triguero D, Buciak J, Pardridge WM (1990) Capillary depletion method for quantification of blood-brain barrier transport of circulating peptides and plasma proteins. J Neurochem 54:1882–1888
Van Houten M, Posner BI, Kopriwa BM, Brawer JR (1979) Insulin binding sites in the rat brain: in vivo localization to the cricumventricular organs by quantitative radioautography. Endocrinology 105:666–673
Verheugen C, Laufer IR, De Facio H, Pardridge WB, Lu JK, Judd HL (1983) Impermeability of the rat blood-brain barrier to exogenously administered gonadotropin-releasing hormone. Neuroendocrinology 36:102–104
Walsh RJ, Slaby F, Posner BI (1987) Prolactin transport from blood to cerebrospinal fluid: a receptor-mediated process. Wiss Z Karl Marx Univ Leipzig Math-Naturwiss R 36:(1), 119
Walter R, Ritzmann RF, Bhargave HN, Flexner LB (1979) Prolyl-leucyl-glycinamide, (cyclo-leucylglycine), and derivatives block development of physical dependence on morphine in mice. Proc Natl Acad Sci USA 76:518–520
Wilson JF, Anderson AG, Snook G, Llewellyn KK (1984) Quantification of the permeability of the blood-CSF barrier to α-MSH in the rat. Peptides 5:681–685
Woods SC, Porte D (1977) Relationship between plasma and cerebrospinal fluid insulin levels of dogs. Am J Physiol 233:E331–E334
Yudilevich DL, Mann GE (1982) Unidirectional uptake of substrates at the blood-tissue interface of secretory epithelia: stomach, salivary gland and pancreas. Fed Proc 41:3045–3053
Zadina JE, Banks WA, Kastin AJ (1986) Central nervous system effects of peptides, 1980–1985: a cross-listing of peptides and their central actions from the first six years of the journal Peptides. Peptides 7:497–537
Zeuthen T, Wright EM (1981) Epithelial potassium transport: tracer and electrophysiological studies in choroid plexus. J Membr Biol 60:105–128
Zloković BV, Begley DJ, Chain DG (1983) Blood-brain barrier permeability to dipeptides and their constituent amino acids. Brain Res 271:66–71
Zloković BV, Begley DJ, Chain-Eliash DG (1985a) Blood-brain barrier permeability to leucine enkephalin, D-alanine2-D-leucine 5-enkephalin and their N-terminal amino acid (Tyrosine). Brain Res 336:125–132
Zloković BV, Segal MB, Begley DJ (1985b) Permeability of the isolated choriod plexus of the sheep to thyrotropin-releasing hormone. In: Yudilevich DL, Mann GE (eds) Carrier mediated transport of solutes from blood to tissue. Longman, London, pp 307–312
Zloković BV, Segal MB, Begley DJ, Davson H, Rakić Lj (1985c) Permeability of the blood-cerebrospinal fluid and blood-brain barriers to thyrotropin releasing hormone. Brain Res 358:191–199
Zloković BV, Begley DJ, Djuričić BM, Mitrović D (1986) Measurement of solute transport across the blood-brain barrier in the perfused guinea pig brain: Method and application to N-methyl-α-aminoisobutyric acid. J Neurochem 46:1444–1451
Zloković BV, Lipovac MN, Begley DJ, Davson H, Rakić LJ (1987) Transport of leucine-enkephalin across the blood-brain barrier in the perfused guinea pig brain. J Neurochem 49:310–315
Zloković BV, Lipovac MN, Begley DJ, Davson H, Rakić Lj (1988a) Slow penetration of thyrotropin releasing hormone across the blood-brain barrier of an in situ perfused guinea pig brain. J Neurochem 51:252–257

Zloković BV, Segal MB, Davson H, Jankov RM (1988b) Passage of delta sleepinducing peptide (DSIP) across the blood-cerebrospinal fluid barrier. Peptides 9:533–538

Zloković BV, Segal MB, Davson H, Mitrović DM (1988c) Unidirectional uptake of enkephalins at the blood-tissue interface of the blood-cerebrospinal fluid barrier; a saturable mechanism. Regul Pept 20:33–44

Zloković BV, Begley DJ, Segal MB, Davson H, Rakić LJ, Lipovac MN, Mitrović DM, Jankov RM (1988d) Neuropeptide transport mechanisms in the central nervous system. In: Rakić LJ, Begley DJ, Davson H, Zloković BV (eds) Peptide and amino acid transport mechanisms in the central nervous system. Macmillan Basingstoke, Stockton, pp 3–19

Zloković BV, Mačkić JB, Djuričić, Davson H (1989a) Kinetic analysis of leucine-enkephalin cellular uptake at the luminal side of the blood-brain barrier of an in situ perfused guinea-pig brain. J Neurochem 53:1333–1340

Zloković BV, Sušić VT, Davson H, Begley DJ, Jankov RM, Mitrović DM, Lipovac MN (1989b) Saturable mechanism for delta sleep-inducing peptide (DSIP) at the blood-brain barrier of the vascularly perfused guinea pig brain. Peptides 10: 249–254

CHAPTER 7

The Movement of Vitamins Into the Brain

O.E. PRATT

The brain needs all of the vitamins, with the possible exception of K and D, a need emphasised by the neurological disturbances associated with deficiencies of almost any of the B-group vitamins. How they move into the brain is thus of some importance. There are different possible routes and various mechanisms by which vitamins cross various barriers between the blood and brain cells.

A. Routes

The anatomical routes by which vitamins reach the brain (Fig. 1) show three alternative paths between the blood and the brain cells. There are permeability barriers or other restrictions which limit the flow along these routes. Such restrictions exclude unwanted solutes and help regulate the concentration of the nutrients, a form of homeostasis to maintain an optimal environment within brain cells. To traverse any of the routes (Fig. 1a–c) nutrients have to cross either the brain capillary endothelium or the cells of the choroid plexus.

I. The Trans-Capillary Route

Nutrients other than vitamins cross the cerebral capillaries by carrier facilitated transport. That the brain's nutritional needs can be supplied mainly, possibly entirely, by this route (Fig. 1b) follows from the facts that the surface area of the capillaries is very much larger than that of the choroid plexuses and the blood flow several orders of magnitude larger than the rate of cerebrospinal fluid formation. It was concluded, therefore, that the trans-capillary route was more important than the choroid plexus–cerebrospinal fluid route (Fig. 1c) (PRATT 1979a; PARDRIDGE 1984) provided that carriers exist to transport the nutrient across the endothelial cell barrier.

II. Other Possible Routes

It has seemed that some vitamins were exceptions to this general rule. In order to establish that the choroid plexus route (Fig. 1c) is an important one

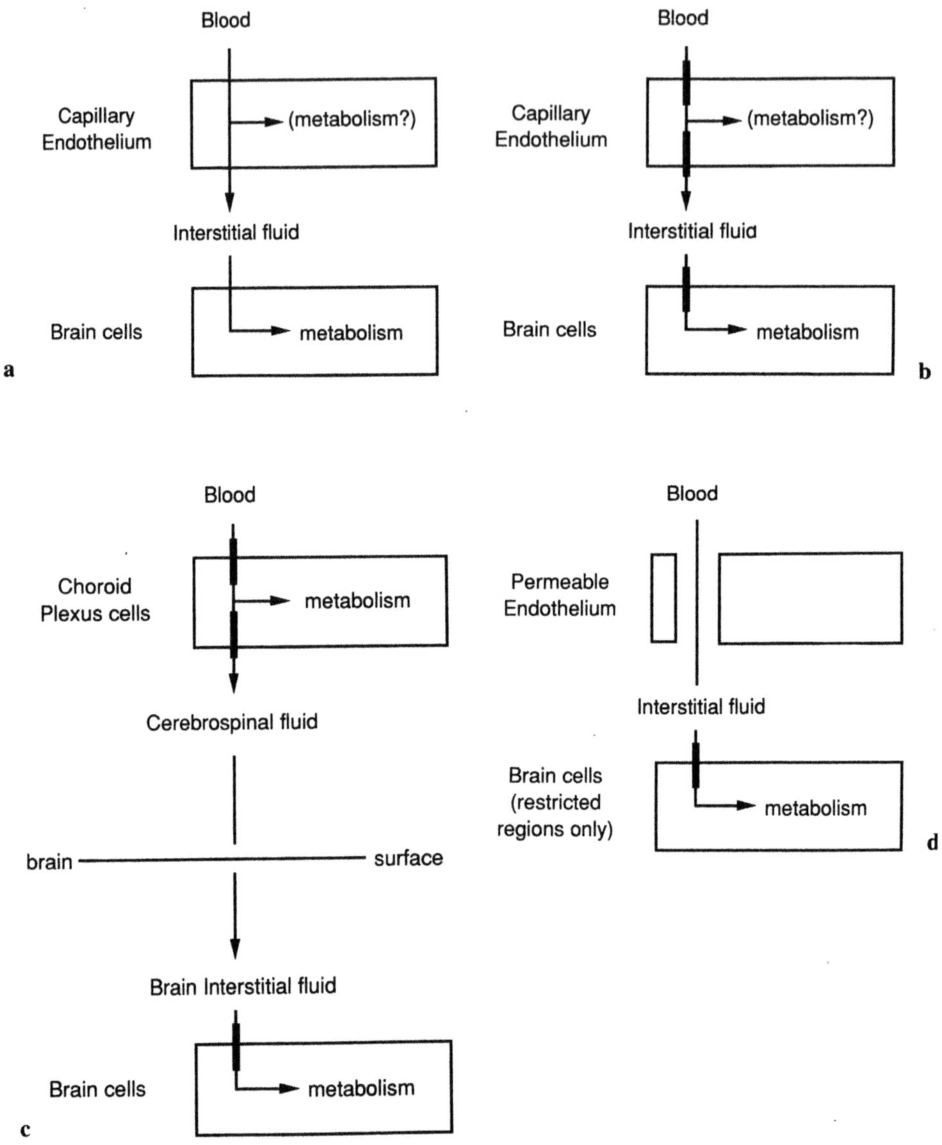

Fig. 1a–d. Anatomical routes by which vitamins reach the brain. **a** The brain capillary route for lipid-soluble nutrients not needing carrier-mediated transport. **b** The brain capillary route with carrier-mediated transport across the capillary endothelium (*thickened lines* represent carrier-mediated steps). **c** The choroid plexus cerobrospinal fluid route which may be important for any vitamins which cannot cross the capillary endothelium by carrier mediated transport (*thickened lines* represent carrier-mediated steps). **d** The rather special route across brain capillaries lacking tight junctions, enabling vitamins to enter a few small special brain regions in the hypothalamus and brain stem. This route is not thought to contribute appreciably to the supply of vitamins to other parts of the brain (*thickened line* indicates carrier-mediated uptake by brain cells)

for the supply of any particular vitamin to the brain it must be shown that the micronutrient actually reaches the cerebrospinal fluid, not just that it is taken up by choroid plexus cells, for the plexus cells may be doing this only to supply their own metabolism. It must be shown not only that the vitamin crosses the choroid plexus but also that transport carriers are lacking in the brain capillaries. Because the capillary surface area is so much larger than that of the choroid plexuses, a relatively low flux across the capillaries may provide the major supply route for a nutrient. To detect the existence of brain capillary transport systems for vitamins, it is essential to use the most sensitive flux measurement methods, namely those which depend upon continuous injection protocols (PRATT 1985).

Another possible pathway (Fig. 1d) is through regions where the blood-brain barrier is defective, especially the median eminence and area postrema, the blood capillaries lacking the usual tight junctions. These capillaries however, supply only a very small part of the total brain and it seems that solutes do not easily diffuse out into other regions of the brain. One function of such regions may be to monitor the concentrations of solutes like glucose in the circulation. This route does not appear to have any importance for vitamin supply to the main bulk of the brain.

B. Mechanisms

There are different ways in which vitamins or other micronutrients cross the permeability barriers in the cerebral capillary endothelium or choroid plexus. Non-carrier mediated processes are only important for solutes with a high lipid/water partition coefficient or when blood concentrations are high: they do not "saturate", that is, the flux increases in proportion, however high the solute concentration.

I. Passive Diffusion

Passive diffusion across the capillary endothelium is the route travelled by the lipid-soluble vitamins A, D and α-tocopherol. They all have high lipid-water partition coefficients, readily crossing the permeability barriers and, by taking advantage of the widespread capillary network, easily reaching brain cells by this route (Fig. 1a).

II. Carrier-Mediated Transport

Carrier-mediated transport across the capillary endothelium enables nutrients other than vitamins (e.g. glucose and amino acids) and most vitamins, too, to reach brain cells. Energy coupling is usually lacking – it is facilitated diffusion, that is, vitamins move down a concentration gradient which is maintained by active transport into brain cells (Fig. 1b). The choroid plexus cells are also well provided with carrier-mediated transport systems (Fig. 1c) which, in contrast to those on the capillary endothelium, are energy driven and can pump against a concentration gradient, resembling in this respect the carriers which transport vitamins into brain cells. The transport at all three sites may be subject to competitive inhibition from chemical analogues of the vitamin.

III. Binding Proteins

Some nutrients are strongly bound in body fluids to specific binding proteins. The B-group vitamins in this category are the cobalamines, the folates, pyridoxal and riboflavin. It is possible that mechanisms exist for moving the specific binding protein, together with the attached vitamin, across the blood-brain barriers but little information is available.

IV. Covalently Bound Carriers

For carrier-mediated transport, an enzyme-like carrier is anchored in the endothelial cell surface and the nutrient interacts with the carrier only transiently. In contrast, a covalently bound carrier is combined chemically with the vitamin on a more enduring basis. In addition to the specific binding proteins (see above) the vitamin may be linked chemically to a lipophilic group and such a process has been used for example to make lipid soluble derivatives of thiamine for therapeutic use (GRAUL et al. 1967; THOMSON et al. 1971).

C. The Individual Vitamins

The vitamins are essential micronutrients, which mostly function as cofactors for enzymes. Considered also are choline, needed in rather larger amounts and nucleic acid precursors which the brain has limited ability to synthesise. The relative importance of the different pathways and mechanisms varies somewhat among the different vitamins and each of them will be considered in turn.

I. Thiamine

There is general agreement that the major route for thiamine supply to the brain is by carrier-mediated transport across the capillary endothelium. Any contribution from the choroid plexus route must be a minor one. The affinity for thiamine of the transport carrier in the choroid plexus (SPECTOR 1976) is of the order of 1000 times lower than that for this carrier in the brain capillaries and other sites (GREENWOOD et al. 1982). It is not clear whether the choroid plexus only contains a low affinity transport carrier which will be far from saturated at normal blood thiamine levels and therefore largely ineffective or whether the apparent low affinity is an artifact due to possible competitive inhibition of the transport carrier (which if it occurred, passed unrecognised) by the pyrithiamine used to prevent phosphorylation of the tracer thiamine. Pyrithiamine is a highly effective inhibitor of thiamine transport systems. Even if the flux across the choroid plexus capillary were of a similar order to that across the brain capillaries the effect would be comparatively negligible since the surface area of brain capillaries is some 5000 times that of choroid plexus capillaries (RAICHLE 1983). Since there are likely to be physical limits to the number of carrier units per unit capillary surface area, it seems reasonable to assume that the alternative route is only likely to be comparatively important for substances of which the transport across the brain capillaries is negligibly small.

Thiamine crosses the endothelial surface of the blood-brain barrier in the free form mainly by a saturable, carrier-mediated process but the transport also has an apparently non-saturable component, as well as a much lower affinity carrier which transports the monophosphate (GREENWOOD et al. 1982; REGGIANI et al. 1984). The blood-brain barrier transport resembles the two component process which carries thiamine across the intestinal border (HOYUMPA et al. 1982). It is likely that the apparently non-saturable minor component of thiamine transport is not so much passive diffusion as transport by carriers not specific for thiamine. Probably, these carriers normally transport other substrates, chemically resembling thiamine in some respects. Such carriers would be expected to have a low affinity, that is a high K_m and a low V_{max} for thiamine, giving an apparently non-saturable kinetic pattern indistinguishable from passive diffusion.

For each carrier the Michaelis equation for the flux, V, of a solute of concentration, S, in the blood plasma is given by:

$$V = \frac{V_{max}}{K_m + S} S$$

For a non-specific carrier with K_m much greater than S, the latter can be neglected and the equation becomes:

$$V = \frac{V_{max}}{K_m} S \quad \text{or} \quad V = K\,S$$

Data for thiamine flux across the blood-brain barrier (GREENWOOD et al. 1982) are found to fit an equation of the form:

$$V = \frac{V_{max}}{K_m + S} S + K\,S$$

which is the sum of the flux due to a specific carrier (the major component) and the second term (the minor component) due to either passive diffusion or, more probably, non-specific carriers.

Kinetic work has shown that the normal flux of thiamine into the brain across the blood-brain barrier (Greenwood et al. 1982) exceeds the rate of turnover of the vitamin in the brain (Rindi et al. 1980) by only a small margin. The carrier-mediated component of the transport can be inhibited competitively by chemical analogues of the vitamin. The inhibition of thiamine transport has been studied for a wide range of thiamine-like substances and gives some clues to how the vitamin is transported.

The range of potential inhibitors is wide, including simple chemical derivatives of thiamine itself, like thiamine monophosphate, acetyl thiamine or thiamine disulphide, as well as fairly close chemical analogues of the vitamin, amprolium, pyrithiamine, oxythiamine. A wider range of substances have only a partial or limited chemical similarity to the thiamine molecule but are widely distributed, e.g. adenosine and the dibasic amino acid L-arginine. Another chemical analogue of thiamine is chlormethiazole edisylate (Heminevrin) used to treat chronic alcoholism under conditions commonly associated with thiamine deficiency. Also, some potential inhibitors are believed to aggravate the effects of thiamine deficiency: delta-pyrrolinium in thiamine deficiency disease of ruminants (Edwin and Jackman 1973) and caffeic acid, present in ferns, berries and various plant products (Hilker and Somogyi 1982). A possible source of such inhibitors is provided by thiaminase, an enzyme which splits apart the two heterocyclic rings of the thiamine molecule and after chemical change can recombine the two parts (Hilker and Somogyi 1982). Thiamine analogues are formed in this way in the digestive systems of ruminants with thiamine deficiency but substances isolated from this source so far do not inhibit the transport of the vitamin.

Both thiamine monophosphate and Heminevrin produce small increases in the thiamine flux of 15 ± 9% (mean + SEM of six experiments) and 9 ± 24% (mean + SEM of three experiments) respectively which are not statistically significant. Other substances with some resemblance to the thiamine molecule, fail to inhibit thiamine flux across the blood-brain barrier significantly, namely adenosine, adenine, arginine, caffeic acid, and delta-pyrrolinium. Substantially complete inhibition of the saturable component of thiamine transport was effected by acetyl thiamine, amprolium, pyrithiamine or thiamine disulphide but neither chronic thiamine deficiency nor chronic administration of ethanol altered appreciably thiamine transport across the blood-brain barrier although acute alcohol intoxication inhibited the apparently non-saturable component, reducing its flux significantly by 42 + 7%, (mean + SEM of six experiments, $P < 0.01$, *t* test) (Greenwood and Pratt 1984, 1985).

Severe inhibition of thiamine transport across the blood-brain barrier appears to be restricted to some simple chemical derivatives, thiamine disulphide and acetyl thiamine and to certain close chemical analogues. Two analogues, amprolium (Greenwood and Pratt 1983) and pyrithiamine (Greenwood and Pratt 1985) stand out as highly competitive inhibitors of thiamine transport across the blood-brain barrier of the rat. The failure of a third chemical analogue, oxythiamine, to inhibit appreciably indicates that the amino group on the pyrimidine ring of thiamine is an essential part of the molecule from the point of view of its affinity for the transport carrier. The failure of thiamine monophosphate to inhibit thiamine transport suggests that the specificities of the different carrier systems for free and phosphorylated thiamine are high enough to preclude cross inhibition. Various claims that both chronic alcohol intake and thiamine deficiency adversely affect thiamine transport across the blood-brain barrier have not been confirmed except for the effect of acute alcoholism described above. The effect upon the supply of the vitamin to the brain of acute alcoholism reducing the non-saturable transport flux will be minimal at normal blood thiamine levels.

II. Ascorbic Acid

Early evidence from autoradiography of the brain after injection of radioactively labelled ascorbate showed that the vitamin was taken up rapidly by the choroid plexuses before entry into the brain, followed by slow penetration into the substance of the brain adjacent to the plexuses and the subarachnoid space (HAMMERSTROM 1966). Later transport work showed that the vitamin was actively accumulated by the choroid plexus and carried across into the cerebrospinal fluid against a concentration gradient by an active, energy-fed transport process (SPECTOR and LORENZO 1973). It is claimed that the predominant route by which ascorbate enters the brain from the blood is via the choroid plexus, the vitamin readily penetrating the brain substance from the cerebrospinal fluid (SPECTOR 1981). In an important paper he showed that tracer ascorbate penetrated the surface region of the rabbit brain from the cerebrospinal fluid more rapidly than mannitol. However, even 2 or 4 h after injection into the cerebrospinal fluid the tracer ascorbate concentration was still considerably higher in regions near cerebrospinal fluid surfaces. Diffusion or bulk flow into the human brain would be considerably slower than that into the rabbit brain for much of the human brain is more than a centimeter from any cerebrospinal fluid-bathed surface. The penetration would be favoured by the relatively high cerebrospinal fluid levels of the vitamin. An important more recent finding is that ascorbate does cross the cerebral capillaries by carrier-mediated transport (LAM and DANIEL 1986). Although ascorbate also crosses the choroid plexus into the cerebrospinal fluid and may enter the brain at an appreciable rate, it is not clearly established that this constitutes an important route of supply to the brain. The main arguments run as follows.

1. The concentrations of ascorbate in the cerebrospinal fluid are higher, often 4 to 5 times, than in the blood plasma but this is the expected result of its active choroid plexus transport and does not prove supply to the brain via the cerebrospinal fluid.

2. When ascorbate is administered, concentrations rise more rapidly in the cerebrospinal fluid than in the brain but this may well be the result of less efficient homeostasis of the vitamin in the cerebrospinal fluid than in the brain. That the brain ascorbate level is better regulated than that in the cerebrospinal fluid is indirect evidence that the brain's supply of the vitamin does not come predominantly from the cerebrospinal fluid route. When the blood concentration of ascorbate is high, a relatively fast, largely non-carrier mediated flux across the choroid plexus can be expected from the kinetic data of SPECTOR and LORENZO (1973).

3. The autoradiographic work shows that administered ascorbate rapidly concentrates in choroid plexuses and then moves into the brain. Rapid uptake by the choroid plexus cells is not disputed and accounts for the intense staining, some of which spreads in vivo or, possibly, during processing, to adjacent brain. What is at issue is how important this movement is under normal conditions.

4. The ascorbate transport system in the choroid plexus has a capacity some 25 times higher than that in the capillaries and an affinity for the ascorbate more than two times higher. However, this advantage is more than offset by the mass of the brain which is some 200 times that of the rat choroid plexus with an even greater disparity in the human and this leads to the conclusion that the choroid plexus route contributes less than 10% of the ascorbate reaching brain cells (LAM and DANIEL 1986). However, this is a mean value and comparison with the more than tenfold difference in tracer distributions 2 h after injecting labelled ascorbate intraventricularly between midbrain slices adjacent to the cerebrospinal fluid and other slices reported by SPECTOR (1981) leads to the conclusion that the choroid plexus route is of negligible importance for most of the human brain. However, this route may supply a substantial amount of the vitamin to brain tissue adjacent to a ventricular surface, presumably in excess of what is actually needed. If so, it is not clear what may be the implications, if any, of this somewhat anomalous circumstance.

III. Pyridoxine

The transport systems carrying this vitamin across the brain capillaries, into choroid plexus or into brain cells, and back into the blood are all carrier-mediated providing homeostasis of vitamin levels (SPECTOR 1978a,b). The rate of its uptake by choroid plexus or brain cells is regulated by pyridoxal kinase which converts the other forms to pyridoxal phosphate and this phosphorylation of the vitamin as it crosses the choroid plexus constitutes the source of the pyridoxal phosphate in the cerebrospinal fluid (SPECTOR and GREENWALD 1978). The latter form acts as the coenzyme but brain cells take up the unphosphorylated forms. The carrier-facilitated diffusion of unphosphorylated forms of the vitamin across the capillary endothelium conveys the vitamin to the fluid surrounding brain cells in the form in which they take it up. Indeed, this uptake into brain cells is inhibited by the phosphorylated form (SPECTOR 1978b). There is no evidence for dephosphorylation of the vitamin in the brain interstitial fluid.

Three conclusions can be drawn. First, carrier-mediated transport efficiently regulates the vitamin B_6 levels. Secondly, the energy-dependent uptake of the vitamin by the choroid plexus cells not only meets the vitamin needs of these metabolically active cells but also supplies phosphorylated vitamin to the cerebrospinal fluid. Finally, the needs of the brain cells for vitamin B_6 are met mainly, if not entirely, by the vitamin crossing the brain capillaries in the unphosphorylated form.

IV. Folates

Deficient supply of folates to the brain impairs the synthesis of the methyl donor *S*-adenosyl methionine which is needed for the synthesis of the central neurotransmitter serotonin (ALGERI et al. 1979). The markedly higher concentrations of folates in the cerebrospinal fluid than in the blood plasma (REYNOLDS et al. 1972) indicate active transport across the choroid plexus. Interpretation of transport is made difficult by metabolism of this vitamin, which is reduced to di- and tetra-hydrofolate, the latter being methylated to provide the active form for one-carbon transfer metabolism. As BRADBURY (1979) points out, absence of the reductase enzyme from brain means that tetrahydrofolate and its metabolites must cross the blood-brain barrier rather than being formed in situ. It is the reduced forms, not folate itself, which enter the cerebrospinal fluid after intravenous injection (LEVITT et al. 1971). Indeed, it is the $N_{(5)}$-methyltetrahydrofolate which is most rapidly taken up both by brain and choroid plexus cells although folate itself is cleared from the cerebrospinal fluid to the blood by a saturable process (SPECTOR and LORENZO 1975a). The transport carrier is about half saturated at the normal blood plasma concentration but subject to inhibition by other folates (SPECTOR and LORENZO 1975b). The choroid plexus carrier is active and energy-using, tetrahydrofolate is probably methylated during transport.

The high cerebrospinal fluid/blood folate ratios provide evidence that the vitamin reaches the cerebrospinal fluid by active transport across the choroid plexus, against the "uphill" concentration gradient. This leaves open, however, the issue of how much of the brain's needs are met by the facilitated diffusion type of transport across the capillary endothelium. The concentration of tetrahydrofolates in the brain interstitial fluid is likely to be well below that in the blood or cerebrospinal fluid due to their rapid uptake into brain cells, helping flux across the brain capillary bed.

V. Vitamin B_{12}

The importance of two specific binding proteins, transcobalamine I and II, for the absorption and transport of vitamin B_{12} is now firmly established (HALL 1975). These are unusually acidic proteins, isoelectric around pH 3.

Congenital deficiency of one of them leads to failure to transport the vitamin from some tissues to others (HALL 1973). The obvious assumption that a specific carrier exist to transport a transcobalamin with attached vitamin into the brain meets certain difficulties. The main problem is that the remarkable slowness of exchange between newly ingested vitamin and that in the rest of the body (more than 30 days needed) suggests that protein-bound B_{12} is not readily transported across cell boundaries. The reported transport of freshly injected B_{12} (presumably not protein-bound) across the blood-brain barrier (BHATT et al. 1980) but this may not have physiological relevance unless sufficient free vitamin is sometimes present in the circulation. Similar considerations apply to choroid plexus transport and cerebrospinal fluid concentrations of the vitamin are quite low. It seems likely that movement into or out of the CNS is very slow, a hypothesis consistent with the chronic nature of CNS lesions in deficiency states.

VI. Nicotinamide, Nicotinic Acid

Deficiency of this vitamins causes pellagra which can include dementia but there has been little interest in its transport into the brain, probably because tryptophan which readily crosses the capillaries of the brain by carrier-mediated diffusion (PRATT 1979b) can be converted into nicotinamide.

VII. Choline

The brain is not able to synthesise choline but needs it for the synthesis of complex lipids. There is a saturable transport carrier in the capillary endothelium with a K_m of 0.22 mM and a V_{max} of 6 nmol min^{-1} per gram of brain (OLDENDORF and BRAUN 1976). Efforts have been made to improve the mental status of demented patients by dietary supplementation with choline (free or combined as lecithin) but with disappointing results (BARTUS et al. 1982), probably because the transport processes taking these substances into or out of the CNS provide an effective homeostatic mechanism so that dietary supplements will only benefit patients suffering from cerebral deficiency of choline.

VIII. Nucleosides and Purine Bases

Although these substances are not usually considered as vitamins, they are micronutrients which the brain does not seem to be able to make. Relatively non-specific transport carriers for uridine (SPECTOR 1985) and other ribonucleosides are found both in the capillary endothelia and choroid plexuses but those for desoxyribonucleosides have been reported so far only at the latter site. Since uridine is not only the predominant nucleoside in blood plasma but also transported across the much larger brain capillary surface, its movement by this route will easily supply the brain's needs.

Since the brain cannot easily make other nucleosides from uridine, other nucleosides must also enter the brain, even if only slowly, possibly with a substantial contribution from the choroid plexus-cerebrospinal fluid route (Spector 1986), if transport carriers are really lacking across the capillaries.

IX. Other B-Vitamins

Both riboflavin (Spector 1980) and pantothenic acid (Spector 1986) enter the brain, cerebrospinal fluid and choroid plexus by saturable transport processes which are, however, rather slow. Brain only slowly phosphorylates pantothenic acid in contrast both to its rapid phosphorylation elsewhere and the rapid phosphorylation of thiamine and pyridoxine in brain. The limited penetration of riboflavin into the cerebrospinal fluid indicates that it is transported mainly across the capillary endothelium. More rapid penetration of pantothenic acid into the cerebrospinal fluid suggests that transport by this route is possible but its relative contribution to brain cell needs has yet to be evaluated.

X. Lipophilic Vitamins

Vitamins with a high lipid/water partition coefficient will freely cross the capillary cells of the brain, despite their tight junctions, and enter the interstitial space to be taken up by brain cells.

D. Clinical Implications

Competitive inhibition of vitamin transport may exacerbate some naturally occurring nutritional deficiencies and malabsorption syndromes or complicate the effects of drugs which are vitamin analogues. In particular, kinetic work has shown that the normal flux of thiamine into the brain across the blood-brain barrier (Greenwood et al. 1982) exceeds the rate of turnover of the vitamin in the brain by only a small margin (Rindi et al. 1980). Clinicians should be aware of the possible effects of thiamine analogues. Thus, transport inhibition by analogues like amprolium or pyrithiamine is severe enough to produce or aggravate deficiencies (Greenwood and Pratt 1983, 1985). Similar considerations hold for other B-vitamins, e.g. methotrexate as a possible inhibitor of folate transport (Horne et al. 1978).

There are examples of congenital defects in the transport of a vitamin. There have been many reports of such defects involving folate transport across the gastrointestinal wall and the blood-brain barrier, which are associated with malabsorption syndromes and mental retardation (Lanzkowsky et al. 1969; Corcino et al. 1970; Lanzkowsky 1970; Makulu et al. 1973; Dougados et al. 1983) and reduced folates in the cerebrospinal fluid (Allan et al. 1983). In an interesting case of nutritional deficiency (Lever et al.

1986), psychiatric symptoms responded readily to methyl tetrahydrofolate, the form transported into the brain from the blood, by whatever route.

There is evidence that cell membrane changes in chronic alcoholism may alter blood-brain barrier transport or permeability properties, e.g. for thiamine. Fuller confirmation is needed but such effects may cause the B-vitamin deficiencies so common in alcoholism to affect the CNS and help explain the role of such deficiencies in much alcohol-related brain damage.

It is conceivable therefore that a deficiency of B-group vitamins specifically localised to the central nervous system may arise either from a defect in the blood-brain barrier transport system or from the effects of ethanol or other of as yet unrecognised substances competing with the vitamin for transport carriers at the blood-brain barrier. The list of analogues of B-group vitamins used in the treatment of various disorders is extensive (see e.g. Waxman et al. 1970). For example, many cytotoxic drugs are thought to interfere with folate metabolism and it would be interesting to know whether the transport of 5-methyl tetrahydrofolate into the brain from the blood is susceptible to inhibition by drugs like methotrexate. Again, it has long been known that treatment of human tuberculosis with cycloserine is likely to lead to a vitamin B-6 deficiency limited to the central nervous system and it would be useful to know whether the drug inhibits the transport of this vitamin across the blood-brain barrier.

Nutritional deficiencies, especially of B-vitamins, are believed to aggravate tissue damage from alcoholism (Thomson 1982) and are frequently found in alcoholics, a large proportion showing either overt or sub-clinical evidence of B-group deficiency (Morgan 1981). The treatment of such deficiencies would be impeded by the existence of any abnormality possibly undetected) affecting B-group vitamin transport across the blood-brain barrier. One possible cause of defective transport is competitive inhibition from vitamin analogues (Greenwood and Pratt 1983, 1985).

E. Homeostasis

Transport processes have a regulatory role as follows:

1. The slowness of vitamin transport across permeability barriers prevents too rapid increase in concentration in the CNS or on the other hand slows loss from the CNS in deficiency states.
2. Saturation of carrier-mediated transport limits movement of a vitamin into the CNS when the blood concentration is abnormally high.
3. Loss of micronutrients from the CNS may be limited not only by permeability barriers but also by active "one-way" transport inwards against a concentration gradient.

Other factors assist in homeostasis. Thus, losses from brain cells may be limited by slow diffusion through the tortuous brain interstitial space and metabolic conversion, especially phosphorylation, to less readily transported forms. For many vitamins, losses are reduced also because brain cells convert them to coenzymes, many of which are also firmly bound to apoenzyme proteins. Almost all of the intracellular thiamine, for example, is combined in the form of thiamine diphosphate with various thiamine-dependent enzymes.

F. Conclusions

All the vitamins appear able to cross the choroid plexus and enter the cerebrospinal fluid (Fig. 1c) although the vitamin fluxes are likely to be somewhat less than the uptake by the choroid plexus for the latter also supplies the metabolic needs of the plexus cells themselves. Because transport carriers for most nutrients are present both on the capillaries and on the choroid plexus, it is tempting to treat the choroid plexus as a continuation of the rest of the vascular system, forming with it the blood-brain barrier, jointly feeding nutrients into a complex compartment: the cerebrospinal fluid in continuity with the brain interstitial fluid. Unfortunately, such a simplification is not helpful. The carriers on the choroid plexus differ in being able to convey solutes against a concentration gradient. Also, three separate lines of evidence show that penetration of solutes into the brain is slow. First, the autoradiographic work shows a slow ascorbate penetration (Hammerstrom 1966); then after injection of protein-bound dyes there is only localised staining of small specially permeable brain regions and, thirdly, solutes entering cerebral tumours diffuse only slowly into adjacent normal brain regions (Levin et al. 1975; Deane et al. 1984). The conclusion to be drawn is that for vitamins which cross the choroid plexus, slow penetration of brain is likely to limit the overall flow by this route, rather than transport across the choroid plexus. In contrast, flux across the cerebral capillary delivers the vitamin much more directly to the interstitial fluid closely surrounding brain cells.

If a vitamin crosses the cerebral capillaries, the presumption must be that major part of the brain's needs are supplied by this route (Fig. 1b) because the surface area of the capillaries is very much larger than that of the choroid plexuses, the blood flow several orders of magnitude larger than the rate of cerebrospinal fluid formation and the vitamin is delivered directly into the fluid immediately surrounding brain cells. Thus, it is important to establish the existence or not of capillary transport systems. A lack of such carriers has not been established for any vitamin. Apart from the difficulty in principle of establishing such a negative, ascorbic acid, at first thought to lack such a carrier, has recently been shown to cross the capillaries by carrier-facilitated transport. The greater surface area and blood flow advantage of the capillary route means that a relatively low flux will be effective. Low fluxes were not always detected by older methods of transport measurement and the more sensitive continuous injection methods must be used to detect low vitamin fluxes across the capillaries. Quantitative kinetic comparisons of the two routes for thiamine and ascorbate show that the choroid plexus-cerebrospinal fluid route makes, at the most, only a minor contribution no more than 10%. Detailed comparisons are lacking for other vitamins but there is no reason to believe that the general picture is any different for any of them.

References

Algeri A, Consolazione A, Colderini G, Achilli G, Canas EP, Garattini S (1979) Effect of administration of [d-ala] methionine-enkephalin on the serotonin metabolism in rat brain. Experientia 34:1488–1489

Allan RJ, DiMauro S, Coulter DL, Papadimitriou A, Rothenburg SP (1983) Kearns-Sayre syndrome with reduced plasma and cerebrospinal fluid folate. Ann Neurol 13:679–682

Bartus RT, Dean RL, Sherman KA, Friedman F, Beer B (1982) Profound effects of combining choline and piracetam on memory enhancement and cholinergic function in aged rats. Neurobiol Ageing 2:105–111

Bhatt H, Daniel PM, Linnell JC, Love ER, Pratt OE (1980) The influx of cyancobalamin into the brain of the rat, in vivo. J Physiol (Lond) 308:88P

Bradbury WB (1979) The concept of the blood-brain barrier. Wiley, Chichester

Corcino JJ, Waxman S, Herbert, V (1970) Absorption and malabsorption of vitamin B-12. Am J Med 48:562–569

Cornford EM, Oldendorf WH (1975) Independent blood-brain barrier transport systems for nucleic acid precursors. Biochem Bio Phys Acta 394(2):211–219

Deane BR, Greenwood J, Lantos PL, Pratt OE (1984) The vasculature of experimental brain tumours: part 4. The quantification of vascular permeability. J Neurol Sci 65:59–68

Dougados M, Zittoun J, Laplane D, Castaigne P (1983) Folate metabolism disorder in Kearns-Sayre syndrome. Ann Neurol 13:687

Edwin EE, Jackman R (1973) Ruminal thiaminase and tissue thiamine in cerebrocortical necrosis. Vet Rec 92:640–641

Graul EH, Ruether W, Kovacs G (1967) Der Einfluβ von Thiamintetrahydrofurfuryldisulfid auf das Strahlensyndrom. Münch Med Wochenschr 109:2192–2197

Greenwood J, Pratt OE (1983) Inhibition of thiamine transport across the blood-brain barrier in the rat by a chemical analogue of the vitamin. J Physiol (Lond) 336:479–486

Greenwood J, Pratt OE (1984) The effect of ethanol upon thiamine transport across the blood-brain barrier in the rat. J Physiol (Lond) 348:61P

Greenwood J, Pratt OE (1985) Comparison of the effects of some thiamine analogues upon thiamine transport across the blood-brain barrier of the rat. J Physiol (Lond) 369:79–91

Greenwood J, Love ER, Pratt OE (1982) Kinetics of thiamine transport across the blood-brain barrier in the rat. J Physiol (Lond) 327:95–103

Greenwood J, Love ER, Pratt OE (1983) The effects of alcohol or of thiamine deficiency upon reproductionin the female rat and fetal development. Alcohol Alcoholism 18:45–51

Greenwood J, Pratt OE, Thomson AD (1985) Thiamine, malnutrition and alcohol-related damage to the central nervous system. In: Parvez S, Burvo Y, Parvez M, Burns E (eds) Progress in alcohol research, vol 1. VNU Science, Utrecht, pp 287–310

Hall CA (1973) Congenital disorders of vitamin B_{12} transport and their contribution to concepts. Gastroenterology 65:684–686
Hall CA (1975) Transcobalamins I and II as natural transport proteins of vitamin B_{12}. J Clin Invest 56:1125–1131
Hammerstrom L (1966) Autoradiographic studies on the distribution of C^{14}-labelled ascorbic acid and dehydroascorbic acid. Acta Physiol Scand Suppl 289:1–70
Hilker DM, Somogyi JC (1982) Antithiamines of plant origin: their chemical nature and mode of action. Ann NY Acad Sci 378:137–144
Horne DW, Briggs WT, Wagner C (1978) Transport of 5-methyltetrahydrofolic acid and folic acid in freshly isolated hepatocytes. J Biol Chem 253:3529–3535
Hoyumpa AM, Strickland R, Sheehan JJ, Yarborough G, Nichols S (1982) Dual system of intestinal thiamine transport in humans. J Lab Clin Med 99:701–708
Lam DK, Daniel PM (1986) The influx of ascorbic acid into the rat's brain. Q J Exp Physiol 71:483–489
Lanzkowsky MD (1970) Congenital malabsorption of folate. Am J Med 48:580–583
Lanzkowsky P, Erlandson ME, Bezan AI (1969) Isolated defect of folic acid absorption associated with mental retardation and cerebral calcification. Blood 34:452–465
Lever EG, Elwes RD, Williams A, Reynolds EI (1986) Subacute combined degeneration of the cord due to folate deficiency; response to methyl folate treatment. J Neurol Neurosurgery Psychiatry 49(10):1203–1207
Levin VA, Freeman-Dove M, Landahl HD (1975) Permeability chacteristics of brain adjacent to tumours in rats, Arch Neurol 32:785–791
Levitt M, Nixon PF, Pincus JL, Bertino JR (1971) Transport characteristics of folates in cerebrospinal fluid: a study utilizing doule labelled 5-methyl tetrohydrofolate and 5-formyltetrohydrofolate. J Clin Invest 50:1301–1308
Makulu DR, Smith EF, Bertino JR (1973) Lack of dihydrofolate reuctase activity in brain tissue of mammalian species: possible implications. J Neurochem 21:241–245
Morgan MY (1981) Enteral nutrition in chronic liver disease. Acta Chir Scand Suppl 507:81–90
Oldendorf WH, Braun LD (1976) [^{3}H]-tryptamine and ^{3}H water as diffusible internal standards for measuring brain extraction of radio-labelled substances following carotid injection. Brain Res 113:219–224
Pardridge WM (1984) Transport of nutrients and hormones through the blood-brain barrier. Fed Proc 43:201–204
Pratt OE (1979a) Adequate nutrition of the developing brain. In: Korobkin R, Guilleminault C (eds) Advances in perinatal neurology I. Spectrum, New York p 23
Pratt OE (1979b) Kinetics of tryptophan transport across the blood-brain barrier. J Neurol Transm [Suppl] 15:29–42
Pratt OE (1985) Continuous injection methods for the measurement of flux across the blood-brain barrier: the steady-state, initial-rate method. In: Marks N, Rodnight R (eds) Research methods in neurochemistry, vol 6. Plenum, New York, pp 117–150
Raichle ME (1983) Neurogenic control of blood-brain barrier permeability, Acta Neuropathol [Suppl] (Berl) 8:75–79
Reggiani C, Patrini C, Rindi G (1984) Nervous tissue thiamine metabolism in vivo: I transport of thiamine and thiamine monophosphate from plasma to different brain regions of the rat. Brain Res 293:319–327
Reynolds EH, Gallagher BB, Mattson RH, Bowers M, Johnson AL (1972) Relationship between serum and cerebrospinal fluid folate. Nature 240:155–157
Rindi G, Patrini C, Comincioli V, Reggiani C (1980) Thiamine content and turnover rates of some rat nervous regions, using labelled thiamine as a tracer. Brain Res 181:369–380
Spector R (1976) Thiamine transport in the central nervous system. Am J Physiol 230:1101–1107

Spector R (1978a) Vitamin B_6 transport in the central nervous system: in vivo studies. J Neurochem 30:881–887
Spector R (1978b) Vitamin B_6 transport in the central nervous system: in vitro studies. J Neurochem 30:889–897
Spector R (1980) Riboflavin homeostasis in the central nervous system. J Neurochem 35:202–209
Spector R (1981) Penetration of ascorbic acid from cerebrospinal fluid into brain. Exp Neurol 72:645–653
Spector R (1982) Thiamine homeostasis in the central nervous system. Ann NY Acad Sci 378:344–353
Spector R (1985) Uridine transport and metabolism in the central nervous system. J Neurochem 45:1411–1418
Spector R (1986) Pantothenic acid transport and metabolism in the central nervous system. Am J Physiol 19:R292–297
Spector R, Lorenzo AV (1973) Ascorbic acid homeostasis in the central nervous system. Am J Physiol 225:757–763
Spector R, Lorenzo AV (1975a) Folate transport by the choroid plexus in vitro. Science 187:540–542
Spector R, Lorenzo AV (1975b) Folate transport in the central nervous system. Am J Physiol 299:777–782
Spector R, Greenwald LL (1978) Transport and metabolism of vitamin B_6 in rabbit-brain and choroid plexus. J Biol Chem 253(7):2373–2379
Thomson AD, Frank H, Levy CM (1971) Thiamine propyl disulphide: absorption and utilisation. Ann Intern Med 74:529–534
Thomson AD, Ryte DR, Shaw GK (1982) Ethaniol, thiamine and brain damage. Alcohol 18:27–43
Waxman S, Corcino JJ, Herbert V (1970) Drugs, toxins and dietary amino acids affecting vitamin B-12 or folic acid absorption or utilization. Am J Med 48:599–608

CHAPTER 8

Electrolyte Transport

G.P. SCHIELKE and A.L. BETZ

A. Brain Electrolyte Homeostasis

I. Stability of the Extracellular Potassium Concentration

A precisely regulated extracellular ionic environment in the central nervous system is vital for normal neuronal function. Thus, the processes that stabilize the concentrations of the principal electrolytes (Na, K, and Cl) in brain are of considerable importance. K is the ion most thoroughly studied, both because of its importance for neuronal function and the availability of methods for determining its concentration in interstitial fluid (ISF) in vivo. The early studies of BITO (1969) indicated that the concentration of K in the cerebrospinal fluid ($[K]_{CSF}$) is maintained slightly lower than its concentration in plasma (approximately 2.8 m*M* and 4.4 m*M*, respectively). Similarly, the concentration of K in the interstitial fluid ($[K]_{ISF}$) of the cortical gray matter is maintained between 2.6 and 3.8 m*M*, a level that is less than that of plasma water (KATZMAN 1976; SOMJEN 1979). Although one study has found regional variations in $[K]_{ISF}$ from 3.35 m*M* in the cortex to 1.95 m*M* in the thalamus (MOGHADDAM and ADAMS 1987) and regional variations in $[K]_{CSF}$ may also exist (BITO 1969), in general, the [K] in the brain's extracellular fluid is lower than its concentration in plasma.

In addition to being lower than plasma [K], $[K]_{ISF}$ and $[K]_{CSF}$ remain stable during acute and chronic alterations in plasma [K] (BRADBURY and KLEEMAN 1967; JONES and KEEP 1987; KEEP et al. 1987). For example, plasma [K] is increased to 5–10 m*M* in rats fed a high K diet for 10 days but CSF and ISF [K] are unchanged (KEEP et al. 1987). This stability exists even though the rate of K uptake from blood to brain is increased in the hyperkalemic state (BRADBURY and KLEEMAN 1967). Since total brain K also does not change significantly (BRADBURY and KLEEMAN 1967), the rate of K

Abbreviations

ANP	Atrial natriuretic pepticle	ISF	Interstitial fluid
CSF	Cerebrospinal fluid	$[K]_{ISF}$	Concentration of potassium in the ISF
ECF	Extracellular fluid		
FITC	Fluorescein isothiocyanate		

clearance from the ISF and/or CSF back into the blood must be increased. This clearance process and the long-term regulation of brain potassium are most likely mediated by the brain capillary endothelium and choroid plexus epithelium (BRADBURY 1979).

II. Spatial Buffering of Potassium

Despite these constant basal values, however, there are conditions in both normal and pathologic brain where $[K]_{ISF}$ is altered. Increases in $[K]_{ISF}$ of 1–8 m*M* have been observed during repetitive neuronal activity elicited by electrical stimulation of the brain (SYKOVÁ 1983). Even a single Purkinje cell, when activated by a depolarizing current, can cause a rise of 0.5–3 m*M* in the $[K]_{ISF}$ near its soma (HOUNSGAARD and NICHOLSON 1983). In addition, $[K]_{ISF}$ increases 5–10 m*M* during seizure activity. In all cases, the $[K]_{ISF}$ rapidly recovers after cessation of the stimulus.

Several processes are likely responsible for this rapid recovery of $[K]_{ISF}$ including diffussion of K away from the source, active and passive glial uptake, and active reuptake by neurons (KATZMAN 1976; SOMJEN 1979). The role of astrocytic glial cells in spatial buffering of K has received considerable attention (KIMELBERG and NORENBERG 1989). Spatial buffering is a proposed mechanism whereby local accumulations of K are cleared by passive uptake through the foot processes of glial cells (NEWMAN 1986; ORKAND 1989; WALZ and HERTZ 1983). This causes an intracellular current in the glial cell which results in release of K at distant sites where $[K]_{ISF}$ is lower. The role of the capillary in the rapid recovery process is probably not major (MUTSUGA et al. 1976; VERN et al. 1979). Rather, it is widely held that the brain capillary is responsible for the long-term stability of K ion homeostasis and maintenance of $[K]_{ISF}$ below plasma [K] (BRADBURY 1979). However, since glial foot processes nearly completely envelop the brain capillaries (WOLFF 1970), it is possible that K is dumped by the glial cell into the pericapillary space and that the endothelial cell then assists in short-term K homeostasis.

III. Extracellular Sodium and Chloride Concentrations

In normal brain, the concentrations of Na and Cl are similar in the CSF and ISF (HANSEN 1985). While [Na] is only slightly higher in CSF than in plasma water, the [Cl] is a approximately 15 m*M* higher in CSF than in plasma (FISHMAN 1980). This difference in [Cl] is believed to result from the relative absence of anionic proteins in the CSF.

The CSF concentrations of Na and Cl are in osmotic equilibrium with plasma and, therefore, changes in the plasma concentrations are mirrored, albeit with some time delay, by changes in the CSF concentrations (FISHMAN 1980). However, if hyponatremia or hypochloremia are induced isomotically through dialysis with mannitol to replace Na and Cl (MELTON and NATTIE

1983) or isothionate to replace Cl (ABBOTT et al. 1971), then substantial concentration gradients can be maintained between CSF and plasma for at least 6 h. Thus, extracellular fluid [Na] and [Cl] appear to be quite sensitive to osmotic gradients between blood and brain but relatively insensitive to concentration gradients.

IV. Pathologic Conditions

In several pathologic states of the central nervous system, ion homeostasis becomes severely disturbed. Neuronal and glial membranes fail to maintain their ionic gradients during spreading depression (HANSEN and ZEUTHEN 1981; VYSKOCIL et al. 1972) and when substrate supply is reduced, as during ischemia (ASTRUP et al. 1980; BRANSTON et al. 1977; HARRIS and SYMON 1984; STRONG et al. 1983), anoxia (HANSEN 1984), and hypoglycemia (HARRIS et al. 1984). Due to massive release of K from the cells in exchange for Na, $[K]_{ISF}$ can increase to values as high as 90 mM while $[Na]_{ISF}$ falls. However, even these severe disturbances are reversible if substrate supply is reestablished in a timely fashion.

A disturbance of the ionic gradients within the brain, for example in ischemia, is associated with changes in the total brain content of ions. This indicates net exchanges of electrolytes between blood and brain. In the early stages of partial ischemia, total brain Na content increases and K content falls (BETZ et al. 1989; LO et al. 1987; YOUNG et al. 1987). These ionic exchanges across the blood-brain barrier (BBB) are the driving force for the cytotoxic or intact-barrier edema which develops at this time (BETZ et al. 1989; YOUNG et al. 1987). The gain in Na is greater than the loss in K and the net gain of cations accounts for most or all of the increased water content (BETZ et al. 1989; LO et al. 1987). These changes occur early after the onset of ischemia, and before the BBB permeability to small hydrophilic compounds is increased. In a study of early ischemic edema formation in the gerbil, we demonstrated that the rate of Na accumulation was the same as the rate of unidirectional ^{22}Na uptake (BETZ et al. 1989). This suggests that BBB Na permeability is the rate limiting factor in early edema formation. Furthermore, the blood to brain flux of ^{22}Na appears to be selectively increased during ischemia (ENNIS et al. 1990; SCHIELKE et al. 1991; SHIGENO et al. 1989) and intra-arterial infusion of the active sodium transport inhibitor ouabain reduces not only brain sodium uptake but brain edema formation as well (SHIGENO et al. 1989).

These examples of normal and altered brain ion homeostasis suggest an important and active role for the brain capillary endothelium. We anticipate that a better understanding of the mechanisms of ion transport at the BBB and the factors which regulate these processes should help explain how the $[K]_{ISF}$ is so tightly controlled and also could lead to new approaches to the therapy of brain edema. Although our current knowledge is far from complete, studies of BBB ion transport during the past decade provide a

Table 1. Blood-brain barrier permeabilities to major electrolytes

Na	K	Cl	Species	Reference
2.5			Gerbil	ENNIS et al. 1990
1.5			Rat	SCHIELKE et al. 1991
2.0			Rat	MURPHY and JOHANSON 1989
1.2	11.3	0.8	Rat	SMITH and RAPOPORT 1986
		1.1	Rat	SMITH and RAPOPORT 1984
	2.8		Rat	HANSEN et al. 1977
	9.2		Cat	BRADBURY and KLEEMAN 1967
1.2	12.2	1.3	Rabbit	DAVSON and WELCH 1971
1.5			Rabbit	DAVSON 1955

Values are expressed as permeability × surface area (*PS*) products in units of $\mu l\, g^{-1}\, min^{-1}$.

framework for the development and testing of new hypotheses. In the following sections, we will discuss the mechanisms by which electrolytes are believed to cross the BBB and then consider how these processes may be regulated.

B. Blood-Brain Barrier Permeability to Electrolytes

I. Measurements of Electrolyte Permeability

Since maintenance of brain water and ion homeostasis is presumed to be a major function of the BBB, it is important that blood to brain uptake of inorganic ions be very restricted. BBB permeability to ions has been measured with a variety of isotopic techniques, as well as with ion-selective microelectrodes (Table 1). In general, the permeability to K is 7–10 times higher than that of Na, while the permeability to Cl is similar to that of Na.

BBB permeability to electrolytes is most commonly estimated using radioisotopes. Since the rates of Na and Cl uptake into CSF are 6–30 times greater than into brain (SMITH and RAPOPORT 1986), care must be taken to ensure that measurements of BBB permeability to ^{22}Na and ^{36}Cl are not affected by movement of isotope from CSF into brain. This artifact can be avoided with short isotope circulation times and by sampling brain most distant from the choroid plexuses (SMITH and RAPOPORT 1986). Alternatively, brain, CSF, and plasma isotope concentrations can be measured and a multicompartmental analysis used to calculate influx rate constants (DAVSON and WELCH 1971). A CSF contribution is less of a problem for measurements of ^{42}K permeability since the rate of blood to brain flux is about the same as the rate of blood to CSF flux (SMITH and RAPOPORT 1986).

A different approach to estimating K permeability was described by HANSEN et al. (1977), who used ion-selective microelectrodes to measure

Table 2. Permeabilities, hydraulic conductivities, and electrical resistances across the blood-brain barrier, muscle capillary, and choroid plexus epithelium

	P_{Na}	P_K (cm/s × 10^{-7})	$P_{sucrose}$	Hydraulic conductivity (cm sec^{-1} dyne × 10^{-12})	Resistance (Ω cm^2)
BBB	1.4[a]	13.5[a]	0.5[b]	0.3[h]	1870[h]
Muscle capillaries	390[c] 350[d]	570[c]	500[b]	25[d]	28[c]
Choroid plexus	380[a]	190[a]	46[f]	2.8[e]	24[g]

[a] Rat (SMITH and RAPOPORT 1986). [b] Rat (OHNO et al. 1978). [c] Frog (OLESEN and CRONE 1983). [d] Cat (RENKIN 1978). [e] Rabbit (WELCH et al. 1966). [f] Rabbit (WELCH and SADLER 1966). [g] Frog (ZEUTHEN and WRIGHT 1981). [h] Frog (CRONE and OLESEN 1982).

brain $[K]_{ISF}$ and the sagittal sinus, muscle, and mesenteric [K] during intra-arterial bolus injections of K. Muscle, mesentery, and sagittal sinus demonstrated an immediate and large increase in [K], whereas brain $[K]_{ISF}$ remained essentially unchanged. This confirms the low blood to brain K permeability as measured isotopically. However, with this method, the upper limit of K permeability was calculated to be 2.8×10^{-7} cm/s, about twice the permeability estimated for Na and lower than most isotopically determined K permeabilities (Table 2). This discrepancy may, in part, be due to assumptions needed to convert measured influx rates into permeabilities.

The large number of measurements of blood to brain electrolyte flux contrasts to the paucity of information on brain to blood fluxes. This is due primarily to the difficulty of accurately measuring the clearance across the BBB of isotopes that are also transported by the choroid plexus and other brain cells. Nevertheless, BRADBURY and STULCOVÁ (1970) were able to demonstrate removal of K from blood to brain via a transport process whose rate varied with the $[K]_{CSF}$. CSERR et al. (1981) measured the rate of clearance of ^{22}Na from brain after injection into the caudate nucleus. They found that the rate of Na clearance was approximately 10 times higher than that of albumin or 900 MW polyethylene glycol. About 20% of the injected Na was cleared into the CSF. The remainder of the Na was cleared from brain, presumably across the BBB, at a rate that approximated the rate of Na influx.

II. Route of Electrolyte Passage Across the BBB

There are several possible routes through which ions might move between blood and brain and the measured flux rates represent the sum of the rates for all of the contributing mechanisms. Movement across the BBB could occur by (1) diffusion through the lipid membranes and cytoplasm of the endothelial cells, (2) diffusion through water-filled pores, either within the membranes of the endothelial cells, within spaces between the tight junctions, or both, or (3) carrier-mediated active and/or passive transport. The first two processes are passive and driven only by ions moving down electrical or concentration gradients. The third process is specific, saturable, and may result in the movement of ions against a gradient.

1. Passive Permeability Pathways

It is unlikely that diffusion through the endothelial cell lipid membranes is a significant route for ion penetration for several reasons. Reports of Na and K permeability coefficients for pure lipid membranes are approximately 6 orders of magnitude lower than the permeability coefficients for the BBB (BRUNNER et al. 1980; PAPAHADJOPOULOS 1971). Also, pure lipid membranes generally exhibit little permselectivity, whereas the BBB is many times more permeable to K than Na when isotopically determined values are compared. Although pure phosphatidylserine vesicles are 5–9 times more permeable to K than to Na, this selectivity is lost when cholesterol or various other lipids that are more representative of biological membranes are added (PAPAHADJOPOULOS 1971).

In systemic continuous capillaries, ions diffuse freely between blood and ISF. This is believed to occur through relatively large water-filled pores (e.g., 70 Å and 240 Å; BUNDGAARD 1980; TAYLOR and GRANGER 1983) in which the charge of an ion has little effect on its permeability. However, as mentioned above, ion and water flux through cerebral capillaries is relatively restricted as compared to systemic vascular beds. This can easily be seen when the permeabilities, hydraulic conductivities, and electrical resistances of brain and systemic capillaries are compared (Table 2). In fact, these values for brain capillaries are more similar to those of tight epithelia, where only a small portion of ion fluxes are thought to occur through a neutral paracellular pathway (FRÖMTER and DIAMOND 1972; POWELL 1981).

One reason for this difference is clearly shown using electron microscopy. Lanthanum hydroxide readily penetrates the space between contiguous endothelial cells of systemic capillaries but not the junctional complex of brain capillaries (BRIGHTMAN and REESE 1969). Nevertheless, it is still possible that a paracellular pathway exists for some ions in the BBB. For example, sulfate and ferrocyanide appear to enter the interjunctional space of brain capillaries following intrathecal injection (CASLEY-SMITH 1969). This raises the possibility that the junctions might be selectively permeable to anions. However, the electrical potential between blood and CSF is not affected by changes in plasma Cl (ABBOTT et al. 1971) and, if also true for the potential difference between blood and ISF, this suggests that electrodiffusive Cl flux across the BBB is insignificant.

CRONE (1984) has suggested that there is no charge selectivity for ions at the BBB. He measured diffusion potentials across surface vessels in the frog brain while varying the ionic composition of the intravascular perfusate and the extravascular superfusate. The results demonstrated that the potentials were symmetrical and that their polarity and magnitude could be explained by the relative mobilities of the ions in water. Crone proposed that, as in systemic capillaries, the brain capillary has a paracellular pathway which consists of neutral pores of approximately 70 Å. The low ionic permeability of the BBB was explained by the relative scarcity of these medium-sized pores. Although well conceived, it is important to consider possible limitations of this study. Since the vessels were penetrated by both recording and perfusion micropipetes, physical damage to the vessel and subsequent leakage may have occurred. A more recent study demonstrated that craniectomy alone results in leakage of Na-fluorescein and fluorescein isothiocyanate (FITC)-albumin from small exposed surface vessels (OLESEN 1987b) and these leaks may have obscured any ion selectivity which exists in unperturbed vessels. Finally, the similarities and differences between amphibian and mammalian BBBs have not been adequately characterized.

In summary, the permselective properties, dimensions, and the relative contribution to total ion flux of the BBB paracellular pathway are not known. However, if the isotopic studies are correct and the BBB exhibits permselectivity of K over Na then, as in tight epithelia, the paracellular pathway may represent a minor route for cations to cross the BBB.

The relative importance of the transcellular pathway is further supported by the high electrical resistance that has been measured across the brain capillary since the electrical resistance of a membrane is a measure of its passive permeability to ions. Using cable analysis, the electrical resistance across brain vessels has been found to be $1800\,\Omega\,cm^2$ (CRONE and OLESEN 1982) or greater (BUTT et al. 1990; CRONE 1984; Table 2). The concerns regarding leakage from exposed, impailed vessels discussed above also apply to measurements of electrical resistance by this method. Therefore, the highest values measured, i.e., $6000\,\Omega\,cm^2$, are likely most accurate. Furthermore, a value of $8000\,\Omega\,cm^2$ was obtained when the resistance was calculated from isotopically determined permeability data for Na, K, and Cl (SMITH and RAPOPORT 1986). These high resistances are in the range of values observed in some tight epithelia (POWELL 1981).

In a developmental study in rats, BUTT et al. (1990) showed that the electrical resistance is low ($300\,\Omega\,cm^2$) until immediately before birth, at which time it increases markedly to its adult value over a period of 2 days. This likely reflects a reduction in the paracellular shunt pathway for ions. Tightening of the BBB is probably necessary for the precise regulation of the brain ionic environment which begins near birth in the rat (JONES and KEEP 1987).

Although the paracellular pathway may not be the primary route for ion permeation under physiologic conditions, its permeability may increase in response to specific mediators. For example, serotonin, bradykinin, phospholipase A_2, arachidonic acid, free radicals, and the Ca ionophore A23187 decrease the electrical resistance of frog cerebral pial vessels

(OLESEN 1986, 1987a; OLESEN and CRONE 1986) and increase passive permeability of the BBB (WAHL et al. 1988). These effects are graded, reversible, and blocked by receptor antagonists.

In summary, the permeability of the BBB to electrolytes is low when compared to systemic capillaries. Similarities to tight epithelia suggest that the paracellular pathway is not a major route for ion movement. This tightness of the barrier to electrodiffusive ion flux could permit the fine regulation of brain ion homeostasis by other, more specific mechanisms.

2. Evidence for Active Transport Pathways

The concept of active electrolyte transport across the BBB developed from early studies of brain K homeostasis, blood to brain potential difference, and radiotracer fluxes between blood and brain. Because of methodologic limitations, it was not possible to clearly distinguish between processes occurring at the choroid plexus and the BBB. Given the major role of the choroid plexus in fluid secretion and active ion transport, the role of the BBB could be questioned.

As already mentioned, CSF and ISF [K] are maintained lower than that of plasma and constant despite acute or chronic changes in plasma [K]. This suggests that the activities of K uptake and/or efflux mechanisms can be changed as needed. BRADBURY and STULCOVÁ (1970) measured the movement of ^{42}K from CSF/ISF into the blood during ventriculocisternal perfusion with ^{42}K and varying concentrations of K in artificial CSF. Barrier clearance increased sigmoidally as the $[K]_{CSF}$ was increased from 2 to 10 mM. The increase was blocked by the Na-K-pump inhibitor ouabain, and was specific for K since [^{14}C]-urea clearance was unaffected. This study strongly supports the idea that K is actively transported from CSF to blood through Na-K-ATPase located either in the choroid plexus or the endothelial cell or both.

In an experiment designed to study K transport without the influence of the choroid plexus, BRADBURY et al. (1972) measured uptake of ^{42}K from blood to brain following subarachnoid perfusion with artificial CSF of varying [K]. ^{42}K uptake from the blood was greatest at 0 mM unlabeled [K] and lowest at 10 mM [K]. These data could be explained either by reduced ISF to blood ^{42}K backflux if low $[K]_{CSF}$ inhibits an active efflux system or by K-K interactions in pores or channels.

Further support for active transport of ions across the BBB is provided by studies of the electrical potential difference which exists between the CSF and the blood. Most types of epithelial tissues produce such a potential difference and active transport of ions provides its basis (BETZ and GOLDSTEIN 1981). In brain, the potential difference is 3–5 mV with the CSF side positive, and it is sensitive to changes in plasma [K] (BRADBURY and KLEEMAN 1967) and pH (HELD et al. 1964). In a recent study of the potential difference across exposed brain surface vessels of the frog, ABBOTT et al.

(1986) found a component of the potential difference which was sensitive to ouabain. Additionally, the ouabain-sensitive component varied sigmoidally with the [K] of the solution bathing the vessel with an apparent K_m of 2.6 mM. This strongly suggests that a portion of the potential difference across the vessel is generated by Na-K-ATPase located in the abluminal membrane of the microvessel.

Active transport of Na between blood and brain may also occur. Go et al. (1979; Go and Pratt 1975) studied brain uptake of ^{24}Na in the rat over 30 min and observed both blood pressure-dependent and -independent uptake components. The blood pressure-independent Na uptake was reduced by severe brain hypothermia (15°C) and by removal of the choroid plexuses from the lateral and fourth ventricles. This effect of choroid plexectomy indicates that some of the blood pressure-independent uptake at this long time point occurred from the CSF. However, hypothermia resulted in an even greater reduction in brain Na uptake and this may indicate the presence of an active Na transport mechanism at the BBB.

Finally, since active transport of Na by epithelial cells usually results in fluid secretion, studies of extrachoroidal CSF production provide further evidence for active Na transport across the BBB. Pollay and Curl (1967) first suggested an intraparenchymal source of CSF when they demonstrated that fluid perfused through the aqueduct of Sylvius was diluted by fluid derived from brain. Milhorat et al. (1971) showed that CSF secretion was reduced but not eliminated by choroid plexectomy. These studies suggest that 25%–30% of CSF production is derived from an intraparenchymal source, presumably the brain capillary. Cserr et al. (1981) has shown that following intraparenchymal injections, several extracellular tracers moved out of the brain tissue at similar rates, even though their molecular weights varied from 900 to 69000. Thus, it appears that the brain's ISF moves by bulk flow through the brain parenchyma to the CSF (Rosenberg et al. 1980), a process that might be explained if brain capillary endothelial cells, like epithelial cells, secrete fluid.

III. Epithelial Properties of the Brain Capillary Endothelium

The studies cited above support the concept of active ion transport across the BBB. Since active transport of water and electrolytes across cellular barriers is a function usually carried out by secretory epithelia, it is of interest to note the similarities between brain capillary endothelium and a typical high-resistance epithelium (Betz 1985; Crone 1986).

Similarities in structural and passive permeability properties between the BBB and tight epithelia have already been described. The continuous tight junctions, high transcellular resistance, and low passive permeability to ions and other polar solutes are hallmarks of both tissues. Epithelia are capable of active transcellular movement of ions because they have a differential distribution of transport proteins between the apical and basolateral membranes (BETZ and GOLDSTEIN 1981). This aspect of the cerebral capillary will be described in more detail in the next section. Active transport of ions requires a large amount of metabolic energy and, therefore, epithelial cells are usually well endowed with mitochondria. Brain capillary endothelial cells are also rich in mitochondria, having about 4 times more than vascular endothelial cells in skeletal muscle (OLDENDORF et al. 1977).

IV. Mechanisms of Ion Transport Across the BBB

Having reviewed the indirect evidence for transport of ions across the BBB and drawn an analogy between the brain capillary endothelium and a typical secretory epithelium, we will review studies designed to elaborate the ion transport mechanisms that are present at the BBB. Most of our knowledge is derived from experiments performed in vivo or with isolated capillaries in vitro. Because of the difficulty of performing unambiguous experiments in vivo and the inability to measure transcellular fluxes in isolated capillaries, there is considerable interest in recreating a BBB in vitro using monolayers of cultured brain capillary endothelial cells (see Chap. 17). Unfortunately, the cultured cells currently available do not reliably retain BBB properties (e.g., high transcellular resistance) and, therefore, results obtained with such systems may be misleading. Still, this approach holds promise for the future as new methods for producing well differentiated cultures are developed.

Overall, studies of the mechanisms of BBB electrolyte transport are few in number and results are not always consistent. Figure 1 summarizes the types and distribution of transporters and channels that have been proposed to be present in the brain capillary endothelial cell; the evidence in support of these systems will be discussed in the following sections. Continuing with the analogy to an epithelial cell, we will consider transport processes at the luminal (blood-facing) and abluminal (brain-facing) membranes separately.

1. Luminal Entry Mechanisms

Blood to brain Na transport has been studied in the rat using a modification of the single-pass carotid injection technique (BETZ 1983a). Brain uptake of ^{22}Na was saturable with increasing [Na] ($K_m \approx 2.5\,\text{m}M$) and was inhibited by both amiloride and furosemide. The K_i for amiloride inhibition was $3 \times 10^{-7}\,M$ and inhibition by amiloride and furosemide was additive. Amiloride but not furosemide inhibited ^{86}Rb uptake as well, suggesting that the amiloride-sensitive pathway is non-selective for Na, Rb and probably K, since Rb generally substitutes well for K in biological systems. These data were interpreted as indicating that two transporters reside on the luminal membrane, an amiloride-sensitive cation channel (Fig. 1a) and a furosemide-sensitive Na-Cl neutral cotransporter (Fig. 1b). However, the endothelial cation channel appears to differ from the typical epithelial Na channel in

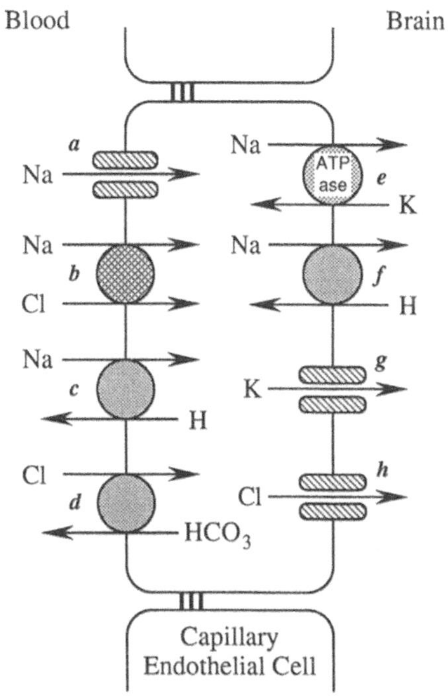

Fig. 1. Electrolyte transporters and channels that may be present in the brain capillary endothelial cell. The transporters and channels that have been proposed and their distribution between the luminal and abluminal membrane of the brain capillary endothelial cell include (*a*) non-selective amiloride-sensitive cation channel, (*b*) Na-Cl cotransporter, (*c*) luminal Na/H exchanger, (*d*) Cl/HCO_3 exchanger, (*e*) Na-K-ATPase, (*f*) abluminal Na/H exchanger, (*g*) K channel, and (*h*) Cl channel

that it is non-selective for Na and K, and the inhibition by amiloride is effective only at low luminal [Na].

An earlier study had failed to demonstrate a clear effect of amiloride on blood to brain ^{22}Na transport (Davson and Segal 1970); however, the method used depended upon measurements of CSF ^{22}Na or brain ^{22}Na at long times after isotope injection (75 min), when flux from CSF into brain can confuse the interpretation of the results. More recently, Murphy and Johanson (1989) measured the rate constant for ^{22}Na uptake into rat cortex 30 min after intraperitoneal injection of isotope. Amiloride and plasma acidosis inhibited ^{22}Na uptake by 22% and 25%, respectively. They suggested that this effect may be due to the existence of a Na/H exchanger (Fig. 1c), although other mechanisms were not excluded and the interrelationship between the effects of acidosis and amiloride were not studied.

In a recent patch clamp study of cultured cerebral microvessel endothelial cells, Vigne et al. (1989) identified a 23-picoseimen, amiloride-sensitive cation channel with little selectivity for Na and K. A similar non-selective cation channel was recently identified in the renal inner medullary collecting duct (Light et al. 1989). This 27-picoseimen channel is not sensitive to membrane potential, Ba^{2+}, and tetraethylammonium, but is inhibited by amiloride ($K_i = 5 \times 10^{-7} M$), atrial natriuretic peptide (ANP), and cyclic guanosine monophosphate. The channel in the brain capillary may also be influenced by ANP since Ibaragi et al. (1989), using isolated microvessels from rat brain, showed that amiloride increases the specific binding of ANP. They also demonstrated that ANP reduces ^{22}Na uptake under conditions where uptake by other systems, i.e., the neutral Na-Cl co-transporter and the Na-H exchanger, was blocked.

Thus, the available evidence supports the existence of a non-selective cation channel in the luminal membrane of the cerebral capillary which is inhibited by amiloride and ANP (Figs. 1a). It is possible that this channel, and the channel observed in the inner medullary collecting duct, are a different form of the more common Na channel in the apical membrane of tight epithelia. This BBB form of the channel may be a specialization found in tissues in which K homeostasis is an important function. Its presence in the luminal membrane of the brain capillary could permit both the entry of Na into brain and the exit of K out of the brain (Fig. 2).

The presence of a Na-Cl cotransport system (Fig. 1b) has not been substantiated so well. Crone (1986) has pointed out that duplicate Na entry systems are usually not observed in the apical membranes of epithelia and neutral Na-Cl cotransport is more often seen in leaky epithelia. Since the Na/H exchanger is also sensitive to amiloride, he suggests that the capillary may have Na/H and Cl/HCO_3 exchangers side by side (Fig. 1c,d), as seen in some epithelia, including the choroid plexus (Spector and Johanson 1989). However, the Na/H exchanger requires a much higher concentration of amiloride (10^{-4}M) for inhibition (Benos 1982). Furthermore, one study using isolated brain capillaries reported the presence of furosemide-sensitive Na uptake (Ibaragi et al. 1989) while another noted furosemide-sensitive K uptake, suggesting Na-K-2Cl cotransport (Lin 1985). These latter studies, however, do not permit the unambiguous localization of transporters between the luminal and abluminal membranes.

The rate of blood to brain ^{36}Cl uptake in rats dialyzed in order to alter their plasma [Cl] was studied by Smith and Rapoport (1984). The rate constant for unidirectional uptake decreased with increasing plasma [Cl] indicating a saturable, carrier-mediated process. ^{22}Na uptake was unaffected by the altered plasma Cl levels. Since the potential difference between blood and brain is not affected by the [Cl] in plasma (Abbott et al. 1971), brain Cl uptake must not occur to a significant degree by electrodiffusion. These data are consistent with either the existence of a Cl/HCO_3 exchanger in the luminal membrane (Fig. 1d) or with a neutral Na-Cl cotransporter (Fig. 1b) if this is not the carrier which is rate limiting for Na uptake.

In summary, a variety of ion transporters and channels have been suggested to be present on the luminal membrane of the brain capillary

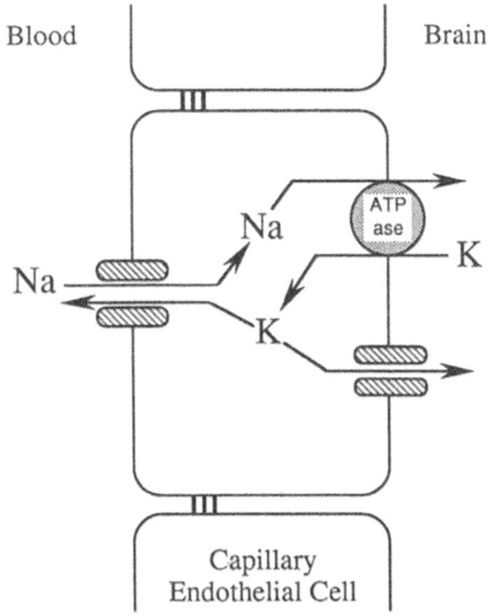

Fig. 2. Simplified model of BBB electrolyte transport. This scheme proposes a distribution of transporters and channels that could explain both the active pumping of Na from blood to brain and the regulation of brain [K] by the capillary endothelial cell. Na-K-ATPase in the abluminal membrane maintains a low intracellular [Na] and a high intracellular [K] by pumping Na from the cell into the brain's ISF in exchange for K. Na enters the endothelial cell from the blood down its concentration gradient through the nonselective cation channel. K taken up from the ISF may either return to brain through the abluminal K channel or leave the brain through the luminal cation channel depending upon whether brain K must be conserved or reduced

endothelium including a nonselective cation channel, a Na-Cl cotransporter, a Na-K-2Cl cotransporter, a Na/H exchanger, and a Cl/HCO_3 exchanger (Fig. 1a–d). It is unlikely that all of these entities are present and, therefore, further study is required.

2. Abluminal Transport Systems

Na-K-ATPase is the primary driving force for all active electrolyte transport in epithelia. That brain capillary endothelium is enriched in Na, K-ATPase compared to other endothelial cells was suggested by the studies of Eisenberg and Suddith (1979). Goldstein (1979) demonstrated that isolated brain microvessels have a ouabain-sensitive K uptake mechanism, the activity of which is dependent on the extracellular [K]. In view of the similarities with the K efflux system described in vivo by Bradbury and Stulcová (1970), Goldstein proposed that the Na-K pumps were located on the abluminal membrane of the brain capillary (Fig. 1e) and that they

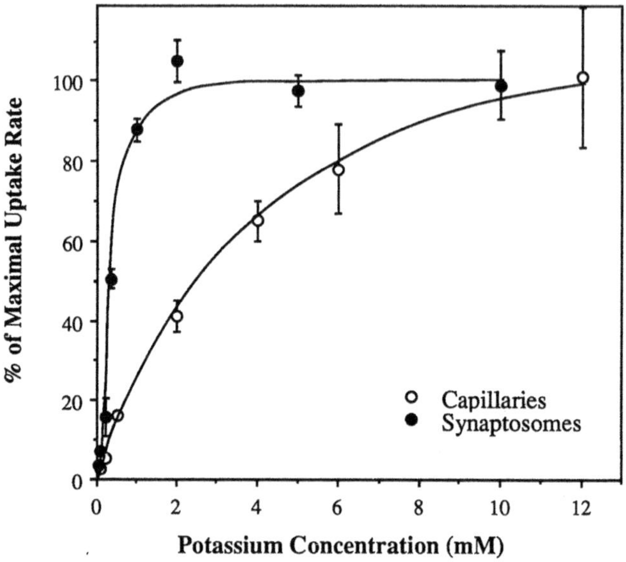

Fig. 3. Comparison of the effect of K on Na-K-ATPase in brain capillaries and synaptosomes. The [K] that results in half-maximal ouabain-sensitive ^{86}Rb uptake is 3.8 m*M* in the capillaries and 0.47 m*M* in the synaptosomes. (Data derived from SCHIELKE et al. 1990)

contributed to ion homeostasis. Direct evidence for the abluminal localization of Na-K-ATPase in the brain capillary was first provided in a cytochemical study by FIRTH (1976). This result was confirmed and expanded using both cytochemical and membrane fractionation techniques (BETZ et al. 1980). A later study by VORBRODT et al. (1982) also found a polar distribution of the transporter in some capillaries, but observed reaction product on both membranes in others. The asymmetric distribution of Na-K-pumps in the brain capillary supports the proposal that they are capable of generating transcellular ion fluxes (BETZ and GOLDSTEIN 1981).

The kinetic properties of BBB Na-K-ATPase have been further studied using isolated brain microvessels. Three studies demonstrated that ouabain-sensitive ^{86}Rb uptake was half-maximal at an extracellular fluid [K] of about 3 m*M* (CHAPLIN et al. 1981; GOLDSTEIN 1979; LIN 1985). More recently, we confirmed this finding and demonstrated that the effect of extracellular fluid [K] on ouabain-sensitive Na efflux from capillaries exhibited similar kinetics (SCHIELKE et al. 1990). Furthermore, the kinetics of K uptake into capillaries differed considerably from those in synaptosomes, a neuronal membrane preparation (Fig. 3). These results suggest that physiological changes in $[K]_{ISF}$ would have little effect on Na-K-ATPase activity in neurons since it is near-maximal in the presence of a normal $[K]_{ISF}$. Na-K-ATPase activity in brain capillary endothelial cells, however, would be affected by changes in $[K]_{ISF}$ in a manner very consistent with the kinetics of K loss from brain during ventriculocisternal perfusion (BRADBURY and STULCOVÁ 1970). Thus, both the kinetics and distribution (Fig. 2) of BBB Na-K-ATPase are well suited for a possible role in regulating the $[K]_{ISF}$.

Other transport systems have been tentatively assigned to the abluminal membrane based on studies with isolated brain microvessels. For example, Na-dependent amino acid transport can be identified using isolated brain capillaries but not when blood to brain amino acid transport is measured in vivo. Therefore, the Na-dependent amino acid transport system was assigned to the abluminal membrane (BETZ and GOLDSTEIN 1978). Similar reasoning was used to assign amiloride-sensitive Na/H exchange to the abluminal membrane (Fig. 1f) (BETZ 1983b). The ability to separate luminal and abluminal transport activities by comparing transport in vivo and in vitro is further supported by the apparent presence of furosemide-sensitive brain sodium uptake in vivo (BETZ 1983a) and its absence in vitro (BETZ 1983b). This is consistent with a luminal localization of the transporter (Fig. 1b) if the capillary lumen is collapsed and inaccessible to ^{22}Na in the isolated micorvessel preparations. However, as noted above, others have identified furosemide-sensitive transporters in isolated brain capillaries (IBARAGI et al. 1989; LIN 1985) and, therefore, this reasoning becomes less tenable unless there is variability in the patency of the capillary lumen when microvessels are prepared in different laboratories. Furthermore, one must be careful in assigning transporters identified in isolated microvessels to the endothelial cells since such preparations also contain pericytes, red blood cells, and occasionally smooth muscle cells. Thus, any proposed localization of transport activities based upon radiotracer fluxes must ultimately be confirmed using more direct approaches such as immunocytochemical techniques.

The presence of additional ion transporters and channels in the abluminal membrane of the brain capillary endothelium can be inferred from the requirements for ion homeostasis and isosmotic fluid flux. For example, most epithelial cells with Na-K-ATPase on the basolateral membrane also have a K channel to permit recycling of K back to the extracellular space (Fig. 1g) (DAWSON and RICHARDS 1990). The barium-inhibitable K efflux from brain capillaries reported by LIN (1985) may represent passage through such a channel. In order to regulate $[K]_{ISF}$ in brain, it seems likely that the endothelial cell would have separate types of K channels in the luminal and abluminal membranes (Fig. 2). Such an arrangement would permit the cell to regulate how much K is released into blood or recycled back to brain. Similarly, if fluid flux across the BBB is linked to isosmotic NaCl transport through the endothelium, then there must be some pathway for Cl to exit the endothelial cell across the abluminal membrane (Fig. 1h).

Thus, our present limited knowledge does not permit a complete and integrated description of how ion transporters and channels mediate BBB electrolyte transport and brain ion homeostasis. However, it appears likely that additional transporters and channels will be identified in the future and this should lead to the formulation of a more complete model. Nevertheless, we have attempted to synthesize the available data into a working model for the distribution of ion transporters within the endothelial cell in a way that most simply explains both the apparent secretion of ISF through active Na and Cl transport into brain and the long-term regulation of $[K]_{ISF}$ through active transport of K out of brain (Fig. 2). It should be emphasized that this is only a hypothetical model and is presented to stimulate further investigation in the field.

V. Regulation of BBB Permeability to Ions

Regulation of BBB electrolyte transport could occur at any of the channels and transporters that are present and through either neural or hormonal mechanisms. In addition, the ambient electrolyte concentration might affect transport rate either directly by an effect on ion binding to transporters or indirectly by causing the release of hormones or neurotransmitters that alter brain capillary transport function.

1. Regulation by Electrolytes and Glucose

The kinetic properties of brain capillary Na-K-ATPase (Fig. 3) indicate that its activity is half-maximal near the normal $[K]_{ISF}$. Thus, if $[K]_{ISF}$ increases transport rate might incease, and if $[K]_{ISF}$ decreases transport rate might decrease. These changes in transport rate would favor the return of $[K]_{ISF}$ to normal. The kinetics of brain capillary Na-K-ATPase with regard to [K] may explain the 50%–70% increase in blood to brain Na transport that occurs when $[K]_{ISF}$ increases during ischemia (ENNIS et al. 1990; SCHIELKE et al. 1991; SHIGENO et al. 1989) and hypoglycemia (G.P. SCHIELKE et al., unpublished results). However, this increase in sodium transport has also been attributed to an alteration in Na-K-ATPase activity as the result of lipid peroxide formation during ischemia (ASANO et al. 1987; KOIDE et al. 1986).

A primary alteration in brain capillary Na-K-ATPase activity has also been postulated to occur in untreated diabetes mellitus. KNUDSEN and JAKOBSEN (1989) noted a 20% reduction in blood to brain sodium transport after either 2 weeks of chronic hyperglycemia or 2 h of acute hyperglycemia in rats. This effect was not related to blood insulin levels or to plasma hyperosmolality, but could be reversed within a few hours following acute insulin treatment of diabetic rats or by dietary supplementation with myoinositol (KNUDSEN et al. 1989). Although Na-K-ATPase activity has not been directly measured in brain capillaries from diabetic rats, it is reduced in many cell types during diabetes and can be normalized with myoinositol (GREENE et al. 1987). Recent studies suggest that this may result from an alteration in regulation of Na-K-ATPase by protein kinase C (LATTIMER et al. 1989).

Alterations in BBB electrolyte transport may also be important during brain volume regulation in response to changes in plasma osmolarity. For example, acute hyperosmolality results in a rapid increase in brain Na and K contents, limiting the amount of brain water that must be lost to reestablish osmotic equilibrium with the plasma (CSERR et al. 1987a). These changes in brain ion content are associated with a marked increase in the influx rate constant for K from blood to brain (CSERR et al. 1987b). Using isolated brain capillaries, LIN (1988) has shown that hyperosmolarity stimulates ouabain-insensitive ^{86}Rb uptake through a Na- and Cl-dependent process. Thus, the increase in brain K uptake observed in vivo might occur through the activation of a Na-K-2Cl cotransporter if such a system is present in the luminal membrane of the brain capillary.

2. Regulation by Nerves

Both physiologic and anatomic data support the hypothesis that brain capillary function is regulated by noradrenergic innervation derived from the locus ceruleus (RAICHLE et al. 1975). Thus, direct stimulation of the locus ceruleus produces a prompt increase in BBB permeability to water while central noradrenergic blockade has the opposite effect. Nevertheless, it is not possible to directly relate changes in water permeability to alterations in electrolyte transport since the BBB permeability to $^{3}H_2O$ is nearly 100 times greater than the net flux of water and ions into brain. However, HARIK (1986) has obtained more direct evidence for a relationship between the locus ceruleus and BBB ion transport. He noted a 40% reduction in the Na-K-ATPase content of capillaries isolated from the cerebral cortex of animals that had received a lesion in the ipsilateral locus ceruleus 2 weeks earlier. The effect of this substantial change in brain capillary pump activity on BBB permeability to sodium has not yet been determined.

3. Regulation by Hormones

That specific aspects of brain capillary function are regulated by hormones is suggested by the presence of receptors for hormones and neurotransmitters on brain capillaries (BETZ 1991). Some of these hormones (e.g., vasopressin, ANP) are known to regulate ion transport and fluid secretion in epithelia including the choroid plexus (FARACI et al. 1990; STEARDO and NATHANSON 1987). As mentioned previously, ANP binding to brain capillaries is modulated by amiloride and it inhibits amiloride-sensitive ^{22}Na uptake by isolated brain capillaries (IBARAGI et al. 1989). Furthermore, intraventricular administration of ANP attenuates brain edema and sodium accumulation during cerebral ischemia; however, this effect has not yet been directly related to an effect on BBB permeability to sodium (NAKAO et al. 1990).

Brain capillaries also contain the V_1 type of vasopressin receptor (PEARLMUTTER et al. 1988). Although intraventricular administration of vasopressin increases the BBB permeability to $^{3}H_2O$ (RAICHLE and GRUBB

1978), as mentioned previously, this is not likely to be mediated through an effect on BBB ion transport. Absence of central and systemic vasopressin as seen in the Brattleboro strain of rat is associated with significantly reduced uptake of sodium during hypernatremia (DEPASQUALE et al. 1989) and ischemia (L.D. DICKINSON and A.L. BETZ unpublished results). However, BBB permeability to sodium in Brattleboro rats is the same as it is in the Long Evans strain from which the Brattleboro is derived (DEPASQUALE et al. 1989). This does not eliminate the possibility that vasopressin is involved in regulating BBB ion transport since its chronic absence in the Brattleboro rat may be balanced by changes in other hormones such as ANP (BAHNER et al. 1990).

Finally, steroid hormones such as progesterone, deoxycorticosterone, androstenedione, and corticosterone inhibit ^{86}Rb uptake by isolated rat brain capillaries in a ouabain-like fashion, suggesting a direct effect on Na-K-ATPase (CHAPLIN et al. 1981). However, very high concentrations are required and acute systemic treatment of rats with progesterone or dexamethasone does not alter the BBB permeability to Na (BETZ and COESTER 1990).

Thus, a variety of preliminary studies suggest the potential for neural and humoral regulation of BBB ion transport. Such systems could provide the means by which the brain is able to control its environment. Although speculative at the present time, it seems likely that these regulatory influences would be activated by changes in the extracellular concentrations of ions or in the intracranial pressure so as to return these important parameters to normal and maintain an optimal environment for brain function.

References

Abbott J, Davson H, Glen I, Grant N (1971) Chloride transport and potential across the blood-CSF barrier, Brain Res 29:185–193

Abbott NJ, Butt AM, Wallis W (1986) The Na-K ATPase of the blood-brain barrier: a microelectrode study. Ann NY Acad Sci 481:390–391

Asano T, Shigeno T, Johshita H, Usui M, Hanamura T (1987) A novel concept on the pathogenetic mechanism underlying ischaemic brain oedema: relevance of free radicals and eicosanoids. Acta Neurochir 41:85–94

Astrup J, Rehncrona S, Siesjö BK (1980) The increase in extracellular potassium concentration in ischemic brain in relation to the perischemic functional activity and cerebral metabolic rate. Brain Res 199:161–174

Bahner U, Geiger H, Palkovits M, Ganten D, Michel J, Heidland A (1990) Atrial natriuretic peptides in brain nuclei of rats with inherited diabetes insipidus (Brattleboro rats). Neuroendocrinology 51:721–727

Benos DJ (1982) Amiloride: a molecular probe of sodium transport in tissues and cells. Am J Physiol 242:C131–C145

Betz AL (1983a) Sodium transport from blood to brain: inhibition by furosemide and amiloride. J Neurochem 41:1158–1164

Betz AL (1983b) Sodium transport in capillaries isolated from rat brain. J Neurochem 41:1150–1157

Betz AL (1985) Epithelial properties of brain capillary endothelium. Fed Proc 44:2614–2615
Betz AL (1991) An overview of the multiple functions of the blood-brain barrier. In: Brown RM, Frankenheim J (eds) NIDA technical review on drug bioavailability and the blood-brain barrier. US Government Printing Office Washington DC
Betz AL, Coester HC (1990) Effect of steroids on edema and sodium uptake of the brain during focal ischemia in rats. Stroke 21:1199–1204
Betz AL, Goldstein GW (1978) Polarity of the blood-brain barrier: neutral amino acid transport into isolated brain capillaries. Science 202:225–227
Betz AL, Goldstein GW (1981) The basis for active transport at the blood-brain barrier. In: Eisenberg HM, Suddith RL (eds) Adances in experimental medicine and biology. Plenum, New York, p 5
Betz AL, Firth JA, Goldstein GW (1980) Polarity of the blood-brain barrier: distribution of enzymes between the luminal and antiluminal membranes of brain capillary endothelial cells. Brain Res 192:17–28
Betz AL, Ennis SR, Schielke GP (1989) Blood-brain barrier sodium transport limits development of brain edema during partial ischemia in gerbils. Stroke 20:1253–1259
Betz AL, Iannotti F, Hoff JT (1989) Brain edema: a classification based on blood-brain barrier integrity. Cerebrovasc Brain Metab Rev 1:133–154
Bito LZ (1969) Blood-brain barrier: evidence for active cation transport between blood and the extracellular fluid of brain. Science 165:81–83
Bradbury M (1979) The concept of a blood-brain barrier. Wiley, Chichester
Bradbury MWB, Kleeman CR (1967) Stability of the potassium content of cerebrospinal fluid and brain. Am J Physiol 213:519–528
Bradbury MWB, Stulcová B (1970) Efflux mechanism contributing to the stability of the potassium concentration in cerebrospinal fluid. J Physiol (Lond) 208:415–430
Bradbury MWB, Segal MB, Wilson J (1972) Transport of potassium at the blood-brain barrier. J Physiol (Lond) 221:617–632
Branston NM, Strong AJ, Symon L (1977) Extracellular potassium activity, evoked potential and tissue blood flow. J Neurol Sci 32:305–321
Brightman MW, Reese TS (1969) Junctions between intimately apposed cell membranes in the vertebrate brain. J Cell Biol 40:648–677
Brunner J, Graham DE, Hauser H, Semenza G (1980) Ion and sugar permeabilities of lecithin bilayers: comparison of curved and planar bilayers. J Membr Biol 57:133–141
Bundgaard M (1980) Transport pathways in capillaries-in search of pores. Ann Rev Physiol 42:325–336
Butt AM, Jones HC, Abbott NJ (1990) Electrical resistance across the blood-brain barrier in anaesthetized rats: a developmental study. J Physiol (Lond) 429:47–62
Casley-Smith JR (1969) An electron microscopical demonstration of the permeability of cerebral and retinal capillaries to ions. Experientia 25/8:845–847
Chaplin ER, Free RG, Goldstein GW (1981) Inhibition by steroids of the uptake of potassium by capillaries isolated from rat brain. Biochem Pharmacol 30:241–245
Crone C (1984) Lack of selectivity to small ions in paracellular pathways in cerebral and muscle capillaries of the frog. J Physiol (Lond) 353:317–337
Crone C (1986) The blood-brain barrier as a tight epithelium: where is information lacking? Ann NY Acad Sci 481:174–185
Crone C, Olesen SP (1982) Electrical resistance of brain microvascular endothelium. Brain Res 241:49–55
Cserr HF, Cooper DN, Suri PK, Patlak CS (1981) Efflux of radiolabeled polyethylene glycols and albumin from rat brain. Am J Physiol 240:F319–F328
Cserr HF, Depasquale M, Patlak CS (1987a) Regulation of brain water and electrolytes during acute hyperosmolality in rats. Am J Physiol 253:F522–F529

Cserr HF, Depasquale M, Patlak CS (1987b) Volume regulatory influx of electrolytes from plasma to brain during acute hyperosmolality. Am J Physiol 253:F530–F537

Davson H (1955) A comparative study of the aqueous humour and cerebrospinal fluid in the rabbit. J Physiol (Lond) 129:111–133

Davson H, Segal MB (1970) The effects of some inhibitors and accelerators of sodium transport on the turnover of ^{22}Na in the cerebrospinal fluid and the brain. J Physiol (Lond) 209:131–153

Davson H, Welch K (1971) The permeation of several materials into the fluids of the rabbit's brain. J Physiol (Lond) 218:337–351

Dawson DC, Richards NW (1990) Basolateral K conductance: role in regulation of NaCl absorption and secretion. Am J Physiol 259:C181–C195

DePasquale M, Patlak CS, Cserr HF (1989) Brain ion and volume regulation during acute hypernatremia in Brattleboro rats. Am J Physiol 256:F1059–F1066

Eisenberg HM, Suddith RL (1979) Cerebral vessels have the capacity to transport sodium and potassium. Science 206:1083–1085

Ennis SR, Keep RF, Schielke GP, Betz AL (1990) Decrease in perfusion of cerebral capillaries during incomplete ischemia and reperfusion. J Cereb Blood Flow Metab 10:213–220

Faraci FM, Mayhan WG, Heistad DD (1990) Effect of vasopressin on production of cerebrospinal fluid: Possible role of vasopressin (V1)-receptors. Am J Physiol 258:R94–R98

Firth JA (1976) Cytochemical localization of the K^+ regulation interface between blood and brain. Experientia 33:1093–1094

Fishman RA (1980) Cerebrospinal fluid in diseases of the nervous system. Saunders, Philadilphia London Toronto

Frömter E, Diamond J (1972) Route of passive ion permeation in epithelia. Nature New Biol 235:9–13

Go KG, Pratt JJ (1975) The dependence of the blood to brain passage of radioactive sodium on blood pressure and temperature. Brain Res 93:329–336

Go KG, Koster-Otte L, Pratt JJ (1979) Brain sodium uptake after choroid plexectomy. Brain Res 170:325–331

Goldstein GW (1979) Relation of potassium transport to oxidative metabolism in isolated brain capillaries. J Physiol (Lond) 286:185–195

Greene DA, Lattimer SA, Sima AA (1987) Sorbitol, phosphoinositides, and sodium-potassium-ATPase in the pathogenesis of diabetic complications. N Engl J Med 316:599–606

Hansen AJ (1984) Ion and membrane changes in the brain during anoxia. Behav Brain Res 14:93–98

Hansen AJ (1985) Effect of anoxia on ion distribution in the brain. Physiol Rev 65:101–148

Hansen AJ, Zeuthen T (1981) Extracellular ion cencentrations during spreading depression and ischemia in the rat brain cortex. Acta Physiol Scand 113:437–445

Hansen AJ, Lund-Andersen H, Crone C (1977) K^+-permeability of the blood-brain barrier, investigated by aid of a K^+-sensitive microelectrode. Acta Physiol Scand 101:438–445

Harik SI (1986) Blood-brain barrier sodium/potassium pump: modulation by central noradrenergic innervation. Proc Natl Acad Sci USA 83:4067–4070

Harris RJ, Symon L (1984) Extracellular pH, potassium, and calcium activity in progressive ischaemia of rat cortex. J Cereb Blood Flow Metab 4:178–186

Harris RJ, Wieloch T, Symon L, Siesjö BK (1984) Cerebral extracellular calcium activity in severe hypoglycemia: relation to extracellular potassium and energy state. J Cereb Blood Flow Metab 4:187–193

Held D, Fend V, Pappenheimer JR (1964) Electrical potential of cerebrospinal fluid. J Neurophysiol 27:942–959

Hounsgaard J, Nicholson C (1983) Potassium accumulation around individual purkinje cells in cerebellar slices from the guinea-pig. J Physiol (Lond) 340:359–388

Ibaragi M-A, Niwa M, Ozaki M (1989) Atrial natriuretic peptide modulates amiloride-sensitive Na^+ transport across the blood-brain barrier. J Neurochem 53:1802–1806
Jones HC, Keep RC (1987) The control of potassium concentration in the cerebrospinal fluid and brain interstitial fluid of developing rats. J Physiol (Lond) 383:441–453
Katzman R (1976) Maintenance of a constant brain extracellular potassium. Fed Proc 35:1244–1247
Keep RF, Jones HC, Cawkwell RD (1987) Effect of chronic maternal hyperkalaemia on plasma, cerebrospinal fluid and brain interstitial fluid potassium in developing rats. J Dev Physiol 9:89–95
Kimelberg HK, Norenberg MD (1989) Astrocytes. Sci Am 260:66–76
Knudsen GM, Jakobsen J (1989) Blood-brain permeability to sodium. Modification by glucose of insulin? J Neurochem 52:174–178
Knudsen GM, Jakobsen J, Barry DI, Compton AM, Tomlinson DR (1989) Myo-inositol normalizes decreased sodium permeability of the blood-brain barrier in streptozotocin diabetes. Neuroscience 29:773–777
Koide T, Asano T, Matsushita H, Takakura K (1986) Enhancement of ATPase activity by a lipid peroxide of arachidonic acid in rat brain microvessels. J Neurochem 46:235–242
Lattimer SA, Sima AAF, Greene DA (1989) In vitro correction of impaired Na^+-K^+-ATPase in diabetic nerve by protein kinase C agonists. Am J Physiol 256:E264–E269
Light DB, Schwiebert EM, Karlson KH, Stanton BA (1989) Atrial natriuretic peptide inhibits a cation channel in renal inner medullary collecting duct cells. Science 243:383–385
Lin JD (1985) Potassium transport in isolated cerebral microvessels from the rat. Jpn J Physiol 35:817–830
Lin JD (1988) Effect of osmolarity on potassium transport in isolated cerebral microvessels. Life Sci 43:325
Lo WD, Betz AL, Schielke GP, Hoff JT (1987) Transport of sodium from blood to brain in ischemic brain edema. Stroke 18:150–157
Melton JE, Nattie EE (1983) Brain and CSF water and ions during dilutional and isosmotic hyponatremia in the rat. Am J Physiol 244:R724–R732
Milhorat TH, Hammock MK, Rall, DP, Levin VA (1971) Cerebrospinal fluid production by the choroid plexus and brain. Science 173:330–332
Moghaddam B, Adams RN (1987) Regional differences in resting extracellular potassium levels of rat brain. Brain Res 406:337–340
Murphy VA, Johanson CE (1989) Acidosis, acetazolamide, and amiloride: effects on ^{22}Na transfer across the blood-brain and blood-CSF barriers. J Neurochem 52:1058–1063
Mutsuga N, Schuette WH, Lewis DV (1976) The contribution of local blood flow to the rapid clearance of potassium from the cortical extracellular space. Brain Res 116:431–436
Nakao N, Itakura T, Hideyoshi Y, Nakai K, Komai N (1990) Effect of atrial natriuretic peptide on ischemic brain edema: changes in brain water and electrolytes. Neurosurgery 27:39–44
Newman EA (1986) High potassium conductance in astrocyte endfeet. Science 233:453–454
Ohno K, Pettigrew KD, Rapoport SI (1978) Lower limits of cerebrovasclar permeability to nonelectrolytes in the conscious rat. Am J Physiol 235:H299–H307
Oldendorf WH, Cornford ME, Brown WJ (1977) The large apparent work capacity of the blood-brain barrier: a study of the mitochondrial content of capillary endothelial cells in brain and other tissues of the rat. Ann Neurol 1:409–417
Olesen S-P (1986) Rapid increase in blood-brain barrier permeability during severe hypoxia and metabolic inhibition. Brain Res 368:24–29

Olesen S-P (1987a) Free oxygen radicals decrease electrical resistance of microvascular endothelium in brain. Acta Physiol Scand 129:181–187
Olesen S-P (1987b) Leakiness of rat brain microvessels to fluorescent probes following craniotomy. Acta Physiol Scand 130:63–68
Olesen S-P, Crone C (1983) Electrical resistance of muscle capillary endothelium. Biophys J 42:31–41
Olesen S-P, Crone C (1986) Substances that rapidly augment ionic conductance of endothelium in cerebral venules. Acta Physiol Scand 127:233–241
Orkand RK (1989) Role of glial cells in the control of the neuronal microenvironment. In: Battaini F, Govoni S, Magnoni MS, Trabucchi M (eds) Regulatory mechanisms of neuron to vessel communication in the brain. Springer, Berlin Heidelberg New York, p 253
Papahadjopoulos D (1971) Na^+-K^+ discrimination by "pure" phospholipid membranes. Biochim Biophys Acta 241:254–259
Pearlmutter AF, Szkrybala M, Kim Y, Harik SI (1988) Arginine vasopressin receptors in pig cerebral microvessels, cerebral cortex and hippocampus. Neurosci Lett 87:121–126
Pollay M, Curl F (1967) Secretion of cerebrospinal fluid by the ventricular ependyma of the rabbit. Am J Physiol 213:1031–1038
Powell DW (1981) Barrier function of epithelia. Am J Physiol 241:G275–G288
Raichle ME, Grubb RL Jr (1978) Regulation of brain water permeability by centrally-released vasopressin. Brain Res 143:191–194
Raichle ME, Hartman BK, Eichling JO, Sharpe LG (1975) Central noradrenergic regulation of cerebral blood flow and vascular permeability. Proc Natl Acad Sci USA 72:3726–2730
Renkin EM (1978) Transport pathways through capillary endothelium. Microvasc Res 15:123–135
Rosenberg GA, Kyner WT, Estrada E (1980) Bulk flow of brain interstitial fluid under normal and hyperosmolar conditions. Am J Physiol 238:F42–F49
Schielke GP, Moises HC, Betz AL (1990) Potassium activation of the Na,K-pump in isolated brain microvessels and synaptosomes. Brain Res 524:291–296
Schielke GP, Moises HC, Betz AL (1991) Blood to brain sodium transport and interstitial fluid potassium concentration during focal ischemia in the rat. J Cereb Blood Flow Metab 11:466–471
Shigeno T, Asano T, Mima T, Takakura K (1989) Effect of enhanced capillary activity on the blood-brain barrier during focal cerebral ischemia in cats. Stroke 20:1260–1266
Smith QR, Rapoport SI (1984) Carrier-mediated transport of chloride across the blood-brain barrier. J Neurochem 42:754–763
Smith QR, Rapoport SI (1986) Cerebrovascular permeability coefficients to sodium, potassium, and chloride. J Neurochem 46:1732–1742
Somjen GG (1979) Extracellular potassium in the mammalian central nervous system. Ann Rev Physiol 41:159–177
Spector R, Johanson CE (1989) The mammalian choroid plexus. Sci Am Nov:68–74
Steardo L, Nathanson JA (1987) Brain barrier tissues: end organs for atriopeptins. Science 235:470–473
Strong AJ, Venables GS, Gibson G (1983) The cortical ischaemic penumbra associated with occlusion of the middle cerebral artery in the cat: 1. Topography of changes in blood flow, potassium ion activity, and EEG. J Cereb Blood Flow Metab 3:86–96
Syková E (1983) Extracellular K^+ accumulation in the central nervous system. Prog Biophys Mol Biol 42:135–189
Taylor AE, Granger DN (1983) Equivalent pore modeling: vesicles and channels. Fed Proc 42:2440–2445
Vern BA, Schuette WH, Mutsuga N, Whitehouse WC (1979) Effects of ischemia on the removal of extracellular potassium in cat cortex during pentylenetetrazol seizures. Epilepsia 20:711–724

Vigne P, Champigny G, Marsault R, Barbry P, Frelin C, Lazdunski M (1989) A new type of amiloride-sensitive cationic channel in endothelial cells of brain microvessels. J Biol Chem 264:7663–7668
Vorbrodt AW, Lossinsky AS, Wisniewski HM (1982) Cytochemical localization of ouabain-sensitive, K^+-dependent P-nitro-phenylphosphatase (transport ATPase) in the mouse central and peripheral nervous systems. Brain Res 243:225–234
Vyskocil F, Kriz N, Bures J (1972) Potassium-selective microelectrodes used for measuring the extracellular brain potassium during spreading depression and anoxic depolarization in rats. Brain Res 39:255–259
Wahl M, Unterberg A, Baethmann A, Schilling L (1988) Mediators of blood-brain barrier dysfunction and formation of vasogenic brain edema. J Cereb Blood Flow Metab 8:621–634
Walz W, Hertz L (1983) Functional interactions between neurons and astrocytes. II. Potassium homeostasis at the cellular level. Prog Neurobiol 20:133–183
Welch K, Sadler K (1966) Permeability of the choroid plexus of the rabbit to several solutes. Am J Physiol 210:652–660
Welch K, Sadler K, Gold G (1966) Volume flow across choroidal ependyma of the rabbit. Am J Physiol 210:232–236
Wolff JR (1970) The astrocyte as link between capillary and nerve cell. Triangle 9:153–164
Young W, Rappaport ZH, Chalif DJ, Flamm ES (1987) Regional brain sodium, potassium, and water changes in the rat middle cerebral artery occlusion model of ischemia. Stroke 18:751–759
Zeuthen T, Wright EM (1981) Epithelial potassium transport: tracer and electrophysiological studies in choroid plexus. J Membr Biol 60:105–128

CHAPTER 9

Secretion and Bulk Flow of Interstitial Fluid

H.F. CSERR and C.S. PATLAK

A. Introduction

Turnover of interstitial fluid (ISF) in most tissues depends on net fluid exchanges with plasma and lymph. ISF is produced by the process of filtration across the semipermeable capillary endothelium. It is then cleared from the interstitium either by filtration back into the capillary or by drainage with escaped colloids into the lymphatics. The situation in the central nervous system can be expected to differ. The highly impermeable cerebral capillary endothelium, or blood-brain barrier, restricts filtration between plasma and cerebral ISF, there are no lymphatics, and the brain and spinal cord are surrounded by another extracellular fluid, the cerebrospinal fluid (CSF).

In this chapter, we review evidence supporting a general model of cerebral ISF production and flow in the normal brain, and discuss the functional significance of these flows, using examples drawn primarily from the authors' work. The proposed model of fluid turnover is based on secretion of ISF at the blood-brain barrier coupled with bulk flow of extracellular fluid between brain and CSF.

For the convenience of the reader, the model is presented briefly prior to a more detailed consideration of its individual components. Relevant characteristics of the blood-brain-CSF system are illustrated in Fig. 1. Cerebral ISF is separated from plasma by the blood-brain barrier, whereas it is connected to CSF by a network of specialized extracellular channels (see Sect. B.III). These channels, represented in Fig. 1 by a single perivascular space, provide a pathway for bulk flow of extracellular fluid between brain and CSF. The direction and rate of bulk flow within the perivascular space will be a function of the hydrostatic pressure gradient between brain ISF and CSF ($P_{isf} - P_{csf}$). In the normal brain, it is proposed that secretion of ISF by the cerebral capillary endothelium generates a driving force for bulk flow of ISF from brain to CSF ($P_{isf} > P_{csf}$), and ISF drains via perivascular spaces around arteries and veins into CSF surrounding the outer surface of the brain. Reversal of this gradient, through an increase in P_{csf} and/or a decrease in P_{isf}, will lead to a reversal of flow. This retrograde flow of CSF into brain provides a novel mechanism for hydration of the cerebral interstitium.

B. Factors Governing Extracellular Fluid Exchanges Within the Brain

I. Blood-Brain Barrier

The cerebral capillary endothelium, the site of the blood-brain barrier in most vertebrates (as reviewed by CSERR and BUNDGAARD 1984), has many of the properties of a tight epithelial membrane (as reviewed by CRONE 1986). This analogy is essential to the proposed role of the barrier as a site of ISF secretion (see Sect. D). As in epithelial membranes, bands of tight junctions between adjacent cerebral endothelial cells impede passive flows across the endothelium, whereas specific transcellular transport mechanisms facilitate the flow of a variety of substances between capillary plasma and ISF. Other features characteristic of secretory epithelia include an increased density of mitochondria (OLDENDORF et al. 1977); the presence of specific transport mechanisms for a number of ions including Na^+ (MURPHY and JOHANSON 1989) and K^+ (as reviewed by KATZMAN and PAPPIUS 1973); polarity in the distribution of Na^+-pumps, to the abluminal border (FIRTH 1977); and a low hydraulic conductivity (FENSTERMACHER 1984). The electrical resistance of the cerebral capillary is so high as to rank it among the "tightest" of epithelia (CRONE and OLESEN 1982).

II. Cerebral ISF and CSF

ISF exists as a thin film surrounding the cellular elements of the nervous tissue, and its volume constitutes 15%–20% of brain volume (LEVIN et al. 1970; NICHOLSON and RICE 1986). CSF fills the ventricular cavities and surrounds the outer pial-glial surface of the brain (Fig. 1). Its volume, relative to brain, shows considerable species variation (CSERR et al. 1980). In man, total CSF volume, 140 ml, approximates 10% of the brain weight. The compositions of ISF and CSF are similar, but both differ markedly from that of the general ISF and lymph. For example, both fluids exhibit marked homeostasis of K^+, Ca^{2+}, and Mg^{2+} concentration and, assuming that ISF protein concentration is as low as that in CSF, have a protein concentration only about 1/400 that of plasma (as reviewed by BRADBURY 1979).

Secretion of CSF by the choroid plexuses accounts for most of CSF production, but drainage of cerebral ISF into CSF constitutes an additional, extrachoroidal source accounting for about 10% of total fluid production (CSERR 1984). CSF leaves the central nervous system by draining into blood, either directly across small valved structures, the arachnoid villi, or indirectly via deep cervical lymph (as reviewed by BRADBURY and CSERR 1985). The hypothesis of ISF turnover discussed in this review views the circulation of cerebral ISF as analogous to that of CSF; specifically, it proposes that ISF is secreted at the blood-brain barrier, through the coupled

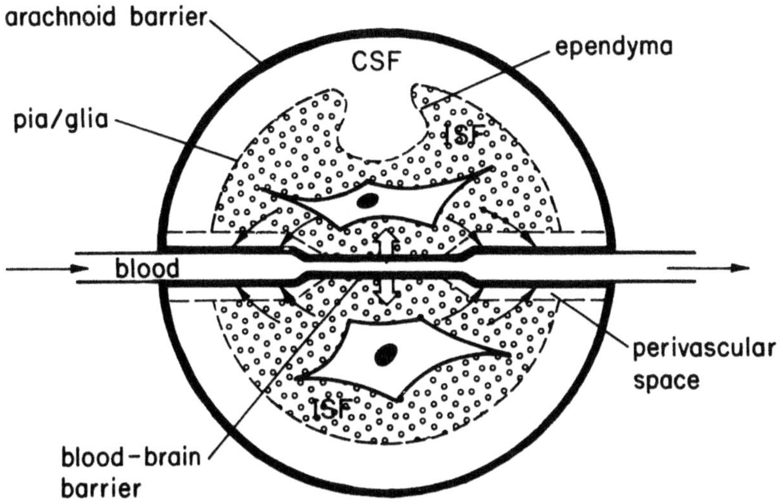

Fig. 1. Schematic diagram illustrating a model of ISF turnover in the brain based on secretion of cerebral ISF by the blood-brain barrier (*open arrows*) and bulk flow of ISF between brain and CSF (*curved arrows*) via perivascular spaces. CSF, cerebrospinal fluid; ISF, interstitial fluid

transport of salts and water, and then drains from the nervous tissue by a pressure-dependent mechanism into the surrounding CSF.

III. Pathways of Flow Between Brain and CSF

The permeable membranes covering the inner (ependymal) and outer (pial-glial) surfaces of the brain (Fig. 1) permit rapid diffusional exchange between CSF and the ISF in superficial portions of the underlying nervous tissue. The resistance to bulk flow through the narrow (20 nm) intercellular clefts which characterize most of the central nervous tissue is too high, however, to accommodate appreciable flow of fluid via this pathway (WELCH 1970). Bulk flow requires spaces of a larger caliber (FENSTERMACHER and PATLAK 1976).

The perivascular space appears to be the main channel for flow (Fig. 1). This space extends along arteries and veins penetrating into the nervous tissue, down to the point at which arterioles and venules merge with capillaries; it also surrounds vessels on the outer, subarachnoid surface of the brain (KRAHN 1982; KRISCH et al. 1984; HUTCHINGS and WELLER 1986). Thus perivascular spaces can be categorized according to their location, as either intracerebral or subarachnoid. Since the outer sheath of the perivascular space is permeable and, further, since blood vessels extend to all portions of the brain, the perivascular space provides a pathway for bulk flow of extracellular fluid between CSF and the entire central nervous system (ZHANG et al. 1990).

Perivascular spaces communicate at the outer and inner surfaces of the brain with a number of other extracellular channels (not illustrated), including the subpial space, tissue spaces within the arachnoid trabeculae and the arachnoid (Krahn 1982; Krisch et al. 1984; Alcolado et al. 1988), and the subependymal zone (Cserr et al. 1977). Expanded ISF spaces between fiber tracts in white matter may also serve as channels of flow (Cserr et al. 1977), especially under pathological conditions (e.g., Reulen et al. 1977).

C. Bulk Flow of ISF

I. Methods of Study

In order to demonstrate bulk flow of ISF it is necessary to distinguish migration of test substances due to diffusion and convection. Methods have been of two types. In the first, the observed distribution of a single test compound within brain tissue is compared with that predicted on the basis of diffusion theory. Bulk flow is indicated by deviation from the predicted pattern of distribution. In the second method, the distribution of test compounds within brain tissue is compared for two or more compounds of markedly different diffusion coefficient. The presence of bulk flow is indicated by components of tracer movement which are independent of diffusion coefficient, i.e., which are the same for compounds of different diffusion coefficient. The second method offers the advantage that the additional tracer(s) serve as controls for differences in experimental factors other than diffusion coefficient, e.g., in tortuosity of the extracellular space.

It is important to appreciate that analysis of tissue diffusion profiles may fail to detect bulk flow, as pointed out by Fenstermacher and Patlak (1976), if the flow of ISF is perpendicular to the direction of tracer diffusion, or if the convective flux of test compound in specialized channels (e.g., perivascular spaces) is small compared to the diffusive flux. Thus, a negative result, based solely on analysis of diffusion profiles, does not necessarily indicate the absence of bulk flow of ISF.

Analyses of the mechanism of tracer distribution within the nervous tissue can be either qualitative or quantitative, depending on the method of tracer detection. Qualitative studies have employed light microscopic evaluation of the pattern of tissue tracer distribution (e.g., Klatzo et al. 1967; Cserr and Ostrach 1974; Reulen et al. 1977), while quantitative studies have typically examined tracer distribution by quantitative autoradiography (e.g., Nakagawa et al. 1987) or radioassay of selected tissue samples (e.g., Rosenberg et al. 1980; Pullen et al. 1987).

II. Drainage of ISF from Brain to CSF

1. Historical Perspectives

His (1865) first ascribed a lymphatic function to the perivascular spaces of the brain. He injected colloidal material into the brain using a multiple intracerebral puncture technique, and found that tracer spread from the injection site along the vast network of perivascular spaces, in the same way that lymphatics are injected by tissue injection in other organs. Subsequent histological studies with other tracers supported the proposed lymphatic function of these channels and, in addition, demonstrated that the normal outward flow of ISF could be reversed under some conditions, such as raised CSF pressure, with retrograde flow of CSF into brain (as reviewed by Davson 1956).

The concept of perivascular drainage of ISF from the normal brain into the subarachnoid space was subsequently abandoned in the 1950s on the basis of light microscopic evidence presented by Woollam and Millen (1954) questioning the continuity between the perineuronal and perivascular spaces. According to these workers, the perivascular space is in the nature of a cul-de-sac, open toward the subarachnoid space but closed toward the brain parenchyma, and consequently cannot serve as a channel for flow.

The conclusion that there is no perivascular flow of cerebral ISF led to questions as to how typical "lymphatic" functions are performed in the central nervous system. Interest focused particularly on the need to explain the removal from the interstitium of proteins, metabolites, and other polar compounds too large to cross the blood-brain barrier. Believing the ISF to be stagnant, Davson et al. (1963) suggested that substances could be cleared from brain by diffusion into CSF. This was referred to as the "sink" action of CSF.

In a 1971 review, Cserr suggested the need to reconsider the possibility of bulk flow of ISF based on (1) a lack of direct evidence that substances are cleared from brain by diffusion, (2) new evidence suggesting that a significant fraction of CSF is produced by the brain, and (3) a theoretical comparison of diffusion and bulk flow. This comparison suggested that, whereas diffusion might be an important mechanism of CSF-brain exchange for tissue located close to CSF, bulk flow would provide a more efficient mechanism for clearing substances from deep within the neuropil, especially in the case of large molecules. Finally, electron microscopy removed the major objection to bulk flow of cerebral ISF by demonstrating continuity between the perivascular space and ISF (Brightman and Reese 1969), i.e., the perivascular space could no longer be viewed as a cul-de-sac.

Studies conducted over the past 20 years have confirmed many of the early concepts of bulk flow between brain and CSF, and have provided the first quantitative estimates of these flows. In addition, appreciation for the unique permeability characteristics of the cerebral endothelium

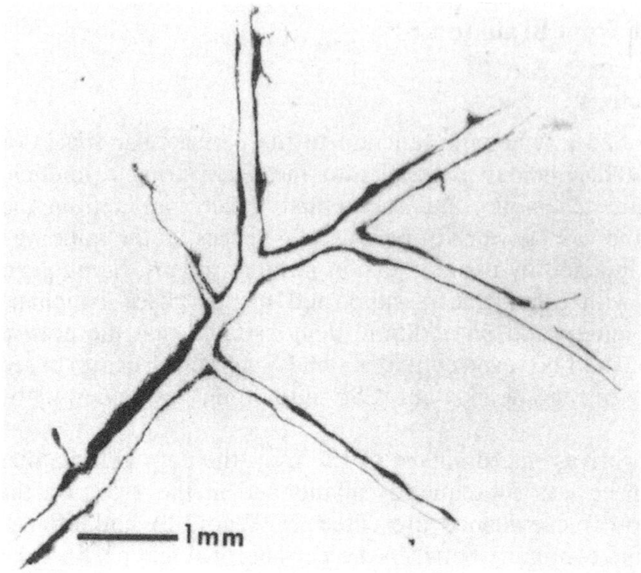

Fig. 2. Perivascular staining of large arteries on the surface of the brain following injection of horseradish peroxidase into the midbrain of a rat. Tracing from a color photograph of a whole-brain preparation 1 h 13 min after tracer injection. (From SZENTISTVANYI et al. 1984)

has demanded special consideration of the mechanism of cerebral ISF production.

2. Recent Studies

CSERR and colleagues reexamined the old question of bulk flow of ISF using a tissue clearance technique, in which extracellular tracers are microinjected slowly into brain tissue through a guide cannula implanted stereotaxically 1 week previously. The 1-week recovery period provides time for return of normal barrier function following cannula placement (CSERR and BERMAN 1978). The characteristics of convection of ISF are then evaluated by following the pathways and kinetics of tracer clearance from the tissue injection site.

Analysis of the pathways of tracer distribution confirmed the importance of perivascular spaces, including their extension along vessels within the subarachnoid space, as conduits for the outward drainage of ISF from brain. Tissue-injected tracers spread extensively within perivascular spaces surrounding blood vessels supplying the tissue injection site, and could be traced in high concentration to large vessels on the surface of the brain (Fig. 2), before they finally leaked across the leptomeningeal sheath surrounding the subarachnoid perivascular space and into CSF. Results also

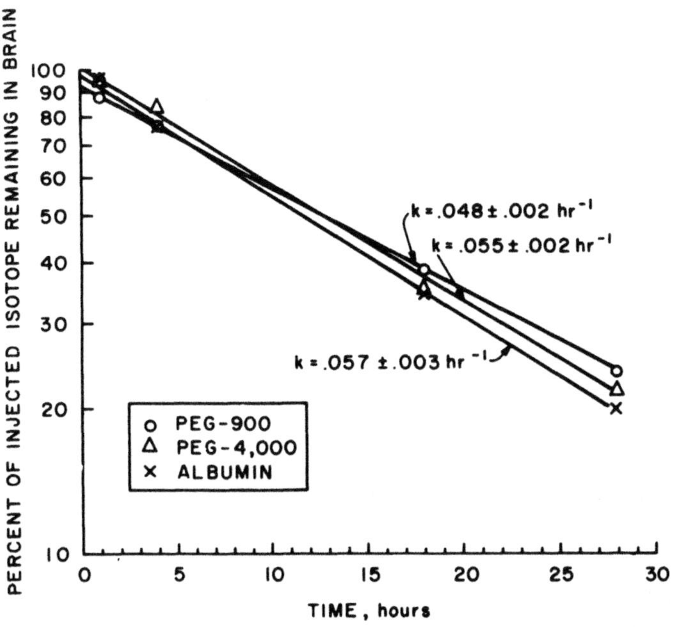

Fig. 3. Disappearance of radiolabelled polyethylene glycols and albumin from brain following injection into rat caudate nucleus. *k* is the first-order rate constant for total efflux from brain. Similarity in efflux rates, despite a fivefold range in diffusion coefficient, is consistent with convective rather than diffusive losses from brain. (From CSERR et al. 1981)

confirmed the importance of other channels of flow, including the subependymal zone of the ventricular ependyma, subpial space, tissue spaces within arachnoid trabeculae and the arachnoid, and spaces between fiber tracts in the white matter (CSERR and OSTRACH 1974; CSERR et al. 1977; BRADBURY et al. 1981; SZENTISTVANYI et al. 1984; ICHIMURA et al. 1991).

Kinetic analyses of tracer efflux provided new evidence for bulk flow of ISF (CSERR et al. 1977, 1981). Specifically, it was found that three extracellular tracers of markedly different diffusion coefficient – two polyethylene glycols (900 and 4000 daltons) and radioiodinated serum albumin (69000 daltons) – are cleared from brain according to a single rate constant, a result inconsistent with diffusion (Fig. 3). The conclusion that tracers are cleared principally by convection meant that this efflux could be used to estimate the rate of ISF drainage from brain. Table 1 compares rates of ISF outflow from different regions of rat and rabbit brain estimated using radioiodinated albumin as a marker of flow.

Results of ROSENBERG et al. (1980) also indicate a volume flow of ISF in normal brain tissue, although only in white matter. They perfused the ventricular system in cats with artificial CSF containing a single tracer, [^{3}H]sucrose, and measured the concentration profiles of radioactivity in

Table 1. Rates of ISF outflow from different regions of rabbit and rat brain

Brain region	ISF outflow rate[a] (μl g brain^{-1} min^{-1})	
	Rabbit	Rat
Caudate nucleus	0.11	0.18
Internal capsule	0.10	0.19
Midbrain	0.15	0.29

[a] Values of ISF outflow for rabbit from YAMADA et al. (1991) and for rat from SZENTISTVANYI et al. (1984).

caudate nucleus and periventricular white matter. Bulk flow is indicated in this technique by a change in the apparent tissue diffusion coefficient for sucrose over time. Apparent diffusion coefficients decreased with time in the white matter, consistent with ISF flow toward the ventricle with an estimated velocity of 0.5 μm/min.

Failure to detect flow within the caudate nucleus, based on the analysis of the sucrose diffusion profile, does not necessarily indicate a lack of flow (as discussed in Sect. C.I). It may be explained by restriction of the area available for bulk flow to specific low-resistance channels, as compared to the broad front available for diffusion, and/or by differences in the relative directions of bulk flow and diffusion between the two test sites. In white matter ISF appears to flow parallel to fiber tracts, whereas in grey matter it follows perivascular spaces and the subependymal zone. These different patterns of flow may yield a detectable vector of flow perpendicular to the ventricular surface only in the case of white matter.

Drainage of ISF from brain to CSF implies a net hydrostatic pressure gradient, favoring bulk flow from the interstitium to the subarachnoid space (i.e., $P_{isf} > P_{csf}$). The existence of this pressure gradient has not been confirmed experimentally (WIIG and REED 1983). This is not surprising, however, assuming that ISF outflow is slow and through low resistance channels, i.e., the pressure gradient is probably too small to detect experimentally.

III. Retrograde Flow of CSF Into Brain

Whereas tissue-injected tracers move centrifugally within perivascular spaces from the interstitium toward CSF (see Fig. 2), CSF-injected tracers normally fail to penetrate in the reverse direction. However, under conditions when it is anticipated that the normal pressure gradient between ISF and CSF is reversed (i.e., when $P_{csf} > P_{isf}$), the direction of perivascular tracer movement is reversed, suggesting retrograde flow from CSF into brain (as reviewed by BRADBURY and CSERR 1985). These conditions include elevation of P_{csf}; plasma hypertonicity, which dehydrates the brain and decreases P_{isf} (WIIG and REED 1983); and a severe reduction in blood pressure, which presumably also decreases P_{isf}.

Quantitative analyses of tissue diffusion profiles or of tracer fluxes from CSF to brain during ventriculocisternal perfusion have confirmed changes in the rate and direction of ISF flow within the nervous tissue indicated by earlier histological studies, both in response to elevation of CSF pressure (ROSENBERG et al. 1982; PULLEN and CSERR 1984) and to plasma hypertonicity (ROSENBERG et al. 1980; PULLEN et al. 1987). These studies have not, however, evaluated the role of the perivascular space, or other pathways, as channels for retrograde flow of CSF into brain.

The magnitude of the volume shift of CSF into brain tissue, in response to elevation of plasma osmolality, is large – 44 µl for a 60 mosmol/kg H_2O increase in osmolality (PULLEN et al. 1987). The time course of this flow is rapid (30 min) compared to the much slower drainage of ISF from brain (half-time of 6–16 h; SZENTISTVANYI et al. 1984; YAMADA et al. 1991), and the presence of an appropriately directed pressure gradient (i.e., $P_{csf} > P_{isf}$) has been confirmed experimentally (WIIG and REED 1983). The osmotically stimulated influx of CSF and contained electrolytes into brain plays a major role in the regulation of brain volume during acute hypernatremia (see Sect. F.II).

A much more limited analysis of the effects of CSF pressure on tracer influx into brain, and on brain electrolyte content, suggests roughly equivalent results; namely, a pressure-induced shift of CSF into brain and an accompanying increase in tissue electrolyte content (PULLEN and CSERR 1984).

IV. Studies Using Horseradish Peroxidase as Tracer

Horseradish peroxidase (HRP) ventriculocisternal perfusion for as little as 6 min results in the labelling of much of the perivascular network in the cat brain. Based on these results RENNELS et al. (1985) have proposed a rapid "paravascular" circulation of extracellular fluid through the brain, with CSF flowing into the brain via arterial perivascular spaces and leaving via venous perivascular spaces a few minutes later. The rapid perivascular influx of HRP into normal brain is at variance with results obtained with all other CSF-administered tracers, as discussed above.

A rapid paravascular circulation of CSF through brain is also not supported by results obtained using a new technique, in which tracers are microinjected directly into the subarachnoid perivascular space. Analysis of subsequent tracer redistribution under intravital microscopy suggested some bulk flow of fluid within the perivascular space, around both arteries and veins. However, this flow was slow (minutes) and its direction varied in an unpredictable way (ICHIMURA et al. 1991); i.e., it was inconsistent with the continuous paravascular circulation of CSF through brain proposed on the basis of results obtained with HRP.

It is possible to explain the discrepancy between results with HRP and other tracers in terms of the very great sensitivity of the tetramethylbenzidine reaction used to detect HRP, especially since a much reduced tracer distribution was found when the less sensitive diaminobenzidine reaction was used (RENNELS et al. 1985). Two mechanisms have been suggested to explain how small amounts of CSF may be carried along the vascular tree, yielding extremely low tracer concentrations that are only detectable with the sensitivity of the assay for HRP. Both mechanisms view oscillations within the perivascular space as the driving force for tracer penetration, but differ with respect to the source of these oscillations. One holds the arterial pulse as responsible, since partial ligation of the cerebral blood supply reduces the perivascular influx of HRP (RENNELS et al. 1985); the other favors a vascular pump, caused by vasoconstriction and dilation of cerebral blood vessels (ICHIMURA et al. 1991). The vascular pump could be associated with normal vasomotion, and also be induced by experimental interventions, such as the administration of toxic HRP type II (COTRAN and KARNOVSKY 1967) or the partial ligation of the carotid arteries used by RENNELS et al. (1985) to damp the vascular pump. It is also consistent with the slow back and forth movement of tracer within the perivascular space seen following tracer injection directly into the subarachnoid perivascular space (ICHIMURA et al. 1991).

D. Secretion of ISF by the Cerebral Endothelium: An Hypothesis

What is the mechanism of cerebral ISF production – metabolic water production, filtration of capillary plasma, or active secretion by the cerebral endothelium? Metabolic water production can be excluded on the basis that estimated rates of ISF production are more than three times total cerebral water production (CSERR 1981). Furthermore, the impermeability of the cerebral endothelium to polar solutes generally, including ions, implies that any free water would be cleared by osmosis into plasma and not be retained within the brain tissue. Filtration can also be excluded as a major mechanism of ISF production based on the barrier's low hydraulic conductivity and impermeability to ions (CSERR and PATLAK 1991).

The failure of other mechanisms to account for estimated rates of ISF production focuses attention on secretion by the cerebral endothelium, through the coupled transport of solutes and water, as the most likely mechanism of ISF production (CSERR 1981). Although there is no direct evidence for secretion, it is supported by the structural and functional similarities of the cerebral endothelium to a secretory epithelium (see Sect. B.I).

Table 2. Recovery of radio-iodinated albumin in deep cerivcal lymph following injection into brain or CSF[a], where recovery in lymph is given as a percentage of total outflow from the central nervous system

Species	Injection site	Duration of lymph collection (h)	Recovery in lymph (%)
Rabbit	Caudate nucleus	25	47
Rabbit	Internal capsule	25	22
Rabbit	Midbrain	25	18
Rabbit	CSF	6	30
Cat	CSF	8	14
Sheep	CSF	26	32

[a] Values from the literature compiled by YAMADA et al. (1991).

E. Outflow of ISF to Blood and Deep Cervical Lymph

Following drainage from the nervous tissue into the surrounding CSF, ISF flows with CSF either into blood or lymph. The main connection to the lymphatics in rabbits (BRADBURY and WESTROP 1983) and rats (HARLING-BERG 1989), the only species in which this has been quantified, is via extensions of the subarachnoid space around the olfactory nerves, to the cribriform plate and nasal submucosa. From submucous spaces of the nose, large molecular weight substances pass into afferent lymphatics.

Analysis of the kinetics of outflow to lymph, using radioiodinated serum albumin as tracer, indicates that protein outflow from the central nervous system to deep cervical lymph is much slower, and the magnitude of this outflow much larger, than previously appreciated. The amount of brain-injected albumin passing through deep cervical lymph in the rabbit increases slowly with time, reaching a peak some 15–20 h after intracerebral injection (YAMADA et al. 1991). A similar delay is seen after CSF-tracer administration, although of shorter duration; specifically, 3–4 h in the rabbit, somewhat later in the cat, and after 8–12 h in the sheep (as reviewed by BRADBURY and CSERR 1985).

A major fraction of the radiolabelled albumin cleared from the brain or CSF (14%–47%) can be recovered from cervical lymph (Table 2). It is important to emphasize that these values are minimal estimates of total protein outflow through lymph, both because of the long delay in passage from the central nervous system to lymph, and because some protein is lost to blood during passage through the nasal submucosa and cervical nodes, i.e., before it is collected for analysis (BRADBURY and WESTROP 1983). The connection to the lymphatic system also extends to the spinal cord (BRIERLEY and FIELD 1948), although this has not been quantified.

F. Functions of ISF Flow

I. Clearance

The normal outflow of ISF provides a mechanism for clearing compounds from the interstitium that cannot be absorbed directly into the blood. Compounds cleared from brain tissue in this way include not only particulate matter and plasma colloids, as in other tissues, but also many smaller substances, such as drugs that have leaked across the blood-brain barrier and polar end-products of brain metabolism. Given the slow rate of ISF turnover, it can be anticipated that this mechanism of tissue clearance is most effective for substances requiring a slow rate of removal.

II. Brain Volume Regulation

The importance of secretion and bulk flow of brain extracellular fluid for brain volume regulation was first proposed by BRADBURY (1979). Figure 4 illustrates the role of these flows in volume regulation in normal brain and during acute hypernatremia in the context of a four-compartment model of electrolyte exchange developed by Cserr and colleagues (as reviewed by CSERR and PATLAK 1991).

Under normal conditions (Fig. 4A), brain volume remains constant despite the continual secretion of fluid at the blood-brain barrier. The pump-mediated influx of Na^+ and Cl^- from plasma is matched in the steady state by an equivalent efflux of Na^+ and Cl^-, due to drainage of extracellular fluid into the surrounding CSF.

During acute hypernatremia (Fig. 4B) the brain shrinks less than predicted on the basis of ideal osmotic behavior and the observed degree of volume regulation after 2 h of this stress appears to be accounted for completely by net tissue uptake of three ions – Na^+, Cl^-, and K^+ (CSERR et al. 1987a). Retrograde flow of CSF into the osmotically dehydrated brain (see Sect. C.III) accounts for most (60%–75%) of the volume regulatory uptake of Na^+ and Cl^-; pump-mediated secretion at the blood-brain barrier for a lesser amount (15%–20%); and, passive movement across the blood-brain barrier, in response to the elevated plasma concentrations of Na^+ and Cl^- that characterize hypernatremia, for the rest (PULLEN et al. 1987). K^+ enters brain from plasma via a single selective pathway stimulated by hypertonicity (CSERR et al. 1987b).

In addition to the fluxes of electrolytes from plasma and CSF into the cerebral interstitium during acute hypernatremia, there is an osmotically stimulated influx of some extracellular Na^+, Cl^-, and K^+ into brain cells. This cellular uptake of ions results in the selective regulation of brain cell volume; i.e., water loss during acute hypernatremia is entirely from the extracellular compartment (CSERR et al. 1991).

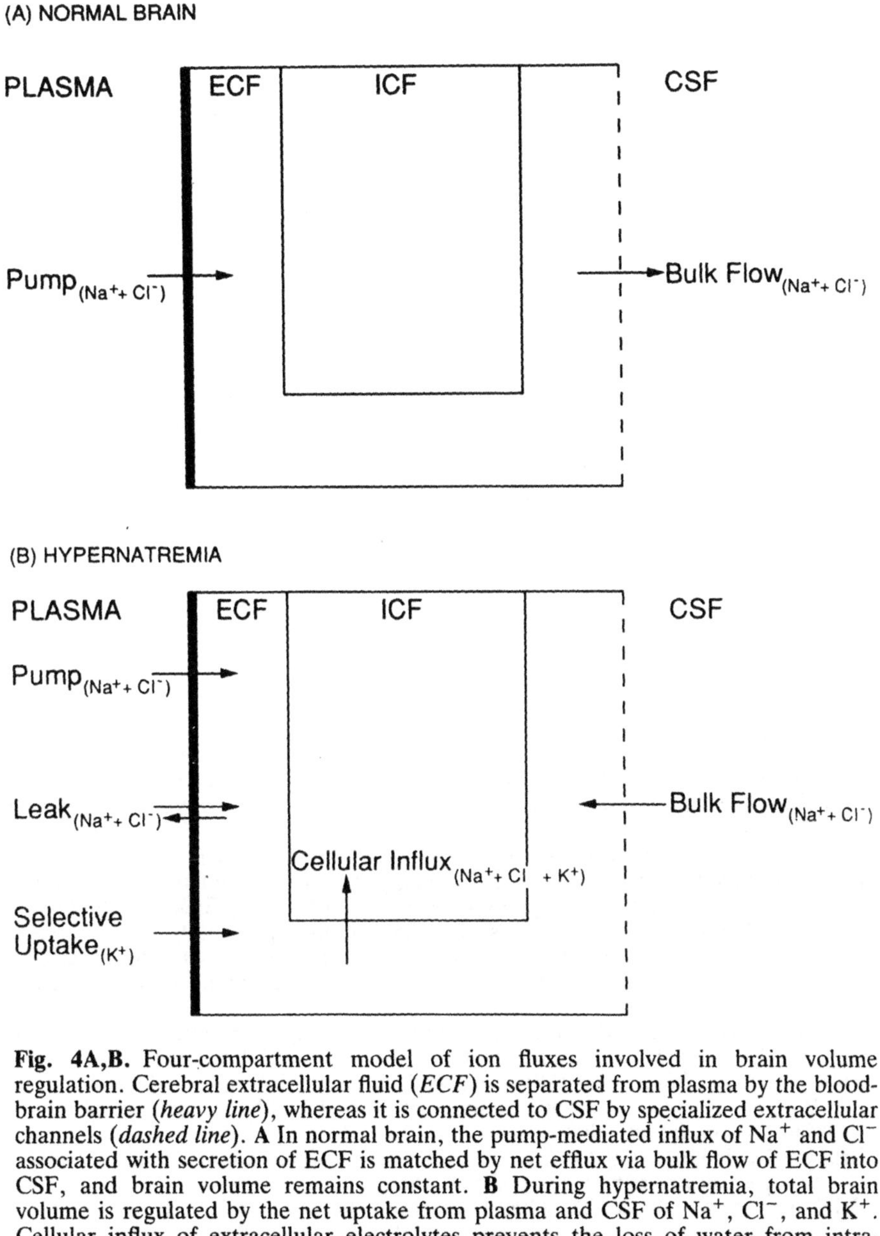

Fig. 4A,B. Four-compartment model of ion fluxes involved in brain volume regulation. Cerebral extracellular fluid (*ECF*) is separated from plasma by the blood-brain barrier (*heavy line*), whereas it is connected to CSF by specialized extracellular channels (*dashed line*). **A** In normal brain, the pump-mediated influx of Na^+ and Cl^- associated with secretion of ECF is matched by net efflux via bulk flow of ECF into CSF, and brain volume remains constant. **B** During hypernatremia, total brain volume is regulated by the net uptake from plasma and CSF of Na^+, Cl^-, and K^+. Cellular influx of extracellular electrolytes prevents the loss of water from intracellular fluid (*ICF*). (From CSERR and PATLAK 1991)

III. Immune Function

Perivascular spaces and other channels of fluid outflow from the central nervous system may be important for filtering and processing of brain-derived antigen (WIDNER 1990), by virtue of their anatomical location and the presence of constitutive antigen-presenting cells, the microglia (HICKEY and KIMURA 1988).

The passage of large molecular weight substances, and of immunocompetent cells (OEHMICHEN 1978), from the central nervous system to lymph provides for the possibility of immune recognition and the generation of antibodies and cell-mediated responses against brain-derived antigens. There have been few studies of the immune response to the lymphatic outflow from brain (as reviewed by BRENDT 1990). However, recent studies of the immune response in cervical nodes have shown that this outflow is effective, both with respect to the generation of antibodies (HOCHWALD et al. 1988; WIDNER et al. 1988; HARLING-BERG et al. 1989) and of $CD8^+$ T cells (DOHERTY et al. 1990) against antigens injected into the central nervous system.

Acknowledgements. We wish to thank Dr. Magnus Bundgaard for assistance in the preparation of Fig. 1. Financial support to H. Cserr is provided by USPHS grant NS 11050.

References

Alcolado R, Weller RO, Parrish EP, Garrod D (1988) The cranial arachnoid and pia mater in man: anatomical and ultrastructural observations. Neuropathol Appl Neurobiol 14:1–17

Bradbury MWB (1979) The concept of a blood-brain barrier. Wiley, New York

Bradbury MWB, Cserr HF (1985) Drainage of cerebral interstitial fluid and of cerebrospinal fluid into lymphatics. In: Johnston MG (ed) Experimental biology of the lymphatic circulation. Elsevier, Amsterdam, pp 355–394

Bradbury MWB, Westrop RJ (1983) Factors influencing exit of substances from cerebrospinal fluid into deep cervical lymph of the rabbit. J Physiol (Lond) 339:519–534

Bradbury MWB, Cserr HF, Westrop RJ (1981) Drainage of cerebral interstitial fluid into deep cervical lymph of the rabbit. Am J Physiol 240:F329–F336

Brent L (1990) Immunologically privileged sites. In: Johansson BB, Owman C, Widner H (eds) Pathophysiology of the blood-brain barrier. Elsevier, Amsterdam, pp 383–402

Brierley JB, Field EJ (1948) The connexions of the spinal sub-arachnoid space with the lymphoid system. J Anat 82:153–166

Brightman MW, Reese TS (1969) Junctions between intimately apposed cell membranes in the vertebrate brain. J Cell Biol 40:648–677

Cotran RZ, Karnovsky MJ (1967) Vascular leakage induced by horseradish peroxidase in the rat. Proc Soc Exp Biol Med 126:557–561

Crone C (1986) The blood-brain barrier as a tight epithelium: where is information lacking? Ann NY Acad Sci 481:174–185

Crone C, Olesen SP (1982) Electrical resistance of brain capillary endothelium. Brain Res 241:49–55

Cserr HF (1971) Physiology of the choroid plexus. Physiol Rev 51;273–311

Cserr HF (1981) Convection of brain interstitial fluid. In: Kovach AGB, Hamar J, Szabo L (eds) Advances in physiological science, vol 7. Pergamon, Budapest, pp 337–341

Cserr HF (1984) Convection of brain interstitial fluid. In: Shapiro K, Marmarou A, Portnoy H (eds) Hydrocephalus. Raven, New York, pp 59–68

Cserr HF, Berman BJ (1978) Iodide and thiocyanate efflux from brain following injection into rat caudate nucleus. Am J Physiol 235:F331–F337

Cserr HF, Bundgaard M (1984) Blood-brain interfaces in vertebrates: a comparative approach. Am J Physiol 246:R277–R288

Cserr HF, Ostrach LH (1974) Bulk flow of interstitial fluid after intracranial injection of Blue Dextran 2000. Exp Neurol 45:50–60

Cserr HF, Patlak CS (1991) Regulation of brain volume under isosmotic and anisosmotic conditions. In: Gilles R, Hoffmann EK, Bolis L (eds) Advances in comparative and environmental physiology, vol 9. Springer, Berlin Heidelberg New York, pp 61–80

Cserr HF, Cooper DN, Milhorat TH (1977) Flow of cerebral interstitial fluid as indicated by the removal of extracellular markers from rat caudate nucleus. Exp Eye Res [Suppl] 25:461–473

Cserr HF, Bundgaard M, Ashby JK, Murray M (1980) On the anatomic relation of choroid plexus to brain: a comparative study. Am J Physiol 238:R76–R81

Cserr HF, Cooper DN, Suri PK, Patlak CS (1981) Efflux of radiolabeled polyethylene glycols and albumin from rat brain. Am J Physiol 240:F319–F328

Cserr HF, DePasquale M, Patlak CS (1987a) Regulation of brain water and electrolytes during acute hyperosmolality in rats. Am J Physiol 253:F522–F529

Cserr HF, DePasquale M, Patlak CS (1987b) Volume regulatory influx of electrolytes from plasma to brain during acute hperosmolality. Am J Physiol 253:F530–F537

Cserr HF, DePasquale M, Nicholson C, Patlak CS, Pettigrew, KD, Rice ME (1991) Extracellular volume decreases while cell volume is maintained by uptake of ions in rat cerebral cortex during acute hypernatraemia. J Physiol (Lond) (in press)

Davson H (1956) Physiology of the ocular and cerebrospinal fluids. Churchill, London, p 50

Davson H, Kleeman CR, Levin E (1963) The blood-brain barrier. In: Hogben AM, Lindgren P (eds) Drugs and membranes. Pergamon, Oxford, vol 4, pp 71–94

Doherty PC, Allan JE, Lynch F, Ceredig R (1990) Dissection of an inflammatory process induced by $CD8^+$ T cells. Immunol Today 11:55–59

Fenstermacher JD (1984) Volume regulation of the central nervous system. In: Staub NC, Taylor AE (eds) Edema. Raven, New York, pp 383–404

Fenstermacher JD, Patlak CS (1976) The movements of water and solutes in the brains of mammals. In: Pappius HM, Feindel W (eds) Dynamics of brain edema. Springer, Berlin Heidelberg New York, pp 87–94

Firth JA (1977) Cytochemical localization of the K^+ regulation interface between blood and brain. Experientia 33:1093–1094

Harling-Berg C (1989) The humoral immune response to human serum albumin infused into the cerebrospinal fluid of the rat. PhD thesis, Brown University, Providence

Harling-Berg C, Knopf PM, Merriam J, Cserr HF (1989) Role of cervical lymph nodes in the systemic humoral immune response to human serum albumin microinfused into rat CSF. J Neuroimmunol 25:185–193

Hickey WF, Kimura H (1988) Perivascular microglial cells of the CNS are bone marrow-derived and present antigen in vivo. Science 239:290–292

His W (1865) Über ein perivasculäres Kanalsystem in den nervösen Central-Organen und über dessen Beziehungen zum Lymphsystem. Z Wiss Zool 15:127–141

Hockwald GM, van Driel A, Robinson ME, Thorbecke GJ (1988) Immune response in draining lymph nodes and spleen after intraventricular injection of antigen. Int J Neurosci 39:299–306

Hutchings M, Weller RO (1986) Anatomical relationships of the pia mater to cerebral blood vessels in man. J Neurosurg 65:316–325
Ichimura T, Fraser PA, Cserr HF (1991) Distribution of extracellular tracers in perivascular spaces of the rat brain. Brain Res (in press)
Katzman R, Pappius HM (1973) Brain electrolytes and fluid metabolism. Williams and Wilkins, Baltimore, p 125
Klatzo I, Wisniewski H, Steinwall O, Streicher E (1967) Dynamics of cold injury edema. In: Klatzo I, Seitelberger F (eds) Brain edema. Springer, Berlin Heidelberg New York, pp 554–563
Krahn V (1982) The pia mater at the site of entry of blood vessels into the central nervous system. Anat Embryol (Berl) 164:257–263
Krisch B, Leonhardt H, Oksche A (1984) Compartments and perivascular arrangement of the meninges covering the cerebral cortex of the rat. Cell Tissue Res 238:459–474
Levin VA, Fenstermacher JD, Patlak CS (1970) Sucrose and inulin space measurements of ceresbral cortex in four mammalian species. Am J Physiol 219:1528–1533
Murphy VA, Johanson CE (1989) Acidosis, acetazolamide, and amiloride: effects on ^{22}Na transfer across the blood-brain and blood-CSF barriers. J Neurochem 52:1058–1063
Nakagawa H, Groothuis D, Owens E, Fenstermacher J, Patlak C, Blasberg R (1987) Dexamethasone effects on (^{125}I) albumin distribution in experimental RG-2 gliomas and adjacent brain. J Cereb Blood Flow Metab 7:687–701
Nicholson C, Rice ME (1986) The migration of substances in the neuronal microenvironment. Ann NY Acad Sci 481:55–68
Oehmichen M (1978) Mononuclear phagocytes in the central nervous system. Springer, Berlin Heidelberg New York, pp 65–82 (Neurology series, vol 21)
Oldendorf WH, Cornford ME, Brown WJ (1977) The large apparent work capability of the blood-brain barrier: a study of the mitochondrial content of capillary endothelial cells in brain and other tissues of the rat. Ann Neurol 1:409–417
Pullen RGL, Cserr HF (1984) Pressure dependent penetration of CSF into brain. Fed Proc 43:2521
Pullen RGL, DePasquale M, Cserr HF (1987) Bulk flow of cerebrospinal fluid into brain in response to acute hyperosmolality. Am J Physiol 253:F538–F545
Rennels ML, Gregory TF, Blaumanis OR, Fujimoto K, Grady PA (1985) Evidence for a "paravascular" fluid circulation in the mammalian central nervous system, provided by the rapid distribution of tracer protein throughout the brain from the subarachnoid space. Brain Res 326:47–63
Reulen HJ, Graham R, Spatz M, Klatzo I (1977) Role of pressure gradients and bulk flow in dynamics of vasogenic brain edema. J Neurosurg 46:24–35
Rosenberg GA, Kyner WT, Estrada E (1980) Bulk flow of brain interstitial fluid under normal and hyperosmolar conditions. Am J Physiol 238:F42–F49
Rosenberg GA, Kyner WT, Estrada E (1982) The effect of increased CSF pressure on interstitial fluid flow during ventriculocisternal perfusion in the cat. Brain Res 232:141–150
Szentistvanyi I, Patlak CS, Ellis RA, Cserr HF (1984) Drainage of interstitial fluid from different regions of rat brain. Am J Physiol 246:F835–F844
Welch K (1970) Discussion. In: Coxon RV (ed) Proceedings of a symposium on the blood-brain barrier. Truex, Oxford, pp 170–171
Widner H (1990) Immunological basis for intracerebral reconstructive transplantations. Tryckt hos Graphic Systems, Malmo, pp 34–35
Widner H, Moller G, Johansson BB (1988) Immune response in deep cervical lymph nodes and spleen in the mouse after antigen deposition in different intracerebral sites. Scand J Immunol 28:563–571
Wiig H, Reed RK (1983) Rat brain interstitial fluid pressure measured with micropipettes. Am J Physiol 244:H239–H246

Woollam DHM, Millen JW (1954) Perivascular spaces of the mammalian central nervous system. Biol Rev 29:251–283
Yamada S, DePasquale M, Patlak CS, Cserr HF (1991) Albumin outflow into deep cervical lymph from different regions of rabbit brain. Am J Physiol (in press)
Zhang ET, Inman CBE, Weller RO (1990) Interrelationships of the pia mater and the perivascular (Virchow-Robin) spaces in the human cerebrum. J Anat 170: 111–123

CHAPTER 10

Trace Metal Transport at the Blood-Brain Barrier

M.W.B. BRADBURY

A. Introduction

A number of trace metals are normally present within the central nervous system (CNS), and may have effects on its function. Several, e.g., zinc, iron, copper and manganese are essential for normal brain development and function. Others, e.g., lead and mercury, have no known essential role, but may be toxic even at low concentrations. Of the essential metals, most may be toxic to the brain at high concentration, e.g., zinc, copper and manganese.

Initial entry into brain must occur from blood and must involve the metal crossing the blood-brain barrier (BBB). In general, one might anticipate this cellular barrier to be relatively impermeable to highly charged metallic ions or to their polar complexes. Hence, the presence of special processes must be suspected. Such mechanisms might be important in abstracting essential metals from blood into brain and in retaining them there, whereas any specific transport of toxic metals at this barrier will obviously have profound implications for neurotoxicology.

Some investigations have been made into transport of trace metals across simpler biological membranes, e.g., the plasma membrane of the red cell, but information is very sparse. There have also been a number of studies of absorption by the gut and transport within the body, but most of these have been made from an interest in general pharmacokinetics rather than from the objective of elucidating transport mechanisms. Primary concern must be with the chemical form in which the metal crosses the BBB. In contrast to the situation of the common cations sodium, potassium and magnesium, there is proportionately very little of the free ionized

---

Abbreviations

BBB	Blood-brain barrier	EDTA	Ethylenediaminetetraacetic acid
CSF	Cerebrospinal fluid	HEPES	Hydroxyethylpiperazine ethanesulphonic acid
DIDS	4,4′-diisothiocyanostilbene-2,2'-disulphonic acid	FCCP	Carbonyl cyanide 4-(trifluoromethoxy) phenylhydrazone

fraction of trace metals present in body fluids (MAY et al. 1977). A consideration of the chemical forms in which trace metals may occur in serum and other extracellular fluids is obviously a first step in assessing what species might actually be transported across the brain endothelium.

Trace metals are present in serum in at least four possible chemical states: (1) as the free metallic ion; (2) as a low molecular weight complex with either another electrolyte ion, e.g., hydroxyl or chloride, or with an amino acid or other organic acid, e.g., cysteine, histidine, citrate; (3) bound reversibly with a protein, often albumin; and (4) tightly bound with a specific binding protein, e.g., iron-transferrin, copper caeruloplasmin and zinc α_2-macroglobulin. A radio-isotopic metallic ion may not readily exchange with the specifically bound metal in serum in vitro, and there are special mechanisms in vivo for releasing the metal from its binding protein. The individual roles of zinc, iron and copper in human biology have been recently reviewed in MILLS (1989), HUEBERS and FINCH (1987), and EVERED and LAWRENSON (1980) respectively.

B. Theoretical and Experimental Approach

The final objective must be to characterize as fully as possible the transport mechanism or mechanisms for the trace metal under consideration. As discussed, each trace metal is often present in serum in all of four general chemical states. Each of these may well be represented by further individual species, e.g., zinc forms several complexes of low molecular weight including at least two with each of L-histidine and L-cysteine (HALLMAN et al. 1971; GIROUX and HENKIN 1972). In order to isolate individual transport systems, it is crucial to study transport under conditions in which it occurs from a medium under the control of the investigator.

One approach is to perfuse the vascular system of the brain or of part of the brain with a fluid of known composition. Another is to study transport into isolated or better cultured cerebral endothelial cells, again from a medium of known composition. Attempts are being made in several laboratories to culture brain endothelium as a tight cellular membrane. Such a preparation, if successful, would permit transport of solutes, including trace metals, to be studies in both directions across a model BBB in vitro.

C. Intravenous Administration in Intact Rat

As a preliminary to use of simplified experimental models such as those suggested above, it is helpful to know what the rate of exchange and flux of a trace metal are across the BBB in the intact animal or man. Most measurements of BBB permeability have been made in the laboratory rat, and those who have looked at trace metals have generally used this species. As with other solutes, the rate of entry into brain may either be estimated after making a continuous intravenous infusion of a radio-isotope of the trace metal or after a single intravenous injection of a bolus of the radiotracer

(OHNO et al. 1978). Intravenous infusion into the anaesthetized animal is best done at a rate which diminishes with time so that the level of the radiotracer remains approximately constant in serum (DAVSON 1955; BRADBURY and DEANE 1986). The diminishing infusion rate can be determined by trial and error, or better by an initial analysis of the decay curve of the radiotracer in serum after single injection into the bloodstream (PRATT 1985). After a set duration of infusion, the rat is decapitated and the radio-isotope estimated in different regions to brain, choroid plexuses and cerebrospinal fluid (CSF) by an appropriate counter or by autoradiography and densitometry.

If it is suspected that the trace metal may bind to the luminal surface of the brain capillaries, it is advisable to wash the circulation arterially with 1 m*M* ethylenediaminetetraacetic acid (EDTA) in isotonic NaCl immediately before decapitation. This also removes residual red cells from the vessels. These may contain very high concentrations of the radiotracer, e.g., as is the case with lead-203. Continuous intravenous infusion has been used by the author and colleagues to estimate flux into brain of ^{203}Pb (BRADBURY and DEANE 1986), ^{65}Zn (ADU et al. 1990) and ^{59}Fe (F. UEDA and M.W.B. BRADBURY, unpublished observations).

The alternative bolus injection method is simpler but depends on the assumption that there is a sufficient reservoir of cold isotope in the brain for there to be no effective back transport of the radio-isotope into blood. The best method of use of this technique is to measure radioactivity in brain at multiple times points after single intravenous injection. If flux is unidirectional, a plot of $C_{br}/C_{s(T)}$ against $\int_0^T C_s \cdot dt/C_{s(T)}$ should be linear, where C_s is the serum concentration of radiotracer and C_{br} the brain concentration after an experiment of duration T (BLASBERG et al. 1983; PATLAK et al. 1983). The slope gives a transfer constant (K_{in}) and any positive intercept with the ordinate, an initial distribution space. It may be noted that the quantity of the abscissa has units of time. Single bolus injection with multiple time point analysis has been used by PULLEN and colleagues to measure brian uptake of ^{67}Ga as a marker for aluminium (PULLEN et al. 1990) and of ^{65}Zn (PULLEN et al. 1991).

A number of points should be considered in undertaking experiments of this general type. The radio-isotope should be of sufficiently high specific activity so that the normal concentration of the metal is not sensibly raised by the radioactive infusion or injection. This can, of course, be readily checked by analysis of the total concentration of the metal in serum with or without infusion or injection. The possibility should be considered that the radiotracer may not be exchanging immediately and completely with all the chemical forms of the metal already present in serum, e.g., although ^{65}Zn exchanges readily with both amino acid- and albumin-complexed zinc, exchange with α_2-macroglobulin-bound zinc is very slow (PARISI and VALLEE 1970) and is thought to require the liver. Finally, it is better not to anticoagulate blood samples. Most anticoagulants, including heparin, may either contain certain trace metals or bind trace metal or both. If arterial samples are centrifuged immediately, plasma may be pipetted at once into a counting vial. Alternatively, the rapidly separated coagulant-free plasma may be allowed to clot to give serum.

D. Short Vascular Perfusion of the Brain etc.

Recently, several relatively simple techniques have been developed for perfusion of either part or the whole of the rat of guinea pig brain (TAKASATO et al. 1984; ZLOKOVIĆ et al. 1986). It appears that the BBB to mannitol remains intact for 20 min or more of perfusion of a physiological saline containing a metabolic inhibitor (LUTHERT et al. 1987). Normal specific transport processes can be demonstrated during perfusion of one hemisphere with simple oxygenated physiological saline (TAKASATO et al. 1984). It seemed that such a technique might be useful for analysing the transport of the more permeable trace metals. The presence of complexing agents in such simple solutions would be largely under the control of the experimenter.

The particular perfusion technique used by the author and colleagues (DEANE and BRADBURY 1990) is a modification of that of TAKASATO et al. (1984). In the anaesthetized rat, the right external carotid, the superior lingual and the thyroid arteries are exposed and ligated. The right common carotid artery is cannulated for infusion at a monitored pressure. After a 5-s saline wash, the circulation of the right hemisphere is perfused with buffered oxygenated saline containing a radio-isotope of the trace metal under study, generally for 1 min. The vasculature is then washed with saline containing 1 m*M* EDTA for 20 s and the rat immediately decapitated. The brain is dissected into various regions for estimation of radio-isotopic uptake by γ-counting. In early experiments, the perfusion pressure was maintained at 10–15 mmHg higher than femoral artery pressure, this artery being cannulated for the comparison. In later experiments, contamination of the perfused vessels with endogenous blood was avoided by section of the heart at the start of the perfusion. The solutions used and results obtained for ^{203}Pb and ^{65}Zn are described in subsequent sections.

It appears that a similar control of the medium from which transport occurs could be achieved by observing uptake of a radio-isotopic trace metal into cultured endothelial cells or better across a membrane of these cells. The author is initiating such experiments with cerebral endothelial cells cultured by the method of HUGHES and LANTOS (1986).

E. Solutions for Brain Perfusion etc.

As indicated above, a trace metal is likely to occur in serum in several general chemical states. Our objective has been to have the trace metal under study present in the perfusion medium or other fluid as a single species which is a (putative) candidate for normal transport across the BBB. An obvious first possibility is to have it as the free ion of the metal – particularly if its normal concentration in serum is not so small as to be insignificant, e.g., Fe^{3+} or Al^{3+}. Pb^{2+} and Zn^{2+} have been investigated in this way.

The trace metal may be added to the physiological salt solution as an inorganic salt, e.g., $PbNO_3$ or $ZnCl_2$. The solubility products of compounds determined by anions in the salt solution must not be exceeded. The normal concentrations of bicarbonate and phosphate in serum limit the lead concentration to about $10^{-8} M$ (T.J.B. SIMONS, personal communication). Again, in the absence of these anions, the free Pb concentration is limited to about $4 \times 10^{-6} M$ by the concentration of OH^- present at a pH of 7.4. Even in the absence of phosphate and bicarbonate, several inorganic complexes may be present in the fluid. The concentration of any one of these may exceed that of the free ion itself. Thus the lead complexes $PbOH^+$ and $PbCl^+$ will occur in a saline solution at pH 7.4, $[Pb^{2+}]$ being less than 0.3 of the total lead concentration (SIMONS 1986). If $FeCl_3$ were to be added to saline at pH 7.4, the maximum $[Fe^{3+}]$ attainable would be about $10^{-18} M$ because of the very low solubility product of ferric hydroxide (MAY et al. 1977). If a pH buffer is to be used in addition to or instead of bicarbonate, it must not bind the trace metal being investigated. Hydroxyethylpiperazine ethanesulphonic acid (HEPES) does not significantly bind either lead (SIMONS 1985) or zinc (KALFAKAKOU and SIMONS 1990) and has been used as a pH buffer at a concentration of 20 mM in bicarbonate-free fluids both for brain perfusion and for studying the uptake of these metals into cultured brain endothelium.

Lead has been added to such simple HEPES-buffered saline at total concentrations of 0.1, 1 and 4 μM for brain perfusion. This allows a high specific activity of ^{203}Pb to be used. The disadvantage of using low concentrations of a metal in saline without metal buffering is that its concentration can readily be diminished by binding either to the syringe or plastic tubing used for the infusion or to the luminal surface of the endothelium of the blood vessels. Binding is also likely to be a problem during in vitro experiments. As uptake occurs into the endothelial cells, the total and radioactive concentration of the trace metal in the medium will diminish rapidly. Further, it is likely to be impossible to achieve concentrations which are equivalent to the very low levels of the free ion which is normally present in blood plasma, e.g., about $10^{-13} M$ for lead or 10^{-10}–$10^{-9} M$ for zinc.

If a specific ion-sensitive electrode is available and is sensitive enough, it is desirable to measure the activity of the free ion directly in the physiological solution being used. Such a solid-state (PbS) electrode is available for lead (Orion) and under the best conditions may be sensitive down to $10^{-12} M$. The calibration of such electrodes has been discussed (KIVALO et al. 1976; SIMONS 1985). A zinc electrode can be constructed but does not have adequate sensitivity at concentrations approaching those in the physiological range.

Because of the above considerations, it may be useful to buffer the concentration of free trace metal. The principle is the same as for buffering hydrogen ion. Whilst this is ideal in theory, practical difficulties may arise. Obviously, the buffering ligand and its complex must be non-toxic. If the experiment is to be of short duration, e.g., saline perfusion of the brain, it may be difficult to incorporate enough radiotracer to both combine with the complexing ligand and give adequate radioactivity for counting in the brain. Also, the very real possibility has to be considered that the complex may penetrate into and across the brain endothelium.

In the case of zinc and the radio-isotope ^{65}Zn, we have used both albumin and the amino acid histidine to buffer $[Zn^{2+}]$. Both of these ligands occur in plasma naturally, a fact which may have both advantages and disadvantages.

If the molar ratio of zinc to albumin is kept below 0.1, one zinc ion binds to a single site of high affinity on the albumin, probably an imidazole group. There are multiple sites of lower affinity (OSTERBERG 1971). The formation constant for zinc-albumin is close to $10^7 M$ (GIROUX and HENKIN 1972; S. BUXANI, unpublished observations). It is highly unlikely that the albumin complex will penetrate significantly into or through the brain endothelium though superficial binding on the abluminal surface is a possibility. Use of ^{65}Zn of high specific activity will permit a $[Zn^{2+}]$ of down to $10^{-9} M$ to be obtained. The normal value in serum is probably between 10^{-10} and $10^{-9} M$. Uptakes into brain during 1 min of perfusion are measurable, but are unsatisfactorily low (<2.5%). Good uptakes into cultured endothelial cells can be measured in vitro over longer time periods. In this case it becomes important to control for possible endocytotic uptake of albumin as well as for superficial binding.

Formation constants for the two zinc complexes with histidine, $ZnHis^+$ and $Zn(His)_2$, are well established (MARTELL and SMITH 1974–1989) – again enabling $[Zn^{2+}]$ to be calculated in a solution containing zinc and either L- or D-histidine. Both L- and D-histidine markedly enhance ^{65}Zn uptake into brain in a manner which is not attributable to $[Zn^{2+}]$ acting on its own. Hence, the use of histidine may give results which are relevant to zinc transport in the intact animal, but is not suitable for buffering $[Zn^{2+}]$ solely without penetration of one or more of the zinc-histidine complexes.

F. Uptake of Various Trace Metals Into Brain After Radiotracer Infusions

Figure 1 shows uptake of four radioactive trace metals into cerebral cortex after an approximately constant blood level had been maintained by intravenous infusion. In each case the total concentration of the metal in serum

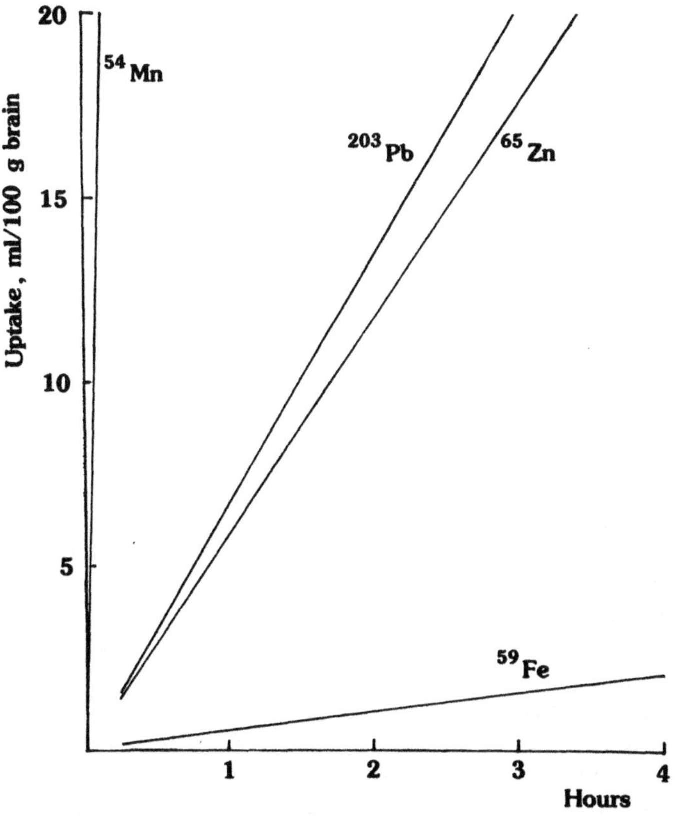

Fig. 1. Rate of uptake of 4 radioactive trace metals into cerebral cortex of rat during maintenance of constant level of the radiotracer in serum by intravenous infusion (slopes from BRADBURY and DEANE 1986; ADU et al. 1990; F. UEDA, J. MICHISON, T.S. PALMA and M.W.B. BRADBURY, unpublished observations)

has not heen increased by more than 25% above its control value. Uptake is linear with time for manganese, lead, zinc and iron, indicating that the BBB is limiting uptake and that there is a sufficient sink for the radiotracer in brain to prevent significant back transport. The differences in rate of uptake are large, the slopes (K_{in}) ranging from $1.7\,ml\,g^{-1}\,h^{-1}$ for manganese to $5.2 \times 10^{-3}\,ml\,g^{-1}\,h^{-1}$ for iron. Gallium enters brain even more slowly than iron (slope $1.5 \times 10^{-4}\,ml\,g^{-1}\,h^{-1}$; PULLEN et al. 1990). The results encouraged us to believe that the mechanism of influx into brain of some of the more permeable trace metals might be studied by short vascular perfusion. As already indicated, this permits control of the composition of the fluid from which transport occurs. Such a method has been applied to the study of ^{203}Pb and ^{65}Zn. Evidence, largely obtained from perfusion, indicates that a major part of lead transport can occur as the free ion or as a simple inorganic complex whereas entry of zinc into brain can involve a low molecular weight complex with histidine. In contrast, much of iron uptake from blood into brain depends on transferrin and its receptors on the endothelium.

I. Lead

Measurements of lead by mass spectrometry indicate normal concentrations in human serum of about 5 n*M* (MANTON and COOK 1984), this concentration being about 1/100 of that in whole blood. Titration of fresh human serum and dialysed serum with lead in the presence of a lead electrode indicate that 20%–25% of serum lead is protein bound, mostly to albumin (AL-MODHEFER et al. 1990). Virtually all the binding of lead to low molecular weight ligands is via SH-groups and part of this binding is to cysteine. Extrapolation of these titration studies to normal serum concentrations suggests that free Pb in human serum is 10^{-13}–10^{-12} *M*.

Uptake of ^{203}Pb into brain after short vascular perfusion with 1 μ*M* lead in HEPES-buffered saline, followed by a 20-s EDTA wash, is shown in Fig. 2 (DEANE and BRADBURY 1990). Uptake under these conditions in which organic binding agents were absent was again approximately linear with time. However, when albumin 5 g dl^{-1} or cysteine 200 μ*M* was added to the perfusion fluid, uptake of ^{203}Pb was virtually abolished (Fig. 2). This is strong evidence that lead can enter the capillary endothelium of the brain as either the lead ion Pb^{2+} or as a simple inorganic complex of it, e.g., $PbOH^+$ or $PbCl^+$. Unidirectional influx of ^{203}Pb was estimated at concentrations of lead in the perfusion fluid between 0.1 and 4 μ*M*. Over this range, flux increased directly with total lead concentration and showed no evidence of saturation. At 4 μ*M*, the flux was about 0.6 μmol kg^{-1} min^{-1}.

Although such rapid movement of a large charged cation is surprising, a high permeability to lead has also been noted for certain cell membranes including that of the red cell (SIMONS 1986). The high permeability of the BBB to lead is compatible with lead exerting its toxic action on the developing CNS through a direct action within the brain rather than via an indirect effect on the transport or regulation of another solute at the BBB.

There are some further clues about the nature of the mechanism by which lead moves from the capillary across the luminal membrane of the endothelium (DEANE and BRADBURY 1990). Increasing potassium in the perfusion fluid, which must depolarize the luminal membrane, reduced the lead flux. This is compatible with lead entering the cell as a positively charged species. Increasing the pH of the fluid enhanced the lead influx to a degree which was proportional to the calculated concentration of $PbOH^+$, suggesting that this may well be the species actually transported.

In addition to the above passive entry of lead, there is evidence for a Ca-ATP-pump which actively moves lead back into blood at the luminal membrane of the endothelium. A series of inhibitors which are known to

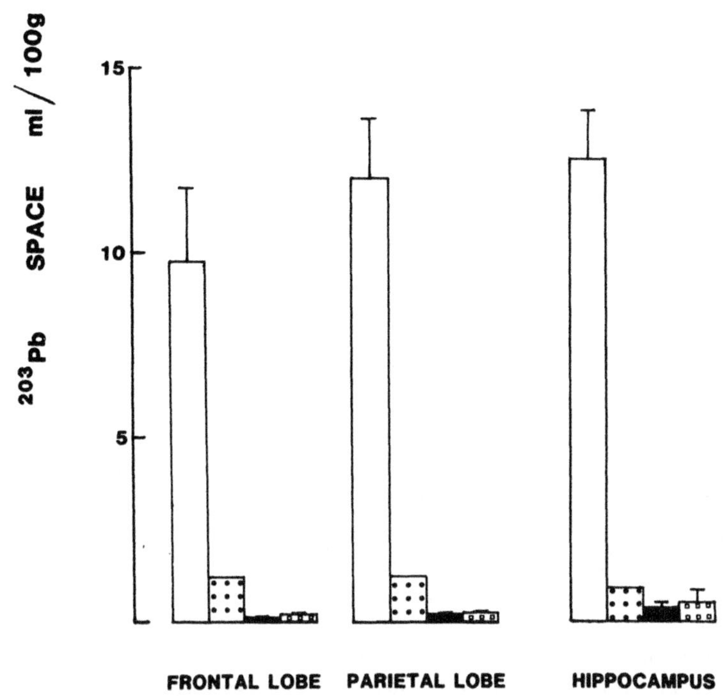

Fig. 2. ^{203}Pb uptake into different regions of rat brain after 1 min cerebrovascular perfusion of ^{203}Pb (1 μM total lead) in HEPES-buffered physiological saline at pH 7.4. *Open blocks* with control saline; *shaded blocks* from *left* to *right* with saline plus 1 mM EDTA, albumin 5 g/dl or 200 μM cysteine (DEANE and BRADBURY 1990)

either act directly on the Ca-ATPase, or to interfere with its energy supply, enhance entry of ^{203}Pb into brain during vascular perfusion (DEANE and BRADBURY 1990). These include vanadate, stannic ions and the mitochondrial uncoupler, carbonyl cyanide 4-(trifluoromethoxy) phenyl hydrazone (FCCP).

II. Zinc

The normal concentration of zinc in human serum is about 15 μM (FOOTE and DELVES 1984) and that in rat serum is similar. In man, 15%–20% is bound to α_2-macroglobulin (FOOTE and DELVES 1984) and is inaccessible to exchange with ^{65}Zn in vitro. A recent study of the distribution of the remaining zinc by computer modelling is Zn-albumin 88%, Zn-transferrin 11% and ≈1% as low molecular weight complexes. These latter comprise various complexes with cysteine and histidine. The free zinc concentration is likely to be between 10^{-10} and 10^{-9} M (GIROUX and HENKIN 1972; MAGNESON et al. 1987).

KASARSKIS (1984) studied the uptake of ^{65}Zn into different regions of brain, optic nerve and choroid plexus but the experimental design did not permit estimation of the BBB to zinc. There is evidence that the concentration of zinc in brain is well maintained in the face of chronic zinc deficiency in young rats but such control is not specific to the CNS, occurring in most cellular tissues, including skeletal muscle, renal cortex and liver (WILLIAMS and MILLS 1970; GIUGLIANO and MILLWARD 1984).

ADU et al. (1990) measured the rate of entry of ^{65}Zn into various brain regions after intravenous infusion to maintain the radiotracer in blood serum constant. The transfer constants for uptake, K_{in}, varied between 0.078 ml $g^{-1} h^{-1}$ (cerebellum) and 0.055 ml $g^{-1} h^{-1}$ (hippocampus). After single bolus injection of ^{65}Zn and multiple time point analysis, PULLEN et al. (1991) considered that there was a fast equilibrating component in brain followed by a linear uptake with time. Their slopes for the second linear phase were rather less than those of ADU et al. (1990), ranging from 0.025 to 0.017 ml $g^{-1} h^{-1}$. It is not certain whether these lower values were due to the different methods or to the fact that Pullen and Franklin used older and larger rats. Although the choroid plexuses accumulated zinc from blood, insignificant ^{65}Zn initially penetrated into CSF.

If assumptions are made about the fraction of serum zinc which is bound to α_2-macroglobulin and hence non-exchangeable, estimations of the normal unidirectional influx of zinc into brain can be made from the uptake slope of ^{65}Zn and the measured concentration total zinc in serum. From the experiments of ADU et al. (1989), this is 20–25 nmol $kg^{-1} min^{-1}$ for cerebral cortex, 20% or zero non-exchangeable serum zinc being assumed respectively.

The mechanism of influx into brain has been investigated by short saline perfusion, using either albumin or histidine as a buffer for zinc. The experiments with albumin indicated a saturable process with a K_m in the nM range (about 20). However, the V_{max} was small (2–3 nmol $kg^{-1} min^{-1}$), indicating that neither albumin-bound zinc nor the free zinc associated with it could account for much of the normal influx into rat brain. A part of this fraction was bicarbonate-dependent and 4,4′-diisothiocyanostilbene-2,2′-disulphonic acid (DIDS) sensitive, as in the red cell, indicating a role for the anion transporter.

When zinc was buffered with L-histidine, larger fluxes were seen. At a constant concentration of Zn^{2+}, increasing histidine between 100 μM and 1 mM caused a linear rise in ^{65}Zn flux into brain. The flux into cerebral cortex increased from 16 nmol $kg^{-1} min^{-1}$ at 100 μM histidine to 40 nmol $kg^{-1} min^{-1}$ at 1 mM. At constant histidine, ^{65}Zn flux tended to saturate with increasing free zinc.

These effects are not likely to be due to transport of either $ZnHis^+$ or $Zn(His)_2$ on an amino acid transporter since there was no inhibition of zinc influx in the presence of 500 μM phenylalanine which almost completely blocked L-[^{14}C]histidine transport into brain. Arginine 500 μM was also

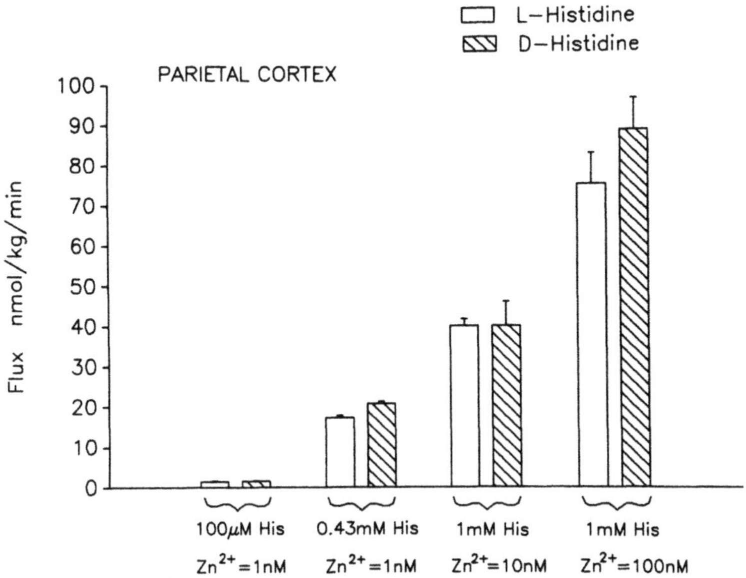

Fig. 3. ^{65}Zn flux into cerebral cortex of rat after 1 min cerebrovascular perfusion of a HEPES-buffered physiological salt solution at pH 7.4 containing different concentrations of L- or D-histidine and of free zinc (BUXANT and ADU, unpublished observations)

without effect on ^{65}Zn transport into brain. D-histidine caused a similar enhancement of ^{65}Zn influx to that caused by L-histidine at the same concentration (Fig. 3).

Some evidence was obtained that zinc fluxes at all L-histidine concentrations could be best fitted by a function containing the sum of separate saturable transports of both $ZnHis^+$ and $Zn(His)_2$. However, it was considered that this might reflect the ability of histidine complexes to diffuse through an unstirred layer at the periphery of the capillary lumen or through a fibre-matrix of glycoprotein at the endothelial surface or both in order for them to present free zinc ions to a saturable transporter at the luminal plasma membrane of the endothelium. Certainly, zinc-albumin would diffuse very slowly or not at all through such initial barriers.

III. Iron

Almost all iron in serum is present as Fe^{3+} in tight combination with the metalloprotein transferrin. Entry of iron into the cells of most tissues is considered to involve initial binding of the iron-transferrin to a membrane receptor followed by endocytosis of the metalloprotein (MAY and CUATRECASAS 1985). The iron is released from the protein by acidification within the vesicular system of the cell.

Monoclonal antibodies against both rat and human transferrin receptors have been used to prove histochemically that the receptors occur in the brain capillaries of these two species respectively (Jefferies et al. 1984). In the case of the rat, they can be labelled after injection into the blood, proof of the presence of the receptor on the luminal surface of the endothelium.

Receptor-mediated uptake of ^{125}I-transferrin has been shown by prolonged vascular perfusion of the rat brain with Ringer's solution (Fishman et al. 1987). If the radioactive perfusion was chased with non-radioactive Ringer, ^{125}I-transferrin was washed out of the capillary fraction of brain, but still increased in the non-vascular fraction. These results indicate passage of transferrin across the endothelium and beyond. Unfortunately, they do not permit calculation of the initial or transendothelial flux of either transferrin or of iron.

Taylor and Morgan (1990) have estimated the uptake of diferric transferrin labelled with both ^{59}Fe and ^{125}I in the developing rat. Iron transport into brain peaked at 15 days post-natal and then declined. ^{59}Fe uptake into the brain of the adult rat and mouse has been measured after intravenous infusion of this radiotracer in the ferrous state to obtain an approximately constant arterial blood level for up to 4 h (F. Ueda and M.W.B. Bradbury, unpublished obsevations). In the rat uptake into various brain regions was linear with time for up to 4 h; in the mouse the rate of uptake rose with time after 2 h (Fig. 4), perhaps due to greater saturation of the transferrin with iron after the longer infusions. The measured flux of iron into cerebellum, as a representative region of brain, was about 3 $nmol\,kg^{-1}\,min^{-1}$ in both rat and mouse.

^{59}Fe uptake into brain regions of both species was inhibited by monoclonal antibodies against the receptors of the specific species. OX26 caused a 50%–70% inhibition in rat and RI7 208 a 70%–80% inhibition in the mouse. The results suggest thst there may be a small component of iron transport across the BBB which is not dependent on transferrin receptors. However, it was not certain that all the receptors were continuously saturated with antibody.

IV. Gallium

Pullen et al. (1990) measured the rate of gallium uptake into brain, using bolus injection of ^{67}Ga and multiple time point analysis. The assumption was that ^{67}Ga is a suitable marker for movement of aluminium. Its linear uptake with time was very slow indeed, having a slope of $0.15 \times 10^{-3}\,ml\,g^{-1}\,h^{-1}$ about 1/35 the rate of transferrin-mediated uptake of iron (Fig. 1). The regional rate of uptake showed a marked similarity to the distribution of ^{125}I-Fe-transferrin receptors. Transport of gallium across plasma membranes is largely transferrin-mediated in other systems, e.g., human leukaemic cells (Chitambar and Zivkovic 1987). It is of interest that the same process can yield such different rates for gallium and iron at the

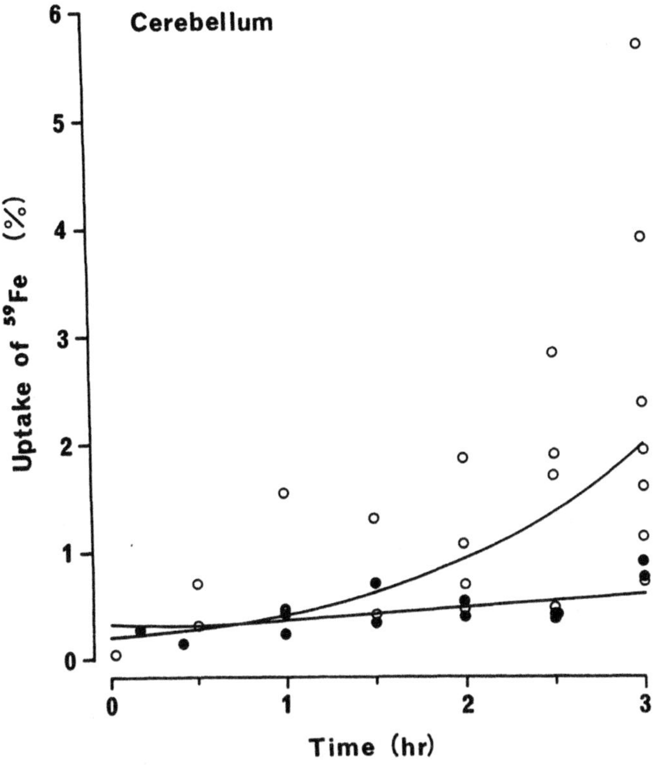

Fig. 4. Uptake of ^{59}Fe into cerebellum of mouse after maintenance of a constant level of ^{59}Fe in serum by intravenous infusion of ^{59}Fe- ferrous chloride. *Open circles* represent control animals and *closed circles* mice treated i.v. with a monoclonal antibody against mouse transferrin receptors, 10 min before infusion of ^{59}Fe (F. UEDA and M.W.B. BRADBURY, unpublished observations)

blood-brain barrier. The experiments of PULLEN et al. (1990) provide no secure evidence concerning the major route for aluminium transport into brain. In this connection, it may be noted that the formation constants for binding of transferrin are close to those for its binding with iron, whereas those for transferrin binding with aluminium are some ten orders of magnitude less (MARTIN et al. 1987).

G. Conclusions

Rates of entry of radioactive trace metals from blood into brain across the barrier vary over four orders of magnitude; manganese-54 having a K_{in} of $ml\,g^{-1}\,h^{-1}$ and gallium-67 having one of only $1.5\,ml\,g^{-1}\,h^{-1}$. Analysis of mechanisms of transport of an individual trace metal is simplified by initial attention to the approximate distribution and concentrations of different species in blood serum under normal and perhaps abnormal conditions. These species may include the free ion, inorganic complexes, low molecular weight complexes with organic molecules, non-specific protein complexes and specific metalloproteins.

Investigation of the properties of each transport process requires control and simplification of the medium from which transport is to take place. This may be achieved by short duration perfusion of the cerebrovascular system of a small laboratory mammal or by use of cultured brain endothelial cells. The medium may be a simple physiological salt solution of maintained pH in which there is not more than one major ligand for the trace metal under consideration. It is useful to measure the activity of the trace metal with an ion-specific electrode, if such is available, or to calculate it from the known formation constant or constants of the complex or complexes present.

Observations are reviewed which indicate for each of four trace metals a major species that is involved in its transport. Lead enters brain as a simple inorganic complex, probably $PbOH^+$; the presence of zinc-histine complexes greatly enhances zinc entry; and entry of iron and almost certainly gallium into brain occurs via binding of transferrin complexes to specific receptors on the brain endothelium and subsequent endocytosis.

References

Adu J, Bradbury MWB, Buxani S (1990) ^{65}Zn transport into brain and other soft tissues of the rat. J Physiol (Lond) 423:40P

Al-Modhefer AJA, Bradbury MWB, Simons TJB (1990) The chemical state of lead in human blood serum. J Physiol (Lond) 422:56P

Blasberg RG, Fenstermacher JD, Patlak CS (1983) Transport of γ-aminoisobutyric acid across brain capillary and cellular membranes. J Cereb Blood Flow Metab 3:8–32

Bradbury MWB, Deane R (1986) Rate of uptake of lead-203 into brain and other soft tissues of the rat at constant radiotracer levels in plasma. Ann NY Acad Sci 481:142–160

Buxani S, Adu J (1991) Histidine-stimulated ^{65}Zn transport at the BBB of the anaesthetized rat. J Physiol (Lond) 438:121P

Chitambar CR, Zivkovic Z (1987) Uptake of gallium-67 by human leukemic cells: demonstration of transferrin receptor-dependent and transferrin-independent mechanisms. Cancer Res 47:3929–3934

Davson H (1955) A comparative study of the aqueous and cerebrospinal fluid in the rabbit. J Physiol (Lond) 129:111–133

Deane R, Bradbury MWB (1990) Transport of lead-203 at the blood-brain barrier during short cerebrovascular perfusion with esaline in the rat. J Neurochem 54:905–914

Evered DC, Lawrenson G (1980) Biological roles of copper. Ciba Found Symp 79

Fishman JB, Rubin JB, Handrahan JV, Fine RE (1987) Receptormediated transcytosis of transferrin across the blood-brain barrier. J Neurosci Res 18:299–304

Foote JW, Delves HT (1984) Albumin bound and α_2-macroglobulin bound zinc concentrations in the sera of healthy adults. J Clin Pathol 37:1050–1054

Giroux EL, Henkin RI (1972) Competition for zinc among serum albumin and amino acids. Biochim Biophys Acta 273:64–72

Giugliano R, Millward DJ (1984) Growth and zinc homeostasis in the severely zinc-deficient rat. Br J Nutr 52:545–560

Hallman PS, Perrin DD, Watt AE (1971) The computed distribution of copper (II) and zinc (II) ions among seventeen amino acids in human plasma. Biochem J 11:549–555

Harris WR, Keen C (1989) Calculations of the distribution of zinc in a computer model of human serum. J Nutr 119:1677–1682

Huebers HA, Finch CA (1987) The physiology of transferrin and transferrin receptors. Physiol Rev 67:520–582

Hughes CCW, Lantos PL (1986) Brain capillary endothelial cells in vitro lack surface IgGFc receptors. Neurosci Lett 68:100–106

Jefferies WA, Brandon MR, Hunt SV, Williams AF, Gatter KC, Mason DY (1984) Transferrin receptor on endothelium of brain capillaries. Nature 312:162–163

Kalfakakou V, Simons TJB (1990) Anionic mechanisms of zinc uptake across the human red cell membrane. J Physiol (Lond) 421:485–497

Kasarskis EJ (1984) Zinc metabolism in normal and zincdeficient rat brain. Exp Neurol 84:114–127

Kivalo P, Virtanen R, Wickstrom K, Wilson M, Pungor E, Horvai G, Toth K (1976) An evaluation of some commercial lead (II)-selective electrodes. Anal Chim Acta 87:401–409

Luthert PJ, Greenwood J, Pratt OE, Lantos PL (1987) The effect of a metabolic inhibitor upon the properties of the cerebral vasculature during a whole head saline perfusion of the rat. Q J Exp Physiol 72:129–141

Magneson GR, Puvathingal JM, Ray WJ (1987) The concentrations of free Mg^{2+} and free Zn^{2+} in equine blood plasma. J Biol Chem 262:11140–11148

Manton WI, Cook JD (1984) High accuracy (stable isotope dilution) measurements of lead in cerebrospinal fluid. Br J Ind Med 41:313–319

Martell AE, Smith RM (1974–1989) Critical stability constants, vols 1–6. Plenum, New York

Martin RB, Savory J, Brown S, Bertholf RL, Wills MR (1987) Transferrin binding of Al^{3+} and Fe^{3+}. Clin Chem 33:405–407

May PM, Linder PW, Williams DR (1977) Computer simulation of metal-ion equilibria in biofluids: models for the low-molecularweight complex distribution of calcium (II), magnesium (II), manganese (II), iron (III), copper (II), zinc (II) and lead (II) ions in human blood plasma. J Chem Soc Dalton: 588–595

May WS, Cuatrecasas F (1985) Transferrin receptor: its biological significance. J Membr Biol 33:205–215

Mills CF (1989) Zinc in human biology. Springer, Berlin Heidelberg New York

Ohno K, Pettigrew KD, Rapoport SI (1978) Lower limits of cerebrovascular permeability to nonelectrolytes in the conscious rat. Am J Physiol 235:H299–H307

Osterberg R (1971) The initial equilibrium steps in the interactions of bovine plasma albumin and Zn (II) ions. A potentiostatic study. Acta Chem Scand 25:3827–3840

Parisi AF, Vallee BL (1970) Isolation of a zinc α_2-macroglobulin in human serum. Biochemistry 9:2421–2426

Patlak CS, Blasberg RG, Fenstermacher JD (1983) Graphical evaluation of blood-to-brain transfer constants from multiple-time uptake data. J Cereb Blood Flow Metab 3:1–7

Pratt OE (1985) Continuous-injection methods for the measurement of flux across the blood-brain barrier: the steady-state, initial rate method. In: Marks N, Rodnight R (eds) Methods in neurochemistry, vol 6. Plenum, New York, pp 117–150

Pullen RGL, Candy JM, Morris CM, Taylor G, Keith AB, Edwardson JA (1990) Gallium-67 as a potential marker for aluminium transport in rat brain: implications for Alzheimer's disease. J Neurochem 55:251–259

Simons TJB (1985) Influence of lead ions on cation permeability in human red cell ghosts. J Membr Biol 84:61–67

Simons TJB (1986) Passive transport and binding of lead by human red blood cells. J Physiol (Lond) 378:267–286

Takasato Y, Rapoport SI, Smith QR (1984) An in situ brain perfusion technique to study cerebrovascular transport in the rat. Am J Physiol 247:H484–H493

Taylor E, Morgan EH (1990) Developmental changes in transferrin and iron uptake by the brain in the rat. Dev Brain Res 55:35–42

Williams RB, Mills CF (1970) The experimental production of zinc deficiency in the rat. Br J Nutr 24:989–1003

Zloković BV, Begley DJ, Djuričić BM, Mitrovic DM (1986) Measurement of solute transport across the blood-brain barrier in the perfused guinea pig brain: method and application to N-methyl-aminoisobutyric acid. J Neurochem 46: 1444–1451

CHAPTER 11
Transport of Drugs

P.J. ROBINSON and S.I. RAPOPORT

A. Introduction

The blood-brain barrier (BBB) at cerebral capillaries consists of a continuous layer of endothelial cells connected by tight junctions (zonulae occludens). The capillary wall lacks water-filled channels of suitable size for aqueous diffusion (except for water itself), and the endothelial cell membranes lack carrier systems of suitable affinity for mediated transport of most drugs [with notable exceptions, such as L-dihydroxyphenylalanine (L-DOPA), melphalan; see below]. Also, it is unlikely that vesicular transport plays a significant role in solute transfer across the BBB (RAPOPORT and ROBINSON 1986). Therefore, most drugs pass between blood and brain extracellular fluid via the lipid membranes of the endothelial cells. To do so, they first must leave the aqueous medium of the blood, then pass into the membrane lipid, diffuse across the membrane, enter and diffuse across the endothelial cell cytoplasm, enter and diffuse across the abluminal membrane, and finally enter the aqueous medium on the other side. Alternatively, the drug could diffuse around the circumference of the endothelial cell via tight junctions and interendothelial cell gaps, thereby bypassing the intracellular and transmembrane diffusion steps.

The permeability of cerebral capillaries to a particular drug which is not specifically transported is therefore directly proportional to its lipid solubility (usually expressed as its octanol/water partition coefficient) and its membrane diffusivity. The latter term is roughly proportional to the square root of its molecular weight. Because octanol/water partition coefficients may vary over a very wide range (about 10^6-fold), whereas the corresponding range of membrane diffusivity is much lower (only about tenfold), BBB permeability is expected to be, to a good approximation, proportional to a

---

Abbreviations

BUI	Brain uptake index	AIB	α-Aminoisobutyric acid
E	Extraction fraction	BBB	Blood-brain barrier
L-DOPA	L-dihydroxyphenylalanine	BSP	Sulfobromophthalein
PA	Permeability–surface area product	BTB	Blood-tumor barrier

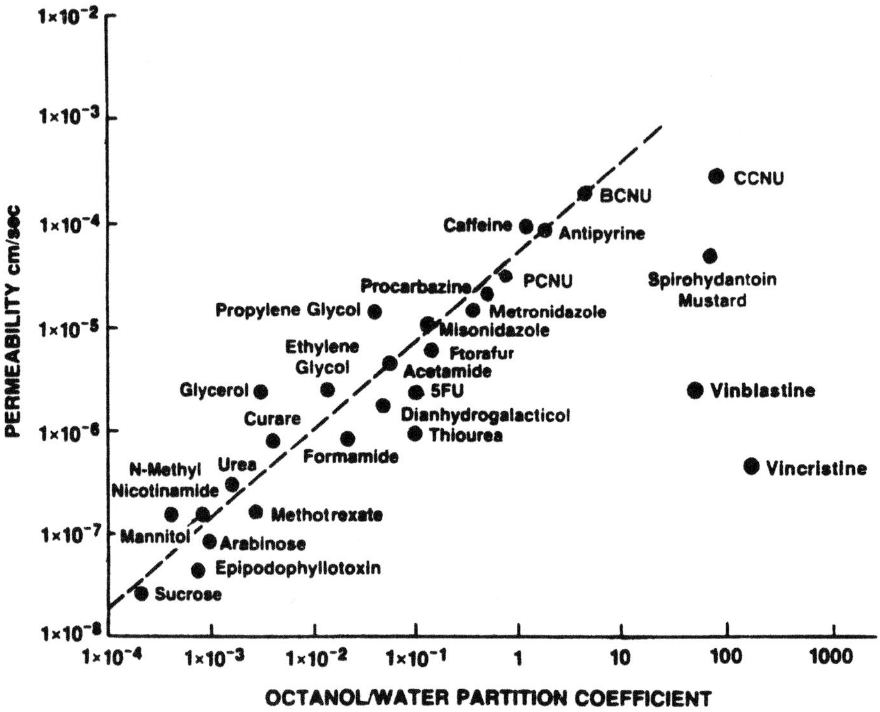

Fig. 1. Permeability coefficients of brain capillaries versus octanol/water partition coefficients for various drugs and other test solutes. (Adapted from RAPOPORT et al. 1979; GREIG, 1989 data for vincristine and vinblastine are from GREIG et al. 1991, in the absence of proteins.)

drug's octanol/water partition coefficient. This is known as the lipid/water partition hypothesis of drug entry into the central nervous system. The ionized form of a drug has much more difficulty penetrating a lipid membrane than the unionized form. A generalized partition coefficient therefore has been introduced that takes into account the fractional concentrations of ionized and unionized species in aqueous solution and their relative lipid solubilities (RAPOPORT and LEVITAN 1974). Figure 1 shows typically good correlation between drug uptake and its octanol/water partition coefficient (RAPOPORT 1976).

Notable deviations from the hypothesized relation have been found for bleomycin, adriamycin, vincristine, epipodophyllotoxin, phenobarbital, dilantin, and methodone (LEVIN 1980; OLDENDORF 1974). Results such as these indicate that factors other than lipid solubility per se may limit drug entry into the brain (FENSTERMACHER 1989). One possible factor is drug size: large drugs with separated hydrophobic and hydrophilic groups (such as vincristine, surfactants) may have to be treated as two separate moieties, with the hydrophilic region retarding movement of the drug as a whole across the BBB (GREIG et al. 1991).

Another important factor is binding to circulating plasma proteins (ROBINSON and RAPOPORT 1986), as the BBB is essentially impermeant to plasma proteins (RAPOPORT 1976). Any compound that binds to proteins is hindered from entering the brain, as only the locally free (unbound) fraction is usually considered available for entry (ROBINSON and RAPOPORT 1986). Many drugs are very highly bound to plasma proteins such as albumin. As protein binding is a major factor in determining a drug's entry into the brain, the underlying mechanism of protein binding forms a major focus for this chapter (see Sect. B). Differences in protein binding due to differences in genetic, environmental, and pathological factors, as well as aging, may markedly affect brain uptake. Protein binding can also prolong the therapeutic effectiveness of a drug by reducing its rate of removal from the blood and by providing a source for long-term action.

As RAPOPORT (1976) has indicated, drug binding to albumin depends on the interaction of lipoidal regions of the drug and the protein; the same kind of interaction occurs between the drug and the membrane. Therefore, as lipid solubility increases, the ability of a free drug to penetrate the BBB increases, but this may be counteracted by increased interaction between drug and protein, reducing the amount of free or exchangeable drug in the plasma. In addition, increase in size may directly affect BBB penetration, as in the case of vincristine (GREIG et al. 1991). These principles have important implications for attempts to increase a drug's entry into the brain by increasing its lipid solubility.

A few therapeutically useful compounds can employ the brain's saturable, carrier mediated transport systems to facilitate their entry into the brain. For example, L-DOPA, which is widely used in the treatment of Parkinson disease, is transported by the large, neutral amino acid system (WADE and KATZMAN 1975). Melphalan, a derivative of L-phenylalanine, can also share the large neutral amino acid carrier system at the BBB (GREIG et al. 1987). By appropriate modification, it may be possible to further synthesize drugs that are able to compete successfully with endogenous compounds for carrier systems at the BBB (GREIG 1989), thereby greatly facilitating their entry into the brain.

Under certain circumstances, the BBB can be made permeable to large molecules, including protein-bound ligands. BBB permeability can be enhanced experimentally by arterial hypertension or hypercapnia, for example (see GREIG 1989 for a review of methods of BBB modification). Experimentally and clinically, however, the carotid infusion of hypertonic solutions such as mannitol, urea, or arabinose has proven most reliable in opening the BBB reversibly (ZIYLAN et al. 1983) and has been most extensively characterized (ZIYLAN et al. 1984; ROBINSON and RAPOPORT 1987a; ROBINSON 1987). The interpretation of these results forms a second focus for this chapter (Sect. C). One of the major applications for osmotic opening of the BBB to both intrinsically large and to protein-bound molecules is in the chemotherapy of primary, malignant brain tumors, and much theoretical and experimental work has been done to establish the efficacy of the osmotic method in this case (see RAPOPORT 1988; RAPOPORT and ROBINSON 1989; ROBINSON and RAPOPORT 1990; NEUWELT and DAHLBORG 1989).

B. Protein Binding and Drug Uptake

Two important and related parameters need to be taken into account in determining the effect of protein binding on brain uptake (ROBINSON and RAPOPORT 1986). Firstly, it is important to know the equilibrium fraction of total drug in plasma that is not bound to protein. This is the equilibrium free (or unbound) fraction (f), which is determined by the ratio of the dissociation and association rate constants (k_{off} and k_{on}) for each binding site (as well as the total concentration of drug and of binding sites; see Eq. 7). If equilibration is achieved in arterial plasma, this fraction determines the amount of drug that is initially (and minimally) available for uptake by the brain.

However, substances with the same equilibrium free fraction may have quite different brain extraction fractions. For example, both bilirubin and palmitate are normally more than 99.99% bound to proteins at equilibrium, but bilirubin has an extraction fraction $E < 0.02$ in a single pass through rat brain, whereas for palmitate $E \simeq 0.058$ (see ROBINSON and RAPOPORT 1986; see also Sect. B.III). In the case of palmitate and many other compounds, some additional drug that is initially bound to protein may be released

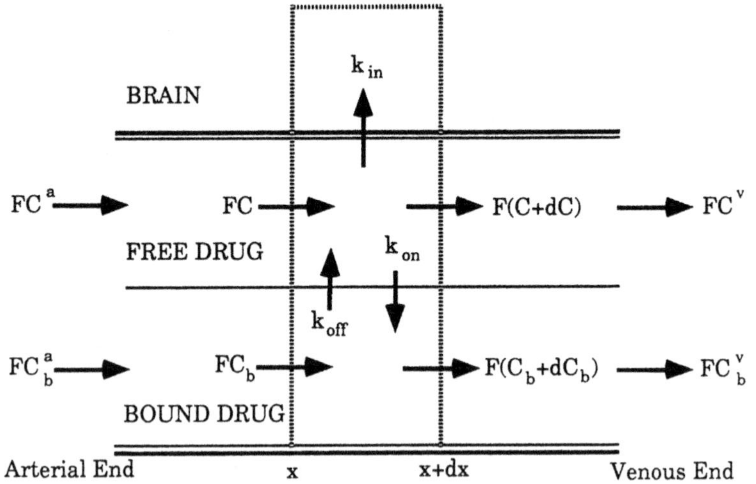

Fig. 2. Schematic diagram of a single capillary showing uptake of free drug, and exchange of drug between bound and free states within a small volume element carried through a capillary by blood flow F. (Adapted from ROBINSON and RAPOPORT 1986)

during passage of blood through the capillary network of the brain, and be subsequently taken up by the brain. The amount of drug taken up therefore depends on the magnitudes of k_{off} and k_{on} for each binding site on the plasma protein, relative to the capillary transit times and the permeability of the BBB to the unbound drug.

In addition to these two primary factors, a number of other factors need to be taken into account in order to form a more complete picture of the effects of protein binding on uptake of drugs by brain (and other perfused organs). These include the interaction of multiple binding sites, the presence of competitors for the binding sites, and the effects of diffusion of both bound and unbound fractions of drug within the unstirred layer adjacent to the capillary endothelial cells on exchange and uptake of free drug (BASS and POND 1988). Direct interaction of the protein-drug complex with the cell surface, resulting in facilitated dissociation and a locally increased free fraction, or uptake of the bound moiety without the intermediate step of release, cannot be ruled out at this stage (PARDRIDGE and LANDAW 1984; WEISIGER et al. 1981).

I. Kinetics of Protein Binding

An initial quantitative description of the effect of binding to plasma proteins on uptake of drugs by the brain can be obtained by following a small volume element of blood as it passes through a typical brain capillary from the arterial to the venous end (Fig. 2; see ROBINSON and RAPOPORT 1986). If we assume that there are no concentration gradients of either bound or free drug perpendicular to the direction of flow, and no significant contribution of diffusion in the direction of the flow, then local mass balance considerations for both bound and free (unbound) forms of the drug (at a distance x from the arterial end of the capillary, reached at a time t) give the following differential equations:

$$\mathrm{d}C_\mathrm{b}/\mathrm{d}t = (F/A)\,(\mathrm{d}C_\mathrm{b}/\mathrm{d}x) = k_\mathrm{on}(C^\mathrm{T} - C_\mathrm{b})C - k_\mathrm{off}C_\mathrm{b} \quad (1)$$

$$\mathrm{d}C/\mathrm{d}t = (F/A)\,(\mathrm{d}C/\mathrm{d}x) = k_\mathrm{off}C_\mathrm{b} - [k_\mathrm{on}(C^\mathrm{T} - C_\mathrm{b}) + k_\mathrm{in}]C \quad (2)$$

where C_b and C are the local concentrations of bound and free drug, respectively, C^T is the concentration of binding sites, and k_{on} and k_{off} are rate constants for association and dissociation of the drug-protein complex. k_{in} is a composite rate constant for uptake of the free drug from the blood, and incorporates the effects of diffusion or transport across the BBB, back-diffusion (or transport) from brain to blood, and metabolism or removal of free drug from the brain extracellular compartment (see ROBINSON and RAPOPORT 1986). F and A are capillary blood flow and capillary cross-sectional area, respectively.

Equations 1 and 2 completely describe the interactions between bound and free forms of the drug at each point along the capillary. To use these equations to describe macroscopic measurable quantities appropriate to the brain (or brain region) as a whole, they must be integrated over the entire length of the capillary, giving concentration profiles of bound and free drug from the arterial to the venous ends of the capillary. The contributions of single capillaries then must be summed, if necessary taking into account possible heterogeneities in capillary properties.

II. Measurable Quantities

A number of the parameters that enter into the model description may either be known or may be measured independently. For example, the association and dissociation rate constants k_{on} and k_{off} for a particular drug-protein complex may be known or measurable in vitro. The number and concentration of protein binding sites may often be known or can be experimentally controlled. Also, the free, or unbound, fraction of the drug at equilibrium can be measured in vitro, or can be calculated from the values of k_{on} and k_{off}.

It has often been assumed that the bound and free forms of a drug equilibrate in arterial plasma. In such a case, the association rate of the drug with the protein equals the dissociation rate of the complex, so that

$$k_{on}(C^T - C^a{}_b)C^a = k_{off}C^a{}_b \tag{3}$$

where the superscript a denotes arterial plasma. The free fraction in arterial plasma is

$$f = C^a/(C^a + C^a{}_b) \tag{4}$$

which, using Eq. 3 for $C^a{}_b$, becomes, for a single binding site:

$$f = 1/[1 + C^T/(C^a + k_{off}/k_{on})] \tag{5}$$

Substituting for C^a in Eq. 5 in terms of the total drug concentration in arterial plasma $C^a{}_{tot}$, where $C^a = fC^a{}_{tot}$, gives, on rearrangement

$$f^2C^a{}_{tot} + f(C^T - C^a{}_{tot} + k_{off}/k_{on}) - k_{off}/k_{on} = 0 \tag{6}$$

Solving for f gives

$$f = \{-[c^T - C^a{}_{tot} + k_{off}/k_{on}] + \sqrt{(C^T - C^a{}_{tot} + k_{off}/k_{on})^2 + 4C^a{}_{tot}k_{off}/k_{on}}\}/(2C^a{}_{tot}) \tag{7}$$

which is a nonlinear increasing function of $C^a{}_{tot}$. Note that as $C^a{}_{tot}$ becomes large, f approaches 1, and as $C^a{}_{tot}$ approaches 0, f approaches $(k_{off}/k_{on})/C^T$.

The extraction fraction E of a drug is defined as the fraction of the arterial total drug that is removed by the brain. Thus:

$$E = [C^a{}_{tot} - C^v{}_{tot}]/C^a{}_{tot} \tag{8}$$

where the superscript v denotes venous plasma.

Other important parameters include the uptake rate V of the brain or other organ:

$$V = F(C^{a}_{tot} - C^{v}_{tot}) = FC^{a}_{tot}E \tag{9}$$

The clearance of a drug is defined as

$$\mathrm{Cl} = V/C^{a}_{tot} \tag{10}$$

while the unbound clearance, or clearance per unbound ligand molecule, is given as

$$\mathrm{Cl}_u = V/C^{a} = V/fC^{a}_{tot} \tag{11}$$

If there is no net uptake of protein or of protein-bound ligand, it is expected that the unbound clearance of a drug will be insensitive to the total protein concentration.

III. Limiting Cases

Equations 1 and 2 together describe local exchanges between bound and free forms of a drug, and uptake of the drug into brain, during passage of blood through a cerebral capillary in the context of the assumptions of our model. In order to relate these processes with measurements of total or regional brain uptake, Eqs. 1 and 2 must be integrated from the arterial to the venous end of the capillary, and then uptake from a single capillary must be related to uptake from an ensemble of capillaries within the brain or brain region.

Equations 1 and 2 can be integrated numerically to give quantitative relations between the arterial and venous concentrations of bound and free drug, total binding site concentration, association and dissociation rate constants, and the uptake parameter k_{in}. If, for simplicity, all capillaries are functionally identical, and the input concentration for all capillaries is the same as in the arterial plasma, then the relations derived for the single capillary are directly applicable to the brain or brain region as a whole.

In many cases, however, we can make certain reasonable assumptions about the rates of interaction between the drug and its protein binding sites that simplify the analysis considerably, and lead to rather simple expressions that are useful for both parameter estimation from the data and for gaining insight into the relative importance of certain parameters for determining brain uptake.

1. Restrictive Elimination

The simpest assumption we can make at this stage is that the dissociation rate of the drug-protein complex is much slower than the time it takes for the complex to pass through a typical capillary. In such a case, only the drug that is initially unbound in the arterial plasma is available for extraction;

there is insufficient time during a single transit for this unbound fraction to be replenished from the circulating pool of bound drug. For such compounds, the bound fraction C_b remains unchanged, so that $dC_b/dt = 0$ in Eq. 1, and $C^a{}_b = C^v{}_b$. Equation 2 becomes, for k_{off} and k_{on} approaching 0:

$$dC/dt = (F/A)\,(dC/dx) = k_{in}C \tag{12}$$

Integrating Eq. 12 from arterial to venous ends of the capillary gives:

$$C^v = C^a e^{-k_{in}\tau} \tag{13}$$

where τ is the capillary transit time. For identical capillaries, the extraction fraction for the brain or brain region becomes (Eq. 8 with $C^a{}_{tot} = C^a{}_b + C^a$ and $C^v{}_{tot} = C^v{}_b + C^v$, and using Eq. 4):

$$E = f(1 - e^{-k_{in}\tau}) \tag{14}$$

The conventional definition of restrictive elimination of a drug by a perfused organ is that $E < f$ (see, for example, WILKINSON and SHAND 1975). Clearly, Eq. 14 satisfies this criterion for all values of the uptake parameter $k_{in}\tau$, and so is an example of restrictive elimination of drug. In this case only the original unbound fraction is available for uptake into the brain. There is no dissociation of bound drug during passage of blood through the capillary to compensate for loss of unbound drug, and thus no bound drug becomes available for uptake. This is true regardless of the magnitude of $k_{in}\tau$ (BBB permeability, binding affinity in brain etc.). There is no competition, or "tug of war" (MACKICHAN 1984; GREIG 1989) between plasma protein binding and brain uptake in this limit: brain uptake is restricted by binding to plasma proteins. However, as we will see below, there are cases in which $E < f$ in which uptake is not restricted by protein binding per se.

For given values of k_{in} and f, Eq. 14 gives a lower bound on the extraction fraction of a protein-bound drug by the brain. It underestimates the true extraction because it neglects possible release of bound drug during the passage of blood through the brain.

2. Instantaneous Equilibration

A second limiting case assumes that exchange of drug between bound and free forms occurs rapidly compared with the capillary transit time, so that there is effectively instantaneous equilibration between bound and free drug at all points along the capillaries. As soon as unbound drug is taken up by brain, bound drug is released to maintain equilibrium. In this limit, concentrations of both bound and free drug decrease along the capillary. Adding Eqs. 1 and 2, we obtain:

$$dC^{tot}/dt = fk_{in}C_{tot} \tag{15}$$

where $C_{tot} = C + C_b$ is the total drug concentration at any point along the capillary. As only unbound drug can be taken up by brain, f can only decrease along the capillary from its initial value given by Eq. 7 (see

ROBINSON and RAPOPORT 1986). Assuming a constant value for f in Eq. 15, given by its initial, or maximum value (Eq. 7), and integrating from arterial to venous ends of the capillary gives the following upper bound on the extraction fraction (for functionally identical capillaries):

$$E = 1 - e^{-fk_{in}\tau} \tag{16}$$

Equation 16 becomes a good approximation for the extraction of a drug when equilibration between bound and free forms of the drug is rapid (compared with the capillary transit time), and when the unbound fraction of the drug does not decrease appreciably along the capillaries. This expression has long been known, and has been used, for example, to describe the uptake of sulfobromophthalein (BSP) sodium by the liver (BAKER and BRADLEY 1966) and the transport of steroid and thyroid hormones across the BBB (PARDRIDGE and LANDAW 1984).

Note that when $k_{in}\tau < 1$, Eq. 16 always gives $E < f$, regardless of the magnitude of f. By definition we therefore again have restrictive elimination of drug, even though in this case both bound and free forms of the drug are available for uptake. Elimination is "restricted," not by binding to proteins, but by small values of $k_{in}\tau$, representing limiting BBB permeability and brain uptake of free drug. The distinction between restrictive and nonrestrictive elimination depends on the absolute magnitude of $k_{in}\tau$; it is independent of the equilibrum free fraction f, and hence of the affinity of the drug for the protein binding sites [there is no "tug-of-war" (MACKICHAN 1984; GREIG 1989) between uptake and protein binding in this case].

For $k_{in}\tau > 1$, on the other hand, there is a transition between nonrestrictive and restrictive elimination as f is increased while $k_{in}\tau$ remains constant. Decreased affinity of the drug for the protein binding sites, relative to its "affinity" for uptake into the brain, in this case brings about the transition between restrictive and nonrestrictive uptake.

3. Dissociation-Limited Uptake

An alternative upper bound to E can be constructed by allowing the drug-protein complex to dissociate during its passage through the capillary, but neglecting possible rebinding during that time (ROBINSON and RAPOPORT 1986):

$$E = [1 - e^{-k_{in}\tau}] - (1 - f)k_{in}(e^{-k_{off}\tau} - e^{-k_{in}\tau})/(k_{in} - k_{off}) \tag{17}$$

Note that for small values of k_{off} (compared with k_{in}), Eq. 17 approaches the lower, bound given by Eq. 14.

For very permeant drugs, k_{in} may become very large. In the limit as $k_{in} \to \infty$,

$$E = 1 - (1 - f)e^{-k_{off}\tau} \tag{18}$$

This is known as dissociation limited uptake, in which the extraction of a drug is determined primarily by its rate of dissociation from the protein binding sites (as well as by the equilibrium free fraction).

IV. Implications and Examples

An important endogenous substance highly bound to plasma proteins is bilirubin, which is normally more than 99.99% bound (FAERCH and JACOBSEN 1975; RAPOPORT 1983), so that $f < 0.0001$. No extraction of bilirubin is observed by the brain uptake index (BUI) method during a single pass through the rat brain (WENNBERG, in RAPOPORT 1983), indicating that $E < 0.02$ (RAPOPORT 1983). This result is consistent with the "free bilirubin hypothesis" (LEVINE 1979), which asserts that only initially unbound bilirubin is available for extraction by brain, and binding to albumin effectively prevents elimination. However, measurements of the dissociation rate of the bilirubin-albumin complex suggest that some of the originally bound bilirubin can be released as blood passes through the capillary bed of the brain. For example, a measured value for k_{off} as high as $0.031\,s^{-1}$ for bilirubin binding to bovine serum albumin (FAERCH and JACOBSEN 1975) suggests that, for a capillary transit time of about 1 s in rat brain, some 3% of bound bilirubin might be released in this way. An extraction fraction of <0.02 therefore suggests that free bilirubin entry at the BBB is restricted, and that uptake is limited to a certain extent by both dissociation of the complex and by the BBB. Brain capillary transit times are about 3.5 s in adult humans (MOLLER and WOLSCHENDORF 1978), and about half this value in children (RAPOPORT 1983), allowing even more bound bilirubin to be released during its passage through the human brain.

LEVITAN et al. (1984) have measured the cerebral uptake of erythrosin B, a widely used food dye (FD and C red no. 3), in rats using the in situ brain perfusion technique (TAKASATO et al. 1984). Uptake of the dye is prevented by its binding to plasma proteins (>99.7% bound: $f < 0.003$), although the BBB is permeable to unbound dye ($PA = 6\text{–}7 \times 10^{-5}\,cm^3\,s^{-1}\,g^{-1}$ in the absence of proteins; LEVITAN et al. 1984). Even if dissociation from proteins is sufficiently rapid that Eq. 16 holds (instantaneous equilibration), $k_{in}\tau = PA/F$ is sufficiently small that E was not measurable.

Another endogenous substance >99.99% bound to proteins is the unsaturated fatty acid palmitate (WOSILAIT et al. 1976). However, palmitate has an extraction fraction of about 0.058 in rat brain (PARDRIDGE and MIETUS 1980), indicating that there is significant release and subsequent uptake of bound palmitate during its passage through cerebral capillaries. A measured dissociation rate of $0.12\,s^{-1}$ with human serum albumin at 37°C (SVENSON et al. 1974) suggests that about half of the palmitate that dissociates during the capillary transit is taken up by the brain, and that both dissociation of the complex and BBB permeability limit brain uptake. However, interpretation of fatty acid uptake studies is made more complicated by binding to plasma proteins other than albumin, to extents depending on their chain length (ROBINSON et al. submitted).

V. Additional Factors

How well do Eqs. 1 and 2, either when integrated numerically or in the form of the approximations given by Eqs. 14, 16, 17, and 18, describe the available experimental data?

Experimentally measured uptake rates of protein-bound substances in the brain and in other perfused organs such as the liver, often differ from predicted values in two ways. Firstly, the magnitudes of these uptake rates tend to be surprisingly high (see, for example BAKER and BRADLEY 1966; PARDRIDGE and LANDAW 1984). Secondly, there often appears to be a dependence on the concentration of protein binding sites that is not explained by the present model. These phenomena are also apparent in studies of uptake by isolated cells, and even by artificial membrane systems (WEISIGER et al. 1989), indicating that they are not a result of the microanatomy of the perfused organ, but are more general. For example, in a hepatocyte monolayer experiment, FLEISCHER et al. (1986) observed that the unbound clearance of palmitate in the presence of 25 μM albumin was about 14 times higher than in the absence of albumin.

These and similar results have led to various "facilitation hypotheses." For example, WEISIGER et al. (1981) postulated specific albumin receptors on the hepatocyte that help to translocate the ligand directly from the albumin molecule into the cell membrane, bypassing the aqueous plasma phase. Others have postulated a catalytic mechanism for the dissociation of the complex at the cell surface (BAKER and BRADLEY 1966; FORKER and LUXON 1981), leading to higher values for the unbound fraction near the cell surface than in bulk solution or in arterial plasma (PARDRIDGE and LANDAW 1984).

BASS and POND (1988) have gone some way towards explaining some of these phenomena in terms of a general theory that also takes into account lateral diffusion of ligand, of ligand-protein complex and of protein in response to equilibrium concentration gradients set up in the unstirred layer in the vicinity of the uptake surface, whether it be a hepatocyte, the membrane of an endothelial cell, or an artificial membrane or interface. This model explains in many cases the enhanced unbound clearances found in the presence of higher protein concentrations, and also the apparent saturation of the unbound clearance with increasing albumin concentration ("pseudo-saturation") without recourse to ad hoc physical structures, such as albumin receptors. Regardless of whether this model explains all of the results so far observed, it seems that diffusion of various moieties in the unstirred layer must be taken into account in any description of uptake of protein-bound ligands, before additional mechanisms are considered.

Additional complicating factors include the likely presence of multiple binding sites of different affinities on the protein molecule (and, in plasma, of different binding proteins). For example, ASHBROOK et al. (1975) computed 12 stepwise equilibrium constants for the binding of palmitate by human serum albumin, ranging in value from 6.15×10^7 to 2.54×10^5.

Ligands would distribute among the binding sites according to their relative affinities (proportional to the ratios k_{on}/k_{off}); as the molar ratio of ligand to protein increases, progressively lower affinity sites would be occupied, and the equilibrium unbound fraction would increase more slowly than in the presence of only high affinity binding sites (see Wosilait et al. 1976). In a complete description, dissociation from, and redistribution among the multiple binding sites on protein would need to be taken into account during passage through the capillary network of the brain.

In addition to the possibility of multiple binding sites for the drug of interest, other substances, both endogenous and exogenous, may compete for the same protein binding sites as the drug. The degree of displacement of a drug by a competitor depends on the relative magnitudes of their respective affinities for the particular binding sites. For example, sulfonamides can compete with bilirubin for specific binding sites on albumin molecules, displacing bound bilirubin and elevating levels of free bilirubin, which then become available to enter the brain (Brodersen 1974). This may explain why premature infants treated with sulfasoxisole suffer a dramatic increase in the incidence of kernicterus (Silverman et al. 1956). In the presence of such a competitor, brain uptake of bilirubin changes from being at least partly dissociation-limited (determined by k_{off}; see, for example, Eq. 18 above) and related to the total bilirubin concentration, to being predominantly membrane-limited (determined by k_{in}; see, for example, Eq. 14) and more highly correlated with the free bilirubin concentration (Robinson and Rapoport 1987b).

C. Modification of the Blood-Brain Barrier

I. Osmotic Blood-Brain Barrier Opening

Hypertonic solutions can disrupt the BBB, both when applied topically to the pia-arachnoid surface of the cortex (Rapoport 1970) and when infused to the blood side of the BBB (Rapoport et al. 1972). Electron microscopy has confirmed that such reversible opening is mediated by increased permeability of interendothelial tight junctions to tracers such as horseradish peroxidase and $La(OH)_3$ (Brightman et al. 1973; Nagy et al. 1979; Dorvini-Zis et al. 1983). It is very likely that reversible BBB opening to water-soluble materials is mediated by osmotically induced shrinkage of cerebrovascular endothelial cells, and consequent reversible widening of interendothelial tight junctions, through which the tracers pass in an aqueous phase (Rapoport 1970; Rapoport et al. 1980).

Radiotracer experiments have been used to quantify the degree and time-course of BBB opening, the critical parameters that determine such opening, and the reversibility of opening. BBB opening (as quantified by the regional cerebrovascular permeability-surface area product, or *PA*), was found to be critically dependent on the osmolality of the solution infused into the carotid circulation. For example, regional [^{14}C]sucrose *PA* between 5 and 15 min following osmotic treatment was found to be seven–tenfold higher than control *PA* values for 30-s infusions of 1.6 and 1.8 *M* arabinose, but showed little increase following a 1.4 *M* arabinose infusion in awake rats (Rapoport et al. 1980). There was a threshold arabinose concentration of 1.6 *M* for barrier opening. A similar threshold for infusion duration, for a fixed infusate concentration, suggested a more general threshold of concentration × duration to produce BBB opening.

The time-course of *PA* following hypertonic arabinose infusion shows that measured *PA* values tend to return to control values by 1 h following osmotic treatment (Rapoport et al. 1980), and lack of histological evidence for brain cell damage or long-term neurological sequelae following osmotic opening (Rapoport and Thompson 1973; Simionescu et al. 1975; Neuwelt et al. 1980) further demonstrates the reversible nature of the phenomenon.

BBB permeability following hypertonic infusion has been most extensively characterized by means of double tracer experiments, with (neutral) tracers of very different molecular sizes (such as sucrose and dextran) injected simultaneously (Ziylan et al. 1983, 1984; Armstrong et al. 1987). If the opened BBB did not discriminate at all on the basis of molecular size, the ratio of the *PA* values for two such substances would equal 1. This would be the case if both molecules crossed the BBB by unrestricted bulk flow. On the other hand, if each molecule passed through the opened BBB by simple (unrestricted) diffusion in an aqueous phase, then the *PA* ratio would equal the ratio of their respective free aqueous diffusion coefficients. If, however, larger molecules were further hindered to a greater extent than smaller molecules (e.g., by passage through relatively small diffusion-limiting pores), the *PA* ratio would be greater than the ratio of the free diffusion coefficients by a factor $\mu(r/r_p) \geq 1$, where r is the molecular radius, and r_p is the pore radius or half slit width (Robinson and Rapoport 1987a). The factor μ takes into account both the steric hindrance and the frictional resistance of the molecule in the pore, relative to free solute (Faxen 1959). The functional form of μ depends on the particular pore geometry (Curry 1984).

Table 1 gives *PA* ratios for sucrose to dextran for 14 brain regions following carotid infusion of 1.6 *M* arabinose in the rat (Ziylan et al. 1984). These data are representative of other pairs of compounds studied in a similar way. Because the *PA* ratios generally are less than the ratio of the free diffusion coefficients for sucrose to dextran (5.3 – Lanman et al. 1971) for the 6-min experiment, there must, at early times after osmotic infusion, be an additional mechanism apart from diffusion that reduces discrimination on the basis of molecular size. It has been proposed that this process is bulk water flow from blood to brain (Ziylan et al. 1984; Robinson and Rapoport 1987a). Such an interpretation is consistent with the independent observation of a transient increase in brain water content by 1%–1.5% of wet

Table 1. Ratio of *PA* products for [^{14}C] sucrose and [^{3}H] dextran (79 000 mol wt) after hypertonic arabinose infusion (from Ziylan et al. 1984)

Brain region	*PA* ratio ([^{14}C] sucrose/[^{3}H] dextran)		
	6-min expt.	35-min expt.	55-min expt.
Olfactory bulb	2.7 ± 0.5	13.0 ± 3.6	7.9 ± 1.2
Caudate nucleus	3.8 ± 0.7	12.3 ± 2.5	10.7 ± 1.5
Hippocampus	4.5 ± 0.7	11.4 ± 1.7	13.8 ± 3.0
Frontal lobe	3.6 ± 0.6	11.4 ± 1.5	8.1 ± 0.6
Occipital lobe	3.7 ± 0.9	16.0 ± 1.2	12.7 ± 1.5
Thalamus and hypothalamus	3.3 ± 0.6	11.7 ± 2.1	9.4 ± 0.9
Superior colliculus	2.5 ± 0.4	10.9 ± 2.0	10.4 ± 1.1
Inferior colliculus	3.1 ± 0.9	8.1 ± 1.2	9.4 ± 1.1
Cerebellum	4.2 ± 0.9	9.2 ± 1.3	9.1 ± 1.5
Pons	3.2 ± 0.4	10.1 ± 1.5	8.9 ± 2.0
Medulla	4.3 ± 0.8	14.9 ± 1.7	13.0 ± 3.5
Midbrain	3.0 ± 0.3	11.0 ± 1.5	8.5 ± 0.7
Parietal lobe	3.3 ± 0.9	9.9 ± 1.2	9.5 ± 1.2
White matter (corpus callosum)	3.5 ± 0.4	30.3 ± 9.0	15.8 ± 1.5

Values are means ± SE; $n = 10$ rats. In all regions 35- and 55-min *PA* ratios are significantly higher ($p < 0.05$) than 6-min values. Values for 35 and 55 min are not significantly different ($p > 0.05$) from each other.

weight within the first 10 min following a hypertonic arabinose injection (Rapoport et al. 1980), and evidence that the brain shrinks during the infusion procedure itself (Fenstermacher and Johnson 1966; Fenstermacher and Rapoport 1984).

A quantitative model describing BBB permeability following osmotic opening in terms of restricted diffusion through cylindrical pores (or narrow slits), together with bulk fluid flow and solute drag, has been developed and applied to data on the size-dependence of BBB permeability (Ziylan et al. 1984; Rapoport and Robinson 1986; Robinson 1987). Assuming a bulk flow value consistent with observed changes in brain water content, such a model suggests an effective open pore radius of about 200 Å or an effective slit width of about 220 Å (Robinson and Rapoport 1986). Changes in *PA* values for sucrose and dextran can be explained on the basis of this model in terms of reductions in bulk water flow from blood to brain, by a factor of about 10, from 6 to 35 min as the barrier recloses, without any changes in the effective pore size (Robinson and Rapoport 1987a). The pore density also seems to remain fairly constant, changing from about 1 pore/200 μm^2 of membrane surface area at 6 and 35 min after osmotic opening, to about 1 pore/300 μm^2 at 55 min (Robinson and Rapoport 1987a).

Once the effective pore size, pore density, and time-course for bulk water flow have been estimated, such a model can be used to predict the entry of neutral, water soluble drugs into the brain following osmotic BBB opening (Robinson 1987). Figure 3 summarizes the results of such an

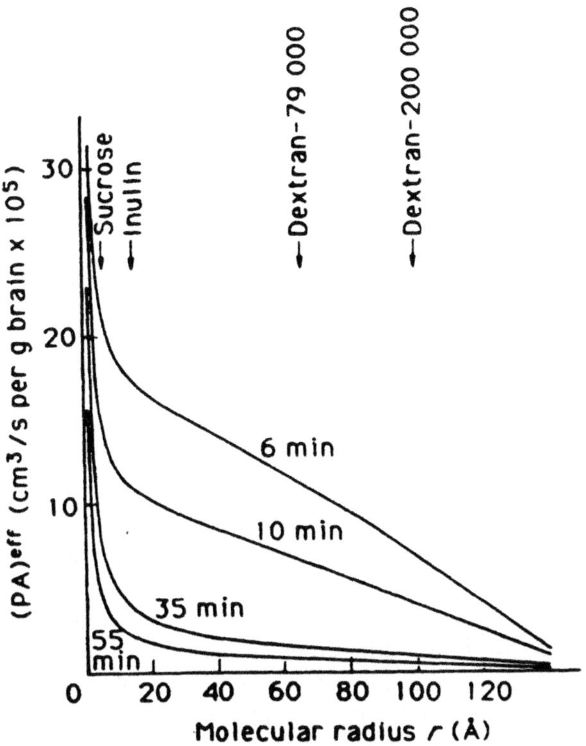

Fig. 3. Predicted variation of effective permeability–surface area product, $(PA)^{eff}$, as a function of molecular radius, using values for cylindrical pore radius (200 Å), pore density and bulk water flow determined from analysis of uptake of sucrose and dextran into rat brain. (From ROBINSON 1987)

analysis. The ordinate gives the effective *PA* value for a neutral, water soluble, and roughly spherical drug as a function of its molecular size at various times after BBB opening (ROBINSON 1987). As the BBB recloses, the effective permeability decreases with time, reaching a particular permeability threshold much sooner for a larger molecule. Larger molecules therefore have a shorter "window" following osmotic opening during which time appreciable quantities of drug can enter the brain for therapeutic effect.

In addition to molecular size, molecular charge may also play an important role in the entry of drugs into the brain following osmotic opening. A negative surface charge associated with interendothelial pores should restrict the passage of negatively charged molecules to a greater extent than their neutral or positively charged counterparts. A number of models have been developed to describe the interaction of charged molecules with charged pores. For example, SMITH and DEEN (1980) solved the Poisson-

Boltzmann equation for a sphere on the axis of a cylindrical pore to evaluate the interaction potential between the sphere and the pore wall. Deen et al. (1980) developed a pore model that describes exclusion and diffusion of solutes in the presence of a uniformly distributed space charge in the pore. There have also been a number of experimental studies in this area (see Curry 1984). Furthermore, Armstrong et al. (1989) observed possible charge effects on BBB permeability to polysaccharides following osmotic BBB opening in rats. Charge modification, in conjunction with osmotic BBB opening, may be a potential technique for further enhancing drug entry into the brain.

As discussed in Sect. B above, an important class of drugs that may be significantly hindered from entering the brain by an intact BBB are those that are highly and tightly bound to circulating plasma proteins. These substances would clearly benefit from the temporary increases in BBB permeability resulting from osmotic treatment, by allowing some bound, as well as free drug to enter the brain. Total drug entry would in this case be made up of two components: both the bound and the free components would follow a time-course for entry into the brain determined by the relevant physicochemical properties of the complex and the free drug, respectively, as well as the time-dependent characteristics of the BBB (see Robinson 1987). It is important to note also that a drug entering the brain, if attached to a plasma protein, would likely have to dissociate from that protein before producing its therapeutic effect. A similar process has been proposed as a model for kernicterus in infants (Levine et al. 1982): hypoxic damage to brain lets in protein-bound bilirubin, which dissociates from protein in the brain and, among other things, inhibits the coupling between oxidation and phosphorylation.

II. Tumor Chemotherapy

In brain tumors of various types, the blood-tumor barrier (BTB) remains a major factor limiting the efficacy of chemotherapy with water-soluble drugs or antibodies. The degree of disruption within a given tumor is very variable and unpredictable, and depends strongly on the type of tumor, its size, history, extent of necrosis, etc. (Hiesiger et al. 1986). For example, transfer coefficients for a nonmetabolizable test substance, α-aminoisobutyric acid (AIB) in a number of brain tumor models in rats range from 1.5 times that for normal cortex (in ethylnitrosourea-induced brain tumors) to 29 times normal cortex (in H-54 transplanted human gliomas); (Blasberg and Groothuis 1986). The extent of disruption of the BTB, in an absolute sense, is greater the lower the initial BTB permeability, and may be insignificant for very highly permeable brain tumors (Rapoport 1988, 1991). Osmotic opening of the BBB has been used clinically to enhance entry of water-soluble drugs into malignant brain tumors with most success in the treatment of primary lymphomas (Neuwelt and Dahlborg 1989).

However, because some chemotherapeutic agents are toxic to normal brain cells, it is critical to limit the exposure of these cells while at the same time maximizing the exposure of the tumor cells. In order to decide whether or not the osmotic method would be clinically useful in achieving this goal, we developed a quantitative model for drug entry into brain and brain tumors, both before and after osmotic treatment (ROBINSON and RAPOPORT 1990; RAPOPORT and ROBINSON 1989). The model incorporates data on the reclosure of the BBB following osmotic opening, as well as on BBB permeability in the tumor of interest, in the brain surrounding the tumor, and in distant brain. It takes into account diffusion of drug within tumor tissue and surrounding brain along concentration gradients set up following any drug infusion schedule, and calculates enhancement factors, defined as the ratio of integrated exposure to drug following osmotic treatment to exposure without osmotic treatment, as a function of distance from the center of the tumor. For example, in untreated brains with a tumor site, drug may diffuse from a higher concentration in the tumor out into the surrounding brain, thereby reducing integrated tumor exposure ("sink effect"). On the other hand, following osmotic treatment, tumor exposure to drug could be enhanced (especially near the tumor edge) both by diffusion and bulk flow from surrounding brain ("source effect"), as well as by increased delivery directly from the blood due to the increased permeability of the BTB.

Tumor exposure, and enhancement of exposure by osmotic treatment, depend on the type and size of tumor, the permeability characteristics of the drug, the drug administration regimen, the rate of movement of the drug in brain and tumor tissue, and the integrity and distribution of blood flow to the tumor. Figure 4 shows the results of applying the model to the exposure to AIB (as a model of a drug) of two kinds of tumor implanted into rat brain, following a continuous infusion of AIB (analysis of data from HIESIGER et al. 1986, by ROBINSON and RAPOPORT 1990). In the Walker-256 carcinoma, which normally has a fairly intact BTB, the effect of osmotic treatment is to greatly increase the exposure of the tumor to AIB, especially near the edge of the tumor, where exposure from blood is supplemented by diffusion from surrounding brain tissue. On the other hand, the osmotic method is less effective for the C-6 glioma, where the BTB is normally much more permeable. Although there is a large increase in BBB permeability in normal brain, increase in BTB permeability is much more modest.

D. Conclusions

In the absence of specific transport mechanisms, passage of drugs across the intact BBB depends in general on their lipid solubility, as measured, for example by the octanol/water partition coefficient. If a drug exists in more than one form in plasma, such as ionized and unionized, then the lipid

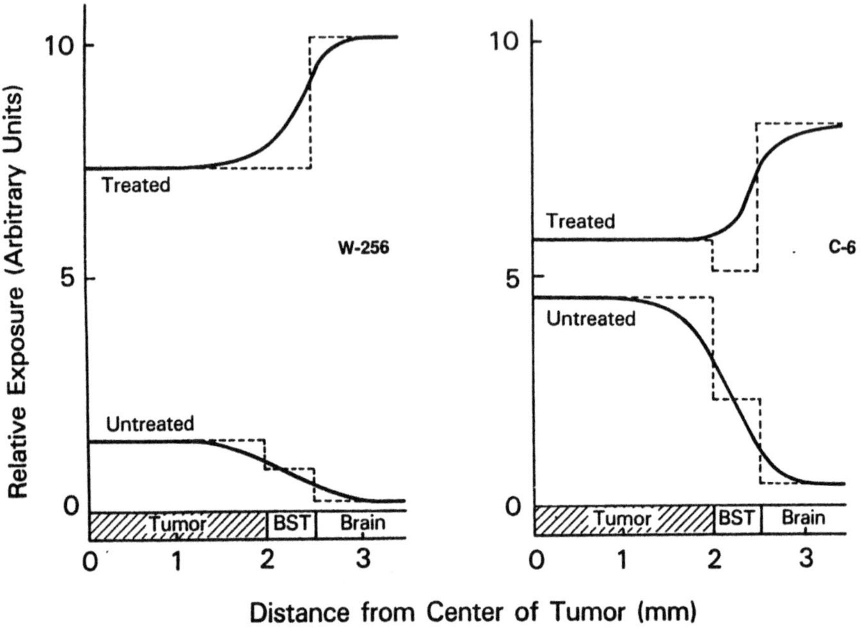

Fig. 4. Calculated 1-h exposures of tumor, brain surrounding tumor (*BST*), and normal brain to α-aminoisobutyric acid (AIB), for Walker-256 cacinoma (*left*) and C-6 glioma (*right*), following a continuous infusion of AIB. *Lower curves* represent untreated brains with intact BBB, while *upper curves* show effect of osmotic treatment concurrently with the onset of the AIB injections. *Broken lines* represent exposure in the absence of diffusion within brain tissue. (From Robinson and Rapoport 1990)

solubility and relative abundance of each must be taken into account, leading to a generalized partition coefficient. An important example of multiple forms of a drug occurs when the drug binds to circulating plasma proteins. The protein-bound forms cannot penetrate the intact BBB, so only the local unbound fraction is able to enter the brain. This unbound fraction is usually measured under equilibrium conditions in vitro, which do not necessarily apply in vivo. Indeed, depending on the absolute magnitudes of the dissociation and association rate constants of the particular drug-protein complex (compared with the capillary transit time), some initially bound drug may be released during passage of blood through the brain ("stripping"), adding to the unbound pool available for entry into the brain. The brain uptake of a drug that is partly protein-bound thus depends on (a) the lipid solubility of the unbound fraction(s); (b) the magnitude(s) of the equilibrium unbound fraction(s), which are determined by the ratios of the dissociation/association rate constants for the relevant binding sites (inversely related to the affinities); and (c) the magnitudes of the dissociation and association rate constants, relative to the capillary transit times. Depending on the relative importance and magnitudes of each of these factors, the kinetics of uptake of a drug that is partly protein bound may assume a particularly simple form, such as in restrictive elimination, instantaneous equilibration, or dissociation limitation. These simple kinetics also furnish upper or lower bounds to uptake in more general cases.

Additional factors that need to be taken into account for a more complete description of drug uptake by the brain include the possibility of multiple affinity binding sites, as well as competition for these binding sites with other endogenous or exogenous substances. Also, in some cases diffusion of ligand, complex, and protein within the unstirred layer adjacent to the endothelial cell surface may play an important role in modulating the uptake of protein-bound ligands. Finally, the rate of direct transfer (if any) from protein to membrane remains to be examined.

In addition to displacement from protein, entry of drugs into the brain may be enhanced either by modifying the drug, or by modifying the BBB. As examples of the former, a drug may be modified to enhance its lipid solubility, while bearing in mind the possibility that such a modification frequently increases its degree of binding to proteins. Alternatively, a drug may be modified to make use of available facilitated transport mechanisms into the brain (e.g., melphalan; GREIG et al. 1987). Reversible osmotic opening of the BBB is perhaps the most important example of BBB modification to enhance drug entry into the brain. It has been extensively characterized in experimental animals, where evidence indicates that it is mediated by the formation of transient aqueous pores between adjacent endothelial cells, resulting from the widening of tight junctions. These pores, which have an estimated radius of about 200 Å (ROBINSON and RAPOPORT 1987a), allow passage of water-soluble substances, including protein, by a combination of restricted diffusion and bulk fluid flow with solute drag.

The osmotic method has also been shown to be clinically effective in humans, and has been used to facilitate the entry of anti-cancer agents into brain tumors in phase II trials; phase III controlled trials for treating primary lymphomas are planned (RAPOPORT 1988; NEUWELT, personal communication).

References

Armstrong BK, Robinson PJ, Rapoport SI (1987) Size-dependent blood-brain barrier opening demonstrated with [14-C]sucrose and a 200 000-Da [3-H]dextran. Exp Neurol 97:686–696

Armstrong BK, Smith QR, Rapoport SI, Strohalm J, Kopecek J, Duncan R (1989) Osmotic opening of the blood-brain barrier permeability to n-(2-hydroxypropyl)-methacrylamide copolymers. Effect of polymer M_w, charge and hydrophobicity. J Controlled Release 10:27–35

Ashbrook J, Spector AA, Santos EC, Fletcher JE (1975) Long chain fatty acid binding to human plasma albumin. J Biol Chem 250:2333–2338

Baker KJ, Bradley SE (1966) Binding of sulfobromophthalein (BSP) sodium by plasma albumin. Its role in hepatic BSP extraction. J Clin Invest 43:281–287
Bass L, Pond SM (1988) The puzzle of rates of cellular uptake of protein-bound ligands. In: Pecile A, Rescigno A (eds) Pharmacokinetics. Plenum, New York
Blasberg RG, Groothuis DR (1986) Chemotherapy of brain tumors: physiological and pharmacokinetic considerations. Semin Oncol 13:70–82
Brightman MW, Hori M, Rapoport SI, Reese TS, Westergaard E (1973) Osmotic opening of tight junctions in cerebral endothelium. J Comp Neurol 152:317–326
Brodersen R (1974) Competitive binding of bilirubin and drugs to human serum albumin studied by enzymatic oxidation. J Clin Invest 54:1353
Curry FE (1984) Mechanics and thermodynamics of transcapillary exchange. In: Renkin EM, Michel CC (eds) Handbook of Physiology, Cardiovascular system, IV. American Physiological Society, Bethesda, pp 309–374
Deen WM, Satvat B, Jamieson JM (1980) Theoretical model for glomerular filtration of charged solutes. Am J Physiol 238:F126–F139
Dorvini-Zis K, Sato M, Goping G, Rapoport SI, Brightman M (1983) Ionic lanthanum passage across cerebral endothelium exposed to hyperosmotic arabinose. Acta Neuropathol (Berl) 60:49–60
Faerch T, Jacobsen J (1975) Determination of association and dissociation rate constants for bilirubin-bovine serum albumin. Arch Biochem Biophys 168: 351–357
Faxen H (1959) About T. Bohlin's paper: on the drag of rigid spheres moving in a viscous liquid inside cylindrical tubes. Kolloidnyi Zh 167:146
Fenstermacher JD (1989) Pharmacology of the blood-brain barier. In: Neuwelt EA (ed) Implications of the blood-brain barrier and its manipulation, vol 1, Basic science aspects. Plenum, New York, pp 137–155
Fenstermacher JD, Johnson JA (1966) Filtration and reflection coefficients of the rabbit blood-brain barrier. Am J Physiol 211:341–346
Fenstermacher JD, Rapoport SI (1984) Blood-brain barrier. In: Renkin EM, Michel CC (eds) Handbook of Physiology, Cardiovascular system IV. American Physiological Society, Bethesda, pp 969–1000
Fleischer AB, Shurmantine WO, Luxon BA, Forker EL (1986) Palmitate uptake by hepatocyte monolayers. J Clin Invest 77:964–970
Forker EL, Luxon BA (1981) Albumin helps mediate removal of taurocholate by rat liver. J Clin Invest 67:1517–1522
Greig NH (1989) Drug delivery to the brain by blood-brain barrier circumvention and drug modification. In: Implications of the blood-brain barrier and its manipulation, vol 1, Basic science aspects. Plenum, New York, pp 311–367
Greig NH, Momma S, Sweeney D et al. (1987) Facilitated transport of melphalan at the rat blood-brain barrier by the large neutral amino acid carrier system. Cancer Res 47:1571–1576
Greig NH, Soncrant TT, Umesha Shetty H, Momma S, Smith QR, Rapoport SI (1991) Brain uptakes and anticancer activities of vincristine and vinblastine are restricted by their low cerebrovascular permeability and binding to plasma constituents in rat. Cancer Chemother Pharmacol (in press)
Hiesiger EM, Voorhies RM, Basler GA, Lipschutz LE, Posner JB, Shapiro WR (1986) Opening the blood-brain and blood-tumor barriers in experimental rat brain tumors: the effect of intracarotid hyperosmolar mannitol on capillary permeability and blood flow. Ann Neurol 19:50–59
Lanman RC, Burton JA, Schanker LS (1971) Diffusion coefficients of some 14-C labeled saccharides of biological interest. Life Sci 10:803–811
Levin VA (1980) Relationship of octanol/water partition coefficient and molecular weight to rat brain capillary permeability. J Med Chem 23:682–684
Levine RL (1979) Bilirubin: worked out years ago? Pediatrics 64:380–385
Levine RL, Fredericks W, Rapoport SI (1982) Entry of bilirubin into the brain due to opening of the blood-brain barrier. Pediatrics 69:255–259

Levitan H, Ziylan YZ, Smith QR et al. (1984) Brain uptake of food dye, erythrosin B, prevented by plasma protein binding. Brain Res 322:131–134
MacKichan JJ (1984) Pharmacokinetic consequences of drug displacement from blood and tissue proteins. Clin Pharmacokinet 9 [Suppl 1] 8:32–41
Moller WD, Wolschendorf K (1978) Dependence of cerebral blood flow on age. Eur Neurol 17, 276–279
Nagy Z, Pappius HM, Mathieson G, Huttner I (1979) Opening of tight junctions in cerebral endothelium. I. Effect of hyperosmolar mannitol infused through the internal carotid artery. J Comp Neurol 185:569–578
Neuwelt EA, Dahlborg SA (1989) Blood-brain barrier disruption in the treatment of brain tumors. Clinical implications. In: Neuwelt EA (ed) Implications of the blood-brain barrier and its manipulation, vol 2, Clinical aspects. Plenum, New York, pp 195–261
Neuwelt EA, Frenkel EP, Diehl J, Vu LH, Rapoport SI, Hill S (1980) Reversible osmotic blood-brain barrier disruption in humans: implications for the chemotherapy of malignant brain tumors. Neurosurgery 7:44–52
Oldendorf WH (1974) Lipid solubility and drug penetration of the blood-brain barrier. Proc Soc Exp Biol Med 147:813–816
Pardridge WM, Landaw EM (1984) Tracer kinetic model of blood-brain barrier transport of plasma protein bound ligands. Empiric testing of the free hormone hypothesis. J Clin Invest 74:745–752
Pardridge WM, Mietus LJ (1980) Palmitate and cholesterol transport through the blood-brain barrier. J Neurochem 34:463–466
Rapoport SI (1970) Effect of concentrated solutions on blood-brain barrier. Am J Physiol 219:270–274
Rapoport SI (1976) Blood-brain barrier in physiology and medicine. Raven, New York
Rapoport SI (1983) Reversible opening of the blood-brain barrier for experimental and therapeutic purposes. In: Levine RL, Maisels MJ (eds) Hyperbilirubinemia in the newborn. Report of the eighty-fifth Ross conference on pediatric research. Ross Laboratories, Columbus, pp 116–124
Rapoport SI (1988) Osmotic opening of the blood-brain barrier. Ann Neurol 24:677–679
Rapoport SI (1991) Osmotic opening of the brain-tumor barrier. J Neurosurg (in press)
Rapoport SI, Levitan H (1974) Neurotoxicity of X-ray contrast media: relation to lipid solubility and blood-brain barrier permeability. AJR 122:186–193
Rapoport SI, Robinson PJ (1986) Tight-junctional modification as the basis of osmotic opening of the blood-brain barrier. Ann NY Acad Sci 481:250–267
Rapoport SI, Robinson PJ (1990) A therapeutic role for osmotic opening of the blood-brain barrier. Re-evaluation of literature and of importance of source-sink relations between brain and tumor. In: Johansson BB, Owman C, Widner H (eds) Pathophysiology of the blood-brain barrier: long-term consequences of barrier dysfunction for the brain. Elsevier, Amsterdam, pp 167–181
Rapoport SI, Thompson HK (1973) Osmotic opening of the blood-brain barrier in the monkey without associated neurological deficits (Abstr). Science 180:971
Rapoport SI, Hori M, Klatzo I (1972) Testing of a hypothesis for osmotic opening of the blood-brain barrier. Am J Physiol 223:323–331
Rapoport SI, Ohno K, Pettigrew KD (1979) Drug entry into brain. Brain Res 172:354–359
Rapoport SI, Fredericks WR, Ohno K, Pettigrew KD (1980) Quantitative aspects of reversible osmotic opening of the blood-brain barrier. Am J Physiol 238:R421–R431
Robinson PJ (1987) Facilitation of drug entry into brain by osmotic opening of the blood-brain barrier. Clin Exp Pharmacol Physiol 14:887–901
Robinson PJ, Rapoport SI (1986) Kinetics of protein binding determine rates of uptake of drugs by brain. Am J Physiol 351:R1212–R1220

Robinson PJ, Rapoport SI (1987a) Size selectivity of blood-brain barrier permeability at various times after osmotic opening. Am J Physiol 253:R459–R466

Robinson PJ, Rapoport SI (1987b) Binding effect of albumin on uptake of bilirubin by brain. Pediatrics 79:553–558

Robinson PJ, Rapoport SI (1990) Model for drug uptake by brain tumors: effects of osmotic treatment and of diffusion in brain. J Cereb Blood Flow Metab 10:153–161

Silverman WA, Anderson DH, Blanc WA et al. (1956) A difference in mortality rate and incidence of kernicterus among premature infants allotted to two prophylactic antibacterial regimens. Pediatrics 18:614

Simionescu N, Simionescu M, Palade GE (1975) Permeability of muscle capillaries to small heme peptides. Evidence for the existence of patent transendothelial channels. J Cell Biol 64:586–607

Smith FG, Deen WM (1980) Electrostatic double-layer interactions for spherical colloids in cylindrical pores. J Colloid Interface Sci 78:444–465

Svenson A, Holmer E, Andersson L-O (1974) A new method for the measurement of dissociation rates for complexes between small ligands and proteins as applied to the palmitate and bilirubin complexes with serum albumin. Biochim Biophys Acta 342:54–59

Takasato Y, Rapoport SI, Smith QR (1984) An in situ brain perfusion technique to study cerebrovascular transport in the rat. Am J Physiol 247:H484–H493

Wade LA, Katzman R (1975) Synthetic amino acids and the nature of L-DOPA transport at the blood-brain barrier. J Neurochem 25:837–842

Weisiger RA, Gollan J, Ockner R (1981) Receptor for albumin on the liver cell surface may mediate uptake of fatty acids and other albumin-bound substances. Science 211:1048

Wilkinson GR, Shand DG (1975) Commentary. A physiologic approach to hepatic drug clearance. Clin Pharmacol Ther 18:377–390

Wosilait WD, Soler-Argilaga C, Nagy P (1976) A theoretical analysis of the binding of palmitate by human serum albumin

Ziylan YZ, Robinson PJ, Rapoport SI (1983) Differential blood-brain barrier permeabilities to 14-C sucrose and 3-H inulin after osmotic opening in the rat. Exp Neurol 79:845–857

Ziylan YZ, Robinson PJ, Rapoport SI (1984) Blood-brain permeability to sucrose and dextran after osmotic opening. Am J Physiol 247:R634–R638

CHAPTER 12

Clinical Assessment of Blood-Brain Barrier Permeability: Magnetic Resonance Imaging

D. BARNES

A. Introduction

Despite the sensitivity of magnetic resonance imaging (MRI) to abnormalities in the central nervous system (CNS), the image appearances lack specificity. Signal intensity in MR images depends predominantly upon the proton density and relaxation times, T_1 and T_2, of tissues. Most pathological processes result in increased tissue water content which increases all three of these parameters. Pathologically dissimilar lesions may, therefore, have similar MRI appearances; for example, it may be impossible to differentiate between tumour tissue and surrounding vasogenic oedema, or between an active and an inactive multiple sclerosis (MS) lesion. It became apparent that as an aid to more accurate diagnosis and for the assessment of treatment efficacy, a contrast agent was required as a marker of abnormal blood-brain barrier (BBB) permeability. Numerous substances act as contrast agents for MRI, but unlike computerised tomography (CT), they produce signal enhancement indirectly via their influence of the relaxation times of neighbouring protons.

In the search for suitable contrast agents, most attention has focused on gadolinium (Gd), a lanthanide which exerts a powerful paramagnetic effect from seven unpaired electrons. When chelated to diethylene-triamine-pentaacetic acid (DTPA) as Gd-DTPA, it is a non-toxic compound with a molecular weight of approximately 550 (WEINMANN et al. 1984; GADIAN

Abbreviations

BBB	Blood-brain barrier
CNS	Central nervous system
CREAE	Chronic relapsing experimental allergic encephalomyelitis
CSF	Cerebrospinal fluid
CT	Computerised tomography
DOTA	10-(2-hydroxypropyl)1,4,7,10-tetraazacyclododecane-1,4,7-triacetic acid
DTPA	Diethylene-triamine-pentaacetic acid
Gd	Gadolinium
HIV	Human immuno-deficiency virus
HLA	Human leucocyte antigen
MRI	Magnetic resonance imaging
MS	Multiple sclerosis
PET	Positron emission tomography
RFLP	Restriction fragment length polymorphism
SPECT	Single photon emission computerised tomography
T_1	Longitudinal relaxation time
T_2	Transverse relaxation time

et al. 1985). Its serum half-life is 90 min, and it is excreted unchanged by the kidney. Apart from occasional reports of hypersensitivity reactions in the USA, it appears to be free from side effects. A number of other Gd compounds such as gadolinium 10-(2-hydroxypropyl)1,4,7,10-tetraazacyclododecane-1,4,7-triacetic acid (Gd-DOTA or Gd-DO3A) have been assessed, but none has so far shown any advantages over the DTPA chelate (PARIZEL et al. 1989; CARVLIN et al. 1990). Gd produces signal enhancement by increasing the efficiency of T_1 relaxation, so shortening T_1 (and increasing the longitudinal relaxation rate, R_1) of tissue water protons to which it is exposed. It does not affect T_2 relaxation significantly at the concentrations normally reached after administration of the usual dose of 0.1 mmol kg^{-1} body weight. For this reason T_1-weighted MRI sequences are used to demonstrate Gd-DTPA enhancement. In normal individuals, brain tissue does not enhance (SZE 1990), but enhancement can be expected in regions unprotected by the BBB, such as the pituitary gland, in vascular structures such as choroid plexus and nasal mucosa, and in blood vessels with *slow* flow such as venous sinuses. Parenchymal brain enhancement is pathological and indicates increased BBB permeability to Gd-DTPA. Experience with Gd-DTPA has now been reported for a variety of CNS diseases, but contrast enhancement has been of particular value in the study of MS and intracranial tumours.

B. Multiple Sclerosis

Although unenhanced MRI is a highly sensitive technique for demonstrating cerebral abnormalities in MS, prior to the introduction of Gd-DTPA it was not possible to determine with confidence the age or biological activity of individual lesions from a single study (YOUNG et al. 1981; ORMEROD et al. 1987). Several lines of evidence suggest that the BBB is abnormal in active MS lesions: first, pathological studies show that acute plaques are usually related to cerebral venules with prominent perivascular inflammation and oedema (ADAMS 1977; ADAMS et al. 1985); secondly, in vivo enhanced CT studies have shown that enhancement is often seen in areas appropriate to recent clinical relapse (SEARS et al. 1981), and enhancing lesions are more often seen during relapse (AITA et al. 1978). MRI is more sensitive for detecting BBB abnormalities in MS lesions than CT (GROSSMAN et al. 1986), and this sensitivity has allowed serial studies to determine the onset and duration of BBB damage in relation to the overall natural history of individual lesions (MILLER et al. 1988). Studies of relapsing/remitting and secondary progressive disease have shown that at least 90% of new MS lesions show abnormal BBB permeability to Gd-DTPA (GONZALEZ-SCARANO et al. 1987; MILLER et al. 1988; THOMPSON et al. 1989, 1990; KERMODE et al. 1990). Of particular interest is the finding that enhancement may be seen up to 2 weeks before a new lesion appears on unenhanced T_2-weighted images

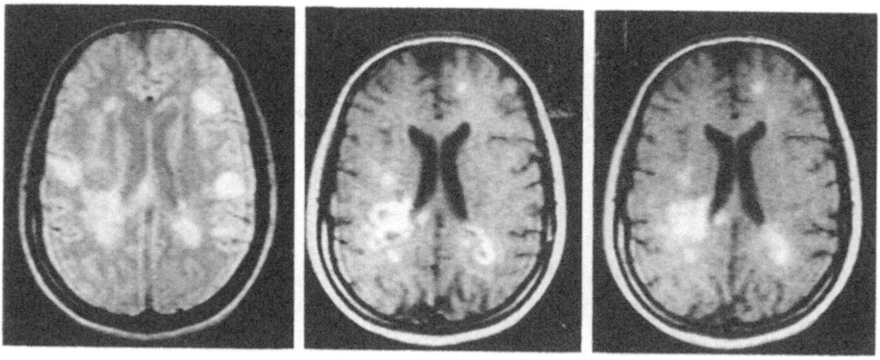

a b c

Fig. 1a–c. Multiple sclerosis. **a** Unenhanced T_2-weighted axial image showing extensive white matter abnormalities. **b** The same slice using T_1-weighted sequence 10 min after injection of Gd-DTPA. Many lesions show ring-like or solid enhancement. **c** 45 min later, enhancement becomes more uniform as Gd-DTPA diffuses into lesions and surrounding white matter

(KERMODE et al. 1990). Since traumatic BBB damage is followed within 1 h by the accumulation of MRI-visible vasogenic oedema in surrounding white matter (BARNES et al. 1988), this finding implies that the alteration in BBB permeability in new MS lesions is initially selective before becoming more generalised.

Serial MRI studies using Gd-DTPA have shown that the overall duration of enhancement in new MS lesions is variable both in brain and spinal cord, but usually lasts between 2 weeks and 2 months (KERMODE et al. 1990; MIKHAEL 1990). Furthermore, previously non-enhancing lesions may develop new enhancement at their margins, often appearing as a ring soon after the administration of Gd-DTPA. Over a period of 1–2 h, the ring gradually fills in and contrast is seen spreading into the white matter beyond the limits of the original ring (Fig. 1). This pattern presumably corresponds to the pathological observation of reactivation of inflammation in older lesions at their edges (PRINEAS and CONNELL 1978).

What is the pathological basis of Gd-DTPA leakage into active MS lesions? This question has been addressed in parallel MRI and ultrastructural studies of the lesions of chronic relapsing experimental allergic encephalomyelitis (CREAE) is guinea pigs (HAWKINS et al. 1990). This model of immune-mediated inflammatory demyelination shows pathological and immunological characteristics closely analogous to those of MS. MRI enhancement in CREAE lesions using Gd-DTPA and much larger Gd-protein complexes was only seen in lesions showing acute inflammatory infiltrates of lymphocytes and macrophages. Ultrastructurally, Gd was identified in endothelial vesicles, perivascular spaces and, sparsely, in spinal cord parenchyma using X-ray dispersive microanalysis. The endothelial transport of Gd was completely inhibited by pre-treatment with dinitrophenol, a metabolic poison, suggesting that the BBB leak was the result of a modification of endothelial cell metabolic activity. It was postulated that this alteration was caused by interaction with activated lymphocytes, possibly mediated by cytokine release. If these observations also apply to MS as seems likely, then Gd-DTPA enhancement in active MS lesions indicates the presence of inflammation, initially with a selective BBB leak, becoming non-selective with more extensive tissue damage.

MRI studies with Gd-DTPA have also provided insight into possible pathogenetic differences between important clinical subgroups of MS, namely (1) relapsing/remitting or secondary progressive (steady clinical progression after an initial relapsing/remitting phase) patients, and (2) primary progressive patients who are progressive from onset without relapses and remissions. Clinically, the latter patients tend to be older at onset, and most commonly present with a progressive paraparesis though other deficits such as visual failure or brainstem syndromes may occur. Furthermore, a Scandinavian restriction fragment length polymorphism (RFLP) analysis of the human leucocyte antigen (HLA) DR and DQ regions has suggested that primary progressive disease is associated with a specific heterozygous *Taq* 1 HLA DQß1 restriction fragment not seen in secondary progressive patients (OLERUP et al. 1989); this finding awaits confirmation. Using strict clinical criteria to categorise their patients, THOMPSON et al. (1991) showed by serial MRI that there were clear differences between the primary and secondary progressive groups in the dynamics of the disease process. Primary progressive patients developed far fewer new lesions (3.3 per patient per year compared to 18.3 for the relapsing and secondary progressive patients), of which only 5% showed enhancement. Overall, much less abnormal white matter was seen in primary progressive MS patients, and confluence of lesions was uncommon. It is unclear why primary progressive patients should show fewer cerebral lesions and less BBB abnormalities to Gd-DTPA on MRI. Indeed, evidence from CSF studies suggests that it is primary progressive patients who show more BBB leakage that relapsing/remitting and secondary progressive patients (who are not in relapse) (B. MCLEAN, unpublished data). This discrepency cannot be explained on the basis of predominantly spinal involvement which was missed on brain imaging, as this form was suspected clinically in only 50% of the primary progressive patients in Thompson's study. A more likely explanation put forward was that lesions are physically smaller or more diffuse than in relapsing or secondary progressive disease, and although BBB leakage did occur, the associated enhancement was beyond the resolution of the imaging technique. Alternatively, BBB leakage in this group may be of the high molecular weight type (WALKER et al. 1985), excluding smaller molecules such as Gd-DTPA. The relationship between cerebrospinal fluid (CSF) abnormalities and BBB damage as assessed by MRI is currently under investigation, and the results should be of interest with respect to lesion pathogenesis in the different clinical groups. Despite the unresolved issues outlined above, these findings highlight real differences in BBB dynamics between different clinical forms of MS, and the need to define clearly patient groups when using MRI to assess treatment aimed at retarding the progression of the disease.

These studies also raise important therapeutic questions: do patients with more marked BBB disturbances benefit in the short or long term from BBB protection with, for example, corticosteroids? Can lesions which enhance before becoming apparent on T_2-weighted images be healed by prompt corticosteroid therapy? Studies with quantitative and Gd-DTPA-enhanced MRI should provide important insights into whether therapy can modify the natural history of individual MS lesions. Many longstanding and apparently inactive lesions are characterised by extensive axonal loss associated with irreversible functional deficit (BARNES et al. 1991). About 20% of such lesions, however, show subtle Gd-DTPA enhancement, possibly as a result of incomplete BBB repair following previous inflammatory insults. If blood-borne factors known to be toxic to myelinated nerve fibres such as cytokines and complement (SCOLDING et al. 1989) are important in lesion progression, then blocking their activity or preventing their access to the CNS should be beneficial.

C. Tumours

As with the lesions of MS, cerebral tumours are well demonstrated by MRI, although it is often impossible to distinguish clearly between tumour tissue and surrounding vasogenic oedema on unenhanced images. CARR et al. (1984), however, showed that Gd-DTPA delineated tumour margin from oedema in a variety of malignant brain lesions. Further studies have confirmed that enhanced MRI is superior to CT for determining tumour extent, particular in regions where CT images are degraded by bone artifact, such as the posterior fossa and chiasmal region (CLAUSSEN et al. 1985; PETERMAN et al. 1985). Subsequently, insight into the BBB characteristics of both experimental and human brain tumours, including degree of disruption and perfusion, have been possible with Gd-DTPA, allowing improved diagnostic accuracy and management (GRAIF and STEINER 1986; WHELAN et al. 1987). Using a canine model of glioma, WHELAN et al. (1987) found that the delineation of tumour margin from surrounding oedema by Gd-enhanced MRI correlated exactly with the postmortem appearances. The sensitivity and specificity of tumour detection may be improved further by the future development for human use of manganese-porphyrin chelates as contrast agents which selectively bind to neoplastic tissue. Experimental studies of these compounds in rats have shown great promise (BOCKHORST et al. 1990).

A detailed discussion of experience with MRI in all intracranial tumours is beyond the scope of this chapter, but certain clear indications for its use are worthy of mention. (1) Metastatic lesions may be isodense with brain on unenhanced MRI. They are, however, well demonstrated by Gd-DTPA with a sensitivity superior to even delayed high dose enhanced CT (Sze et al. 1990), and there is evidence to suggest that doubling the normal dose of Gd-DTPA to $0.2\,\mathrm{mmol\,kg^{-1}}$ body weight increases the detection rate of small lesions still further (Niendorf et al. 1987). (2) Small extra-axial lesions such as pituitary adenomas and acoustic neuromas may only be apparent after Gd-DTPA administration, but it is important to realise that up to 33% of microadenomas causing Cushing's disease will be missed (Doppman et al. 1988). (3) Some indication of the nature of a tumour can be gained from semi-quantitative measurement of the time course of T_1 shortening after contrast administration. This technique improves the differentiation between meningiomas and gliomas or metastases because the former shows maximum enhancement approximately twice as quickly as the latter due to differences in blood-tumour and blood-brain barrier dynamics (Yoshida et al. 1989). (4) Enhanced MRI is a valuable adjunct to the follow-up of patients after tumour resection (Schoerner et al. 1990). Although MRI improves the early detection of recurrent intrinsic tumour, it must be borne in mind that in the first 6–8 months after surgery, brain at the edge of the resection may show enhancement, often in a nodular fashion, in the absence of recurrent tumour (Sze 1990). Similarly, following removal of a meningioma, irregular gyral and/or dural enhancement may be seen as a normal post-operative phenomenon: progressive changes on serial MRI are more definite indicators of recurrence (Ernst et al. 1990).

Future developments with contrast agents should improve diagnostic specificity. One approach is the use of monoclonal antibody-labelled contrast agents directed against tumour-specific antigens. Provided such chelates can cross the BBB and diffuse in the extracellular space like Gd-DTPA, then specific MRI-visible labelling should be possible. Preliminary results using paramagnetic and superparamagnetic (Fe_3O_4) chelates are encouraging (Macri et al. 1988; Cerdan et al. 1989).

D. Cerebrovascular Disease

Cerebral infarcts are MRI-visible by virtue of increased T_1 and T_2 within the ischaemic brain (DeWitt et al. 1984; Sipponen 1984), and several studies have shown that the characteristics of MRI enhancement in infarcts mirrors that been by CT (Hesselink et al. 1988; Schwaighofer et al. 1990; Cordes et al. 1989). MRI enhancement is consistently seen in subacute infarcts up to about 6 weeks old, whereas it is absent in older lesions. In subcortical and lacunar infarcts, BBB breakdown leads to focal Gd-DTPA leakage, but cortical infarcts usually show a gyral pattern of enhancement as with CT. The earliest abnormality seen with Gd-DTPA in a fresh infarct is arterial enhancement (Crain et al. 1990): since only slow-flowing vessels such as venous sinuses are normally visualised, this finding is thought to indicate arterial hypoperfusion as the first MRI-detectable event. Parenchymal enhancement is not usually apparent until a few days later, possibly due to complete vascular occlusion prior to reperfusion. CORDES et al. (1989) studied the correlation between single photon emission computerised tomography (SPECT) and MRI in 26 ischaemic lesions, and found that the presence of Gd-DTPA enhancement was associated with an increase in regional cerebral blood volume in all lesions. This finding suggested that enhancement in infarcts reflects paretic or dilated vessels within hypoxic but viable tissue. In addition to indicating the age of the lesion, therefore, MRI may act as a guide to the use of treatment aimed at salvaging viable brain.

For the assessment of perfusion in cerebrovascular and other CNS disease, specific motion-sensitive sequences have been developed which detect changes in slow flow velocity within voxels (LEBIHAN et al. 1986). Modifications to LEBIHAN's original method allow contributions from perfusion (coherent) and diffusion (incoherent) motions within regions of interest to be visualised separately (LEBIHAN et al. 1988; DORAN and BYDDER 1990). These techniques remain very sensitive to motion artefacts of all kinds, but the use of more rapid imaging sequences should overcome this difficulty. Finally, MRI angiography, where available, is likely to supplant current angiographic techniques for the study of large vessel disease, and future refinements should allow clear visualisation of smaller intracranial vessels.

E. Other Cerebral Diseases

Gd-enhanced MRI is proving of diagnostic value in a variety of other CNS diseases where the BBB may be abnormal. Infectious processes of the brain and meninges, including human immuno-deficiency virus (HIV) infection (HENKES et al. 1989), and head trauma (LANG et al. 1991) are obvious applications. Interestingly, enhancement has been seen at the edges of abnormal white matter in adrenoleukodystrophy (HENKES et al. 1990), and preliminary studies suggest the same might apply to Krabbe's disease. Discerning the time course of BBB abnormalities in these diseases has important implications for attempts at enzyme replacement (YEAGER et al. 1984).

F. Quantification of BBB Permeability

Methods of mathematical modelling of quantitative Gd-DTPA enhancement data have been described by TOFTS and KERMODE (1989) and by LARSSON et al. (1990) which allow the calculation of BBB permeability coefficients from regions of interest. In MS, the almost invariable association between new lesion formation and BBB damage suggests that such information should provide important insights into the natural history and response to treatment of individual lesions at the pathological level.

These techniques use "dynamic" Gd scanning, in which a pre- and then repeated post-contrast T_1-weighted sequences are performed for 1–2 h after administration of Gd-DTPA. A curve of signal change with time in the region of interest is then constructed. Multi-exponential analysis of this curve then yields values for BBB permeability of Gd-DTPA in the selected region, assuming that the T_1-weighted signal changes linearly with local tissue Gd-DTPA concentration (Larsson et al. 1990; Tofts and Kermode 1991).

Both methods require knowledge of Gd concentration in arterial blood to calculate the input function. Larsson et al. measured this parameter directly by brachial artery sampling whereas Tofts and Kermode used reference values from ten normal subjects. The need for arterial cannulation in the former method may, however, be obviated by flow-sensitive imaging of the carotid arteries combined with venous sampling. A further difference between the two techniques is that by assuming a value for relaxivity of Gd-DTPA (the increase in R_1 per unit concentration), Tofts and Kermode were able to derive a value for the distribution space of Gd-DTPA within the lesion. Errors would result, however, if the relaxivity is significantly affected by tissue factors such as macromolecules.

These techniques require a number of assumptions to be made with regard to the transport of Gd-DTPA in brain which are implicit to the definition of permeability: first, that delivery to the tissues is uniform; secondly, that tracer flow is linearly related to concentration differences across the BBB; and thirdly, that at equilibrium, tissue and plasma concentrations will be equal. These assumptions may lead to errors, however, as studies of postmortem MS lesions (Broman 1964) and experimental allergic encephalitis (Hawkins et al. 1990) suggest that endothelial cells control egress of tracer from the vascular space and concentrates it in the perivascular space using active, energetic processes. Parenchymal penetration of tracer may be irregular and dictated by the physical characteristics of the surrounding tissue. These difficulties require clarification in animal models where MRI results can be compared with independent methods of BBB quantification and ultrastructure.

Nevertheless, the techniques of Larsson et al. and Tofts and Kermode provided BBB permeability coefficients of $4–8 \times 10^{-4} s^{-1}$ and $4–40 \times 10^{-4} s^{-1}$ respectively for a variety of MS lesions, implying good agreement between the methods and with independent values derived by positron emission tomography (PET). Tofts and Kermode found the distribution (presumed extracellular) space to be 20%–50% of tissue volume which would be compatible with the extracellular space size in other forms of vasogenic oedema (Barnes et al. 1987).

It seems likely that further development of, and experience with Gd-DTPA-enhanced and other MRI techniques will provide both qualitative and precise quantitative information about BBB permeability in a variety of brain diseases, with spatial resolution superior to that of other imaging modalities.

References

Adams CWM (1977) Pathology of multiple sclerosis: progression of the lesion. Br Med Bull 33:15–20

Adams CWM, Poston RN, Buk SJ, Sidhu YS, Vipond H (1985) Inflammatory vasculitis in multiple sclerosis. J Neurol Sci 69:269–283

Aita JF, Bennett DR, Anderson RE, Ziter F (1978) Cranial CT appearance of acute multiple sclerosis. Neurology (NY) 28:251–255

Barnes D, McDonald WI (1988) A magnetic resonance imaging study of experimental cerebral edema and its response to dexamethasone. Magn Reson Med 7:125–131

Barnes D, McDonald WI, Johnson G, Tofts PS, Landon DN (1987) Quantitative nuclear magnetic resonance imaging: characterisation of experimental cerebral oedema. J Neurol Neurosurg Psychiatry 50:125–133

Barnes D, Munro PMG, Youl BD, Prineas JW, McDonald WI (1991) The longstanding MS lesion: a quantitative MRI and electron microscopic study. Brain 114:1271–1280

Bockhorst K, Höhn-Berlage M, Kocher M, Hossmann K-A (1990) Proton relaxation enhancement in experimental brain tumors – in vivo NMR study of manganese (III) TPPS in rat brain gliomas. Magn Reson Imaging 8:499–504

Broman T (1964) Blood-brain barrier damage in multiple sclerosis: supravital dye observations. Acta Neurol Scand [Suppl 10] 40:21–24

Carr DH, Brown J, Bydder GM, Weinmann H-J, Speck U, Thomas DJ, Young IR (1984) Intravenous chelated gadolinium as a contrast agent in NMR imaging of cerebral tumours. Lancet 1:484–486

Carvlin M, Rosa L, Schellinger D, Francisco J, DeSimone D (1990) Report on clinical trials of ProHance: efficacy and safety evaluation of a new low osmolar MR contrast agent. In: Book of Abstracts, Society of Magnetic Resonance in Medicine, New York, p 731

Cerdan S, Lötscher HR, Künnecke B, Seelig J (1989) Monoclonal antibody-coated magnetite particles as contrast agents in magnetic resonance imaging of tumors. Magn Reson Med 12:151–163

Claussen C, Laniado M, Kazner E, Schörner W, Felix R (1985) Application of contrast agents in CT and MRI (NMR): their potential in imaging of brain tumors. Neuroradiology 27:164–171

Cordes M, Henkes H, Roll D, Eichstädt H, Christe W, Langer M, Felix R (1989) Subacute and chronic cerebral infarctions: SPECT and gadolinium-DTPA enhanced MR imaging. J Comput Assist Tomogr 13:567–571

Crain MR, Yuh WTC, Greene GM, Ryals TJ, Sato Y, Loes DJ (1990) Application of Gd-DTPA in acute ischaemic stroke. In: Book of Abstracts Society of Magnetic Resonance in Medicine, New York, p 6

DeWitt LD, Buonanno S, Kistler JP, Brady TJ, Pykett IL, Goldman MR, Davis KR (1984) Nuclear magnetic resonance imaging in evaluation of clinical stroke syndromes. Ann Neurol 16:535–545

Doppman JL, Frank JA, Dwyer AJ, Oldfield EH, Miller DL, Nieman LK, Chrousos GP, Cutler GB, Loriaux DL (1988) Gadolinium DTPA enhanced MR imaging of ACTH-secreting microadenomas of the pituitary gland. J Comput Assist Tomogr 12:728–735

Doran M, Bydder GM (1990) Magnetic resonance: perfusion and diffusion imaging. Neuroradiology 32:392–398

Ernst RJ, Weingarten K, Frissora CL, Zimmerman RD, Deck MDF (1990) Postoperative meningiomas: assessment with gadolinium-enhanced MR imaging. In: Bood of Abstracts, Society of Magnetic Resonance in Medicine, New York, p 262

Gadian DG, Payne JA, Bryant DJ, Young IR, Carr DH, Bydder GM (1985) Gadolinium-DTPA as a contrast agent in MR imaging – theoretical projections and practical observations. J Comput Assist Tomogr 9:242–251

Gonzalez-Scarano F, Grossman RI, Galetta S, Atlas SW, Silberberg DH (1987) Multiple sclerosis disease activity correlates with gadolinium-enhanced magnetic resonance imaging. Ann Neurol 21:300–306

Graif M, Steiner RE (1986) Contrast-enhanced magnetic resonance imaging of tumours of the central nervous system: a clinical review. Br J Radiol 59:865–873

Grossman RI, Gonzalez-Scarano F, Atlas SW, Galetta S, Silberberg DH (1986) Multiple sclerosis: gadolinium enhancement in MR imaging. Radiology 161:721–725

Hawkins CP, Munro PMG, Mackenzie F, Kesselring J, Tofts PS, DuBoulay EPGH, Landon DN, McDonald WI (1990) Duration and selectivity of blood-brain barrier breakdown in chronic relapsing experimental allergic encephalomyelitis studied by gadolinium-DTPA and protein markers. Brain 113:365–378

Henkes H, Schörner W, Sander B, Felix R (1989) Gd-DTPA enhanced MRI in cerebral infections, inflammations and AIDS. In: Book of Abstracts, Society of Magnetic Resonance in Medicine, Amsterdam, p 7

Henkes H, Sperner J, Sander B (1990) Magnetic resonance tomography of adrenoleukodystrophy. Rontgenblatter 43:7–10

Hesselink JR, Healy ME, Press GA, Brahme FJ (1988) Benefite of Gd-DTPA for MR imaging of intracranial abnormalities. J Comput Assist Tomogr 12:266–274

Kermode AG, Tofts PS, Thompson AJ, MacManus DG, Rudge P, Kendall BE, Kingsley DPE, Moseley IF, DuBoulay EPGH, McDonald WI (1990) Heterogeneity of blood-brain barrier changes in multiple sclerosis: an MRI study. Neurology 40:229–235

Lang DA, Hadley DM, Teasdale GM, Macpherson P, Teasdale E (1991) Gadolinium enhanced MRI following acute head injury. Acta Neurochir (Wien) (in press)

Larsson HBW, Stubgaard M, Frederiksen JL, Jensen M, Henriksen O, Paulson OB (1990) Quantitation of blood-brain barrier defect by magnetic resonance imaging and gadolinium-DTPA in patients with multiple sclerosis and brain tumors. Magn Reson Med 16:117–131

LeBihan D, Breton E, Lallemand D, Grenier P, Cabanis E, Laval-Jeantet M (1986) MR imaging of intravoxel incoherent motions: application of diffusion and perfusion in neurological disorders. Radiology 161:401–407

LeBihan D, Breton E, Lallemand D, Ubin ML, Vignaud J, Laval-Jeantet M (1988) Separation of diffusion and perfusion in intravoxel incoherent motion MR imaging. Radiology 168:497–505

Macri MA, de Luca F, Maraviglia B, Polizio F, Stella A, Cavallo S, Natali PJ (1988) Relaxation study of Gadolinium labelled monoclonal antibody. In: Book of Abstracts, Society of Magnetic Resonance in Medicine, San Fransisco, p 523

Mikhael MA (1990) Serial enhanced MR of the spinal cord and the evolution of multiple sclerosis plaques. In: Book of Abstracts, Society of Magnetic Resonance in Medicine, New York, p 148

Miller DH, Rudge P, Johnson G, Kendall BE, MacManus DG, Moseley IF, Barnes D, McDonald WI (1988) Serial gadolinium enhanced magnetic resonance imaging in multiple sclerosis. Brain 111:927–939

Niendorf HP, Laniado M, Semmler W, Schörner W, Felix R (1987) Dose administration of Gd-DTPA in MR imaging of intracranial tumors. AJNR 8:803–815

Olerup O, Hillert J, Fredrikson S, Olsson T, Kam-Hansen S, Möller E, Carlson B, Wallin J (1989) Primarily chronic progressive and relapsing/remitting multiple sclerosis: two immunogenetically distinct disease entities. Proc Natl Acad Sci USA 86:7113–7117

Ormerod IEC, Miller DH, McDonald WI, DuBoulay EPGH, Rudge P, Kendall BE, Moseley IF, Johnson G, Tofts PS, Halliday AM, Bronstein AM, Scaravilli F, Harding AE, Barnes D, Zilkha KJ (1987) The role of NMR imaging in the assessment of multiple sclerosis and isolated lesions: a quantitative study. Brain 110:1579–1616

Parizel PM, Degryse HR, Gheuens J, Martin J-J, van Vyve M, de La Porte C, Selosse P, van de Heyning P, de Schepper AM (1989) Gadolinium-DOTA enhanced MR imaging of intracranial lesions. J Comput Assist Tomogr 13: 378–385

Peterman SB, Steiner RE, Bydder GM, Thomas DJ, Tobias JS, Young IR (1985) Nuclear magnetic resonance imaging (NMR), (MRI), of brain stem tumours. Neuroradiology 27:202–207

Prineas JW, Connell F (1978) The fine structure of chronically active multiple sclerosis plaques. Neurology (Minneap) 28:68–75

Schoerner W, Henkes H, Mitrovics T, Heim T, Iglesias J, Lanksch W, Felix R (1990) Gd-DTPA enhanced MR imaging in the postoperative follow-up of recurrenet brain tumor. In: Book of Abstracts, Society of Magnetic Resonance in Medicine, New York, p 260

Schwaighofer BW, Klein MV, Wesbey G, Hesselink JR (1990) Clinical experience with routine Gd-DTPA administration for MR imaging of the brain. J Comput Assist Tomogr 14:11–17

Scolding NJ, Morgan BP, Houston A, Campbell AK, Linington C, Compston DAS (1989) Normal rat serum cytotoxicity against syngeneic oligodendrocytes. J Neurol Sci 89:289–300

Sears ES, Hayman LA, Bigelow R (1981) Emerging patterns of lesion activity during multiple sclerosis exacerbations. Trans Am Neurol Soc 106:259–261

Sipponen JT (1984) Visualization of brain infarction with nuclear magnetic resonance imaging. Neuroradiology 26:387–391

Sze G (1990) New applications of MR contrast agents in neuroradiology. Neuroradiology 32:421–438

Sze G, Milano E, Johnson C, Heier L (1990) Intraparenchymal metastases: contrast MR versus non-contrast MR versus contrast CT. Am J Neuroradiol 11:785–791

Thompson AJ, Kermode AG, MacManus DG, Kingsley DPE, Kendall BE, Moseley IF, McDonald WI (1989) Pathogenesis of progressive multiple sclerosis. Lancet 1:1322–1323

Thompson AJ, Kermode AG, MacManus DG, Kendall BE, Kingsley DPE, Moseley IF, McDonald WI (1990) Patterns of disease activity in multiple sclerosis: clinical and magnetic resonance imaging study. Br Med J 300:631–634

Thompson AJ, Kermode AG, Wicks D, MacManus DG, Kendall BE, Kingsley DPE, McDonald WI (1991) Major differences in the dynamics of primary and secondary progressive multiple sclerosis. Ann Neurol 29:53–62

Tofts PS, Kermode AG (1989) Measurement of blood brain barrier permeability using Gd-DTPA scanning. Magn Reson Imaging 7 (Suppl 1):150

Tofts PS, Kermode AG (1991) Measurement of the blood-brain barrier permeability and leakage space using dynamic MR imaging – 1 fundamental concepts. Magn Reson Med 17:357–367

Walker RHW, Thompson EJ, McDonald WI (1985) CSF in multiple sclerosis: relationships between immunoglobulins, leucocytes and clinical features. J Neurol 232:250–259

Weinmann HJ, Brasch RC, Press WR, Wesbey GE (1984) Characteristics of gadolinium-DTPA complex: a potential NMR contrast agent. AJR 142:619–624

Whelan HT, Clanton JA, Moore PM, Tolner DJ, Kessler RM, Whetsell WO Jr (1987) Magnetic resonance brain tumor imaging in canine glioma. Neurology 37:1235–1239

Yeager AM, Brennan S, Tiffany C, Moser HW, Santos GW (1984) Prolonged survival and remyelination after hematopoietic cell transplantation in the twitcher mouse. Science 225:1052–1054
Yoshida K, Furuse M, Kaneoke Y, Saso Y, Inao S, Motegi Y, Ichihara K, Izawa A (1989) Assessment of T_1 time course changes and tissue-blood ratios after Gd-DTPA administration in brain tumors. Magn Reson Imaging 7:9–15
Young IR, Hall AS, Pallis CA, Legg NJ, Bydder GM, Steiner RE (1981) Nuclear magnetic resonance imaging of the brain in multiple sclerosis. Lancet 2: 1063–1066

CHAPTER 13
Clinical Assessment of the Blood-Brain Barrier: Positron Emission Tomography

D.J. Brooks

A. Introduction

Currently, neuroimaging uses two main approaches for studying human blood-brain barrier (BBB) integrity and function in vivo. The first approach is structural imaging employing contrast agents that only penetrate the BBB at sites of damage. Computed tomography (CT) scanning with both ionic and non-ionic radiodense iodine-based contrast agents has been in use for this purpose for 15 years, but is relatively insensitive. More recently the highly sensitive technique of magnetic resonance imaging (MRI) of brain water proton relaxation times with the paramagnetic contrast agent gadolinium diethylenetriaminepentaacetic acid (Gd-DTPA) has become available. The second approach is to use functional imaging. Positron emission tomography (PET) enables the transport of electrolytes, sugars, amino acids, albumin and other biological substrates across intact and damaged BBB to be quantitated in vivo. It can also quantitate cerebral uptake of therapeutic agents such as cytotoxic drugs and monoclonal antibodies. Single photon emission tomography (SPECT) is less versatile than PET, but can provide semi-quantitative measurements of BBB leakage of albumin or red cells.

Abbreviations

ACPC	Aminocyclopentanecarboxylic acid
BBB	Blood-brain barrier
BCNU	Bis-chloroethylnitrosourea
rCBF	Regional cerebral blood flow
rCBV	Regional cerebral blood volume
CT	Computed tomography
DTPA	Diethylenetriaminepentaacetic acid
EDTA	Ethylenediaminetetraacetic acid
E	Extraction
FDG	2-[^{18}F]fluoro-2-deoxy-D-glucose
K_i	Influx constant
MA	[^{11}C]-methylalbumin
MeG	3-*O*-[^{11}C]methyl-D-glucose
MRI	Magnetic resonance imaging
MS	Multiple sclerosis
P	Permeability coefficient
PET	Positron emission tomography
PS	Permeability–surface area product
SPECT	Single photon emission tomography
SSVS	Superior sagittal venous sinus
$t_{1/2}$	Half-life

The purpose of this chapter is to review the ways in which PET can be used to provide biological and clinical information about BBB function. It should be borne in mind that the resolution of current commercial PET scanners ranges from 3 to 8 mm and so one is reduced to studying the BBB at a macroscopic rather than capillary level. In spite of this limitation, using PET it has proved possible to obtain a considerable amount of interesting data about BBB function in both normal subjects and patients with neurological disorders.

In order to perform PET studies it is necessary to tag suitable biological substrates with a short-lived positron emitting isotope. ^{15}O ($t_{1/2}$ 2 min) can be used to label $C^{15}O_2$, $H_2{}^{15}O$, and $C^{15}O$, tracers which are in common use for measuring cerebral blood flow and volume. ^{11}C ($t_{1/2}$ 20 min) can be used to label most sugars and amino acids. There is no positron emitting isotope of hydrogen, but ^{18}F ($t_{1/2}$ 110 min) bound covalently to carbon behaves chemically like C-H. ^{15}O, ^{11}C and ^{18}F isotopes all require a cyclotron for their production. The tracer is administered intravenously, or by inhalation, and the regional cerebral distribution of the isotope is measured with PET in axial tomographic slices. Knowledge of the kinetics of regional cerebral uptake of the tracer and its arterial plasma input function enable first order rate constants describing the transport of the tracer across the BBB to be computed. If the isotope used is very short-lived, such as $^{82}Rb^+$ ($t_{1/2}$ 75 s), continuous intravenous infusion of the isotope results in a steady cerebral level of activity within 3–5 half-lives as cerebral uptake and washout of the tracer are balanced by its rapid radioactive decay. In this situation steady-state kinetics can be used to compute regional cerebral extraction of the tracer.

The majority of studies on regional cerebral function using PET have been directed at determining glucose or amino acid metabolism rather than BBB permeability. Nevertheless, valuable information about transport of these substrates can be derived from regional cerebral uptake data. Table 1 details PET tracers that have been used to determine regional BBB integrity, cerebral metabolism, and cerebral haemodynamics.

B. $^{82}Rb^+$ Transport

$^{82}Rb^+$ is a positron-emitting tracer with behaves chemically and biologically like K^+ and can therefore be used to study transport of this cation (LOVE et al. 1954; KILPATRICK et al. 1956). It is easily obtained by passing saline through a $^{82}Sr/^{82}Rb$ generator. The intact BBB is highly impermeable to K^+, animal studies having yielded permeability coefficients (P) of $2–20 \times 10^{-7}$ cm s^{-1} (BRADBURY 1979; DAVSON and WELCH 1971; HANSEN et al. 1977). Normal plasma levels of K^+ range from 3.5 to 5 mM. The function of the Na/K-ATPase in the endothelium constituting the BBB is to maintain the K^+ level in the cerebral extracellular compartment at 2–3 mM irrespec-

Table 1. PET tracers used to determine blood-brain barrier and cerebral function

Application	Labelled tracers
Cerebral blood flow	$C^{15}O_2$, $H_2{}^{15}O$, ^{15}O-butanol, ^{11}C-butanol, $^{13}NH_3$, $CH_3{}^{18}F$
Cerebral blood volume	^{11}CO, $C^{15}O$, ^{68}Ga-EDTA, ^{68}Ga-transferrin
Cerebral oxygen utilisation	$^{15}O_2$
Cerebral glucose utilisation	2-[^{18}F]fluoro-2-deoxyglucose(^{18}FDG) 2-[^{11}C]deoxyglucose, [^{11}C]glucose
Aminoacid utilisation	[^{11}C]methionine, [^{11}C]leucine, [^{11}C]valine [^{11}C]tryptophan, [^{11}C]alanine
BBB integrity	$^{82}Rb^+$, ^{68}Ga-EDTA 3-O-[^{11}C]methylglucose (^{11}C-MeG) [^{11}C]methylalbumin (^{11}C-MA) [^{11}C]1-aminocyclopentanecarboxylic acid
Drug uptake	^{11}C-BCNU

tive of swings in the plasma cation concentration. Using PET, Yen and Budinger (1981) showed qualitatively that the BBB in primates was impermeable to $^{82}Rb^+$ unless opened with an intracarotid infusion of hyperosmolar 3 *M* urea. Yen et al. (1982) also demonstrated qualitatively with PET that $^{82}Rb^+$ could be used to detect the BBB disruption associated with cerebral tumours.

The first approach to quantitating BBB permeability to $^{82}Rb^+$ with PET used a steady-state model (Lammertsma et al. 1984a; Brooks et al. 1984). These workers assumed that the tracer was distributed between plasma and brain extracellular compartments. By measuring plasma and cerebral ^{82}Rb activity under steady-state conditions, regional cerebral blood flow (rCBF) with $C^{15}O_2$, and blood volume (rCBV) with ^{11}CO, regional cerebral extraction (E) of Rb could be determined. Use of the Renkin-Crone formula (Crone 1963; Renkin 1959) enabled permeability–surface area (PS) products to be derived from E values. This approach yielded a mean E of 2% under steady-state conditions for normal human grey and white matter. As within its 75-s half-life cerebral efflux of ^{82}Rb is negligible, an extraction of 2% is equivalent to a P value of 11×10^{-7} cm s^{-1} (assuming a value of 100 cm ^{2}/g for S, the exchanging surface area of the brain). This P value is in good agreement with animal studies on BBB K^+ permeability.

With this steady-state approach, Brooks et al. (1984) were able to demonstrate that conditions causing up to fourfold increases in intracranial pressure, such as obstructive hydrocephalus and benign intracranial hypertension, were not associated with any increase in BBB permeability to Rb^+. This suggests that regulation of electrolyte balance remains intact in these patients, and explains why mentation is generally preserved. Figure 1a shows PET scans of $^{82}Rb^+$ extraction, rCBF, and rCBV, for a patient with a grade 3 left temporo-occipital glioma. The contrast-enhanced CT is shown in

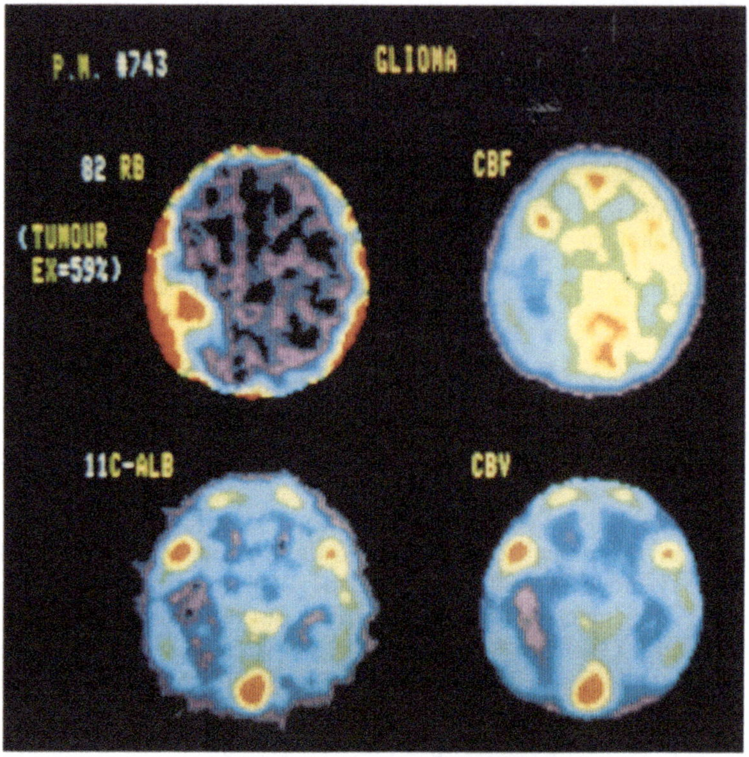

a

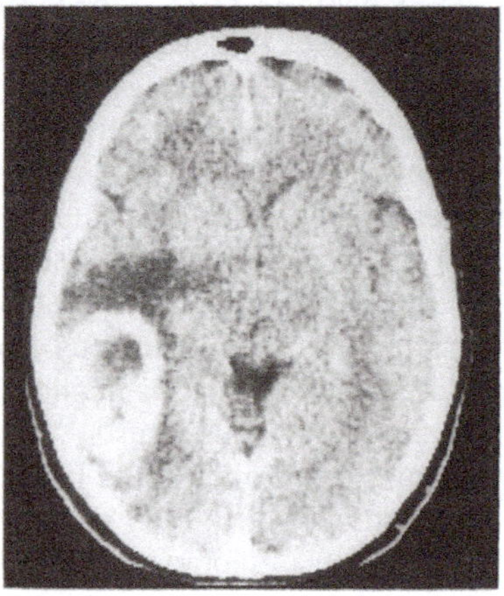

b

Fig. 1. a PET scans of $^{82}Rb^{+}$ extraction, rCBF measured with steady-state $C^{15}O_2$ inhalation, $[^{11}C]$methylalbumin uptake 45 min after tracer administration, and rCBV measured with ^{11}CO inhalation, in a subject with a grade 3 left temporo-occipital glioma. **b** The corresponding CT scan to **a** enhanced with an intravenous radiodense contrast agent

Table 2. Regional cerebral extraction of $^{82}Rb^+$ in conditions associated with raised intracranial pressure

Clinical condition	Rb extraction % (mean ± SD)	$P \times 10^7 cm\,s^{-1}$ (mean ± SD)
CT enhancing tumours (11)	29 ± 13	109 ± 86
Non-enhancing tumours (3)	1.3 ± 1.1	6.6 ± 5.8
Perifocal tumour oedema (5)	1.4 ± 0.3	3.8 ± 1.3
Contralateral brain (10)	2.4 ± 1.6	9.1 ± 7.1
Obstructive hydrocephalus (5)	1.9 ± 0.7	7.9 ± 4.5
Benign intracranial hypertension (3)	2.0 ± 0.7	9.5 ± 1.1
Normal controls (3)	2.0 ± 0.1	11 ± 2.4

Fig. 1b. It can be seen that $^{82}Rb^+$ extraction by the glioma is greatly elevated, but there is no increased tracer uptake in the perifocal tumour oedema. In general, when patients with cerebral tumours were studied with $^{82}Rb^+$, increased tracer extraction by tumours was associated with enhancement on CT with contrast agents. This suggests that increased $^{82}Rb^+$ uptake by cerebral tumours is due to the presence of abnormal vasculature, rather than a consequence of focally increased intracranial pressure. Table 2 details Rb extraction values in patients with brain tumours and other causes of raised intracranial pressure.

Regional cerebral $^{82}Rb^+$ extraction can also be used to determine the effects of therapy on BBB function. Brooks et al. (1984) found that after a 2-week course of radiotherapy the Rb permeability of a cerebral metastasis from the colon doubled, but that there were no long-term effects of fractionated radiation doses up to 5500 rad on normal BBB integrity. Jarden et al. (1985) studied the effects of an intravenous dose of 100 mg dexamethasone, followed by 24 mg orally 4× daily for 24–72 h, on BBB integrity in brain tumour patients. They also examined the effects of 200–600 rad of radiotherapy on tumour Rb uptake within hours of its administration. These workers used a dynamic rather than steady-state approach and fitted their Rb uptake data to a two-compartment model to obtain regional cerebral $^{82}Rb^+$ influx constants (K_i). They found that dexamethasone led to a significant decrease in mean tumour K_i for $^{82}Rb^+$ from 0.056 to 0.040 $ml\,min^{-1}\,g^{-1}$, but had little effect on contralateral brain permeability. This finding would fit with animal studies which have suggest the beneficial effects of dexamethasone on perifocal tumour oedema are exerted via its ability to decrease BBB water and electrolyte permeability. Radiotherapy, within hours of treatment, had no significant effect on tumour permeability to $^{82}Rb^+$. This is not surprising as tumour vasculature in animal models can take several days to fragment following exposure to radiation (Rubin and Casarett 1966). Our tumour patient was studied 2 weeks after exposure to 3000 rad of radiation when extensive disintegration of her blood-tumour barrier had resulted.

C. ^{68}Ga-EDTA Transport

A problem with $^{82}Rb^+$ as a tracer is that its short half-life to 75 s limits its value for detecting subtle changes in BBB integrity. One way around this problem is to generate ^{81}Rb ($t_{1/2}$ 4 h) with a cyclotron. Unfortunately, this isotope of Rb decays to another positron emitter, ^{82m}Rb ($t_{1/2}$ 6 h), resulting in high radiation doses and making the kinetics of Rb^+ uptake complex to analyse. ^{68}Ga-EDTA is a positron emitting tracer ($t_{1/2}$ 68 min) which is similar to sucrose in size and is easily obtained from a ^{68}Ge/^{68}Ga generator. It can be used to detect BBB disruption that may not be evident within the short half-life of $^{82}Rb^+$.

Hawkins et al. (1984) were the first to demonstrate that brain tumour uptake of ^{68}Ga-EDTA could be described by a two-compartment model representing tracer distribution between plasma and the cerebral extracellular space. Tracer uptake was significantly increased in CT enhancing tumours. Iannnotti et al. (1987) also noted increased ^{68}Ga-EDTA uptake in tumours. These authors developed a graphical model to determine tumour ^{68}Ga-EDTA influx constants (K_i) which assumed negligible tracer efflux. They also used superior sagittal venous sinus (SSVS) activity as an input function to avoid the necessity for arterial cannulation, assuming that low tracer extraction would lead to similar arterial and venous sinus tracer levels. Their graphical model worked well for normal brain, but non-linearity of tracer uptake by tumours suggested that the model's assumptions of low ^{68}Ga-EDTA efflux and extraction were probably invalid in tumours.

Using $^{82}Rb^+$, Brooks et al. (1984) studied nine patients with multiple sclerosis (MS) in remission and were unable to demonstrate BBB disruption in any of the demyelinating plaques evident on CT. This finding is in line with pathological reports that endothelial tight junctions remain intact in old MS lesions. Pozzilli et al. (1988) studied 15 MS patients (eight in acute relapse, two in remission, and five in a chronic progressive phase) with ^{68}Ga-EDTA. None of the patients in remission or in a chronic progressive phase of their illness showed evidence of BBB disruption. Four of the eight MS patients in relapse, however, showed evidence of focal intracerebral leakage of ^{68}Ga-EDTA over 80 min of serial PET scanning, K_i values of acute plaques being on average fourfold greater than normal. Leakage of radiodense contrast into these same acute plaques could also be demonstrated by delaying CT scanning by 1 h. While PET enabled the BBB disruption in the MS plaques to be quantitated, it was no more sensitive than delayed CT in identifying acute demyelinating lesions. In chapter 12 it has been shown that Gd-DTPA-enhanced MRI is the most sensitive method for qualitatively detecting BBB disruption in demyelinating lesions.

D. Glucose Transport

The PET tracer *O*-3-[^{11}C]methyl-D-glucose (^{11}C-MeG) is a glucose analogue that is transported by the D-hexose carrier with a similar affinity to glucose, but has a very low affinity for hexokinase. It is excreted unchanged in the urine and so this tracer is ideal for studying glucose transport across the BBB under normal and pathological conditions. 2-[^{18}F]fluoro-2-deoxyglucose (FDG) is also transported by the hexose carrier, but is then phosphorylated by hexokinase to ^{18}FDG-6-phosphate. FDG can also be used to study glucose transport, though it has a greater affinity for the carrier than glucose and traditionally has been used to derive information about cerebral glucose phosphorylation.

VYSKA et al. (1985) first proposed the use of MeG for studying glucose transport but their approach had certain drawbacks. Tracer activity in the cerebral blood pool was ignored although its low cerebral extraction leads to a significant vascular signal. SSVS tracer activity measured tomographically was taken as being equivalent to an arterial plasma input function. The authors studied four normal subjects under conditions of normo- and hyperglycaemia and obtained glucose transport Michaelis-Menten constants of 6.4 m*M* and 2.5 μmol g^{-1} min^{-1} for K_m and V_{max}, respectively. These are in reasonable agreement with those obtained from animal studies (PARDRIDGE and OLDENDORF 1977). The same group (FEINENDEGEN et al. 1986) performed further studies using arterialised venous blood drawn from the vein of their subjects' hands warmed to 37°C, rather than SSVS activity, as an input function, and obtained K_m and V_{max} values for glucose transport in normal controls of 3.8 m*M* and 2.0 μmol g^{-1} min^{-1}, respectively. The mean cerebral glucose level was computed to be 1.1 m*M*.

GJEDDE et al. (1985) compared the kinetics of cerebral uptake of MeG and FDG in four stroke patients. They used a graphical approach to derive cerebral volumes of distribution of the free tracer, and first order rate constants K^M and K^D describing irreversible accumulation of the two tracers in the brain. The authors stated that K^M for MeG was zero, that is rapid equilibration of MeG occurred between plasma and brain, but inspection of their data shows that in fact MeG slowly entered a second cerebral compartment. By setting K^M to zero, and assuming a value of 3.7 m*M* for K_m, these workers computed cerebral glucose levels of 1–1.5 m*M* in normal brain tissue and 0.1–1.1 m*M* in infarcted tissue. They found that transport of FDG into infarcted tissue was invariably depressed. In normal brain the rate of glucose metabolism is determined by hexokinase activity. In two of these authors' four stroke cases glucose metabolism in the infarcted tissue was in fact transport limited due to the low levels of substrate influx.

BROOKS et al. (1986a) studied the kinetics of cerebral uptake of ^{11}C-MeG in four control and seven glioma patients. They used an arterial plasma input function, and found that a model with both rapidly and slowly exchanging cerebral compartments was necessary to describe tracer uptake kinetics. By concommitantly measuring rCBF these workers were able to determine grey and white matter extractions of 14% and 17% for MeG. Glioma extractions of MeG ranged from 14%–29%, while transport of the tracer was depressed in the contralateral cortex of the tumour patients. It has recently been shown that gliomas are deficient in the human glucose transporter (GUERIN et al. 1990). The raised glioma MeG extractions found by BROOKS et al. are therefore likely to represent increased free diffusion of the tracer through abnormal endothelial junctions rather than increased carrier-mediated transportation.

The same group also examined cerebral uptake of MeG in five subjects with long-standing insulin diabetes that had proved difficult to control (BROOKS et al. 1986b). Subjects were studied when hyperglycaemic (mean plasma glucose level 13 m*M*), and when clamped at a mean normoglycaemic plasma glucose level of 4 m*M*. At normoglycaemic levels the diabetic subjects had normal grey and white matter MeG extractions of 15% and 16%, implying that no significant down-regulation of their cerebral glucose transporters had occurred as a consequence of chronic hyperglycaemia. A mean K_m of 4.8 m*M* was computed for the human glucose transporter, in agreement with other workers' findings.

HERHOLZ et al. (1989) compared the kinetics of cerebral uptake of MeG and FDG in four stroke and three tumour patients. These workers used two- and three-compartment models to describe MeG and FDG uptake kinetics, respectively, and used an arterialised venous blood input function. They found that individual subjects had MeG influx and efflux constants values that were consistently 50% of those obtained for FDG in both normal and pathological tissue. These data would support the existence of a single symmetrical hexose carrier in the BBB with double the affinity for FDG than MeG. It is interesting that this MeG:FDG relationship held even for metastatic brain tumours, which are known to be deficient in cerebral glucose transporters (GUERIN et al. 1990). Like GJEDDE et al. (1985), HERHOLZ and coworkers found that glucose transport was impaired in infarcted tissue, glucose metabolism being determined by rate of substrate transport in one out of their four stroke cases.

E. Albumin Diffusion and Microvascular Haematocrit

Serum albumin diffusion across the BBB can be followed with the PET tracer [^{11}C]methylalbumin (MA). A problem with this tracer is that the short 20-min half-life of ^{11}C only enables cerebral uptake of albumin to be monitored for about 1 h in practice. BROOKS et al. (1986c) followed regional cerebral uptake of MA in seven patients with brain tumours and in one normal control. In all subjects brain:plasma ^{11}C activity ratios became constant 5 min after intravenous tracer administration, indicating that the MA had become uniformly distributed throughout the vascular compartment. This ratio remained constant up to 1 h after tracer administration in five out of seven brain tumours scanned, indicating no appreciable leakage of albumin into the tumour was occurring. Figure 1a shows absence of MA uptake by a grade 3 glioma scanned 45 min after tracer administration. The other two brain tumours scanned showed a small increase in MA signal over 1 h.

While MA is not a useful tracer for quantitating BBB disruption in practice, it does enable regional cerebral intravascular haematocrit to be determined. The haematocrit of blood in microvessels is an indirect measure of vessel calibre due to the Fahreus effect (GAEHTGENS et al. 1978). By measuring cerebral volumes of distribution of both red cells and albumin with the tracers ^{11}CO-haemoglobin and MA, respectively, the haematocrit of cerebral microvasculature can be computed (LAMMERTSMA et al. 1984b). Such PET haematocrit measurements primarily reflect venular calibre. BROOKS et al. (1986c) found that both grey and white matter microvasculature had a mean haematocrit value 70% that of large vessels, though rCBF of grey matter was twice that of white matter. Such a finding implies that vessel calibre stays uniform throughout brain tissue, irrespective of local levels of cerebral perfusion. The mean haematocrit value of a group of brain tumours was also 70% that of large vessels, but there was a far wider range of tumour haematocrit values than found for normal brain, reflecting the abnormal vasculature associated with neoplasia.

F. Amino Acid Transport

There are a number of problems associated with the use of PET to study amino acid transport in man. Firstly, both active and facilitated carrier systems exist to transport amino acids. Secondly, amino acids have a complex metabolic fate both peripherally and in the brain, making kinetic modelling of their cerebral uptake kinetics difficult. One way of simplifying the situation is to use an amino acid PET tracer that cannot be metabolised, such as [^{11}C]aminocyclopentanecarboxylic acid (ACPC), to study amino acid transport. Rat models have shown that this tracer is taken up well by tumours (HAYES et al. 1976; WASHBURN et al. 1978), but to date no human data have been reported.

The most widely used amino acid tracer to date, due to its ease of synthesis, has been [^{11}C]methionine. The ^{11}C isotope is situated on the methyl group, and cerebral ^{11}C activity following intravenous administration of this tracer reflects both incorporation of methionine into proteins and its products of catabolism. A theoretical way of determining the cerebral ^{11}C activity due to protein incorporation alone would be to perform a study with [^{11}C]methionine labelled on its carboxyl group, all catabolised ^{11}C then being exhaled as $^{11}CO_2$ (PHELPS et al. 1984). To date no such studies have been reported in man.

Bustany et al. (1985) used a simplified three-compartment model to describe cerebral [^{11}C]methionine uptake kinetics. These workers assumed that the tracer was either taken up into a free cerebral pool or incorporated into protein, and that catabolism was insignificant within the 1 h of PET scanning. Measurements of levels of plasma [^{11}C]methionine metabolites were performed with paper chromotography. Using this approach, the authors found that both [^{11}C]methionine influx and protein incorporation rate were globally depressed in patients with senile dementia of Alzheimer's type, but only frontally depressed in young schizophrenic patients. It is likely that the low [^{11}C]methionine influx observed in Alzheimer's disease and schizophrenia was a consequence of low metabolic requirements rather than decreased numbers of BBB carriers, but in the absence of measurements of rCBF by these authors tracer extractions could not be computed.

The same group studied [^{11}C]methionine uptake by gliomas (Bustany et al. 1986) and showed that tracer influx and protein incorporation rates correlated with the grade of tumour malignancy on the Kernohan scale. They also demonstrated that cerebral influx of [^{11}C]methionine was impaired by competition for the neutral amino acid transporter from the high levels of plasma phenylalanine present in patients with phenylketonuria until they were placed on phenylalanine restriction diets (Bustany et al. 1981).

More recently Hawkins et al. (1989) have used a four-compartment model to describe cerebral uptake kinetics of L-[1-^{11}C]leucine in humans. Catabolism of the tracer was quantitated by measuring plasma $^{11}CO_2$ levels. With this approach the authors were able to compute rate constants describing leucine influx and efflux across the BBB and its rate of incorporation into brain proteins. To date only data on normal subjects have been reported with this tracer. Hübner et al. (1982) have reported the use of ^{11}C-labelled DL-valine and tryptophan for imaging brain tumours. Their approach was, however, semi-quantitative and provides no direct information on BBB function. Until an amino acid tracer is developed that is trapped without being catabolised it is likely that PET studies on cerebral amino acid transport and protein incorporation will continue to be complex to model and difficult to interpret accurately.

G. Conclusions

In this chapter it has been shown how PET can be used to study BBB transport of electrolytes, hexoses, and amino acids under normal and pathological conditions. Studies on BBB K^+ permeability in man using the PET analogue $^{82}Rb^+$ give similar PS products to animal studies, and fourfold increases in intracranial pressure do not appear to impair control of K^+ transport. The Michaelis-Menten constants K_m and V_{max} for glucose transport, measured using the PET tracer 3-*O*-[^{11}C]methylglucose, have similar values in man to those obtained for animals. Tumours with an abnormal vasculature show a generally increased permeability to Rb^+, glucose and amino acids. An interesting future problem to tackle with PET would be measurement of changes in BBB permeability to water under different pathological situations, such as focally and diffusely raised intracranial pressure. $H_2{}^{15}O$ is the tracer routinely used to measure CBF, but can be used to assess water permeability if a separate CBF measurement is performed with a flow tracer such as [^{11}C]butanol (HERSCOVITCH et al. 1987). In normal human cortex the *PS* product for water has been estimated to be 1.27 ml min^{-1} g^{-1}. It would be of interest to know whether the *PS* product is increased in hydrocephalus and in regions of perifocal tumour oedema, and whether such *PS* increases, if present, can be reversed with steroids or hyperosmolar agents.

A problem with the PET approach is the short half-lives of positron-emitting isotopes and the limited number of isotopes in current use. The 20-min half-life of ^{11}C makes the slow cerebral uptake of tracers such as ^{11}C-labelled MA methylalbumin and monoclonal antibodies impractical to follow. ^{124}I is a positron emitter with a half-life of 4 days. Methods of labelling macromolecules with this isotope are currently being developed so that their cerebral uptake can be followed over days. In the future it is to be hoped that interest in the BBB will expand in the PET field, and that this exciting technique will be used to learn more about the effects of pathology on BBB function.

References

Bradbury M (1979) The concept of a blood-brain barrier. Wiley, Chichester

Brooks DJ, Beaney RP, Lammertsma AA et al. (1984) Quantitative measurement of blood-brain barrier permeability using rubidium-82 and positron emission tomography. J Cereb Blood Flow Metab 4:535–545

Brooks DJ, Beaney RP, Lammertsma AA et al. (1986a) Glucose transport across the blood-brain barrier in normal human subjects and patients with cerebral tumours studied using [^{11}C]3-O-methyl-D-glucose and positron emission tomography. J Cereb Blood Flow Metab 6:230–239

Brooks DJ, Gibbs JSR, Sharp P et al. (1986b) Regional cerebral glucose transport in insulin-dependent diabetic patients studied using [^{11}C]3-O-methyl-D-glucose and positron emission tomography. J Cereb Blood Flow Metab 6:240–244

Brooks DJ, Beaney RP, Lammertsma AA et al. (1986c) Studies on regional cerebral haematocrit and blood flow in patients with cerebral tumours using positron emission tomography. Microvasc Res 31:267–276

Bustany P, Sargent T, Saudubray JM, Hanry JF, Comar D (1981) Regional human brain uptake and protein incorporation of ^{11}C-L-methionine studied in vivo with PET. J Cereb Blood Flow Metab 1:S17–S18

Bustany P, Henry JF, de Rotrou J et al. (1985) Correlations between clinical state and positron emission tomography measurement of local brain protein synthesis in Alzheimer's dementia, Parkinson's disease, schizophrenia, and gliomas. In: Greiz T et al. (eds) The metabolism of the human brain studied with positron emission tomography. Raven, New York, pp 241–249

Bustany P, Chatel M, Derlon JM et al. (1986) Brain tumour protein synthesis and histological grades: a study by positron emission tomography with C-11-L-methionine. J Neurooncol 3:397–404

Crone C (1963) Permeability of capillaries in various organs as determined by use of the indicator diffusion method. Acta Physiol Scand 58:292–305
Davson H, Welch K (1971) The permeation of several materials into the fluids of the rabbit's brain. J Physiol (Lond) 218:337–351
Feinendegen LE, Herzog H, Wieler H et al. (1986) Glucose transport and utilisation in the human brain: model using carbon-11 methyl glucose and positron emission tomography. J Nucl Med 27:1867–1877
Gaehtgens F, Albrecht KH, Kreutz F (1978) Fahreus effect and cell screening during tube flow of human blood. I. Effect of variation of flow rate. Biorheology 15:147–154
Gjedde A, Wienhard K, Heiss WD et al. (1985) Comparative regional analysis of 2-fluorodeoxyglucose and methylglucose uptake in brain of four stroke patients. With special reference to the regional estimation of the lumped constant. J Cereb Blood Flow Metab 5:163–178
Guerin C, Laterra J, Hruban RH et al. (1990) The glucose transporter and blood-brain barrier of human brain tumours. Ann Neurol 28:758–765
Hansen AJ, Lund-Anderson H, Crone AG (1977) K^+ permeability of the BBB investigated by aid of K^+ sensitive microelectrode. Acta Physiol Scand 101: 438–445
Hawkins RA, Phelps ME, Huang SC et al. (1984) A kinetic evaluation of blood-brain barrier permeability in human brain tumours with [^{68}Ga-EDTA] and positron computed tomography. J Cereb Blood Flow Metab 4:507–515
Hawkins RA, Huang SC, Barrio JR et al. (1989) Estimation of local protein synthesis rates with L-[1-^{11}C]leucine and PET: methods, model, and resuts in animals and humans. J Cereb Blood Flow Metab 9:446–460
Hayes RL, Washburn LC, Wieland BW et al. (1976) Carboxyl-labelled ^{11}C-1-aminocyclopentanecarboxylic acid, a potential agent for cancer detection. J Nucl Med 17:748–751
Herholz K, Wienhard K, Pietrzyk U, Pawlik G, Heiss WD (1989) Measurement of blood-brain hexose transport with dynamic PET: comparison of [^{18}F]2-fluoro-2-deoxyglucose and [^{11}C]O-methylglucose. J Cereb Blood Flow Metab 9:104–110
Herscovitch P, Raichle ME, Kilbourn MR, Welch MJ (1987) Positron emission tomography measurement of cerebral blood flow and permeability-surface area product of water using [^{15}O]water and [^{11}C]butanol. J Cereb Blood Flow Metab 7:527–542
Hübner KF, Purvis JT, Mahaley SM Jr et al. (1982) Brain tumour imaging by positron emission computed tomography using ^{11}C-labelled aminoacids. J Comput Assist Tomogr 6:544–550
Iannotti F, Fieschi C, Alfano B et al. (1987) Simplified, non-invasive PET measurement of blood-brain barrier permeability. J Comput Assist Tomogr 11:390–397
Jarden JO, Dhawan V, Poltorak A, Posner JB, Rottenberg DA (1985) Positron emission tomographic measurement of blood-to-brain and blood-to-tumour transport of ^{82}Rb: the effect of dexamethasone and whole-brain radiation therapy. Ann Neurol 18:636–646
Kilpatrick R, Reuschler HE, Munro DS, Wilson GM (1956) A comparison of the distribution of ^{42}K and ^{86}Rb in rabbit and man. J Physiol (Lond) 133:194–201
Lammertsma AA, Brooks DJ, Frackowiak RSJ et al. (1984a) A method to quantitate the fractional extraction of rubidium-82 across and blood-brain barrier using positron emission tomography. J Cereb Blood Flow Metab 4: 523–534
Lammertsma AA, Brooks DJ, Beaney RP et al. (1984b) In vivo measurement of regional cerebral haematocrit using positron emission tomography. J Cereb Blood Flow Metab 4:317–322
Love WD, Romney RB, Burch GE (1954) A comparison of the distribution of potassium and exchangeable rubidium in the organs of the dog using ^{82}Rb. Circ Res 2:112–122

Pardridge WM, Oldendorf WH (1977) Transport of metabolic substrates through the blood-brain barrier. J Neurochem 28:5–12
Phelps ME, Barrio JR, Huang SC et al. (1984) Criteria for the tracer kinetic measurement of cerebral protein synthesis in humans with positron emission tomography. Ann Neurol 15:S192–S202
Pozzilli C, Bernardi S, Mansi L et al. (1988) Quantitative measurement of blood-brain barrier permeability in multiple sclerosis using ^{68}Ga-EDTA and positron emission tomography. J Neurol Neurosurg Psychiatry 51:1058–1062
Renkin EM (1959) Transport of potassium-42 from blood to tissue in isolated mammalian skeletal muscles. Am J Physiol 197:1205–1210
Rubin P, Casarett G (1966) Microcirculation of tumours. II. The supervascularised state of irradiated regressing tumours. Clin Radiol 17:346
Vyska K, Magloire JR, Freundlieb C et al. (1985) In vivo determination of the kinetic parameters of glucose transport in the human brain using ^{11}C-methyl-D-glucose (CMG) and dynamic positron emission tomography (dPET). Eur J Nucl Med 11:97–106
Washburn LC, Sun TT, Anon JB, Hayes RL (1978) Effect of structure on tumour specificity of alicyclic alpha-aminoacids. Cancer Res 38:2271–2273
Yen CK, Budinger TF (1981) Evaluation of blood-brain barrier permeability changes in rhesus monkeys and man using ^{82}Rb and positron emission tomography. J Comput Assist Tomogr 5:792–799
Yen CK, Yano Y, Budinger TF et al. (1982) Brain tumour evaluation using Rb-82 and positron emission tomography. J Nucl Med 23:532–537

CHAPTER 14

Ontogenetic Development of Brain Barrier Mechanisms

N.R. Saunders

A. Introduction

The state of development of "the" blood-brain barrier in the immature brain remains a muddled and rather contentious matter, in spite of many decades of study. This is probably mainly because of a rather general preconception that a barrier present in the adult brain would necessarily be immature or absent in the fetal brain. For example, Barcroft (1938) used the following teleological argument: "There is no reason why the brain of the embryo should require an environment of very great chemical constancy. It will of course require a certain minimum of the various materials necessary for growth, but otherwise on first principles we must suppose that the good things of the life may exist in and may vary in the foetal blood to an extent much greater than in the maternal." This misconception is compounded by the fact that the term "blood-brain barrier" has come to describe a wide range of barrier mechanisms that contribute to the overall control of the stable internal environment of the adult brain. Often the term is used without clear specification of which barriers are being considered. Some of these barriers are undoubtedly immature in the developing brain. However, the original (and in many ways the fundamental) blood-brain barrier which was first described is the barrier that excludes or largely excludes protein in the blood from entering the brain and cerebrospinal fluid (CSF). It is formed very early indeed during brain development. Probably, as will be discussed in Sect. B this occurs as the vessels first penetrate into the brain substance (blood-brain barrier) and as the choroid plexus epithelial cells first differentiate (blood-CSF barrier).

Abbreviations

BUI	Brain uptake index	P	Postnatal age in days
cAMP	Cyclic adenosine monophosphate	E	Embryonic age in days
		CRL	Crown–rump length
CSF	Cerebrospinal fluid	IP	Intraperitoneal
ECF	Extracellular fluid	IV	Intravenous
HRP	Horseradish peroxidase		

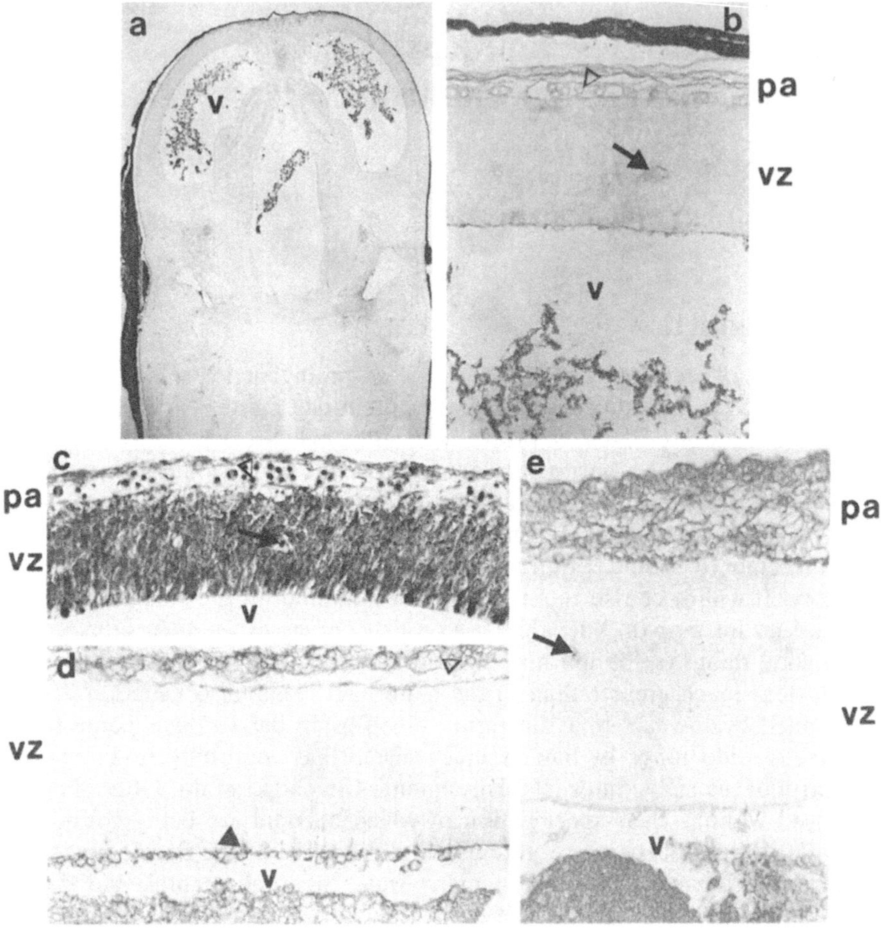

Fig. 1a–e. Immunocytochemistry of plasma proteins in fetal brain of different species, indicating integrity of blood-brain barrier to protein very early in brain development. **a** Low power E14 fetal rat stained for plasma proteins by the peroxidase-antiperoxidase method (see MØLLGÅRD et al. 1988 for details of method). **b** High power view of same section as **a** showing very immature telencephalic wall, also stained for plasma proteins. **c** Adjacent section to **b** at same magnification stained with toluidine blue to show structure. **d** High power of dorsal telencephalic wall of E28 sheep fetus stained for α-fetoprotein (PAP method). **e** High power of lateral telencephalic wall of E28 sheep fetus stained for albumin (PAP method). *v*, lateral ventricle; *pa*, pia arachnoid; *vz*, ventricular zone. *Arrows* indicate vessels within telencephalic wall. *Open arrowheads* indicate red cells in vessels in pia arachnoid. *Filled arrowhead* indicates ventricular zone cells at CSF interface that contain α-fetoprotein (probably taken up from CSF). The telencephalic wall (which will become the neocortex of the adult) is a simple undifferentiated structure at this stage (see **c**) consisting of a pseudostratified layer of ventricular zone cells. Even the primordial plexiform layer (the first stage of differentiation of the neocortex) is barely apparent between the upper layer of ventricular zone cells and the inner surface of the pia arachnoid. There are few vessels within the telencephalic wall at this stage; their endothelial cells stain for plasma proteins (**b**,**e**) but there is no evidence of a "leak" of protein from these vessels or from the pial vessels into the

The barrier to protein, which also restricts (but does not entirely prevent) the penetration from blood into brain and CSF of other smaller lipid insoluble molecules such as sucrose and inulin, is the fundamental basis of all barrier mechanisms. This is because it provides a diffusion restraint upon which other mechanisms such as ion pumps can work. Thus it is, that in the adult the internal environment of the brain has a composition which is different in important respects from that of the general extracellular environment of the body, as represented by plasma. More importantly, because of the presence of this diffusion restraint between blood on the one hand and brain and CSF on the other, the internal environment of the brain is much more stable than that of the rest of the body.

The nature of the diffusion restraint between blood and brain lies in the presence of tight junctions between the cerebral endothelial cells (and between the epithelial cells of the choroid plexus in the case of the blood-CSF barrier). Numerous cellular mechanisms (described in other chapters in this volume) are responsible for moving metabolically important materials into and out of the brain. Such mechanisms are only able to function adequately and to maintain a stable environment within the brain because of the presence of tight junctions at the interfaces between blood and brain extracellular fluid (ECF) and blood and CSF (see Chap. 1).

In the fetus, as will be described below, tight junctions that are effective enough to exclude protein from the blood and CSF, are present very early in brain development. CSF, in spite of the presence of tight junctions between the choroid plexus epithelial cells, contains a high concentration of protein early in fetal life, as will be discussed later. Studies involving the in vitro culture of endothelial cells (see Sect. E) suggest that although low resistance junctions develop very rapidly, they do not necessarily achieve their final tightly restricted permeability properties immediately. As will be discussed later, there is increasing evidence suggesting that the permeability properties of tight junctions depend partly upon external influences from developing astrocytes. If these properties develop sequentially, this would explain the experimental observation that the fetal blood-brain barrier appears to be tight to protein from very early in development (e.g. Fig. 1) yet is much more permeable to small molecular weight compounds such as sucrose and inulin (see Dziegielewska et al. 1979 and Sect. G). The various cellular mechanisms that contribute to the specific properties of the brain's internal environment and their control, develop subsequent to the initial formation of tight junctions.

The traditional way of demonstrating the blood-brain barrier phenomenon has been to inject dyes (which bind to plasma proteins, Tschirgi

dorsal surface of the developing forebrain vesicle. Note that before vessels invade the nervous tissue (**d**) protein from plasma is present in CSF (*v*) and within pia arachnoid vessels, but does not penetrate into the brain ECS; there is some intracellular uptake of protein from CSF into *vz* cells (**d**) which is probably specific for different proteins at different stages of brain development.

1950) intravenously or intraperitoneally. In more recent years horseradish peroxidase (HRP) has been used; HRP can be visualized in the electron microscope because it is an enzyme, the reaction product of which is electron dense. As will be reviewed in the next section, the results of these experiments are contradictory, although it is generally only those that appear to show barrier immaturity that are cited. This is such a fundamental point for the understanding of the development of the whole range of barrier mechanisms in the immature brain that these experiments will be summarized before proceeding to outline briefly what is known of the morphology of tight junctions and other intercellular junctions in the developing brain. This will be followed by sections on permeability to lipid insoluble molecules, metabolically important materials such as amino acids and glucose, as well as sections on CSF secretion and pressure and the composition of CSF (particularly proteins and electrolytes) during brain development. The pharmacology of barriers in the developing brain has not been much studied. However, understanding of the various barrier mechanisms in the immature brain is a fundamental basis for future studies, as will be considered in Sect. N. Several of these topics have been reviewed in more detail elsewhere (DZIEGIELEWSKA and SAUNDERS 1988, 1991; JOHANSON 1989; SAUNDERS 1991). Other related topics (e.g. kernicterus, pathophysiological effects on barrier mechanisms, viral entry in the immature central nervous system) are not considered here because of lack of space. They have been reviewed recently (DZIEGIELEWSKA and SAUNDERS 1991; SAUNDERS 1991, 1992).

B. Barriers to Dyes and Proteins in the Developing Brain

Some of the more recent experiments designed to test the integrity of the blood-brain barrier in the immature brain are summarised in Table 1. WISLOCKI (1920) appears to have been the first to inject a dye (trypan blue, which binds to plasma proteins, particularly albumin) into a fetus. His colour lithograph shows the same appearance reported by others on injecting trypan blue into adult animals (e.g. EHRLICH 1885). The dye stained almost all of the tissues of the body with the striking exception of the brain. One of the most commonly cited of the early studies of barrier permeability in the developing animal is that of BEHENSEN (1927) in postnatal rats. However, BEHENSEN injected large amounts of dye and his published evidence shows drawings of rat brain indicating where the dye penetrated. This was generally in the areas in which dye penetrates in the adult (e.g. area postrema, hypothalamus and choroid plexuses). These areas of dye staining were reported as being larger in the immature brain. However, this may have been because BEHENSEN used outline drawings of sagittal sections of *adult* brain for both postnatal and adult results. The neocortex and

Table 1. Injection experiments used as test of blood-brain barrier integrity in fetal and neonatal animals

Authors	Species	Age (days)	Body weight[c] (g)	Injection[a] (vol)	Effect on protein concentration[b]	"Immature" barrier claimed	Injected material[d]
Grazer and Clemente (1957)	Rat	E10.5+	<0.05	Not stated	–	No	Trypan blue IP/IV[b]
Olsson et al. (1968)	Rat	E15+	0.14	<5%	–	No	Fluorescein labelled albumin Umb. A.
Delorme et al. (1970)	Chick	E4.5–10	0.1–2.39	Not stated	3% Increase	Yes	HRP i.v.
Wakai and Hirokawa (1979, 1981)	Chick	E9	1.56	5%–10%	2 × Increase	Yes	HRP i.v.
Dziegielewska et al. (1979)	Sheep	E60	60	7.5%	20%	No	Human protein i.v.
Roncali et al. (1986)	Chick	E6–12	0.34–5.04	40%	<10%	Yes	HRP intracardiac[e]
Risau et al. (1986)	Mouse	E13	0.08	1–3 ×	1% Increase	Yes	HRP intracardiac[e]
Vorbrodt et al. (1986)	Mouse	Newborn	1.4	= total blood volume	2–3 × Increase	Yes	HRP i.v.
Stewart and Hayakawa (1987)	Mouse	E15	0.26	Not stated	Up to 4 × Increase	Yes	HRP i.p.[f]
Dziegielewska et al. (1988)	Tammar wallaby	Newborn+	0.50	10%–23% Blood vol	24%–48% Increase	No	HRP/HSA i.v.
Hulsebosch and Fabian (1989)	Rat	Newborn	5–7	Up to 40% Increase	Up to 3.3 × Increase	Yes	IgG i.p.[b]

[a] The authors' injection volume when stated has been compared to estimates of circulating blood volume (10% of body weight and allowing a factor of 2–3 for placental blood volume in the case of fetuses, although it is doubtful if under the conditions of some of these experiments, much mixing with fetal placental blood would have occurred. The actual circulating blood volume in sheep fetuses has been measured (Dziegielewska et al. 1991).

[b] The effect of the amount of injected protein on the plasma protein concentration is also estimated when sufficient information is available. Plasma protein concentrations from Birge et al. (1974, chick embryo) and Dziegielewska et al. (1981, rat fetus).

[c] Body weights and circulating blood and plasma volumes for chick embryos from Romanoff (1967). Body weights of fetal rats from unpublished data.

[d] Information is not usually available about the speed of injection nor about the likely distribution volume (e.g. in experiments involving[e] intracardiac injection in which the normal circulation may not have continued or[f] i.p. injection in which it is not clear what proportion of the volume and protein will have entered the circulation).

E is embryonic age from day of conception.

cerebellum are of course relatively much smaller in the neonatal period so the apparently greater staining of other areas of the brain may have been a visual impression rather than real. Also, BEHENSEN did not report dye throughout the brain as would be expected if the barrier were generally permeable (for more detailed comments see JANAS et al. 1991).

The importance of using only moderate quantities of dye was stressed by Stern (e.g. STERN et al. 1929). She commented that dye staining of the neonatal brain was only obtained if substantially greater amounts of dye were used that was customary in adult experiments. More recent experiments using HRP that are often quoted as evidence of barrier immaturity are those of WAKAI and HIROKAWA (1979, 1981) in the chick embryo and RISAU et al. (1986) in the mouse embryo. As can be seen from Table 1, the former experiments involved a very large increase in plasma protein concentration whereas the latter involved a very large volume of injected material. Experiments in which the barrier appears intact generally seem to have involved smaller injection volumes and protein amounts.

In fact it is possible to demonstrate a barrier to protein in the developing brain without injecting anything! The circulating plasma contains various proteins, e.g. α-fetoprotein, transferrin, that can be visualized in tissue sections by immunocytochemistry. Such sections are illustrated in Fig. 1. The immunoreaction product is confined to precipitated plasma within vessels and to the endothelial cells. There is no protein apparent in the ECF around the vessels as would be expected if they were "leaking". There is, however, clear evidence of plasma proteins in the CSF, which, as will be described in Sect. L, are present in very high concentrations compared with the adult, especially early in brain development.

C. Formation of the Fundamental Blood-Brain and Blood-CSF Barriers in the Embryonic and Fetal Brain

The basis of the various blood-brain barrier mechanisms described in Sect. A has been defined above as the presence of tight junctions between cerebral endothelial cells (and for the blood-CSF barrier, tight junctions between choroid plexus epithelial cells). The formation of tight junctions at these two sites has been studied directly in vivo and in vitro using electron-microscopical techniques in a variety of preparations. It has also been studied indirectly using macromolecules such as dyes bound to proteins or HRP in permeability studies, again both in vivo and in vitro. In addition, measurement of transendothelial resistance in vivo (BUTT et al. 1990) and in vitro (RUBIN et al. 1991) has been used as an indirect measure of the diffusion restraint that is fundamental to all barrier mechanisms. In many studies HRP has been injected i.v. or i.p. as a test of barrier integrity. However, as pointed out in Sect. B, unless care is taken to limit the volume and protein concentration of the solution injected and also the pressure, in the case of i.v. injections, an apparently greater permeability to HRP in immature brains needs to be interpreted with caution. One aspect of tight junction formation which does not seem to have been fully considered is that it may not be an all or nothing process. Indeed, it will be argued below, that the results of measurements of electrical resistance both in vivo and in vitro and the strikingly different permeabilities for large macromolecules compared with smaller ones in the immature brain, suggest that tight junctions may form initially as a relatively weak intercellular connection that is effective in excluding protein but not smaller molecules from intercellular passage. Subsequent "maturation" of the junctions then results in increasing transendothelial resistance and additional "barrier" properties such as decreased permeability to smaller molecules, in addition to carrier properties determined by the endothelial cells themselves.

D. Ultrastructure of Tight Junctions in Embryonic and Fetal Brain

It has been clearly shown in the adult that the fundamental basis for the blood-brain barrier (to protein) is the presence of tight junctions between adjacent cerebral endothelial cells (REESE and KARNOVSKY 1967); in the case of the blood-CSF barrier, it is tight junctions between the choroid plexus epithelial cells that are important (BRIGHTMAN and REESE 1969). Since it had for long been the assumption that the blood-brain barrier (to protein) is immature in the developing brain, it is perhaps less than surprising that most of the initial reports on tight junctions claimed that they were incomplete. As was studied extensively by BRIGHTMAN and REESE (1969) in the adult brain, the electron-microscopical appearance of membrane apposition depends very much upon the fixation conditions. In addition, the fetal brain has generally been found to be much more difficult to fix adequately. Thus many of the earlier (and some of the more recent) studies need to be evaluated with care. Also, transmission electron microscopy of thin sections allows only a very small part of a junction to be examined. The work involved in reconstructing a whole junction from serial sections is considerable and not more than a few junctions in any brain can therefore be studied. The freeze fracture technique has the advantage that a large proportion of a single tight junction can be examined for its completeness. However, double replicas are needed in order to be sure whether an apparent gap in one fracture face is really a gap or if the membrane particles separated into the complementary fracture face (e.g. see MØLLGÅRD et al. 1979). Cerebral endothelial cell tight junctions are difficult to fracture because of their spiral configuration; also, in the very early stages of development the brain is poorly vascularized. In so far as it has been possible to obtain successful fractures of endothelial tight junctions in the developing brain, they appear to be well formed (Fig. 2; MØLLGÅRD and SAUNDERS

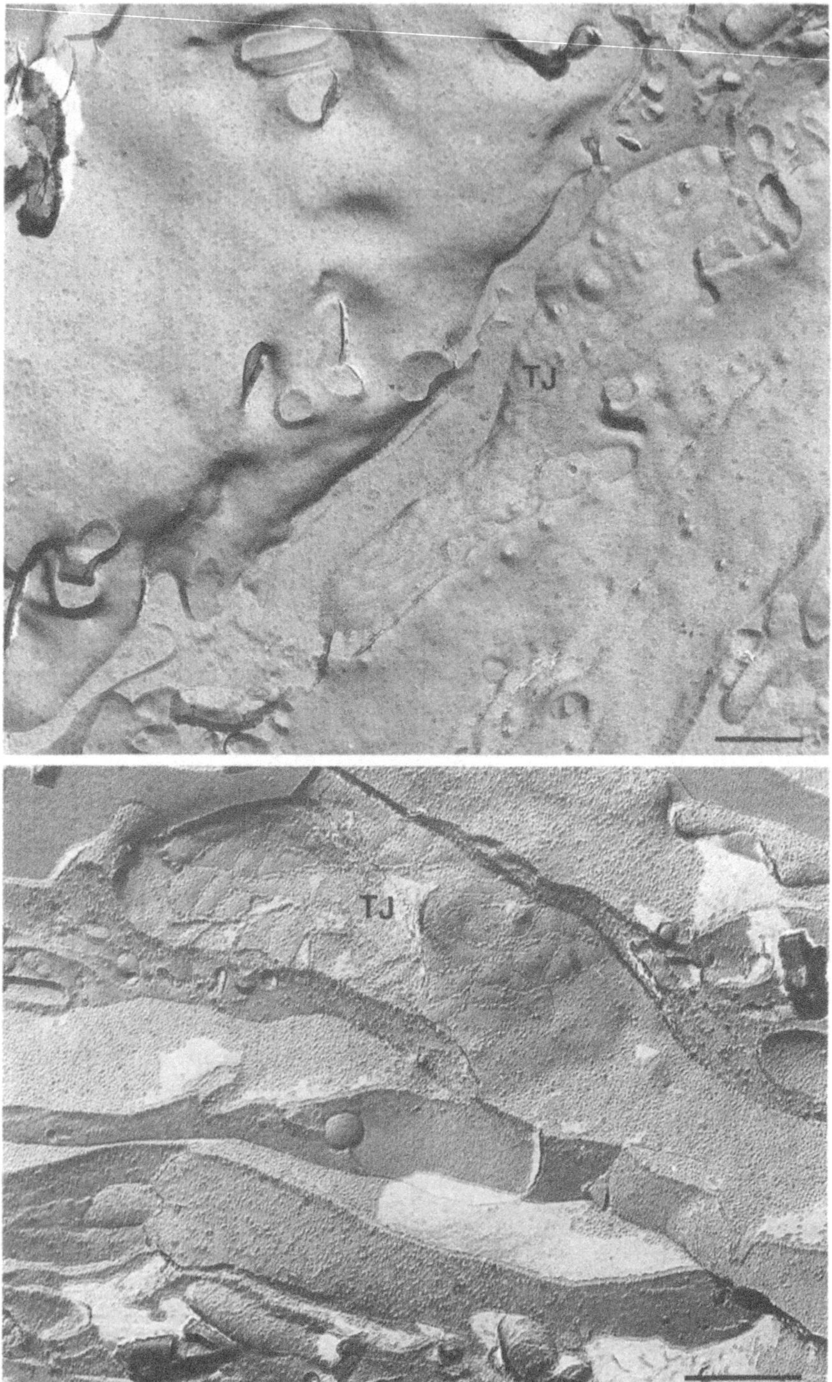

Fig. 2. Freeze fracture electron micrographs of cerebral capillary endothelial cell tight junctions from human fetal brain (crown-rump length 141 mm, *above*) and sheep fetal brain (125 days gestation, *below*). *Bars*, 0.5 μm; *TJ*, tight junction. Note that strands within junctional complex are continuous and that in both replicas a large area of tight junction has been exposed. (From MØLLGARD and SAUNDERS 1975); original micrographs kindly supplied by Professor K. MØLLGÅRD)

1975, 1986; SAUNDERS and MØLLGÅRD 1984). However, the most convincing evidence for the "tightness" of immature cerebral blood vessels comes from the more generalized observation that naturally occurring plasma proteins remain within the vessel lumina (although they may be taken up into endothelial cells) as is illustrated in Fig. 1.

It has been possible to study the ultrastructure of tight junctions of the developing choroid plexuses more systematically in fetal sheep, in parallel with permeability studies (cf. MØLLGÅRD et al. 1976; DZIEGIELEWSKA et al. 1979). Table 2 shows that there are only marginal changes (although statistically significant) in the complexity of choroid plexus tight junctions during development. This is in contrast to the considerable decline in blood-CSF permeability to protein and smaller molecules such as sucrose and inulin (DZIEGIELEWSKA et al. 1979, 1980). These observations in fetal sheep appear to be inconsistent with CLAUDE and GOODENOUGH's (1973) hypothesis that junctional resistance (and thus permeability) of epithelia is proportional to junctional structural complexity. The high concentration of proteins in CSF is accounted for by transfer from plasma across choroid plexus epithelial cells (see Sect. L). However, the apparently greater permeability to smaller molecules (inulin and sucrose) of the blood-CSF barrier may be due to indirect exchange of these materials across the blood-brain barrier and equilibration via the brain-CSF interface, rather than to "leakiness" of the choroid plexus itself.

E. Cerebral Endothelial Cells In Vitro

Over the last 15 years numerous methods for isolating endothelial cells from brain have been published, together with reports on a variety of their properties (e.g. BOWMAN et al. 1983; GOLDSTEIN and BETZ 1986). Most of these studies involved endothelial cells from adult brain although a few developmental studies have been published (see BETZ and GOLDSTEIN 1981; JOÓ 1985 for review). Initial investigations involved study of the presence of various enzymes and uptake systems. However, the polarity of transport systems could not be determined. In recent years these studies have been

Table 2. Freeze fracture tight junction morphology at 40 and 125 days gestation in fetal sheep (from MØLLGÅRD et al. 1976)

Gestational age[a] (days)	Minimum strand no.[b]	Junctional depth (μm)	*n*
40	3.64 ± 1.12	0.34 ± 0.17	121
125	3.45 ± 1.13	0.28 ± 0.14[c]	49

[a] Term is 150 days.
[b] Minimum strand number and junctional depth were measured within a standard grid. Mean ± SD. *n*, number of areas of freeze fracture replicas counted.
[c] Significantly different from 40 days.

pursued in two important areas: (1) development of techniques for culturing cerebral endothelial cells as monolayers between two chambers, so that the transport properties of the monolayer can be studied (e.g. RUTTEN et al. 1987; DEHOUCK et al. 1990; RUBIN et al. 1991); and (2) studies of the development of tight junctions between cultured cerebral endothelial cells and the possible influence of astroglia on their development (e.g. ARTHUR et al. 1987; TAO-CHENG and BRIGHTMAN 1988). This second problem has also been studied in vivo in experiments in which tissues are implanted into foreign sites and the properties of the resulting blood vessels within the graft studied in order to determine whether it is the tissue of origin of the vessels or the surrounding milieu which determines the "tightness" of vessels subsequently developing (e.g. STEWART and WILEY 1981; BROADWELL et al. 1991). Although some of the results reported appear contradictory it seems clear that providing the experimental circumstances are properly defined, the clear-cut result is obtained that it is the milieu which is important rather than the origin of the endothelial cells (see also Chap. 15).

Culture of monolayers of cerebral endothelial cells has been reported by a number of groups (see also Chap. 17). One of the more successful is that of RUBIN et al. (1991), who cultured bovine brain endothelial cells, initially on collagen-fibronectin with subsequent trypsinization and transfer to Costar Transwells. The culture medium included conditioned medium from cultured neonatal rat brain type 1 astrocytes and agents that raise cellular levels of cyclic adenosine monophosphate (cAMP). This protocol resulted in the growth of single monolayers of cerebral endothelial cells with transmonolayer electrical resistances in excess of $1000\,\Omega\,cm^2$ in some preparations. This is considerably higher than obtained previously and approaches that of cerebral vessels in vivo (OLESEN and CRONE 1983; BUTT et al. 1990). The high resistance cultures were shown to be much less leaky, for example to molecules as small as sucrose, than previous culture systems that did not use conditioned medium or increased levels of intracellular cAMP. RUBIN et al. (1991) have also shown that modification of cell adhesion molecules or reduction in cAMP levels in these cultures reduced the transmonolayer resistance to low levels. Preliminary in vivo experiments suggested that agents which either decrease cAMP or inhibit protein kinase A may facilitate entry of impermeant molecules across the blood-brain barrier (RUBIN et al. 1991). Monolayers of brain endothelial cells obtained by culture in the presence of astrocyte conditioned medium without raised levels of cellular cAMP, have relatively lower trans-monolayer electrical resistances, but they were able to exclude the passage of albumin whilst still allowing that of sucrose (L. Rubin, personal communication). This correlates with the observation that cerebral vessels in the developing brain are impermeable to protein from the earliest stages of vascularization of the embryonic/fetal brain (see Sects. B, L) but are much more permeable to sucrose than in the adult (see Sect. G). These in vitro observations also correlate with the dramatic increase in vessel resistance that occurs in vivo around the time of birth in the rat (BUTT et al. 1990; ABBOTT and REVEST 1991 and Chap. 15). Brightman and his colleagues (e.g. TAO-CHENG et al. 1987) have provided important ultrastructural evidence of the effect of astrocytes on tight junction formation between cerebral endothelial cells when the two cell types were co-cultured.

F. Suitable Preparations for Studies of Barrier Mechanisms in the Developing Brain In Vivo

It will be apparent from Sect. B that one of the deficiencies in the qualitative experiments demonstrating or refuting the presence of a barrier to protein or dyes bound to protein in the developing brain is that the preparation used was often subjected to extremely unphysiological conditions. This limitation was often compounded by the fact that the fetuses or neonates used were generally too small for any attempt to be made to assess their physiological condition. This is particularly important in quantitative permeability studies because it is known that a high level of carbon dioxide (hypercapnia) in the blood causes increased exchange of a variety of materials from blood to brain and CSF (e.g. EVANS et al. 1976; HABGOOD and SINCLAIR 1988; HABGOOD 1990), especially in the immature brain. A further difficulty of small preparations is that their circulating blood volume is too low for it to be possible to take several blood samples in order to check the plasma level of a marker introduced into the circulation when measuring its penetration into brain and CSF. Blood-brain barrier results are usually expressed as the ratio brain (or CSF) concentration of marker/plasma concentration of marker. It is therefore essential to achieve a steady plasma level or to obtain sufficient plasma samples so that a time-weighted mean can be calculated. If small animals are used, such as neonatal rats, only single terminal samples of blood can be obtained. HABGOOD (1989, 1990) has shown recently that it is possible to use a litter of neonatal rats as the experimental model. Having injected all of the litter with a standard dose of permeability tracer, the blood concentrations are sufficiently similar for a blood concentration curve against time to be plotted as shown in Fig. 3; from Fig. 3 it can be seen that a steady level in plasma is achieved at 4–6h after an intraperitoneal injection. Subsequently the plasma level falls, which of course will give rise to progressively increasing CSF or brain/plasma ratios until the marker concentrations in CSF and brain also begin to fall. Thus it is likely that the studies of FERGUSON and WOODBURY (1969) in neonatal rats and of STASTNY (1983) in chick embryos will have considerably overestimated the blood/brain and blood/CSF permeability to the markers inulin and sucrose. A similar concern probably also applies to the work of AMTORP (1976) who studied blood-brain and blood-CSF exchange of ^{125}I-labelled human albumin in postnatal rats. In fact in order to study albumin or other plasma protein permeability, particularly in response to changing physiological conditions,

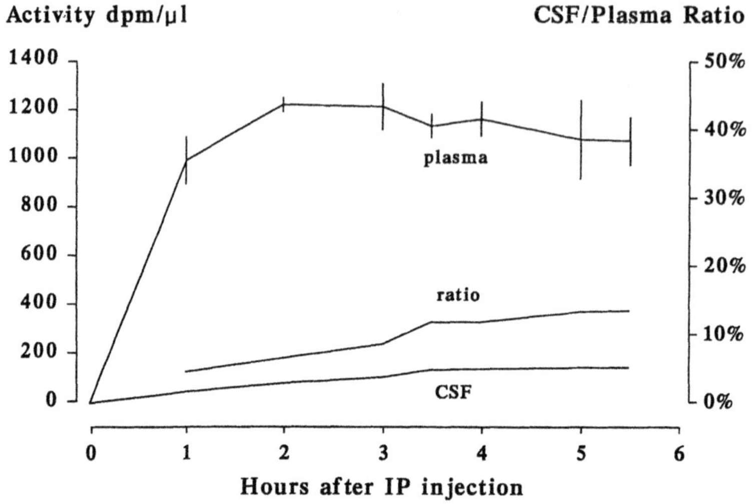

Fig. 3. Time course of uptake of [^{14}C]sucrose into plasma and CSF (cisternal) following intraperitoneal injection into a litter of bilaterally nephrectomized 2-day-old rats. All littermates were injected at time zero; individual animals were sampled at 1–5 h later. Note stable plasma level of [^{14}C]sucrose between 1 and 5 h after injection. CSF levels and CSF/plasma ratios reach stable levels by about 5 h after injection. *Bars*, SEM. (From HABGOOD 1990)

e.g. hypercapnia, it is not necessary to inject any foreign or radioative protein: the endogenous proteins occurring in CSF and plasma can be measured directly using an immunoassay and the CSF/plasma ratio thus obtained (see for example DZIEGIELEWSKA et al. 1980; HABGOOD and SINCLAIR 1988; HABGOOD 1990). Such an approach assumes that all of the protein in CSF originates from plasma. In view of the high concentration of proteins in fetal CSF (see DZIEGIELEWSKA and SAUNDERS 1988 and below) it is necessary first to show whether or not this is the case.

To date the most satisfactory preparation for physiologically controlled, quantitative barrier permeability studies during brain development appears to be the fetal sheep (see EVANS et al. 1974; DZIEGIELEWSKA et al. 1979, 1991). It has two main advantages: (1) Size; that is, down to about 50 days gestation (term is 150 days) the fetal circulating blood volume is large enough for several samples to be taken so that the blood level of markers can be measured and the physiological state of the fetus (blood gas and pH together with continuous monitoring of arterial blood pressure and heart rate) can be assessed. (2) The placenta is made up of a large number, up to 100, of small placentae. An individual placenta can be everted and its arterial and venous supply cannulated to give access to the fetal circulation with minimal disturbance. The fetus can generally be left within the intact amniotic sac (DZIEGIELEWSKA et al. 1979). Results from this preparation are reviewed in Sect. G and L.

The use of neonatal rats for barrier permeability studies has been extensively reviewed by JOHANSON (1989) and HABGOOD (1990). If they are unanaesthetized for most, if not all, of the time course of a steady state experiment, it is likely that they would be in a reasonable physiological state. The use of a whole litter to estimate blood levels of marker in neonatal rats has been discussed above, but a further difficulty is that for small molecules such as inulin and sucrose the kidneys have to be tied off or too much of the marker is excreted in the urine. This was found not to be necessary in immature fetal sheep (e.g. DZIEGIELEWSKA et al. 1979); fortunately, the placenta in this species is so impermeable to lipid insoluble molecules that there is a negligible loss of molecules such as inulin or sucrose to the maternal circulation (BOYD et al. 1976).

A more recent and promising preparation for permeability studies at immature stages of brain development is the newborn marsupial. Studies have been carried out in two species, the tammar wallaby (*Macropus eugenii*, DZIEGIELEWSKA et al. 1988) and the grey short-tailed opossum (*Monodelphis domestica*, KNOTT et al. 1991). These animals, although rather small, are very robust and are born at a very immature stage of brain development (see REYNOLDS et al. 1985; SAUNDERS et al. 1989).

The use in immature animals of Oldendorf's (e.g. BRAUN et al. 1980) Brain Uptake Index (BUI) for rapidly penetrating molecules has been reviewed recently by JOHANSON (1989) and will be considered briefly in a later section (K) dealing with permeability to metabolically important materials.

G. Barrier Permeability to Lipid Insoluble Molecules in the Fetal/Neonatal Brain

The earliest such studies are probably those of STERN and her colleagues (e.g. STERN and PEYROT 1927; STERN et al. 1929) who showed that in various species more sodium ferrocyanide (304 daltons) penetrates into the newborn brain than in the adult. These qualitative studies were followed much later by the quantitative studies of FERGUSON and WOODBURY (1969) in neonatal rats and of STASTNY (1983) in chick embryos. These authors used sucrose or inulin or both, labelled radioactively with ^{14}C or ^{3}H. These experiments confirmed the much earlier observations of STERN and colleagues, that the blood-brain and blood-CSF barriers appear to be much more permeable to small molecules (such as inulin and sucrose) than in the adult. The methodological limitations of the studies of FERGUSON and WOODBURY (1969) and of STASTNY (1983) have been outlined in the previous section.

The results of a series of permeability studies in fetal sheep carried out under well controlled physiological conditions are illustrated in Fig. 4. These results show that blood-CSF exchange for a wide range of molecular sized markers is much greater in the immature brain (60 days gestation) than in

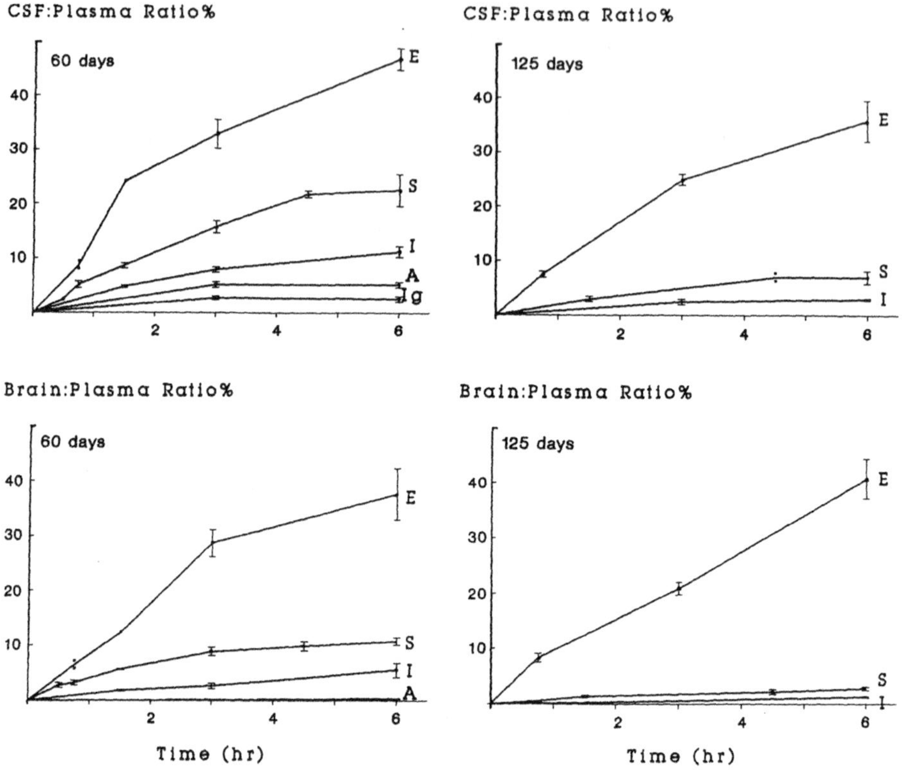

Fig. 4. Permeability of blood-brain and blood-CSF barriers in anaesthetized fetal sheep at 60 and 125 days gestation (term is 150 days from conception). Fetuses were given i.v. infusions or intermittent injections to maintain approximately constant blood levels of a wide range of molecular size markers: erythritol (*E*, mol. radius 0.35 nm), sucrose (*S*, 0.51 nm), inulin (*I*, 1.3 nm), human albumin (*A*, 3.5 nm), human IgG (*Ig*, 5.3 nm). Experiments were run for 20 min to 6 h. Three to six experiments were performed for each time point except where individual points are shown. *Bars* ± standard error of mean. Results are expressed as the concentration ratio CSF: plasma × 100 or brain to plasma × 100. The ratios for erythritol at both ages and for CSF and brain are similar probably because erythritol is small enough to pass through cell membranes. The ratios for brain are smaller than for CSF at each age because the markers are distributed in brain within the extracellular space. Note the decline in ratios with age, which is thought to reflect a decline in passive permeability with age. The values for human albumin in 60 day fetal brain are not significantly above background blood contamination of the brain samples, i.e. the blood-brain barrier to protein is well formed (cf. qualitative barrier experiments in Fig. 1 and data of Fig. 5). (From DZIEGIELEWSKA and SAUNDERS 1991)

the more mature brain (125 days gestation). The results confirm that protein penetration from blood into brain could not be detected above the level in blood contaminating the brain samples, but blood-brain exchange for the smaller molecules was clearly greater in the less mature brain. Other authors with similar if less quantitatively secure results have suggested that such a result is due to a reduced rate of CSF secretion in the younger brain (e.g. FERGUSON and WOODBURY 1969; AMTORP 1976). However, as will be discussed below, the "sink effect" of CSF at different ages is probably similar and if related to the size of the brain it may even be that the sink effect of CSF secretion is greater rather than smaller in the immature brain (see Table 3). Thus it seems likely that the greater blood-brain and blood-CSF exchange of small molecular weight compounds (<5000 daltons) seen in younger fetuses is due to a genuinely greater permeability of the immature cerebral vessels and of the choroid plexus. In the case of the vessels their individual permeability is probably considerably greater than is indicated by the change in brain/plasma ratio for the markers used, because the vascularity of the immature brain is considerably less than in the more mature brain (see e.g. CALEY and MAXWELL 1970; DZIEGIELEWSKA et al. 1979; JOHANSON 1989).

Permeability between blood and brain and between blood and CSF in the adult has generally been discussed in terms of pore-restricted diffusion (e.g. BRADBURY 1979; RAPOPORT 1976; RAPOPORT and PETTIGREW 1979) and it has often been assumed that in the developing brain the "pores" responsible for the greater permeability of the immature brain are larger and reduce during development (e.g. BRADBURY 1979). However, FELGENHAUER (1974) has put forward a simple analysis for blood-CSF exchange in the adult human, especially for proteins, and has shown that unrestricted diffusion can satisfactorily account for the steady state CSF/plasma ratios of many proteins in CSF. A similar approach in fetal sheep shows that the permeability data for a wide range of molecular sizes can be accounted for by unrestricted diffusion through large pores, and that with age it is the number of pores rather than their diameter which decreases (DZIEGIELEWSKA et al. 1979). An inverted "Felgenhauer plot" for fetal sheep data compared with the adult ewe and adult human is shown in Fig. 5. The parallel slope of the lines at different ages suggests that with increasing age the declining permeability is due to a reduction in the surface area for unrestricted diffusional exchange rather that to a reduction in pore size (which would be expected to change the slope of the line).

H. CSF-Brain Barrier in the Immature Brain

A novel feature of barriers in the developing brain is the presence of a CSF-brain barrier. This was described by FOSSAN et al. (1985) who perfused the ventricular systems of 60- and 125-day fetal sheep with artificial CSF containing horseradish peroxidase as is illustrated in Fig. 6. At 125 days, as has previously been described in adult mice by BRIGHTMAN and REESE (1969), the HRP penetrated across the CSF-brain interface to an extent that depended upon the duration of perfusion. In contrast, at 60 days, no HRP passed into the brain extracellular space although some was taken up by the neuroependymal cells lining the ventricles. The ultrastructural basis for

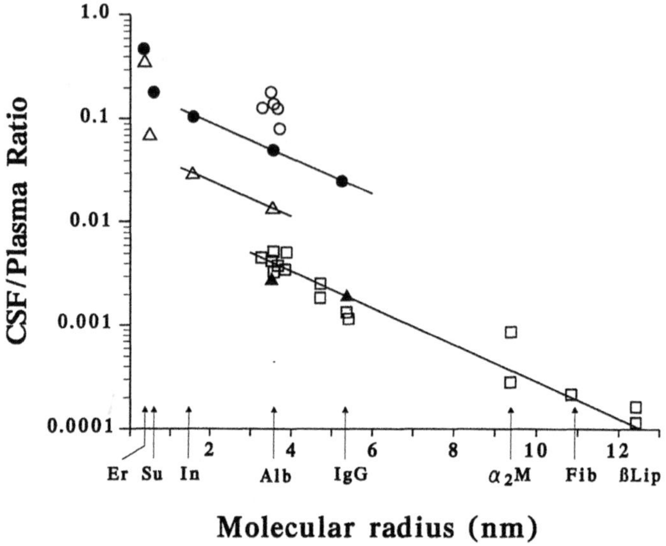

Fig. 5. Inverted FELGENHAUER (1974) plot of CSF/plasma ratio (log scale) against molecular radius (nm) for different molecular sized compounds. *Er*, erythritol; *Su*, sucrose; *In*, inulin; *Alb*, albumin; *IgG*, immunoglobulin G; $\alpha_2 M$, α_2-macroglobulin; *Fib*, fibrinogen; βLip, β-lipoprotein. Sheep data from Fig. 4 and 11 (*filled circles* and *open triangles*). Points above 60 day line (*open circles*) are sheep protein data from Fig. 11 and unpublished data. Adult sheep data from DZIEGIELEWSKA et al. (1991) and unpublished (*filled triangles*). Human data (*open squares*) from FELGENHAUER (1974). Note parallel log-linear lines with decreasing CSF/plasma ratios as age increases. This relationship suggests unrestricted diffusion through very large pores, with a reduction in the number of pores rather than their size during development. Values for sucrose and especially for erythritol are above the line, suggesting an additional population of small pores. Erythritol is probably small enough to pass through cell membranes. Note protein points above 60 days fetal sheep line. These are values for the five main proteins in CSF and plasma of fetal sheep: α-fetoprotein, α_1-antitrypsin, albumin, fetuin and transferrin; their CSF/plasma ratios are all higher than can be accounted for by diffusion and may be explained by a specific transfer of proteins from blood to CSF probably via the choroid plexuses (especially III and IVth ventricular) (cf. data in Fig. 11). (From SAUNDERS 1991; with additional unpublished data from DZIEGIELEWSKA and MATTHEWS)

this exclusion of protein at the CSF-brain interface in the immature brain is a membrane specialization, the "strap junction", which is present at regions of contact between adjacent neuroependymal cells. These junctions have been described in humans (MØLLGÅRD and SAUNDERS 1986), sheep (MØLLGÅRD et al. 1987a) and tammar wallaby (DZIEGIELEWSKA et al. 1988) brain early in development. Examples from a 19-day sheep embryo are illustrated in Fig. 7. They disappear during brain development and at least in the sheep fetus this disappearance seems to correlate with cessation of neurogenesis and a decline in the protein concentration in CSF (see Sect. L). The functions of the strap junctions that form the CSF-brain barrier to protein would appear to be in the control of the protein environment of

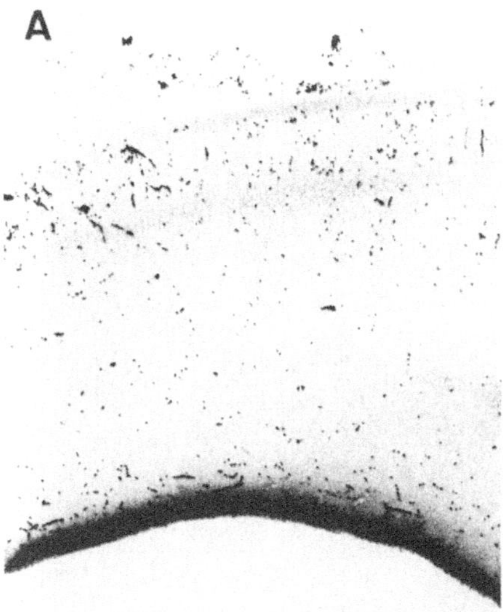

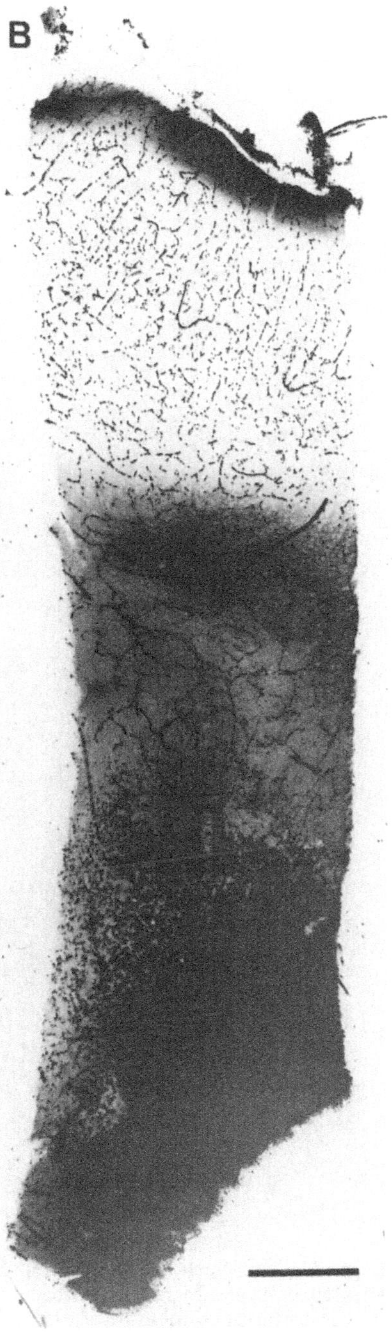

Fig. 6A,B. Coronal sections of telencephalic wall of 60-day (**A**) and 125-day (**B**) sheep fetuses after 5 h ventriculocisternal perfusion with HRP in artificial CSF. The pial surface is *above* and the lateral ventricle is *below*. Peroxidase can be localized up to 3 mm away from the CSF-brain interface at 125 days, whereas at 60 days the peroxidase was confined to the normal ventricular zone and was largely intracellular. Note filling of the subpial layer at 125 days but not at 60 days. Peroxidase activity within vessels was associated with erythrocytes. *Bar*, 0.5 mm. (From FOSSAN et al. 1985)

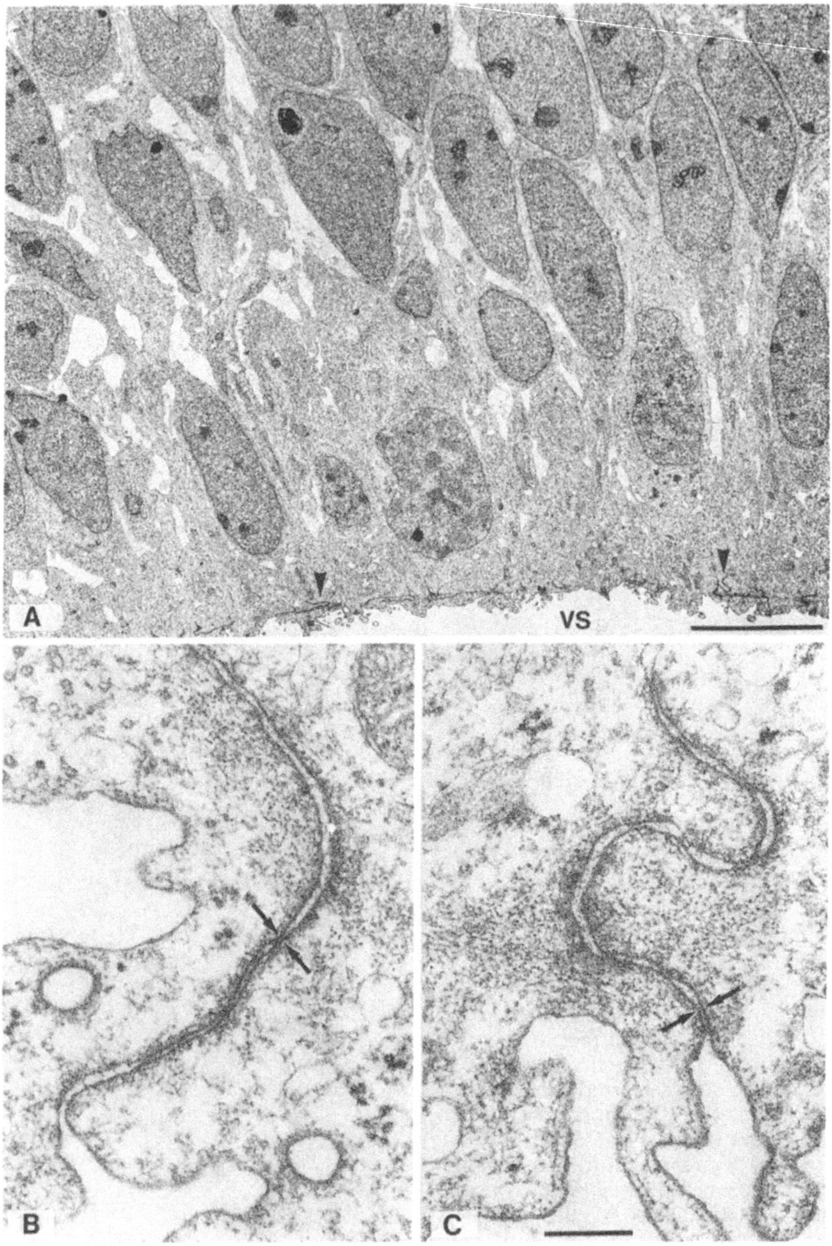

Fig. 7A–C. Thin Section electron micrographs of the cerebral vesicle from a 19-day-old sheep embryo. In **A** the ventricular surface (*vs*) is seen in low magnification. Note the tortuous configuration of the intercellular clefts (*arrowheads*); *bar*, 10 µm. The junctional zone is shown in high magnification in **B** and **C**. It exhibits very narrow intercellular celfts which at places seem to be totally occluded (*arrows*); *bar*, 0.5 µm. (From MØLLGÅRD et al. 1987a; micrographs kindly supplied by Professor K. MØLLGÅRD)

developing neuroblasts and neurons. They may also have a mechanical function in stabilizing the rapidly dividing cells of the neuroependyma.

I. CSF Secretion and Pressure

The formation of CSF in the adult is firmly established to be due largely to active secretion by the choroid plexus epithelium together with a contribution from brain interstitial fluid (for review see DAVSON et al. 1987). Quantitative estimates of CSF secretion rate in the developing brain have been reported for neonatal rats (BASS and LUNDBORG 1973; JOHANSON and WOODBURY 1974; WOODBURY et al. 1974). EVANS et al. (1974) and FOSSAN et al. (1985) have published data on CSF secretion rates in different ages of fetal sheep and in the adult. The data from the two species are summarised in Table 3. Clearly, the lowest secretion rates occur in the smallest brains, whether because of the size of the species or because of immaturity. If secretion rates are related to the weight of the choroid plexus, then there is evidence of an increased secretory capacity with age in both species. The turnover of CSF in the sheep does not change much with age in spite of enormous developmental increases in CSF volume, secretion rate and brain size. The data for postnatal rats are less complete. Also, the values obtained for CSF secretion rate by FERGUSON and WOODBURY (1969) and by BASS and LUNDBORG (1973) although similar in the neonatal and adults rats were found to be substantially higher by BASS and LUNDBORG (1973) at 10 and 30 days postnatal (P10, P30). Both used an inulin clearance method, but in the case of FERGUSON and WOODBURY (1969) this involved ventriculo-cisternal perfusion in the anaesthetized rat, which is therefore directly comparable to the sheep experiments, whereas BASS and LUNDBORG (1973) used spinal subarachnoid infusion in unanaesthetized rats. Although BASS and LUNDBORG (1973) claimed on the basis of the visually detected distribution of dye included in the perfusion fluid that the CSF in ventricular and subarachnoid systems was adequately mixed with the perfusion fluid, this seems doubtful. Inadequate mixing would lead to an overestimate of CSF secretion rate. For these reasons the P10 and P30 results of BASS and LUNDBORG (1973) are not included in Table 3. It seems from the data in Table 3 that the turnover of CSF increases in the rat between the neonatal period and late postnatal adult, in contrast to the sheep fetus. However, the age ranges in Table 3 for the two species may not be comparable and the turnover of CSF in fetus sheep earlier in gestation than 60 days may well be less in both absolute and proportional terms. If turnover rates are related to brain size, then in both the rat and the sheep the "sink effect" (DAVSON et al. 1987) would appear to *decline* with age (much more so in the sheep than in the rat). This is perhaps not surprising when one compares the relative size of choroid plexuses and brain at different ages. The choroid plexus is disproportionately much larger than the brain early in development. In contrast, the CSF volume as a proportion of brain weight does not change much (Table 3). Earlier this

Table 3. CSF secretion in postnatal and adult rats and fetal and adult sheep

Age	Brain wt.[b] (g)	Choroid plexus wt.[c] (mg)	CSF secretion (μl/min)	Secretory capacity (μl min^{-1} mg^{-1} choroid plexus)	CSF vol. (ml)	CSF vol/brain wt. (%)	Turnover (%/min)	"Sink" (% min^{-1} g^{-1} brain wt.)
Rats[a]								
P3	0.49	0.60	0.2	0.33	<0.05	~10	~0.40	~0.82
P5	0.54	0.65	0.34	0.52	0.12	22	0.28	0.52
P8	1.05	1.14	0.44	0.39	0.24	23	0.18	0.17
P13–14	1.18	1.36	0.53	0.39	(0.24)[b1]	20	0.22	0.19
P21–23	1.28	1.52	0.86	0.57	(0.25)[b2]	20	0.34	0.27
Adult	1.92	2.5	2.1	0.84	0.28	15	0.75	0.39
Sheep[d]								
E60	1.91	66	2.8	0.04	0.45	24	0.62	0.32
E125	36.7	111	62.5	0.56	7.12	19	0.88	0.02
Adult	78.4	195[e]	118	0.60	14.2	18	0.83	0.01

[a] From JOHANSON and WOODBURY (1974), WOODBURY et al. (1974), DAVSON et al. (1987) and BASS and LUNDBORG (1973).
[b] Some brain weights are from ZEMAN and INNES (1963), BASS and LUNDBORG (1973): [b1] P10, [b2] P20.
[c] Choroid plexus weights at P3 and P5 are derived from JOHANSON et al. (1976).
[d] From EVANS et al. (1974), DZIEGIELEWSKA et al. (1979) and FOSSAN et al. (1985).
[e] Adult choroid plexus weight from SEGAL (personal communication).
P is postnatal age in days from birth.
E is embryonic age in days from conception.

century it was considered by some authors that, because of the effectiveness of the blood-brain barrier, the choroid plexuses and CSF might be a dominant route for the supply of nutrients to the brain (e.g. STERN and GAUTIER 1921, 1922; STERN and RAPOPORT 1928). Subsequently, for the adult brain this has largely been discounted (DAVSON 1967, 1989). However, the idea has resurfaced in relation to the developing brain. Because of the disproportionately larger size of the choroid plexuses in the developing brain it has sometimes been suggested that the choroid plexus may play a more important role in nutrition of the brain early in development than in the adult (e.g. KLOSOVSKI 1963; JOHANSON 1989).

CSF pressure in the developing brain is lower than in the adult; presumably this is at least partly due to the distensibility of the immature skull compared with the adult. JONES et al. (1987) estimated the pressure to be 17 ± 1.5 mm H_2O at E19 (days of embryonic life) in the rat fetus, rising to 21 ± 1.8 mm H_2O at E21 with further increases up to P10 (34 ± 3.6 mm H_2O). Similar values were reported for postnatal rats by BASS and LUNDBORG (1973). In the rabbit fetus, CSF pressure was estimated to be 25 ± 16 mm H_2O with a progressive rise in the postnatal period to an adult value of 129 ± 54 mm H_2O (HORNIG et al. 1987).

DAVSON et al. (1987) have reviewed studies of CSF pressure in human neonates and children. As in the animal studies, pressure is low in early life (10–14 mm H_2O in neonates) and rises progressively: 40–100 mm H_2O in children and 150 mm H_2O in adults.

The CSF pressure is a balance between production and drainage, a relationship that appears to be important for normal brain development – at least early on. Thus, DESMOND and JACOBSEN (1977) showed that decompression of the CSF space in chick embryos shortly after neural tube closure resulted in severe disturbances of the developing brain within 24 h of intubation. CARNAN and GILLESPIE (1988) showed that ion transport inhibitors that would be expected to reduce CSF secretion resulted in a decrease in ventricular volume and brain size in chick embryos.

It is possible that the very high concentration of protein in CSF early in brain development (see Sect. L) contributes to the formation of CSF and expansion of the brain early in its growth. The presence of membrane specializations ("strap junctions"; MØLLGÅRD et al. 1987a) between adjacent neuroependymal cells at this stage of brain development (see Sect. H) has been shown to largely prevent the penetration of protein from CSF to brain ECF (FOSSAN et al. 1985), unlike the adult in which the ependyma is freely permeable to protein (BRIGHTMAN and REESE 1969).

The colloid osmotic pressure exerted by the huge concentration of protein in CSF early in brain development must be almost as large as that in the circulating plasma since the protein concentration in plasma is much less that in the adult, e.g. around 1600 mg/100 ml at E31 in fetal sheep and 2100 mg/100 ml in E31 fetal pigs, at which age the CSF total protein concentration is about 1150 mg/100 ml in the sheep and 960 mg/100 ml in the pig

(DZIEGIELEWSKA 1982). It is not known how active the choroid plexuses are in secreting CSF so early in brain development. WEED (1917) studied the drainage pathway for CSF in the developing pig embryo by injecting a mixture of sodium ferrocyanide and ferric ammonium citrate (which gives the well known Prussian blue reaction on acidification) into the CSF spaces to replace the CSF. These experiments showed that there was no escape of material from the neural tube until 14 mm crown–rump length (CRL; = E24) and later. Thus, the high concentration of protein in CSF very early in brain development may well be within a "closed" system in terms of bulk flow and probably rises rapidly once the neural tube closes (see Fig. 10 and Sect. L). Recent studies of the isolated central nervous system of the South American opossum (*Monodelphis domestica*) maintained in culture over 7–10 days (NICHOLLS et al. 1990; STEWART et al. 1991) have shown abnormal inward growth of regions of the ventricular zone lining the cerebral ventricles (MØLLGÅRD, NICHOLLS and SAUNDERS, unpublished observations). This is probably due to the lack of pressure within the ventricular system as a result of isolating the brain.

J. Electrolytes in CSF in the Developing Brain

The ionic composition of CSF (and thus, it is usually assumed of brain ECF) in the adult is remarkably stable (see DAVSON et al. 1987 and Chap. 8). This stability is one example of barrier mechanisms that depend upon the combination of a diffusion restraint and active transport of specific ions into or out of the CSF and brain ECF, thus maintaining characteristic ion gradients between CSF and plasma. In the developing brain, if such gradients are absent, we cannot immediately distinguish between an inadequate diffusion restraint or the lack of appropriate ion pumps. However, the occurence of even one such gradient indicates the presence of an adequate diffusion restraint (i.e. presence of effective tight junctions). However, some ions are bound to protein to a significant extent, for example Ca^{2+} and Mg^{2+}. Early in gestation there is a high concentration of protein in fetal CSF (see Sect. L). Thus, if only total magnesium or calcium are estimated, it may be more difficult to decide to what extent a gradient is due to an active ion pump, to a diffusion restraint to protein or to a diffusion restraint that is effective for small ions. Determination of both total and ionic forms of a particular electrolyte would be needed to interpret the distribution of an ion between plasma and CSF. The absence of some ion gradients at a stage of brain development when others are present would clearly indicate that ion pumps develop sequentially. Available experimental evidence suggests that this is the case.

FLEXNER (1938) made pioneering studies of electrolytes and urea in the fetal pig, followed by permeability studies using radioactive sodium in the fetal guinea pig (FLEXNER and FLEXNER 1949). Subsequently, limited studies of the monkey fetus (BITO and MYERS 1970) and sheep fetus (BRADBURY et al. 1972) and more comprehensive studies of the fetal and neonatal rat (JOHANSON et al. 1974, 1976; JONES and KEEP 1987, 1988) have been published.

Interspecies comparisons are complicated by different lengths of the gestational period and stage of brain development at the time of birth. The chloride ion seems to be the most widely studied in different species; it has the advantage of not being significantly bound to protein. In pig (FLEXNER 1938) and sheep (BRADBURY et al. 1972) there is a clear CSF/plasma Cl^- gradient (of about 1.2) by around 60 days gestation. In the pig term is at 115 days and in the sheep it is at 150 days. In the rat (gestation 22 days) the Cl^- gradient is not present at E18 but is apparent by the day of birth (FERGUSON and WOODBURY 1969). In the rabbit fetus (gestation is 31 days) the CSF Cl^- concentration is rather stable from 6 days before birth until at least 10–12 days postnatal. This is in spite of changes in plasma Cl^- concentrations, which implies the presence of Cl^- homeostasis within the developing brain from well before birth in this species (AMTORP and SØRENSEN 1974).

Development of a CSF/plasma gradient for K^+ in the rabbit (AMTORP and SORENSEN 1974) has a similar time course to that for Cl^- in the rat (FERGUSON and WOODBURY 1969; JONES and KEEP 1987) but in other species the gradient for K^+ develops more slowly and rather later in gestation. For example, in sheep fetuses it is about 0.9 around 90 days gestation and 0.78 shortly before birth (BRADBURY et al. 1972); in the monkey the K^+ ratio is about 0.9 at 70 days gestation (term is 170 days) and 0.75 shortly before birth. JONES and KEEP (1987) have investigated the stability of CSF [K^+] in the face of imposed changes in plasma [K^+]. Their studies showed clearly that although a CSF/plasma [K^+] gradient is established by E21, there is very little CSF [K^+] regulation at this age but it begins to be apparent by 1 day after birth (Fig. 8). PARMALEE and JOHANSON (1989) have shown that lateral ventricular choroid plexus of neonatal rats has only limited Na^+-K^+ ATPase activity but that this matures rapidly during the first 2–3 weeks after birth.

The estimations of calcium and magnesium in fetal and neonatal CSF and plasma are more difficult to interpret because of the developmental changes that occur in proteins in these fluids. The concentration of total protein increases very early in brain development to reach a peak and then decline (see Fig. 10). In plasma there is a steady increase in total protein concentration. The situation is further complicated by the fact that the individual proteins contributing to the total protein concentration change differently with age. In the adult it is known that albumin is the major plasma calcium and magnesium binding protein (PETERS 1975) but it is not clear whether this is the case during development. JONES and KEEP (1988) suggested that CSF from the immature brain may contain proteins or other substances that complex calcium to a higher degree than in the adult, in addition to the overall effect of the higher concentration of protein in fetal

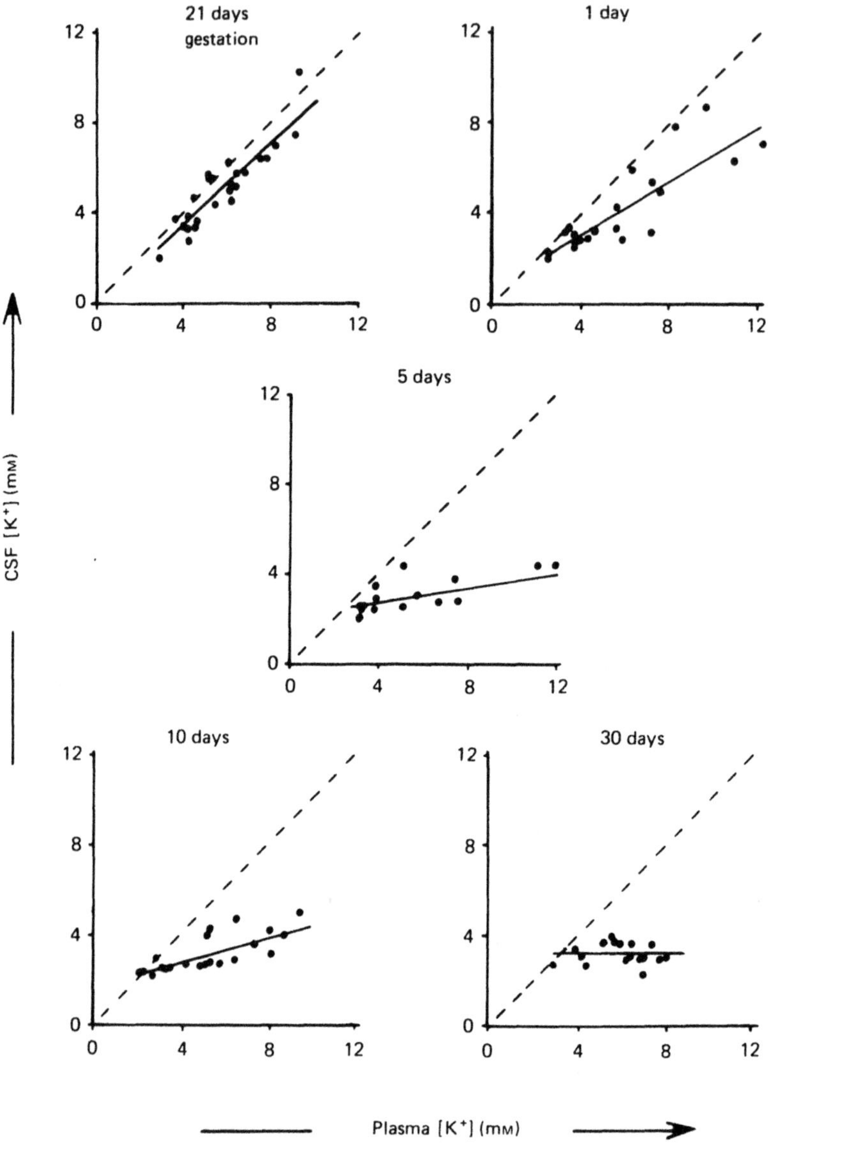

Fig. 8. Relation between steady-state subarachnoid CSF potassium concentration [K^+] and plasma [K^+] at different ages in fetal and postnatal rats. In each group the *broken line* represents the isionic line and the *continuous line* is the regression of CSF [K^+] upon plasma [K^+]. Note that there is no evidence of CSF [K^+] control in the fetus but that it begins to be apparent in the early neonatal period. (From JONES and KEEP 1987)

and neonatal CSF. JONES and KEEP (1988) measured both total and ionic calcium concentrations in CSF from the newborn period to adulthood in rats. They also measured calcium in brain interstitial fluid. They present convincing evidence for the onset of calcium regulation during the first week of life; they found that it was more effective for brain interstitial fluid than for CSF (Fig. 9). The characteristic adult CSF/plasma gradient for $[Ca^{2+}]$ did not appear until after P30 in the rat, probably because of the confounding effects of changes in protein concentration in CSF and plasma during the first few weeks of postnatal life in the rat (see DZIEGIELEWSKA et al. 1981). The ratio CSF/plasma for total calcium declines progressively through the first month of postnatal life in the rat. In the sheep (BRADBURY et al. 1972) and monkey (BITO and MYERS 1970) a similar decline has been reported but in those species with their long gestational periods the decline occurs during fetal life.

The only comprehensive study of magnesium in CSF during development is that of BRADBURY et al. (1972) in sheep. They found that the CSF/plasma ratios for the youngest fetuses studied (44–50 days gestation) and adults were similar: 1.25 and 1.28 respectively. However, during the rest of the fetal period the ratio declined to approach 1.0 by 100 days gestation and it did not recover to the adult value until 4–7 weeks postnatal. The early gradient is probably a reflection of the high concentration of protein in fetal CSF (see Fig. 10) early in brain development. By late in gestation CSF protein concentration is only about 2–3 times the adult value (Fig. 10). The subsequent recovery of a CSF/plasma gradient for magnesium in the postnatal period suggests that development of magnesium homeostasis is a function that does not occur until late compared with other ions. The more limited data on CSF magnesium for the monkey would appear to support this conclusion (BITO and MYERS 1970).

K. Glucose and Amino Acid Transport in the Developing Brain

This subject has been reviewed recently by JOHANSON (1989) and will only be dealt with briefly here. It has been known for some time that many metabolically important materials are taken up by the developing brain at a greater rate than in the adult. Several authors have interpreted this as meaning that the blood-brain barrier for these materials is immature (e.g. BAKAY 1953; SESSA and PEREZ 1972). However, as pointed out by DOBBING (1969), it could as well reflect a high rate of metabolism of the materials by the developing brain. The application of Oldendorf's single pass BUI method to neonatal animals (e.g. CORNFORD et al. 1982a) has shown that the transfer of a range of materials such as lactate, arginine, tryptophan and choline is indeed greater in the developing brain. Since the adult mechanisms for this transfer are known to be cellular and specialized (see Chap.

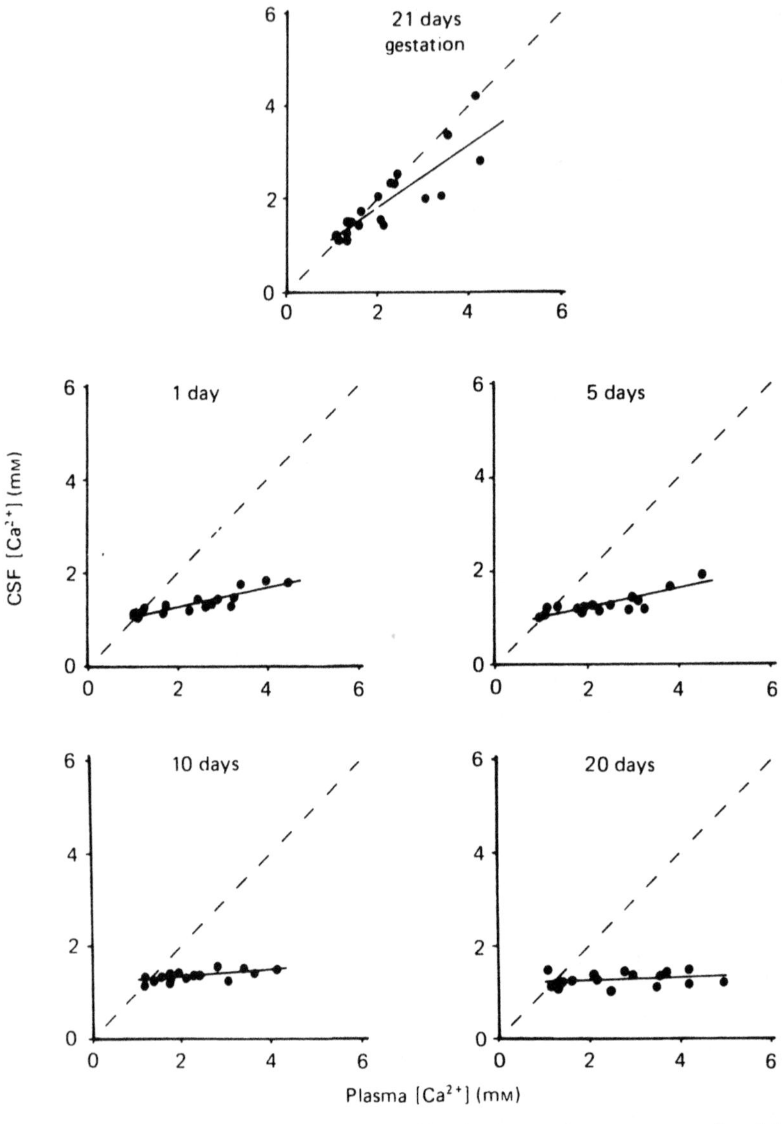

Fig. 9. Relation between steady-state CSF ionic calcium concentration [Ca^{2+}] and plasma [Ca^{2+}] at different ages in fetal and postnatal rats. In each group the *broken line* indicates the isionic line and the *continous line* is a linear regression of CSF [Ca^{2+}] upon plasma [Ca^{2+}]. Similar results were obtained for brain interstitial fluid [Ca^{2+}] except that the slope of the regression line at 5 days was flatter, suggesting an earlier approach to adult control of [Ca^{2+}] in brain interstitial fluid than in CSF. (From JONES and KEEP 1988)

5), it seems perverse to interpret the greater uptake in the fetus or in the neonate as an indication of barrier immaturity. Rather, it would seem to reflect the greater metabolic demands of a rapidly growing tissue (BANOS et al. 1978; CORNFORD et al. 1982a). However, the actual molecular mechanisms of these transport systems even in the adult are only beginning to be understood (e.g. the glucose transporter, see PARDRIDGE et al. 1990). It will require similar studies in developing brain to determine whether the higher rate of transport is indeed due to a developmental specialization rather than to immaturity.

L. Proteins in CSF in the Fetus and Neonate

There have been several recent reviews of the data on the protein composition of CSF during brain development in a wide range of species (DZIEGIELEWSKA 1982; DAVSON et al. 1987; DZIEGIELEWSKA and SAUNDERS 1988, 1991). In all species so far investigated, the concentrations of total protein in CSF are characteristically high early in brain development, as was originally reported by KLOSOVSKI (1963). Data from most of the species studied are summarized in Fig. 10. There are only a limited amount of

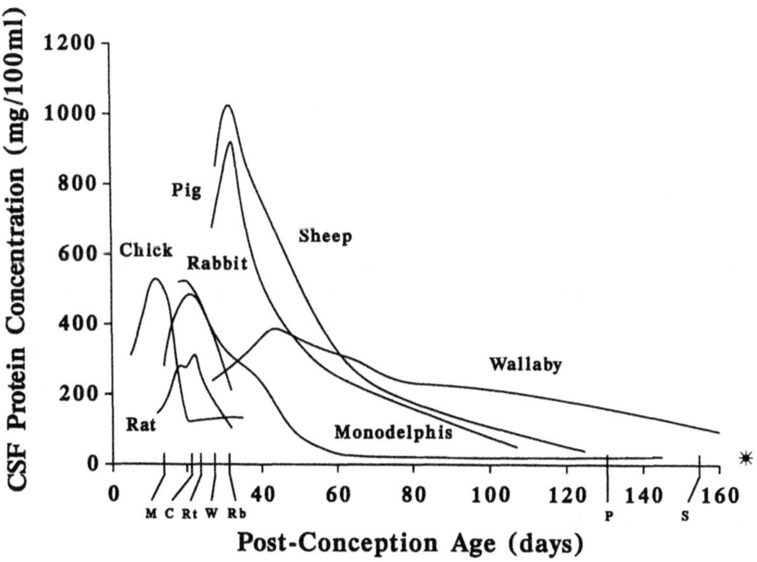

Fig. 10. Total protein concentration in CSF from *cisterna magna* of various species during development estimated by the LOWRY et al. (1951) or BRADFORD (1976) methods. *Abscissa*, age from conception in days; *ordinate*, protein concentration in mg/100 ml. Letters along *abscissa* indicate time of birth: *M*, Monodelphis; *C*, chick; *Rt*, rat; *W*, tammar wallaby; *Rb*, rabbit; *P*, pig; *S*, sheep, *, adult cisternal CSF protein concentration in mammalian species. (Redrawn from DZIEGIELEWSKA and SAUNDERS 1991; with additional data from DZIEGIELEWSKA 1982)

data on the human fetus (see SAUNDERS 1991). Both qualitative immunoelectrophoretic and quantitative immunoassay studies have shown that the proteins in fetal CSF are immunologically indistinguishable from plasma proteins. Extensive studies on 60 day gestation fetal sheep have shown that most if not all of the steady state CSF levels of at least the major proteins in CSF at this age can be accounted for by transfer from plasma as is summarized in Fig. 11. Immunocytochemical evidence indicates that the route of transfer is an intracellular one via the choroid plexus epithelial cells; cell counts of different choroid plexuses for different plasma proteins suggest that transport occurs to a different extent in different choroid plexuses (lateral ventricular, 3rd ventricular or 4th ventricular) for different proteins (JACOBSEN et al. 1982a,b, 1983; MØLLGÅRD et al. 1987b; DZIEGIELEWSKA et al. 1991). Different proteins of rather similar molecular size have different steady state CSF plasma ratios early in brain development (see Fig. 11 and Table 4). This suggests that there may be a specific (receptor mediated) transfer mechanism across the choroid plexus. The best evidence for such a mechanism comes from studies of albumin permeability in 60 day gestation fetal sheep (DZIEGIELEWSKA et al. 1991). In these experiments the natural steady state CSF/plasma ratios for endogenous sheep albumin in three CSF compartments (lateral ventricle, dorsal cortical subarachnoid space and *cisterna magna*) were compared with those obtained after 3–5 h exposure to intravenous radiolabelled sheep albumin or albumin from different species (goat, cow, human and chicken), as is illustrated in Fig. 12.

These experiments showed that only labelled sheep albumin and goat albumin approached the natural steady state for endogenous sheep albumin in all three compartments by 3 h after starting the albumin injection. Bovine albumin approached the sheep albumin steady state more slowly (by 5 h, Fig. 11) but human (and chicken) remained at only about 40% of the steady state CSF/plasma concentration for sheep albumin. Immunocytochemical studies showed that human albumin was transferred in the same choroid plexus cells (predominantly in the 3rd ventricular choroid plexus at 60 days gestation) as the sheep albumin; fewer cells were positive for human albumin than for sheep albumin, which is consistent with the lower transport of human albumin. Structural differences between sheep and human albumins have been identified which may be the molecular basis for this difference in transport of the two albumins (DZIEGIELEWSKA et al., in preparation). Discrimination in the transfer of different species of albumins between plasma and CSF has also been found for neonatal rats (DZIEGIELEWSKA et al. 1990) and opossums (KNOTT et al. 1991). Species differences in the cellular uptake of transferrins have been described (LIM et al. 1987) which are clearly due to differences in effectiveness of the molecular recognition between different species of transferrin and transferrin receptors. Albumin receptors are much less well characterized (see DZIEGIELEWSKA et al. 1991 for discussion) than are transferrin receptors (HUEBERS and FINCH 1987). Further work is

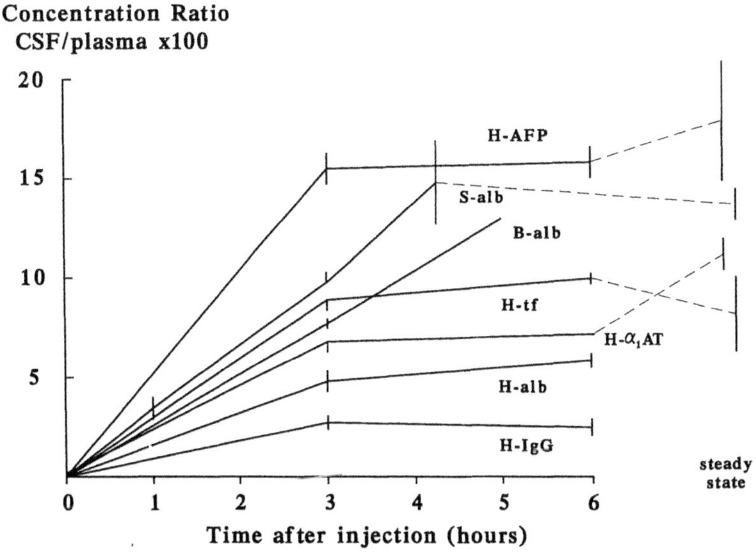

Fig. 11. Penetration of proteins from plasma into CSF of 60-day fetal sheep. Proteins were injected intravenously and blood was sampled to give an estimate of mean plasma concentration. At times indicated CSF was sampled from *cisterna magna*. Concentrations of protein in CSF and plasma were estimated by radial immunodiffusion assay. *Abscissa*, time in hours following i.v. injection; *ordinate*, CSF concentration/plasma concentration × 100. *Steady state* indicates CSF/plasma ratio for naturally occurring sheep proteins. Mean ± SEM for three to six experiments, except at early time point (where there is no error bar, $n = 2$). All injected proteins were human (*H-*) except for *S-alb* ($^{125}I/^{35}S$-sheep albumin) and *B-alb* (bovine albumin measured using sheep anti-bovine albumin antiserum). Experimental details are in Dziegielewska et al. (1980, 1991). The data are a summary of results from Dziegielewska (1982) and Dziegielewska et al. (1980, 1991). *AFP*, α-fetoprotein; *alb*, albumin; *tf*, transferrin; $\alpha_1 AT$, α_1-antitrypsin. These four proteins together with fetuin quantitatively account for the total protein concentration in fetal sheep CSF at 60 days gestation. The sheep fetus does not possess any IgG of its own, hence no steady state ratio is shown. *Note*: (1) There is a relation between molecular size and permeability (the largest molecule, IgG, has the lowest ratio and the smallest molecule, AFP, has the largest ratio). (2) Proteins of similar size may have different ratios; this is even more apparent earlier in gestation (see Table 4). (3) Albumin from different species may have different ratios (see also Fig. 12). These results, together with those of Fig. 12, suggest that there is a selective mechanism that transports proteins from plasma to CSF in addition to the greater passive permeability of the blood-CSF barrier in fetal sheep early in gestation, as illustrated in Fig. 4. The route of protein transfer appears to be via the epithelial cells of the choroid plexus (Jacobsen et al. 1983; Dziegielewska et al. 1991). The specific transport mechanism is not present after about 70 days gestation in this species

needed to identify the nature of the albumin transfer system present in the immature choroid plexus. Later in development (e.g., 70 days gestation in sheep, P20 in the rat) the level of albumin penetration from blood into CSF in much less and there does not appear to be any species discrimination.

Table 4. CSF/plasma ratios (%) for proteins early in brain development

Species	Age	Albumin	AFP	Fetuin/ α_2HS	Transferrin	α_1AT	IgG	Total protein
Sheep	E31d	30.0 ± 5.2	86.3 ± 8.7	64.0 ± 4.5	25.3 ± 4.2	64.1 ± 2.4	0	70.2 ± 8.7
Pig	E31d	0	43.1 ± 11.6	70.0 ± 16.5	45.9 ± 10.4	nd	0	44.4 ± 6.1
Rat	E22d	5.0 ± 0.6	18.2 ± 1.9	nd	10.4[a]	6.2 ± 0.9	nd	11.7 ± 0.9
Tammar	P15d	16.4 ± 3.0	nd	31.2 ± 2.1	10.0[a]	9.6 ± 0.4	0	32 ± 4.0
Opossum	P19d	16.4 ± 1.7	nd	nd	60.0[a]	65.0[a]	nd	46.6 ± 3.6
Human	E20–24w	9.7 ± 1.0	13.1 ± 2.2	13.7 ± 2.5	nd	nd	11.1 ± 1.6	nd

CSF/plasma ratios (%) of plasma proteins for different species at the peak of total protein concentration in CSF (mean ± SEM).
[a] Derived from crossed immunoelectrophoretic plates.
nd, not determined; 0, not detectable; AFP, α-fetoprotein; α_1AT, α_1-antitrypsin. Age is embryonic/fetal (E) or postnatal (P) in days (d) or weeks (w).
Data from Adinolfi and Haddad (1977, human); Dziegielewska (1982, sheep and pig); Dziegielewska et al. (1981, rat; 1987, human; 1988, tammar; 1989 opossum).

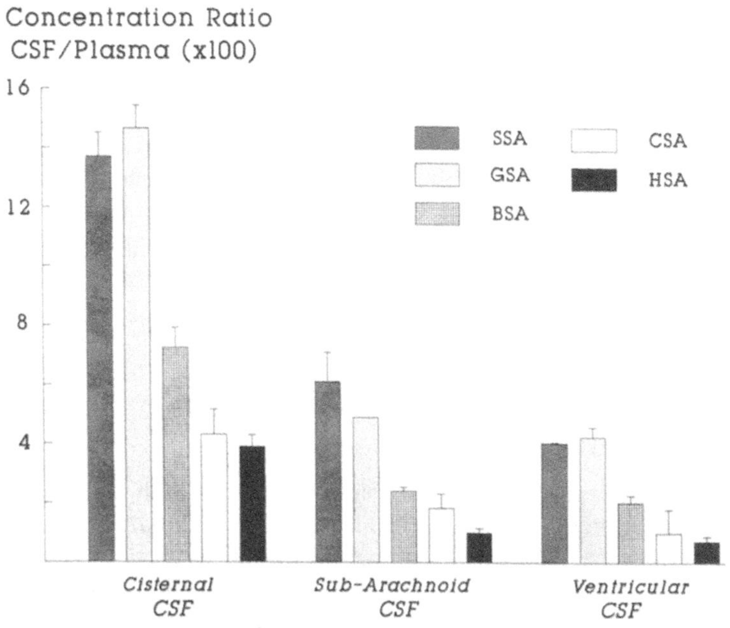

Fig. 12. CSF/plasma ratios for several different species of albumin 3 h after i.v. injection, compared with the naturally occurring sheep albumin ratio. CSF samples were collected from the *cisterna magna*, from the dorsal subarachnoid space (*SSA*) and from within the lateral ventricles as indicated in the figure. *Error bars* shown are standard errors of the mean. Note that the concentrations of albumins in lateral ventricular CSF are only about 20%–30% of those in cisternal CSF. *GSA*, goat; *BSA*, bovine; *CSA*; chicken; *HSA*, human serum albumin. ^{35}S-labelled sheep albumin (obtained from hepatocytes cultured in vitro) reached the same CSF/plasma ratio as for endogenous SSA and exogenous GSA. (Data from DZIEGIELEWSKA et al. 1991 and unpublished observations)

The functions of these high concentrations of proteins in CSF in the immature brain are not known. Two general possibilities are: (1) The overall high concentration of protein may exert a significant colloid osmotic pressure and provide an expanding force around which brain grows. As mentioned in Sect. I, decompression of the ventricles or inhibition of CSF secretion results in malformed or retarded brain brain growth. (2) The proteins may have general carrier and nutritive functions as in plasma. The immature brain at the time of peak levels of protein in CSF is poorly vascularized. Indeed, in the earliest stages of brain development, just after closure of the neural tube, there are no vessels within the neural tissue.

At least some of the proteins in CSF are taken up into the ventricular zone cells surrounding the ventricular spaces (e.g., CAVANAGH and WARREN 1985). Immunocytochemical evidence suggests that different proteins may be taken up at different times (e.g., REYNOLDS and MØLLGÅRD 1985). As indicated in Sect. H, a CSF-brain barrier in the immature brain prevents the passage of CSF protein into brain ECF. Thus the cells of the neuroependyma are presented with a protein-rich solution on the surface that is in contact with CSF and a protein-poor solution on their surfaces within the brain. This asymmetry occurs at a time when the neuroependyma of the neocortex is producing predominantly neurons, at least in the sheep fetus (cf. ÅSTROM 1967; MØLLGÅRD et al. 1987a). Thus the prsence of high concentrations of plasma proteins in CSF of the immature brain may be important for the generation of neurons in general or of specific populations of neurons.

M. Trace Metals and Brain-Liver Glycoproteins in the Developing Brain

Little is known about the uptake and distribution of metals such as zinc and iron in the developing brain. Zinc has been shown to be essential for normal brain development, especially in early stages. The deprivation of zinc in the critical period E9–11 in pregnant rats results in a high incidence of fetal malformations, including several involving the nervous system such as ancencephaly or hydrocephalus. The mechanism of this effect is not known (for details and references see HURLEY 1981; KEEN and HURLEY 1989). FREDERICKSON (1989) has reviewed the involvement of zinc in brain structure and function. He points out the very large concentration gradient for zinc between CSF and plasma on the one hand and intraneuronal zinc on the other. This implies an active cellular uptake mechanism, but as indicated by FREDERICKSON (1989) nothing is known of the possible nature of such a mechanism. Only a few reports of zinc distribution in and uptake into the brain of developing animals have been published and these have mainly concerned the hippocampus (e.g., DREOSTI 1984). There is good evidence for the involvement of zinc in synaptic transmission in some brain regions, especially in the hippocampus. The recent work of XIE and SMART (1991) suggests that zinc may have an important functional role in the developing hippocampus. It has an inhibitory effect on both pre- and postsynaptic $GABA_B$ receptors in this brain region which could, for example, explain the occurrence of postnatal seizures in young children with acute zinc deficiency (GOLDBERG and SHEEHY 1982). FREDERICKSON (1989) has summarized the large number of enzymes that contain zinc and proteins which bind zinc in the brain. However, he does not mention zinc binding globulin, a liver protein (SCHWICK and HAUPT 1981) which is probably an example of a whole series of liver proteins that are also synthesized by the developing brain (see MØLLGÅRD et al. 1988). Zinc binding globulin has a striking distribution in the developing neocortex (as well as the hippocampus) within new fibres on either side of the developing cortical plate (DZIEGIELEWSKA et al. 1992). This is a known region of synaptogenesis (e.g., MOLLIVER et al. 1973). XIE and

SMART (1991) suggest that their findings of the important neuromodulating effects of zinc at GABA synapses in the developing hippocampus may be a more general phenomenon for zinc containing fibres throughout the developing telencephalon. The distribution of zinc binding globulin in the immature neocortex supports this proposal.

Iron and its transport by transferrin is the metal-carrier protein system to have been most studied in the brain and in other tissues. Both transferrin (BLOCH et al. 1985; MØLLGÅRD et al. 1987b, 1988) and iron (HILL and SWITZER 1984) are present within cells in the adult and developing brain. Transferrin receptors are present on the luminal surface of cerebral endothelial cells (JEFFERIES et al. 1984) and also on the surface membranes of cells within the brain, particularly neurons (GIOMETTO et al. 1990). There is good evidence for uptake of iron and the involvement of transferrin in the process in both the adult and developing brain (TAYLOR and MORGAN 1990), but it is not clear what the relationship is between intracellular transferrin in the brain which appears to be synthesized in vitro (BLOCH et al. 1985; MØLLGÅRD et al. 1988) and transferrin taken up from plasma by cerebral endothelial cells. It seems to be assumed by some authors (e.g., PARDRIDGE et al. 1987) that transferrin taken up into cerebral endothelial cells will necessarily pass into brain extracellular fluid. However, there is no direct evidence for this. FISHMAN et al. (1987) report the use of adult rat brain perfused with ^{125}I-transferrin, with subsequent cell fractionation. They compared the distribution of labelled transferrin in cerebral endothelial cells and supernatant from brain homogenate after different times of infusion. However, the results are given as counts per minute in "microvessels" and supernatant of brain homogenate without it being clear whether this was on a tissue weight or whole brain basis. This makes it difficult to assess the quantitative and likely functional significance of labelled transferrin recovered in brain homogenate, as an index of penetration beyond the cerebral endothelium. TAYLOR and MORGAN (1990) have reported that more radiolabelled transferrin than albumin could be recovered from rat brain following injection. TAYLOR and MORGAN (1990) used albumin to account for blood contamination and assumed that any excess of transferrin in brain was a reflection of brain uptake. However, this was some 45 times less than the uptake of iron in P15 rats. This suggests a considerable degree of recycling of transferrin at the blood-brain interface with only a minor amount being transferred into brain. Presumably the intracellular transferrin, predominantly in neurons in the developing brain (MØLLGÅRD et al. 1987b, 1988) and in oligodendrocytes in the adult (BLOCH et al. 1985; CONNOR and FINE 1986), is important for the binding and trapping of intracellular iron in the cells where it is functionally important. The mechanism of the step between export of iron from endothelial cells to uptake into specific cells in the brain remains unclear. Numerous other brain-liver proteins have been identified in different cell populations of the developing brain (MØLLGÅRD et al. 1988) but their function remains to be elucidated.

N. Drugs and Barriers in the Developing Brain

Relatively few studies of the penetration into brain and CSF of pharmacologically active molecules or naturally occurring neurotransmitters and related substances have been carried out in developing animals. This may be a reflection of the lack of any perceived therapeutic indication for drug treatment of fetuses. However, it is of importance to know whether drugs administered to pregnant women will reach the fetal brain and also, in the case of drugs administered to neonates, especially when premature, whether the drugs will reach the developing brain.

The general principles determining the penetration and steady state levels in brain and CSF of materials present within plasma are dealt with in several chapters of this volume and have been outlined elsewhere (SAUNDERS 1991b). However, these can at best allow only a qualitative prediction of whether or not a particular drug or other neuroactive substance will penetrate into the brain and CSF. The methodological considerations outlined in this chapter, particularly Sect. F, should allow adequate design of experiments for the study of the pharmacology of barriers in the developing brain. Overriding considerations are the lipid solubility of a drug, its degree of protein binding and whether or not it will attach itself to naturally occurring transport systems, into or out of the brain. Specific design of drugs that will enter the brain even for the adult is in its infancy (see Chap. 20), so it is hardly surprising that still less is known for the immature brain, particularly since a lack of clinical application makes this an unattractive field to the drug industry. However, it is a cause for concern that drugs are used in patients with immature nervous systems, particularly in the neonatal period. There is little experimental evidence to give guidance on their likely degree of penetration into the nervous system, let alone whether their effects on the developing nervous system are similar to those in the adult.

The study of KALARIA and HARIK (1983) is an example of a series of investigations of monoamine oxidases in brain tissue and isolated cerebral microvessels at different stages of development. They show differential developmental regulation of monoamine oxidases A and B in postnatal rat brain microvessels. This is presumably important for control of monoamine levels during brain development, but this seems not to have been directly studied.

CORNFORD et al. (1982b) have provided a model study of blood-brain barrier penetration of the anticonvulsant diphenylhydantoin in newborn and adult rabbits. They applied the BUI method (BRAUN et al. 1980) and showed that uptake of the drug was about three times higher in the newborn than in the adult rabbit. They showed important effects of plasma proteins on the solubility and uptake of the drug; they also showed that the uptake was not due to a saturable carrier system. The authors were careful not to conclude that the greater uptake in the newborn period was necessarily due to immaturity of the blood-brain barrier. Rather they pointed to the finding that for several other drugs (e.g., caffeine, ethanol and heroin) there was no difference in uptake for neonatal and adult brain. The mechanism of the greater uptake of phenylhydantoin (and also of phenobarbitone) is not known, but it is clearly not due to a generalized immaturity of the blood-brain barrier in the neonatal period (in rabbits).

O. Conclusions

The evidence presented in this review supports the proposition that the state of development of barrier mechanisms in the immature brain is much more complex than is described by the common view that "the" blood-brain barrier is immature in the developing brain. The fundamental underlying blood-brain barrier in the adult brain is the barrier that prevents the penetration of protein from blood into brain (and largely prevents its penetration into CSF). In the fetus this barrier (which is principally due to the presence of tight junctions between adjacent cerebral endothelial cells) also appears to be well formed. Naturally occurring proteins in plasma do not appear to "leak" across even the most immature cerebral vessels and the demonstration of apparent leakiness of the barrier on injection of dye or HRP seems to result from the use of excessive volumes or protein concentrations. However, this is only one aspect of barrier mechanisms in the developing brain. Although proteins seem to be excluded from the ECF of the brain, the blood-brain barrier in the fetus (and newborn of altricial species) is much more permeable to smaller molecules such as inulin in sucrose. In addition, barrier mechanisms which depend upon cellular carriers or enzymes do appear later in brain development and available evidence suggests that this is a maturational process with different systems becoming apparent at different stages of brain development. It seems likely that the underlying cellular and molecular basis for all these developmentally regulated changes is as follows: the cerebral endothelial cells appear to have an intrinsic ability to form tight junctions when these cells first invade the developing brain. However, these may only be effective in excluding the passage of protein. Their subsequent ability to exclude or largely exclude smaller molecules (which does not appear until much later) may depend upon further maturation of the tight junctions which, at least in many areas of the brain, may be importantly influenced by materials secreted by astrocytic end feet. It may also be that the astrocytes are important for expression of the various specific mechanisms within the cerebral endothelial cells.

There is an important aspect of the barrier systems in the developing brain which is unexpectedly different from those in the adult brain. CSF in the immature brain contains very high concentrations of proteins that appear to originate from plasma by a transcellular, probably receptor mediated transport system across the choroid plexus. However, unlike the adult brain, protein in fetal CSF is prevented from crossing the CSF-brain interface (the neuroependyma in the fetus). This restriction is confined to the earlier stages of brain development and probably corresponds to the period of neurogenesis rather than to the later period of gliogenesis, by which time the CSF protein concentration has fallen considerably.

These observations suggest that the brain grows and develops within a local environment that is carefully controlled at least with respect to its protein composition. Proteins present in the general extracellular environment of the organisms do not appear to be present in anything like such large concentrations in the extracellular environment of the developing brain except at the specialized neuroependymal interface. The presence of a high concentration of protein at one surface of differentiating neuroblasts may be important for their normal development. Similarly, the apparently low extracellular protein concentration in the rest of the brain's extracellular space may be important for subsequent maturation and connectivity of neurons once they have migrated from the neuroependyma. One possible biological significance of this "isolation" of developing neurons from extracellular (plasma) proteins may be that some neuron populations appear to synthesize some of these proteins (e.g., transferrin, fetuin). If these proteins are being secreted at low levels as trophic or recognition signals for other neuron populations, such a mechanism could not work in the face of high concentrations of extracellular protein flooding in from the plasma.

Thus it seems that the blood-brain barrier to protein, far from being immature in early brain development is well established and displays features that are unique to the fetal brain and essential for its normal development.

Acknowledgements. I should like to thank the numerous friends and colleagues from laboratories in different parts of the world who have contributed to much of the work described in the review, in particular the contributions of my wife Katarzyna Dziegielewska and of our mutual friend Kjeld Møllgård. Without their particular contributions and support, none of this work would have been possible. I should also like to thank Mark Habgood for preparing the figures and Geraldine Cole for endlessly retyping the manuscript.

Many of the studies described here were supported by grants from Action Research, AFRC, MRC, NATO, Nuffield Foundation and The Wellcome Trust over many years. I am most grateful for their generous support.

References

Abbott NJ, Revest PA (1991) Control of brain endothelial permeability. Cerebrovasc Brain Metab Rev 3:39–72

Adinolfi M, Haddad SA (1977) Levels of plasma proteins in human and rat fetal CSF and the development of the blood-CSF barrier. Neuropädiatrie 8:345–353

Amtorp O (1976) Transfer of I^{125} albumin from blood into brain and cerebrospinal fluid in newborn and juvenile rats. Acta Physiol Scand 96:399–406

Amtorp O, Sørensen SC (1974) The ontogenetic development of concentration differences for protein and ions between plasma and cerebrospinal fluid in rabbits and rats. J Physiol (Lond) 243:387–400

Arthur FE, Shivers RR, Bowman PD (1987) Astrocyte-mediated induction of tight junctions in brain capillary endothelium: an efficient in vitro model. Dev Brain Res 36:155–159

Åstrom KE (1967) On the early development of the isocortex in fetal sheep. Prog Brain Res 26:1–59

Bakay L (1953) Studies on blood-brain barrier with radioactive phosphorus. III. Embryonic development of the barrier. Arch Neurol Psychiatry 70:30–39

Baños G, Daniel PM, Pratt OE (1978) The effect of age upon the entry of some amino acids into the brain, and their incorporation into cerebral protein. Dev Med Child Neurol 20:335–346

Barcroft J (1938) The Brain and its Environment. Yale University Press, New Haven

Bass NH, Lundborg P (1973) Postnatal development of bulk flow in the cerebrospinal fluid system of the albino rat: clearance of carboxyl [^{14}C] inulin after intrathecal infusion. Brain Res 52:323–332

Behensen G (1927) Über die Farbstoffspeicherung in Zentralnervensystem der weißen Maus in verschiedenen Alterszuständen. Z Zellforsch 4:515–572

Betz AL, Goldstein GW (1981) Developmental changes in metabolism and transport of properties of capillaries isolated from rat brain. J Physiol (Lond) 312:365–376

Birge WJ, Rose AD, Haywood JR, Doolin PF (1974) Development of the blood-cerebrospinal fluid barrier to proteins and differentiation of cerebrospinal fluid in the chick embryo. Dev Biol 41:245–254

Bito LZ, Myers RE (1970) The ontogenesis of haematoencephalic cation transport processes in the rhesus monkey. J Physiol (Lond) 208:153–170

Bloch B, Popovici T, Levin MJ, Tuil D, Kahn A (1985) Transferrin gene expression visualized in oligodendrocytes of the rat brain by using in situ hybridization and immunohistochemistry. Proc Natl Acad Sci USA 82:6706–6710

Bowman PD, Ennis SR, Rarey KE, Betz AL, Goldstein GW (1983) Brain microvessel endothelial cells in tissue culture: a model for study of blood-brain barrier permeability. Ann Neurol 14:396–402

Boyd RDH, Haworth C, Stacey TE, Ward RHT (1976) Permeability of the sheep placenta to unmetabolized polar non-electrolytes. J Physiol (Lond) 256:617–634

Bradbury MWB (1979) The concept of a blood-brain barrier. Wiley, Chichester

Bradbury MWB, Crowder J, Desai S, Reynolds JM, Reynolds ML, Saunders NR (1972) Electrolytes and water in the brain and cerebrospinal fluid of the foetal sheep and guinea-pig. J Physiol (Lond) 227:591–610

Bradford MM (1976) A rapid sensitive method for the quantitation of microgram quantities of protein utilizing the principle of protein-dye binding. Anal Biochem 72:248–254

Braun LD, Cornford EM, Oldendorf WH (1980) Newborn rabbit blood-brain barrier is selectively permeable and differs substantially from the adult. J Neurochem 34:147–152

Brightman MW, Reese TS (1969) Junctions between intimately apposed cell membranes in the vertebrate brain. J Cell Biol 40:648–677

Broadwell RD, Charlton HM, Ebert PS, Hickey WF, Shirazi Y, Villegas J, Wolf AL (1991) Allografts of CNS tissue possess a blood-brain barrier. II. Angiogenesis in solid tissue and cell suspension grafts. Exp Neurol 112:1–28

Butt AM, Jones HC, Abbott NJ (1990) Electrical resistance across the blood-brain barrier in anaesthetized rats: a developmental study. J Physiol (Lond) 429:47–62

Caley WD, Maxwell DS (1970) Development of the blood vessels and extracellular spaces during postnatal maturation of rat cerebral cortex. J Comp Neurol 138:31–48

Carnan E, Gillespie JI (1988) The formation of intra-ventricular fluid in the developing nervous system of the early chick embryo. J Physiol (Lond) 403: 66P
Cavanagh ME, Warren A (1985) The distribution of native albumin and foreign albumin injected into lateral ventricles of prenatal and neonatal rat forebrains. Anat Embryol (Berl) 172:345–351
Claude P, Goodenough D (1973) Fracture faces of zonulae occludentes from "tight" and "leaky" epithelia. J Cell Biol 58:390–400
Connor JE, Fine RE (1986) The distribution of transferrin immunoreactivity in the rat central nervous system. Brain Res 368:319–328
Cornford EM, Braun LD, Oldendorf WH (1982a) Developmental modulations of blood-brain barrier permeability as an indicator of changing nutritional requirements in the brain. Pediatr Res 16:324–328
Cornford EM, Braun LD, Oldendorf WH (1982b) Age-related alterations in blood-brain barrier penetration of diphenylhydantoin. In: Akimoto H, Kazamatsuri H, Seino M, Ward A (eds) Advances in epileptology. 13th Epilepsy International Symposium. Raven, New York, pp 285–288
Davson H (1967) Physiology of the cerebrospinal fluid. Churchill, London
Davson H (1989) History of the blood-brain barrier concept. In: Neuwelt EA (ed) Implications of the blood-brain barrier and its manipulation. Plenum, New York, pp 27–52
Davson H, Welch K, Segal MB (1987) The physiology and pathophysiology of the cerebrospinal fluid. Livingstone, Edinburgh
Dehouck M-P, Méresse S, Delorme P, Torpier G, Fruchart J-C, Cecchelli R (1990) The blood-brain barrier in vitro: co-culture of brain capillary endothelial cells and astrocytes. Circ Metab Cerveau 7:151–162
Delorme P, Gayet J, Grignon G (1970) Ultrastructural study on transcapillary exchanges in the developing telencephalon of the chicken. Brain Res 22: 269–283
Desmond ME, Jacobsen AG (1977) Embryonic brain enlargement requires cerebrospinal fluid pressure. Dev Biol 57:188–198
Dobbing J (1969) The development of the blood-brain barrier. Progr Brain Res 29:417–425
Dreosti IE (1984) Zinc in the central nervous system: the emerging interactions. In: Fredrickson CJ, Howell GA, Kasarkis EJ (eds) The neurobiology of zinc, Liss, New York, pp 1–26
Dziegielewska KM (1982) Proteins in fetal CSF and plasma. PhD thesis, University of London
Dziegielewska KM, Saunders NR (1988) The development of the blood-brain barrier: proteins in fetal and neonatal CSF, their nature and origins. In: Meisami E, Timiras PJ (eds) Handbook of human growth and developmental biology, vol 1A. CRC, Boca Raton, pp 169–191
Dziegielewska KM, Saunders NR (1991) The internal environment of the developing brain. In: Mednick SA (ed) Developmental neuropathology of schizophrenia. Plenum, New York (NATO ASI series) (in press)
Dziegielewska KM, Evans CAN, Malinowska DH, Møllgård K, Reynolds JM, Reynolds ML, Saunders NR (1979) Studies of the development of brain barrier systems to lipid insoluble molecules in fetal sheep. J Physiol (Lond) 292: 207–231
Dziegielewska KM, Evans CAN, Malinowska DH, Møllgård K, Reynolds ML, Saunders NR (1980) Blood-cerebrospinal fluid transfer of plasma proteins during fetal development in the sheep. J Physiol (Lond) 300:457–465
Dziegielewska KM, Evans CAN, Lai PCW, Lorscheider FL, Malinowska DH, Møllgård K, Saunders NR (1981) Proteins in cerebrospinal fluid and plasma of fetal rats during development. Dev Biol 83:193–200

Dziegielewska KM, Møllgård K, Reynolds ML, Saunders NR (1987) A fetuin-related glycoprotein (α_2HS) in embryonic and fetal development. Cell Tissue Res 248:33–41

Dziegielewska KM, Hinds LA, Møllgård K, Reynolds ML, Saunders NR (1988) Blood-brain, blood-cerebrospinal fluid and cerebrospinal fluid-brain barriers in a marsupial (*Macropus eugenii*) during development. J Physiol (Lond) 403: 367–388

Dziegielewska KM, Habgood MD, Jones SE, Reader M, Saunders NR (1989) Proteins in cerebrospinal fluid and plasma of postnatal *Monodelphis domestica* (grey short-tailed opossum). Comp Biochem Physiol [B] 92:569–576

Dziegielewska KM, Habgood MD, Saunders NR, Sedgewick JE (1990) A specific albumin transfer mechanism at the blood-CSF barrer in the rat. J Physiol (Lond) 423:35P

Dziegielewska KM, Habgood MD, Møllgård K, Stagaard M, Saunders NR (1991) Species specific transfer of plasma albumin from blood into different cerebrospinal fluid compartments in the immature fetal sheep. J Physiol 439:215–237

Dziegielewska KM, Matthews N, Møllgård K, Saunders NR (1992) Zinc binding globulin a passive marker for blood-brain carrier permeability in the human fetal brain. J Physiol 446, 6069

Ehrlich P (1885) Das Sauerstoff-Bedürfnis des Organismus. Eine farbenanalytishche Studie. Hirschwald, Berlin, pp 69–72

Evans CAN, Reynolds JM, Reynolds ML, Saunders NR, Segal MB (1974) The development of a blood-brain barrier mechanism in foetal sheep. J Physiol (Lond) 238:371–386

Evans CAN, Reynolds JM, Reynolds ML, Saunders NR (1976) The effect of hypercapnia on a blood-brain barrier mechanism in foetal and newborn sheep. J Physiol (Lond) 255:701–714

Felgenhauer K (1974) Protein size and cerebrospinal fluid composition. Klin Wochenschr 52:1158–1164

Ferguson RK, Woodbury DM (1969) Penetration of ^{14}C-inulin and ^{14}C-sucrose into brain, cerebrospinal fluid and skeletal muscle of developing rats. Exp Brain Res 7:181–194

Fishman JB, Rubin JB, Handrahan JV, Connor JR, Fine RE (1987) Receptor-mediated transcytosis of transferrin across the blood-brain barrier. J Neurosci Res 18:299–304

Flexner LB (1938) Changes in the chemistry and nature of cerebrospinal fluid during fetal life in the pig. Am J Physiol 124:131–135

Flexner LB, Flexner JB (1949) Biochemical and physiological differentiation during morphogenesis. IX. The extracellular and intracellular phases of the liver and cerebral cortex of the foetal guinea-pig as estimated from distribution of chloride and radiosodium. J Cell Comp Physiol 34:115–127

Fossan G, Cavanagh ME, Evans CAN, Malinowska DH, Møllgård K, Reynolds ML, Saunders NR (1985) CSF-brain permeability in the immature sheep fetus: a CSF-brain barrier. Dev Brain Res 18:113–124

Frederickson CJ (1989) Neurobiology of zinc and zinc-containing neurons. Int Rev Neurobiol 31:145–238

Giometto B, Bozza F, Argentiero V, Gallo P, Pagni S, Piccinno MG, Tavolato B (1990) Transferrin receptors in rat central nervous system – an immunocytochemical study. J Neurol Sci 98:81–90

Goldberg J, Sheehy EM (1982) Fifth day fits: an acute zinc deficiency syndrome? Arch Dis Child 57:633–635

Goldstein GW, Betz AL (1986) Blood vessels and the blood-brain barrier. In: Ashbury AK, McKhann GM, McDonald WI (eds) Diseases of the nervous system, vol 1. W.B. Saunders, Philadelphia, pp 172–184

Grazer FM, Clemente CD (1957) Developing blood brain barrier to trypan blue. Proc Soc Exp Biol Med 94:758–760

Habgood MD (1989) Blood-CSF barrier permeability in very immature rats. J Physiol (Lond) 417:31P

Habgood MD (1990) Barriers in the developing brain. PhD thesis, University of Southampton

Habgood MD, Sinclair JD (1988) The effects of hypercapnia on the transfer of proteins between plasma and CSF in neonatal and adult rats. J Physiol (Lond) 400:47P

Hill JM, Switzer RC (1984) The regional distribution and cellular localization of iron in the rat brain. Neuroscience 11:595–603

Hornig GW, Lorenzo AV, Zavala LM, Welch K (1987) Brain tissue pressure measurements in perinatal and adult rabbits. Z Kinderchir 42, Suppl 1:23–26

Huebers HA, Finch CA (1987) The physiology of transferrin and transferrin receptors. Physiol Rev 67:520–587

Hulsebosch CE, Fabian RH (1989) Penetration of IgGs into the neuraxis of the neonatal rat. Neurosci Lett 98:13–18

Hurley LS (1981) Teratogenic aspects of manganese, zinc and copper nutrition. Physiol Rev 61:249–295

Jacobsen M, Clausen PP, Jacobsen GK, Saunders NR, Møllgård K (1982a) Intracellular plasma proteins in human fetal choroid plexus during development. I. Developmental stages in relation to the number of epithelial cells which contain albumin in telencephalic, diencephalic and myelencephalic choroid plexus. Dev Brain Res 3:239–250

Jacobsen M, Jacobsen GK, Clausen PP, Saunders NR, Møllgård K (1982b) Intracellular plasma proteins in human fetal choroid plexus during development. II. The distribution of prealbumin, albumin, alpha-fetoprotein, transferrin, IgG, IgA, IgM, and Alpha$_1$-antitrypsin. Dev Brain Res 3:251–262

Jacobsen M, Møllgård K, Reynolds ML, Saunders NR (1983) The choroid plexus in fetal sheep during development with special reference to intracellular plasma proteins. Dev Brain Res 8:77–88

Janas MS, Moos T, Møllgård K (1991) Is the differential localization of specific plasma proteins in the developing neo- and archicortex a reflection of differences in trans-barrier transport? Possible implications for the neuropathology of schizophrenia. In: Mednick SA (ed) Developmental neuropathology of schizophrenia. Plenum, New York (NATO ASI series) pp 61–73

Jefferies WA, Brandon MR, Hunt SV, Williams AF, Gatter KC, Mason DY (1984) Transferrin receptor on endothelium of brain capillaries. Nature 312:162–163

Johanson CE (1989) Ontogeny of the blood-brain barrier. In: Neuwelt EA (ed) Implications of the blood-brain barrier and its manipulation, vol 1. Plenum, New York, pp 157–198

Johanson CE, Woodbury DM (1974) Changes in CSF flow and extracellular space in the developing rat. In: Vernadakis A, Weiner N (eds) Drugs and the developing brain. Plenum, New York, pp 281–287

Johanson CE, Reed DJ, Woodbury DM (1974) Active transport of sodium and potassium by the choroid plexus of the rat. J Physiol (Lond) 241:359–372

Johanson CE, Reed DJ, Woodbury DM (1976) Developmental studies of the compartmentalization of water and electrolytes in the choroid plexus of the neonatal rat brain. Brain Res 116:35–48

Jones HC, Keep RF (1987) The control of potassium concentration in the cerebrospinal fluid and brain interstitial fluid of developing rats. J Physiol (Lond) 383:441–453

Jones HC, Keep RF (1988) Brain fluid calcium concentration and response to acute hypercalcaemia during development in the rat. J Physiol (Lond) 402:579–593

Jones HC, Deane R, Bucknall RM (1987) Developmental changes in cerebrospinal fluid pressure and resistance to absorption in rats. Dev Brain Res 33:23–30

Joó F (1985) The blood-brain barrier in vitro: ten years of research on microvessels isolated from the brain. Neurochem Int 7:1–25

Kalaria RN, Harik SI (1987) Differential postnatal development of monoamine oxidases A and B in the blood-brain barrier of the rat. J Neurochem 49: 1589–1594

Keen CL, Hurley LS (1989) Zinc and reproduction: effects of deficiency on foetal and postnatal development. In: Mills CF (ed) Zinc in human Biology. Springer, Berlin Heidelberg New York, pp 183–220

Klosovski BN (1963) The development of the brain and its disturbance by harmful factors. Pergamon, Oxford, p 94

Knott G, Dziegielewska KM, Habgood MD, Saunders NR (1991) Species specific blood-CSF transfer of albumins in the grey short-tailed opossum (*Monodelphis domestica*). Eur J Neurosci Suppl 4:4158

Lim BC, McArdle HJ, Morgan EH (1987) Transferrin-receptor interaction and iron uptake by reticulocytes of vertebrate animals – a comparative study. J Comp Physiol 157:363–371

Lowry OH, Rosebrough NJ, Farr AL, Randall RJ (1951) Protein measurement with the folin phenol reagent. J Biol Chem 193:265–275

Møllgård K, Saunders NR (1975) Complex tight junctions of epithelial and of endothelial cells in early foetal brain. J Neurocytol 4:453–468

Møllgård K, Saunders NR (1977) A possible transepithelial pathway via endoplasmic reticulum in fetal sheep choroid plexus. Proc R Soc Lond [Biol] 199:321–326

Møllgård K, Saunders NR (1986) The development of the human blood-brain and blood-CSF barriers. Neuropathol Appl Neurobiol 12:337–358

Møllgård K, Malinowska DH, Saunders NR (1976) Lack of correlation between tight junction morphology and permeability properties in developing choroid plexus. Nature 264:293–294

Møllgård K, Lauritzen B, Saunders NR (1979) Double replica technique applied to choroid plexus from early foetal sheep: completeness and complexity of tight junctions. J Neurocytol 8:139–149

Møllgård K, Balslev Y, Lauritzen B, Saunders NR (1987a) Cell junctions and membrane specializations in the ventricular zone (germinal matrix) of the developing sheep brain: a CSF-brain barrier. J Neurocytol 16:433–444

Møllgård K, Stagaard M, Saunders NR (1987b) Cellular distribution of transferrin immunoreactivity in the developing rat brain. Neurosci Lett 78:35–40

Møllgård K, Dziegielewska KM, Saunders NR, Zakut H, Soreq H (1988) Synthesis and localization of plasma proteins in the developing human brain. Dev Biol 128:207–221

Molliver ME, Kostovic I, van der Loos H (1973) The development of synapses in cerebral cortex of the human fetus. Brain Res 50:403–407

Nicholls JG, Stewart RR, Erulkar SD, Saunders NR (1990) Reflexes, fictive respiration and cell division in the brain and spinal cord of the newborn opossum, *Monodelphis domestica*, isolated and maintained in vitro. J Exp Biol 152:1–15

Olesen SP, Crone C (1983) Electrical resistance of muscle capillary endothelium. Biophys J 42:31–41

Olsson Y, Klatzo I, Sourander P, Steinwall O (1968) Blood-brain barrier to albumin in embryonic new born and adult rats. Acta Neuropathol (Berl) 10:117–122

Pardridge WM, Eisenberg J, Yang J (1987) Human blood-brain barrier transferrin receptor. Metabolism 26:892–895

Pardridge WM, Boado RJ, Farrell CR (1990) Brain-type glucose transporter (GLUT-1) is selectively localized to the blood-brain barrier. J Biol Chem 265:18035–18040

Parmalee JT, Johanson CE (1989) Development of potassium transport capability by choroid plexus of infant rats. Am J Physiol 256:R786–R791

Peters T Jr (1975) Serum albumin. In: Putnam FW (ed) The plasma proteins, vol 1. Academic, New York, pp 133–183

Rapoport SI (1976) Blood-brain barrier in physiology and medicine. Raven, New York

Rapoport SI, Pettigrew KD (1979) A heterogenous, pore-vesicle membrane model for protein transfer from blood to cerebrospinal fluid at the choroid plexus. Microvasc Res 18:105–119

Reese TS, Karnovsky MJ (1967) Fine structural localization of a blood-brain barrier to exogenous peroxidase. J Cell Biol 34:207–217

Reynolds ML, Møllgård K (1985) The distribution of plasma proteins in the neocortex and early allocortex of the developing sheep brain. Anat Embryol (Berl) 171:41–60

Reynolds ML, Cavanagh ME, Dziegielewska KM, Hinds LA, Saunders NR, Tyndale-Biscoe CH (1985) Postnatal development of the telencephalon of the tammar wallaby (*Macropus eugenii*). Anat Embryol (Berl) 173:81–94

Risau W, Hallman R, Albrecht U (1986) Differentiation-dependent expression of proteins in brain endothelium during development of the blood-brain barrier. Dev Biol 117:537–545

Romanoff AL (1967) Biochemistry of the avian embryo, Wiley, New York

Roncali L, Nico B, Ribatti D, Bertossi M, Mancini L (1986) Microscopical and ultrastructural investigations on the development of the blood-brain barrier in the chick embryo optic tectum. Acta Neuropathol (Berl) 70:193–201

Rubin LL, Barbu K, Bard F, Cannon C, Hall DE, Horner H, Janatpour M, Liaw C, Manning K, Morales J, Porter S, Tanner L, Tomaselli K, Yednock T (1991) Differentiation of brain endothelial cells in cell culture. Ann NY Acad Sci 633, 420–425

Rutten MJ, Hoover RL, Karnovsky MJ (1987) Electrical resistance and macromolecular permeability of brain endothelial monolayer cultures. Brain Res 425:301–310

Saunders NR (1992) The development of the blood-brain barrier to macromolecules. In: Segal MB (ed) The fluids and barriers of the eye and the brain. Macmillan, Basingstoke pp 129–155

Saunders NR (1992) Development of the blood-brain barrier and properties of CSF in the developing brain. In: Crockard A, Haywood R, Hoff JT (eds) Neurosurgery, the scientific basis of clinical practice. Blackwell, Oxford pp 22–37

Saunders NR, Møllgård K (1984) Development of the blood-brain barrier. J Dev Physiol 6:45–57

Saunders NR, Adam E, Reader M, Møllgård K (1989) *Monodelphis domestica* (grey short-tailed opossum): an accessible model for studies of early neocortical development. Anat Embryol (Berl) 180:227–236

Schwick HG, Haupt H (1981) Purified human plasma proteins of unknown functions. In: Bing DH, Rosenbaum RA (eds) Plasma and cellular modulatory proteins. Centre for Blood Research, Boston

Sessa G, Perez MM (1972) Biochemical changes in rat brain associated with the development of the blood-brain barrier. J Neurochem 25:779–782

Stastny F (1983) Glucocorticoids and brain development. Avicenum, Prague

Stern L, Gautier G (1921) Rapports entre le liquide céphalorachidien et la circulation sanguine. Arch Int Physiol. 17:138–192

Stern L, Gautier G (1922) Les rapports entre le liquide céphalorachidien et les éléments nerveux de l'axe cérébrospinal. Arch Int Physiol 17:391–448

Stern L, Peyrot R (1927) Le fonctionnement de la barrière hémato-encéphalique aux divers stades de développement ches les diverses espèces animales. C R.Soc Biol (Paris) 96:1124–1126

Stern PL, Rapoport JL (1928) Les échanges entre le liquide céphalorachidien et les éléments nerveux cérébro-spinaux. C R Soc Biol (Paris) 98:1518–1519

Stern L, Rapoport JL, Lokschina ES (1929) Le fonctionnement de la barrière hémato-encéphalique chez les nouveau-nés. C R Soc Biol (Paris) 100:231–233
Stewart PA, Hayakawa EM (1987) Interendothelial, functional changes underlie the developmental 'tightening' of the blood-brain barrier. Dev Brain Res 32:271–281.
Stewart PA, Wiley MJ (1981) Developing nervous tissue induces formation of blood-brain barrier characteristics in invading endothelial cells: a study using quail-chick transplantation chimeras. Dev Biol 84:183–192
Stewart RR, Zou DJ, Treherne JM, Møllgård K, Saunders NR, Nicholls JG (1991) The intact nervous system of the newborn oppossum in long-term culture: fine structure and GABA-mediated inhibition of electrical activity. J Exp Biol 161:25–41
Tao-Cheng JH, Brightman MW (1988) Development of membrane interactions between brain endothelial cells and astrocytes in vitro. Int J Dev Neurosci 6:25–37
Tao-Cheng JH, Nagy Z, Brightman MW (1987) Tight junctions of brain endothelium in vitro are enhanced by astroglia. J Neurosci 7:3293–3299
Taylor EM, Morgan EH (1990) Developmental changes in transferrin and iron uptake by the brain in the rat. Dev Brain Res 55:35–42
Tschirgi RD (1950) Protein complexes and the impermeability of the blood-brain barrier to dyes. Am J Physiol 163:756
Vorbrodt AW, Lossinsky AS, Wisniewski HM (1986) Localization of alkaline phosphatase activity in endothelia of developing and mature mouse blood-brain barrier. Dev Neurosci 8:1–13
Wakai S, Hirokawa N (1979) Development of the blood-brain barrier to horseradish peroxidase in the chick embryo. Cell Tissue Res 195:195–203
Wakai S, Hirokawa N (1981) Development of blood-cerebrospinal fluid barrier to horseradish peroxidase in the avian choroidal epithelium. Cell Tissue Res 214:271–278
Wislocki GB (1920) Experimental studies on fetal absorption. I. The vitally stained fetus. Contrib Embryol Carnegie Inst 5:45–52
Weed LH (1917) The development of the cerebrospinal fluid spaces in pig and man. Contrib Embryol Carnegie Inst 4:41–52
Woodbury DM, Johanson C, Brønsted H (1974) Maturation of the blood-brain and blood-cerebrospinal fluid barriers and transport systems. In: Zimmermann E, George R (eds) Narcotics and the hypothalamus. Raven, New York, pp 225–247
Xie X, Smart TG (1991) A physiological role for endogenous zinc in rat hippocampal synaptic neurotransmission. Nature 349:521–524
Zeman W, Innes JRM (1963) Craigie's Neuroanatomy of the rat, Academic, New York, p 45

CHAPTER 15

Comparative Physiology of the Blood-Brain Barrier*

N.J. ABBOTT

A. Introduction

I. History of Blood-Brain Barrier Studies

The earliest studies of blood-brain exchange were done on mammals. In his classic experiments on dyes injected into the vascular system, EHRLICH (1885) showed that many dyes distributed widely into body tissues, but failed to stain brain parenchyma; the choroid plexus was stained, but the cerebrospinal fluid (CSF) was not. GOLDMANN later showed (1909, 1913) that the brain was stained by trypan blue introduced into the CSF, but not by the same dye injected into the blood. Electron microscopic studies using the tracer horseradish peroxidase (HRP) showed that restriction by the endothelium (REESE and KARNOVSKY 1967), but not by the bulk of the ependymal lining of the ventricles (BRIGHTMAN and REESE 1969), was responsible for the earlier dye observations. It is now well established that the mammalian blood-brain barrier (BBB) is formed by tight junctions between the endothelial cells of the brain parenchyma, and between the epithelial cells of the choroid plexus (Chap. 1). Tight junctions in the arachnoid layer restrict penetration between the dura and subarachnoid space (NABESHIMA et al. 1975). The mammalian studies have concentrated on the characterisation of the cellular processes occurring at each of these barrier sites, their development in the embryo, and the ways in which they can be modulated by physiological, pathological and experimental conditions. However, additional insights into the function and evolution of the barrier have come from

Abbreviations

ACE	angiotensin converting enzyme	CSF	cerebrospinal fluid
BBB	blood-brain barrier	γ-GTP	γ-glutamyl transpeptidase
CM	conditioned medium	HRP	horseradish peroxidase
		ISF	interstitial fluid

* Our comparative studies have been generously supported by the SERC, Wellcome Trust, Royal Society, Nuffield Foundation, NSF and Danish Natural Science Research Council.

work on non-mammalian species. This chapter discusses the insights to be gained from the comparative approach, and presents some of the key conclusions.

II. Value of Comparative Studies

1. Comparisons Between Animal Groups

There are advantages in studying the BBB in more than one animal group. This approach helps to establish whether an observed mechanism is a general phenomenon, or specific to one group only; i.e., it helps to separate basic and incidental properties of the barrier layer. If a common problem is resolved by different mechanisms in different animal groups, it implies that the end result is more important than the means by which it is achieved. If a mechanism appears in "advanced" representatives of a group, but not in their more "primitive" relatives, deductions can be made about the evolutionary sequence and processes at work. And if a mechanism appears to have evolved independently in different animal groups, this is strong evidence that the mechanism confers selective advantage.

2. Selection of Convenient Experimental Preparation

Mammalian studies have used rat, mouse, rabbit, guinea-pig, dog, sheep and human subjects, and although quantitative differences have been found, there is an overall similarity of mechanisms, so that the investigator is free to choose a particular preparation for experimental convenience. Small animals are cheaper and require smaller amounts of drugs and radioisotopes. Rats and mice, particularly inbred strains, offer advantages of reproducibility, ease of dietary control, and availability of a number of specific antibodies against receptors and transport mechanisms. Rabbits have easily accessible ear veins and arteries, which can be tapped with minimum disturbance to the animal, ideal for studies where anaesthesia would have undesirable effects on blood flow and kinetics. Guinea-pigs have an unusual cerebrovascular anatomy resulting in a functional separation of vertebral and carotid artery circulations, providing favourable conditions for perfusion of the forebrain (Zloković et al. 1988). Small brains are more convenient for tissue uptake studies (terminal brain sampling); large brains (e.g., dog, sheep) give better resolution for single-pass extraction from the blood (Crone 1965; Yudilevich et al. 1988), and for manoeuvres requiring cannulation of the choroid plexus and fetus (Segal and Zloković 1988; Saunders, Chap. 14).

As for in vivo studies, so for culture; cultured bovine brain endothelial cells have been used where large numbers of cells are an advantage (Chap. 17), while use of rat cultures permits exploitation of the immunocytochemical markers developed for this species (Hughes and Lantos 1986; Abbott et al. 1992b).

Among other vertebrate groups, the frog (amphibia) has proved particularly advantageous for microelectrode study of brain microvessels, as its pial vessels are readily accessible in an auto-perfused or saline-perfused preparation, and the brain is relatively resistant to hypoxia (CRONE and OLESEN 1982; CRONE 1984; ABBOTT et al. 1986c; OLESEN 1989). The chick has proved useful for study of barrier induction in endothelium, using the chorio-allantoic membrane, accessible in the egg, as an implantation site for brain-derived cells (astrocytes) (JANZER and RAFF 1987).

Among invertebrates, most studies have been done on animal groups with large complex brains and central nervous system (CNS) ganglia, including some that have rich vascularisation, permitting direct comparison with vertebrates (ABBOTT et al. 1986b).

3. Preparations Offering Unique Advantages

Some non-mammalian species, as a result of an unusual anatomy or physiology, offer a uniquely suitable preparation for the study of a key topic. Questions concerning the role of the choroid plexus can be tackled by comparing the lamprey and the hagfish, jawless vertebrates (cyclostomes) in which the choroid plexus is respectively large (lamprey) and absent (hagfish) (CSERR and BUNDGAARD 1984). Putative sites of interstitial fluid (ISF) secretion can be examined in cephalopod molluscs, which have a rich intracerebral blood supply but no choroid plexus or ventricular system to complicate the analysis (ABBOTT et al. 1985). Substrate transport by glial cells of the brain can be studied by looking at brain uptake of tracer substrates from blood in elasmobranch fish (ABBOTT et al. 1988a,b), since here the blood-brain interface is formed by glial cells (BRIGHTMAN et al. 1971; BUNDGAARD and CSERR 1981b). As the eminent Danish physiologist August KROGH pointed out, for every physiological question, there is one species above all others most appropriate for its investigation (KROGH 1929; KREBS 1975).

B. Evolutionary Pressures Leading to Development of a Blood-Brain Barrier

I. Why Do We Need a Blood-Brain Barrier?

The mammalian BBB combines low passive permeability with specialised transport mechanisms for ions, glucose, amino acids, purines, nucleosides, thiamine, peptides and proteins (BRADBURY 1979, 1989). It is clear that the transport mechanisms became necessary as a result of the tightening of the barrier layer, but studies confined to the fully developed barrier of mammals provide little information on the evolutionary pressures leading to formation of a barrier. In other words, why does the brain need a barrier? What is the key advantage bestowed by the presence of a barrier? Here studies on lower animals can give useful information.

II. Ionic Homeostasis and Neural Function

Fluctuations in the ionic composition of the plasma, particularly for K^+ and H^+, result from natural physiological events such as absorption from the gut, ventilation, tissue respiration and exercise. Axons conduct all-or-none action potentials, which behave like digital signals, with information being carried in the frequency and pattern of action potentials (Fig. 1). Thus small variations in action potential amplitude, arising from ionic fluctuations in the medium, will have little effect on the information-carrying abilities of the axons (ABBOTT et al. 1986a). By contrast, synaptic potentials are graded signals, analogue events, with information contained in the amplitude of the signal (Fig. 1); quite small fluctuations in the ionic environment of synapses can cause changes in the size of post-synaptic potentials, which will be interpreted as arising from changes in the input signal. However, some kinds of synaptic activity, such as the 1:1 "slave relay" of the peripheral nerve-muscle junction, have such a large safety factor (size of end-plate potential $\gg$ threshold for production of post-synaptic action potential) that ionic fluctuations here would cause little functional disturbance. Similarly, the control of smooth muscle tone (gastrointestinal tract, blood vessels) by autonomic ganglia is a relatively imprecise activity, often overidden by local tissue events; it is therefore also relatively tolerant of changes in synaptic function arising through ionic fluctuations. These considerations lead to the prediction that the central synapses integrating detailed quantitative information (especially in the visual and motor systems) should need better local ionic homeostasis than axons, peripheral ganglia and neuromuscular

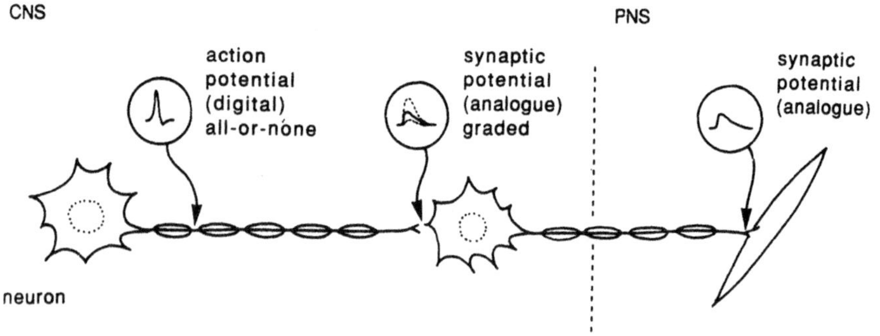

Fig. 1. Schematic diagram of types of electrical signal used for communication within the central (CNS) and peripheral (PNS) nervous systems. The action potential is an all-or-none (digital) signal, where information is coded by signal frequency. Synaptic potentials are graded (analogue) signals, where information is coded by signal amplitude. The latter should be more vulnerable to disturbance by ionic fluctuations

synapses (ABBOTT et al. 1986a). As a blood-tissue barrier is the simplest way of protecting the local microenvironment from fluctuations in the plasma, it would be predicted that the barrier would be tightest around central synapses. Moreover, it could be argued that the development of a BBB in evolution occurred because of the selective advantage achieved by animals with a tighter barrier, reflected in an improved efficiency and speed of CNS integrative activity.

These ideas lead to the hypothesis that a BBB is primarily a mechanism for achieving ionic homeostasis around integrating synapses, and that other benefits follow (i.e., are secondary adaptations). This hypothesis is hard to test in mammals, because although the barrier in the brain appears to be tighter than that of autonomic ganglia and peripheral nerve, central synapses and axons are both well protected (ABBOTT et al. 1986a). However, strong evidence comes from invertebrate studies. Most invertebrate groups have no BBB – this includes most animals with a simple behavioural repertoire (annelids, lower molluscs, primitive chelicerates such as ticks and the horseshoe crab *Limulus*). The CNS of insects and decapod crustaceans has an efficient barrier to small ions, whereas the peripheral nerves are leakier (LANE and TREHERNE 1972, 1973; TREHERNE and PICHON 1972; ABBOTT et al. 1975; ABBOTT and PICHON 1987). The cephalopod mollusc *Sepia* (cuttlefish) has a BBB tight to HRP and ionic lanthanum in the optic lobe of the brain, where visual input is processed, but the axon tracts connecting the retina with the optic lobe are leaky to HRP (ABBOTT et al. 1986b). Thus a barrier is associated with the development of higher functions of the brain that depend on precise synaptic activity and integration. In animal groups where only part of the nervous system is protected by a barrier, the observations support the hypothesis.

III. Isolation for Chemical Signalling

It has been pointed out that the presence of a barrier not only protects the brain microenvironment from fluctuations in plasma, but also isolates the environment so that endogenous signals released by neurons (ions, transmitters, neuromodulators) are not rapidly lost to blood, so can be used as efficient intercellular signals, perhaps acting over large distances (CSERR and BUNDGAARD 1984, 1986). Movement through the brain would occur by a combination of local diffusion (NICHOLSON and RICE 1986), and bulk flow along pathways of reduced resistance (see below, discussion of relation of ISF and CSF). This is an important idea, and there are now several examples of situations within the CNS where the local accumulation of ions or neuroactive agents can cause (often long-lasting) changes in neural activity (reviewed in SPIRA et al. 1984; SYKOVÁ 1987; LENG et al. 1988). There is no doubt that once the barrier is present, the isolation it confers on the brain interstitium can be exploited for this additional level of neurochemical signalling.

It is worth asking whether the need to achieve isolation for neurochemical signalling could have been the primary evolutionary pressure giving advantage to animals with a BBB. CSERR and BUNDGAARD (1986) have pointed out that most species with a large enough neural mass to require ingrowth of vessels also show a BBB; if the ingrowing vessels did not have a barrier, they would dissipate the local gradients of signalling molecules. This argues in favour of a barrier being required primarily for neurochemical isolation.

The argument is somewhat weakened by the observation that insects, which have large ganglia but no invading vessels (oxygen is supplied via tracheoles), do show a barrier at the blood-tissue surface; as the neuropil, the region of synaptic interaction, lies deeper within the ganglion, a barrier would not be predicted from the "isolation hypothesis". Conversely, the primitive chelicerate *Limulus* (horseshoe crab) has vessels growing into the CNS, but no BBB (HARRISON and LANE 1981). Moreover, many lower molluscs with sizeable ganglia, and neurons sensitive to complex combinations of neuroactive agents, appear to manage modulation of neural activity without the strict isolation of a barrier (SATTELLE and LANE 1972; SATTELLE 1973; MIROLLI and GORMAN 1973). On balance it seems more likely that the barrier was initially developed to protect the synapses from ion fluctuations in blood, rather than to conserve local signalling chemicals.

IV. Regulation of the Periaxonal Environment

The above discussion on digital and analogue signalling leads to the prediction that the peripheral nervous system of invertebrates should not need a blood-nerve barrier. There are some complications here, because although this prediction is generally borne out (ABBOTT et al. 1975; VILLEGAS and SANCHEZ 1991), a barrier *is* present around larger peripheral nerves of some insects (LANE and TREHERNE 1972, 1973). However, many insects have a haemolymph (blood) whose ionic composition is either unsuitable for nerve conduction (high in K^+, low in Na^+; TREHERNE and PICHON 1972) or experiences large fluctuations in ions such as K^+ (LETTAU et al. 1977). Hence it could be argued that the insect barrier is necessary to enable the peripheral nervous system to develop and maintain a local milieu more appropriate for action potential generation and conduction; indeed, studies on the interstitial microenvironment of the insect nervous system show that it can be dramatically different from that of haemolymph (TREHERNE and PICHON 1972), and may possess useful ion exchange properties providing a reservoir of extracellular ions (TREHERNE et al. 1982).

It was, however, somewhat surprising to find that in the squid, the nerve supplying the mantle muscle appeared to be protected by a relatively efficient blood-nerve barrier (ABBOTT and PICHON 1987). Thus axonal action potentials recorded from in situ giant axons in the mantle muscle were relatively insensitive to raising the $[K^+]$ or lowering the $[Na^+]$ in the vascular perfusate, manoeuvres causing rapid nerve block in dissected preparations. The ionic composition of squid plasma resembles that of sea water, and is a suitable medium for axonal function, so here a barrier is not needed to keep the plasma out. While action potential conduction and frequency in unbranched axons is relatively insensitive to fluctuations in local ions such as K^+, this is not necessarily the case for axons branching into the target tissue (SMITH 1980a,b), especially when the branch point is associated with a dramatic drop in the diameter of the axon (and hence safety factor for cable conduction). Failure of even one action potential in a burst of three (as could follow depolarisation caused by local K^+ accumulation) can cause as much as a 20% drop in force generated by the mantle muscle (BROWN 1991). As the nerve runs through active muscle, which releases large amounts of K^+, significant local K^+ accumulation could occur, particularly in the poorly vascularised anaerobic muscle. The blood-nerve barrier here may be needed to ensure ionic homeostasis around axonal branch points.

It has been known for approaching 20 years that the squid giant axon communicates with its surrounding Schwann cell sheath by chemical signals, triggered by repetitive axonal stimulation. The process involves an neurotransmitter cascade which appears to include glutamate, acetylcholine and an endogenous peptide, and can be modulated by agents such as octopamine (VILLEGAS 1984; VILLEGAS et al. 1988). The activation results in elevation of intracellular cyclic adenosine monophosphate (cAMP) and a long-lasting hyperpolarisation of the Schwann cell membrane; the physiological function of this process is uncertain, but it could be a way of "switching on" the Schwann cells so that they become more efficient in K^+ removal and metabolic support of the axons during and following repetitive axonal activity (BROWN et al. 1991).

As with the argument above concerning "isolation" and neuromodulatory signals, it is unlikely that the advantage resulting from this axon-Schwann cell signalling was the primary evolutionary pressure that resulted in development of the blood-nerve barrier. However, once the barrier was formed to control the ionic environment around axon branch points, the greater stability of the periaxonal microenvironment thus created would have permitted development of axon-Schwann cell signalling. As hyperpolarisation of the Schwann cell can be detected to application of $10^{-9}\,M$ glutamate and is maximal at ca. $10^{-6}\,M$ (LIEBERMAN et al. 1989), while the plasma glutamate concentration is reported to be as high as $2\,mM$, the signalling system would be largely saturated/desensitised if the plasma compartment were in free communication with the periaxonal space.

C. Site of the Barrier Layer

I. Endothelial Barrier of Vertebrates

An endothelial BBB is universal among all mammalian species studied. This is also the pattern in birds, amphibia, reptiles, teleost fish and cyclostomes (lamprey, hagfish) (reviewed in Cserr and Bundgaard 1984; Abbott et al. 1986b; Fig. 2c). Microelectrode studies of the transendothelial resistance of the accessible pial vessels show that those of the frog and rat have a similar tightness, with resistance in the range 1000–5000 $\Omega\,cm^2$ (Crone and Olesen 1982; Butt et al. 1990b). Similar junctional structures (*zonulae occludentes*) appear to be responsible for the barrier in each group (Brightman and Reese 1969; Bundgaard 1982b; Nagy et al. 1984), although there may be species differences in the extent to which the barrier is rendered leaky by inflammatory mediators and experimental manipulation (Olesen 1989; Abbott and Revest 1991).

II. Glial Barriers in Higher Invertebrates

Most invertebrates with simple brains (annelids, lower molluscs, lower chelicerates) have no BBB (Abbott et al. 1986b). A barrier is found in higher chelicerates (spiders, scorpions), crustaceans and insects (reviewed in Abbott et al. 1986b), all "intelligent" invertebrates capable of complex CNS activity; examples are the prey-catching strategies of spiders and scorpions, and the social interactions and communication skills of insects. This pattern led to the proposal that a barrier is associated with higher integrative activities of central synapses (Abbott et al. 1986a, see above). In all these invertebrate groups with a barrier, the restriction occurs at a glial, not an endothelial layer (Fig. 2a). In many cases the endothelium or equivalent is poorly developed or discontinuous, so if a barrier is needed, it has to be

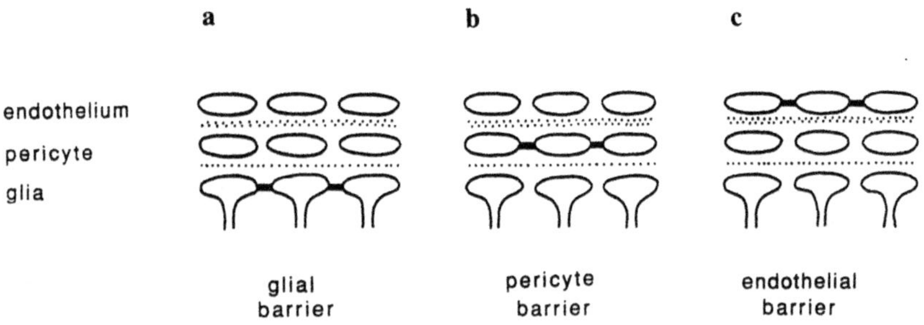

Fig. 2a–c. The site of the barrier layer at the blood-brain interface. The barrier is shown by *bars* linking adjacent cells. In most invertebrates with a barrier the restriction is formed by glia (**a**); in most vertebratres it is formed by the endothelium (**c**). The cuttlefish *Sepia* (cephalopod mollusc) shows an intermediate condition (glial barrier in capillaries, pericyte barrier in arterioles; **b**)

glial. However, in some groups a reasonably complete layer of mesodermal origin is also present at the blood-tissue interface (fat body sheath and mesodermal layer in insect nerve cord), yet the barrier is still formed by the deeper glial cells (LANE and TREHERNE 1972; EDWARDS et al. 1991). This raises the interesting possibility that a glial barrier is the primitive condition, and a mesodermal/endothelial barrier came later. In *Peripatus*, the segmented invertebrate thought to be close to the ancestral origin of uniramian arthropods from annelids, there is no barrier (LANE and CAMPIGLIA 1987), confirming that a barrier is a more advanced evolutionary feature.

III. Transitional Glial/Pericyte Barrier of Cephalopod Molluscs

At the level of the intracerebral microvessels, cephalopod molluscs such as cuttlefish, octopus and squid have a glial barrier (ABBOTT and BUNDGAARD 1992; Fig. 2a). On the luminal side this is lined by a discontinuous layer of endothelial cells and pericytes. In the smallest vessles, the distinction between endothelial cells and pericytes breaks down, only a single class of endothelial/pericyte being seen (BUNDGAARD and ABBOTT 1992). In the venous part of the cerebral vasculature, the barrier is still formed by glia, even though the endothelial and pericyte layers may be relatively complete. However, the arterial vessels show a different pattern – although endothelial, pericyte and glial layers are present, intravascular tracers such as HRP and ionic lanthanum are blocked by the pericyte layer (ABBOTT and BUNDGAARD 1992; Fig. 2b). This suggests strongly that *Sepia* is demonstrating a transitional evolutionary condition, in which the primitive glial barrier is retained in the capillaries, but is superseded by a pericyte barrier when this layer becomes sufficiently continuous. The final shift to the endothelial layer, as in higher vertebratcs, is then seen to be a later evolutionary specialisation.

IV. Apparently Anomalous Glial Blood-Brain Barrier of Elasmobranch Fish

Elasmobranchs have a well developed endothelium, but the barrier is formed by the end feet of perivascular glia (BRIGHTMAN et al. 1971; BUNDGAARD and CSERR 1981b; Fig. 2a). The holocephalan *Chimaera monstrosa*, a modern cartilaginous fish thought to be related to the ancestors of elasmobranchs, has an endothelial barrier (BUNDGAARD 1982c). This, and the presence of an endothelial barrier in another primitive fish group, the cyclostomes (BUNDGAARD et al. 1979; BUNDGAARD and CSERR 1981a; BUNDGAARD 1982a), led to the idea that the earliest vertebrate ancestors already possessed an endothelial BBB, and the glial barrier of elasmobranchs must therefore be a secondary development (CSERR and BUNDGAARD 1984; Fig. 3a). It was not clear when or why the elasmobranchs lost their endothelial barrier.

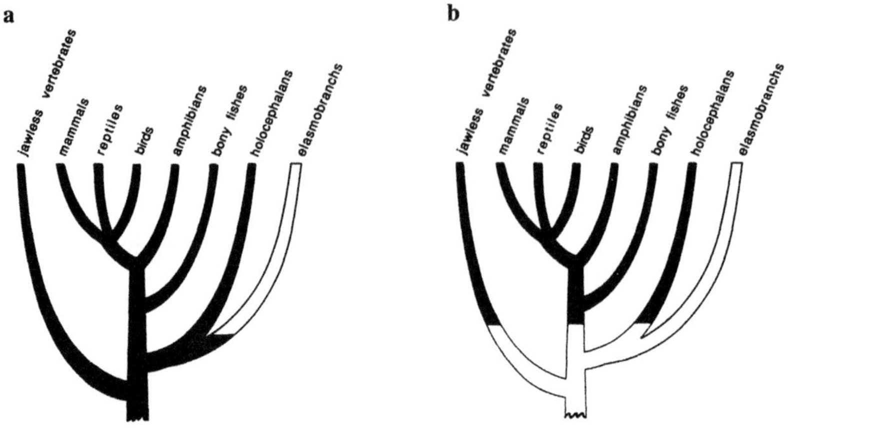

Fig. 3a,b. Schematic family trees of vertebrates including all extant groups. *Tips* of branches represent present time; the *base* is approximately 500 million years ago. *Branching points* represent common ancestors. *White area*, glial BBB; *shaded area*, endothelial barrier. **a** Scheme based on Cserr and Bundgaard (1984); elasmobranchs have a glial barrier, regarded as a secondary anomaly. **b** New scheme proposed here; a glial barrier is regarded as the primitive condition, still seen in modern elasmobranchs. By this scheme, an endothelial barrier would have evolved more than once

The recent studies on invertebrate barriers prompt an alternative explanation; they suggest that a glial barrier was the primitive condition, and an endothelial barrier came later. According to this scheme, the elasmobranchs, which preserve many other primitive features, are only anomalous in that they have retained the primitive glial barrier. In view of the likely evolutionary tree (Fig. 3b), this means that the ancestral vertebrate had a glial barrier, and that an endothelial barrier arose independently several times in vertebrate evolution. The finding of a pericyte barrier in *Sepia*, while lower molluscs have no barrier, confirms that a barrier may arise independently in different animal groups, and that the barrier layer is likely to have shifted with progressive evolutionary development, from glia to pericytes to endothelium.

V. Residual Barrier-Forming Properties of Vertebrate Glia and Ependyma

The ependyma shows many features of an undifferentiated glial layer. Traces of the barrier-forming potential of glial cells (tight junctions) can be seen in ependymal derivatives in the adult mammalian brain including the ependymal lining of the floor of the IIIrd ventricle over neurosecretory zones, tanycytes of the median eminence/arcuate nucleus border, choroid plexus epithelium, and pigment epithelium of the eye (reviewed in Abbott et al. 1986a). In spite of the appearance of electron-dense tracers in the perivascular space soon after intravenous injection (BALIN et al. 1986), apparently via leaks across large vessel walls into the subarachnoid space and thence into the Virchow-Robin spaces, tracer penetration between glial end feet into the interstitium is negligible. Glial sheaths around some tumours can restrict penetration of tracers from tumour tissue into normal brain parenchyma (DEANE et al. 1984). In all these situations the glial cells may be demonstrating their residual barrier-forming potential.

VI. Role of Pericytes in Barrier Function

Most studies of the vertebrate BBB have concentrated on the endothelial cells, and where the perivascular glia (astrocytic end feet) have been considered, it is usually in the context of induction of barrier properties in endothelium. Pericytes are conspicuously missing from the discussion. However, the demonstration that modified proteins that are transported across the endothelium are largely prevented from entering the brain interstitium by endocytosis into pericytes (BROADWELL et al. 1988) suggests that pericytes are active as a second line of defence. The pericytes and glial end feet also demarcate the perivascular space, which by its continuity with the Virchow-Robin spaces and subarachnoid space may have a significant role in channelling the flow of interstitial fluid (RENNELS et al. 1985). The BBB should be regarded as a multicellular layer, in which the three cell types cooperate (BROADWELL and SALCMAN 1981).

D. Inductive Signals in Barrier Formation

Comparative studies have proved extremely useful in studying the interactions between the brain tissue and vascular elements, and the inductive signals responsible for formation of barrier properties at a particular cell layer.

I. Grafting Studies: Brain Versus Non-Brain

SVENDGAARD et al. (1975) showed that when brain explants were grafted into the anterior chamber of the eye in rats, vessels within the graft showed barrier-associated properties, whereas in the opposite type of graft (iris into brain), the vessels showed characteristics typical of leaky non-brain vessels (Fig. 4a,b). It was assumed that the vessels had grown into the explants from the host tissue, and it was concluded that brain tissue exerted an inductive influence on the microvessel properties. However, it was possible that the vessels had been transferred in the graft and become reconnected; they would then be simply expressing their normal phenotype. This issue was resolved by STEWART and WILEY (1981), who confirmed the inductive effect in chick-quail grafts, where differences in nuclear morphology permit

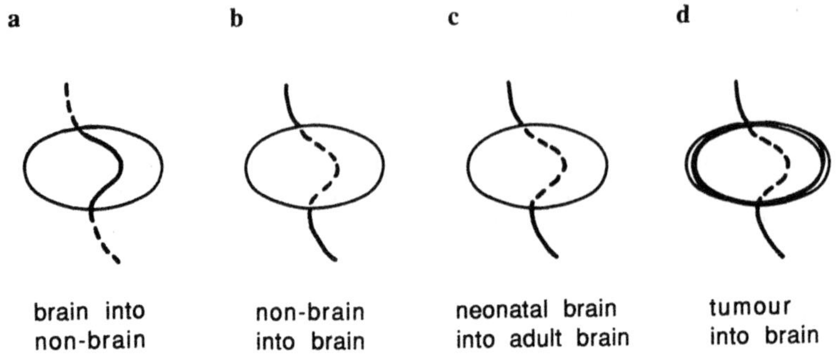

Fig. 4a–d. Summary of grafting experiments and observations in tumours. Vessels are shown outside and inside the graft or tumour, tight vessels indicated by a *solid line*, leaky vessels by a *dashed line*. The experiments provide evidence that the tissue environment around the vessels determines vessel permeability

unequivocal distinction between host and donor endothelial cells. In recent studies, cerebellar grafts into the anterior chamber of the eye in the rat were invaded by iris vessels which developed tight junctions impermeable to HRP, while adjacent iris tissue had no barrier (Nagy and Martinez 1991). As the tight vessels had a complete perivascular ensheathment of astrocytes, an inductive factor from glia was the most likely explanation for the development of tight vessels. However, the presence of occluding junctions in frog pial vessels which lack a glial ensheathment (Bundgaard 1982b) suggests that close contact by glia is unnecessary for induction. The fact that an endothelial barrier fails to develop fully in some grafts of immature into mature CNS (Rosenstein 1987; Fig. 4c) and in some brain tumours (Greig 1987; Fig. 4d) suggests some defect in tissue-endothelial signalling in these situations. However, other studies have reported a relatively normal BBB in some neonatal brain grafts (Broadwell et al. 1991) and some brain tumours (Greig 1987). It would be of great interest to establish whether the tissues that do and do not develop a BBB differ in the putative signals secreted into the interstitial fluid.

II. Implanted Astrocytes

Vessels invading astrocytic clusters implanted into the anterior chamber of the rat eye and on to the chorio-allantois of the chicken developed impermeability to albumin as assessed by light microscopic immunocytochemistry (Janzer and Raff 1987). This confirmed that glial factors alone could have a significant inductive effect on barrier properties. Further physiological and fine-structural studies are needed to establish the tightness of the vessels to small molecules and the extent to which transport properties are also induced.

III. Induction in Culture

1. Differences Between Cultured and In Situ Endothelial Cells

Several groups have successfully cultured brain endothelial cells and determined the extent to which they retain BBB characterisitics. Cultures that express von Willebrand Factor (factor VIII-related antigen) and angiotensin converting enzyme (ACE), reliable markers of endothelial cells, have also been shown to retain the L-system amino acid carrier, tight junctions and transferrin receptor, all characteristic of in situ brain microvessels (Hughes and Lantos 1986; Abbott et al. 1992b). However, cultured preparations generally show reduced expression of γ-glutamayl transpeptidase (γ-GTP) and the glucose carrier (Risau 1991), and the tight junctions are leakier than in situ. These features are candidates for induction by astrocytes.

2. Effects of Astrocytes on the Endothelium

Co-culture of endothelium with astrocytes has been reported to cause reappearance of barrier enzymes (Méresse et al. 1989) and increased complexity and tightness of tight junctions (Arthur et al. 1987; Dehouck et al. 1990; Rubin et al. 1991). Conditioned medium (CM) from astrocytes can at least partly mimic the effect, although the growth substrate (plastic or extracellular matrix) may be an additional determinant. Elevation of cAMP together with addition of CM produces the tightest layer (Rubin et al. 1991), implicating intracellular messengers in the tightening effect.

3. Effects of Endothelial Cells on Astrocytes

Astrocytes at certain locations, particularly the end feet facing the perivascular space, are characterised by orthogonal arrays of intramembraneous particles or "assemblies" (Landis and Reese 1981); their role is unknown, but they have been proposed as the site of high densities of K^+ channels, thought to be important in the "siphoning" of K^+ away from active neurons (Newman 1987) and in the mediation of vasodilatation of cerebral arterioles (Paulson and Newman 1987). Co-culture of astrocytes with endothelial cells causes appearance of a greater density of assemblies on the glial end feet (Tao-Cheng et al. 1990), suggesting inductive influences from the endothelial cells to astrocytes.

IV. Induction in Development

The gradual tightening of the BBB to small solutes during embryonic development is associated with an increasingly complete perivascular astrocytic ensheathment. However, the barrier in the rat shows a further dramatic tightening at the time of birth (from ca. 300 to $>1200\,\Omega\,cm^2$; Butt et al. 1990b) which is not associated with any detectable change in astrocytic morphology. It is possible that barrier induction occurs in at least two stages, an initial stage associated with astrocytic ensheathment, to a resistance of 300–600 $\Omega\,cm^2$ (mimicked in culture), and a second stage to the 1000–5000 $\Omega\,cm^2$ characteristic of the adult. The final tightening may require as yet unidentified signals released at the time of birth.

V. Global Hypothesis of Induction

Vertebrate studies have concentrated on inductive effects between astrocytes and endothelial cells. However, the glial barriers described above pose some problems for any simple inductive model. If glial cells can produce inductive factors, as the mammalian studies suggest, and glial cells in the lower groups have the right receptors to respond by forming a barrier (as in insects, cephalopods), why is a barrier not found throughout the glial population in these groups, but only at the blood-tissue interface? The main distinguishing feature of glial cells at the blood-tissue boundary is that they are exposed to different cellular/fluid environments on their two faces, and their apical/basal polarity is marked. This leads to the suggestion that only the interface glia are able to respond to glial inductive signals, either because a plasma signal is simultaneously required, or because the glial signal is only able to elicit a response in glial cells expressing strong apical/basal polarity (Fig. 5a). Once the endothelial layer becomes sufficiently confluent to develop strong polarity itself, and to reduce access of plasma proteins to the glial layer, the barrier will shift to the endothelium (Fig. 5b). A pericyte barrier (cephalopod arterioles) could represent an intermediate condition. An endothelial barrier may fail to form in the circumventricular organs of the vertebrate brain (e.g., choroid plexus, neurosecretory areas), either because the endothelium here has not developed the usual polarity, or because plasma factors are able to penetrate the leaky vessel walls to act at the next polarised layer, whether choroid plexus epithelium, ependyma, or tanycytes. The wider perspective from comparative studies leads to formulation of additional hypotheses to test in mammalian co-culture experiments.

E. Comparative Approach in Developmental Studies

I. Maturity at Birth

There are major benefits to be gained from studying a variety of species in determining the key features that govern development of the blood-brain barrier. Mammals are born showing a wide range of CNS maturity, from more or less fully mature (sheep, guinea-pig) to deficient in sensory and motor capabilities (rat). Marsupials show an interesting mixture of maturity of forelimb control with rudimentary hind limb control. Some species are particularly suitable for studies in utero (sheep, rat; see Chap. 14).

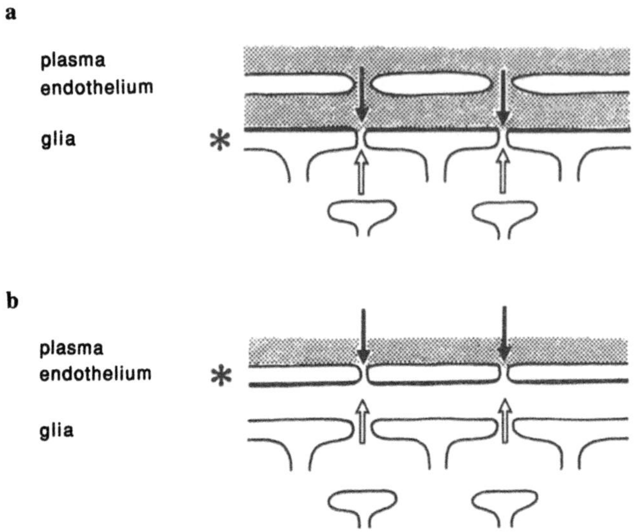

Fig. 5a,b. Global hypothesis for induction of BBB properties. It is proposed that a barrier will form (*) at the layer with most polarised properties (strong apical-basal polarity, basal membrane shown as *thicker line*) under the influence of factors coming from glial cells (*open arrows*) and plasma (*closed arrows*). The barrier will form at the glial layer in invertebrates without a complete endothelium (**a**), but once the endothelium becomes sufficiently confluent to develop strong polarity and restrict penetration of plasma factors, the barrier will shift to the endothelial layer (**b**). A similar argument could explain a pericyte barrier as a transitional stage between **a** and **b**

II. Blood-Brain Barrier Development and Ion Homeostasis

Electrical recording has proved a sensitive way of assessing barrier development in the rat, and the sudden improvement in intracerebral ion homeostasis on the day of birth (shown with ion-sensitive electrodes, JONES and KEEP 1987, 1988) is well correlated with a rise in transendothelial resistance (as measured by cable analysis of pial vessels; BUTT et al. 1990b; Fig. 6). As fine-structural studies show an endothelial mitochondrial volume fraction before birth similar to that of the adult (KEEP and JONES 1990), the improved homeostasis is likely to result from the reduction in the paracellular (tight junctional) shunt, rather than from activation of cellular metabolism and ion transport.

III. Relation of Ontogeny and Phylogeny

Karl Ernst VON BAER (1828) introduced the idea that the general anatomical features seen in all members of a group (e.g., segmentation, limbs) develop early in the embryo, whereas the more special features that distinguish the members of the group (e.g., digits, skin, feathers) develop later. MÜLLER in 1864 and HAECKEL in 1868 added an evolutionary perspective (the "biogenic

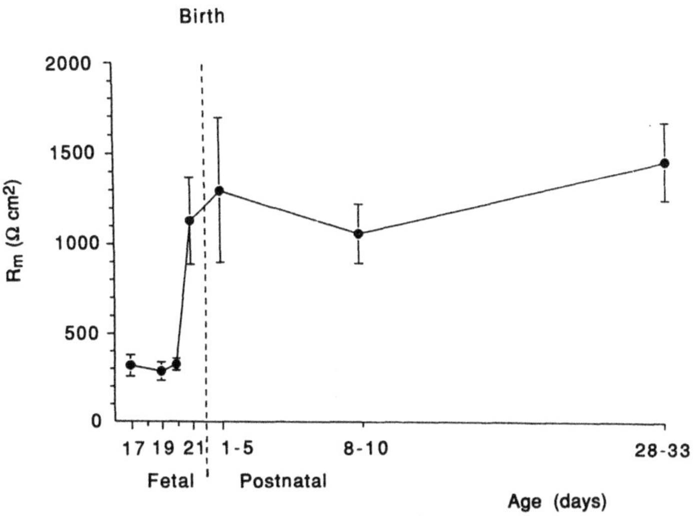

Fig. 6. Transendothelial resistance ($\Omega\,cm^2$) as a function of developmental age in the rat. There is an abrupt increase at the time of birth. (From BUTT et al. 1990b)

law") and proposed that ontogenetic development presents the features of an animal's organisation in the same sequence as they evolved during phylogenetic development; i.e., ontogeny is a recapitulation of phylogeny (BALINSKY 1960). While undoubtedly too simple, the concept has proved useful in explaining the presence of features in the embryo (e.g., gill arches, segmental myotomes) appropriate for our aquatic ancestors, but not used by the modern adult form. There has been little attempt to apply the concept at the cellular level. However, in view of the evidence that a glial barrier was the primitive condition, and an endothelial barrier came later, it is worth examining embryonic forms to see if the same sequence applies.

K. Møllgård and co-workers have studied the development of the neural tube in a range of vertebrates. The superficial glial layer of the developing neural tube, and the ependymal lining of the ventricles, pass through a stage when the glial-ependymal cells are linked first by a rather open striated junctional structure (the "linker" junction), and subsequently by a single-stranded tight junction relatively impermeable to protein (MØLLGÅRD et al. 1987; ABBOTT et al. 1988b; Fig. 7a–d). The linker junction resembles the "matrix" junction between perivascular glial cells in the cephalopod *Sepia*, apparently responsible for the restriction to penetration of electron-dense tracers such as lanthanum and HRP (ABBOTT and BUNDGAARD 1992; ABBOTT et al. 1992a). Thus the early embryo displays a primitive type of junctional stucture, coupling glio-ependymal cells. Only with vascularisation of the brain does the glial junction disappear (except in specialised parts of the ependyma, see above, where it is superseded by the full adult *zonula occludens*), to be replaced by tight junctions between endothelial cells (Fig. 7e). In BBB development, ontogeny appears to repeat phylogeny.

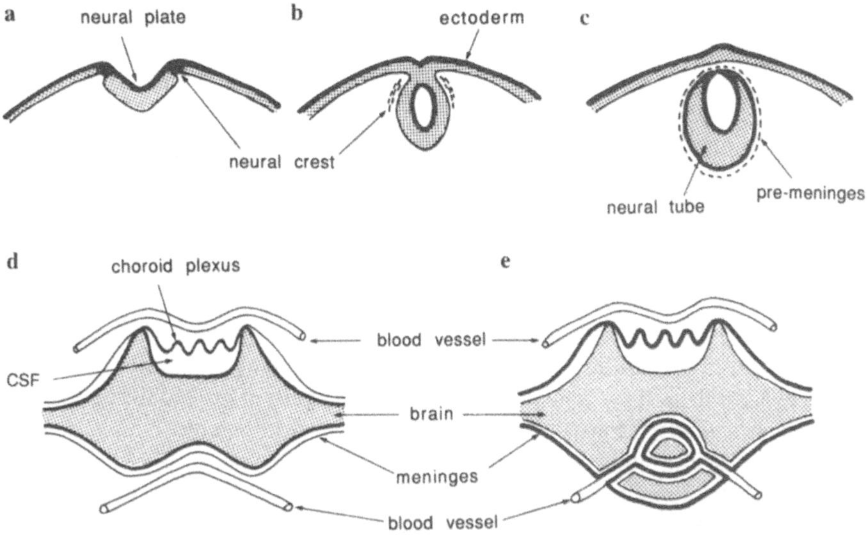

Fig. 7a–e. Diagram to summarise the shift in barrier layers during mammalian neural tube ontogeny. Barrier layers are drawn as *heavy lines*, leaky layers as *thin lines*. **a** Neural plate; the *dark stippled areas* will form neural crest. **b** Closure of neural tube and neural crest migration. **c** Separation of neural tube from remaining ectoderm. Neural tube develops glial/glioependymal barrier layers at outer and inner surfaces. Neural crest contributes to developing meninges (*outer dotted line*). **d** Leaky blood vessels associated with the surface of the neural tube. Outer glial layer and glio-ependyma (including choroid plexus) are relatively tight. CSF high in protein. **e** Stage of vascularisation: vessels entering brain become tight, glial and glio-ependymal barrier lost except in choroid plexus. CSF low in protein. At brain surface, barrier shifts to meninges (arachnoid). The figure shows that in ontogeny, as in phylogeny, the barrier is first glial, later endothelial. (Fig. 7e modified from CSERR and BUNDGAARD 1984)

F. Permeability and Transport Properties of Barrier Layers

I. The Vertebrate Endothelial Barrier

The endothelial barrier of the mammal has been very thoroughly characterised, as described in this volume. Briefly, the luminal membrane possesses a carrier for glucose, at least three types of amino acid carrier (L-system, basic and ASC), carriers for nucleosides, and for certain peptides. The abluminal membrane contains an electrogenic Na^+-K^+-ATPase, carboxylic acid carrier, and Na^+-dependent A system amino acid carrier.

II. The Vertebrate Choroid Plexus

The choroid plexus shows a glucose transporter that is Na^+-independent on the blood side (DEANE and SEGAL 1985), Na^+-dependent on the brain side, and an apical (brain facing) Na^+-K^+-ATPase (Wright and Saito 1986). The luminal (basal) membrane also shows amino acid transporters including the L-system carrier, no A-system carrier, but significant uptake of glycine (Segal and Zloković 1988; Preston et al. 1989). The choroid plexus may also be the primary site for brain entry of vitamins (Spector 1986).

III. The Elasmobranch Glial Barrier

The brain uptake index (BUI) technique of Oldendorf (1970) has been applied to the glial blood-brain barrier of the dogfish. A Na^+-independent glucose carrier is present on the abluminal membrane (Abbott et al. 1988a), similar, but not identical to that of the mammalian endothelial barrier (e.g., less inhibition by phloretin; cytochalasin B no effect). Microelectrode experiments demonstrate a K^+-selectivity and electrogenic Na^+-pump on the brain-facing membrane (Abbott and Butt 1986a; Fig. 8), and isotope studies show a luminal permeability to K^+ that is ca. 17 times greater than that to Na^+ (Abbott and Butt 1986b), compared to 5–10 times greater for mammalian brain endothelium). In addition to luminal L-system and probably abluminal A-system carriers, the glial barrier shows prominent uptake of glycine (Abbott et al. 1988c), resembling that other glio-ependymal derivative, the choroid plexus (see above). Unlike the vertebrate endothelium, there is marked uptake of glutamate and γ-aminobutyric acid (GABA), showing that the glial barrier retains some glial characteristics (Abbott and Hart 1988). It is likely that these neurotransmitter amino acids are prevented from entering into brain by metabolism within the barrier glia (Abbott 1991).

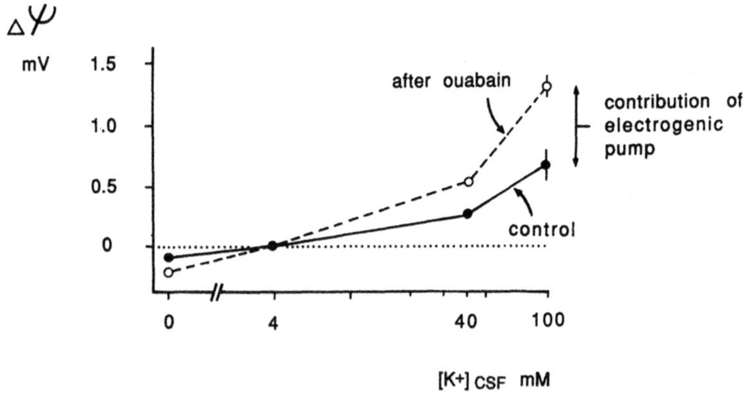

Fig. 8. Microelectrode study of subependymal vessels in elasmobranch brain, where barrier is at glial level. The change in vessel wall potential (ψ, mV) with variation in bathing potassium concentration $[K^+]_{CSF}$ reflects the K^+ selectivity of the glial membrane; the effect of ouabain provides evidence for an abluminal electrogenic Na^+-K^+-ATPase. (From Butt 1986)

IV. The Cephalopod Glial Barrier

At the capillary level in *Sepia* brain, low permeability is achieved by a combination of a "seamless" perivascular glial layer covering ca. 90% of the microvascular tree, together with tortuous clefts and "matrix junctions" in the remainder (BUNDGAARD and ABBOTT 1992; ABBOTT and BUNDGAARD 1992). The luminal membrane has a Na^+-independent glucose carrier which, like that of the elasmobranch glial barrier, is less sensitive to phloretin than phloridzin (ABBOTT and PICHON 1987). There is measurable uptake of glycine, taurine and β-alanine, again glial properties.

V. Crustacean and Insect Glial Barriers

The glial perineurium forming the blood-nervous tissue barrier in these groups shows a transepithelial resistance of ca. 300–900 $\Omega\,cm^2$, and selective permeability of the outwardly facing membrane that may be switched on by depolarisation – by either a calcium-dependent or voltage-dependent rise in potassium conductance (BUTT et al. 1990a; BUTT 1991). Cooling and inhibitor studies in crustaceans provide evidence for a brain-facing electrogenic Na^+-K^+-ATPase, and a blood-facing Na^+-K^+-Cl^- co-transporter (ABBOTT and PICHON 1987). Tight junctions have not been described in the crustacean glial BBB (LANE and ABBOTT 1975; LANE et al. 1977), but may comprise discontinuous strands as in the peripheral nerves (VILLEGAS and SANCHEZ 1991), together with a relatively restricting fibre matrix. The tight junctions of the insect barrier are relatively simple networks of intramembranous particle strands (LANE 1981).

VI. Comparison of Endothelial and Glial Barriers

1. Properties Common to Barrier Layers

From this survey of barriers in different animal groups, it can be seen that the blood-facing membrane of the BBB is characterised by glucose and L-system amino acid carriers, and the brain-facing membrane by an electrogenic Na^+-K^+-ATPase (Fig. 9a,b). The latter may provide the driving force for secretion of interstitial fluid. An abluminal A-system amino acid carrier may subserve (Na^+-dependent) amino acid efflux. The barrier permeability is low, corresponding to a resistance of 300–3000 $\Omega\,cm^2$, and a permeability to small non-electrolytes like sucrose of ca. $1 \times 10^{-8}\,cm\,s^{-1}$.

2. Differences of Endothelial and Glial Barriers

The low permeability of the barrier is achieved in vertebrate endothelial and glial barriers by multistranded *zonula occludens*, but in invertebrates by more varied means including simple tight junctions, seamless glial layers, tortuous clefts and "matrix" junctions. Both endothelial and glial barriers

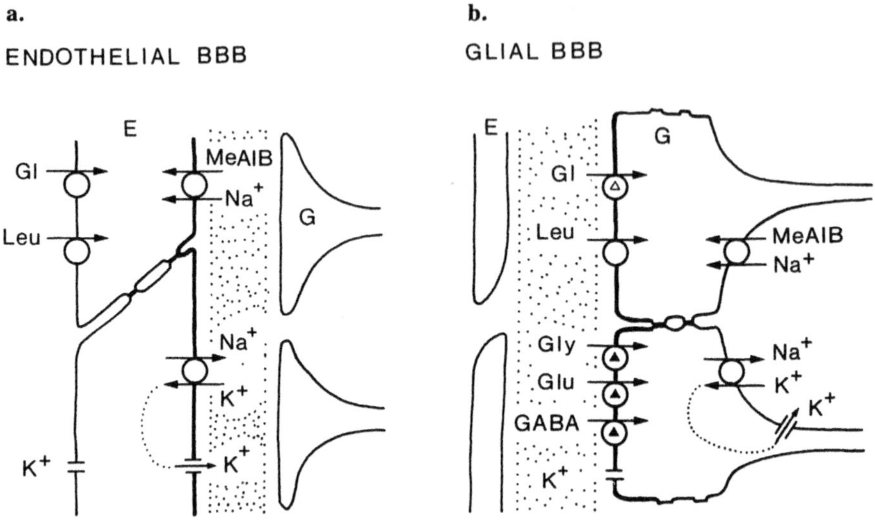

Fig. 9a,b. Comparison of transport mechanisms at vertebrate **a** endothelial (*E*) and **b** glial (*G*) barriers. Both have carriers for glucose (*Gl*) and for neutral amino acids (Leu, L-system) on the luminal membrane, Na^+-K^+-ATPase and A-system amino acid carrier (*MeAIB*) on the abluminal membrane. The glial barriers show additional glial-type carriers for glycine (*Gly*), GABA and glutamate (*Glu*)

appear able to restrict penetration of neuroactive agents such as transmitters. For glutamate and GABA this is achieved by a low density of luminal carrier molecules on endothelial barriers, but by cellular uptake and metabolism in glial barriers (Fig. 9b). The glucose carrier of the glial barriers shows some differences in substrate and inhibitor specificity, which could point to interesting differences in molecular configuration of the carrier.

G. Relation of Interstitial Fluid and Cerebrospinal Fluid

I. Vertebrate Brain

H.F. Cserr has reviewed the evidence for bulk flow of vertebrate ISF (Chap. 9), and although the presence of the ventricles and CSF compartment makes quantitation difficult, a secretion of ISF at approximately $0.1\,\mu l\,g^{-1}\,min^{-1}$ has been calculated for the rat, around one tenth of the rate of CSF secretion. The vertebrate brain interstitial space seems to be organised as a system of low resistance fluid pathways in which bulk flow occurs (predominantly in perivascular spaces, and in the subependymal zone), in parallel with higher resistance routes through the interstitium, which, at least at the microscopic level (few hundred micrometers), behaves relatively isotropically (i.e., indistinguishable from diffusion, NICHOLSON and RICE 1986). In the presence of oedema, additional routes for bulk flow may open up. The drainage of ISF in vertebrates is partly into the CSF system, partly along cranial nerve sheaths into lymphatics.

II. Cephalopod Brain

In the cuttlefish *Sepia*, with no ventricles or CSF, so that all ISF must be derived from blood vessels, ^{14}C-labelled polyethylene glycol (4 kDa) and ^{51}-Cr-labelled ethylenediaminetetraacetic acid (EDTA) (340 Da) introduced into the blood reach steady-state concentrations in brain interstitium of 50% and 65% of their plasma concentrations respectively (ABBOTT et al. 1985). This distribution is difficult to explain if ISF is static, but is that expected if ISF is secreted at a rate of ca. $0.1 \mu l\, g^{-1} min^{-1}$, similar to the vertebrate figure. Drainage of ISF in *Sepia* is likely to be into the veins in the low pressure side of the system. "Gliovascular" channels, prominent in the optic lobe and retinal nerves, contain myocyte-like cells and apparent ciliary profiles (BUNDGAARD and ABBOTT, unpublished); these channels may represent the invertebrate equivalent of lymphatics for drainage of ISF, the motile cells having the same function as the pulsations produced by vertebrate vessels in massaging fluid along the extracellular pathways.

H. Summary and Conclusions

Comparative studies offer a range of novel experimental possibilities for tackling questions concerning the brain microenvironment. The survey of anatomical arrangements and mechanisms, in the adult and in the embryo, permits deductions about phylogenetic and ontogenetic development. When it can be demonstrated that different cellular arrangements produce the same functional result (e.g., endothelial versus glial barriers), this underlines the generality of the phenomenon. Systems that are geometrically simpler than in mammals (e.g., brain fluids in cephalopods) permit separation of subcomponents of a regulatory process. A comparative approach helps in the development and testing of hypotheses concerning the mechanisms at work, and leads to insights that would not be possible from the study of mammalian forms alone.

References

Abbott NJ (1991) Permeability and transport of glial blood-brain barriers. Ann NY Acad Sci 633:378–393

Abbott NJ, Bundgaard M (1992) Electron-dense tracer evidence for a blood-brain barrier in the cuttlefish *Sepia officinalis*. J Neurocytol 21:276–303

Abbott NJ, Butt AM (1986a) A microelectrode study of K^+ transport at the dogfish blood-brain barrier. J Physiol (Lond) 374:39P

Abbott NJ, Butt AM (1986b) Permeability of the dogfish glial blood-brain barrier to ions and small molecules. J Physiol (Lond) 377:103P
Abbott NJ, Hart J (1988) Glutamate and GABA transport by the glial blood-brain barrier of the dogfish. J Physiol (Lond) 407:141P
Abbott NJ, Pichon Y (1987) The glial blood-brain barrier of Crustacea and cephalopods: a review. J Physiol (Paris) 82:304–313
Abbott NJ, Revest PA (1991) Control of brain endothelial permeability. Cerebrovasc Brain Metab Rev 3:1–34
Abbott NJ, Moreton RB, Pichon Y (1975) Electrophysiological analysis of potassium and sodium movements in crustacean nervous system. J Exp Biol 63:85–115
Abbott NJ, Bundgaard M, Cserr HF (1985) Tightness of the blood-brain barrier and evidence for brain interstitial fluid flow in the cuttlefish, *Sepia officinalis*. J Physiol (Lond) 368:213–226
Abbott NJ, Bundgaard M, Cserr HF (1986a) Comparative physiology of the blood-brain barrier. In: Suckling AJ, Rumsby MG, Bradbury MWB (eds) The blood-brain barrier in health and disease. Horwood, Chichester, pp 52–72
Abbott NJ, Lane NJ, Bundgaard M (1986b) The blood-brain interface in invertebrates. Ann NY Acad Sci 481:20–41
Abbott NJ, Butt AM, Joels S, Pitchford S, Wallis W (1986c) Na^+-K^+-transport at the blood-brain barrier – a microelectrode study in the frog. J Physiol (Lond) 372:68P
Abbott NJ, Zloković BV, Butt AM (1988a) Methods for study of the blood-brain barrier in non-mammalian species. In: Rakic L, Begley DJ, Davson H, Zloković BV (eds) Peptide and amino acid transport mechanisms in the central nervous system. Macmillan, London, Serbian Academy of Sciences and Arts, Belgrade, pp 289–303
Abbott NJ, Bundgaard M, Lane NJ, Møllgård K (1988b) Parallels between junctions in invertebrate brain and embryonic mammalian brain. J Physiol (Lond) 400:72P
Abbott NJ, Zloković BV, Taylor M, Hart J, Rogac L (1988c) Amino acid transport by a glial blood-brain barrier: studies in an elasmobranch fish. In: Rakic L, Begley DJ, Davson H, Zloković BV (eds) Peptide and amino acid transport mechanisms in the central nervous system. Macmillan, London, Serbian Academy of Sciences and Arts, Belgrade, pp 241–244
Abbott NJ, Lane NJ, Bundgaard M (1992a) A fibre-matrix model for the restricting junction of the blood-brain barrier in a cephalopod mollusc: implications for capillary and epithelial permeability. J Neurocytol 21:304–311
Abbott NJ, Hughes CCW, Revest PA, Greenwood J (1992b) Development and characterization of a rat brain capillary endothelial culture: towards an in vitro blood-brain barrier. J Cell Sci (in press)
Arthur FE, Shivers RR, Bowman PD (1987) Astrocyte-mediated induction of tight junctions in brain capillary endothelium: an efficient in vitro model. Dev Brain Res 36:155–159
Balin BJ, Broadwell RD, Salcman M, El-Kalliny M (1986) Avenues of entry of peripherally administered protein to the CNS in mouse, rat, and squirrel monkey. J Comp Neurol 251:260–280
Balinsky BI (1960) An introduction to embryology. Saunders, Philadelphia
Bradbury MWB (1979) The concept of a blood-brain barrier. Wiley, Chichester
Bradbury MWB (1989) Transport across the blood-brain barrier. In: Neuwelt EA (ed) Implications of the blood-brain barrier and its manipulation, vol 1, Basic science aspects. Plenum, New York, pp 119–136
Brightman MW, Reese TS (1969) Junctions between intimately apposed cell membranes in the vertebrate brain. J Cell Biol 40:648–677
Brightman MW, Reese TS, Olsson Y, Klatzo I (1971) Morphologic aspects of the blood-brain barrier to peroxidase in elasmobranchs. Prog Neuropathol 1: 146–161
Broadwell RD, Salcman M (1981) Expanding the definition of the blood-brain barrier to protein. Proc Natl Acad Sci USA 78:7820–7824

Broadwell RD, Balin BJ, Salcman M (1988) Transcytosis pathway for blood-borne protein through the blood-brain barrier. Proc Natl Acad Sci USA 85:632–636
Broadwell RD, Charlton HM, Ebert P, Hickey WF, Shirazi Y, Villegas J, Wolf AL (1991) Allografts of CNS tissue possess a blood-brain barrier. I. Angiogenesis in solid tissue and cell suspension grafts. Exp Neurol 112:1–28
Brown ER (1991) Axon-Schwann cell interaction in the squid. PhD thesis, University of London
Brown ER, Bone Q, Abbott NJ, Ryan KP (1991) Morphology and electrical properties of Schwann cells around the giant axon of the squids *Loligo forbesi* and *Loligo vulgaris*. Proc R Soc Lond B 243:255–262
Bundgaard M (1982a) Brain barrier systems in the lamprey. I. Ultrastructure and permeability of cerebral blood vessels. Brain Res 240:55–64
Bundgaard M (1982b) Ultrastructure of frog cerebral and pial microvessels and their permeability to lanthanum ions. Brain Res 241:57–65
Bundgaard M (1982c) The ultrastructure of cerebral blood capillaries in the ratfish *Chimaera monstrosa*. Cell Tissue Res 226:145–154
Bundgaard M, Abbott NJ (1992) Fine structure of the blood-brain interface in the cuttlefish *Sepia officinalis* (Mollusca, Cephalopoda). J Neurocytol 21:260–275
Bundgaard M, Cserr HF (1981a) Impermeability of hagfish cerebral capillaries to radiolabeled polyethylene glycols and to microperoxidase. Brain Res 206:71–81
Bundgaard M, Cserr HF (1981b) A glial blood-brain barrier in elasmobranchs. Brain Res 226:61–73
Bundgaard M, Cserr HF, Murray M (1979) Impermeability of hagfish cerebral capillaries to horseradish peroxidase. Cell Tissue Res 198:65–77
Butt AM (1986) Comparative physiology of the blood-brain barrier in the frog, *Rana pipiens* and *Rana temporaria*, and the dogfish, *Scyliorhinus canicula*. PhD thesis, University of London
Butt AM (1991) Modulation of a glial blood-brain barrier. Ann NY Acad Sci 633: 363–377
Butt AM, Hargittai PT, Lieberman EM (1990a) Calcium-dependent regulation of potassium permeability in the glial perineurium (blood-brain barrier) of the crayfish. Neuroscience 38:175–185
Butt AM, Jones HC, Abbott NJ (1990b) Electrical resistance across the blood-brain barrier in anaesthetized rats: a developmental study. J Physiol (Lond) 429:47–62
Crone C (1965) Facilitated transfer of glucose from blood to brain. J Physiol (Lond) 181:103–113
Crone C (1984) Lack of selectivity to small ions in paracellular pathways in cerebral and muscle capillaries of the frog. J Physiol (Lond) 353:317–337
Crone C, Olesen S-P (1982) Electrical resistance of brain capillary endothelium. Brain Res 241:49–55
Cserr HF, Bundgaard M (1984) Blood-brain interfaces in vertebrates: a comparative approach. Am J Physiol 246:R277–R288
Cserr HF, Bundgaard M (1986) The neuronal microenvironment: a comparative view. Ann NY Acad Sci 481:1–6
Deane BR, Greenwood J, Lantos PL, Pratt OE (1984) The vasculature of experimental brain tumours. Part 4. The quantification of vascular permeability. J Neurol Sci 65:59–68
Deane R, Segal MB (1985) The transport of sugars across the perfused choroid plexus of the sheep. J Physiol (Lond) 362:245–260
Dehouck M-P, Méresse S, Delorme P, Fruchart J-C, Cecchelli R (1990) An easier, reproducible, and mass-production method to study the blood-brain barrier in vitro. J Neurochem 54:1798–1801
Edwards JS, Swales LS, Bate CM (1991) Development of non-neural elements in the central nervous system of *Drosophila*. Ann NY Acad Sci 633:617–618
Ehrlich P (1885) Das Sauerstoffbedürfnis des Organismus. Eine farbenanalytische Studie. Hirschwald, Berlin

Goldmann EE (1909) Die äussere und innere Sekretion des gesunden und kranken Organismus im Lichte der "vitalen Färbung." Beitr Klin Chir 64:192–265
Goldmann EE (1913) Vitalfärbung am Zentralnervensystem. Abh Preuss Akad Wiss Phys Math Kl 1:1–60
Greig NH (1987) Optimizing drug delivery to brain tumours. Cancer Treat Rev 14:1–28
Harrison JB, Lane NJ (1981) Lack of restriction at the blood-brain-interface in *Limulus* despite atypical junctional arrangements. J Neurocytol 10:233–250
Hughes CCW, Lantos PL (1986) Brain capillary endothelial cells lack surface IgG Fc receptors. Neurosci Lett 68:181–186
Janzer RC, Raff MC (1987) Astrocytes induce blood-brain barrier properties in endothelial cells. Nature 325:253–257
Jones HC, Keep RF (1987) The control of potassium concentration in the cerebrospinal fluid and brain interstitial fluid of developing rats. J Physiol (Lond) 383:441–453
Jones HC, Keep RF (1988) Brain fluid calcium concentration and response to acute hypercalcaemia during development in the rat. J Physiol 402:579–593
Keep RF, Jones HC (1990) Cortical microvessels during brain development: a morphometric study in the rat. Microvasc Res 40:412–426
Krebs HA (1975) The August Krogh principle: "For many problems there is an animal on which it can be most conveniently studied". J Exp Zool 194:221–226
Krogh A (1929) Progress of physiology. Am J Physiol 90:243–251
Landis DMD, Reese TS (1981) Membrane structure in mammalian astrocytes: a review of freeze-fracture studies on adult, developing, reactive and cultured astrocytes. J Exp Biol 95:35–48
Lane NJ (1981) Tight junctions in arthropod tissues. Int Rev Cytol 73:243–318
Lane NJ, Abbott NJ (1975) The organization of the nervous system in the crayfish *Procambarus clarkii*, with emphasis on the blood-brain interface. Cell Tissue Res 156:173–187
Lane NJ, Campiglia SS (1987) The lack of a structured blood-brain barrier in the onycophoran *Peripatus acacioi*. J Neurocytol 16:93–104
Lane NJ, Treherne JE (1972) Studies on the perineurial junctional complexes and the sites of uptake of microperoxidase and lanthanum in the cockroach central nervous system. Tissue Cell 4:427–436
Lane NJ, Treherne JE (1973) The ultrastructural organization of peripheral nerves in two insect species (*Periplaneta americana* and *Schistocerca gregaria*). Tissue Cell 5:703–714
Lane NJ, Swales L, Abbott NJ (1977) Lanthanum penetration in crayfish nervous system; observations on intact and "desheathed" preparations. J Cell Sci 23:315–324
Leng G, Shibuki K, Way SA (1988) Effects of raised extracellular potassium on the excitability of, and hormone release from, the isolated rat neurohypophysis. J Physiol (Lond) 399:591–605
Lettau J, Foster WA, Harker JE, Treherne JE (1977) Diel changes in potassium activity in the haemolymph of the cockroach *Leucophaea maderae*. J Exp Biol 71:171–186
Lieberman EM, Abbott NJ, Hassan S (1989) Evidence that glutamate mediates axon-to-Schwann cell signaling in the squid. Glia 2:94–102
Méresse S, Dehouck M-P, Delorme P, Bensaïd M, Tauber JP, Delbert C, Fruchart J-C, Cecchelli R (1989) Bovine brain endothelial cells express tight-junctions and monoamine oxidase activity in long-term culture. J Neurochem 53: 1363–1371
Mirolli M, Gorman ALF (1973) The extracellular space of a simple molluscan nervous system and its permeability to potassium. J Exp Biol 58:423-435
Møllgård K, Balslev Y, Lauritzen B, Saunders NR (1987) Cell junctions and membrane specializations in the ventricular zone (germinal matrix) of the developing sheep brain: a CSF-brain barrier. J Neurocytol 16:433–444

Nabeshima S, Reese TS, Landis DMD, Brightman MW (1975) Junctions in the meninges and marginal glia. J Comp Neurol 164:127–170

Nagy Z, Martinez K (1991) Astrocytic induction of endothelial tight junctions. Ann NY Acad Sci 633:395–404

Nagy Z, Peters H, Hüttner I (1984) Fracture faces of cell junctions in cerebral endothelium during normal and hyperosmotic conditions. Lab Invest 450: 313–322

Newman EA (1987) Distribution of potassium conductance in mammalian Müller (glial) cells: a comparative study. J Neurosci 7:2423–2432

Nicholson C, Rice ME (1986) The migration of substances in the neuronal microenvironment. Ann NY Acad Sci 481:55–71

Oldendorf WH (1970) Measurement of brain uptake of radiolabeled substances using a tritiated water internal standard. Brain Res 24:372–376

Olesen S-P (1989) An electrophysiological study of microvascular permeability and its modulation by chemical mediators. Acta Physiol Scand [Suppl 579] 136: 1–28

Paulson OB, Newman EA (1987) Does the release of potassium from astrocytic endfeet regulate cerebral blood flow? Science 237:896–898

Preston JE, Segal MB, Walley GJ, Zloković BV (1989) Neutral amino acid uptake by the isolated perfused sheep choroid plexus. J Physiol (Lond) 408:31–43

Reese TS, Karnovsky MJ (1967) Fine-structural localization of a blood-brain barrier to exogenous peroxidase. J Cell Biol 34:207–217

Rennels ML, Gregory TF, Blaumanis OR, Fujimoto K, Grady PA (1985) Evidence for a "paravascular" fluid circulation in the mammalian central nervous system, provided by the rapid distribution of tracer protein throughout the brain from the subarachnoid space. Brain Res 326:47–63

Risau W (1991) Induction of blood-brain barrier endothelial cell differentiation. Ann NY Acad Sci 633:405–419

Rosenstein JM (1987) Neocortical transplants in the mammalian brain lack a blood-brain barrier to macromolecules. Science 235:772–774

Rubin LL, Barbu K, Bard F, Cannon C, Hall DE, Horner H, Janatpour M, Liaw C, Manning K, Morales J, Porter S, Tanner L, Tomaselli K, Yednock T (1991) Differentiation of brain endothelial cells in cell culture. Ann NY Acad Sci 633: 420–425

Sattelle DB (1973) Potassium movements in a central nervous ganglion of *Lymnaea stagnalis* (L.) (Gastropoda: Pulmonata). J. Exp Biol 58:15–28

Sattelle DB, Lane NJ (1972) Architecture of gastropod central nervous tissue in relation to ionic movements. Tissue Cell 4:253–270

Schofield PK, Treherne JE (1984) Localization of the blood-brain barrier of an insect: electrical model and analysis. J Exp Biol 109:255–267

Segal MB, Zloković BV (1988) Factors which influence the concentration of amino acids in cerebrospinal fluid. In: Rakić L, Begley DJ, Davson H, Zloković BV (eds) Peptide and amino acid transport mechanisms in the central nervous system. Macmillan, London, Serbian Academy of Sciences and Arts, Belgrade, pp 229–239

Smith DO (1980a) Mechanisms of action potential propagation failure at sites of axon branching in the crayfish. J Physiol (Lond) 301:243–259

Smith DO (1980b) Morphological aspects of the safety factor for action potential propagation at axon branch points in the crayfish. J Physiol (Lond) 301:261–269

Spector R (1986) Nucleoside and vitamin homeostasis in the mammalian central nervous system. Ann NY Acad Sci 481:221–229

Spira ME, Yarom Y, Zeldes D (1984) Neuronal interactions mediated by neurally evoked changes in the extracellular potassium concentration. J Exp Biol 112:179–197

Stewart PA, Wiley MJ (1981) Developing nervous tissue induces formation of blood-brain barrier characteristics in invading endothelial cells: a study using quail-chick transplantation chimeras. Dev Biol 84:183–192

Svendgaard NA, Bjorklund A, Hardebo JE, Stenevi U (1975) Axonal degeneration associated with a defective blood-brain barrier in cerebral implants. Nature 255:334–336

Syková E (1987) Modulation of spinal cord transmission by changes in extracellular K^+ activity and extracellular volume. Can J Physiol Pharmacol 65:1058–1066

Tao-Cheng J-H, Nagy Z, Brightman MW (1987) Tight junctions of brain endothelium in vitro are enhanced by astroglia. J Neurosci 7:3293–3299

Tao-Cheng J-H, Nagy Z, Brightman MW (1990) Astrocytic orthogonal arrays of intramembranous particle assemblies are modulated by brain endothelial cells in vitro. J Neurocytol 19:143–153

Treherne JE, Pichon Y (1972) The insect blood-brain barrier. Adv Insect Physiol 9:257–313

Treherne JE, Schofield PK, Lane NJ (1982) Physiological and ultrastructural evidence for an extracellular anion matrix in the central nervous system of an insect *Periplaneta americana*. Brain Res 247:255–267

Villegas G, Sanchez F (1991) The periaxonal ensheathment of lobster giant nerve fibers as revealed by freeze-fracture and lanthanum penetration. J Neurocytol 20:504–517

Villegas J (1984) Axon-Schwann cell relationship. Curr Top Membr Transp 22: 547–571

Villegas J, Evans PD, Reale V (1988) Electrophysiology of Schwann cell receptors. In: Kimelberg HK (ed) Glial cell receptors. Raven, New York

Von Baer KE (1828) Ueber Entwicklungsgeschichte der Tiere, Beobachtung und Reflexion. Königsberg

Wright EM, Saito Y (1986) The choroid plexus as a route from blood to brain. Ann NY Acad Sci 481:214–219

Yudilevich DL, Wheeler CPD, Bustamante JC (1988) A comparative view of amino acid transport across the blood-brain barrier (endothelium) and the placenta (trophoblast). In: Rakic L, Begley DJ, Davson H, Zloković BV (eds) Peptide and amino acid transport mechanisms in the central nervous system. Macmillan, London, Serbian Academy of Sciences and Arts, Belgrade, pp 245–266

Zloković BV, Begley DJ, Segal MB, Davson H, Rakic L, Lipovac N, Mitrović DM, Kankov RM (1988) Peptide interactions with the blood-brain-CSF interfaces. In: Rakić L, Begley DJ, Davson H, Zloković BV (eds) Peptide and amino acid transport mechanisms in the central nervous system. Macmillan, London, Serbian Academy of Sciences and Arts, Belgrade, pp 3–19

CHAPTER 16

Immunology of Brain Endothelium and the Blood-Brain Barrier

D.K. MALE

A. Introduction

The central nervous system (CNS) is partly shielded from the effector arms of the immune system. For example, a proportion of allogeneic grafts will survive in the CNS, whereas they would be rapidly rejected if implanted into skin. This protection is largely due to functions of the specialised cerebral endothelium, and the critical factors in this respect are:

1. Very low levels of leucocyte migration into normal CNS.
2. Low levels of immunologically important molecules, such as immunoglobulin and complement, in interstitial fluids and cerebrospinal fluid (CSF).
3. Limited access of antigens from the brain to the general immune system.

Although these conditions prevail in normal circumstances, the shielding is only partial. Consequently, immune reactions may develop in the brain in diseases such as multiple sclerosis (MS). In these cases there is usually an increased flux of serum molecules across the endothelium and this is often accompanied by a greatly increased migration of cells, predominantly lymphocytes and mononuclear phagocytes. This chapter reviews the factors which control access of leucocytes and serum molecules to the brain dur-

---

Abbreviations

APC	Antigen presenting cell
CSF	Cerebrospinal fluid
CNS	Central Nervous System
CREAE	Chronic relapsing experimental allergic encephalomyelitis
EAE	Experimental allergic encephalomyelitis
ELAM	Endothelial leucocyte adhesion molecule
ICAM-1	Intercellular adhesion molecule 1
IFN	Interferon
IgG	Immunoglobulin G
IL	Interleutin
LFA-3	Lymphocyte functional antigen 3
MBP	Myelin basic protein
MHC	Major histocompatibility complex
MS	Multiple sceerosis
PAF	Platelet activating factor
SMC	Smooth muscle cell
SSPE	Subacute sclerosing panencephalitis
TNF	Tumour necrosis factor
VCAM	Vascular cell adhesion molecule

ing disease. Once lymphocytes have entered the brain, immune reactions develop by their interaction with microglia and other parenchymal cells, but the initial point of lymphocyte contact and entry is the endothelium.

B. Molecular Permeability

Important molecules which control the development of immune reactions are immunoglobulins, cytokines, and molecules of the plasma enzyme systems – complement, kinin, plasmin and clotting systems. The great majority of these have molecular weights >15 kDa and therefore are present at very low levels in normal brain parenchyma. For example, the level of immunoglobulin G (IgG) is normally much less than 1% of the level in corresponding sera, and this is also true of most complement system molecules (Table 1). In the absence of any specific transport system for these molecules, the permeability of the normal barrier is (approximately) inversely related to the molecular weight of the protein. In encephalitis there is an increase in barrier permeability to these large serum molecules, but even in the most severe cases of encephalitis, immunoglobulin levels rarely exceed 3% of the level found in serum (LEIBOWITZ and HUGHES 1983).

I. Antibodies

Antibodies in cerebral interstitial fluids deserve special consideration, since they may come either from the serum, or be produced by local synthesis. Locally synthesised antibodies are derived from B cells and plasma cells which have migrated into the CNS during inflammatory reactions, and can continue to produce their specific antibodies over an extended period. This occurs in diseases as diverse as MS, viral encephalitis and subacute sclerosing panencephalitis (SSPE). When these CSF immunoglobulins are examined by isoelectric focusing, they are found to be oligoclonal, i.e. the progeny of a small number of B-cell clones. Within a single individual the profile of antibodies in the CSF can be remarkably stable over a course of months or even years. This suggests that particular B-cell clones remain

Table 1. Immunoglobulins and complement components in CSF[a]

Molecule	Mol. wt.	CSF conc. (μg/ml)	CSF/serum (%)
IgG	160000	20–40	0.16–0.32
IgA	170000	0.8–2.0	0.03–0.77
IgM	960000	0.05–0.25	0.005–0.025
C3	185000	4.6–14	0.41–1.07
C4	205000	0.9–4.0	0.15–0.67

[a] Based on LEIBOWITZ and HUGHES (1983) and references therein.

active within the brain parenchyma over an extended period. Antibodies of the same clonotype are also found in the serum of the individual patients, which suggests that the B cells which have penetrated the CNS are a random, small selection from those available in the circulation. In diseases initiated by viral infection the antibodies are usually directed to the specific virus, but the specificities in MS are diverse and uninformative.

It is possible to distinguish the contribution of local antibody synthesis from that due to altered barrier permeability, because there is a defined relationship between permeability of the barrier to different serum molecules. Serum albumin distribution (CSF/serum ratio) can be used as a measure of barrier permeability. If it is found that the IgG CSF/serum ratio is higher than can be accounted for by this degree of permeability, then this is evidence of local antibody synthesis.

II. Routes of Molecular Movement

There are four potential ways in which large serum molecules, such as antibody, can enter CNS during inflammatory reactions (Fig. 1):

1. By an increase in vesicular shuttling across the endothelium,
2. by the opening of continuous channels through the endothelium,
3. by separation of tight junctions between the cells,
4. by toxic or cytotoxic damage to the cells.

Although, it is clear that barrier permeability does alter in immunologically mediated diseases of the brain, the relative contributions of each of these mechanisms is still unclear (Broadwell 1989).

Large molecules, such as horseradish peroxidase, are taken up at the serosal side of the endothelium in vivo and are found in endosomes some

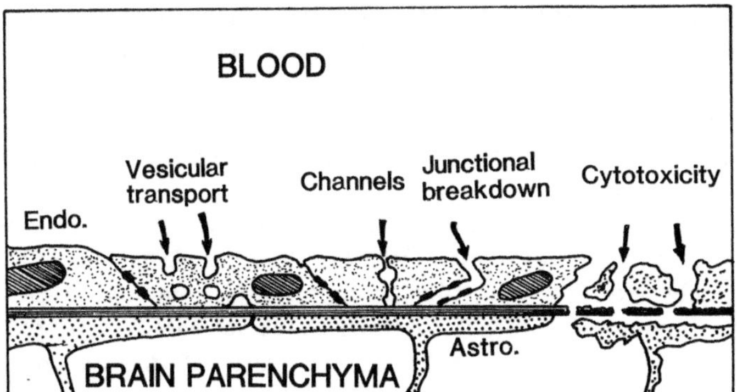

Fig. 1. Four potential mechanisms for increased movement of serum molecules across brain endothelium during immune reactions. In most cases, even in severe immune reactions, tight junctions remain closed, and currently vesicular transport is thought to be of greatest importance

hours later (BALIN et al. 1986). In experimental allergic encephalomyelitis (EAE), electron microscopic studies show that thorotrast is transported from the luminal to the abluminal surface of the endothelium, and this is thought to be mediated by vesicular transport. There is also evidence that increased vesicular transport occurs during EAE in rats. For example, there is a fivefold increase in the numbers of vesicles seen in the endothelium at the height of the encephalitis, and this is restricted to areas of increased permeability and cellular infiltration (CLAUDIO et al. 1990).

It has been proposed that the vesicles seen in electron micrographs are shuttling directly from the luminal to the abluminal surface. However others consider that movement occurs via a Golgi-associated intracellular compartment (BROADWELL et al. 1988), and there is also some evidence for reverse transport of material in the vesicles (see below). The evidence for transendothelial channels is more tenuous, and based largely on transmission electron microscopy of transverse sections of brain endothelium during immune reactions. However, it is difficult to distinguish a potential channel from a series of vesicles. On balance, the evidence favours vesicular transport as being the primary route for movement of immunoglobulins and other large serum molecules into the brain during disease.

The immunological signals for changes in vascular permeability include histamine, platelet activating factor (PAF) and eicosanoids, all of which have receptors on brain endothelium. Notably histamine and PAF are chemokinetic, while leukotriene B4 is chemotactic. Nevertheless, these reagents do not increase leucocyte adhesion to endothelium themselves. Since the interaction between leucocytes and endothelium is the first step in cell migration, it appears that the mediators controlling endothelial permeability and cell traffic are different.

In more extreme conditions, there is also the possibility for barrier opening via disruption of interendothelial cell tight junctions, or by direct damage to the endothelium. Antibodies to cultured brain endothelium have been detected in MS, type 1 human T-cell leukemia virus infection and neuro-Behcet syndrome (TSUKADA et al. 1989). These could theoretically induce complement mediated damage to the endothelium, although the supposition that any sublethal damage caused to the endothelium would increase permeability is unproven.

Brain endothelial cells may become infected by a number of viruses, as detected by viral particles or nucleic acid (WISNIEWSKI et al. 1983; HO et al. 1984). Since the cells express major histocompatibility complex (MHC) class I molecules, they become targets for virus-specific, MHC-restricted, cytotoxic T cells (RODRIGUEZ et al. 1987). In severe cases of viral encephalitis, lymphocytes do accumulate in cerebral vessels, and the endothelial cells appear disrupted. Similarly, in tissue culture, brain endothelium can be destroyed by antigen-specific T cells, although in this case the interactions are MHC class II restricted (RISAU et al. 1990; SEDGWICK et al. 1990). Cytokines released by activated T cells (such as tumour necrosis factor,

TNF) alone appear to be unable to destroy the endothelium in this way, which indicates the importance of direct cell-mediated cytotoxicity. Total destruction of the endothelial cells allows platelets access to basement membranes, with release of further vasoactive mediators. It could also lead to localised intracranial haemorrhage. However, in vivo, such severe damage to the vasculature is very uncommon and is only really seen in severe cases of fulminating MS, and in certain forms of EAE where the animals are irradiated before disease induction.

C. Movement of Antigens out of the CNS

Since the CNS lacks (leucocyte) dendritic cells, which are important in antigen presentation, it has been debated how antigens originating in the CNS could reach the immune system to sensitise lymphocytes. There is clear evidence that antigens directly injected into the CNS can stimulate antibody production in serum (WIDNER et al. 1988). There are thought to be three potential routes by which antigens might move out of the brain. These are:

1. By reverse transcytosis across cerebral endothelium,
2. by passage into the CSF and thence via arachnoid granulations into the venus sinuses,
3. draining via lymphatics to the deep and superficial cervical lymph nodes.

The first two routes would lead to antigen entering the blood stream, while the third causes it to pass to secondary lymphoid organs, where it may be trapped by professional antigen presenting cells (APCs). It has been possible to detect movement of antigens across brain endothelium, from the basal to the serosal surface. By immunogold staining and electron microscopy, VASS and colleagues (1984) have shown that a xenogeneic protein injected into the brain parenchyma, appears associated with endothelial vesicles, and at the serosal surface of the endothelium. Despite this, it appears that the most important route of egress is via the cervical lymphatics: following antigen injection into brain, antigen-sensitised lymphocytes are preferentially found located in the cervical lymph nodes, in comparison with e.g. spleen (HARLING-BERG et al. 1989; Fig. 2). Furthermore, a potential route of migration for antigen (and possibly also cells) has been identified from the brain to these nodes (BRADBURY and CSERR 1985).

D. Antigen Presentation by Brain Endothelial Cells

Following the observation that the brain normally lacks APCs, it was proposed that either brain endothelium or astrocytes could act as the brain's APC (FONTANA and FIERZ 1985). The rationale behind this proposal was that the brain endothelium is the first potential site at which circulating lymphocytes could contact CNS antigens. However, in view of the findings

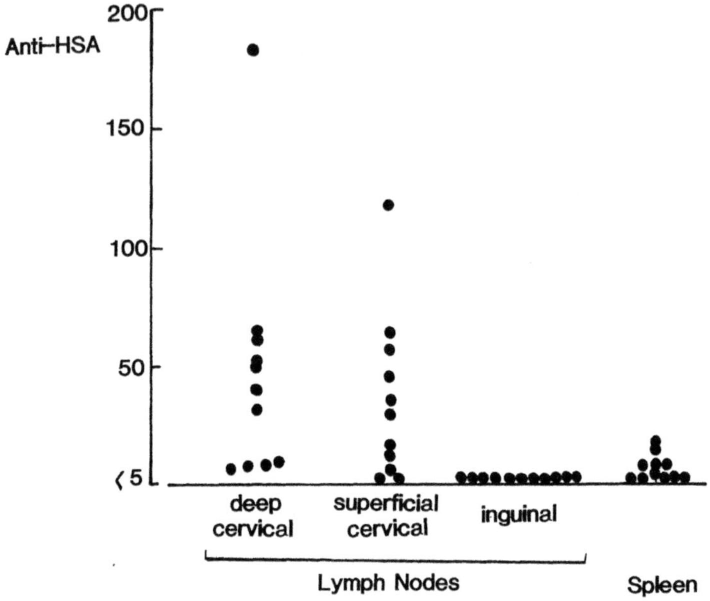

Fig. 2. Production of anti-human serum albumin (anti-HSA) antibodies in cultures of lymphocytes from different sources following an intra-CSF antigen injection. (Based on data of HARLING-BERG et al. 1989)

above, this argument has lost much of its force. Nevertheless, there is a considerable amount of work which examines the potential of brain endothelium to present antigens. One should first state, the necessary requirements of an APC. These are:

1. Expression of MHC class II molecules,
2. the ability to endocytose and process antigens which become associated with the MHC molecules,
3. the expression of surface molecules, which can deliver costimulatory signals to lymphocytes, to trigger their proliferation,
4. the secretion of cytokines such as interleukin-1 (IL-1) and TNF-α, which can modulate lymphocyte function and receptor expression.

I. MHC Expression and Induction

There appear to be genuine species-specific differences in the expression and induction of MHC molecules by brain endothelium. Like all other nucleated cells, endothelium expresses MHC class I molecules which are involved in interactions with cytotoxic T cells. In most species, normal brain endothelium lacks the MHC class II molecules, which are involved in interactions with helper T cells and are crucial for the initiation of immune responses. In man and guinea pig, small vessels occasionally express class II molecules at very low levels in normal CNS (PARDRIDGE et al. 1989). The debate has centred around whether class II molecules are induced during immune reactions. Early studies by TRAUGOTT et al. (1985) suggested that endothelium in the periphery of active MS plaques does express class II MHC, but it is difficult to distinguish endothelial cell staining from that on other cells associated with the vessels. Most subsequent studies have not confirmed the initial findings.

In EAE, the findings vary according to the species used. In rats, class II is not expressed at significant levels during acute EAE, even during the most severe phases of the disease (MATSUMOTO et al. 1986; VASS et al. 1986). However, in chronic relapsing EAE (CREAE) in guinea pigs class II expression on endothelium does occur and seems to precede relapse.

Studies on cerebral endothelium in vitro indicate that resting cells express class I, but not class II MHC molecules (MALE et al. 1987). Class I expression is enhanced on rat endothelium by TNF-α or interferon-γ (IFN-γ) over a period of hours, and this may be affected by low levels of the cytokines. IL-1 is ineffective in this respect. IFN-γ, but not TNF-α is able to induce class II, but this requires high levels of IFN-γ over a period of at least 48 h (Table 2) (MALE and PRYCE 1988a). This indicates that the endothelial cells will make different responses depending on the levels of cytokines in their environment. Interestingly, in the rat, TNF-α synergises with IFN-γ in MHC induction, while IL-1 has a slight inhibitory effect (MALE and PRYCE 1988b; TANAKA and MCCARRON 1990). However, in SJL/J mice TNF had an antagonistic effect on IFN-γ actions.

It has been noted that MHC class II molecules are more readily induced by IFN-γ in some strains than in others. This applies to a group of cells which normally do not express MHC class II molecules, including astrocytes and brain endothelium (MASSA et al. 1987; MALE and PRYCE 1988c). The strains which are most readily induced (LEW rat and SJL mouse) are also the ones which are susceptible to EAE. However, it is not yet certain whether this is due to the MHC regulatory genes of these strains, or whether the endothelium is just more responsive to cytokine activation. Preliminary evidence suggests that the latter is true, since intracellular adhesion molecule-1 (ICAM-1) is also more readily induced on brain endothelium in LEW rats than in the EAE-resistant PVG strain.

Table 2. Expression of MHC molecules on brain endothelium in vitro

	MHC class I	MHC class II
Resting cells	++	−
TNF-α stimulated	+++	−
IFN-γ stimulated	+++	++
IFN-γ level needed to induce MHC	Low	High
Induction time(h)	<8	>48

II. Stimulation of T-Cell Proliferation

The first in vitro studies on antigen presentation by mouse brain endothelium indicated that they were able to induce T-cell proliferation (McCARRON et al. 1985). Guinea pig endothelium in vitro has also been able to induce a low level of T-cell activation (WILCOX et al. 1989). However, two careful studies using highly purified rat brain endothelium showed that the cells were absolutely unable to induce significant T-cell proliferation (PRYCE et al. 1989; RISAU et al. 1990). The differences in these various reports may be related to species-specific variation, as occurs with MHC class II expression. However, another explanation may lie in the degree of purity of the cell populations. For example, other studies comparing endothelium with microvessel-associated smooth muscle cells (SMCs) have shown that the SMCs can express class II MHC molecules and can induce T-cell proliferation (HART et al. 1987; FABRY et al. 1990).

Although the endothelial cells do not appear to be efficient inducers of T-cell proliferation, there is no doubt that they do express antigen which can be recognised by T cells. For example, two studies have shown that brain endothelium can become the target for antigen-specific MHC class II-restricted cytotoxic cells (RISAU et al. 1990; SEDGWICK et al. 1990). This indicates that the endothelium must express something recognisable by the T cells, although it cannot induce their proliferation. The presumption is that the endothelium is unable to deliver the necessary costimulatory signals to the T cells to initiate division. Studies on extracerebral endothelium indicate that molecules such as lymphocyte functional antigen 3 (LFA-3) on endothelium, interacting with CD2 on T cells, and ICAM-1 on endothelium, interacting with LFA-1 on lymphocytes, may contribute to the T-cell activation. It is known that ICAM-1 is present on brain endothelium (see below) and it is presumed that LFA-3 is present because of its wide distribution on endothelium throughout the body. Endothelium can also stimulate lymphocytes by the release of cytokines, including IL-1, IL-6 and TNF-α. Unfortunately, no definitive studies are yet available to determine whether *cerebral* endothelium can also produce these molecules.

The most probable conclusion from these studies is that brain endothelium is not an important APC for the brain. Opinion is now moving towards the view that the microglial cell (which belongs to the mononuclear phagocyte lineage) acts as the brain's resident APC, and that cells which enter the brain during immune reactions such as B cells are also efficient APCs. Furthermore, most antigens such as viruses will sensitise the immune system before they reach the brain. Consequently, there is now no particular reason for looking to cerebral endothelium to perform this particular immunological function.

E. Cellular Migration Across Brain Endothelium

I. The Role of Lymphocytes

Under normal circumstance the levels of leucocyte migration across brain endothelium are very low. In studies where labelled lymphocytes are injected into the periphery, it has been difficult to detect any infiltraing cells at all. However, cell traffic is considerably increased even in low grade inflammatory reactions such as MS. For example, HAFLER and WEINER (1987) injected anti-CD2 into individuals with MS to label peripheral T cells, and then mesured the incidence of labelled cells within the CSF over a period of 1–10 days. Within 4 days virtually all the CSF lymphocytes were labelled, indicating a rapid turnover in this disease: the antibody itself did not cross into CSF to label cells which were already present.

The characteristics of cell migration have been extensively examined in graft rejection (Fig. 3) and in EAE. In the rat, lymphocytes and mononuclear phagocytes start to enter the CNS at approximately 8 days after immunisation with myelin basic protein (MBP) in complete Freund's adjuvant. There is also some migration of neutrophil polymorphs in the earliest stages of the disease. Cellular infiltration precedes the onset of the first signs

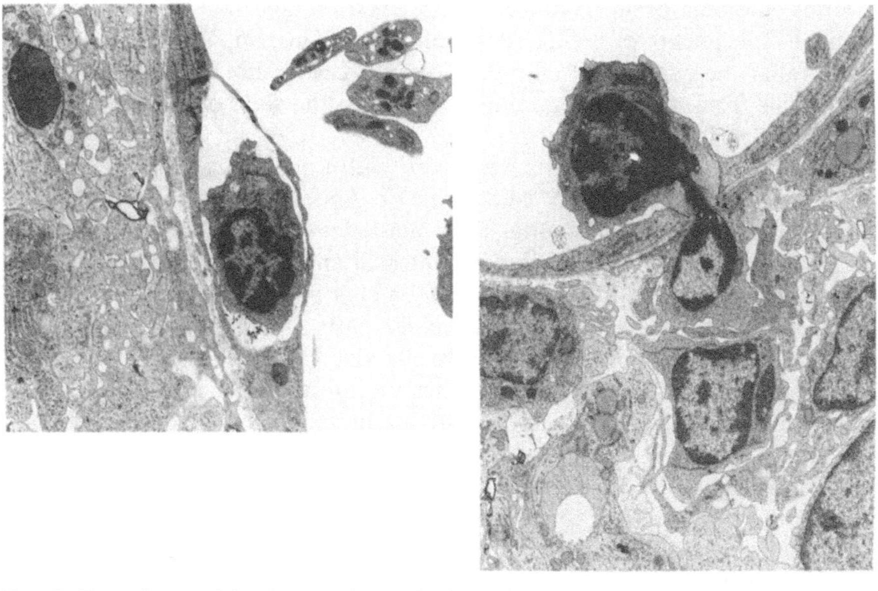

Fig. 3. Lymphocyte lying in a pocket of brain endothelium 9 days after transplanting an allograft (embryonal hippocampal primordium) into hippocampus of adult rats (*left*). Leucocyte undergoing diapedesis into the perivascular space, which already contains several lymphocytes. (From LAWRENCE et al. 1990, courtesy of Dr. G. RAISMAN, with permission)

of paralysis by around 1 day. The paralysis develops over the following 4–5 days and then resolves, leaving the animals refractory to reinduction. The course of the acute disease is essentially similar in mice and guinea pigs, albeit with different induction times (Traugott 1989). Modifications of the injection schedule and a judicious choice of animals has allowed the development of CREAE (guinea pigs and mice, but not rats), in which the animals undergo a series of relapses and remissions at intervals after the acute disease. It is notable that the paralysis of EAE develops in these animals before considerable demyelination has occurred, and this has led to the suggestion that modification of the blood-brain barrier is responsible for the initial symptoms (Kelero de Rosbo et al. 1985). Indeed, barrier breakdown has been detected in EAE, by perfusion of postmortem brains and by direct gadolinium-enhanced nuclear magnetic resonance measurements. Although it appears that the modulators of barrier permeability and those which control cell movement are different, these events usually happen concommitantly and with the same regions of the vasculature (Lossinsky et al. 1989).

It is also possible to induce EAE by transfer of MBP-specific T cell lines (passive EAE), in which case the paralysis develops more quickly than following active immunisation with MBP. One crucial requirement for the development of passive EAE is that the cells must be activated with antigen and cytokines in vitro, before injection. It appears that non-activated cells do not enter the brain to induce EAE, even though they have the correct antigen specificity to induce demyelination. This important observation shows that the control of cell migration into CNS is not initially dependent on signals from the brain endothelium, but on the state of activation of the lymphocytes.

Numerous pieces of evidence accord with this finding. For example, in CREAE in guinea pigs, it is possible to find MBP-specific cells in the periphery, but they do not enter the brain unless triggered to do so. Studies in vitro have shown that mitogen-activated or antigen-activated lymphocytes adhere to cerebral endothelium at much higher levels than resting cells, and there is a specific stage during the early part of the cell cycle when the binding to endothelium is particularly effective (Male et al. 1990a; Fig. 4). This explains some older studies indicating the importance of the stage of the cell cycle in lymphocyte homing to brain. It also accords with the observation that many of the cells found bound to the endothelium in acute EAE (mice) have centrioles, indicating that they are in cell cycle at around the period of mitosis (Raine et al. 1990). This effect is not unique to brain endothelium, since activated lymphocytes tend to preferentially migrate into inflammatory sites, while resting cells tend to move into secondary lymphoid tissues. The adhesion molecules controlling these patterns of traffic are different (see below).

Despite the importance of $CD4^+$ MBP-specific cells in the induction of EAE, it is found that only a small proportion of the cells which enter the

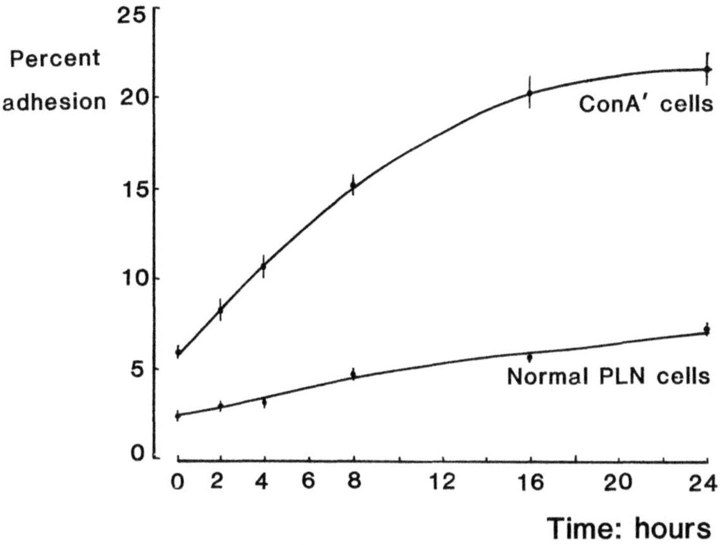

Fig. 4. Percent adhesion of normal peripheral lymph node cells (*PLN*) and mitogen activated cells (*Con-A*), −16 h) to brain endothelial monolayers activated for different amounts of time with TNF-α 50 ng/ml

brain are specific for MBP. In passive EAE, using radiolabelled MBP-specific cells, less than 4% of the lymphocytes localised in the acute lesion were from the injected population: the remainder were of host origin (CROSS et al. 1990). In lesions of relapsed animals, radiolabelled cells were almost undetectable. This confirms studies using chimaeric animals, where it was noted that the great majority of the brain infiltrating cells were of host origin. The implication of this work is that the disease is initivated by activated $CD4^+$ MBP-specific T cells entering the brain. These now interact with APCs in the brain, causing the release of cytokines which can activate the endothelium to express adhesion molecules, causing the recruitment of other lymphocytes and mononuclear phagocytes. A second conclusion from these studies is that the interaction between the brain endothelium and lymphocytes is primarily antigen non-specific. Otherwise, one could not explain the accumulation of the large numbers of cells which are not specific for CNS antigens.

Various studies have examined whether there is a differential interaction of different lymphocyte subpopulations with the brain endothelium, and whether particular subsets migrate across more or less effectively. In EAE, it is noted that $CD4^+$ cells enter the lesion site first, but tend to remain in the perivascular region. CD8 cells and small numbers of B cells enter the lesion later, but these tend to move away from the vasculature. In MS, however, most studies have shown there to be a surprising preponderance of $CD8^+$ cells. Using ^{51}Cr-labelled lymphocytes. TRETTER and STEINMAN showed that Lyt^{2+} ($CD8^+$) and Lyt^{2-} cells (presumably mostly $CD4^+$) entered the brain during acute EAE equally: however, B cells did not migrate to any degree. In contrast, SIMMONS and WILLENBORG (1991) injected the cytokines TNF-α or IFN-γ directly into spinal cord and found surprisingly large numbers of B cells and $CD8^+$ cells. This latter finding mirrors studies in vitro which have shown that B cells tend to bind better to cytokine-activated endothelium than T cells, and CD8 T cells tend to bind more effectively than CD4 cells (PRYCE et al. 1991). It is possible to resolve these discrepancies in the light of the original proposal that both the state of lymphocyte activation and the state of the endothelium control cell traffic. For instance, activated $CD4^+$ cells enter the lesion first, an event controlled by lymphocyte activation in the periphery. Then at a later stage these cells can activate the endothelium, leading to an influx of CD8 cells and small numbers of B cells. The switch in the migration of cell types may correspond to the switch from the development to remission phases of the acute disease (HICKEY and GONATAS 1984).

Finally, PRYCE et al. (1991) have examined the possibility that there is a population of cells which preferentially homes through the brain, and might thus preferentially accumulate in cervical lymph nodes. These studies have shown that the population of lymphocytes present in the cervical nodes does interact more strongly with brain endothelium, but this can be explained by the higher proportions of B cells and $CD8^+$ T cells which are often present in these nodes.

II. Control by Brain Endothelium

Lymphocyte migration into brain occurs primarily across small venular endothelium. In the lesions of EAE and the plaques of MS, lymphocytes are seen to be clustered around the venules, and these regions selectively suffer from the characteristic disease pathologies. There are three possible explanations for the selective migration of cells across these vessels:

1. Difference in surface charge on capillary endothelium and venular endothelium.
2. Selective expression of adhesion molecules on venular endothelium.
3. Haemodynamic constraints on adhesion.

Brain endothelium usually carries a high negative charge, which tends to limit non-specific adhesion of blood cells. In scrapie-infected mouse brains, it has been shown that the surface charge is reduced on the endothelium, as demonstrated using binding of cationized ferritin (VORBRODT et al. 1990). Moreover, capillary endothelium has a higher negative charge than adjoining venular endothelium, providing supporting evidence for the first mechanism above.

Studies are now appearing which demonstrate the expression of intercellular adhesion molecules on brain endothelium by immunostaining of sections from normals and in disease (see below). Although none of these has directly addressed the question of which vessels express the adhesion molecules, there have been no reports of differences between capillary and venular endothelium.

The third reason explaining traffic across venules is that haemostatic pressure is lowest in these regions. Leucocytes have a very limited period during which they must make multi-point contacts with the endothelium which are sufficiently strong to prevent them from being pushed through into the venular circulation. Theoretical calculations indicate that the interaction with endothelium involves at least several hundred molecules on either cell. Adhesion would therefore be favoured in venules where the flow slows. In EAE it is has been noted that lymphocytes often stick near the bifurcations of vessels, where flow may be turbulent and the cells have longer to contact the endothelium. In addition, in practical terms, when leucocytes bind to venules they do not block the plasma flow, as they would if attached to capillary endothelium. On balance, then, the selective migration of cells across venules appears to relate to haemodynamic and charge effects.

Brain endothelium in vitro has been shown to respond to the cytokines TNF-α and IFN-γ by increased adhesion for lymphocytes (HUGHES et al. 1988). The increased adhesion occurs rises over 24 h and requires new protein synthesis. The enhanced adhesion is seen within 1–2 h of cell stimulation and the endothelium is sensitive to low levels of cytokines in this respect. The effects of lymphocyte activation and endothelial activation are independent, such that the binding when both cell types are activated may be 5–10 times that of resting cells. Antigen-specific effects seem to be relatively insignificant, since the presence of antigen produces only marginal increases in adhesion to endothelial cells which have been induced to express MHC class II by high levels of IFN-γ. This accords with the observation that most cells entering inflammatory sites in the brain are antigen non-specific.

III. Molecules Controlling Cell Migration into Brain

Much of the work on the molecules which control endothelial migration has been carried out on large vessel endothelium. It is now clear that large vessel endothelium (which would not normally support leucocyte migration) differs considerably in its immunological properties from cerebral endothelium. For example, the base level of lymphocyte adhesion to aortic endothelium is approximately 4 times greater than the binding to brain endothelium in equivalent conditions (MALE et al. 1990b). However, the work on large vessel endothelium has provided a starting point to examine the adhesion molecules which might be expressed by brain endothelium and control cell traffic (Table 3). Amongst these, important candidates are ICAM-1 and VCAM for lymphocytes, since their time course of induction is

Table 3. Potential adhesion molecules for control of leukocyte migration

Molecule	Induction[a] (h)	Ligand	Comment[b]
ICAM-1	8–48	LFA-1	Present at low levels, induced in severe immune responses
VCAM	8–48	VLA-4+?	No studies available
ELAM	2–16	Neutrophil	No studies available
Addressins	>48	Homing receptors	Induced in chronic EAE

[a] Based on studies using large vessel endothelium, particularly from human umbilical vein, stimulated with, e.g., TNF.
[b] Studies on expression of adhesion molecules on brain endothelium by immunohistochemistry.

similar to the kinetics of enhanced cellular adhesion, following cytokine activation of the endothelium. At sites of chronic immune responses, vascular addressins may be induced on the endothelium. These molecules would normally be located on the high endothelial venules in lymphoid tissues, and their expression is associated with a change in the morphology of the endothelium from a flattened to columnar form (CANNELLA et al. 1990; CROSS et al. 1990).

Cytokines which have been implicated in the control adhesion molecules include IFN-γ, lymphotoxin, TNF-α, IL-1 and IL-4. In addition, IL-8, which is important in chemotaxis and adhesion of neutrophils and mononuclear cells, can be directly produced by stimulated endothelium.

Although neutrophil infiltration is not a major feature of CNS immune reactions in man, these cells do enter the brain during the earliest phases of EAE. Studies on large vessel endothelium suggest that GMP-140 and ELAM expression on endothelium may be important in neutrophil adhesion. GMP-140 is released to the endothelial surface from intracellular stores, while ELAM is synthesised and expressed over 4–12 h following endothelial cell activation and declines by 24 h.

The data on the expression of these molecules on brain endothelium is incomplete. In normal human brain the levels of ICAM-1 expression are low on the endothelium. They are greatly enhanced during viral encephalitis (herpes simplex), but in MS the increase in expression is comparatively modest, and several other cell types including astrocytes are also induced to express ICAM-1 (SOBEL et al. 1990).

Studies in vitro suggest that ICAM-1 is expressed on unstimulated rat brain endothelium: in this, brain endothelium differs from the standard human umbilical vein endothelial preparation, which does not express ICAM-1 in the resting state. IFN-γ and TNF-α enhance ICAM-1 expression over 24 h. The rate of induction of ICAM-1 and the levels of cytokines required to induce it are very similar to the kinetics of increased lymphocyte adhesion induced by these cytokines. However, anti-ICAM-1 only blocks lymphocyte adhesion to a limited degree, suggesting that other adhesion molecules are involved. Although VCAM induction is thought to be similar to that of ICAM-1, there are no studies as yet to show whether it has a role in lymphocyte migration into CNS.

The vascular addressins are recognised by the MECA series of antibodies in mice and the HECA series in man. These receptors are normally only expressed at chronic inflammatory sites and in secondary lymphoid tissues. They are absent from normal brain endothelium, but in mouse CREAE MECA-325 is induced at the same time as ICAM-1 in both the acute phase of the disease and in relapse. These adhesion molecules disappear again during remission. MECA-325 is found on the peripheral lymph node high endothelium, and it is notable that the endothelium in remission of CREAE is often plumper than normal. A separate study on expression of adhesion molecules in CREAE has indicated that the mucosal addressin (MECA-79) is preferentially expressed at the inflammatory sites, while the mucosal addressin is absent. Resolution of this discrepancy will require further work.

A number of studies have considered the possibility that MHC class II molecules, expressed on brain endothelium, might contribute to lymphocyte adhesion (and activation) by a CD4-dependent mechanism. Several studies have suggested that the expression of these molecules on brain endothelium is associated with the onset of relapses in EAE (Sobel et al. 1984; Sobel and Colvin 1985; Rose et al. 1988). However, most studies have been unable to block lymphocyte/endothelial cell adhesion with anti-CD4. It therefore seems probable that MHC expression on the endothelium at the time of relapse is a marker of endothelial activation, and other more relevant adhesion molecules are also expressed at this time.

The processes which cells use to detach themselves from the endothelium are not well understood. It is notable that infusion of enzyme inhibitors can block the development of EAE. In addition, blocking the expression of cell surface proteases (e.g. with canastospermine) also reduces EAE. These enzymes are synthesized and the functional avidity of adhesion molecules such as LFA-1 are induced by cellular activation. These observations suggest an important role for lymphocyte proteases in penetration of the basement membrane and release of the migrating cells into the tissues. It is often found that migrating cells lie within pockets in the endothelium, or under the endothelium adjoining the basement membrane (Lawrence et al. 1990). It has been hypothesised that cells in such a protected environment would be able to release their enzymes in a region protected from serum enzyme inhibitors. In the final stages of the migration lymphocytes lose the adhesion molecules which allow them to interact with endothelium and start to express new molecules (such as the β_1-integrins which allow them to interact with extracellular matrix components in the tissues).

F. Summary and Synthesis

Brain endothelium normally supports low levels of leucocyte migration and excludes the majority of immunologically important molecules. In the initial stages of immune reactions, activation of lymphocytes in the periphery may induce an increase in cellular migration. Alternatively, local activation of endothelium by, for example, TNF-α from microglia can induce adhesion molecules on the endothelium and also enhance migration. In the second stage, lymphocytes may encounter their specific antigen on APCs in the CNS and release further cytokines to further induce endothelial adhesion molecules. Transcytosis of serum molecules may also be enhanced at this stage by mediators such as histamine, PAF and leukotrienes. Termination of immune reactions is normally determined by the availability of the antigen. Where an external antigen is present, the immune response may be able to eliminate it to end the reaction. However, if the response is to an autoantigen, or where the effector actions cannot eliminate the antigen (e.g. SSPE), there is persistent cell traffic across the endothelium and chronic immune responses develop.

References

Balin BJ, Broadwell RD, Salcman M, El-Kalliny M (1986) Avenues of entry of peripherally administered protein to the CNS in mouse, rat and monkey. J Comp Neurol 251:260–280

Bradbury MW, Cserr HF (1985) Drainage of cerebral interstitial fluid and of cerebrospinal fluid into lymphatics. In: Johnston MG (ed) Experimental biology of the lymphatic circulation. Elsevier, London

Broadwell RD (1989) Transcytosis of macromolecules through the blood-brain barrier: a cell biological perspective and critical appraisal. Acta Neuropathol (Berl) 79:117–128

Broadwell RD, Balin BJ, Salcman M (1988) Transcytosis of blood-bourne protein through the blood-brain barrier. Proc Natl Acad Sci USA 85:632–636

Cannella B, Cross AH, Raine CS (1990) Upregulation and coexpression of adhesion molecules correlating with relapsing autoimmune demyelination in the central nervous system. J Exp Med 172:1521–1531

Claudio L, Kress Y, Norton WT, Brosnan CF (1990) Increased vesicular transport and decreased mitochondrial content in blood-brain barrier endothelial cells during experimental autoimmune encephalomyelitis. Am J Pathol 135: 1157–1168

Cross AH, Cannella B, Brosnan CF, Raine CS (1990) Homing to central nervous system vasculature by antigen-specific lymphocytes. I. Localization of ^{14}C-labeled cells during acute, chronic and relapsing experimental allergic encephalomyelitis. Lab Invest 63:162–170

Fabry Z, Waldschmidt MM, Moore SA, Hart MN (1990) Antigen presentation by brain microvessel smooth muscle and endothelium. J Neuroimmunol 28:63–71

Fontana A, Fierz W (1985) The endothelium-astocyte immune control system of the brain. Springer Semin Immunopathol 8:57–70

Hafler DA, Weiner HL (1987) T cells in multiple sclerosis and inflammatory central nervous system diseases. Immunol Rev 100:307–332

Harling-Berg C, Knopf PM, Merriam J, Cserr HF (1989) Role of cervical lymph nodes in the systemic humoral immune response to human serum albumin microinfused into rat CSF. J Neuroimmunol 25:185–193
Hart MN, Waldschmidt MM, Hanley-Hyde JM, Moore SA, Kemp JD, Schelper RL (1987) Brain microvascular smooth muscle expresses class II antigens. J Immunol 138:2960–2963
Hickey WF, Gonatas NK (1984) Suppressor T lymphocytes in the spinal cord of Lewis rats recovered from acute experimental allergic encephalomyelitis. Cell Immunol 85:284–288
Ho DD, Rota TR, Hirsch MS (1984) Infection of human endothelial cells by human T-lymphocytotropic virus type I. Proc Natl Acad Sci USA 81:7588–7590
Hughes CCW, Male DK, Lantos PL (1988) Adhesion of lymphocytes to cerebral microvascular cells: effects of interferon-γ, tumour necrosis factor and interleukin-1. Immunology 64:677–682
Kelero de Rosbo N, Bernard CC, Simmons RD, Carnegie PR (1985) Concommitant detection of changes in myelin basic protein and permeability of blood spinal cord barrier in experimental allergic encephalomyelitis by immunoblotting. J Neuroimmunol 9:349
Lawrence JM, Morris RJ, Wilson DJ, Raisman G (1990) Mechanisms of allograft rejection in the rat brain. Neuroscience 37:431–462
Leibowitz S, Hughes RAC (1983) Immunology of the nervous system Arnold, London
Lossinsky AS, Badmajew V, Robson JA, Moretz RC, Wisniewski HM (1989) Sites of egress of inflammatory cells and horse radish peroxidase transport across the blood brain barrier in a murine model of chronic relapsing experimental allergic encephalomyelitis. Acta Neuropathol (Berl) 78:359–371
Male DK, Pryce G (1988a) Kinetics of MHC gene expression and mRNA synthesis in brain endothelium. Immunology 63:37–42
Male DK, Pryce G (1988b) Synergy between interferons and monokines in MHC induction on brain endothelium. Immunol Lett 17:267–272
Male DK, Pryce G (1988c) Induction of Ia molecules on brain endothelium is related to susceptibility to experimental allergic encephalomelitis. J Neuroimmunol 21:87–90
Male DK, Pryce G, Hughes CCW (1987) Antigen presentation in brain: MHC induction on brain endothelium and astrocytes compared. Immunology 60: 453–459
Male DK, Pryce G, Hughes CCW, Lantos PL (1990a) Lymphocyte migration into brain modelled in vitro: control by lymphocyte activation, cytokines and antigen. Cell Immunol 127:1–11
Male DK, Pryce G, Rahman J (1990b) Comparison of the immunological properties of rat cerebral and aortic endothelium. J Neuroimmunol 30:161–168
Massa PT, Ter Meulen V, Fontana A (1987) Hyperinducibility of Ia on astrocytes correlates with strain-specific susceptibility to experimental allergic encephalomyelitis. Proc Natl Acad Sci USA 84:4219–4223
Matsumoto Y, Naoyuki H, Tanaka R, Fujiwara M (1986) Immunochemical analysis of the rat central nervous system during experimental allergic encephalomyelitis with special reference to Ia-positive cells with dendritic morphology. J Immunol 136:3668–3676
McCarron RM, Kempsi O, Spatz M, McFarlin DE (1985) Presentation of myelin basic protein by murine cerebral vascular endothelial cells. J Immunol 134:3100–3103
Pardridge WM, Yang J, Buciak J, Tourtellotte WW (1989) Human brain microvascular DR antigen. J Neurosci Res 23:337–341
Pryce G, Male DK, Sedgwich J (1989) Antigen presentation in brain: brain endothelial cells are poor stimulators of T cell proliferation. Immunol 66:207

Pryce G, Male DK, Sarkar C (1991) Control of lymphocyte migration into brain: selective interaction of lymphocyte subpopulations with brain endothelium. Immunology (in press)

Raine CS, Cannella B, Duijvestijn AM, Cross AH (1990) Homing to central nervous system vasculature by antigen-specific lymphocytes. II. Lymphocyte/endothelial cell adhesion during the intial stages of autoimmune demyelination. Lab Invest 63:476–489

Risau W, Engelhardt B, Wekerle H (1990) Immune function of the blood brain barrier: incomplete presentation of protein (auto-) antigens by rat brain microvascular endothelium in vitro. J Cell Biol 110:1757–1766

Rodriguez M, Pierce ML, Howie EA (1987) Immune response gene products on glial and endothelial cells in virus-induced demyelination. J Immunol 138:3438–3442

Rose LM, Petersen R, Mehra R, Alvord EC (1988) Endothelial cell Ia increases before inflammatory cell infiltration in EAE induced in long-tailed macaques. Ann NY Acad Sci 540:315–318

Sedgwick J, Hughes CCW, Male DK, MacPhee IAM, Ter Meulen V (1990) Antigen-specific damage to brain vascular endothelial cells mediated by encephalitogenic and non-encephalitogenic $CD4^+$ T cell lines in vitro. J Immunol 145:2474–2481

Simmons RD, Willenborg DO (1991) Direct injection of cytokines into the spinal cord causes autoimmune encephalitis-like inflammation. J Neurol Sci (in press)

Sobel RA, Colvin RB (1985) The immunopathology of experimental allergic encephalomyelitis (EAE). Differential in situ expression of strain 13 Ia on endothelial and inflammatory cells of (strain 2 × 13) F1 guinea pigs with EAE. J Immunol 132:2382–2401

Sobel RA, Blanchette BW, Bahn AK, Colvin RB (1984) The immunopathology of experimental allergic encephalomyelitis. II. Endothelial cell Ia increases prior to inflammatory cell infiltration. J Immunol 132:2402–2407

Sobel RA, Mitchell ME, Fondren G (1990) Intercellular adhesion molecule-1 (ICAM-1) in cellular immune reactions in the human central nervous system. Am J Pathol 136:1309

Tanaka M, McCarron RM (1990) The inhibitory effect of tumour necrosis factor and interleukin-1 on Ia induction by interferon-gamma on endothelial cells from murine central nervous system microvessels. J Neuroimmunol 27:209–215

Traugott U (1989) Detailed analysis of early immunopathologic events during lesion formation in acute experimental allergic encephalomyelitis. Cell Immunol 119:114–129

Traugott U, Scheinberg LC, Raine CS (1985) On the presence of Ia-positive endothelial cells and astrocytes in multiple sclerosis lesions and its relevance to antigen presentation. J Neuroimmunol 8:1–14.

Tretter J, Steinman L (1984) Homing of $Lyt2^+$ and $Lyt2^-$ T cell subsets and B lymphocytes to the CNS of mice with acute experimental allergic encephalomyelitis. J Immunol 132:2919–2923

Tsukada N, Tanaka Y, Yanagisawa N (1989) Autoantibodies to brain endothelial cells in the sera of patients with human T-lymphocytotropic virus type I associated myeolpathy and other demyelinating disorders. J Neurol Sci 90:33–42

Vass K, Lassmann H, Wisniewski HM, Iqbal K (1984) Ultracytochemical distribution of myelin basic protein after injection in the CSF. Evidence fro transport through the blood brain barrier and binding to the luminal surface of cerebral vessels. J Neurol Sci 63:423–433

Vass K, Lassmann H, Wekerle H, Wisniewski HM (1986) The distribution of Ia antigens in the lesions of rat acute experimental allergic encephalomyelitis. Acta Neuropathol (Berl) 70:149–160

Vorbrodt AW, Dobrogowska DH, Lossinsky AS, Wisniewski HM (1990) Changes in the distribution of anionic sites in brain micro-blood vessels with and without amyloid deposits in scrapie-infected mice. Acta Neuropathol (Berl) 79:353–363

Widner H, Moller G, Johansson BB (1988) Immune response in deep cervical lymph nodes and spleen after antigen deposition in different intra-cerebral sites. Scand J Immunol 28:563–571
Wilcox CE, Healey DG, Baker D, Willoughby DA, Turk JL (1989) Presentation of myelin basic protein by normal guinea pig brain endothelial cells and its relevance to experimental allergic encephalomyelitis. Immunology 67:435–440
Wisniewski HM, Brown HR, Thormar H (1983) Pathogenesis of viral encephalitis: demonstration of viral antigens in the brain endothelium. Acta Neuropathol (Berl) 60:107

CHAPTER 17

The Blood-Brain Barrier In Vitro and in Culture*

J.J. LATERRA and G.W. GOLDSTEIN

A. Introduction

Since the first qualitative descriptions by EHRLICH (1885, 1902) and then GOLDMAN (1913), numerous in vivo studies established a foundation from which in vitro studies have expanded our understanding of blood-brain barrier development and function. OLDENDORF (1971) pioneered the application of intracarotid injection single-pass techniques in defining the unique transport properties of cerebral capillaries in vivo. The ultrastructural studies of REESE and KARNOVSKY (1967) and BRIGHTMAN and REESE (1969) localized the diffusion barrier to the microvascular endothelial cells and the quail-chick chimera studies of STEWART and WILEY (1981) addressed the importance of endothelial interactions with brain parenchyma in the expression of the barrier phenotype. Recent developments in the isolation of metabolically active barrier microvessels and the culture of their constituent cell types have allowed us to experimentally address questions in vitro that are difficult or impossible to answer in vivo. For instance, in vitro studies support what had been a reasonable assumption that the extraction by brain of substances injected into the carotid artery represented transport properties of endothelium within the cerebral microvasculature. Studies of isolated brain microvessels which access the abluminal vessel surface led to descriptions of novel asymmetric abluminal endothelial transport systems. Likewise, studies using isolated retinal microvessels demonstrate endothelial contributions to the blood-retinal barrier since both microvessels and pigmented epithelium constitute this barrier in vivo. The development of methods for the maintenance and growth of microvessel endothelial cells in vitro confirmed the importance of intercellular relationships for barrier expression. Most recently, experiments that reconstitute the relationships of cultured cerebral microvessel cells in vitro demonstrate the role of the perivascular astrocyte in endothelial differentiation.

* This work was supported by National Institutes of Health Research Grants ES-02380 and EY-03772 (G.W.G.) and by a grant from The Juvenile Diabetes Foundation International (J.L.). J. Laterra is a Clinical Investigator Development Awardee of the National Institute of Neurological Disorders and Stroke (NS-01329).

We acknowledge that this chapter does not exhaustively review all relevant reports and experimental systems. We hope that those discussed provide a useful historical and experimental perspective of the contributions the in vitro approach has made to the understanding of brain capillary formation and blood-brain barrier function.

B. Isolated Microvessels

The first in vitro analysis of purified barrier microvessels was described by SIAKOTOS et al. (1969). Microvessels were isolated from bovine and human brain by a series of fractionation steps consisting of tissue homogenization, nylon mesh filtration, differential and density centrifugations, Sephadex G-25 chromatography, and glass bead chromatography. Purity was demonstrated by phase contrast and electron microscopy and their phospholipid content examined. This demonstrated the feasibility of using such microvascular preparations for detailed biochemical analysis. Joó and KARNUSHINA (1973), apparently unaware of this previous report, developed a somewhat simpler method of rat brain microvessel isolation consisting of nylon mesh filtration followed by differential and density centrifugation. These authors, by applying the fact that brain microvessels are rich in certain enzymes, were the first to demonstrate alkaline phosphatase histochemically in microvessels in vitro. In addition to using microscopic techniques to assess the quality of microvessel preparations, brain microvessel enrichment is now routinely checked by demonstrating either histologically or enzymatically the relative concentrations of alkaline phosphatase or gamma-glutamyl transpeptidase (Table 1). BRENDEL et al. (1974) and GOLDSTEIN et al. (1975) introduced the possibility of performing detailed metabolic and transport studies on brain microvessels in vitro. The major oxidative energy-producing pathways (i.e., glycolysis, fatty acid oxidation, and the citric acid cycle) were shown by BRENDEL et al. (1974) to be maintained for up to 3 h in bovine brain microvessels in vitro. Using rat brain microvessels consisting of endothelial cells and pericytes, with their associated basement membranes free of abluminal astrocytes, GOLDSTEIN et al. (1975) identified linear oxidative glucose metabolism and a saturable, temperature dependent uptake of 2-deoxy-D-glucose. MRSULJA et al. (1976) identified similar properties of 2-deoxy-D-glucose uptake and metabolism in isolated rabbit cerebral microvessels. We have found bovine retina to be a particularly reliable source of highly purified barrier microvessels (BETZ and GOLDSTEIN 1980; Figs. 1, 2).

C. Transport

Transport studies using isolated microvessels require special considerations from those performed in vivo. Isolated microvessels typically remain viable for a limited time of at most 2–3 h. Since both luminal and abluminal

Table 1. Enzyme Activity in Brain, Retina and Purified Microvessels

Enzyme	Rat cortex[a]		Bovine cortex[b]		Bovine retina[b]	
	Microvessel	Cortex	Microvessel	Cortex	Microvessel	Retina
Alkaline phosphatase	202	13.4	147	22	94	47
Glucose-6-phosphatase	20	3.5	—	—	—	—
Gamma-glutamyl transpeptidase	156	7.7	186	4	8.5	20
Choline acetyltransferase	0.01	0.6	—	—	—	—

Activities are expressed as nmol product mg protein^{-1} min^{-1}. Cortical microvessels are enriched in alkaline phosphatase, glucose-6-phosphatase and gamma-glutamyl transpeptidase. In contrast, retinal microvessels contain little gamma-glutamyl transpeptidase relative to intact retina or cortical vessels.

[a] From GOLDSTEIN et al. (1975).

[b] From BRESSLER (personal communication).

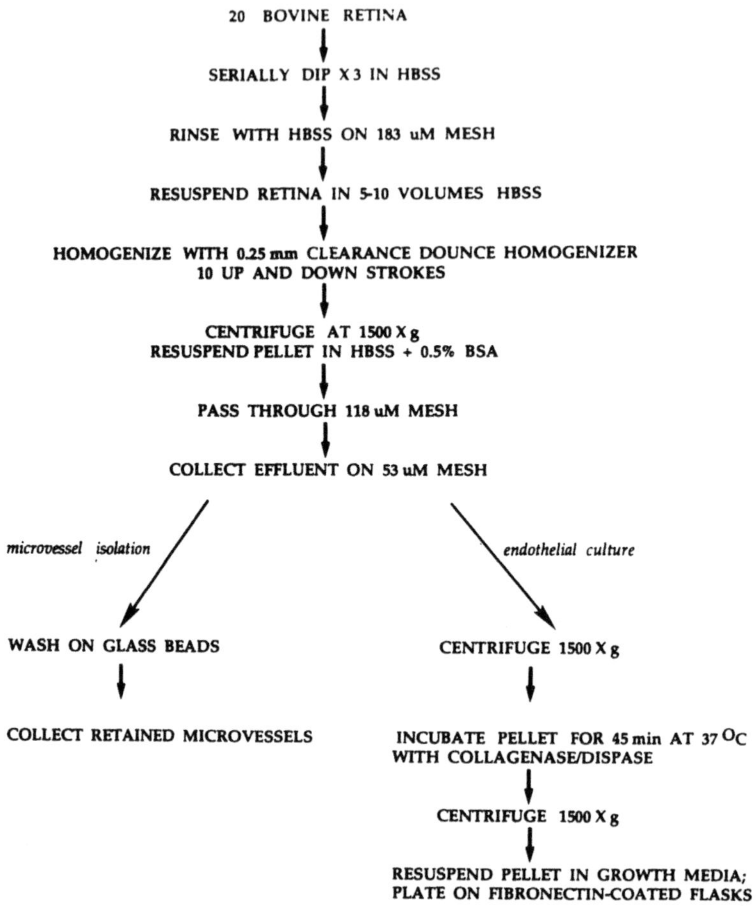

Fig. 1. Procedure for isolating bovine retinal microvessels and establishing microvascular endothelial cell cultures. Material from 20 retinas routinely yields 700 µg capillary protein or two 25-cm^2 flasks of cells (see Fig. 2). *HBSS*, Hanks balanced salt solution

surfaces are exposed, substrate transport into microvessel suspensions (i.e., uptake) substitutes for unidirection transcellular transport. Furthermore, it is technically difficult to obtain time points before equilibrium is complete for certain kinetic analyses since the final distribution space for water soluble substrates is very small in isolated capillaries compared to whole brain. MURTA et al. (1990) have developed a technique for microperfusing individual isolated microvessels that partially circumvents these problems. Despite and in certain cases because of these differences, transport studies in vitro have augmented in vivo analyses in creating a model of barrier transport systems (Fig. 3).

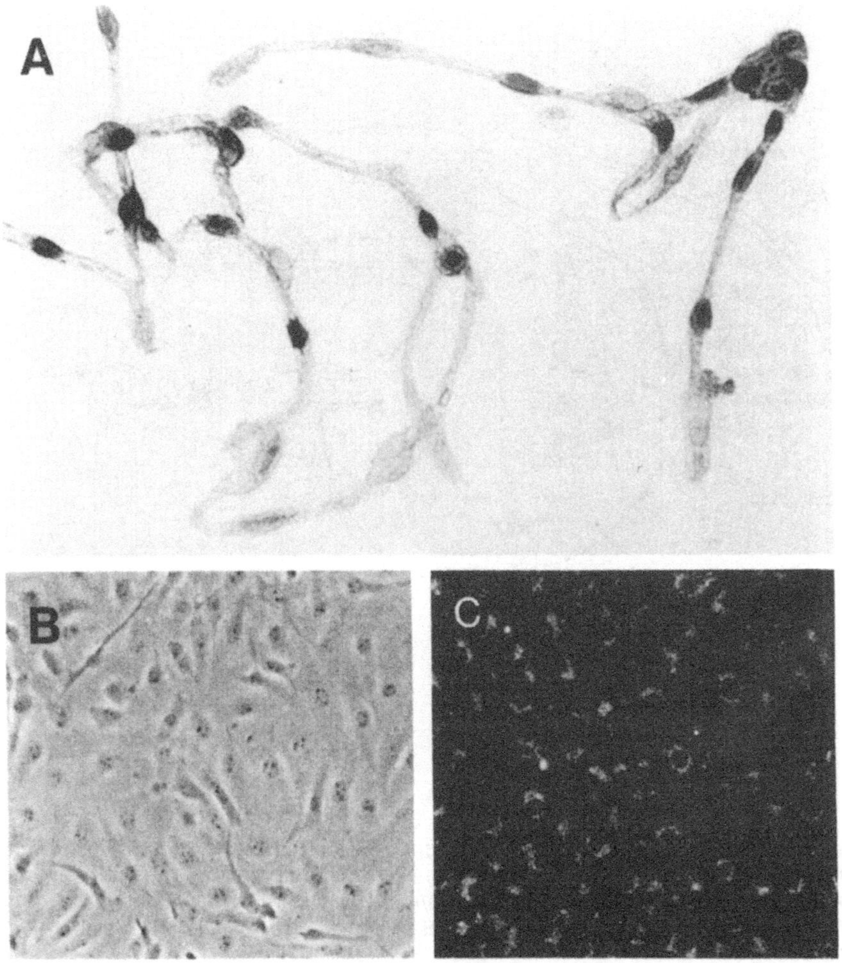

Fig. 2A–C. Isolated bovine retinal microvessels (**A**) and passaged microvascular endothelial cells (**B**, **C**). The endothelial character of these cultured cells is evident from their typical cobblestoned morphology (**B**) and their ability to rapidly accumulate the fluorescent probe DiI-acyl-LDL (**C**)

I. Glucose

Quantitation of glucose transporter density in various fractions of brain using D-glucose-displaceable cytochalasin B binding indicates that cerebral microvessels are markedly enriched in this transporter in comparison to the intact tissue from which they were obtained (DICK and HARIK 1986). This is consistent with the large glucose requirements of brain, since glucose must traverse the microvessel endothelium to enter brain and since microvessel endothelial cells comprise less than 0.1% of brain by weight. Regional differences in brain of microvessel glucose transporter density have also

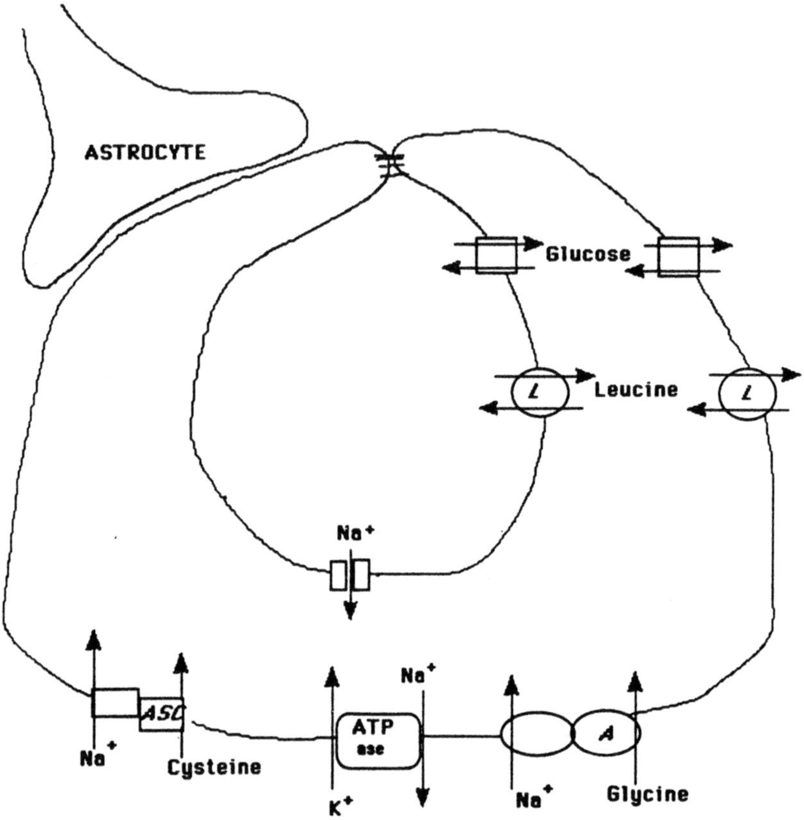

Fig. 3. Schematic representation of brain endothelial transport systems. Glucose and large neutral amino acid (*L*) facilitative transporters are symmetrically located at the luminal and abluminal endothelial membranes. The small neutral amino acid (*A*) and alanine/serine/cysteine (*ASC*) active transporters are at the abluminal surface only and linked to Na^+-K^+-ATPase. (Modified from GOLDSTEIN and BETZ 1983)

been described (DICK and HARIK 1986) but the physiological significance of these differences remains obscure. BETZ et al. (1978) examined the uptake of 3-*O*-methyl-D-glucose into isolated rat brain microvessels. Uptake was found to be non concentrative, stereospecific, sodium-independent, and inhibitable with cytochalasin B but unaffected by insulin, ouabain, or 2,4-dinitrophenol. These results are consistent with data from in vivo transport studies. Definition of transport kinetics and determination of K_m at 37 °C were not technically possible due to the very rapid initial transport velocities and rapid equilibration between extracellular and intracellular compartments. In order to perform detailed kinetic analyses of hexose transport, KOLBER et al. (1979) analyzed 3-*O*-methyl-D-glucose and 2-deoxy-D-glucose uptake in rat brain microvessels at 20 °C to slow the time course of uptake. Despite correcting for temperature based on the isotherm for the K_m of the erythrocyte glucose transporter, which is identical to the brain endothelial transporter (BIRNBAUM et al. 1986), the K_m for microvessel transport was four times higher than that determined in vivo (PARDRIDGE and OLDENDORF 1975). This difference was attributed to a significant diffusional component of hexose uptake due to leakiness of damaged microvessel membranes.

The glucose transporter appears to be altered in certain pathological conditions. Transporter density of brain microvessels isolated from animals with streptozocin-induced diabetes mellitus is reduced in comparison to control animals (MATTHAEI et al. 1986). Incubating microvessels from streptozocin-treated animals with insulin resulted in a return to normal of transporter density, thus implicating insulin depletion and not solely hyperglycemia as a primary mechanism. Although these findings have been contradicted (HARIK et al. 1988), they are consistent with in vivo studies demonstrating decreased V_{max} of glucose uptake in experimental diabetes (GJEDDE and CRONE 1981) as well as with the sensitivity of chronically hyperglycemic diabetics to rapid drops in plasma glucose concentrations. Microvessels isolated from neocortex and hippocampus of Alzheimer diseased brains have been found to contain reduced glucose transporter density (KALARIA and HARIK 1989). In contrast, brain regions not commonly affected in Alzheimer disease demonstrated normal transporter density. The phosphorylation of 2-deoxy-D-glucose is reduced in microvessels from Alzheimer diseased brains, which is consistent with an additional alteration in endothelial hexokinase activity (MARCUS et al. 1989). Although it is unclear whether these alterations represent primary or secondary pathogenic events, the decrease in transporter density and hexokinase activity may both contribute to the regional glucose hypometabolism that is characteristic of Alzheimer disease.

II. Amino Acids

Three transport systems account for most neutral amino acid transport of isolated brain microvessels (Fig. 3). Similar systems are present in a variety of cells. These consist of the L-system (preferring neutral amino acids with branched or ringed side chains, sodium independent, and sensitive to 2-aminobicycloheptane-2-carboxylic acid, BCH), the A-system (preferring neutral amino acids with short polar or linear side chains, sodium-dependent, and sensitive to alpha-methylaminoisobutyric acid, MeAIB) and the ASC system (preferring alanine, serine, and cysteine, sodium-dependent, and insensitive to both BCH and MeAIB) (CHRISTENSEN 1975). SERSHEN and LAJTHA (1976) and HJELLE et al. (1978) showed that the L-system is present in isolated microvessels from brain and retina. Insensitivity to ouabain, energy independence, and stereospecificity were demonstrated. Since the rate of amino acid transport is relatively slow, K_m values could be calculated. Those for leucine and valine (133 μM and 500 μM, respectively) agree with in vivo determinations of blood-to-brain uptake (PARDRIDGE and

OLDENDORF 1975). In contrast to the L-system, in vivo evidence suggested that the A-system was not present in brain microvessels. However, these in vivo studies which evaluated blood-to-brain uptake accessed transporters only on the luminal microvascular surface. Using isolated brain microvessels, BETZ and GOLDSTEIN (1978) identified a sodium-dependent small neutral amino acid transporter that was sensitive to MeAIB, consistent with A-system transport. This apparent discrepancy is consistent with a localization of the A transporter solely at the abluminal microvessel surface since both the luminal and abluminal surfaces of isolated microvessels are accessible in vitro. A similar conclusion has been made with regard to the ASC system since it functions in microvessels isolated from the newborn rat (TAYARANI et al. 1987) but not in amino acid uptake from blood to brain in the adult rat in vivo (WADE and BRADY 1981). The abluminal localization of Na^+-K^+-ATPase supports the asymmetric localization of both the A and ASC systems whose sodium dependence is linked to this enzyme (BETZ et al. 1980).

There is significant overlap between these different brain microvessel neutral amino acid transport systems (TAYARANI et al. 1987). The ASC system accounts for 50%–70% of alanine, serine, and cysteine uptake into microvessels isolated from adult rats. The L-system accounts for approximately 25%–30% of their uptake and the A-system for 15% of alanine and 28% of serine. Leucine is predominantly transported by the L-system (80%) with a much smaller component contributed by ASC (18%). Microvascular transport of proline occurs predominantly via the A-system (TAYARANI et al. 1987; HWANG et al. 1983). In interpreting the contribution of these systems to the transport of particular amino acids, one must keep in mind their asymmetric distribution.

There is evidence from in vitro studies for cooperation between different transport systems. L-system transport in brain microvessels is linked, in part via the A-system, to intracellular glutamine levels (CANGIANO et al. 1983; CARDELLI-CANGIANO et al. 1984). Preloading of isolated rat brain microvessels with glutamine via a sodium-dependent and MeAIB-inhibitable (i.e., A-system) mechanism enhances L-system uptake of neutral amino acids. Similarly, ammonia stimulates L-system transport by increasing intracellular glutamine concentrations, presumably via endothelial glutamine synthetase activity. Enhanced L-system transport in response to both A-system glutamine loading and ammonia result from an increased V_{max} without significant alteration in K_m. These responses may be relevant to conditions associated with elevated brain levels of ammonia and glutamine such as hepatic encephalopathy. In fact, brain microvessels isolated from rats with portacaval shunts have increased uptake of large neutral amino acids (CARDELLI-CANGIANO et al. 1984).

III. Potassium

BRADBURY et al. (1972) and KATZMAN (1976) characterized K^+ transport of cerebral microvessels in vivo. Based on the quantification of ouabain-inhibitable release of ATP-derived phosphate, EISENBERG and SUDDITH (1979) demonstrated that the density of the potassium transporter Na^+-K^+-ATPase in isolated rat brain microvessels is 500-fold higher than in peripheral large vein endothelial cells and 60% higher than in choroid plexus. Using ouabain binding as a marker of Na^+-K^+-ATPase, HARIK et al. (1985) concluded that Na^+-K^+-ATPase density in isolated microvessels from rat and pig is 25%–30% of that in whole brain and 50% of that in choroid plexus. Microvessel uptake of rubidium, a K^+ analog, is ouabain sensitive (consistent with Na^+-K^+-ATPase mediation) and the kinetics of Na^+-K^+-ATPase activity has revealed a K_m of 0.23 mM and V_{max} of 14.77 mmol/h (EISENBERG and SUDDITH 1979). The K_m for K^+ uptake has been reported by GOLDSTEIN (1979) to be 2.8 mM, nearly identical to the interstitial K^+ concentration in brain. Microvessel membrane fractionation in conjunction with whole brain cytochemistry showed that Na^+-K^+-ATPase is asymmetrically localized at the abluminal endothelial membrane, consistent with the primarily brain-to-blood transport of potassium in vivo (BETZ et al. 1980).

D. The Blood-Brain Barrier in Cell Culture

Many endothelial functions that comprise the diffusion and transport barrier of brain microvessels have been defined by studies in whole animals and in isolated capillaries in vitro. These systems have been less useful in defining the biochemical and molecular mediators of barrier regulation due to the complexity of whole animal manipulations and the limited viability of isolated microvessels in vitro. The ability to grow central nervous system microvascular endothelial cells in culture (BOWMAN et al. 1980, 1982) opens the door to many new experimental approaches in elucidating the mechanisms of blood-brain barrier expression and regulation (Fig. 2B,C).

Under routine culture conditions, endothelial cells derived from brain or retinal barrier microvessels retain endothelial markers but rapidly cease to express certain barrier-specific characteristics. Two barrier characteristics lost in culture are the expression of gamma-glutamyl transpeptidase (MERESSE et al. 1989; FUKUSHIMA et al. 1990) and the formation of very complex continuously interdigitating interendothelial tight junctions (TAO-CHENG et al. 1987). The role of gamma-glutamyl transpeptidase in barrier function remains unknown. Although there is an abundance of evidence indicating that the endothelial cell is the source of microvessel gamma-glutamyl transpeptidase, this has not to our knowledge been definitively proven. Thus the significance of its absence in cultured cells is unclear. On the other hand, the decreased complexity of the tight junctions expressed in cultured endothelial monolayers (BOWMAN et al. 1983; TAO-CHENG et al. 1987; HORNER et al. 1991) accounts for their inability to block the intercellular diffusion of small polar molecules and ions that is the hallmark of barrier in vivo (CRONE and OLESEN 1982). Despite these significant limitations, microvascular endothelial cells grown on solid tissue culture substrates or permeable membranes have been frequently used to model the blood-brain barrier.

Environmental signals induce endothelial cells to express tissue-specific properties in vivo and in vitro. Transplant studies in murine (SVENDGAARD et al. 1975) and avian systems (STEWART and WILEY 1981; RISAU et al. 1986) indicate that certain signals derived from brain parenchyma are required for endothelial cell expression of the blood-brain barrier. The perivascular astrocyte which maintains a most intimate anatomic association with barrier microvascular endothelial cells appears to be the most likely source of this inductive influence. JANZER and RAFF (1987) supplied direct experimental evidence for this hypothesis in their dimonstration that astrocytes specifically induced vessels of the chick chorioallantoic membrane to develop a barrier to the diffusion of Evan's blue.

Application of endothelial culture systems has the potential of defining how brain-derived signals influence endothelial differentiation at the biochemical and molecular levels. DE BAULT and CANCILLA (1980) demonstrated by histochemical methods that a brain-derived endothelial cell line is induced to express gamma-glutamyl transpeptidase when cocultured with C_6 astroglial cells. Heterologous cell-cell contact appeared to be required and the possibility of enzyme transfer rather than true induction was not addressed. The analysis of gamma-glutamyl transpeptidase mRNA levels should resolve this issue. In a later study, MAXWELL et al. (1987) demonstrated that media previously, conditioned with astrocytes increases endothelial-associated gamma-glutamyl transpeptidase levels. The magnitude of the effect (30%) was much less than expected, as based on the amount of enzyme expressed by capillaries in vivo or isolated microvessels in vitro. Exposing cloned brain microvessel endothelial cells to astroglial plasma membranes resulted in a 12-fold increase in endothelial-associated gamma-glutamyl transpeptidase (BAUER et al. 1990). The possibility for enzyme transfer in this study was reduced by the observation that enzyme levels increased only in endothelial isolates with cobblestoned morphology and not in those with spindled morphologies. The in vitro expression of interendothelial tight junctions is also influenced by astrocytic cells. TAO-CHENG et al. (1987) demonstrated this quantitatively in a system that required astrocyte-endothelial cell contact. SHIVERS et al. (1988), based on qualitative findings, concluded that astrocyte conditioned media induced tight junction expression as well. The relative importance of soluble and insoluble factors in barrier induction remains unclear.

For cultured endothelial cells to adequately model the blood-brain barrier, they must form extensive enough interendothelial tight junctions in conjunction with minimal transcellular diffusion pathways such that cell monolayers block the passage of small hydrophilic molecules and generate a high electrical resistance. Primary cultures retain some of these properties and have been used as models of blood-brain barrier (BOWMAN et al. 1983; BETZ et al. 1983; AUDUS and BORCHARDT 1986). Unfortunately, as discussed above, they rapidly lose their barrier characteristics with time and certainly after passage under routine culture conditions. Two long-term monolayer culture systems have recently been described that retain high electrical resistance as well as other blood-brain barrier characteristics in the presence of rat astrocytes or astrocytic cell products. Bovine brain microvessel endothelial cells, grown on collagen-coated Millicell-CM filters containing astrocytes on the other side, generated an average of $660\,\Omega\,cm^2$ that was stable for up to 10 days (DEHOUCK et al. 1990). For comparison, these brain endothelial cells generated a resistance of $416\,\Omega\,cm^2$ when cultured alone, and bovine aortic endothelial cells yielded $257\,\Omega\,cm^2$ in the presence of astrocytes. Under these coculture conditions brain microvessel and not aortic endothelial cells blocked the passive diffusion of inulin. Leucine but not sucrose crossed this in vitro barrier, consistent with a functioning large neutral amino acid transport system, but no specific evidence for a receptor-mediated process was provided. HORNER et al. (1991) have developed a system in which astrocyte conditioned media in conjunction with agents that raise intracellular cyclic adenosine monophosphate (AMP) levels increased the electrical resistance generated by bovine brain endothelial cell monolayers from $100\,\Omega\,cm^2$ to $834\,\Omega\,cm^2$. These high-resistance monolayers exhibit the tight junction-associated protein ZO-1 in a pattern consistent with enhanced tight junction formation, and block the diffusion of fluorescein, dextran of 4kDa, and sucrose. In contrast to these properties that were dependent upon increased intracellular cyclic AMP, reduced fluid phase pinocytosis in peripheral endothelial cell monolayers was not dependent upon or enhanced by cyclic AMP manipulations. Agents that increased cyclic GMP reduced electrical resistance by 40%. Whether the astrocyte conditioned media functions to improve the growth and homogeneity of the endothelial cell cultures or acts more specifically to induce barrier is unclear. These experimental systems, if easily replicable, should lead to detailed characterization of blood-brain barrier regulatory mechanisms.

E. Microvessel Morphogenesis In Vitro

The development of the blood-brain barrier is closely linked to microvessel formation and the establishment of physiological interactions with perivascular astrocytes (BÄR and WOLFF 1972; EVANS et al. 1974; BÄR 1983; STEWART and HAYAKAWA 1987). Therefore, the cellular and biochemical events involved in the organization of neural endothelial cells into tubular networks within brain are fundamental to our understanding of endothelial-parenchymal interactions relevant to vessel behavior and blood-brain barrier expression. Most organs undergo a significant degree of development in the absence of blood vessels. Vascularization occurs as the tissues enlarge and develop morphogenic complexity. Microvessel formation within the central nervous system appears to be distinct from that in systemic tissues. Using avian chimeras, NODEN (1989) showed that tissues of mesodermal origin contain cells with angiogenic potential that coalesce in situ to form segments of vessels. In contrast, the central nervous system which is derived from neural ectoderm was vascularized exclusively by ingrowth of exogenous vessels. Developing vessels within the central nervous system are both anatomically and proliferatively linked to the developing astroglia (BÄR and WOLFF 1972). Evidence from retinal studies suggest that astrocytes play a role in regulating angiogenesis. LING and STONE (1988) found that the ingrowth of retinal vessels is temporally and spatially coordinated with the ingrowth of astrocytes from optic nerve. Furthermore, capillaries are restricted to retinal regions populated by astrocytes in mammals (e.g., monkey, horse) whose fully developed retinae contain avascular zones (SCHNITZER 1987).

Microvascular proliferation (i.e., angiogenesis) is a multistep process consisting of endothelial activation, basement membrane dissolution, endothelial migration and proliferation leading to first cord and then tube formation, capillary maturation, and basement membrane deposition (D'AMORE and THOMPSON 1987; FOLKMAN 1986). At a biochemical level these steps involve a series of alterations in cell-cell and cell-matrix interactions such as degradation of abluminal extracellular matrix (i.e., basement membrane), dissociation of endothelial cells via changes in interendothelial adhesive interactions, endothelial interaction with parenchymal pericellular and extracellular matrix, regeneration of interendothelial adhesive interactions, and cytoskeletal and cell surface rearrangement to reestablish the mature capillary morphology. The dissection of these events at the biochemical and molecular levels is aided by the use of in vitro systems of capillary-like structure formation (i.e., in vitro angiogenesis).

Isolated endothelial cells are capable of forming capillary-like structures or "tubes" in culture following the prolonged withdrawal of growth factors (FEDER et al. 1983; FOLKMAN and HAUDENSCHILD 1980; MACIAG et al. 1982), in response to certain well-defined matrices such as fibronectin and reconstituted basement membrane (GERHART et al. 1988; KUBOTA et al. 1988; INGBER and FOLKMAN 1989) or soluble signals such as fibroblastic growth factor and transforming growth factor-beta (MONTESANO and ORCI 1985; HOCKEL et al. 1987; MADRI et al. 1988). Thus, endothelial interactions with extracellular matrix operate in conjunction with soluble factors to regulate endothelial growth and morphogenic behavior. These in vitro studies also demonstrate that endothelial cells contain the intracellular mechanisms necessary for endothelial organization into capillaries and support the hypothesis that environmental signals regulated not only the expression of tissue-specific endothelial properties (e.g., blood-brain barrier) but also mechanisms of microvessel formation.

We recently described a coculture system that permits the study of astrocyte-endothelial interactions relevant to neural microvessel formation. Rat brain glial fibrillary acidic protein-positive astrocytes and C_6 astroglial cells induce neural microvascular endothelial cells to form capillary-like structures (Figs. 4, 5) (LATERRA et al. 1990). These in vitro capillaries mimic true capillaries in size, shape, their accumulation of the basement membrane-associated protein laminin, and their "ensheathment" by surrounding astroglia (LATERRA et al. 1990). In addition, central lumens occur in an uncertain percentage of these structures (LATERRA and GOLDSTEIN 1991). This endothelial morphogenic response requires very close endothelial-astroglial interactions and probably direct heterologous cell-cell contact consistent with a cell surface, extracellular matrix, or an extremely labile soluble inducer.

We investigated neural microvessel formation in response to astroglial cells and reconstituted basement membrane – two complementary experimental systems that model different but overlapping steps in microvessel morhogenesis. Microvessel formation in response to astroglial cells can be conceptually divided into three steps: – (1) the generation of the morphogenic signal, (2) the reception of this signal by endothelial cells, and (3) the final endothelial morphogenic response. Microvessel morphogenesis in response to reconstituted basement mambrane (GERHART et al. 1988; KUBOTA et al. 1988) requires only the last two steps since the inductive signal is present in the exogenously added basement membrane material. This may account for why microvessel morphogenesis occurs much more rapidly on reconstituted basement membrane than in response to astroglia, 12 h versus 48–72 h. Despite the potential differences between these two systems, such as the nature of the angiogenic signals and their reception by endothelial cells, the final endothelial responses appear to be nearly identical by light and electron microscopic criteria (LATERRA et al. 1990; KUBOTA et al. 1988; LATERRA and GOLDSTEIN 1991). Thus, differences in the response of these systems to experimental manipulation may be attributed to these differences and not to the endogenous endothelial events involved in the final common response (e.g., migratory events, homologous cell-cell interaction, cytoskeletal rearrangements, final shape change).

The morphological differentiation of endothelial cells into capillaries may either require the expression of new genes or occur exclusively by means of post-translational events. To answer this question we examined the effects of RNA and protein synthesis inhibition on in vitro microvessel morphogenesis. Neither actinomycin D (0.01–0.025 μg/ml), cycloheximide (0.01–0.1 μg/ml), or puromycin (0.1–0.25 μg/ml) was found to inhibit neural microvessel morphogenesis in response to reconstituted basement membrane (LATERRA and GOLDSTEIN 1991). In contrast, astroglial-induced

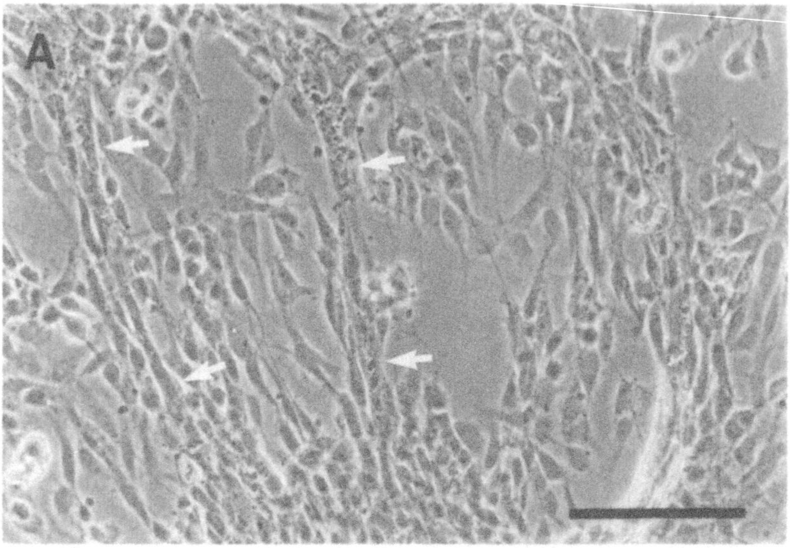

A

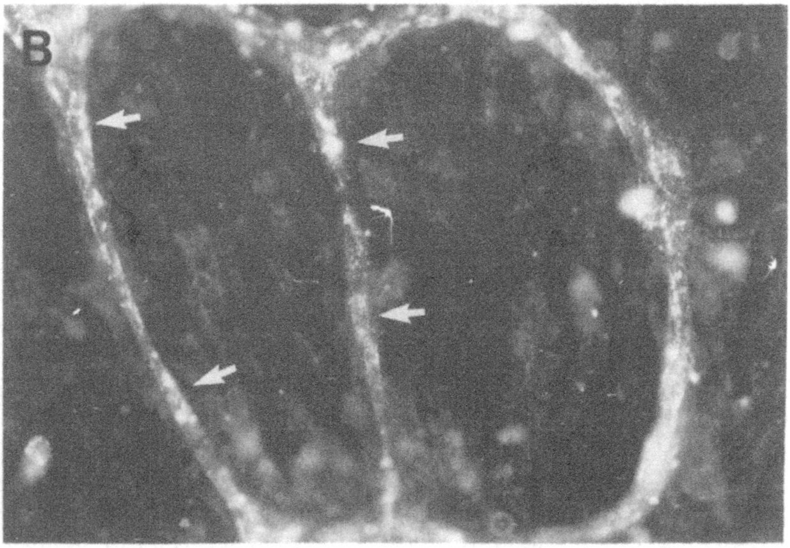

B

Fig. 4A,B. Astroglial-induced microvessel morphogenesis. Bovine retinal microvessel endothelial cells differentiate into capillary-like structures when cocultured with C_6 astroglial cells. **A** Phase contrast photomicrograph demonstrates endothelial-astroglial coculture with linearly associated endothelial cells (*arrows*) forming vessel-like structures. **B** Fluorescent photomicrograph of same field identifies endothelial cells (*arrows*) which have been labeled with the fluorescent marker DiI-acyl-LDL. Endothelial cells cultured alone (Fig. 2B,C) or in the presence of fibroblastic cells (LATERRA et al. 1990) fail to form these structures. *Bar*, 125 μm

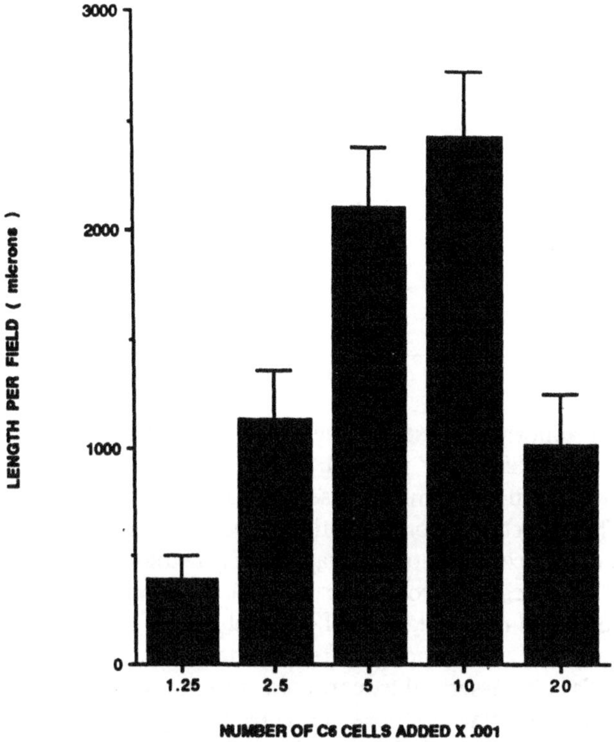

Fig. 5. Astroglial dependence of microvessel morphogenesis. The length of astroglial-induced capillary-like structures per low powered microscopic field as a function of C_6 cell density was quantitated using computer-assisted image analysis. The decrease in microvessel morphogenesis seen with 20000 astroglial cells is attributed to the rapid overgrowth of cocultures by C_6 at this cell density. (From LATERRA et al. 1990)

tube formation was markedly reduced by these metabolic inhibitors (Table 2). These findings suggest that the astroglial-dependent inductive mechanism and not the final endothelial response requires transcription and translation. This dependence on gene transcription implies a role for specific astroglial or endothelial regulatory genes. Furthermore, the endogenous endothelial mediators that generate microvessel morphology (e.g., cell surface adhesion molecules, cytoskeletal proteins) despite RNA and protein synthesis inhibition probably are constitutively synthesized, have relatively long functional half-lives, and can be activated exclusively by post-translational mechanisms.

It is well known that glucocorticoids regulate gene transcription via receptor-mediated mechanisms and also enhance blood-brain barrier function (LONG et al. 1966; YAMADA et al. 1983; JARDEN et al. 1989). The mechanism of this effect is unknown. Steroids are believed to tighten the barrier of preexisting vessels but may additionally inhibit vessel proliferation that is

Table 2. Inhibition of Astroglial-Induced Microvessel Morphogenesis

Drug	Inhibition %
Dexamethasone (10^{-8} *M*)	80
Dexamethasone (10^{-8} *M*) + cortexolone	25
Hydrocortisone (10^{-8} *M*)	40
Progesterone (10^{-8} *M*)	35
Tetrahydrocortisone (10^{-8} *M*)	0
17-α-hydroxyprogesterone (10^{-8} *M*)	5
Cycloheximide (0.1 μg/ml)	80
Actinomycin D (0.025 μg/ml)	70

typically associated with abnormal vessel permeability (e.g., brain tumors, inflammation). We explored this latter possibility in vitro. Dexamethasone at 10^{-8} *M* inhibited astroglial-induced microvessel formation by 70%–100% (WOLFF et al. 1992). This effect was concentration dependent and receptor mediated, since the glucocorticoid receptor antagonist cortexolone reversed the effect, and specific since cholesterol at up to 100 μ*M* as well as other non glucocorticoid steroids had minimal or no effect (Table 2). The cell upon which the glucocorticoids acted was not identified since both endothelial cells and astroglial cells expressed glucocorticoid receptors. Progesterone, which regulates genes similar to those influenced by glucocorticoids (STRAHLE et al. 1989), inhibited astroglial-induced microvessel formation by 30% at 10^{-8} *M* and 60% at 10^{-6} *M*. Only the astroglial cells used in these studies expressed functional progesterone receptors consistent with inhibition of astroglial-derived inductive signals. In support of this conclusion were the findings that none of the steroids affected astroglial-independent microvessel formation in response to reconstituted basement membrane. We have proposed that steroids modulate neural microvessel function, in part, indirectly via a primary effect upon perivascular astrocytes (Fig. 6).

F. Conclusion

Experimental systems using isolated microvessels have both confirmed many conclusions of in vivo studies and produced novel findings relevant to the formulation of a model of blood-brain barrier. Microvessel enrichment in certain enzymes, transporters, and junctional complexes is associated with the barrier phenotype. Transporters for glucose and large neutral amino acids are symmetrically located on both the luminal and abluminal endothelial membranes and facilitatively transport substrate down concentration gradients. The small neutral amino acid transporter and the ASC transport system exist predominantly on the abluminal endothelial surface and are capable of pumping substrate up a concentration gradient by virtue of

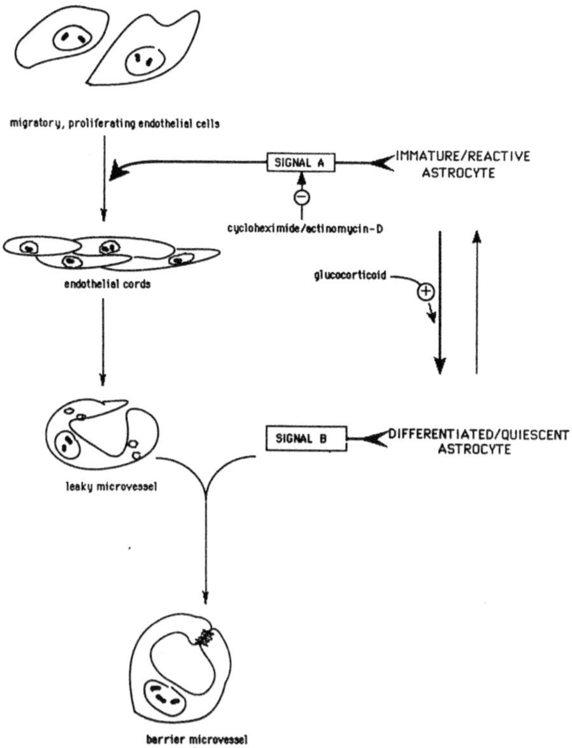

Fig. 6. Model of the balance between angiogenic and barrier-inducing astrocyte-endothelial interactions. Immature, undifferentiated, and reactive astrocytes influence microvascular proliferation. Differentiated and nonreactive astrocytes induce microvessels to express the barrier phenotype. Experiments summarized in the text demonstrate that angiogenic astrocyte-endothelial interactions require RNA and protein synthesis. Glucocorticoids, which are known to increase astroglial expression of differentiated markers (FRESHNEY 1984; KUMAR and DE VELLIS 1988) and enhance blood-brain barrier function, inhibit astroglial-induced microvessel formation

linkage to Na^+-K^+-ATPase. Most of these components of barrier microvessels have not been characterized at the molecular biological level. As a step toward this goal, brain microvessel preparations have recently been used as a source of barrier-specific cDNA. These probes are likely to modify concepts of barrier and help define regulatory mechanisms. Cultured microvascular endothelial cells are being used to study brain microvessel morphogenesis and formation of the diffusional barrier in vitro. These culture systems are proving to be effective in defining the cellular and molecular interactions associated with these events. A combined experimental approach that takes advantage of in vivo preparations as well as these in vitro model systems will be useful in revealing mechanisms of brain microvascular injury, microvessel response to neoplasia, and in creating novel approaches for the therapeutic manipulation of brain microvessels and barrier.

References

Audus KL, Borchardt RT (1986) Bovine brain microvessel endothelial cell monolayers as a model system for the blood-brain barrier. Ann NY Acad Sci 481:9–18

Bär TH (1983) Patterns of vascularization in the developing cerebral cortex. Ciba Found Symp 100:20–36

Bär TH, Wolff JR (1972) The formation of capillary basement membranes during internal vascularization of the rat's cerebral cortex. Z Zellforsch 133:231–248

Bauer HC, Tontsch U, Amberger A, Bauer H (1990) Gamma-glutamyl-transpeptidase (DDTP) and $Na\text{-}K^+$-ATPase activities in different subpopulations of cloned cerebral endothelial cells: responses to glial stimulation. Biochem Biophys Res Commun 168:358–363

Betz AL, Goldstein GW (1978) Polarity of the blood-brain barrier: neutral amino acid transport into isolated brain capillaries. Science 202:225–227

Betz AL, Goldstein GW (1980) Transport of hexoses, potassium and neutral amino acids into capillaries isolated from bovine retina. Exp Eye Res 30:593–605

Betz AL, Csejtey J, Goldstein GW (1978) Hexose transport and phosphorylatin by capillaries isolated from rat brain. Am J Physiol 236:C96–C102

Betz AL, Firth JA, Goldstein GW (1980) Polarity of the blood-brain barrier: distribution of enzymes between the luminal and antiluminal membranes of the brain capillary endothelial cell. Brain Res 192:17–28

Betz AL, Bowman PD, Goldstein GW (1983) Hexose transport in microvascular endothelial cells cultured from bovine retina. Exp Eye Res 36:269–277

Birnbaum MJ, Haspel HC, Rosen OM (1986) Cloning and characterization of a cDNA encoding the rat brain glucose-transporter protein. Rroc Natl Acad Sci USA 83:5784–5788

Bowman PD, Betz AL, Ar D, Wolinsky JS, Penney JB, Shivers RR, Goldstein GW (1980) Primary culture of capillary endothelium from rat brain. In Vitro 17: 353–362

Bowman PD, Betz AL, Goldstein GW (1982) Primary culture of microvascular endothelial cells from bovine retina: selective growth using fibronectin-coated substrate and plasma-derived serum. In Vitro 18:626–632

Bowman PD, Ennis RR, Rarey KE, Betz AL, Goldstein GW (1983) Brain microvessel endothelial cells in tissue culture: a model for study of blood-brain barrier permeability. Ann Neurol 14:396–402

Bradbury MW, Segal MP, Wilson J (1972) Transport of potassium at the blood-brain barrier. J Physiol (Lond) 221:617–632

Brendel K, Meezan E, Carlson EC (1974) Isolated brain microvessels: a purified, metabolically active preparation from bovine cerebral cortex Science 185: 953–955

Brightman MW, Reese TS (1969) Junctions between intimately apposed cell membranes in the vertebrate brain. J Cell Biol 40:648–677

Cangiano C, Cardelli-Cangiano P, James JH, Rossi-Fanelli F, Patriz MA, Brackett KA, Strom R, Fischer JE (1983) Brain microvessels take up large neutral amino acids in exchange for glutamine. J Biol Chem 258:8948–8954

Cardelli-Cangiano P, Cangiono C, James JH, Ceci F, Fischer JE, Strom R (1984) Effect of ammonia on amino acid uptake by brain microvessels. J Biol Chem 259:5295–5300

Christensen HN (1975) Recognition sites for material transport and information transfer. In: Bonner F, Kleinzeller A (eds) Current topics in membranes and transport. Academic, New York, pp 227–258

Crone C, Oleson SP (1982) Electrical resistance of brain mivrovascular endothelium. Brain Res 241:49–55

D'Amore PA, Thompson RW (1987) Mechanisms of angiogenesis. Annu Rev Physiol 49:452–464

DeBault LE, Cancilla PA (1980) Gamma-glutamyl transpeptidase in isolated brain endothelial cells: induction by glial cells in vitro. Science 207:653–655
Dehouck MP, Meresse S, Delorme P, Fruchart JC, Cecchelli R (1990) An easier, reproducible, and mass-production method to study the blood-brain barrier in vitro. J Neurochem 54:1798–1801
Dick APK, Harik SI (1986) Distributin of the glucose transporter in the mammalian brain. J Neurochem 46:1406–1411
Ehrlich P (1885) Das Sauerstoff-Bedürfnis des Organismus: eine farbenanalytische Studie. Hirschwald, Berlin
Ehrlich P (1902) Über Beziehungen von chemischer Constitution, Vertheilung, und pharmakologischer Wirkung. (Reprinted and translated in: Collected studies in immunity, 1906. Wiley, New York, pp 567–595)
Eisenberg HM, Suddith RL (1979) Cerebral vessels have the capacity to transport sodium and potassium. Science 206:1083–1085
Evans CAN, Reynolds JM, Reynolds ML, Saunders NR, Segal MB (1974) The development of a blood-brain barrier mechanism in foetal sheep. J Physiol (Lond) 238:371–386
Feder J, Marasa JC, Olander JV (1983) The formation of capillary-like tubes by calf aortic endothelial cells grown in vitro. J Cell Physiol 116:1–6
Folkman J (1986) How is blood vessel growth regulated in normal and neoplastic tissue? Cancer Res 46:467–473
Folkman J, Haudenschild CC (1980) Angiogenesis in vitro. Nature 288:551–556
Freshney RI (1984) Effects of glucocorticoids on glioma cells in culture. Exp Cell Biol 52:286–292
Fukushima H, Fujimoto M, Ide M (1990) Quantitative detection of blood-brain barrier-associated enzymes in cultured endothelial cells of porcine brain microvessels. In Vitro Cell Dev Biol 26:612–620
Gerhart DZ, Broderius MA, Drewes LR (1988) Cultured human and canine endothelial cells from brain microvessels. Br Res Bull 21:785–793
Gjedde A, Crone C (1981) Blood-brain glucose transfer: repression in chronic hyperglycemia. Science 214:456–457
Goldman E (1913) Vitalfärbung am Zentralnervensystem. Abh Preuss Akad Wiss Phys Math Kl 1:1–60
Goldstein GW (1979) Relation of potassium transport to oxidative metabolism in isolated brain capillaries. J Physiol (Lond) 286:185–195
Goldstein GW, Betz AL (1983) Recent advances in understanding brain capillary function. Ann Neurol 14:389–395
Goldstein GW, Wolinsky JS, Csejtey J, Diamond I (1975) Isolatin of metabolically active capillaries from rat brain. J Neurochem 25:715–717
Harik SI, Doull GH, Dick APK (1985) Specific ouabain binding to brain microvessels and choroid plexus. J Cereb Blood Flow Metab 5:156–160
Harik SI, Gravian SA, Kalaria RN (1988) Glucose transporter of the blood-brain barrier and brain in chronic hyperglycemia. J Neurochem 51:1930–1934
Hjelle JT, Baird-Lambert J, Cardinale G, Spector S, Udenfriend S (1978) Isolated microvessels: the blood-brain barrier in vitro. Proc Natl Acad Sci USA 75: 4544–4548
Hockel M, Sasse J, Wissler JH (1987) Purified monocyte-derived angiogenic substance (angiotropin) stimulates migration, phenotypic changes, and "tube formation" but not proliferation of capillary endothelial cell in vitro. J Cell Physiol 133:1–13
Horner HC, Barbu K, Bard F, Hall D, Janatpour M, Manning K, Morales J, Porter S, Setler PE, Tanner L, Rubin LR (1991) Permeation of the blood-brain barrier for drug delivery to the brain. Natl Inst Drug Abuse Res Monogr Ser (in press)
Hwang SM, Miller M, Segal S (1983) Uptake of L-[14C]proline by isolated rat brain capillaries. J Neurochem 40:317–323

Ingber DE, Folkman J (1989) Mechanochemical switching between growth and differentiation during fibroblast growth factor-stimulated angiogenesis in vitro: role of extra-cellular matrix. J Cell Biol 109:317–330

Janzer RC, Raff MC (1987) Astrocytes induce blood-brain barrier properties in endothelial cells. Nature 325:253–257

Jarden JO, Dhawan V, Moeller JR, Strother SC, Rottenberg DA (1989) The time course of steroid action on blood-to-brain and blood-to-tumor transport of ^{82}Rb: a positron emission tomographic study. Ann Neurol 25:239–245

Joó F, Karnushina I (1973) A procedure for the isolation of capillaries from rat brain. Cytobios 8:41–48

Kalaria RN, Harik SI (1989) Reduced glucose transporter at the blood-brain barrier and in cerebral cortex in Alzheimer's disease. J Neurochem 53:1083–1088

Katzman R (1976) Maintenance of a constant brain extracellular potassium. Fed Proc 35:1244–1247

Kolber AR, Bagnell CR, Krigman MR, Hayward J, Morell P (1979) Transport of sugars into microvessels isolated from rat brain: a model for the blood-brain barrier. J Neurochem 33:419–432

Kubota Y, Kleinman HK, Martin GR, Lawley TJ (1988) Role of laminin and basement membrane in the morphological differentiation of human endothelial cells into capillary-like structures. J Cell Biol 197:1589–1598

Kumar S, de Vellis J (1988) Glucocorticoid-mediated functions in glial cells. In: Kimelberg HK (ed) Glial cell receptors. Raven, New York, pp 243–264

Laterra J, Goldstein GW (1991) Astroglial induced in vitro angiogenesis: requirements for RNA and protein synthesis. J Neurochem (in press)

Laterra J, Guerin C, Goldstein GW (1990) Astrocytes induce neural microvascular endothelial cells to form capillary-like structures in vitro. J Cell Physiol 144:205–215

Ling T, Stone J (1988) The development of astrocyte in the cat retina: evidence of migration from the optic nerve. Dev Brain Res 44:73–85

Long DM, Hartmann JF, French LA (1966) The response of human cerebral edema to glucosteroid administration. An electron microscopic study. Neurology (Minncap) 16:521–528

Maciag T, Kadish J, Wilkins L, Stemerman MB, Weinstein R (1982) Organizational behavior of human umbilical vein endothelial cells. J Cell Biol 94:511–520

Madri JA, Pratt BM, Tucker AM (1988) Phenotypic modulation of endothelial cells by transforming growth factor-B depends upon the composition and organization of the extracellular matrix. J Cell Biol 106:1375–1384

Marcus DL, de Leon MJ, Goldman BS, Logan J, Christman DR, Wolf AP, Fowler JS, Hunter K, Tsai J, Pearson J, Freedman ML (1989) Altered glucose metabolism in microvessels from patients with Alzheimer's disease. Ann Neurol 26:91–94

Matthaei S, Horuk R, Olefsky JM (1986) Blood-brain glucose transfer in diabetes mellitus. Diabetes 35:1181–1184

Maxwell K, Berliner JA, Cancilla PA (1987) Induction of γ-glutamyl transpeptidase in cultured cerebral endothelial cells by a product released by astrocytes. Brain Res 410:309–314

Meresse S, Dehouck MP, Delorme P, Bensaid M, Tauber JP, Delbart C, Fruchart JC, Cecchelli R (1989) Bovine brain endothelial cells express tight-junctions and monoamine oxidase activity in long-term culture. J Neurochem 53:1363–1371

Montesano M, Orci L (1985) Tumor-promoting phorbol esters induce angiogenesis in vitro. Cell 42:469–477

Mrsulja BB, Mrsulja BJ, Fujimoto T, Klatzo I, Spatz M (1976) Isolation of brain capillaries: a simplified technique. Brain Res 110:361–365

Murta JN, Cunha-Vaz JG, Sabo CA, Jones CW, Laski ME (1990) Microperfusin studies on the permeability of retinal vessels. Invest Ophthalmol Vis Sci 31: 471–480

Noden DM (1989) Embryonic origins and assembly of blood vessels. Am Rev Respir Dis 140:1097–1103

Oldendorf WH (1971) Brain uptake of radiolabelled amino acids, amines and hexoses after arterial injection. Am J Physiol 221:1629–1639

Pardridge WM, Oldendorf WH (1977) Transport of metabolic substrates through the blood-brain barrier. J Neurochem 28:5–12

Reese TS, Karnovsky MJ (1967) Fine structural localization of a blood-brain barrier to exogenous peroxidase. J Cell Biol 34:207–217

Risau W, Hallmann R, Albrecht U, Henke-Fahle S (1986) Brain induces the expression of an early cell surface marker for blood-brain barrier-specific endothelium. EMBO J 5:3179–3183

Schnitzer J (1987) Retinal astrocytes: their restriction to vascularized parts of mammalian retina. Neurosci Lett 78:29–34

Sershen H, Lajtha A (1976) Capillary transport of amino acids in the developing brain. Exp Neurol 53:465–474

Shivers RR, Arthur FE, Bowman PD (1988) Induction of gap junctions and brain endothelium-like tight junctions in cultured bovine endothelial cells: local control of cell specialization. J Submicrosc Cytol Pathol 20(1):1-14

Siakotos AN, Rouser G, Fleischer S (1969) Isolation of highly perified human and bovine endothelial cells and nuclei and their phospholipid composition. Lipids 4:234–239

Stewart PA, Hayakawa EM (1987) Interendothelial junctional changes underlie the developmental "tightening" of the blood-brain barrier. Dev Brain Res 32: 271–281

Stewart PA, Wiley MJ (1981) Developing nervous tissue induces formation of blood-brain barrier characteristics in invading endothelial cells: a study using quail-chick transplantation chimeras. Dev Biol 84:183–192

Strahle U, Boshart M, Klock G, Stewart F, Schutz G (1989) Glucocorticoid- and progesterone-specific effects are determined by differential expression of the respective hormone receptors. Nature 339:629–632

Svendgaard N, Bjorklund A, Hardebo JE, Stenevi U (1975) Axonal degeneration associated with a defective blood-brain barrier in cerebral implants. Nature 255:334–337

Tao-Cheng JH, Nagy Z, Brightman MW (1987) Tight junctions of brain endothelium in vitro are enhanced by astroglia. J Neurosci 7:3293–3299

Tayarani I, Lefauconnier JM, Roux F, Bourre JM (1987) Evidence for an alanine, serine, and cysteine system of transport in isolated brain capillaries. J Cereb Blood Flow Metab 7:585–591

Wade LA, Brady HM (1981) Cysteine and cystine transport at the blood-brain barrier. J Neurochem 37:730–734

Wolff JE, Laterra J, Goldstein GW (1992a) Astroglial induced in vitro angiogenesis: inhibition by steroids. (submitted)

Wolff JE, Laterra J. Goldstein GW (1992b) Steroid inhibition of neural microvessel morphogenesis in vitro: receptor-mediation and astroglial dependence. J Neurochem (in press)

Yamada K, Ushio Y, Hayakawa T, Arita N, Yamada N, Mogami H (1983) Effects of methylprednisolone on peritumoral brain edema. A qualtitative autoradiographic study. J Neurosurg 59:612–619

CHAPTER 18

Opening of the Barrier in Cerebral Pathology*

P.J. LUTHERT

A. Introduction

It has long been appreciated that many cerebral disorders are associated with deficits in blood-brain barrier (BBB) function and the interest in barrier pathology is as intense now as it has ever been. There are a number of reasons why this should be the case.

1. The increasing understanding of the nervous system and its disorders is generating a wealth of potential therapeutic agents which fail to cross the barrier and have yet to be modified in such a way that they can (see Chaps. 12 and 21). The study of barrier opening in pathological states is providing new ideas and suggesting new approaches to experimental (Chap. 20), and ultimately therapeutic (Chap. 21) barrier manipulation.
2. The morbidity and mortality associated with cerebral oedema, the major clinical complication of barrier disruption, provides impetus to those interested in controlling pathological barrier opening. The increasing exploitation of therapeutic barrier opening makes all the more urgent an increased appreciation of the potential complications of barrier disruption.
3. Despite recent major advances in the understanding of many neuropathological states the pathogenesis of most remains obscure. There are, however, several pointers to barrier pathology being an integral component of conditions such as Alzheimer's disease (AD) and demyelinating states.

It would be impracticable to review the nature of barrier opening in every instance of cerebral pathology where it has been studied. Therefore, a more general approach will be taken and two widely differing disorders, namely

Abbreviations

AD	Alzheimer's disease
BBB	Blood-brain barrier
CNS	Central nervous system
CSF	Cerebrospinal fluid
HRP	Horseradish peroxidase
MHC	Major histocompatibility complex

* The author's barrier studies were supported by the Wellcome Trust.

AD and cerebral tumours, will be used to illustrate broader principles. The elements involved in the pathology of the barrier will be discussed on an anatomical basis. It should, however, be emphasised that such an approach is largely arbitrary and certainly there are very few, if any, diseases that affect a single component of the nervous system in isolation. Indeed, the close anatomical and physiological integration of tissue elements within the brain makes elucidation of the nature of many of the processes to be considered extremely difficult.

B. Demonstration of Barrier Breakdown in Pathological States

I. Technical Aspects

The earliest demonstrations of barrier breakdown in pathological conditions (BROMAN 1944) were made along the lines of the original visualisation of the BBB, that is, by the use of intravascular dyes which stained areas of pathology and left normal brain uncoloured (for discussion and references see BRADBURY 1979). Since then, the underlying principle has remained more or less identical but different tracers detectable by different means have been employed. Which method is most appropriate in a given circumstance depends on a variety of factors. The focal nature of many cerebral disorders, however, makes regional techniques particularly valuable.

II. Barrier Breakdown in Cerebral Tumours

Innumerable investigations have demonstrated barrier breakdown in a wide range of cerebral tumours (e.g., NEUWELT et al. 1986; BULLARD et al. 1986; DEANE et al. 1984; GROOTHUIS et al. 1984; VICK and KHANDEKAR 1977). The disruption is often severe and extensive but kinetic studies have shown that despite this, the blood-tumour barrier may offer a significant barrier to non-transported, polar therapeutic agents (GROOTHUIS et al. 1984) and the earlier view that there was effectively no barrier in tumours (VICK and KHANDEKAR 1977) now seems over-optimistic. An additional complication is that this degree of barrier pathology is not universal and some tumours seem to have a relatively intact barrier (PIRES et al. 1987; SCHACKERT et al. 1988) and others demonstrate a mixture of relatively permeable and relatively tight regions (GROOTHUIS et al. 1984). What determines the degree of barrier disruption in a given part of a given tumour is unclear but type, grade, site and size of the tumour all have some influence (GROOTHUIS et al. 1984; SEITZ and WECHSLER 1987). The transport of drugs is further compromised by the nature of tumour growth. Particularly in gliomas, single cells spread from the main tumour mass into adjacent, oedematous brain. Here the majority of studies suggest that the barrier remains intact (DEANE et al. 1984; LEVIN et al. 1975; however, see STEWART et al. 1987b), and so the most rapidly growing, advancing edge of the tumour, which is also the portion least readily removed surgically, is protected by the barrier from blood-borne drugs.

III. Barrier Breakdown in Alzheimer's Disease

AD is the commonest cause of dementia and the commonest neurodegenerative disease. It is charaterised by nerve cell loss/shrinkage, intracellular accumulations of paired helical filaments and plaque formation. The observation that the latter contain a 42/43 amino acid protein (A4 or beta protein) that can form amyloid and which is also found in amyloid deposits in vessel walls (congophilic angiopathy) has stimulated great interest in AD research (for review see MASTERS and BEYREUTHER 1989).

Attempts to determine whether there is a barrier defect in AD or not have proved to be fraught with difficulties. Not only is any barrier dysfunction likely to be subtle in comparison with that seen in most destructive cerebral lesions but individual focal events may be transient. Three main approaches have been adopted. Firstly, cerebrospinal fluid (CSF)/serum ratios of a variety of proteins have been measured. Unfortunately, these studies have yielded no consensus (for references see ROZEMULLER et al. 1988). In fact, CSF studies may not be appropriate in this circumstance. Although CSF is often considered to be equivalent to cerebral extracellular space, factors other than BBB function have a profound influence on its composition. Secondly, imaging studies in patients have failed to show a defect (SCHLAGETER et al. 1987) but sensitivity may be an issue here. Finally, there have been a considerable number of immunohistochemical studies demonstrating plasma proteins in relation to plaques, around vessels (particularly if affected by amyloid deposition – see below) and in neurons and glia (for methodological critique and references see ROZEMULLER et al. 1988). Although albumin, immunoglobulin, fibrinogen and complement components are found in AD they can also be found to a lesser degree (ROZEMULLER et al. 1988) in controls. So, however, can plaques and tangles, the histological hallmarks of AD. Recently, a rather more specific association between serum amyloid P, a component of most systemic amyloidoses, and plaques and tangles has been reported (e.g., KALARIA and GRAHOVAC 1990). This is of barrier interest for, as far as is known, this protein is only synthesised in the liver. The relative importance of circulating and intracerebrally produced A4 in AD remains uncertain.

IV. Barrier Breakdown in Other Pathological States

Without exception, any destructive CNS lesion will produce a barrier defect, even if only temporary and highly localised. Many other disorders are associated with barrier disruption even though they are less obviously primarily destructive. A comprehensive description of all pathologies associated with barrier opening would be impracticable but many different disorders are discussed in the following sections.

C. An Anatomical Approach to Barrier Opening in Cerebral Pathology

When considering barrier pathology it is of value to take into account not only the different components of the endothelial cell, the major determinant of barrier function (see Chaps. 1 and 2) but also the basal lamina, pericytes, astrocytes, vascular innervation and components of the immune system. Each of these elements, individually or multiply, can be altered in a disease state and in turn interact with each other. Nevertheless, the main focus of attention will inevitably be the endothelial cell and the way in which the other components interact with it.

I. Luminal Factors

The BBB presents a large surface area to the circulation and cerebral blood flow rates are high. It is therefore not surprising that blood-borne factors are of fundamental interest in barrier pathology. These factors may be physical or chemical.

1. Loss of Autoregulation

Cerebral blood flow is normally maintained within narrow limits over a relatively wide range of perfusion pressures. Outside these limits, or when the autoregulatory mechanisms fail, life-threatening pathology can ensue.

a) Increased Vessel Wall Tension

One "final common pathway" resulting in barrier opening is increased circumferential tension in vessel walls. From Laplace's law, an increase in pressure or vessel size may lead to opening (BRADBURY 1979). This phenomenon may be seen following head injury, particularly in children, in altitude sickness, cerebral ischaemia, acute hypertension (chronic hypertension may actually be protective), epilepsy and following neurosurgical procedures involving large-scale intracranial decompression. The most extensively studied of these disorders have been acute hypertension (NAG and HARIK 1987; NAG 1986) and hypercapnia (SØRENSEN 1974). In the former, the capacity of the circulation to autoregulate is overcome and in the latter the reflex arteriolar dilatation presents excessive pressure to the capillary bed. If autoregulatory capacity of the cerebrovascular bed is pharmacologically impaired acute hypertensive insults produce greater barrier damage

(HOLLERHAGE et al. 1988). Conversely, attempts to reduce post-ischaemic hyperaemia reduce leakage and tissue damage (SEIDA et al. 1988).

The close inter-relationship between increased flow, vasodilatation and barrier opening makes it important to measure flow in parallel to barrier permeability when attempting barrier manipulation (e.g., SHARMA 1987) if direct endothelial effect is to be distinguished from changes in flow (GREENWOOD et al. 1989; HOLLERHAGE et al. 1988). Barrier opening by hyperosmolar solutions also operates by this mechanism (see Chaps. 20 and 21).

b) Cerebral Ischaemia/Hypoxia

At the other extreme lies cerebral ischaemia where perfusion drops below a critical level. Before any barrier defect becomes apparent there must be sufficient reperfusion and, as ischaemia interferes with autoregulation, there may be an early hyperperfusion event (see Sect. C.I.1.a) followed later by barrier breakdown secondary to the cerebral ischaemia per se (KUROIWA et al. 1985). Unlike the brain the endothelial cell barrier is relatively resistant to ischaemia/hypoxia and there is mounting evidence that energy deprivation actually inhibits barrier opening (BROMAN 1944; HOSSMANN and OLSSON 1971; LUTHERT et al. 1987; KUROIWA et al. 1988). An exception is an investigation of the energy-dependence of barrier integrity in the frog (OLESEN 1986) where species differences may be of importance.

2. Circulating Factors

There are few investigations in which circulating chemical agents have been shown to contribute to barrier dysfunction. One exception is an elegant study where DIETRICH et al. (1988) have used a cross-circulation system to demonstrate that circulating factor(s) generated during cerebral ischaemia can provoke barrier opening. Heat and immobilisation stress (SHARMA and DEY 1987, 1988) in the rat has been shown to produce barrier opening apparently related to the production of high circulating levels of serotonin. It is entirely possible that in man, particularly in multiple organ failure and during other systemic disturbances, barrier changes take place through similar mechanisms. Finally, some of the mediators discussed below and in Chap. 20 possibly gain access to the endothelium from the circulation as well.

II. Reorganisation of Angioarchitecture

Several pathological states demonstrate striking remodelling of the vasculature. There may be changes in the area available for exchange and, even if the overall vascularity is maintained, should the normally highly ordered arrangement of vessels be destroyed, flow may become sluggish and heterogeneous from area to area. Loss of the normal relationships between endothelium and other cells may also have profound consequences on the reactivity and functional properties of the barrier.

1. Angioarchitecture of Cerebral Tumours

The complex pattern of organisation of tumour vasculature is the result of several processes. Tumours produce angiogenesis factors (FOLKMAN and KLAGSBRUN 1987) and new vessel formation may be discernible at the tumour's edge (WELLER et al. 1977; DEANE and LANTOS 1981a). Tumour also appears to be able to take over existing vessels as it infiltrates (LUTHERT and LANTOS 1985). The uncoordinated growth of tumour between established vessels can lead to a reduction in vascularity and the resultant necrotic portions of the tumour, which may become massive, receive no blood flow and effectively have no barrier as a result.

Generally tumour vessels are larger (see Sect. C.I.1.a) and more irregular (Fig. 1) in cross section when compared to normal vessels (LUTHERT and LANTOS 1985). They show a tortuous path through the tumour and flow is often lower than that of the "host" tissue. Such flow as is present seems to be more readily compromised than normal (LUTHERT and GREENWOOD, submitted) but some responses to vasoactive agents remain (PANTHER et al. 1985).

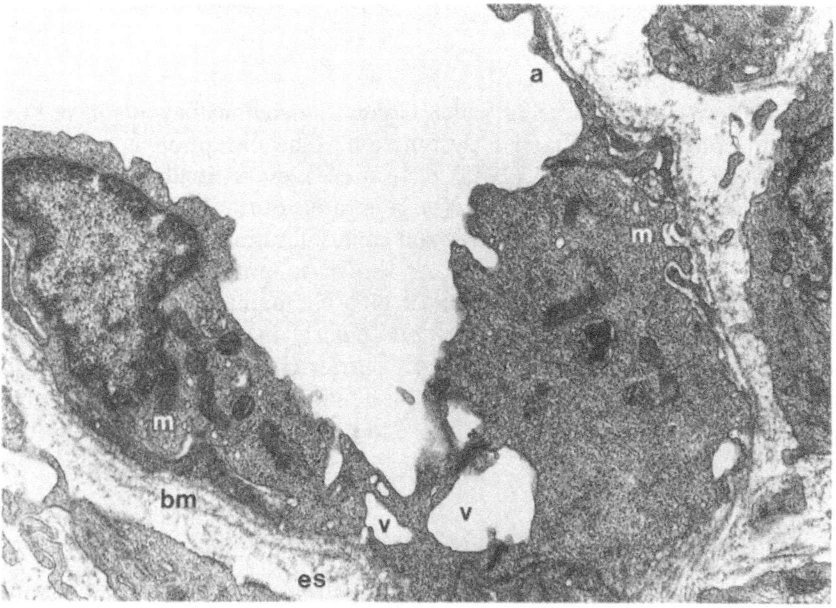

Fig. 1. Electron micrograph of experimental glioma vessel. Note the irregularity of the lumen, the just-discernible microvesicular profiles (*m*), duplicated basement membrane (*bm*), large vacuoles (*v*), widened extracellular space (*es*) and attenuated cytoplasm approaching fenestration (*a*). Magnification ×12 500

2. Angioarchitecture in Alzheimer's Disease

A number of studies have reported a reduction in vascularity which largely parallels the loss of neuronal elements (e.g., FISCHER et al. 1990). Whilst it seems probable that the vascular changes are secondary to loss of neuronal elements a more direct involvement of vessel elimination in AD cannot be excluded (LUTHERT and WILLIAMS, submitted). A discussion of the possibility that a barrier defect may lie at the centre of the pathogenesis of AD (HARDY et al. 1986) is beyond the scope of this chapter.

III. The Endothelial Cell

The endothelial cell's pivotal role in normal barrier function makes it no surprise that it has received great attention from barrier pathologists. However, although there is a consensus that the barrier is lost in many conditions, many controversies remain.

1. Ultrastructural Correlates of Barrier Breakdown

In this section some of the electron microscopical features of pathological cerebral endothelium are reviewed. To date, assigning a particular structural abnormality to a transendothelial transport role has proved to be fraught with difficulties: The number of "channels" required to produce the observed degree of leakage is very low and the sampling problems of electron microscopy make it difficult to know whether what is seen is specifically relevant to barrier breakdown or whether it is simply a manifestation of an "activated" endothelium. Possible fixation artefacts (NAGY et al. 1988) complete the confusion.

a) Glycocalyx

The luminal aspect of the endothelium possesses a glycoprotein coat that bears a negative charge which can be demonstrated with cationic electron dense markers. The abluminal surface and basement membrane also bear a negative charge but the molecules responsible differ (VORBRODT 1989). Stripping of this charge with agents such as protamine is associated with barrier opening (STRAUSBAUGH 1987), and parallel loss of charge and barrier dysfunction have been shown both in tumours (VINCENT et al. 1988) and in scrapie (LOSSINSKY et al. 1987; however, see EIKELENBOOM et al. 1987). Components of the glycocalyx bind lectins and studies of lectin binding have shown changes in acute hypertension (NAGY et al. 1988) and human tumours (DEBBAGE et al. 1988).

b) Cytoplasmic Changes

Normal cerebral endothelium shows few cytoplasmic membranous profiles (see Chap. 1) and the presence of any such feature in pathological endothelium makes it a likely candidate for a transendothelial transport system. The case for its involvement is strengthened by showing that it contains an electron dense polar marker such as lanthanum or horseradish peroxidase (HRP) following its intravascular or intraventricular injection. The most commonly reported abnormality is the microvesicular profile which is almost universally found in pathological cerebral endothelium (CERVOS-NAVARRO et al. 1983). For instance, they are seen in association with trauma (OLSSON et al. 1990) and tumours (Fig. 1) (LONG 1970; DEANE and LANTOS 1981a; NISHIO et al. 1983). They are also found in association with the administration of mediators of barrier opening such as histamine (DUX et al. 1988) and dibutyryl cyclic nucleotides (JOÓ 1987). A recent study has shown that gliomas found to have an open barrier as assessed by ^{99m}technetium-pertechnetate uptake had a higher number of cytoplasmic vesicles (KOHN et al. 1989; however, see also NIR et al. 1989). Whether these vesicular profiles can be interpreted in functional terms as a shuttle operating in one (DUX et al. 1988) or both directions is far from clear. Their number may be increased by aldehyde fixatives (NAGY et al. 1988) and serial sectioning studies (see Chap. 1 and COOMBER and STEWART 1986) have demonstrated that many apparently "free" vesicles are interconnected with others to form a cluster which invaginates from the luminal or abluminal aspect of the endothelium. Transient fusions between opposing clusters may provide a route across.

Another candidate is the transendothelial tubule. These have been most clearly demonstrated using high voltage electron microscopy with HRP as an electron dense tracer designed to demonstrate continuity between luminal and abluminal membranes. With this technique thick sections can be examined to demonstrate the entirety of the tubule but the resulting pictures can be difficult to interpret. A full discussion with references is given in a recent study by LOSSINSKY et al. (1989) where a clear distinction is made between transendothelial tubules and those running into the endosomal system.

A less frequently seen finding is attenuation of the cytoplasmic space to the point of fenestrae formation (Fig. 1) (DEANE and LANTOS 1981a; STEWART et al. 1985).

The most severe type of damage which clearly occurs in traumatic lesions (MAXWELL et al. 1988) is endothelial disruption. There are also a number of reports showing this degree of injury following other insults.

Great emphasis has been given to the opening of tight junctions (for example in tumours; see NIR et al. 1989). Again, sampling issues and the improbability of a tracer-filled open junction lying entirely within the plane of an ultrathin, or moderately thick section complicate matters. In many pathological conditions, for each study suggesting increased vesicular transport to be the mechanism of barrier disruption there will be another implicating junctional changes; perhaps the two tend to go together and other confounding factors determine which is the most apparent in a given investigation. Tight junctional abnormalities are generally believed to be of major importance in hyperosmolar opening (NAGY et al. 1988; Chaps. 20 and 21).

IV. The Basement Membrane

The basement membrane lies between the endothelium and subjacent brain (or tumour) and does not appear to offer a significant barrier to proteins. It also surrounds pericytes and, in larger vessels, smooth muscle cells. It can be visualised at the light microscope level by traditional reticulin stains or, more recently, by immunohistochemistry to its constituent protein elements.

In a number of conditions, including epilepsy (CORNFORD and OLDENDORF 1986), there is thickening of the basement membrane and characteristically it becomes fragmented and multi-layered in tumours (Fig. 1; LONG 1970; WELLER et al. 1977; DEANE and LANTOS 1981a). In AD the vascular deposition of A4 protein in relation to vessels (Fig. 2) leads to obliteration and destruction of the basement membrane and it is possible that interference with basement membrane function is responsible for the barrier deficits reported in relation to congophilic angiopathy. A general deterioration in the basement membrane in AD has also been reported (SCHEIBEL and DUONG 1988).

Basement membrane may have an important structural role in providing an anchor for cell adhesion molecules and growth factors arising from the endothelial and astrocytic aspects. The basement membrane may provide an important substrate for chemical signals between astrocytes and endothelium and in in vitro studies a suitable substrate has been shown to be

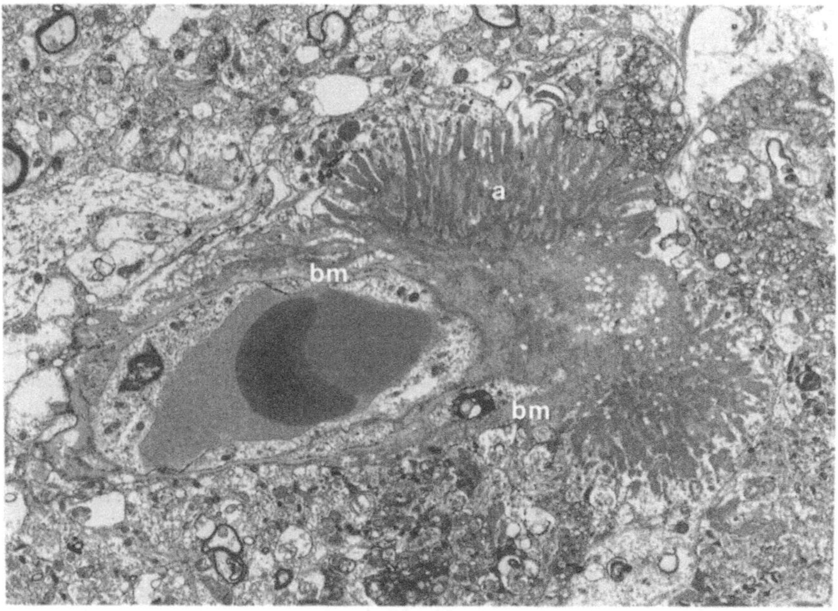

Fig. 2. Electron micrograph of vessel from cortex of a case of Alzheimer's disease. Note the amyloid (*a*) extending from the basement membrane (*bm*). Magnification ×4350 (with permission from Dr. I. JANOTA)

necessary for astrocyte-conditioned culture medium to induce tight junctions between rat cerebral endothelial cells (ARTHUR et al. 1987).

There have been relatively few attempts to investigate basement membrane role in barrier function directly although it has been shown that intravascularly delivered proteases lead to barrier opening (GODEAU and ROBERT 1979). Whether this is a direct abluminal effect on the basement membrane or a more non-specific effect on the cerebral endothelium is debatable.

V. Pericytes

Pericytes form an extensive yet incomplete layer around the cerebral microvasculature in a position within the basement membrane that is analogous to the smooth muscle coat of larger vessels and, indeed, there is evidence relating the pericyte to smooth muscle. The pericyte also appears to have a macrophagic role. It accumulates material with ageing and it has been suggested that an age-related loss of pericytes might reduce the capacity of the aged brain to cope with transient barrier leaks (STEWART et al. 1987a). More recently, pericytes have been shown to express major histocompatibility complex (MHC) class II and myelomonocytic antigens following cerebral injury, confirming their macrophage-like role (GRAEBER et al. 1990).

VI. Astrocytes

1. The Role of the Astrocyte in Barrier Maintenance

There is now increasing evidence from transplant, grafting, anatomical and in vitro (ARTHUR et al. 1987; JANZER and RAFF 1987; MAXWELL et al. 1987) studies that astrocytes, and probably a particular subset (RAFF et al. 1987), play a central role in the induction of barrier properties. It is therefore to be expected that astrocyte malfunction might lead to loss of barrier integrity. A reciprocal effect of cultured endothelium upon astrocyte membrane assemblies has also been demonstrated (TAO-CHENG et al. 1990). There are, however, instances where tight endothelium is not intimately associated with astrocytes but other barrier inducing mechanisms could operate in these circumstances. Similarly, close apposition of reactive glial processes to endothelial cells is seen in some grafting experiments where the barrier remains leaky (KRUM and ROSENSTEIN 1989).

2. Astrocyte Reactions to Cerebral Injury

Few if any pathological states selectively damage astrocytes but a relatively specific astrocyte swelling is seen in response to their extreme metabolic load in liver failure. Here, however, the barrier appears to remain largely intact to large (Lo et al. 1987), but not necessarily small molecules (KATO et al. 1989), at least until secondary effects such as raised intracranial pressure add another dimension to the pathology.

In tumours, and at later stages of a wide range of pathologies, the endothelium's astrocytic investment is either separated or completely lost. Before loss of astrocyte factors can be implicated in these circumstances it will be necessary to reverse the barrier opening with appropriate astrocyte-derived factors. Neoplastic astrocytes actually produce an endothelial permeability factor (CRISCUOLO et al. 1988) which opens the barrier. This may well, with other factors (see below), have an important bearing on barrier opening in glial tumours and other tumours. On the other hand, induction of gamma-glutamyl transpeptidase (a barrier-specific enzyme) in vitro has been demonstrated with media from C6 glioma cells (MAXWELL et al. 1987).

VII. Blood-Brain Barrier Innervation

Recent interest has focused upon the innervation of the barrier and its potential role in cerebral pathology. Whilst the large intracranial vessels receive innervation from the sympathetic chain there is increasing evidence that the locus coeruleus has a role in barrier function. Lesions of the locus coeruleus increased the susceptibility of the arteriolar barrier to insults such as hypertension (NAG and HARIK 1987). This may be a direct effect or perhaps more likely by modulating the autoregulatory system. Cerebral microvessels possess a wide range of receptors for neurotransmitters (O'NEILL et al. 1990) and it seems likely that they respond to circulating and freely diffusing extracellular transmitters as well as to more specific local innervation. The situation in tumours is, however, complex. On the one hand, tumour circulation may be abnormally susceptible to some vasoactive agents because they can gain access to the abluminal side and, on the other, at least some tumour microvessels lack adrenergic receptors (MAGNONI et al. 1988). Clearly any destructive lesion, including a tumour, will interfere with any fine perivascular nerve terminals. Several aspects of the innervation of cerebral vasculature have been reviewed by MACKENZIE and SCATTON (1987).

The locus coeruleus is of especial interest in neurodegenerative disease as it is not infrequently the site of severe damage. It might even play a crucial role in the pathogenesis of AD where neocortical plaque formation leads to retrograde locus damage which, in turn, could lead to focal barrier and further plaque formation and so the vicious circle continues (HARDY et al. 1986). SCHEIBEL and DUONG (1988) have reported changes in the innervation of the microcirculation in AD. The locus coeruleus is also damaged in Parkinson's disease but, to date, there appears to be no compelling evidence for barrier pathology in that condition.

VIII. Immune Aspects

Immunological interactions with the cerebral endothelium are covered in detail elsewhere in this volume (Chap. 17). Here it is worth emphasising that elements of the immune system are involved in most neuropathological conditions. Perivascular lymphocyte cuffs are not infrequently seen in tumours and T cells and microglia/macrophages are features of AD. Evidence from clinical and experimental studies clearly points to immune aspects to the pathogenesis of barrier dysfunction in certain demyelinating disorders.

IX. Parenchymal Factors

Any destructive lesion, whether it involves the endothelium or not, will yield a large number of substances that may contribute to further barrier breakdown. Lymphokines associated with an immune response have already been mentioned. Of particular recent interest has been the liberation of free radicals and their role in tissue damage, barrier breakdown and oedema formation. It seems probable that membrane damage liberates, via phospholipase action, arachidonic acid which directly or by metabolism into prostaglandins and thromboxanes, or leukotrienes increases barrier permeability (CHAN and FISHMAN 1984; UNTERBERG et al. 1987; VILLACARA et al. 1989). From the point of view of the pathogenesis of barrier permeability in tumours, it is of interest that leukotrienes have been shown to be liberated by glioma cells in culture (BLACK et al. 1986). It has been proposed that this conversion of arachidonic acid liberates free radicals (KONTOS and WEI 1986), but this has been questioned (UNTERBERG et al. 1988). Nevertheless, interference with free radical generation appears to be a useful therapeutic approach (for instance MARTZ et al. 1990).

Other likely locally generated mediators of barrier disruption include bradykinin, the kallikrein-kinin system and glutamate (BAETHMANN et al. 1988), polyamines (KOENIG et al. 1989), which have been shown to be released from increasing malignant gliomas in increasing amounts (MARTON et al. 1976), and platelet activating factor (see Chap. 20 for references). An interesting area for further study will be that of growth factors which are produced by tumours and injured tissue.

Finally, it has been suggested that the disruption of parenchymal cells leads to an increased osmotic load in the extracellular space which, in turn, will supplement the osmotic elements leaking from the circulation and contribute to oedema formation (KUROIWA et al. 1988).

D. Consequences of Barrier Opening

I. Benefits of Barrier Opening

Outside the CNS it is generally accepted that oedema formation is an active and potentially beneficial component of the inflammatory response. Given that barrier breakdown appears to be active it is reasonable, at least in principle, to consider what benefits it might bring.

The barrier clearly presents a significant obstacle to circulating immunoglobulin and this may contribute to the late CNS damage seen in some viral infections. If the barrier is osmotically opened after administration of anti-herpes simplex immunoglobulin G the CNS consequences of a prior virus challenge are significantly diminished (KENT and MCKENDALL 1988). Failure of the barrier to open once cerebral damage has begun to take place might lead to even more severe pathology in a range of infectious conditions. On a similar note, despite all the problems of drug delivery to the brain, it is undoubtedly enhanced when the barrier is damaged. This is, however, unlikely to have been a significant evolutionary influence until rather recently!

CNS regeneration is generally absent and at best minimal. It has, however, been argued that the opening of the barrier and resultant oedema and enlarged extracellular space might provide an environment similar to that in development through which some processes might be able to regenerate (IKUTA et al. 1983).

II. Complications of Barrier Opening

Space does not permit a detailed discussion of the adverse consequences of barrier opening in pathological states. Briefly, however, by far the most major clinical problem is cerebral oedema with the associated life-threatening raised intracranial pressure. Several aspects of the pathogenesis of oedema have been reviewed elsewhere (KLATZO 1987; JOÓ 1987). Following severe disruption small numbers of lymphocytes cross (KAJIWARA et al. 1990) and, as would be expected, the brain becomes susceptible to neurotoxic agents including adriamycin (KONDO et al. 1987) and convulsants (REMLER and MARCUSSEN 1986). Similarly, vasoactive substances like histamine whose receptors are situated abluminally are able to act from the luminal side of the endothelium (VESELY et al. 1987; VEZZANI et al. 1989). That barrier opening per se may produce neuronal damage has been suggested by a number of recent studies (SOKRAB et al. 1988; SUZUKI et al. 1988), but it is extremely difficult to be certain that the process employed to open the barrier is not causing damage by some other mechanism.

E. Manipulation of the Pathologically Disturbed Barrier

Strategies for closing the opened barrier and for additional opening of the closed barrier would provide material for an entire separate review. The major approach to the latter has been hyperosmolar opening (see Chap. 22). The rapid increase in the understanding of increasingly subtle aspects of cerebral endothelial cell physiology and pathophysiology makes probable the development of novel, non-toxic, well-targeted techniques for barrier opening in the next few years. As part of an integrated approach to barrier manipulation it will be equally important to be able to close the barrier after opening it and in ready opened situations. This will not only reduce the likelihood of life-threatening cerebral oedema but will reduce the efflux of any non-irreversibly bound therapeutic agent from brain into the circulation. There are already a number of approaches to this.

The major clinical strategy for the reduction of cerebral oedema remains the use of glucocorticoids, especially dexamethasone. These undoubtedly work in a wide variety of ways (LUTHERT and GREENWOOD 1990) and in some instances a reduction of barrier permeability seems to play a part. In others it does not (LUTHERT et al. 1986). Steroids may modulate specific transport processes; they produce cerebral vasoconstriction and possibly alter the spread and resolution of oedema fluid as well as its formation. Interference with the cascades of mediators that generate permeability-promoting species is another area where steroids may operate (BLACK et al. 1986; CHAN and FISHMAN 1984; JOÓ 1987; CRISCUOLO et al. 1988; REICHMAN et al. 1986). These systems may also be targeted by other agents and inhibition of any of the "damage response" systems may directly (by action on cerebral endothelium) or indirectly (by reducing tissue damage) lead to a reduction in oedema by reducing vascular permeability. Effects have been demonstrated experimentally with non-steroidal anti-inflammatory agents (REICHMAN et al. 1986), inhibitors of serotonin (OLSSON et al. 1990), inhibitors of ornithine decarboxylase (KOENIG et al. 1989) and inhibitors of free radical generation (SØRENSEN 1974).

F. Conclusions

The opening of the BBB in cerebral pathology plays a pivotal role in the pathogenesis of many conditions. It may lead to death through the formation of cerebral oedema and it may be a major determinant in the efficacy or otherwise of drug therapy. These disorders, and the endothelial changes occurring within them, provide an invaluable insight into the repertoire of cellular mechanisms available within cerebral endothelium that may be exploited in other situations where control of barrier function is of prime importance. The next few years will almost certainly show the BBB to be an even more vital feature of CNS therapeutics than it is already.

References

Arthur FE, Shivers RR, Bowman PD (1987) Astrocyte-mediated induction of tight junctions in brain capillary endothelium: an efficient in vitro model. Brain Res 433:155–159

Baethmann A, Maier-Hauff K, Kempski O, Unterberg A, Wahl M, Schurer L (1988) Mediators of brain edema and secondary brain damage. Crit Care Med 16:972–978

Black KL, Hoff JT, McGillicuddy JE, Gebarski SS (1986) Increased leukotriene C4 and vasogenic edema surrounding brain tumors in humans. Ann Neurol 19: 592–595

Bradbury MWB (1979) The concept of a blood-brain barrier. Wiley, Chichester

Broman T (1944) Supravital analysis of disorders in the cerebral vascular permeability in man. Acta Med Scand 118:76–83

Bullard BE, Adams CJ, Coleman RE, Bigner DD (1986) In vivo imaging of intracranial human glioma xenografts comparing specific with nonspecific radiolabelled monoclonal antibodies. J Neurosurg 64:257–262

Cervos-Navarro J, Artigas J, Mrsulja BJ (1983) Morphofunctional aspects of the normal and pathological blood-brain barrier. Acta Neuropathol (Berl) [Suppl] 8:1–19

Chan PH, Fishman RA (1984) The role of arachidonic acid in vasogenic brain edema. Fed Proc 43:210–213

Coomber BL, Stewart PA (1986) Three-dimensional reconstruction of vesicles in endothelium of blood-brain barrier versus highly permeable microvessels. Anat Rec 215:256–261

Cornford EM, Oldendorf WH (1986) Epilepsy and the blood-brain barrier. Adv Neurol 44:787–812

Criscuolo GR, Merrill MJ, Oldfield EH (1988) Further characterization of malignant glioma-derived vascular permeability factor. J Neurosurg 69:254–262

Deane BR, Lantos PL (1981a) The vasculature of experimental brain tumours. I. A sequential light and electron microscopic study of angiogenesis. J Neurol Sci 49:55–66

Deane BR, Lantos PL (1981b) The vasculature of experimental brain tumours. II. A quantitative assessment of morphological abnormalities. J Neurol Sci 49:67–77

Deane BR, Papp MI, Lantos PL (1984) The vasculature of experimental brain tumours. III. Permeability studies. J Neurol Sci 65:47–58

Debbage PL, Gabius HJ, Bise K, Marguth F (1988) Cellular glycoconjugates and their potential endogenous receptors in the cerebral microvasculature of man: a glycohistochemical study. Eur J Cell Biol 46:425–434

Dietrich WD, Prado R, Watson BD (1988) Photochemically stimulated blood-borne factors induce blood-brain barrier alterations in rats. Stroke 19:857–862

Dux E, Doczi T, Joo F, Szerdahelyi P, Siklos L (1988) Reverse pinocytosis induced in cerebral endothelial cells by injection of histamine into the cerebral ventricle. Acta Neuropathol (Berl) 76:484–488

Eikelenboom P, Scott JR, McBride PA, Rozemuller JM, Bruce ME, Fraser H (1987) No evidence for involvement of plasma proteins or blood-borne cells in amyloid plaque formation in scrapie-affected mice. An immunohistoperoxidase study. Virchows Arch [B] 53:251–256

Ellison MD, Povlishock JT, Hayes RL (1986) Examination of the blood-to-brain transfer of alpha-aminoisobutyric acid and horseradish peroxidase: regional alterations in blood-brain barrier function following acute hypertension. J Cereb Blood Flow Metab 6:471–480

Fischer VW, Siddiqi A, Yusufaly Y (1990) Altered angioarchitecture in selected areas of brains with Alzheimer's disease. Acta Neuropathol (Berl) 79:672–679

Folkman J, Klagsbrun M (1987) Angiogenic factors. Science 235:442–447

Godeau G, Robert AM (1979) Mechanism of action of collagenase on the blood-brain barrier permeability. Increase of endothelial cell pinocytotic activity as shown with horse-radish peroxidase as a tracer. Cell Biol Int Rep 3:747–751
Graeber MB, Streit WJ, Kiefer R, Schoen SW, Kreutzberg GW (1990) New expression of myelomonocytic antigens by microglia and perivascular cells following lethal motor neuron injury. J Immunol 27:121–132
Greenwood J, Hazell AS, Luthert PJ (1989) The effect of a low pH saline perfusate upon the integrity of the energy-depleted rat blood-brain barrier. J Cereb Blood Flow Metab 9:234–242
Groothuis DR, Molnar P, Blasberg RG (1984) Regional blood flow and blood-to-tissue transport in five brain tumor models. Prog Exp Tumor Res 27:132–153
Hardy JA, Mann DMA, Wester P, Winblad B (1986) An interactive hypothesis concerning the pathogenesis and progression of Alzheimer's disease. Neurobiol Aging 7:489–502
Hollerhage HG, Gaab MR, Zumkeller M, Walter GF (1988) The influence of nimodipine on cerebral blood flow autoregulation and blood-brain barrier. J Neurosurg 69:919–922
Hossmann K-A, Olsson Y (1971) Influence of ischaemia on the passage of protein tracers across capillaries in certain blood-brain barrier injuries. Acta Neuropathol (Berl) 18:113–122
Ikuta F, Yoshida Y, Ohama E, Oyanagi K, Takeda S, Yamazaki K, Watabe K (1983) Revised pathophysiology on BBB damage: the edema as an ingeniously provided condition for cell motility and lesion repair. Acta Neuropathol (Berl) [Suppl] 8:103–110
Janzer RC, Raff MC (1987) Astrocytes induce blood-brain barrier properties in endothelial cells. Nature 325:253–257
Joó F (1987) A unifying concept on the pathogenesis of brain oedemas. Neuropathol Appl Neurobiol 13:161–176
Kajiwara K, Ito H, Fukumoto T (1990) Lymphocyte infiltration into normal rat brain following hyperosmotic blood-brain barrier opening. J Immunol 27:133–140
Kalaria RN, Grahovac I (1990) Serum amyloid P immunoreactivity in hippocampal tangles, plaques and vessels: implications for leakage across the blood-brain barrier in Alzheimer's disease. Brain Res 516:349–353
Kato M, Sugihara J, Nakamura T, Muto Y (1989) Electron microscopic study of the blood-brain barrier in rats with brain edema and encephalopathy due to acute hepatic failure. Gastroenterol Jpn 24:135–142
Kent TA, McKendall RR (1988) The effect of modification of the blood-brain barrier to immunoglobulin in the course of herpes simplex myelitis. Ann NY Acad Sci 529:272–274
Klatzo I (1987) Pathophysiological aspects of brain edema. Acta Neuropathol (Berl) 72:236–239
Koenig H, Goldstone AD, Lu CY, Trout JJ (1989) Polyamines and Ca^{2+} mediate hyperosmolal opening of the blood-brain barrier: in vitro studies in isolated rat cerebral capillaries. J Neurochem 52:1135–1142
Kohn S, Front D, Nir I (1989) Blood-brain barrier permeability of human gliomas as determined by quantitation of cytoplasmic vesicles of the capillary endothelium and scintigraphic findings. Cancer Invest 7:313–321
Kondo A, Inoue T, Nagara H, Tateishi J, Fukui M (1987) Neurotoxicity of adriamycin passed through the transiently disrupted blood-brain barrier by mannitol in the rat brain. Brain Res 412:73–83
Kontos HA, Wei EP (1986) Superoxide production in experimental brain injury. J Neurosurg 64:803–807
Krum JM, Rosenstein JM (1989) The fine structure of vascular-astroglial relations in transplanted fetal neocortex. Exp Neurol 103:203–212

Kumar R, Harvey SA, Kester M, Hanahan DJ, Olson MS (1988) Production and effects of platelet-activating factor in the rat brain. Biochim Biophys Acta 963:375–383

Kuroiwa T, Ting P, Martinez H, Klatzo I (1985) The biphasic opening of the blood-brain barrier to proteins following temporary middle cerebral artery occlusion. Acta Neuropathol (Berl) 68:122–129

Kuroiwa T, Shibutani M, Okeda R (1988) Blood-brain barrier disruption and exacerbation of ischemic brain edema after restoration of blood flow in experimental focal cerebral ischemia. Acta Neuropathol (Berl) 76:62–70

Levin VA, Freeman-Dove M, Landahl HD (1975) Permeability characteristics of brain adjacent to tumors in rats. Arch Neurol 32:785–791

Lo WD, Ennis SR, Goldstein GW, McNeely DL, Betz AL (1987) The effects of galactosamine-induced hepatic failure upon blood-brain barrier permeability. Hepatology 7:452–456

Long DM (1970) Capillary ultrastructure and the blood-brain barrier in human malignant brain tumors. J Neurosurg 32:127–144

Lossinsky AS, Moretz RC, Carp RI, Wisniewski HM (1987) Ultrastructural observations of spinal cord lesions and blood-brain barrier changes in scrapie-infected mice. Acta Neuropathol (Berl) 73:43–52

Lossinsky AS, Song MJ, Wisniewski HM (1989) High voltage electron microscopic studies of endothelial cell tubular structures in the mouse blood-brain barrier following brain trauma. Acta Neuropathol (Berl) 77:480–488

Luthert PJ, Lantos PL (1985) A morphometric study of the microvasculature of a rat glioma. Neuropathol Appl Neurobiol 11:461–473

Luthert PJ, Greenwood J (1990) Experimental studies and the blood-brain barrier. In: Capildeo R (ed) Steroids in disease of the central nervous system. Wiley, Chichester

Luthert PJ, Greenwood J, Lantos PL, Pratt OE (1986) The effect of dexamethasone on vascular permeability of experimental brain tumours. Acta Neuropathol (Berl) 69:288–294

Luthert PJ, Greenwood J, Pratt OE, Lantos PL (1987) The effect of a metabolic inhibitor upon the properties of the cerebral vasculature during a whole-head saline perfusion of the rat. Q J Exp Physiol 72:129–141

MacKenzie ET, Scatton B (1987) Cerebral circulatory and metabolic effects of perivascular neurotransmitters. CRC Crit Rev Clin Neurobiol 2:357–419

Magnoni MS, Frattola L, Piolti R, Govini S, Kobayashi H, Trabucchi M (1988) Glial brain tumors lack microvascular adrenergic receptors. Eur Neurol 28:27–29

Marton LJ, Heby O, Levin VA, Lubich WP, Crofts DC, Wilson CB (1976) The relationship of polyamines in cerebrospinal fluid to the presence of central nervous system tumors. Cancer Res 36:973–977

Martz D, Beer M, Betz AL (1990) Dimethylthiourea reduces ischemic brain edema without affecting cerebral blood flow. J Cereb Blood Flow Metab 10:352–357

Masters CL, Beyreuther KT (1989) The pathology of the amyloid precursor of Alzheimer's disease. Ann Med 21:89–90

Maxwell K, Berliner JA, Cancilla PA (1987) Induction of gamma-glutamyl transpeptidase in cultured cerebral endothelial cells by a product released by astrocytes. Brain Res 410:309–314

Maxwell WL, Irvine A, Adams JH, Graham DI, Gennarelli TA (1988) Response of cerebral microvasculature to brain injury. J Pathol 155:327–335

Nag S (1986) Cerebral endothelial plasma membrane alterations in acute hypertension. Acta Neuropathol (Berl) 70:38–43

Nag S, Harik SI (1987) Cerebrovascular permeability to horseradish peroxidase in hypertensive rats: effects of unilateral locus coeruleus lesion. Acta Neuropathol (Berl) 73:247–253

Nagy Z, Pettigrew KD, Meiselman S, Brightman MW (1988) Cerebral vessels cryofixed after hyperosmosis or cold injury in normothermic and hypothermic frogs. Brain Res 440:315–327
Neuwelt EA, Specht HD, Hill SA (1986) Permeability of human brain tumor to 99mTc-gluco-heptonate and 99mTc-albumin. Implications for monoclonal antibody therapy. J Neurosurg 65:194–198
Nir I, Levanon D, Iosilevsky G (1989) Permeability of blood vessels in experimental gliomas: uptake of 99mTc-glucoheptonate and alteration in blood-brain barrier as determined by cytochemistry and electron microscopy. Neurosurgery 25: 523–31
Nishio S, Ohta M, Abe M, Kitamura K (1983) Microvascular abnormalities in ethylnitrosourea (ENU)-induced rat brain tumors: structural basis for altered blood-brain barrier function. Acta Neuropathol (Berl) 59:1–10
O'Neill C, Fowler CJ, Winblad B (1989) Alpha 1-adrenergic receptor binding sites in postmortal human cerebral microvessel preparations: preservation in multi-infarct dementia and dementia of the Alzheimer type. J Neurol Transm 1:308–310
Olesen SP (1986) Rapid increase in blood-brain barrier permeability during severe hypoxia and metabolic inhibition. Brain Res 368:24–29
Olsson Y, Sharma HS, Pettersson CAV (1990) Effects of p-chlorophenylalanine on microvascular permeability changes in spinal cord trauma. Acta Neuropathol (Berl) 79:595–603
Panther LA, Baumbach GL, Bigner DD, Piegors D, Groothuis DR, Heistad DD (1985) Vasoactive drugs produce selective changes in flow to experimental brain tumors. Arch Neurol 18:712–715
Pires MM, Pilkington GJ, Lantos PL (1987) Vascular permeability in transplantable murine gliomas: morphological correlation with tracer studies. Neuropathol Appl Neurobiol 12:251–262
Raff MC, Ffrench-Constant C, Miller RH (1987) Glial cells in the rat optic nerve and some thoughts on remyelination in the mammalian CNS. J Exp Biol 132:35–41
Reichman HR, Farrell CL, del Maestro RF (1986) Effects of steroids and nonsteroid anti-inflammatory agents on vascular permeability in a rat glioma model. J Neurosurg 65:233–237
Remler MP, Marcussen WH (1986) Systemic focal epileptogenesis. Epilepsia 27: 35–42
Rozemuller JM, Eikelenboom P, Kamphorst W, Stam FC (1988) Lack of evidence for dysfunction of the blood-brain barrier in Alzheimer's disease: an immunohistochemical study. Neurobiol Aging 9:383–391
Schackert G, Simmons RD, Buzbee TM, Hume DA, Fidler IJ (1988) Macrophage infiltration into experimental brain metastases: occurrence through an intact blood-brain barrier. JNCI 80:1027–1034
Scheibel AB, Duong T (1988) On the possible relationship of cortical microvascular pathology to blood-brain barrier changes in Alzheimer's disease. Neurobiol Aging 9:41–42
Schlageter NL, Carson RE, Rapoport SI (1987) Examination of blood-brain barrier permeability in dementia of Alzheimer type with [^{68}Ga] EDTA and positron emission tomography. J Cereb Blood Flow Metab 7:1–8
Schurer L, Temesvari P, Wahl M, Unterberg A, Baethmann A (1989) Blood-brain barrier permeability and vascular reactivity to bradykinin after pretreatment with dexamethasone. Acta Neuropathol (Berl) 77:576–581
Seida M, Wagner HG, Vass K, Klatzo I (1988) Effect of aminophylline on postischemic edema and brain damage in cats. Stroke 19:1275–1282
Seitz RJ, Wechsler W (1987) Immunohistochemical demonstration of serum proteins in human cerebral gliomas. Acta Neuropathol (Berl) 73:145–152
Sharma HS (1987) Effect of captopril (a converting enzyme inhibitor) on blood-brain barrier permeability and cerebral blood flow in normotensive rats. Neuropharmacology 26:85–92

Sharma HS, Dey PK (1987) Influence of long-term acute heat exposure on regional blood-brain barrier permeability, cerebral blood flow and 5-HT level in conscious normotensive young rats. Brain Res 424:153–162
Sharma HS, Dey PK (1988) EEG changes following increased blood-brain barrier permeability under long-term immobilization stress in young rats. Neurosci Res 5:224–239
Sokrab TE, Johansson BB, Tengvar C, Kalimo H, Olsson Y (1988) Adrenaline-induced hypertension: morphological consequences of the blood-brain barrier disturbance. Acta Neurol Scand 77:387–396
Sørensen SC (1974) The permeability to small ions of tight junctions between cerebral endothelial cells. Brain Res 70:174–178
Stewart PA, Hayakawa K, Hayakawa E, Farrell CL, del Maestro RF (1985) A quantitative study of blood-brain barrier permeability ultrastructure in a new rat glioma model. Acta Neuropathol (Berl) 67:96–102
Stewart PA, Magliocco M, Hayakawa K, Farrell CL, del Maestro RF, Girvin J, Kaufmann JC, Vinters HV, Gilbert J (1987a) A quantitative analysis of blood-brain barrier ultrastructure in the aging human. Microvasc Res 33:270–282
Stewart PA, Hayakawa K, Farrell CL, del Maestro RF (1987b) Quantitative study of microvessel ultrastructure in human peritumoral brain tissue. Evidence for a blood-brain barrier defect. J Neurosurg 67:697–705
Strausbaugh LJ (1987) Intracarotid infusions of protamine sulfate disrupt the blood-brain barrier of rabbits. Brain Res 409:221–226
Suzuki M, Iwasaki Y, Yamamoto T, Konno H, Kudo H (1988) Sequelae of the osmotic blood-brain barrier opening in rats. J Neurosurg 69:421–428
Tao-Cheng JH, Nagy Z, Brightman MW (1990) Astrocytic orthogonal arrays of intramembranous particle assemblies are modulated by brain endothelial cells in vitro. J Neurocytol 19:143–153
Unterberg A, Dautermann C, Baethmann A, Muller-Esterl W (1986) The kallikrein-kinin system as mediator in vasogenic brain edema. III. Inhibition of the kallikrein-kinin system in traumatic brain swelling. J Neurosurg 64:269–276
Unterberg A, Wahl M, Hammersen F, Baethmann A (1987) Permeability and vasomotor response of cerebral vessels during exposure to arachidonic acid. Acta Neuropathol (Berl) 73:209–219
Unterberg A, Wahl M, Baethmann A (1988) Effects of free radicals on permeability and vasomotor response of cerebral vessels. Acta Neuropathol (Berl) 76:238–244
Vesely R, Hoffman WE, Gil KS, Albrecht RF, Miletich DJ (1987) The cerebrovascular effects of curare and histamine in the rat. Anesthesiology 66:519–523
Vezzani A, Stasi MA, Wu HQ, Castiglioni M, Weckermann B, Samanin R (1989) Studies on the potential neurotoxic and convulsant effects of increased blood levels of quinolinic acid in rats with altered blood-brain barrier permeability. Exp Neurol 106:90–98
Vick NA, Khandekar JD (1977) Chemotherapy of brain tumors. The "blood-brain barrier" is not a factor. Arch Neurol 34:523–526
Villacara A, Spatz M, Dodson RF, Corn C, Bembry J (1989) Effect of arachidonic acid on cultured cerebromicrovascular endothelium: permeability, lipid peroxidation and membrane "fluidity". Acta Neuropathol (Berl) 78:310–316
Vincent S, DePace D, Finkelstein S (1988) Distribution of anionic sites on the capillary endothelium in an experimental brain tumor model. Microcirc Endothelium Lymphatics 4:45–67
Vorbrodt AW (1989) Ultracytochemical characterization of anionic sites in the wall of brain capillaries. J Neurocytol 18:359–368
Weller RO, Foy M, Cox S (1977) The development and ultrastructure of the microvasculature in malignant gliomas. Neuropathol Appl Neurobiol 3:307–322

CHAPTER 19

Experimental Manipulation of the Blood-Brain ”and Blood-Retinal Barriers

J. GREENWOOD

A. Introduction

There exists two major areas of research in which experimental manipulation of the blood-brain barrier (BBB) and blood-retinal barrier (BRB) are routinely employed. Firstly, there are a broad range of investigations undertaken to elucidate the various cellular mechanisms involved in precipitating both the structural and fuctional alterations of the barrier often found in diseases of the central nervous system (CNS). Secondly, there are those studies involved in the discovery and characterisation of agents capable of disrupting the integrity of the BBB and BRB with the ultimate aim of developing a safe method for enhancing the delivery of therapeutic agents to the CNS.

The capillary endothelial cells of the brain that constitute the BBB severely restrict the movement of non-transported polar molecules from the capillary lumen to the extracellular fluid of the brain. The BBB's ability to precisely control the passage of substances is a function of specialized characteristics such as tight junctions, paucity of vesicular activity and lack of fenestrations and transcapillary channels (Fig. 1; also see Chap. 1). The BRB, however, consists of two morphologically distinct sites: the retinal capillary endothelium and the retinal pigment epithelium, although their function is identical to that of the BBB and many of the factors that influence one also affect the other.

Breakdown of both the BBB and BRB is associated with many diseases of the CNS. Consequently, a great deal of research has been undertaken to

---

Abbreviations

ADP	Adenosine diphosphate	5-HT	5-hydroxytryptamine
AMP	Adenosine monophosphate	IL	Interleukin
ATP	Adenosine triphosphate	LAK	Lymphokine-activated killer
BBB	Blood-brain barrier	MIP	Macrophage inflammatory protein
BRB	Blood-retinal barrier		
cAMP	Cyclic AMP	PAF	Platelet activating factor
CNS	Central nervous system	TCDC	Taurochenodeoxycholate
DMSO	Dimethyl sulphoxide	TNF	Tumour necrosis factor
HRP	Horseradish peroxidase	VPF	Vascular permeability factor

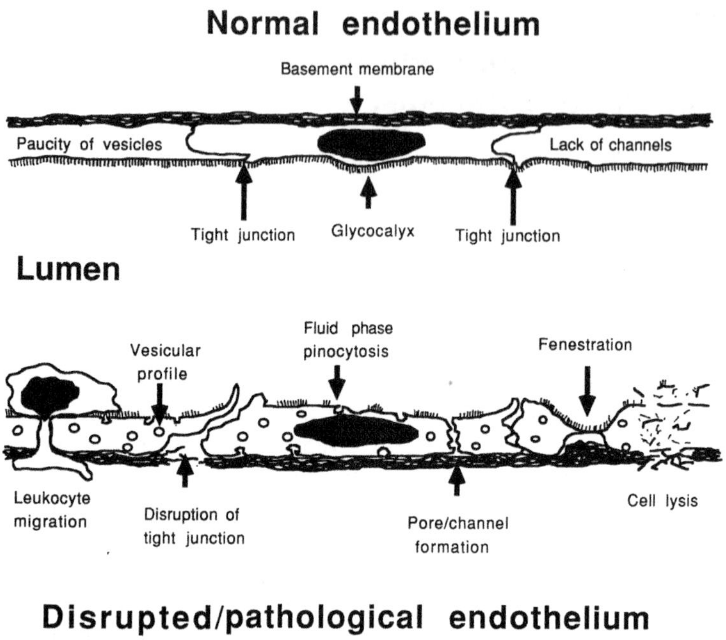

Fig. 1. Diagramatic representation of normal and pathological endothelium from blood-brain or blood-retinal barrier. A variety of potential routes through which extravasation of tracer may occur are shown

develop animal models of diseases in which the barrier is involved and to investigate the potential mediators of barrier dysfunction. This review, however, will only deal with studies in which the role of specific pathologically derived mediators upon barrier integrity have been investigated. The considerable literature relating to dysfunction of the barriers in experimentally induced animal models of disease such as ischaemia, hypoxia, hypercapnia, hypertension, tumours, and experimental allergic encephalopathy are covered in part elsewhere in this book (see Chap. 18) and will not be covered here. Similarly, radiation induced damage to the BBB and the use of cerebral grafts as a means of circumventing the barrier will not be within the scope of this review. The other major area of research that will be covered will be the studies in which substances are introduced into the circulation with the aim of inducing a reversible opening of the barrier. A vast number of different compounds have been screened for barrier effects and as such only those that have been studied in more detail will be referred to. Despite these two philosphically different approaches to manipulating the BBB and BRB, there remain many fundamental similarities in the cellular mechanisms of barrier disruption.

Almost without exception, techniques used to determine the integrity of the BBB in vivo rely upon the detection of extravasated ions or molecules that, under normal conditions, are excluded from the extracellular space of the brain. Experimental methods for studying the BBB are covered extensively elsewhere in this volume (see Chaps. 2 and 17) and readers should familiarize themselves with the different approaches and tracers used to study experimental manipulation of the barrier.

B. Manipulation of the Blood-Brain and Blood-Retinal Barriers

I. Disruptive Agents of Non-Pathological Origin

1. Hyperosmolar Solutions

The discovery that raising the molarity of the fluids surrounding the brain capillary endothelium causes a reversible disruption of the BBB has led to exhaustive investigations. The prime motive for such extensive research has been its potential as a method for enhancing the delivery of therapeutic agents to the brain and cerebral neoplasms (see Chap. 21). In most cases the hypertonic solution is delivered as a bolus or short infusion into the carotid circulation followed by the tracer or drug under investigation. The most common solutes used are the inert sugars mannitol and arabinose which are usually injected at concentrations of between $1.4\,M$ and $1.8\,M$ (RAPOPORT and ROBINSON 1986; GREENWOOD et al. 1988; COSOLO et al. 1989), although other compounds such as urea have also been used at concentrations of up to $3\,M$ (BRIGHTMAN et al. 1973). Despite intensive studies there still remains some disagreement over the mechanism by which plasma proteins and tracers become extravasated. A number of reports suggest that induction of vesicular transport (HANNSSON and JOHANSSON 1980; HOUTHOFF et al. 1982; LEHTOSALO et al. 1982; MAYHAN and HEISTAD 1985) or formation of transendothelial channels (FARRELL and SHIVERS 1984) may play a role in hypertonia-induced barrier disruption. However, the majority of experimental evidence strongly suggests that opening is brought about by osmotic shrinkage of the endothelium and pulling apart of the tight junctions (BRIGHTMAN et al. 1973; NAGY et al. 1979; RAPOPORT and ROBINSON 1986; ROBINSON and RAPOPORT 1987; GREENWOOD et al. 1988). An alternative mechanism recently reported proposes that hyperosmotic opening is mediated by increased polyamine synthesis and Ca^{2+} uptake which induces vesicular transport and opening of tight junctions (KOENIG et al. 1989a,b). By irreversibly inhibiting ornithine decarboxylase, the enzyme responsible for putrescine production, hypertonia-induced BBB opening was inhibited, an effect which could overcome by administering exogenous putrescine. In the frog, however, OLESEN and CRONE (1986) were unable to open the BBB with either intra- or extravascular administration of the polyamine putrescine.

There are many reports that attempt to define the threshold concentration required to disrupt the BBB. Although reasonably consistent values are given, various factors must be taken into consideration such as the size and lipid solubility of the tracer used, the duration over which it is presented to the vascular bed and the rate and route of administration. Furthermore, with some techniques the actual concentration of hypertonic solution at the microvascular bed of the brain may be lower than that injected into the carotid artery. With this caveat in mind, previous studies have reported a relatively small increase in opening following administation of hypertonic solutions of between 0.9 and 1.4 *M* with a more dramatic increase occurring between 1.4 and 1.8 *M* (RAPOPORT et al. 1980; NAKAGAWA et al. 1984; GREENWOOD et al. 1988). Other studies, however, have demonstrated that an increase in osmolarity of only 100 m*M* above normal is capable of opening the BBB (CSERR et al. 1987; FRASER and DALLAS 1990). Although it appears that hypertonic opening of the BBB is not an "all or nothing" effect it is not clear whether the dose-related opening is due to an increase in either the number or size of the "openings". It is likely that as the osmolarity of the fluid perfusing the brain is increased the greater will be the tension generated between the endothelial cells, and the greater the number of breached junctions. In addition, as fluid is drawn into the lumen from the tissue a more concentrated solution would be able to maintain an effective osmotic gradient further down the vascular tree. It is clear that the extravasation of visual tracer following hyperosmolar opening does not occur uniformly along the length of a vessel but only at particular points, giving a punctate multifocal pattern of leakage (Fig. 2). It is likely that these disrupted regions represent the weakest point along the vessel, and may occur preferentially at the capillary vessel bifurcation (LUTHERT and GREENWOOD, unpublished observations).

There have been many studies investigating the time course of hyperosmolar opening and the nature of the "holes" induced in the BBB. The BBB has been shown to close more rapidly to larger molecular weight tracers than to smaller ones (ZIYLAN et al. 1983, 1984), a result that is consistent with the formation of paracellular routes at the site of the tight junctions. Pores with a width of approximately 200 Å have been calculated (ROBINSON and RAPOPORT 1987), occurring at an average of approximately 1/200 μm^2, 6 min after opening. As the integrity of the barrier begins to recover there is only a minimal decrease in the number of these pores. The main factor responsible for the reduction in opening between 6 and 35 min is thought to be a dramatic reduction in bulk flow (ROBINSON and RAPOPORT 1987; ROBINSON 1987). Consistent with this is the finding that disruption of the barrier is maximal within the 5 min following an intracarotid hypertonic bolus (COSOLO et al. 1989; NEUWELT and BARNETT 1989).

The duration of intracarotid delivery of hyperosmolar solutions will also determine the extent of BBB disruption and the mortality rate of the animal. Using a 25% solution of mannitol (1.37 *M*), COSOLO et al. (1989)

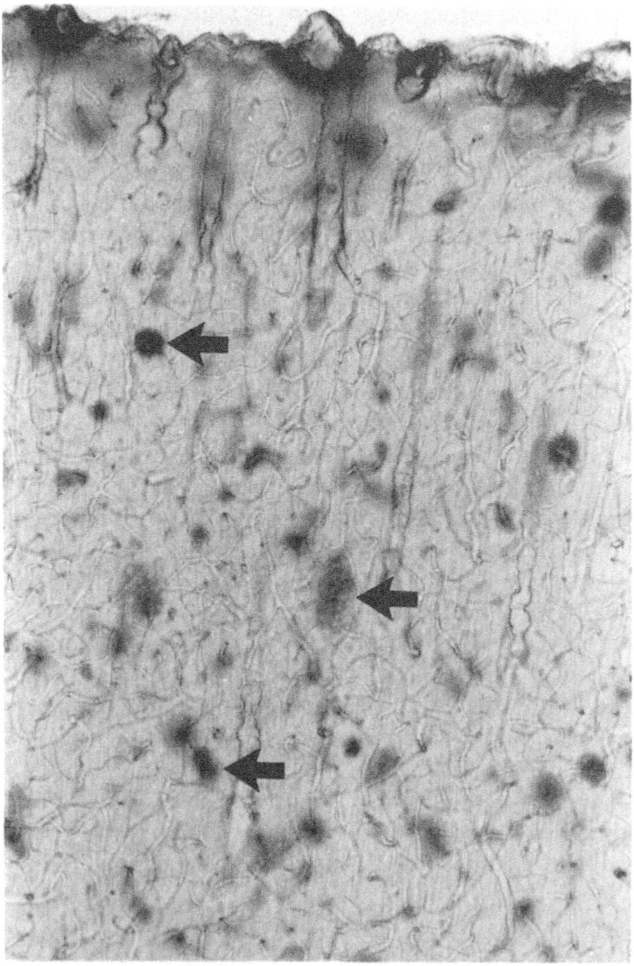

Fig. 2. Light micrograph of 100-μm-thick section through rat cerebral cortex following 1.8 *M* arabinose infusion for 1 min in the supravitaly perfused rat brain (see GREENWOOD et al. 1988). Horseradish peroxidase is used as a marker of barrier disruption. Small punctate areas of reaction product can be seen (e.g. *arrows*) but only in a small percentage of the vessels. (From LUTHERT and GREENWOOD, unpublished data)

found that a rate of $0.25\,\mathrm{ml\,s^{-1}\,kg^{-1}}$ for 20 s provided an optimal rate of delivery of tracer without producing any signs of neurological deficit. By infusing for 30 s at greater rates, the proportion of deaths amongst the animals was greatly increased.

As a method of drug delivery, hyperosmolar disruption of the barrier is far from satisfactory, but it is preferable to other reported techniques. The small degree of oedema that is sometimes reported following hypertonic opening is readily reversible upon repair of the barrier. However, even with thorough removal of particulate matter from the hypertonic bolus this procedure may still give rise to a small degree of neuronal cell death, affecting up to 25% of animals (SUZUKI et al. 1988), and occasionally mild nerological and behavioural changes (RAPOPORT et al. 1977; GREENWOOD, unpublished observations). Following a single episode the damage is likely to be small and "clinically" undetectable in the experimental animal, especially if sensitive behavioural tests are not performed. It is, however, important to consider that such damage is irreversible and, with subsequent openings, cumulative. In addition to the possibility of mild neuronal damage, hyperosmolar infusions have also been associated with an increase in cerebral blood flow and a rise in systemic blood pressure (PICKARD et al. 1977; HARDEBO 1980; HARDEBO and NILSSON 1980; GULATI et al. 1985a), both of which may independently lead to alterations in BBB permeability. Furthermore, hypertonic infusions may induce ictal activity (FIESCHI et al. 1980), which has also been reported to cause BBB disruption (for review see BOLWIG 1989) and possibly neuronal damage. The vascular delivery of contrast media for CNS imaging can also lead to a hypertonic disruption of tight junctions and induction of vesicular activity (WALDRON and BRYAN 1975; BRINKER 1979; GREENWOOD 1991a). High concentrations, especially of the ionic species of imaging agents, are injected in order to achieve sufficient opacity of the area of interest, which can lead to problems such as a reversible cortical blindness (LANTOS 1989).

In the eye, the intracarotid delivery of hyperosmolar solutions has also been shown to disrupt the BRB in a dose-related manner (Fig. 3; RAPOPORT 1977; ENNIS and BETZ 1986; FRANK et al. 1986; LI et al. 1987). As with the BBB, the prime mechanism of opening is believed to be by osmotic shrinkage of the retinal capillary endothelium and the retinal pigment epithelium with the subsequent disruption of the tight intercellular junctions.

2. Bile Salts

Another group of compounds to be evaluated as potentially safe compounds for reversible disruption of the BBB are the bile salts. The intracarotid administration of dehydrocholate disrupts the BBB of the rat for periods of up to 3 days without any reported sign of neuronal injury or parenchymal damage (SPIGELMAN et al. 1983). The mechanism leading to this increase in permeability was not elucidated although modification of the tight junctions, stimulation of vesicular transport or formation of transcellular channels were all regarded as being equally possible (SPIGELMAN et al. 1983). Bile salts, however, are strong detergents and at concentrations of 1.5 m*M* and above result in lysis of cerebral endothelial cells both in situ and in vitro (Fig. 4; GREENWOOD et al. 1991). When applied at lower concentrations it has been shown that the perfused BBB can be opened without any apparent damage to the cerebral endothelium (Fig. 4) and that this could operate in the absence of energy-producing metabolism (GREENWOOD et al. 1991). It was

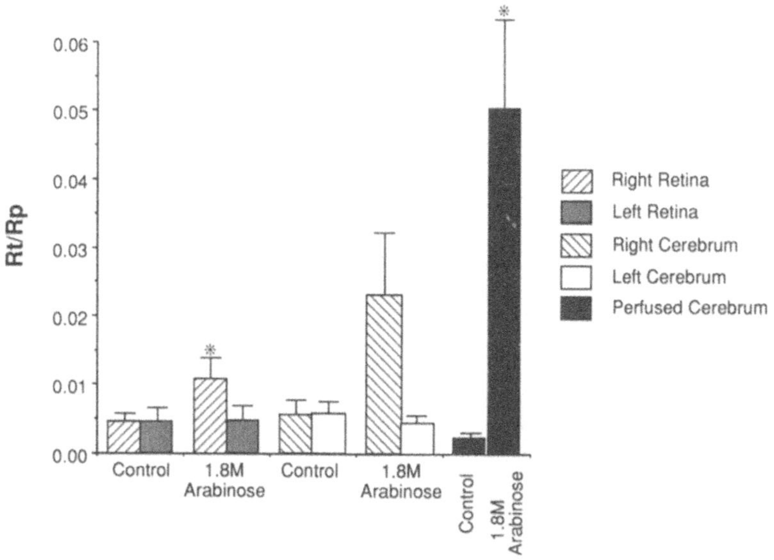

Fig. 3. Permeability of the blood-brain and blood-retinal barrier of the rat following a 20-s infusion of 1.8 M arabinose into the right carotid artery. The radiotracer ^{14}C-mannitol was maintained at a steady level within the circulation and the permeability is expressed as the ratio of tissue to plasma radioactivity (R_t/R_p; for procedure see LUTHERT et al. 1986). Control animals were infused with saline for 20 s (unpublished data from PATEL and GREENWOOD). The increase in permeability of the cerebral cortex following a 1-min perfusion of 1.8 M arabinose in the perfused rat brain are also included for comparison (data from GREENWOOD et al. 1988). * Significant difference from controls

proposed that the disruptive actions of the bile salts deoxycholate and taurochenodeoxycholate (TCDC) at low concentrations were unlikely to operate via energy-requiring processes such as pinocytosis, but interfere with the arrangement of tight junctions. Indeed, in support of this it has been demonstrated that bile salt disruption of the gut epithelium can be inhibited by the addition of 20 μM of the cytoskeletal poison, cytochalasin D (BLEAKMAN and NAFTALIN 1990). It is, however, inconceivable that such a method could ever be used for drug delivery in man.

3. Drugs and Anaesthetics

The antineoplastic agent etoposide causes a reversible, dose-related opening of the BBB following intracarotid administration (SPIGELMAN et al. 1984a,b, 1986). The dose required to disrupt the barrier for periods of up to 4 days was readily tolerated by the animals without the appearance of any severe behavioural abnormalities (SPIGELMAN et al. 1986). The extended duration of opening, however, suggests that etoposide may operate by causing severe structural damage to the vascular endothelium. Histological investigations

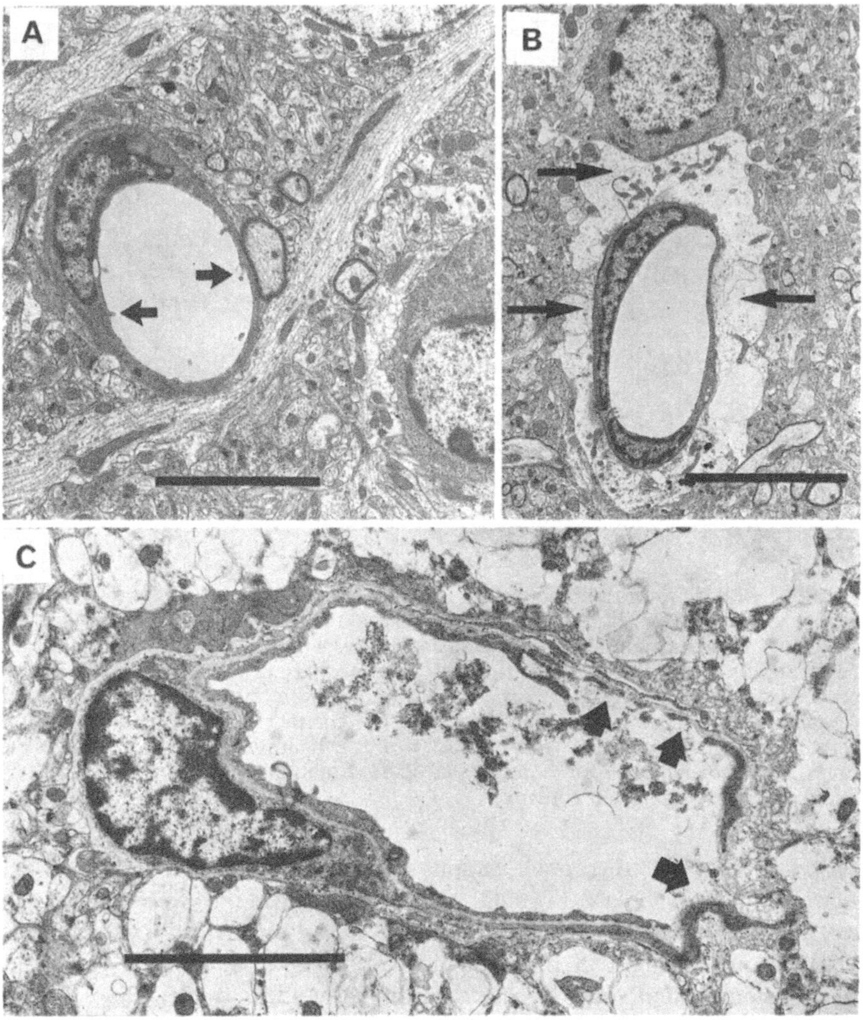

Fig. 4A–C. Electron micrographs from a rat brain perfused with the bile salt taurochenodeoxycholate (TCDC) for 1 min (see Greenwood et al. 1991b). **A** 0.2 m*M* TCDC causes little alteration of the ultrastructure but does open the barrier to the tracer ^{14}C-mannitol (Greenwood et al. 1991). Possible increase in luminal membrane microvilli (*arrows*). *Bar*, 5 μm. **B** 1-min perfusion with 2 m*M* TCDC. Considerable pericapillary oedema is present (*arrows*). *Bar*, 10 μm. **C** Severe lysis of endothelial cell (*arrow*) and destruction of parenchyma following 1-min perfusion with 2 m*M* TCDC. Lumen filled with cell debris. *Bar*, 5 μm

failed to identify any parenchymal or vascular damage at light microscope level but did reveal perivascular lymphocyte infiltration 4 days after the etoposide administration which disappeared within 3 weeks (Spigelman et al. 1986). What initiates the leucocyte infiltration is of considerable importance as active cellular migration may well cause a degree of non-specific and reversible opening of the BBB and BRB. An ultrastructural study of the BBB following etoposide administration may ultimately provide important information as to the route of extravasation. One must also consider that etoposide can only be delivered in a complex solvent that has itself been shown to have some effect upon the BBB (SPIGELMAN et al. 1984b).

The reduction of cerebral oedema following steroid treatment is thought by many to be a consequence of its ability to reduce the permeability of a damaged BBB (LUTHERT et al. 1986; LUTHERT and GREENWOOD 1989). Indeed, it has been reported that the permeability of the normal BBB can be further reduced with chronic administration of the steroid dexamethsone (ZIYLAN et al. 1989), and discontinuation of the steroid reversed the effect. In contrast to this, glioma-bearing rats treated with dexamethsaone did not show any reduction in permeability of either the blood-tumour barrier or the surrounding BBB (LUTHERT et al. 1986). Similarly, NEUWELT et al. (1990) were unable to detect any alteration in barrier permeability of normal animals following dexamethasone treatment. Furthermore, in animals in which the BBB was rendered permeable with an osmotic shock it was only at the highest steroid dose that a modest decrease in permeability could be recorded. It is clear that apart from possibly affecting barrier permeability steroids may also operate by acting upon the cerebral endothelium in other ways, such as by interfering with calcium influx (see under tumour permeability factors below).

The anaesthetic agents pentobarbital and ketamine appear to reduce the permeability of the BBB (SAIJA et al. 1989) although whether this is the result of a direct effect upon the cerebral vasculature or due to alterations in blood flow, blood pressure or glucose utilization is still unclear. Indeed, some anaesthetics may influence hyperosmolar barrier disruption by inducing cardiovascular changes (GUMERLOCK and NEUWELT 1990). In combination with acute ethanol treatment the barbiturate pentobarbital brought about extravasation of horseradish peroxidase (HRP) by structuraly damaging the capillary endothelium (STEWART et al. 1988). It was postulated by the authors that barrier leakage under these conditions is due to the cumulative effect upon membranes of the two lipophilic agents. Moreover, these compounds are believed to influence the handling of cellular calcium which may affect barrier function.

When administered intravascularly the adrenergic agents epinephrine and phenylephrine are thought to cause BBB disruption by inducing hypertension which in turn disrupts the endothelial cell tight junctions (MURPHY and JOHANSON 1985). This hypothesis is supported by the finding that the adrenergic agents isoproterenol and amphetamine, which do not increase blood pressure, do not alter the permeability of the BBB. Another drug which causes hypertension, metaraminol, also disrupts both the BBB and BRB to both small (sucrose) and large (microperoxidase) tracers (LIGHTMAN et al. 1987). Hypertension as a major cause of BBB damage has recently been reviewed by JOHANSSON (1989) and the evidence suggests that this is the mode of action of many of these drugs. That these compounds may act upon the barrier in a process independent of blood pressure, however, remains a possibility. It has been reported that chronic amphetamine intoxication opens the BBB in a reversible manner that is independent of hypertension (RAKIC et al. 1989). A similar degree of extravasation of two different molecular weight tracers suggested to the authors that the route through the BBB was unlikely to be due to simple mechanical disruption of the tight junctions and the formation of pores.

Amitriptyline, a commonly used antidepressant, can disrupt the rat's BBB following an intraperitoneal injection of $34\,mg\,kg^{-1}$. Ultrastructural analysis revealed a significant increase in pinocytotic vesicles in cortical capillary endothelial cells with no other obvious morphological alterations (SARMENTO et al. 1990). It was concluded that vesicular transport was likely to be responsible, at least in part, for the disruption of the BBB. It is important to consider, however, that this antidepressant is also associated with a dramatic increase in the plasma concentration of 5-hydroxytryptamine (5-HT), a compound with considerable vasoactive properties (see below).

A great deal of interest has been shown in the barrier disrupting properties of the CNS stimulant Metrazol. Intracarotid administration is reported to lead to a reversible (<4 h) and apparantly innocuous opening of the BBB (GREIG and HELLMANN 1983; GREIG et al. 1984), although its action on the cerebral vasculature is far from being understood. There is some concern that the dose required to induce BBB dysfunction can also lead to seizure activity, which is known to disrupt the cerebral vasculature (for review see BOLWIG 1989). Metrazol-induced seizures have been shown to disrupt the barrier by altering the capillary endothelial tight junctions (LORENZO et al. 1972, 1975). Under such conditions it is unclear whether barrier opening is brought about by seizure activity, by an increase in cerebral blood flow or by a mechanism independent of these. In a recent review, GREIG (1989) proposes that the mechanism of Metrazol-induced BBB opening may possibly be independent of its seizure activity and that this is worthy of further investigation. If, however, the mechanism of opening is found to be related to its seizure activity then there are serious implications in using this as a method for enhancing drug delivery as epileptic seizures may lead to a degree of permanent neuronal damage (SOKRAB et al. 1990).

4. Enzymes

The intravascular infusion of a variety of enzymes specific for particular substrates at the BBB and BRB produces a dramatic increase in the permeability of the cerebral and ocular vasculature. Intravenously injected collagenase significantly increases BBB permeability (ROBERT et al. 1977). This enzyme will digest basement membrane and is routinely used in the separation of microvessel fragments from brain for tissue culture (GREENWOOD 1991b). Its effect upon the BBB in vivo also appears to be through its enzymatic action upon the basement membrane although how it reaches the abluminal aspect of the barrier is unclear. ROBERT et al. (1977) also reported induction of pinocytosis, which may account for some of the dysfunction of the barrier as well as the delivery of the enzyme to the basement membrane. The effect of collagenase upon the BBB could be attenuated by pretreatment with the drug chromocarb diethylamine (GODEAU et al. 1987), which is thought to operate by acting directly upon the collagen substrate.

Another enzyme, trypsin, which is also used for cell separation, rapidly disrupts the BBB of the frog following intravascular administration (OLESEN and CRONE 1986). Similarly, intravascular injection of neuraminidase also leads to barrier breakdown, possibly by stripping off the sialic acid residues of the glycocalyx and altering the surface charge, a phenomenon which is known to affect BBB permeability (see below). Neither thrombin nor hyaluronidase, however, opened the BBB when applied either intravascularly or extravascularly (OLESEN and CRONE 1986).

The effect of intravascular delivery of enzymes upon the integrity of the BRB has also been studied. Unlike studies on the BBB, neuraminidase did not cause any extravasation of macromolecular tracer into the surrounding tissue, although there was some evidence of tracer-filled vesicles at the luminal aspect of the endothelium (PINO 1987). Similar findings were found for heparinase. The enzymes heparitinase and pronase, on the other hand, led to extensive extravasation of tracer with tracer-filled channels and vesicles evident, but no apparant junctional abnormalities. These results strengthen the case for the role of the glycocalyx, and in particular heparin sulphate proteoglycan, in maintaining barrier integrity.

5. Cations and Endothelial Cell Surface Charge

The glycocalyx that coats the luminal surface of the BBB and BRB is negatively charged, principally by virtue of the large number of sialic acid residues, and appears to play an important part in the function of the barrier (VORBRODT 1987; SAHAGUN et al. 1990). In hypertensive animals the negative luminal surface charge is found to be greatly reduced with a concomitant increase in BBB permeability (NAG 1984; MAYHAN et al. 1989; for review see JOHANSSON 1989). A return of the arterial blood pressure to within normal limits leads to an increase in the negative charge of the endothelial cell glycocalyx and repair of the BBB. Such a correlation, however, does not necessarily indicate that an alteration in surface charge is the mechanism by which hypertension damages the barrier. Although significant evidence does exist that alteration of surface charge disrupts the BBB and BRB, this involves the introduction of positively charged molecules into the cerebral circulation. If, therefore, surface charge is involved in hypertensive barrier disruption the mechanism by which it operates is still unclear.

By introducing cations, such as protamine sulphate and poly-L-lysine, into the cerebral circulation it is possible to neutralize the anionic endothelial cell surface and disrupt the BBB (HARDEBO and KAHRSTROM 1985;

STRAUSBAUGH 1987; OLESEN and CRONE 1986). By using extravasated endogenous albumin as a measure of barrier disruption the duration of opening has been shown to be relatively short term (WESTERGREN and JOHANSSON 1990) with extravasated albumin levels returning to normal within 72 h. This suggests that the endothelial cell is not severely damaged, although how permeability is increased requires further investigation. Some evidence suggests that removal of the surface negative charge induces pinocytosis (NAGY et al. 1981, 1983) and that certain cations may induce non-specific absorptive endocytosis (PARDRIDGE et al. 1987). Disruption, however, may also operate independently of surface charge and be related to the direct toxic action of some of these molecules.

Infusion of protamine sulphate and poly-L-lysine into the carotid artery of rats also leads to the extravasation of protein tracer across the BRB (LIN 1988). However, ultrastructural analysis revealed that protamine sulphate caused some degree of endothelial cell damage allowing tracer to flood the cellular cytoplasm and from there diffuse into the subendothelial and perivascular areas. Contrary to this, poly-L-lysine can induce leakage of tracer but is not associated with endothelial cell damage. This supports the idea that in addition to its effect upon cell surface charge, the cytotoxic action of protamine may also be responsible for damaging the BBB and BRB.

In a related study, NAGY et al. (1985) infused a low pH solution into the carotid artery of the rat and demonstrated extensive extravasation of HRP into the cerebral parenchyma. It was found that the negatively charged surface of the cerebral endothelium was greatly reduced by this procedure. In a supravital brain perfusion system (GREENWOOD et al. 1985; LUTHERT et al. 1987) in which primary energy metabolites are severely depleted, the perfusion of a saline solution of pH 5.5 was not able to disrupt the BBB (GREENWOOD et al. 1989). Lack of opening under these conditions may be due to inhibition of energy-requiring processes such as vesicular transport. In the in vivo frog brain, however, where the cerebral energy status is presumably normal, neither intravascular nor extravascular perfusion of a low pH solution was able to increase vessel permeability (OLESEN and CRONE 1986).

II. Disruptive Agents of Pathological Origin

1. Arachidonic Acid and the Eicosanoids

There are an increasing number of pathologically derived compounds being implicated either directly or indirectly in barrier disruption (for review see WAHL et al. 1988; GREENWOOD 1991a). One of the first group of compounds to be implicated in the inflammation and pathogenesis of the BBB was arachidonic acid and its metababolic products, the eicosanoids and oxygen-derived free radicals. The release of arachidonic acid and its metabolites from pathological tissue, leucocytes and platelets has stimulated considerable interest. This fatty acid is liberated from membrane phospholipids of damaged cells of the CNS by the activation of phospholipase A_2 and C, possibly brought about by the release of free radicals or a rise in intracellular free calcium ($[Ca^{2+}]_i$). The concentration of arachidonic acid in extracellular fluid increases substantially in cerebral and ocular lesions and, along with its metabolites the prostaglandins and leukotrienes, it has been implicated in endothelial cell damage and disruption of the BBB (KONTOS et al. 1980; BLACK and HOFF 1985) and BRB (NAVEH and WEISSMAN 1990). Moreover, cerebral endothelial cells possess the capacity to metabolize arachidonic acid and produce further potentially damaging metabolites (HAMBRECHT et al. 1987; MOORE et al. 1989).

Reports that the disruptive action of arachidonic acid upon the BBB could be prevented by inhibition of its metabolism suggested that its action was mediated in part by its degradation products (KONTOS et al. 1980; BLACK and HOFF 1985; PAPADOPOULOS et al. 1989). Other studies, however, suggest that this may not be the complete story. The superfusion of arachidonic acid on to the cerebral cortex of the cat in the presence of inhibitors of cyclooxygenase and lipoxygenase was found to promote extravasation of fluorescein tracers (UNTERBERG et al. 1987). Ultrastructural analysis failed to identify any junctional abnormalities or vesicular activity at the BBB, but did record alterations at the venous endothelium, including penetration of polymorphonuclear granulocytes. They suggested that arachidonic acid may be a chemotactic agent and that cellular infiltration of these leucocytes across the BBB is likely to be responsible for the vascular leakage. In an in vitro study on cultured cerebral endothelium and in the absence of leucocytes, arachidonic acid in the presence of inhibitors of its metabolism increased the permeability of monolayers grown on microcarrier beads (KEMPSKI et al. 1987). Unfortunately, the ultrastructural appearance of these monolayers was not investigated. In frogs, superfusion of arachidonic acid on to the surface of the brain elicited an increase in vascular permeability within seconds, suggesting once again that arachidonic acid on its own damages the BBB (OLESEN and CRONE 1986).

A further mode of action suggested is by damage of membrane lipids by peroxidative and lipolytic processes. Indeed, arachidonic acid-induced opening of the BBB can be reduced by treating the animal with a steroid with antioxidant and antilipolytic activity (ZUCCARELLO and ANDERSON 1989).

The role of arachidonic acid degradation products in the pathogenesis of barrier dysfunction is unclear. Both leukotrienes and prostaglandins (KONTOS et al. 1980; BLACK 1984; BLACK and HOFF 1985; OLESEN and CRONE 1986; MAYHAN et al. 1986; NAVEH and WEISSMAN 1990) as well as free radicals (KONTOS 1985; WEI et al. 1986; OLESEN and CRONE 1986; OLESEN 1987b; CHAN et al. 1987) have been reported to be associated with some degree of BBB disruption. Contrary to these findings, Unterberg and colleagues were unable to detect an increase in BBB permeability with either leukotrienes (WAHL et al. 1988) or free radicals (UNTERBERG et al. 1988), although both were reported to possess vasomotor properties. Similarly, OLESEN and CRONE (1986) were unable to disrupt the BBB of the frog with either intravascular or extravascular administration of prostaglandins.

There is considerably more evidence to suggest that leukotrienes are stronger chemotactic agents than arachidonic acid. Indeed, if leukotrienes are able to induce cellular infiltration of leucocytes across the BBB this is likely to lead to a degree of delayed barrier disruption independent of any immediate direct action upon the barrier.

Further elucidation of the role of arachidonic acid and its metabolites in BBB dysfunction are clearly required. There are considerable problems involved in dissecting out the role of action of an individual metabolite from others in the metabolic cascade. Indeed, arachidonic acid and its products may stimulate other vasoactive mediators such as the kinins (BAETHMANN et al. 1988) and may alter the physiochemical properties and function of membranes (MOORE et al. 1988). Exacerbating the problems of experimental interpretation is the diversity of experimental approach. The mode of application of these compounds varies considerably. The agent may be superfused onto the cortical surface or injected into the parenchyma, where its action will be upon the abluminal aspect of the BBB, or conversly into the vascular compartment, where it will act upon the luminal plasma membrane. In addition, there may be considerable differences between the response of capillary endothelium and those of arterioles and venules as well as between pial and parenchymal vessels. The concentrations applied may also differ by several orders of magnitude and may exceed concentrations found in pathological tissue. Furthermore, the wide variety of species used may explain some of the differences reported. Naturally, these comments are also pertinant to other studies reviewed in this chapter.

2. Histamine

Histamine, like many other vasoactive compounds, can be released from a variety of cell types including those of the brain. Both intravascular (GROSS et al. 1981, 1982; DUX and JOÓ 1982; GULATI et al. 1985b) and cortical superfusion (OLESEN 1987a) of histamine have been reported to open the BBB. Contrary to these findings, other studies were unable to evoke extravasation of tracer following administration of histamine (GABBIANI et al. 1970; MARTINS et al. 1980; SARIA et al. 1983; KILZER et al. 1985; OLESEN and CRONE 1986). It is clear that further studies are required to clarify these contradictions. As with investigations into the vasoactive actions of other potential mediators, the considerable diversity in experimental approach may be the source of many of these discrepancies.

In those studies in which barrier permeability was reported, a number of alternative explanations have been given for the route of tracer extravasation. DUX and JOÓ (1982) and GROSS et al. (1982) suggest that extravasation is due largely to the induction of vesicular transport and that this is mediated via the H_2 receptor and stimulation of cyclic nucleotides (Joó 1987). They also reported a small degree of tight junction abnormalities (Dux and Joó 1982), which, considering the difficulty in detecting ultrastructural alterations in tight junctions, suggests that this may well be a predominant route of extravasation. To support this view, evidence has been given that histamine gives rise to the opening of pores, presumably at the tight junctions (Olesen 1987a). Furthermore, histamine-induced BBB breakdown is more pronounced to small molecular weight tracers than to large ones (Domer et al. 1983), a finding consistent with a junctional abnormality. As with many of these mediators, careful monitoring of blood pressure is required so that any direct effect upon the BBB and BRB can be isolated from any hypertensive damage.

In the eye, histamine H_1 receptors are implicated in BRB dysfunction in experimental diabetes (Enea et al. 1989). Subsequent to streptozocin-induced diabetes, animals that are given the H_1 receptor antagonist diphenhydramine show significantly less extravasated macromolecular tracer than either controls or animals treated with the H_2 receptor antagonist cimetidine. In contrast to these findings, the same group also reported that they could inhibit diabetes-induced BRB permeability with another H_2 receptor antagonist, ranitidine (Hollis et al. 1988). It must be concluded from this that opening may also operate via this receptor and that the difference in results is likely to be due to differences in the effectiveness of the H_2 receptor antagonists used. Following intravascular administration of histamine with large molecular weight tracers, Kilzer et al. (1985) recorded extravasation into the eye but not the brain. This difference, however, may have resulted from a sampling artefact caused by tracer leaking into the eye from the more permeable vasculature of the ciliary body. Large tracers, such as those used in this study, are generally not recommended for the detection of what may be small increases in permeability. Intravascular administration of histamine appears to elicit an H_1 receptor-mediated leakage at the BRB (Dull et al. 1986), unlike the predominantly H_2 receptor-mediated dysfunction of cerebral capillaries (Gross et al. 1981, 1982; Dux and Joó 1982; Gulati et al. 1985b). Extravascular application of histamine to the retina in normal animals, on the other hand, does not lead to extravasation of circulating macromolecules (Ashton and Cunha-Vaz 1965; Shakib and Cunha-Vaz 1966).

3. Bradykinin

The role of kinins as mediator substances in cerebral damage is well known. Bradykinin increases the permeability of the BBB to small tracers following cortical superfusion in the cat (Unterberg et al. 1984). The opening of small functional pores of an approximate diameter of 11–15 Å at tight junctions brought about by contraction of the endothelial cells is postulated as the mechanism of disruption. This process is likely to be initiated via B_2

kininergic receptors (UNTERBERG et al. 1984), which can be induced with either intra- or extravascular application of bradykinin (OLESEN and CRONE 1986), suggesting the presence of these receptors on both the luminal and abluminal aspect of the cerebral endothelium. RAYMOND et al. (1986) also demonstrated extravasation of HRP following intracarotid infusion with increased evidence of endothelial vesicular activity. In addition, they found that extravasation of HRP could be blocked by inhibiting arachidonic acid metabolism, suggesting a role for the eicosinoids in bradykinin-induced barrier breakdown. Inhibition of phospholipase A_2 with dexamethasone, however, did not prevent opening of the BBB following cerebral administration of bradykinin (SCHURER et al. 1989). This does not rule out the possibility that potentially active metabolites of arachidonic acid were being produced via phospholipase C, which is unaffected by dexamethasone. A number of studies have been unable to demonstate bradykinin-mediated BBB disruption (GABBIANI et al. 1970; SARIA et al. 1983).

The potent vasomotor activity of bradykinin may play a part in inducing barrier disruption although this is unlikely as dexamethasone pretreatment can block the vasodilatory action of bradykinin but not prevent BBB disruption (WAHL et al. 1985; SCHURER et al. 1989).

4. 5-Hydroxytryptamine

As with many vasoactive compounds there is considerable disagreement in the literature over the action that the bioactive amine 5-hydroxytryptamine (5-HT; serotonin) has upon the BBB. In a number of reports both intraventricular and intravascular administration of 5-HT precipitated opening of the BBB (WESTERGAARD 1977; SHARMA et al. 1990), with evidence that extravasation occurred via an increase in vesicular transport. SHARMA et al. (1990) found that 5-HT-induced opening of the BBB could be prevented by pretreatment with a variety of specific agents: cyproheptadine, a 5-HT_2 antagonist; indomethacin, a cyclooxygenase inhibitor; and vinblastine, which interferes with vesicular transport. They postulated that 5-HT acts by binding to endothelial cell 5-HT_2 receptors, which stimulates prostaglandin synthesis, leading to induction of vesicular transport either directly or via cyclic nucleotides. Using frog pial vessels, OLESEN (1985) found that intravascular but not extravascular administration of 5-HT causes a rapid and reversible increase in vessel permeability that can be blocked by the 5-HT_2 antagonist ketanserin and the calcium entry blocker verapamil. He also suggested that the mechanism of barrier opening was operating through binding to luminal 5-HT_2 receptors and induction of Ca^{2+} influx. Unlike the study of SHARMA et al. (1990), however, it was proposed that an influx of Ca^{2+} leads not to vesicular transport but to endothelial cell shrinkage and opening of tight junctions. The fact that 5-HT also posesses vasomotor activity may also account for some of its reported disruptive action upon the BBB and BRB. Other studies have failed to demonstrate extravasation of tracer, irrespective of the route of 5-HT administration (GABBIANI et al. 1970; HARDEBO et al. 1981; SARIA et al. 1983).

5. Cytokines, Platelet Activating Factor and Complement

Endothelial cells are known to be a prime site for cytokine action (POBER and COTRAN 1990) eliciting a number of diverse cellular responses, in particular, the induction of cellular adhesion molecules (see Chap. 16). The potential role of cytokines in directly affecting BBB permeability, however, is poorly understood and remains to be resolved.

Many cytokines will induce adhesion and migration of leucocytes across the vascular bed of the brain and eye, a process which may itself increase barrier permeability. There is growing evidence, however, that some of these compounds may act directly upon the endothelium to induce barrier disruption. Interleukin-1 (IL-1) and tumour necrosis factor (TNF), when injected into the vitreous of the rabbit eye, have been shown to lead to a transient increase in endothelial cell pinocytosis (BROSNAN et al. 1990), which may play a part in retinal capillary disruption (ESSNER 1987). In addition, IL-1 also causes cellular infiltration of neutrophils and macrophages, although further evidence is required to establish whether these cytokines can increase the permeability of the barrier independently of cell migration. Using extravasated plasma proteins in the cerebrospinal fluid as a marker of barrier integrity, SAUKKONEN et al. (1990) were able to demonstrate that the cytokines IL-1α, IL-1β, TNF, and macrophage inflammatory proteins 1 and 2 (MIP-1, MIP-2) could disrupt the BBB even in the absence of significant leucocyte infiltration. It would be of great interest if this study were repeated with a more sensitive method for detecting barrier opening and with careful temporal assessment of leucocyte migration. Furthermore, it must also be noted that IL-1 can induce endothelial cells to synthesize platelet activating factor (PAF), which also posseses vasoactive properties (see below). Indeed, many of these compounds will promote the production of a host of other potentially active substances which must be taken into consideration.

The development of interleukin-2 (IL-2) and adoptive transfer of lymphokine-activated killer (LAK) cells as a method of treating cerebral neoplasms has led to the discovery that IL-2 administration alone can induce breakdown of the BBB. ELLISON et al. (1987, 1990) found that in non-tumour-bearing animals treated with either single or multiple IL-2 infusions, both HRP and autologous immunoglobulin G became extravasated, particularly in white matter. Alterations in cerebrovascular morphology was evident as early as 6 h after a single dose of IL-2 and ultimately led to more widespread parenchymal damage. In a separate study on tumour-bearing rats, increases in permeability were only recorded following chronic IL-2 treatment in the tumour and surrounding region but not in the normal brain (ALEXANDER et al. 1989).

Unlike a number of other cytokines, there is no evidence for IL-2 acting directly upon endothelial cells. It is likely that barrier disruption with IL-2 is operating via IL-2 activation of leucocytes (WATTS et al. 1989) or by IL-2-stimulated release of other cytokines with vasoactive properties.

PAF is a product of the sequential activities of the enzymes phospholipase A and acetyl transferase and is a potent lipid mediator with a variety of metabolic roles. Along with leucocytes and platelets, endothelial cells may also be stimulated to synthesize PAF. KUMAR et al. (1988) have demonstrated that PAF can be synthesized in rat brain especially following seizure activity. By perfusing the brain with a synthetic version of PAF, they were also able to show significant disruption of the barrier despite its poor penetration across the BBB. Further evidence that PAF has a direct role in barrier dysfunction was demonstrated by the effectiveness of a PAF antagonist to reduce BBB disruption following induction of focal cortical lesions (FRERICHS et al. 1990).

In the peripheral vasculature the complement cleavage peptides C3a and C5a are thought to posses vasoactive and chemotactic properties. In a study in which the complement-derived polypeptide C3adesArg was introduced into the ventricles, a concomitant increase in BBB permeability has been reported (HOLLERHAGE et al. 1989). The increase in barrier permeabilty coincided with a considerable perivascular and meningeal infiltration of granulocytes and an increase in cerebrospihal fluid prostanoid levels. Unlike C3a and C5a, C3adesArg does not stimulate histamine release, which indicates that it is the complement factor alone that possesses the strong chemotactic activity for granulocytes. The levels of complement used, however, were very high and there was no comment on the mechanism of barrier disruption, especially whether the leakage was occurring independently of leucocyte infiltration.

6. Tumour-Secreted Vascular Permeability Factor

The permeability of the vascular bed of many cerebral tumours is often greater than that of normal brain. Furthermore, it has been observed that microvessels in the brain immediately adjacent to a tumour may also show increased permeability. This has led to the search for a vascular permeability factor (VPF) secreted by neoplastic cells. It has been reported that cultured human malignant glioma cells produce a factor that is capable of increasing the permeability of guinea-pig skin and that this action can be inhibited with cyclohexamide and dexamethasone but not by antihistamines (BRUCE et al. 1987; CRISCUOLO et al. 1988). This factor was identified as a cationic peptide with a molecular weight of between 41 000 and 56 000 and apparently binds to heparin residues on capillary endothelium. When applied to cultured endothelial cells from various sources, including rat brain, it has been found to induce a rapid and transient elevation of cytosolic calcium that resulted from an influx of extracellular Ca^{2+}. This effect can be blocked with non-specific calcium channel blockers but not with verapamil, an antagonist for voltage-gated calcium channels (CRISCUOLO et al. 1989). In addition, dexamethasone blocked the VPF-induced rise in intracellular calcium even though it was unable to affect cyclic nucleotide-induced calcium influx. The authors proposed that the cationic VPF may operate by interacting with the negatively charged glycocalyx in a manner similar to that reported for other cations such as protamine (see above). In a separate study in which conditioned medium from cultured neoplastic cells was injected into the cerebral parenchyma, a significant increase in vascular permeability was recorded (OHNISHI et al. 1990). This effect could be prevented by pretreating the animal with either dexamethasone or a lipoxygenase inhibitor but not by inhibiting cyclooxygenase, thus implicating products of arachidonic acid metabolism in this process.

III. Miscellaneous Techniques

Initial reports that the chemical solvent dimethyl sulphoxide (DMSO) causes a reversible opening of the BBB with no apparent parenchymal damage (BROADWELL et al. 1982; WRIGHT et al. 1985) have not been corroborated (NEUWELT et al. 1983; GREIG et al. 1985; ZIYLAN et al. 1988). In view of these findings it appears unlikely that infusion of DMSO at subtoxic concentrations will alter the permeability of the BBB by affecting tight junctions or by inducing vesicular transport and transendothelial channels (BALIN et al. 1987). High concentrations of this solvent, however, may cause severe structural damage to the endothelium and parenchyma of the brain and are likely to be incompatible with life.

In recent studies by RUBIN et al. (1991) the problem of circumventing the BBB to deliver therapeutic agents has been approached in a novel and potentially rewarding manner. By increasing the levels of cyclic adenosine monophosphate (cAMP) in cultured brain endothelial cells with agents such as forskolin, they reported a concomitant increase in the expression of tight junction proteins and an increase in transmonolayer resistance. The suggestion that cAMP, operating via a kinase, brought about phosphorylation of proteins involved in tight junction integrity was also tested in vivo. Reducing cAMP levels or inhibiting protein kinase A in living animals led to an enhancement of uptake of tracer molecules across the BBB. Furthermore, by applying synthesized amino acid sequences from a group of cell adhesion molecules associated with tight junctions, the cadherins (LIAW et al. 1990), they were able to reversibly open tight junctions in vitro (RUBIN et al. 1991). These investigations may prove to be of substantial importance in developing physiological methods of opening the BBB. Such agents, however, must operate in a reversible manner, and the agents used to induce the internal cellular alterations must not induce significant harmful side effects.

Photothrombosis of the middle cerebral artery of the rat leads to subsequent opening of the BBB to HRP (DIETRICH et al. 1988a). Release of an active substance into the blood from either damaged endothelium or platelets can be demonstrated by removing blood from animals in which a common carotid artery thrombosis had been photochemically induced and injecting this into the carotid circulation of a normal recipient rat. Extensive ipsilateral extravasation of HRP can be induced in the brains of recipient animals, demonstrating the presence of a potent vasoactive component in the donor blood (Dietrich et al. 1988b). Ultrastructural analysis of the vasculature revealed many tracer-filled vesicular profiles with no evidence of tight junction abnormalities. The question of which active component is being released remains unanswered but it may well be one of the vasoactive agents commented on above.

Finally, the purine nucleotide adenosine, which has been postulated to be a mediator in ischaemic damage, has been shown to disrupt the BRB when applied intra-ocularly (Sen and Campochiaro 1989). Indeed, adenosine agonists could also cause a dose-dependent opening of the BRB, an effect that could be attenuated by applying adenosine antagonist. These results strongly implicate adenosine receptors in barrier dysfunction. Both intra- and extravascular administration of adenosine to the frog brain, however, failed to disrupt the BBB (Olesen and Crone 1986) although the adenine nucleotides ATP, ADP and AMP all succeeded in increasing the permeability of the barrier.

C. Cellular Mechanisms of Increased Barrier Permeability and Route of Leakage

It is clear that an increasing number of investigations are concentrating on the intracellular mechanisms involved in barrier disruption. As indicated above, there are many problems associated with elucidating the mechanism of a particular chemical mediator as many are involved in complex metabolic pathways, the products of which may also be vasoactive. Isolating the action of a particular compound from the surrounding plethora of pathophysiological processes, therefore, must remain a priority for future research. A number of underlying trends, however, are begining to emerge. Alterations in intracellular free calcium ($[Ca^{2+}]_i$) appear to be a frequent phenomenon in endothelial cell dysfunction (Olesen 1985, 1987c; Criscuolo et al. 1989; Koenig et al. 1989a,b), although how such changes lead to increases in barrier permeability remain to be resolved. The proposal that an increase in cyclic nucleotide production leads to pinocytosis and vesicular transport (Joó et al. 1983; Joó 1987; Koenig et al. 1989a,b) has not been supported by other groups (Olesen 1987c; Kempski et al. 1987; Rubin et al. 1991). Indeed, many of these studies suggest that alterations in second messenger levels leads to opening of tight junctions. Although our current understanding of these processes is limited, the recent advent of convincing in vitro cultures of both brain and retinal endothelium, as well as retinal pigment epithelium, should allow us to investigate in greater detail the cellular mechanisms involved in barrier dysfunction.

The ultrastructural correlate of barrier opening, in most cases, also remains to be elucidated. A distinction can often be made, however, between endothelium of the normal barrier, in which changes are induced with exposure to vasoactive mediators, and the endothelium from neovascularized regions, such as in tumours and following ischaemia. The latter often structurally resemble the peripheral microvasculature (GREENWOOD 1991b), whereas the former may appear relatively unchanged despite increases in permeability. In most cases of barrier opening, there still remains no absolutely compelling or irrefutable evidence that allows us to dismiss any of the purported routes of leakage. These include the paracellular route through disrupted tight junctions, the formation of pores or channels through the endothelium and the induction of vesicular transport (Fig. 1). Certainly, some routes would appear to be more likely than others but because of the immense difficulty in visually demonstrating these we often have to rely upon circumstantial evidence to provide evidence for the route of passage.

Few topics in the study of the BBB and BRB cause more controversy than the existence of vesicular transport. This contentious issue has occupied a significant proportion of the literature and has caused a degree of polarization within the field. It is beyond any doubt that there is often a dramatic increase in so-called vesicular profiles both in stimulated and pathological cerebral and retinal endothelium (see above and Chap. 18), although their true three-dimensional structure has caused considerable debate. Indeed, even if pinocytosis does occur there is little evidence for net transfer of tracer from capillary lumen to brain. However, the use of more elaborate studies does suggest that vesicular transfer of plasma proteins across the barrier may occur in some pathological conditions. When tackling the problem of which route tracers take when passing from the lumen to the extracellular space, investigators should approach the subject with an open mind and take great care in their experimental design.

References

Alexander JT, Saris SC, Oldfield EH (1989) The effect of interleukin-2 on the blood-brain barrier in the 9L gliosarcoma rat model. J Neurosurg 70:92–96

Ashton N, Cunha-Vaz JG (1965) Effect of histamine on the permeability of the ocular vessels. Arch Ophthalmol 73:211–223

Baethmann A, Maier-Hauff K, Kempski O, Unterberg A, Wahl M, Schurer L (1988) Mediators of brain edema and secondary brain damage. Crit Care Med 16:972–978

Balin BJ, Broadwell RD, Salcman M (1987) Tubular profiles do not form transendothelial channels through the blood-brain barrier. J Neurocytol 16: 721–735

Black KL (1984) Leukotriene C_4 induces vasogenic cerebral edema in rats. Prostaglandins Leukotrienes Med 14:339–340

Black KL, Hoff JT (1985) Leukotrienes increase blood-brain barrier permeability following intraparenchymal injections in rats. Ann Neurol 18:349–351

Bleakman D, Naftalin RJ (1990) Hypertonic fluid absorption from rabbit descending colon in vitro. Am J Physiol 258:G377–G390

Bolwig TG (1989) Epileptic seizures and the blood-brain barrier. In: Neuwelt EA (ed) Implications of the blood-brain barrier and its manipulations, vol 2. Plenum, New York, pp 567–574

Brightman MW, Hori M, Rapoport SI, Reese TS, Westergaard E (1973) Osmotic opening of tight junctions in cerebral endothelium. J Comp Neurol 152:317–326

Brinker RA (1979) Neuroangiographic contrast agents. In: Miller RE, Skucas J (eds) Radiographic contrast agents. University Park Press, Baltimore, pp 365–374

Broadwell RD, Salcman M, Kaplan SR (1982) Morphological effect of dimethyl sulfoxide on the blood-brain barrier. Science 217:164–166

Brosnan CF, Claudio L, Tansey FA, Martiney J (1990) Mechanisms of autoimmune neuropathies. Ann Neurol [Suppl] 27:S75–S79

Bruce JN, Criscuolo GR, Merrill MJ, Moquin RR, Blacklock JB, Oldfield EH (1987) Vascular permeability induced by protein product of malignant brain tumors. J Neurosurg 67:880–884

Chan PH, Longar S, Fishman RA (1987) Protective effects of liposome-entrapped superoxide dismutase on posttraumatic brain edema. Ann Neurol 21:540–547

Cosolo WC, Martinello P, Louis WS, Christophidis N (1989) Blood-brain barrier disruption using mannitol: time course and electron microscopy studies. Am J Physiol 256:R443–R447

Criscuolo GR, Merrill MJ, Oldfield EH (1988) Further characterization of malignant glioma-derived vascular permeability factor. J Neurosurg 69:254–262

Criscuolo GR, Lelkes PI, Rotrosen D, Oldfield EH (1989) Cytosolic calcium changes in endothelial cells induced by a protein product of human gliomas containing vascular permeability factor activity. J Neurosurg 71:884–891

Cserr HF, DePasquale M, Patlak CS (1987) Volume regulatory influx of electrolytes from plasma to brain during acute hyperosmolarity. Am J Physiol 253: F530–F537

Dietrich WD, Prado R, Watson BD, Nakayama H (1988a) Middle cerebral artery thrombosis: acute blood-brain barrier consequences. J Neuropathol Exp Neurol 47:443–451

Dietrich WD, Prado R, Watson BD (1988b) Photochemically stimulated blood-borne factors induce blood-brain barrier alterations in rats. Stroke 19:857–862

Domer FR, Boertje SB, Bing EG, Reddix I (1983) Histamine- and acetylcholine-induced changes in the permeability of the blood-brain barrier of normotensive and spontaneously hypertensive rats. Neuropharmacology 22:615–619

Dull R, Vergis GJ, Hollis TM (1986) Effect of chronic histamine infusion on the permeability of the blood-retinal barrier. Fed Proc 45:462

Dux E, Joó F (1982) Effects of histamine on brain capillaries. Fine structural and immunohistochemical studies after intracarotid infusion. Exp Brain Res 47: 252–258

Ellison MD, Povlishock JT, Merchant RE (1987) Blood-brain barrier in cats following recombinant interleukin-2 infusion. Cancer Res 47:5765–5770

Ellison MD, Krieg RJ, Povlishock JT (1990) Differential central nervous system responses following single and multiple recombinant interleukin-2 infusions. J Neuroimmunol 28:249–260

Enea NA, Hollis TM, Kern JA, Gardner TW (1989) Histamine H_1 receptors mediate increased blood-retinal barrier permeability in experimental diabetes. Arch Ophthalmol 107:270–274

Ennis SR, Betz AL (1986) Sucrose permeability of the blood-retinal and blood-brain barriers. Effects of diabetes, hypertonicity, and iodate. Invest Ophthalmol Vis Sci 27:1095–1102

Essner E (1987) Role of vesicular transport in breakdown of the blood-retinal barrier. Lab Invest 56:457–460

Farrell CL, Shivers RR (1984) Capillary junctions of the rat are not affected by osmotic opening of the blood-brain barrier. Acta Neuropathol (Berl) 63: 179–189

Fieschi C, Lenzi GL, Zanette E, Orzi F, Passero S (1980) Effects on EEG of the osmotic opening of the blood-brain barrier in rats. Life Sci 27:239–243
Frank JA, Dwyer AJ, Girton M, Knop RH, Sank VJ, Gansow OA, Magerstadt M, Brechbiel M, Doppman JL (1986) Opening of blood-ocular barrier demonstrated by contrast-enhanced MR imaging. J Comput Assist Tomogr 10:912–916
Fraser PA, Dallas AD (1990) Measurement of filtration coefficient in single cerebral microvessels of the frog. J Physiol (Lond) 423:343–361
Frerichs KU, Lindsberg PJ, Hallenbeck JM, Feuerstein GZ (1990) Platelet-activating factor and progressive brain damage following focal brain injury. J Neurosurg 73:223–233
Gabbiani G, Badonnel MC, Majno G (1970) Intra-arterial injections of histamine, serotonin, or bradykinin: a topographic study of vascular leakage. Proc Soc Exp Biol Med 135:447–452
Godeau G, Gavignet C, Groult N, Robert AM (1987) Effect of chromocarb diethylamine on the permeability of the blood-brain barrier. Clin Physiol Biochem 5:15–26
Greenwood J (1991a) Mechanisms of blood-brain barrier breakdown. Neuroradiology 33:95–100
Greenwood J (1991b) Astrocytes, cerebral endothelium and cell culture: the pursuit of an in vitro blood-brain barrier. Ann NY Acad Sci 633:426–431
Greenwood J, Luthert PJ, Pratt OE, Lantos PL (1985) Maintenance of the integrity of the blood-brain barrier in the rat during an in situ saline-based perfusion. Neurosci Lett 56:223–227
Greenwood J, Luthert PJ, Pratt OE, Lantos PL (1988) Hyperosmolar opening of the blood-brain barrier in the energy-depleted rat brain. I. Permeability studies. J Cereb Blood Flow Metab 8:9–15
Greenwood J, Hazell AS, Luthert PJ (1989) The effect of a low pH saline perfusate upon the integrity of energy-depleted rat blood-brain barrier. J Cereb Blood Flow Metab 9:234–242
Greenwood J, Adu J, Davey AJ, Abbott NJ, Bradbury MWB (1991) The effect of bile salts upon the permeability and ultrastructure of the perfused, energy-depleted, rat blood-brain barrier. J Cereb Blood Flow Metab 11:644–654
Greig N, Hellmann K (1983) Enhanced cerebrovascular permeability by metrazol: significance for brain metastases. Clin Exp Metastasis 1:83–95
Greig N, Newell D, Hellmann K (1984) Metrazol enhances brain penetration and therapeutic efficacy of some anticancer agents: implications for brain metastases. Clin Exp Metastasis 2:55–59
Greig NH (1989) Drug delivery to the brain by blood-brain barrier circumvention and drug modification. In: Neuwelt EA (ed) Implications of the blood-brain barrier and its manipulations, vol 1. Plenum, New York, pp 311–367
Greig NH, Sweeney DJ, Rapoport SI (1985) Inability of dimethyl sulfoxide to increase brain uptake of water-soluble compounds: implications to chemotherapy for brain tumors. Cancer Treat Rep 69:305–312
Gross PM, Teasdale GM, Angerson WJ, Harper AM (1981) H_2-receptors mediate increases in permeability of the blood-brain barrier during arterial histamine infusion. Brain Res 210:396–400
Gross PM, Teasdale GM, Graham DI, Angerson WJ, Harper AM (1982) Intra-arterial histamine increases blood-brain transport in rats. Am J Physiol 243:H307–H317
Gulati A, Agarwal SK, Shukla R, Srimal RC, Dhawan BN (1985a) The mechanism of opening of the blood-brain barrier by hypertonic saline. Neuropharmacology 24:909–913
Gulati A, Dhawan KN, Shukla R, Srimal RC, Dhawan BN (1985b) Evidence for the involvment of histamine in the regulation of blood-brain barrier permeability. Pharmacol Res Commun 17:395–404
Gumerlock MK, Neuwelt EA (1990) The effects of anesthesia on osmotic blood-brain barrier disruption. Neurosurgery 26:268–277

Hambrecht GS, Adesuyi SA, Holt S, Ellis EF (1987) Brain 12-HETE formation in different species, brain regions, and in brain microvessels. Neurochem Res 12:1029–1033

Hansson HA, Johansson BB (1980) Induction of pinocytosis in cerebral vessels by acute hypertension and by hyperosmolar solutions. J Neurosci Res 5:183–190

Hardebo JE (1980) A time study in rat on the opening and reclosure of the blood-brain barrier after hypertensive or hypertonic insult. Exp Neurol 70: 155–166

Hardebo JE, Kahrstrom J (1985) Endothelial negative surface charge areas and blood-brain barrier function. Acta Physiol Scand 125:495–499

Hardebo JE, Nilsson B (1980) Hemodynamic changes in brain caused by local infusion of hyperosmolar solutions, in particular relation to blood-brain barrier opening. Brain Res 181:49–59

Hardebo JE, Owman C, Wiklund L (1981) Influence of neurotransmitter monoamines and neurotoxic analogues on morphologic blood-brain barrier function. In: Cervos-Navarro J, Fritschka E (eds) Cerebral microcirculation and metabolism. Raven, New York, pp 177–180

Hollerhage H-G, Walter GF, Stolke D (1989) Complement-derived polypeptide C3adesArg as a mediator of inflammation at the blood-brain barrier in a new experimental cat model. Acta Neuropathol (Berl) 77:307–313

Hollis TM, Gardner TW, Vergis GJ, Kirbo BJ, Butler C, Dull RO, Campos MJ, Enea NA (1988) Antihistamines reverse blood-ocular barrier breakdown in experimental diabetes. J Diabetic Complications 2:47–49

Houthoff HJ, Go KG, Gerrits PO (1982) The mechanism of blood-brain barrier impairment by hyperosmolar perfusion. Acta Neuropathol (Berl) 56:99–112

Johansson BB (1989) Hypertension and the blood-brain barrier. In: Neuwelt EA (ed) Implications of the blood-brain barrier and its manipulations, vol 2. Plenum, New York, pp 389–410

Joó F (1987) A unifying concept on the pathogenesis of brain oedemas. Neuropathol Appl Neurobiol 13:161–176

Joó F, Temesvari P, Dux E (1983) Regulation of the macromolecular transport in the brain microvessels: the role of cyclic GMP. Brain Res 278:165–174

Kempski O, Villacara A, Spatz M, Dodson RF, Corn C, Merkel N, Bembry J (1987) Cerebromicrovascular endothelial permeability. In vitro studies. Acta Neuropathol (Berl) 74:329–334

Kilzer P, Chang K, Marvel J, Kilo C, Williamson JR (1985) Tissue differences in vascular permeability changes induced by histamine. Microvasc Res 30:270–285

Koenig H, Goldstone AD, Lu CY (1989a) Polyamines mediate the reversible opening of the blood-brain barrier by the intracarotid infusion of hyperosmolal mannitol. Brain Res 483:110–116

Koenig H, Goldstone AD, Lu CY, Trout JJ (1989b) Polyamines and Ca^{2+} mediate opening of the blood-brain barrier: in vitro studies in isolated rat cerebral capillaries. J Neurochem 52:1135–1142

Kontos HA (1985) Oxygen radicals in cerebral vascular injury. Circ Res 57:508–516

Kontos HA, Wei EP, Povlishock JT, Dietrich WD, Magiera CJ, Ellis EF (1980) Cerebral arteriolar damage by arachidonic acid and prostaglandin G_2. Science 209:1242–1245

Kumar R, Harvey SAK, Kester M, Hanahan DJ, Olson MS (1988) Production and effects of platelet-activating factor in the rat brain. Biochim Biophys Acta 963:375–383

Lantos G (1989) Cortical blindness due to osmotic disruption of the blood-brain barrier by angiographic contrast material: CT and MRI studies. Neurology 39:567–571

Lee JC, Olszewski J (1961) Increased cerebrovascular permeability after repeated electroshocks. Neurology (Minneap) 11:515–519

Lehtosalo J, Panula P, Laitinen LA (1982) The permeability alteration of brain and spinal cord vasculature to horseradish peroxidase during experimental

decompression sickness as compaired to the alteration in permeability induced by hyperosmolar solution. Acta Neuropathol (Berl) 57:179–187
Li V, Turski PA, Levin AB, Weinstein J, Rozental J, Strother CM (1987) Osmotic disruption of the blood-ocular barriers. AJNR 8:347–348
Liaw CW, Cannon C, Power MD, Kiboneka PK, Rubin LL (1990) Identification and cloning of two species of cadherins in bovine endothelial cells. EMBO J 9: 2701–2708
Lightman S, Rechthand E, Latker C, Palestine A, Rapoport S (1987) Assessment of the permeability of the blood-retinal barrier in hypertensive rats. Hypertension 10:390–395
Lin WL (1988) Leakage of blood-retinal barrier due to damaging effect of protamine sulfate on the endothelium. Acta Neuropathol (Berl) 76:427–431
Lorenzo AV, Shirahige I, Liang M, Barlow CF (1972) Temporary alteration of cerebrovascular permeability to plasma proteins during drug-induced seizures. Am J Physiol 223:268–277
Lorenzo AV, Hedley-Whyte ET, Eisenberg HM, Hso DW (1975) Increased penetration of horseradish peroxidase across the blood-brain barrier induced by metrazol seizures. Brain Res 88:136–140
Luthert PJ, Greenwood J (1989) Experimental studies and the blood-brain barrier. In: Capildeo R (ed) Steroids in diseases of the central nervous system. Wiley, New York, pp 47–57
Luthert PJ, Greenwood J, Lantos PL, Pratt OE (1986) The effects of dexamethasone on vascular permeability of experimental brain tumours. Acta Neuropathol (Berl) 69:288–294
Luthert PJ, Greenwood J, Pratt OE, Lantos PL (1987) The effect of a metabolic inhibitor upon the properties of the cerebral vasculature during a whole-head saline perfusion of the rat. Q J Exp Physiol 72:129–141
Martins AN, Doyle TF, Wright SJ, Bass B (1980) Response of cerebral circulation to topical histamine. Stroke 11:469–476
Mayhan WG, Heistad DD (1985) Permeability of blood-brain barrier to various sized molecules. Am J Physiol 248:H712–H718
Mayhan WG, Sahagun G, Spector R, Heistad DD (1986) Effects of leukotriene C_4 on the cerebral microvasculature. Am J Physiol 251:H471–H474
Mayhan WG, Faraci FM, Siems JL, Heistad DD (1989) Role of molecular charge in disruption of the blood-brain barrier during acute hypertension. Circ Res 64:658–664
Moore SA, Prokuski LJ, Figard PH, Spector AA, Hart MN (1988) Murine cerebral microvascular endothelium incorporate and metabolize 12-hydroxyeicosatetraenoic acid. J Cell Physiol 137:75–85
Moore SA, Figard PH, Spector AA, Hart MN (1989) Brain microvessels produce 12-hydroxyeicosatetraenoic acid. J Neurochem 53:376–382
Murphy VA, Johanson CE (1985) Adrenergic-induced enhancement of brain barrier system permeability to small non-electrolytes: choroid plexus versus cerebral capillaries. J Cereb Blood Flow Metab 5:401–412
Nag S (1984) Cerebral endothelial surface charge in hypertension. Acta Neuropathol (Berl) 63:276–281
Nagy Z, Pappius HM, Mathieson G, Huttner I (1979) Opening of tight junctions in cerebral endothelium. I. Effect of hyperosmolar mannitol infused through the internal carotid artery. J Comp Neurol 185:569–578
Nagy Z, Peters H, Huttner I (1981) Endothelial surface charge: blood-brain barrier opening to horseradish peroxidase induced by the polycation protamine sulfate. Acta Neuropathol (Berl) 7[Suppl]:7–9
Nagy Z, Peters H, Huttner I (1983) Charge-related alterations of the cerebral endothelium. Lab Invest 49:662–671
Nagy Z, Szabo M, Huttner I (1985) Blood-brain barrier impairment by low pH buffer perfusion via the internal carotid artery in the rat. Acta Neuropathol (Berl) 68:160–163

Nakagawa H, Groothuis D, Blasberg RG (1984) The effect of graded hypertonic intracarotid infusions in drug delivery to experimental RG-2 gliomas. Neurology (NY) 32:1571–1581
Naveh N, Weissman C (1990) The correlation between excessive vitreal protein levels, prostaglandin E2 levels, and the blood-retinal barrier. Prostaglandins 39:147–156
Neuwelt EA, Barnett PA (1989) Blood-brain barrier disruption in the treatment of brain tumours: animal studies. In: Neuwelt EA (ed) Implications of the blood-brain barrier and its manipulations, vol 2. Plenum, New York, pp 107–194
Neuwelt EA, Barnett P, Barranger J, McCormick C, Pagel M, Frenkel E (1983) Inability of dimethyl sulfoxide and 5-fluorouracil to open the blood-brain barrier. Neurosurgery 12:29–34
Neuwelt EA, Horaczek A, Pagel MA (1990) The effect of steroids on gentamicin delivery to brain after blood-brain barrier disruption. J Neurosurg 72:123–126
Ohnishi T, Sher PB, Posner JB, Shapiro WR (1990) Capillary permeability factor secreted by malignant brain tumor. J Neurosurg 72:245–251
Olesen SP (1985) A calcium-dependent reversible increase in microvessels in frog brain induced by serotonin. J Physiol (Lond) 361:103–113
Olesen SP (1987a) Leakeness of rat brain microvessels to fluorescent probes following craniotomy. Acta Physiol Scand 130:63–68
Olesen SP (1987b) Free oxygen radicals decrease electrical resistance of microvascular endothelium in brain. Acta Physiol Scand 129:181–188
Olesen SP (1987c) Regulation of ion permeability in frog brain venules. Significance of calcium, cyclic nucleotides and protein kinase C. J Physiol (Lond) 387:59–68
Olesen SP, Crone C (1986) Substances that rapidly augment ionic conductance of endothelium in cerebral venules. Acta Physiol Scand 127:233–241
Papadopoulos SM, Black KL, Hoff JT (1989) Cerebral edema induced by arachidonic acid: role of leukocytes and 5-lipoxygenae products. Neurosurgery 25:369–372
Pardridge WM, Kumagai AK, Eisenberg JB (1987) Chimeric peptides as a vehicle for peptide pharmaceutical delivery through the blood-brain barrier. Biochem Biophys Res Commun 146:307–313
Pickard JD, Durity F, Welsh FA, Langfitt TA, Harper AM, MacKenzie ET (1977) Osmotic opening of the blood-brain barrier: value in pharmacological studies on the cerebral circulation. Brain Res 122:170–176
Pino RM (1987) Perturbation of the blood-retinal barrier after enzyme perfusion. A cytochemical study. Lab Invest 56:475–480
Pober JS, Cotran RS (1990) Cytokines and endothelial cell biology. Physiol Rev 70:427–451
Rakic LM, Zlokovic BV, Davson H, Segal MB, Begley DJ, Lipovac MN, Mitrovic UM (1989) Chronic amphetamine intoxication and the blood-brain barrier permeability to inert polar molecules in the vascularly perfused guinea pig brain. J Neurol Sci 94:41–50
Rapoport SI (1977) Osmotic opening of blood-brain and blood-ocular barriers. Exp Eye Res 25[Suppl]:499–509
Rapoport SI, Robinson PJ (1986) Tight-junctional modification as the basis of osmotic opening of the blood-brain barrier. Ann NY Acad Sci 481:250–266
Rapoport SI, Matthews K, Thompson HK, Pettigrew KD (1977) Osmotic opening of the blood-brain barrier in the rhesus monkey without measurable brain edema. Brain Res 136:23–29
Rapoport SI, Fredericks WR, Ohno K, Pettigrew KD (1980) Quantitative aspects of reversible omotic opening of the blood-brain barrier. Am J Physiol 238: R421–R431
Raymond JJ, Robertson DM, Dinsdale HB (1986) Pharmacological modification of bradykinin induced breakdown of the blood-brain barrier. Can J Neurol Sci 13:214–220

Robert AM, Godeau G, Miskulin M, Moati F (1977) Mechanism of action of collagenase on the permeability of the blood-brain barrier. Neurochem Res 2:449–455

Robinson PJ (1987) Facilitation of drug entry into brain by osmotic opening of the blood-brain barrier. Clin Exp Pharmacol Physiol 14:887–901

Robinson PJ, Rapoport SI (1987) Size selectivity of blood-brain barrier permeability at various times after osmotic opening. Am J Physiol 253:R459–R466

Rubin LL, Barbu K, Bard F, Cannon C, Hall DE, Horner H, Janatpour M, Liaw C, Manning K, Morales J, Porter S, Tanner L, Tomaselli K, Yednock T (1991) Differentiation of brain endothelial cells in cell culture. Ann NY Acad Sci 633: 420–425

Sahagun G, Moore SA, Hart MN (1990) Permeability of neutral vs. anionic dextrans in cultured brain microvascular endothelium. Am J Physiol 259:H162–H166

Saija A, Princi P, de Pasquale R, Costa G (1989) Modifications of the permeability of the blood-brain barrier and local cerebral metabolism in pentobarbital- and ketamine-anaesthetized rats. Neuropharmocology 28:997–1002

Saria A, Lundberg JM, Skofitsch G, Lembeck F (1983) Vascular protein leakage in various tissues induced by substance P, capsaicin, bradykinin, serotonin, histamine and by antigen challenge. Naunyn Schmiedebergs Arch Pharmacol 324:212–218

Sarmento A, Albino-Teixeira A, Azevedo I (1990) Amitriptyline-induced morphological alterations of the rat blood-brain barrier. Eur J Pharmacol 176:69–74

Saukkonen K, Sande S, Cioffe C, Wolpe S, Sherry B, Cerami A, Tuomanen E (1990) The role of cytokines in the generation of inflammation and tissue damage in experimental gram-positive meningitis. J Exp Med 171:439–448

Schurer L, Temesvari P, Wahl M, Unterberg A, Baethmann A (1989) Blood-brain barrier permeability and vascular reactivity to bradykinin after pretreatment with dexamethasone. Acta Neuropathol (Berl) 77:576–581

Sen HA, Campochiaro PA (1989) Intravitreous injection of adenosine or its antagonists causes breakdown of the blood-retinal barrier. Arch Ophthalmol 107:1364–1367

Shakib M, Cunha-Vaz JG (1966) Studies on the permeability of the blood-retinal barrier permeability. IV. Junctional complexes of the retinal vessels and their role in the permeability of the blood-retinal barrier. Exp Eye Res 5:229–234

Sharma HS, Olsson Y, Dey PK (1990) Changes in blood-brain barrier and cerebral blood flow following elevation of circulating serotonin level in anesthetized rats. Brain Res 517:215–223

Sokrab TE, Kalimo H, Johansson BB (1990) Parenchymal changes related to plasma protein extravasation in experimental seizures. Epilepsia 31:1–8

Spigelman MK, Zappulla R, Malis LI, Holland JF, Goldsmith SJ, Goldberg JD (1983) Intracarotid dehydrocholate infusion: a new method for prolonged reversible blood-brain barrier disruption. Neurosurgery 12:606–612

Spigelman MK, Zappulla RA, Goldberg JD, Goldsmith SJ, Marotta D, Malis LI, Holland JF (1984a) Effect of intracarotid etoposide on opening the blood-brain barrier. Cancer Drug Deliv 1:207–211

Spigelman MK, Zappulla RA, Johnson J, Goldsmith SJ, Malis LI, Holland JF (1984b) Etoposide induced blood-brain barrier disruption – effect of drug compared with that of solvents. J Neurosurg 61:674–678

Spigelman MK, Zappulla RA, Strauchen JA, Feuer EJ, Johnson J, Goldsmith SJ, Malis LI, Holland JF (1986) Etoposide induced blood-brain barrier disruption in rats: duration of opening and histological sequelae. Cancer Res 46:1453–1457

Stewart PA, Hayakawa EM, Carlen PL (1988) Ethanol and pentobarbital in combination increase blood-brain barrier permeability to horseradish peroxidase. Brain Res 443:12–20

Strausbaugh LJ (1987) Intracarotid infusions of protamine sulfate disrupt the blood-brain barrier of rabbits. Brain Res 409:221–226

Suzuki M, Iwasaki Y, Yamamoto T, Konno H, Kudo H (1988) Sequelae of the osmotic blood-brain barrier opening in rats. J Neurosurg 69:421–428
Unterberg A, Wahl M, Baethmann A (1984) Effects of bradykinin on permeability and diameter of pial vessels in vivo. J Cereb Blood Flow Metab 4:574–585
Unterberg A, Wahl M, Hammersen F, Baethmann A (1987) Permeability and vasomotor response of cerebral vessels during exposure to arachidonic acid. Acta Neuropathol (Berl) 73:209–219
Unterberg A, Wahl M, Baethmann A (1988) Effects of free radicals on permeability and vasomotor response of cerebral vessels. Acta Neuropathol (Berl) 76:238–244
Vorbrodt AW (1987) Demonstration of anionic sites on the luminal and abluminal fronts of endothelial cells with poly-L-lysine gold complex. J Histochem Cytochem 35:1261–1266
Wahl M, Unterberg A, Baethmann A (1985) Intravital fluorescence microscopy for the study of blood-brain barrier function. Int J Microcirc Clin Exp 4:3–18
Wahl M, Unterberg A, Baethmann A, Schilling L (1988) Mediators of blood-brain barrier dysfunction and formation of vasogenic brain edema. J Cereb Blood Flow Metab 8:621–634
Waldron RL, Bryan RN (1975) Effect of contrast agents on the blood-brain barrier: an electron microscopic study. Radiology 116:195–198
Watts RG, Wright JL, Atkinson LL, Merchant RE (1989) Histopathological and blood-brain barrier changes in rats induced by an intracerebral injection of human recombinant interleukin-2. Neurosurgery 25:202–208
Wei EP, Ellison MD, Kontos HA, Povlishock JT (1986) O_2 radicals in arachidonate-induced increased blood-brain barrier permeability to proteins. Am J Physiol 251:H693–H699
Westergaard E (1977) The blood-brain barrier to horseradish peroxidase under normal and experimental conditions. Acta Neuropathol (Berl) 39:181–188
Westergren I, Johansson BB (1990) Albumin content in brain and CSF after intracarotid infusion of protamine sulfate: a longitudinal study. Exp Neurol 107:192–196
Wright DC, McCormick PC, Lawrence P, Owens E, Blasberg RG (1985) Infusions of dimethyl sulfoxide (DMSO): blood-brain barrier effects. J Cereb Blood Flow Metab 5 [Suppl 1]:75–76
Ziylan YZ, Robinson PJ, Rapoport SI (1983) Differential blood-brain barrier permeabilities to 14-C sucrose and 3-H inulin after osmotic opening in the rat. Exp Neurol 79:845–857
Ziylan YZ, Robinson PJ, Rapoport SI (1984) Blood-brain barrier permeabilities to sucrose and dextran after osmotic opening. Am J Physiol 247:R634–R638
Ziylan YZ, Korkmaz G, Bernard G, Lefauconnier JM (1988) Effect of dimethyl sulfoxide on blood-to-brain transfer of a-aminoisobutyric acid: examination of regional blood-brain barrier function. Neurosci Lett 89:74–79
Ziylan YZ, Lefauconnier JM, Bernard G, Bourre JM (1989) Regional alterations in blood-to-brain transfer of alpha-aminoisobutyric acid and sucrose, after chronic administration and withdrawal of dexamethasone. J Neurochem 52:684–689
Zuccarello M, Anderson DK (1989) Protective effect of a 21-aminosteroid on the blood-brain barrier following subarachnoid hemorrhage in rats. Stroke 20: 367–371

CHAPTER 20

Drug Entry Into the Brain and Its Pharmacologic Manipulation

N.H. GREIG

A. Introduction

During recent years much progress has been made in pharmacology which has resulted in the introduction of several important new drugs into clinical medicine. Even greater progress has been made in the arena of biopharmaceutical technology, with the development of a number of peptides and genetically engineered agents which interact, with high specificity, with intra- and/or extracellular targets at a molecular level. It is probable that many of these new peptide and protein therapeutics will undergo clinical assessment during this decade. For the vast majority of drugs, peptides, and proteins the relation between dose and effect is critical. Most therapeutic agents produce their required pharmacologic response by interfering with some specific aspect of cell function or structure in a concentration-dependent and reversible manner. To obtain a required pharmacologic response, the administration, absorption, and transport of a threshold amount of drug to the target is essential. Further, this often must occur at the correct time for a pharmacologic effect to be elicited and maintenance of drug at the target may be required for a specific time to achieve a clinically valuable response. Thus, for an optimal pharmacologic response, control of drug input and time-dependent knowledge of drug

---

Abbreviations

AAD	Amino acid decarboxylase
ADTN	2-Amino-6,7-dihydroxy-tetrahydronaphthalene
AIDS	Acquired immunodeficiency syndrome
AZT	Azidothymidine
BChE	Butyrylcholinesterase
CDS	Chemical delivery system
COMT	Catechol-*O*-methyl-transferase
DOPA	3,4-Dihydroxyphenylalanine
Enk	Enkephalin
GABA	γ-Aminobutyric acid
5-HT	5-Hydroxytryptamine
5-HTP	5-Hydroxytryptophan
IGF	Insulin-like growth factor
MAO	Monoamine oxidase
MCPP	*m*-Chlorophenylpiperazine
MPP^+	1-Methyl-4 phenylpyridinium
MPTP	1-Methyl-4-phenyl-1,2,3,6-tetrahydropyridine
NAM	DL-2-amino-7-bis-[(2-chloroethyl) amino]-1,2,3,4-tetra-hydro-2-napthoic acid
PA	Permeability – surface area (product)
TRH	Thyrotrophin releasing hormone

disappearance, due to distribution, metabolism, and excretion processes, is required.

In general, improvements in drug delivery and their implementation have lagged achievements in medicinal chemistry and biochemistry. All too often drug delivery is considered only in the late stages of drug development, if, indeed, it is remembered at all. Ideally, it should be included as an integral part of drug design.

For many therapeutic agents, depending on the location of their target and mechanism of action, the conventional systemic administration by intravenous or oral routes results in the achievement of therapeutic levels at their site or sites of action and a useful clinical activity. For others, and in particular therapeutic agents whose target is sequestered within the brain, their efficacy in vivo is minimal despite remarkable in vitro promise. Brain entry and efficacy of numerous potentially valuable agents is often reduced due to the presence of a blood-brain barrier, even during pathologic conditions when the integrity of this barrier may become reduced.

B. The Presence and Function of a Blood-Brain Barrier

I. The Structural Barrier

The historical conception of a blood-brain barrier and its anatomical localization have recently been extensively reviewed by Davson and his colleagues (DAVSON 1989; DAVSON et al. 1987) and BRIGHTMAN (1989). The capillary endothelium of most nonneural tissues allows free diffusion of ions and water-soluble agents up to the size of approximately 40000 molecular weight, from the vasculature into the interstitial space. Larger compounds, such as serum albumin, have restricted passage as the endothelial cells are linked by tight junctions. However, unlike the brain, these junctions are not circumferential and therefore do not occlude the paracellular spaces or openings, allowing paracellular transport, as described. In addition, some non-neural capillaries possess tight junctions that are interrupted by gap junctions. Cerebral capillary endothelial cells are joined by continuous belts of tight junctions and are devoid of fenestrations and gap junctions. The critical zonularity of the tight junctions of cerebral vasculature sets them apart from that of other organs and imparts to them a low conductivity and with a high electrical resistance, approximately $2000\,\Omega\,cm^2$, which is similar to that of frog skin. Other capillary systems have dramatically lower electrical resistances. Mesenteric capillaries have a resistance of only $1-2\,\Omega\,cm^2$, whereas that of muscle capillary endothelium is about $20\,\Omega\,cm^2$.

The blood-brain barrier can be simplistically considered as a system of layers of cells that are interconnected by continuous tight junctions at the level of (1) the cerebral capillary endothelium, (2) the choroid plexus epithelium, and (3) the arachnoid membranes. These collectively separate the brain and cerebrospinal fluid from the blood (Fig. 1). The presence of continuous tight junctions confers on each of the described layers the properties of an extended cell membrane, as paracellular exchange is all but obliterated (RAPOPORT 1976). The cell membranes are lipid, and hence the property that generally governs passive diffusion/penetration of a compound into brain, via the layers, is its lipid solubility – often measured as its octanol/water partition coefficient (RAPOPORT and LEVITAN 1974; LEVIN 1980; GREIG 1989a).

In addition to forming the structural elements of the barrier that can be visualized in morphological studies (KARNOVSKY 1967; REESE and KARNOVSKY 1967), cerebral capillary endothelium are metabolically highly active, containing large numbers of mitochondria. These are involved in both the maintenance and functioning of (1) selective, stereospecific, and saturable transport mechanisms to regulate and supply the brain with essential amino acids, monosaccharides, and nucleosides, and of (2) a degradative enzymatic barrier.

II. The Enzymatic Barrier

A variety of neurotransmitters and peptides possess differential actions in the periphery and systemic circulation, compared to the brain. Monoamines (dopamine, epinephrine, norepinephrine, and serotonin), for example, are essential neurotransmitters in brain but, additionally, act as hormones in the periphery. These dual roles obviously must be separated, requiring

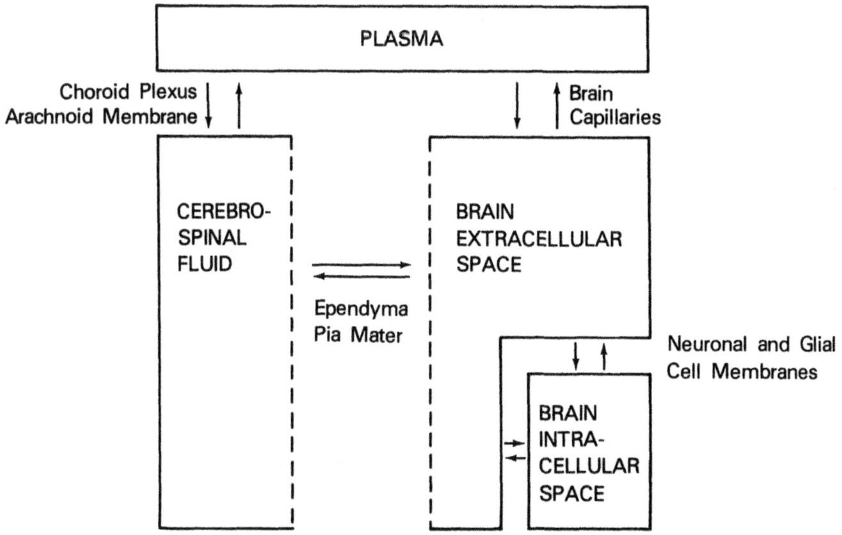

Fig. 1. Model of solute exchange between plasma and the compartments of the central nervous system

that monoamines in blood be restricted from entering the brain. In addition to the described physical barrier, this also is efficiently achieved by the presence of degrading enzymes. These exist both in the blood, for example unspecific esterases, and at the level of the capillary endothelial cells to restrict the brain uptake of not only endogenous neuroactive agents, such as monoamines, but also potentially neurotoxic exogenous compounds. Unfortunately, this is not the case for 1-methyl-4-phenyl-1,2,3,6-tetrahydropyridine (MPTP) which is oxidized by monoamine oxidase-B (MAO-B) at the enzymatic barrier to eventually form the highly potent and selective dopaminergic toxin 1-methyl-4-phenylpyridinium (MPP^+) (Kalaria et al. 1987).

The high activity of alkaline phosphatase in brain capillaries has been known for some 50 years (Gomori 1941), and visualization of this enzyme is commonly used for demonstrating brain microvasculature. However, studies by Bertler et al. (1966) first clearly demonstrated the importance of the enzymatic barrier by showing that the amine precursors L-3,4-dihydroxyphenylalanine (L-DOPA) and 5-hydroxytryptophan (5-HTP) were prevented from entering the brain by enzymatic mechanisms (Fig. 2). Biochemical studies have identified a variety of degradative enzymes associated with cerebral capillary endothelium, including L-amino acid decarboxylase (AAD) and the monoamine degrading enzymes MAO (subtypes A and B) and catechol-*O*-methyl transferase (COMT). Butyrylcholinesterase (BChE) and 4-aminobutyrate aminotransferase, involved in the degradation of γ-aminobutyric acid (GABA) by transamination, have also been found. Other degradative enzymes undoubtedly exist, including peptidases and unspecific esterases. In addition, the presence of enzyme systems such as γ-glutamyl transpeptidase have been implicated in the facil-

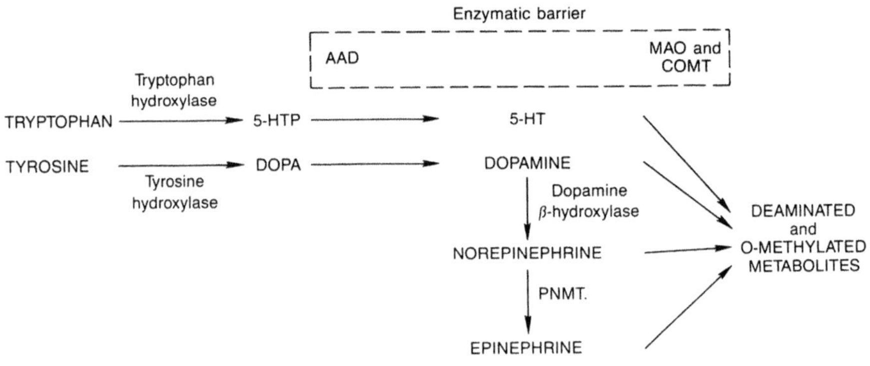

Fig. 2. Enzymes systems related to the formation and degradation of neurotransmitter monoamines, with *hatched area* representing the influence of the enzymatic barrier. *DOPA*, 3,4-dihydroxyphenylalanine; *5-HTP*, 5-hydroxytryptophan; *5-HT*, 5-hyroxytyrptamine; *AAD*, aromatic L-amino acid decarboxylase; *MAO*, monoamine oxidase; *COMT*, catechol-*O*-methyltransferase; *PNMT*, phenylethanolamine-*N*-methyl transferase

itated transport of amino acids from blood to brain (DURIČIĆ and MRŠULJA 1988; BRIGHTMAN 1989).

From a drug delivery perspective, both types of enzyme systems, degradative and transport related, are of importance and, in certain circumstances, can be manipulated. The classical example is that of L-DOPA, which shares an uptake site for neutral amino acids at the blood-brain barrier – the sodium-dependent "L-system." As described, the presence of intracellular AAD in cerebral capillary endothelium decarboxylates L-DOPA to dopamine (Fig. 2). The latter is not a substrate for the L-system, has a low passive cerebrovascular permeability and is subject to further potential metabolism by MAO and COMT. Decarboxylation is a rapid process in brain and, due to its similarly efficient degradation at the level of the gastrointestinal tract, circulating L-DOPA levels are only about 30 nM. The capacity of AAD to decarboxylate L-DOPA has an upper limit, and hence some exogenous L-DOPA enters brain following its administration for the treatment of Parkinson's disease. However, high doses are required to produce plasma concentrations in the micromolar range, at which level debilitating side effects are commonly induced (RINNE et al. 1973). The decarboxylase inhibitor carbidopa (α-methyl-DOPA hydrazine), however, lowers the upper limit. Carbidopa is charged, and thus possesses relatively selective actions against peripheral decarboxylases, including those within the cerebral capillary endothelium, without inhibiting neuronal enzyme, due to its low cerebrovascular permeability (BARTHOLINI and PLETSCHER 1969). Additional administration of the MAO-B inhibitor deprenyl can result in a further clinical improvement in the treatment of Parkinson's disease (BIRKMAYER et al. 1977). Dopamine is a substrate for MAO-B, whereas epinephrine, norepinephrine, and 5-HTP are substrates for MAO-A. Deprenyl may therefore have its effect by not only reducing dopamine breakdown at the postsynaptic level, but by also inhibiting its metabolism at the level of the enzymatic barrier following L-DOPA administration.

III. Barrier Permeability

Quantitative techniques to determine the cerebrovascular permeability (P) – surface area (A) product, PA, of an agent at the blood-brain barrier have been extensively reviewed in Chap. 2. As illustrated in Fig. 3, which shows a series of nonelectrolytes that includes several widely used anticancer drugs, the cerebrovascular permeability of an agent is roughly proportional to its lipid solubility (FENSTERMACHER and RAPOPORT 1984; GREIG 1987, 1989a). In general, transfer across the blood-brain barrier of a lipophilic agent having an octanol/water partition coefficient that exceeds 1.0 (log P value >0), i.e., equal partition, is rapid and rate limited by the supply of drug – the rate of local blood flow. Conversely, a lipid-insoluble polar drug whose octanol/water partition coefficient is less that 0.1 (log P value <−1.0) is restricted from entering brain as a consequence of its low permeability, irrespective of blood flow. For compounds whose octanol/water partition coefficients lie between these two limits, a combination of brain capillary permeability and cerebral blood flow will determine brain uptake.

Exceptions to the proportionality relation illustrated in Fig. 3 have been demonstrated, and occur for agents that share a facilitated transport mechanism at the cerebral capillaries, i.e., large neutral L-amino acids and related drugs, as well as certain compounds of large molecular weight. As a consequence of the latter, a molecular weight cut-off of between 800 and 1000 has been suggested to occur (Levin 1980).

Studies by Greig et al. (1990a) with the lipophilic anticancer alkaloids vincristine and vinblastine (molecular weights 825 and 814 and log *P* values 2.14 and 1.68, respectively) suggest how a molecular weight cut-off phenomena may occur. By using the isolated, in situ brain perfusion technique of Takasato et al. (1984) (described in detail in Chap. 2), and controlling for the effects of drug binding to plasma proteins, metabolism, and spontaneous degradation, the *PA* values of both alkaloids were determined and were approximately 500-fold less than that predicted from their octanol/water partition (Greig et al. 1990a). Although lipophilic when partitioned between octanol and water, both agents are endowed with several highly polarized

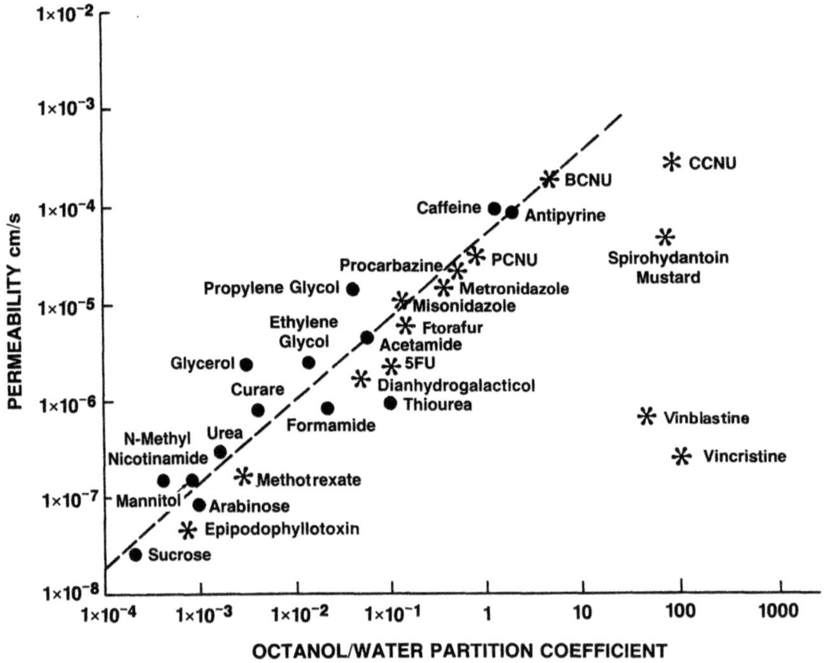

Fig. 3. Relation between octanol/water partition coefficient and cerebrovascular permeability. *Asterisk*, an anticancer drug; *BCNU*, 1,3-bis-chloro(2-chloroethyl)-1-nitrosourea; *CCNU*, 1-(2-chloroethyl)-3-cyclohexyl-1-nitrosourea; *PCNU* 1-(2-chloroethyl)-3-(2,6-dioxo-3-piperidyl)-1-nitrosourea; *5FU*, 5-fluorouracil

functional groups that include two basic tertiary amines and several hydroxyl moieties. It is probable that repulsive forces occur between these charged and regionally distributed functional groups and the lipid component of the cerebral capillary membranes. These forces, together with the unusual geometry (STEIN 1986) of both agents, may restrict their cerebrovascular permeability. Similarly, the lipophilic immunosuppressive agent cyclosporin (log *P* value 2.99), which is a cyclic endecapeptide with a large number of lipophilic alkyl groups but with several amino and carboxylic acid moieties, does not readily enter brain (CEFALU and PARDRIDGE 1985). Together, these studies suggest that for large compounds, the additive rules that govern log *P* values and bulk phase octanol/water partitioning (LEO et al. 1971; HANSCH 1972) break down when the distances between regionally distributed hydrophilic and hydrophobic groups comprising the compound become significant relative to their respective distances to the lipophilic cell membrane/aqueous plasma interphase. The size of an agent alone will not restrict its passive diffusion across the cerebral capillary endothelium and transfer from blood to brain, as, for example, unbound erythrosin B (red dye no. 3), molecular weight 880, readily enters brain (LEVITAN et al. 1984).

C. Factors Determining Time-Dependent Brain Drug Levels

The cerebrovascular permeability of an agent, as described above, is a key determinant of its brain entry. Although important, this physicochemical characteristic is not the sole factor that determines the time-dependent concentration integral of a compound in brain and its pharmacologic activity.

I. Ionization

The physicochemical characteristics of a compound derive from its chemical structure. Simplistically, most drugs can be considered as organic molecules formed from carbon-containing superstructures with different functional groups. The nature of the functional groups, the number of hydrocarbons, and their combined structural arrangement impart on the agent its mechanism of action (i.e., receptor or subsite fit to stimulate a neuron or inhibit an enzyme) and its ability to penetrate tissues or cross biological barriers. Drug-induced enzyme inhibition or receptor stimulation generally involves the interaction of ionized groups between the drug and target. Consequently, most pharmacologically active agents are either weak acids or weak bases, coexisting in ionized and unionized forms. Due to the hydration of the ion and the hydrophobic nature of cell membranes, the ionic species of a drug is both absorbed and carried across cell membranes to a far lesser degree than the unionized species, with up to a 1000-fold difference (RAPOPORT 1976). The equilibrium between ionized and unionized species therefore is a further important factor that codetermines the brain penetrability of a compound. The ionized/unionized fraction can be determined from the pH of the local environment and from the pK_a of the compound, by the Henderson-Haselbalch equation (GREIG 1989a). Differences in the pH of two compartments sometimes can be exploited to modulate drug distribution, as weak acids will accumulate in an environment with a relatively low pH and weak bases in one of a relatively high pH.

A pH gradient exists between arterial blood, brain cells, and cerebrospinal fluid with mean pH values of 7.45, 7.0 and 7.35, respectively, in adults (JOHANSON 1989). This pH difference is too small to dramatically alter the relative distribution between blood and brain of most drugs, apart from weak bases or acids that possess a pK_a close to that of physiological pH. One specific example is that of morphine, with a pK_a of 7.93 at 37°C. Slight alterations in blood pH, between 7.2 and 7.6, due to metabolic acidosis and alkalosis, will change the unionized fraction of morphine and thereby alter its brain uptake by up to threefold (SCHULMAN et al. 1984). Such an example is relatively uncommon in healthy individuals, but can occur pathologically. Indeed, it is probable that a similar or larger pH differences may occur between arterial blood or brain and a malignant brain tumor. It should be possible to exploit such a difference to deliver more drug to the tumor than normal brain, particularly if the local pH of the tumor could be artificially further lowered by glucose shock.

II. Drug Binding to Plasma Proteins

Many systemically administered therapeutics become bound, in varying degrees and with various affinities, to plasma constituents, and in particular to serum albumin and γ-globulin. GREIG (1989a), KRAGH-HANSEN (1981), and ROBINSON and RAPOPORT (1986) have reviewed the molecular aspects of drug binding to plasma proteins, and how such interactions affect drug uptake into brain. Additionally, the binding of drugs to plasma proteins can dramatically alter their pharmacokinetics by protecting them from elimination and metabolism, and by slowly releasing them to enter tissues (JUSKO and GRETCH 1976; VALNER 1977; MACKICHAN 1984).

Although receptor-mediated transport of proteins such as transferrin and transcytosis of positively charged, as compared to neutral and negatively charged, albumin have been demonstrated to occur across cerebral capillaries (PARDRIDGE 1988; PARDRIDGE et al. 1990), and although proteins gain a limited access to the cerebrospinal fluid via the choroid plexus (DAVSON 1967; DAVSON et al. 1987), drugs that are restrictively bound to plasma proteins are generally restricted from entering the brain. The extent to which drug/protein binding occurs is dependent on the concentration of the drug, which is time dependent, the concentration of protein, and the affinity of the interaction. As plasma proteins such as serum albumin are present in large amounts (approximately 5 g/dl) and possess a multitude of ionized groups and lipophilic domains on their surface, they can specifically bind both water-soluble and lipophilic drugs. As a consequence, the physicochemical characteristic of lipophilicity alone does not guarantee a high brain uptake. Indeed, SCHOLTAN (1968) demonstrated a positive linear relation between binding constants to human albumin and log P values of a variety of drugs, including derivatives of steroids, sulfapyrimidine, tetracycline, sulfonamide, and penicillin. In each case, the addition of an alkyl or phenyl substituent increased the binding capacity by a constant that was specific to the series of drugs studied. Accordingly, several highly lipophilic agents such as Evans blue and erythrosin B are restricted from entering brain due to their extensive binding to serum albumin (FREEDMAN and JOHNSON 1969; RAPOPORT 1976; LEVITAN et al. 1984).

Recent studies have challenged the "free drug concept" that only the free dialyzable fraction of a drug is available to enter brain (GREIG 1989a; GREIG et al. 1990b). Studies with the benzodiazepines, in particular diazepam, with palmitate, and with phenytoin have shown that these agents, unlike others, such as bilirubin and phenobarbital, are nonrestrictively bound to plasma proteins (LEVINE et al. 1982; CORNFORD 1984; JONES et al. 1988). Their extraction from blood into brain is greater than and independent of their free plasma concentration. Simplistically, this restrictive/nonrestrictive binding phenomenon occurs during the transit of a drug through the microvasculature of an organ when the binding affinity between the drug and plasma protein is greater or less, respectively, than the affinity between the drug and some component within the organ, such as a receptor, an enzyme, or a transport system (MACKICHAN 1984; GREIG 1989a). In the event of restrictive binding, only the free plasma concentration is available to enter the organ. For nonrestrictive binding, by contrast, drug stripping from plasma protein will occur during the passage of the complex through the organ as rapidly as the binding off-rate, between drug and plasma protein, will allow (ROBINSON and RAPOPORT 1986). This phenomenon is well described for the hepatic and renal clearance of a wide variety of drugs, including propranolol, lidocaine, and penicillins (ROWLAND 1984).

III. Time-Dependent Plasma Concentration

The time required for an agent to achieve equilibrium between blood and brain when steady-state levels are maintained in plasma, and in the absence of restrictive plasma protein binding, is proportional to its cerebrovascular permeability. Ethyl alcohol, for example, achieves an equilibrium within seconds, whereas approximately an hour is required for ethyl thiourea, and for agents such as salicylic acid, sucrose, and sulfate, brain levels can be predicted to be lower than those of plasma at infinite time (HOLLINGSWORTH and DAVSON 1973; DAVSON et al. 1987). Accordingly, for water-soluble agents and for those with equal partition in octanol and water, the maintenance of high plasma levels optimizes drug delivery and retention in brain (GREIG 1987, 1989a).

The brain concentration of a water-soluble drug is invariably lower than its concomitant plasma level after systemic administration, in the absence of a facilitated transport system and except in the late phase of elimination when plasma levels often decline more rapidly than those in brain. Conversely, the brain concentration of a lipophilic drug can be substantially higher than its concomitant plasma level almost immediately after the plasma peak, in the absence of restrictive plasma protein binding. Indeed, plasma and cerebrospinal fluid concentrations often provide little valuable information concerning the brain concentration of a lipophilic drug. For some, such as phenobarbital and the cholinergic drug arecoline, plasma and brain concentrations are similar throughout their time-course, with a ratio of approximately unity (KAPETANOVIĆ et al. 1982a; SONCRANT et al. 1989a). For these agents brain levels appear to be dependent on those in plasma, and as plasma levels decline due to metabolism and elimination back-diffusion of drug from brain occurs.

The concept that all lipophilic drugs readily enter brain but quickly back-diffuse as plasma levels decline (BODOR and BREWSTER 1982) (i.e., "easy in – easy out") is a misconception. Many lipophilic agents maintain high brain levels despite low plasma concentrations. Examples of many such agents are the centrally acting dopamine antagonist haloperidol and 5-hydroxytryptamine (5-HT) (1B) agonist *m*-chlorophenylpiperazine (MCPP) which possess brain/plasma concentration integral ratios of approximately 30 (KAPETANOVIĆ et al. 1982b; GREIG et al. 1991a) (Fig. 4). For these, sequestration probably occurs within the intracellular compartment of the brain (Fig. 1) and both their brain levels and pharmacologic activities are largely independent of their time-dependent plasma concentration.

It is possible to manipulate the time-dependent plasma concentration, and thereby the amount of drug that the brain has potential access to, for both water-soluble and lipophilic compounds by controlling their rate of administration. Continuous infusion schedules can optimize the pharmacokinetics of agents that possess a short half-life and, accordingly, a short duration of action. As an example, the lipophilic muscarinic agonist arecoline (log P value 0.3) readily enters brain following its systemic administration. However, it undergoes rapid ester hydrolysis to arecaidine, which has no cholinergic activity, in plasma and brain with a half-life of between 3 and 5 min in rats and humans (SONCRANT et al. 1989a; SHETTY et al. 1991). Administration of arecoline by continuous intravenous infusion maintains steady-state plasma levels (Fig. 5), and hence brain levels, and has proved of benefit in improving short-term memory in patients with dementia of the Alzheimer's type (RAFFAELE et al. 1989; SONCRANT et al. 1989b, SONCRANT et al. 1991). Similarly, the water-soluble anticancer drug cytosine arabinoside is rapidly inactivated to uracil arabinoside, with a half-life of 12 min, by the hepatic enzyme cytidine deaminase following its administration to cancer patients (FINKLESTEIN et al. 1970). As a consequence, minimal drug enters cerebrospinal fluid or brain. Its continuous

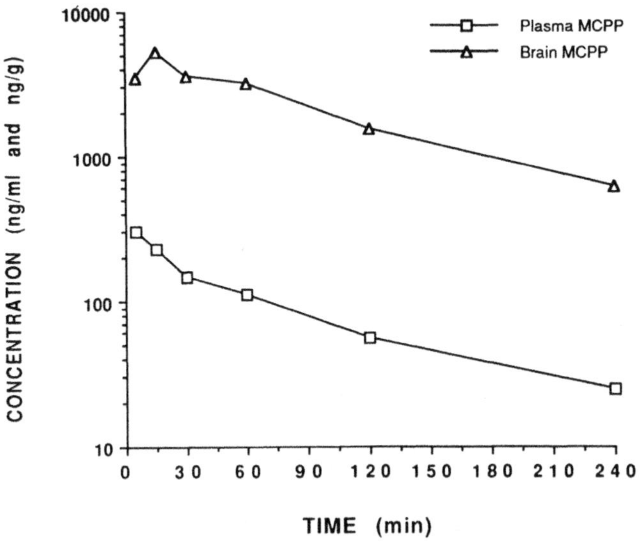

Fig. 4. Brain (*triangles*) and plasma (*squares*) time-dependent concentration profile of *m*-chlorophenylpiperazine (MCPP) in rats. Similar to many lipophilic agents, MCPP brain levels remain high despite the disappearance of drug from plasma, with a brain/plasma concentration integral ratio of 30

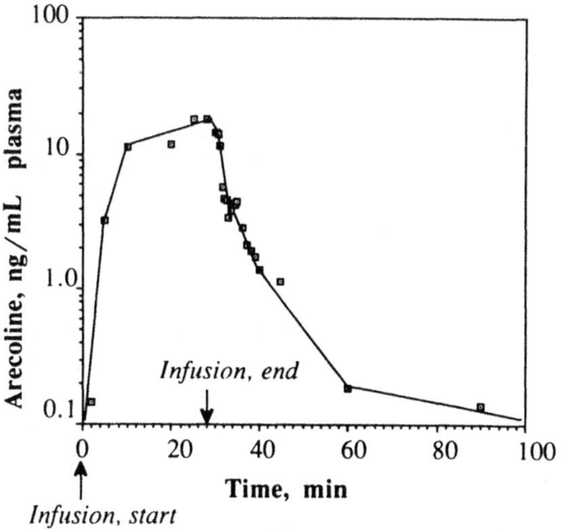

Fig. 5. Time-dependent plasma concentration profile of the short-acting muscarinic cholinergic agonist arecoline following 28-min continuous intravenous infusion of 5 mg into a patient with dementia of the Alzheimer's type. This agent has a disappearance half-life of between 3 and 5 min (see end of infusion), but steady-state levels can be achieved by continuous 24-h infusion. (From SHETTY et al. 1991)

infusion, however, probably saturates its hepatic metabolism and has resulted in elevated cerebrospinal fluid/serum ratios of between 0.4 and 0.58 (HO and FRIE 1971; WEINSTEIN et al. 1982).

IV. Cerebral Blood Flow

Cerebral blood flow is an additional factor that codetermines the amount of a systemically administered drug that eventually reaches and enters brain. As described, this is of particular importance for lipophilic drugs, and a critical analysis of the interaction between cerebrovascular permeability, drug extraction into brain, and cerebral blood flow is that of FENSTERMACHER and RAPOPORT (1983). From a drug delivery perspective, the rate of cerebral blood flow is high compared to many peripheral organs, and accounts for approximately 20% of cardiac output. The arterial administration of a drug, however, allows its 100% delivery into the ipsilateral cerebral vasculature, and this can dramatically increase the brain delivery of specific agents (FENSTERMACHER and COWLES 1977; HOCHBERG et al. 1985; GREIG 1989a). The advantage of arterial administration occurs only during the initial passage of the compound through the cerebral vasculature. Thereafter, the drugs enters the systemic circulation and its pharmacokinetics are similar to those pertaining to its intravenous administration. Ideal drugs for arterial administration are lipophilic, are not restrictively plasma protein bound, bind or enter the intracellular compartment to reduce back-diffusion from brain, possess a rapid systemic clearance, and are not neurotoxic. The relative advantage of this technique for the brain delivery of a variety of drugs of differing pharmacokinetics and physicochemical characteristics has been reviewed by GREIG (1989a). With knowledge of the total body clearance of an agent, Cl (ml/min), derived from its intravenous pharmacokinetics, as well as of blood flow through the artery into which the drug would be administered, F (ml/min), an estimate of the relative advantage of the intraarterial versus the intravenous route is given as

$$\text{Relative advantage} = 1 + Cl/F.$$

The intraarterial route has been used successfully with a variety of agents in three ways. First, to increase target concentrations of the drug where a maximal pharmacologic effect has not been achieved, without the side effect of concomitantly increasing peripheral tissue levels; second, a variation of the first, to decrease peripheral tissue levels of a drug whose toxicity is a target other than the brain, without reducing brain levels; and third, in combination with the hyperosmotic blood-brain barrier disruption technique (NEUWELT and DAHLBORG 1989). By combining arterial administration with transient barrier opening one can deliver and sequester water-soluble agents up to the size of immunoglobulin G in brain, provided that they are administered before barrier rehydration occurs.

D. Strategies for Increasing the Brain Concentration of Drugs

Several techniques can be undertaken to augment drug delivery to brain, and a number of these have recently been reviewed (PARDRIDGE 1985; GREIG 1987, 1989a; NEUWELT and DAHLBORG 1989). These can be broadly divided into two major categories, namely "chemical/pharmacologic" techniques and "physical" techniques. The former includes modification of water-soluble agents by the development of lipophilic analogues and prodrugs, as well as developing agents to utilize naturally occurring blood-brain barrier transport systems. The latter includes the direct administration of drugs into the cerebrospinal fluid (RICARDI et al. 1983) and occasionally into brain (HARBAUGH et al. 1988; BREM 1990), and transient innocuous opening of the blood-brain barrier by the hyperosmotic method (NEUWELT and DAHLBORG 1989; ROBINSON and RAPOPORT 1990). Both broad categories of techniques have proved highly valuable in specific circumstances. This chapter, however, will briefly review only the principles and examples of chemical/pharmacologic techniques.

I. Development of Lipophilic Analogues and Prodrugs

If a potentially valuable therapeutic is restricted from entering the brain by the presence of an ionized group, rendering it water-soluble, it may be possible to mask that group and increase the compound's cerebrovascular permeability. Provided that such a modification is not too rapidly reversed with regeneration of the original ionized agent in the blood prior to its reaching brain microvasculature, and provided that the masked compound does not bind restrictively to plasma constituents or widely distribute to peripheral organs, it may enter brain substantially. For drugs that require the interaction of the masked ionized group with the target (i.e., receptor or enzyme subsite), then regeneration/unmasking is required at the target before a pharmacodynamic effect can be elicited. Such agents are termed prodrugs, and there are a number of extensive reviews on their development (HIGUCHI and STELLA 1975; SINKULA and YALKOWSKY 1975; NOTARI 1981; WERMUTH 1984; STELLA et al. 1985; BODOR and KAMINSKI 1987; BUNDGAARD 1987).

Prodrugs are normally inactive and generate active drug, hopefully at their target, by enzymatically or chemically mediated cleavage of their lipophilic transport moiety. Their therapeutic activity, therefore, not only relies on their delivery to their target, but generation of active drug must occur with sufficiently rapid kinetics to provide effective drug levels. Furthermore, the prodrug and released transport moiety must not be neurotoxic. If an ionized group is not required to elicit pharmacologic activity, its regeneration is not required and a lipophilic analogue of the compound can be developed (GREIG 1989a).

In prodrug design, use is generally made of one or multiple endogenous enzymes to transform it into its active form in vivo. A rational approach to initial prodrug design is to determine the relative concentrations and activities of brain enzyme systems and compare them to those in peripheral organs. In addition to the brain, high concentrations of degradative enzymes are present in the gastrointestinal tract, liver, and blood (AUGUSTINSSON 1961; BOYER 1971; HIGUCHI 1987). Relatively few such studies have been undertaken to quantitate and compare the susceptibilities of series of widely differing substrates to differentiate enzyme systems that are preferentially localized in brain for which prodrugs can be specifically designed. Should such systems exist, it is probable that their role is related to neuronal function and the synthesis or degradation of neurotransmitters.

The most popular prodrugs are esters, since by appropriate esterification of molecules containing carboxylic or hydroxyl functions (or thiol groups) it is feasible to obtain derivatives with almost any desired lipophilicity or hydrophilicity. By varying steric and electrostatic factors, it is possible to control the lability of such prodrugs. Consequently, many alcoholic and carboxylic drugs have been modified for a variety of reasons. Most, however, have not been designed for delivery to the brain, as esterases, required for regeneration of active drug, are ubiquitous, abundant, and relatively unspecific with regard to their substrate affinity. Despite this, there are examples of "brain directed" ester-type prodrugs of varying value. None can truly be described as "site specific" because prodrug enters and regenerates active drug in organs other than the brain. In the event that brain selectivity of action does occur, it is generally because the pharmacologic target (i.e., receptor or enzyme), rather than the regenerating prodrug, is absent from other organs.

Based on extensive structure-activity studies that demonstrated release of epinephrine in mouse cardiac muscle from phenolic derivatives (DALY et al. 1966), CREVELLING et al. (1969) studied the brain delivery and action of 3,4,β-triacetyl-L-norepinephrine and a 3,4,β-trimethysilyl derivative (Fig. 6). Although both readily entered the brain, norepinephrine release was extremely slow compared to cardiac tissue, and neither agent demonstrated dramatic pharmacologic action. Conversely, similar structure-activity studies by HORN et al. (1978a, 1979) resulted in the development of several interesting derivatives of the long-acting dopaminergic agonist 2-amino-6,7-dihydroxytetraphydronaphthalene (ADTN). ADTN possesses a structure similar to dopamine and hence has poor brain penetrability and is subject to extensive metabolism by MAO-B at the level of the enzymatic barrier. Administration of dibenzoyl-ADTN (Fig. 7) to rat, however, led to the delivery and maintenance of increased ADTN levels in brain over several hours. Comparison between the above described prodrugs of structurally similar compounds, norepinephrine and ADTN (CREVELLING et al. 1969; HORN et al. 1979), demonstrates that for two series of lipophilic agents, relatively small differences in the ester moiety can lead to differences in

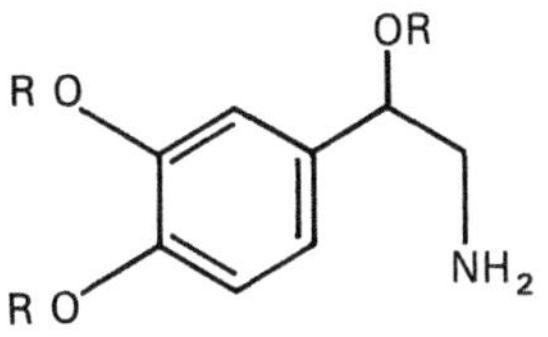

R = H or Ac or $(CH_3)_3$ Si

Fig. 6. Chemical structures of 3,4,β-triacetyl and 3,4,β-trimethysilyl derivatives of norepinephrine

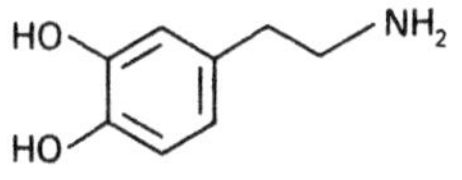

Dopamine

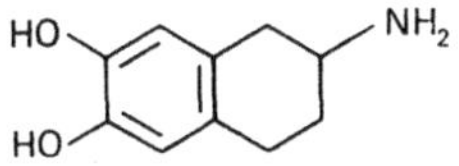

ADTN

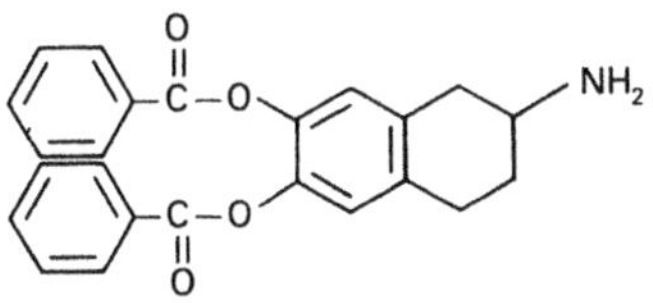

Dibenzoyl-ADTN

Fig. 7. Chemical structures of dopamine, its water-soluble agonist analogue ADTN, and the lipophilic ADTN prodrug dibenzoyl-ADTN

lability that are sufficient to determine whether the compound is active or inactive. Dibenzyl-ADTN proved to be more stable than simple aliphatic ADTN esters (Horn et al. 1978b), allowing it sufficient systemic stability to enter brain before regenerating ADTN. The epinephrine prodrugs, however, proved too stable.

Extensive studies have been undertaken to assess how, for carboxylic acid and alcohol drugs, structural alterations in their ester derivatives alter their rate of hydrolysis in plasma (Glick 1941; Sturge and Whittaker 1951; Chong 1970). Two interesting series of studies have applied these principles to brain and provide an example of how, for different agents, two alternative approaches, (1) to stabilize an ester, and (2) to make it more labile, have culminated in the development of several valuable compounds.

In the first series, Greig et al. (1990c) developed aliphatic and aromatic esters of the widely used, water-soluble, anticancer alkylating agent chlorambucil, an agent restricted from entering brain by the presence of a single carboxylic acid group, pK_a 4.6, that is highly ionized at physiologic pH (Greig et al. 1988). Brain and plasma pharmacokinetic studies in rat, together with binding and activity studies, demonstrated that short-chain aliphatic (methyl and *n*-propyl) esters and aromatic (methylphenyl, ethylphenyl, and prednisolone) esters were too rapidly hydrolyzed in plasma, whereas long-chain aliphatic (hexyl and octyl) esters were too restrictively plasma protein bound to dramatically enter brain. Accordingly, their brain/plasma concentration ratios were not different from that of chlorambucil, 0.02. As the alkylating activity and hence anticancer action of chlorambucil is largely independent of its carboxylic acid moiety, Greig and colleagues (GREIG et al. 1990d,e; GENKA et al. 1991a) developed sterically hindered esters to provide the resulting derivatives with sufficient stability in plasma, without the lipophilicity (and hence restrictive protein binding) associated with a long-chain ester, so they would readily enter brain. The studies culminated in the development and preclinical testing of the chlorambucil-tertiary butyl ester, which is sufficiently stable and lipophilic (log P value 3.2) that it readily enters and maintains high brain concentrations, despite quickly decreasing plasma levels. Its brain levels far exceeded those derived from administration of equimolar chlorambucil (GREIG et al. 1990d; GENKA et al. 1991a). The comparative time-dependent plasma and brain pharmacokinetics of tri-, di-, and mono-substituted esters of chlorambucil are shown in Fig. 8, and illustrate the importance of steric effects on ester hydrolysis rates, and hence on brain delivery. The brain and plasma pharmacokinetics of chlorambucil mirrored those of the methyl ester, which almost instantaneouly generates the former, and hence chlorambucil's pharmacokinetics are not shown. As also illustrated in Fig. 8, chlorambucil-tertiary butyl ester represents a further example of how a lipophilic agent will not necessarily flow out of brain immediately as plasma levels decline, but maintains high brain levels (GREIG et al. 1990d,e). Chlorambucil-tertiary butyl ester has demonstrated dramatic activity against human malignant gliomas in vitro, removed from patients that previously failed routine chemotherapy (ALI-OSMAN et al. 1991). The compound was specifically designed to work on two levels, due to the variable nature of the blood-brain barrier in brain tumors (GREIG 1984, 1989b). As the ester possesses intrinsic activity, not requiring hydrolysis, it can readily enter and kill tumor cells sequestered behind an intact blood-brain barrier. Furthermore, as a consequence of high systemic esterase levels, the agent releases chlorambucil in the circulation so that this can enter central tumor regions where localized areas of barrier breakdown occur (GREIG 1989b). Chlorambucil-tertiary butyl ester is dramatically less toxic than chlorambucil, LD_{50} values being approximately 150 and 15 mg/kg, respectively, in the rat, and is not associated with neurotoxicity (GREIG et al. 1991b). The same principles could prove of value if applied to other compounds. For example, SUZUKI et al. (1987) recently reported the improved brain delivery of prostaglandin D_2 in the form of its methyl ester. However, both ester and active drug rapidly disappeared from brain with a half-life of 9 s, and minimal amounts were detected minutes later. Addition of steric hindrance may overcome this problem.

In the second series approach, an example of which is the development of "double ester" prodrugs of GABA (KROGSGAARD-LARSEN and CHRISTENSEN 1979), increased lability can be designed into a compound.

(1) (2) (3) (4) (5)

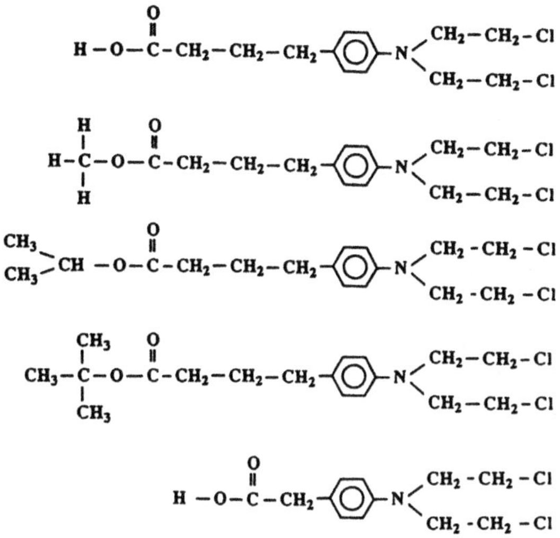

a

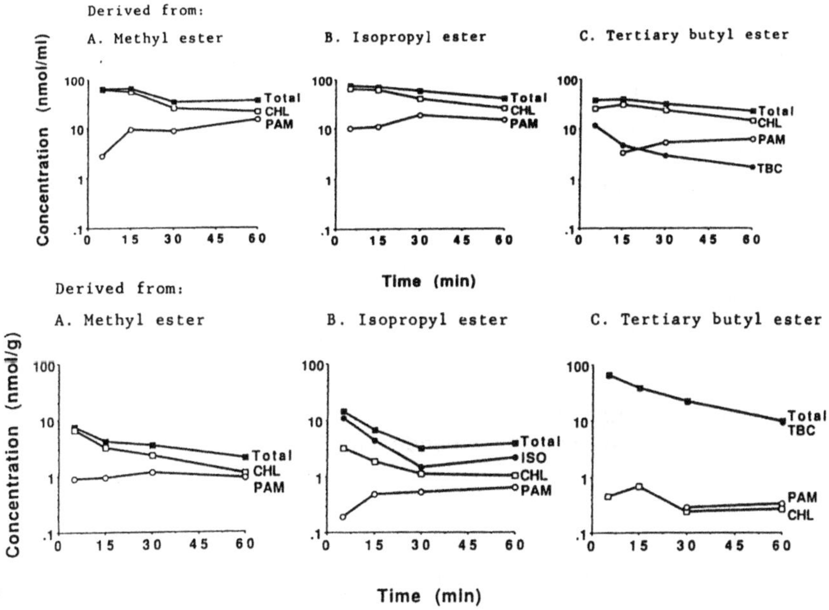

b

Fig. 8. **a** Chemical structures of chlorambucil (*1*), its methyl ester (*2*), isopropyl ester (*3*), and tertiary butyl ester (*4*), and the active water-soluble metabolite of chlorambucil β-oxidation, phenylacetic mustard (*5*). **b** Time-dependent plasma (*top*) and brain (*bottom*) concentration profiles of total active agents (*Total*), chlorambucil derivatives, chlorambucil (*CHL*), and phenylacetic mustard (*PAM*) after equimolar intravenous administration of chlorambucil-methyl ester (10.8 mg/kg), chlorambucil-isopropyl ester (*ISO*, 12.6 mg/kg) and chlorambucil-tertiary butyl ester (*TBC*, 13 mg/kg) to rats. *N.B.* For chlorambucil-tertiary butyl ester in brain, concentrations of the compound are superimposed with the profile for total active agents in brain. Additionally, pharmacokinetics of equimolar chlorambucil alone, although undertaken, are not shown as they were similar to those pertaining to the administration of the methyl ester

Occasionally, simple aliphatic and aromatic esters are too stable to provide the required generation of active drug at their target, particularly in the event that the esterified group is already highly sterically hindered. A solution to this problem was found by JANSEN and RUSSELL (1965) who developed a special double ester (acycloxymethyl ester) of benzylpenicillin that was rapidly hydrolyzed in blood and peripheral tissues of rodents and man to improve the oral bioavailability of the drug. Such esters of ampicillin, including the pivaloyloxymethyl ester (pivampicillin), the carbonyloxyethyl ester (bacampicillin), and the phthalidyl ester (talampicillin), are in wide clinical use. Double esters have similarly been applied to develop brain-directed prodrugs as they are lipophilic and provide a means to tailor the release of drug for optimal pharmacologic activity. The first step in the hydrolysis of double ester prodrugs (Fig. 9) is the enzymatic cleavage of the terminal ester bond with the formation of a highly unstable hydroxymethyl ester. This then rapidly dissociates to the parent drug and formaldehyde. Rate of hydrolysis, therefore, is strongly dependent on the acyl moiety in the terminal ester.

KROGSGAARD-LARSEN and CHRISTENSEN (1979; KROGSGAARD-LARSEN et al. 1982) developed a series of heterocyclic GABA analogues with desirable toxicologic and pharmacokinetic properties, which included the potent specific GABA agonist isoguvacine (1,2,3,6-tetrahydropyridine-4-carboxylic acid; Fig. 10). However, similar to GABA, isoguvacine is a zwitterion and

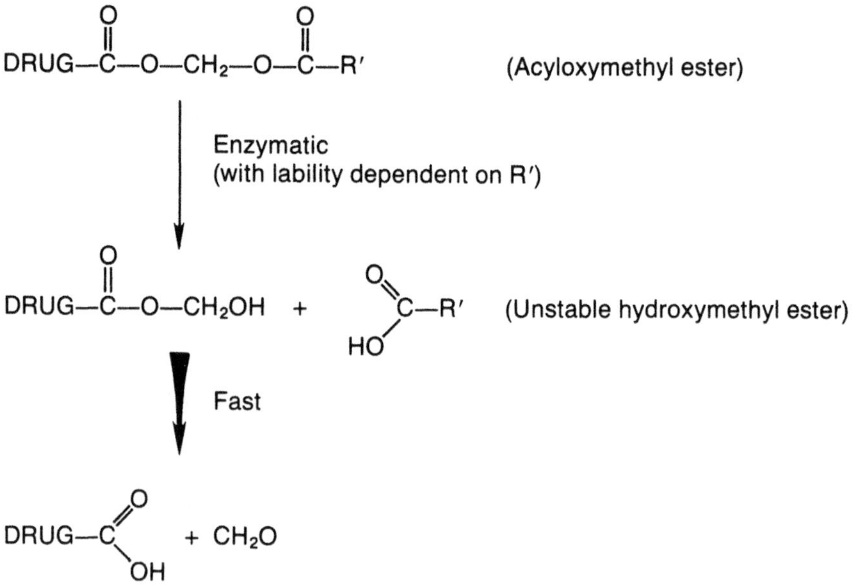

Fig. 9. Scheme for the degradation of acyloxymethyl "double ester" prodrugs to release parent drug

hence is predominantly ionized at physiologic pH; pK_a values are 4.0 and 10.7, and 3.6 and 9.8, respectively. Accordingly, neither compound readily enters brain. As simple ester derivatives of isoguvacine such as the *n*-butyl ester (Fig. 10) proved highly stable in human serum albumin in vitro and in mouse in vivo, more labile acycloxymethyl esters were developed. By varying the structure of the terminal ester, and its steric hindrance, the rate of regeneration of isoguvacine by these esters could be controlled (Fig. 10). These rates, determined in vitro in 10% human serum, correlated reasonably well with their antagonism of electroshock-induced seizures in mice after systemic administration (FALCH et al. 1981; KROGSGAARD-LARSEN et al. 1982). This double ester concept has been applied to many drugs to transiently alter their physicochemical properties; a further example is the brain delivery of the GABA uptake inhibitor nipecotic acid as its pivalyloxymethylester (CROUCHER et al. 1983) (a similar ester to compound 3, Fig. 10). Indeed, such esters can mask several types of ionized moiety. The double ester and sterically hindered single ester provide two extremes of lability that can be designed into a compound to optimize its specific brain delivery and action. Other approaches to these have been undertaken to deliver GABA and chlorambucil as well as many other water-soluble agents to brain, and a selected few will be reviewed.

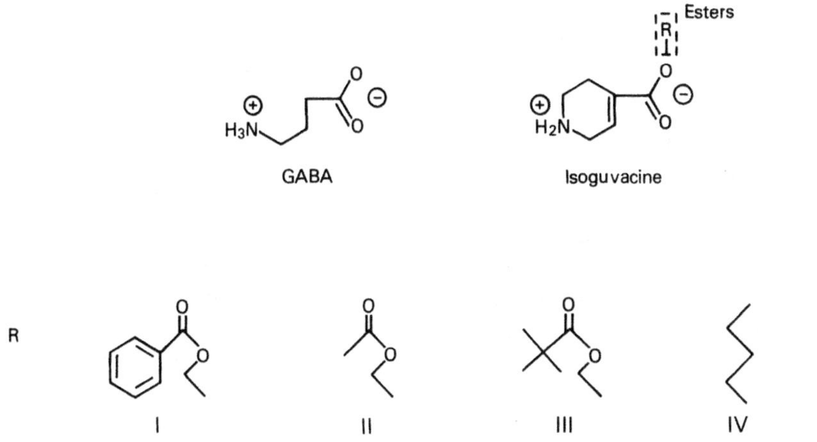

	I	II	III	IV
T½	4 min	36 min	4.6 hours	>100 hours
Activity (post administration)	None	1 to 2 hours	4 to 6 hours	17 to 24 hours

Fig. 10. Chemical structure of GABA and of its water-soluble analogue isoguvacine (*top*); *R* indicates placement of ester moiety of prodrugs. Chemical structures of isoguvacine prodrugs (*bottom*), with compounds *I–III* being double esters containing a different terminal ester, and the fourth compound (*IV*) being a single *n*-butyl ester of isoguvacine. The ability of the compounds to release isoguvacine in 10% human serum and the duration prior to initiation of neuropharmacologic action in mice are shown

CH_2—OCO—$(CH_2)_7(CH{=}CH\ CH_2)_3\ CH_3$
CH—OCO$(CH_2)_3\ NH_2$
CH_2—OCO$(CH_2)_3\ NH_2$

CH—OCO—$(CH_2)_7(CH{=}CH\ CH_2)_3\ CH_3$
CH—OCO—$(CH_2)_7(CH{=}CH\ CH_2)_3\ CH_3$
CH_2—OCO$(CH_2)_3\ NH_2$

Fig. 11. Chemical structures of mono- (*left*) and di-GABA linolenoyl esters of glycerol (*right*)

Two lipid esters of GABA were synthesized to assess whether or not natural lipid analogues that resemble normal components of lipid bilayer membranes could penetrate and transport GABA into brain. The differential uptakes of mono- and di-GABA linolenoyl esters of glycerol (Fig. 11) into mouse brain and liver were compared to those following administration of GABA and of its simple water-soluble glycerol ester (Jacob et al. 1985). Brain uptakes of both GABA linolenoyl esters of glycerol were modestly increases over GABA-alone administration, and delivery to liver was dramatically reduced. Additionally, both esters resulted in neurophamacologic activity, unlike GABA and its simple glycerol ester. Interestingly, recent studies have indicated that endogenous fatty acids, such as palmitate and arachidonate, are not only readily taken up by brain (Noronha et al. 1990) but differentially distribute into intracerebral tumors up to six fold more than into brain (Greig et al. 1991c; Nariai et al. 1991a). They are, therefore, potentially valuable imaging probes (Genka et al. 1991b; Nariai et al. 1991b). Attachment of a pharmacologically active moiety or a drug to a fatty acid that is highly utilized by brain, or that is differentially utilized by a brain sequestered pathologic process (i.e., a malignant brain tumor), represents a potentially useful approach for delivering drugs to brain or to the pathological site. A further example of targeting a disease within brain is in a recent report by Barachi et al. (1991) that describes a series of lipophilic 2′,3′-dideoxynucleoside antiviral prodrugs that are activated by adenosine deaminase (Fig. 12). Although studies on the brain pharmacokinetics of these interesting agents are yet to be completed, they are of great potential in the treatment of acquired immunodeficiency syndrome (AIDS) encephalopathy as brain levels of adenosine deaminase are reported increased in AIDS patients (Raiteri et al. 1989). Cerebrospinal fluid levels in patients with AIDS are approximately 15-fold higher than those in healthy individuals.

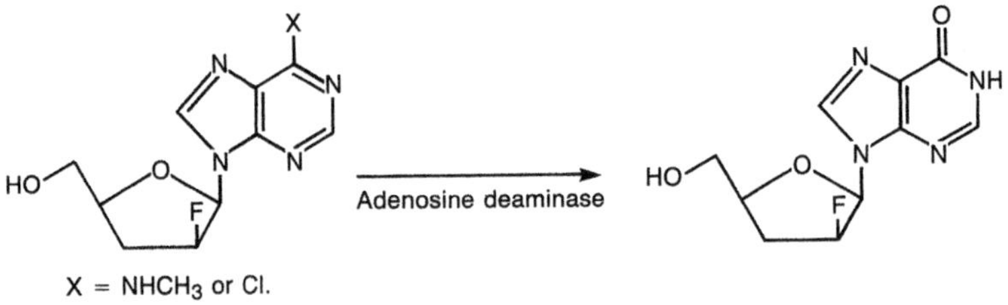

Fig. 12. Chemical structure of adenosine deaminase-activated antiviral lipophilic adenine nucleoside derivatives

A chemical delivery system (CDS) for brain that has been widely studied (BODOR and BREWSTER 1982; BODOR and KAMINSKI 1987; BODOR 1987) involves the attachment of a drug via an ester or amide link to a pyridine ring capable of participating in dihydropyridine ⇌ pyridinium salt redox-type reactions. The dihydropyridine form renders the compound lipophilic for brain entry and, thereafter, becomes oxidized to a charged pyridinium product (Fig. 13). If this occurs, the latter is hydrophilic and has restricted diffusion from brain. Regeneration of active drug from carrier is dependent on the kinetics of enzyme cleavage versus carrier loss from brain. A variety of predominantly water-soluble but also lipophilic agents have been attached to this system and studied for improved brain delivery or reduced systemic delivery, respectively (BODOR and KAMINSKI 1987; BODOR 1987; DIETZEL et al. 1990). There have been variable successes. Amongst the many compounds, both GABA and chlorambucil have been assessed. In studies with GABA (ANDERSON et al. 1987), neuropharmacologic activity was observed following administration of GABA CDS to rat. Unfortunately, measurements of brain GABA levels in treated and control animals were not reported. Concentrations of free chlorambucil of up to 8 µg/g brain after 60 mg drug/kg were reported following administration of chlorambucil CDS to rats, demonstrating clear release from carrier (BODOR et al. 1989). A comparison to equimolar chlorambucil cannot be made; however, comparison with the previously described sterically hindered ester approach, illustrated in Fig. 8 with administration of a lower dose, can be made. Conversion of concentrations from nmol/g into µg/g indicates that peak levels of active drug were 23 µg/g brain and after administration of chlorambucil-tertiary butyl ester 13 mg/kg.

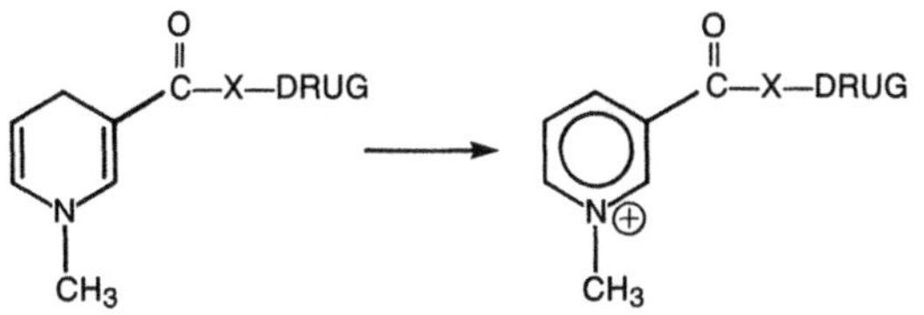

Fig. 13. Chemical structure and scheme of the dihydropyridine ⇌ pyridinium chemical delivery system

Interestingly, administration of isonicotinoyl- and nicotinoyl-GABA to rodents resulted in moderate increases in brain GABA levels and neuropharmacologic activity. These agents (Fig. 14) bear structural similarity to dihydropyridine-linked GABA, but are unable to undergo redox reaction. As recently discussed by BODDY et al. (1991), the utility of the dihydropyridine CDS is critically dependent on the structure of the drug that it is linked to. Minor modifications in the drug can dramatically affect the ability of the

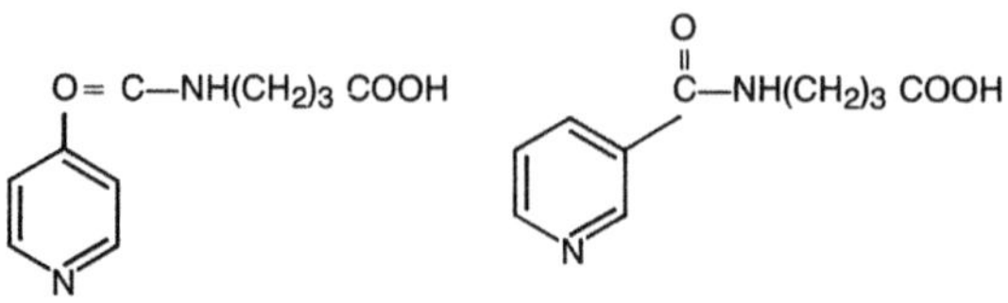

Fig. 14. Chemical structures of isonicotinoyl-GABA (*left*) and nicotinoyl-GABA (*right*)

carrier not only to reach brain but also to enzymatically release drug there. This may partly explain the variable results that many find when attempting to utilize the dihydropyridine CDS for drug delivery to brain. Indeed, the importance of drug selection for drug-targeting systems has been consistently stressed in the literature (HUNT et al. 1986; BODDY et al. 1989). Finally, improvement in drug delivery of any system can only be truly ascertained when active drug concentrations in brain are compared to those achieved following the equimolar administration of the parent compound. Differences in neuropharmacologic activity alone may derive from increases in available drug at the target, i.e., delivery, or improved/altered activity of the prodrug/analogue versus parent compound at the cellular level, i.e., mechanism of action.

Derivatives of the cholinesterase inhibitor physostigmine provide an example of how improved neuropharmacologic activity of a drug can be derived from enhanced interactions between drug and target at the cellular level, i.e., drug/enzyme subsite interactions, rather than by altered brain delivery. Physostigmine readily enters brain following its systemic administration and causes both systemic and central inhibition of acetyl- and butyrylcholinesterase. The duration of enzyme inhibition as well as the disappearance half-life of physostigmine are relatively short, approximately 20–30 min. Physostigmine undergoes hydrolysis to water-soluble eseroline, which has poor brain entry and no anticholinesterase activity (ATACK et al. 1989). Difficulties in maintaining long-term, steady-state inhibition of brain cholinesterase may account for the reported variable therapeutic value of physostigmine in treating the cholinergic deficit in patients with dementia of the Alzheimer's type (POMPONI et al. 1990). BROSSI (1990) and POMPONI et al. (1990; DE SARNO et al. 1989) developed a wide series of physostigmine analogues, including many with altered carbamate groups such as *n*-heptylcarbamoyl eseroline (heptyl-physostigmine; DE SARNO et al. 1989; POMPONI et al. 1990; Fig. 15). Heptyl-physostigmine readily enters brain and causes cholinesterase inhibition for over 6 h. This extended duration of action probably derives from its binding with enzyme to form a carbamylated enzyme intermediate which is more stable than physostigmine (FREEDMAN et al. 1991), rather than to a greater delivery or longer half-life.

Such drugs provide an important example of how the pharmacologic activity of certain agents, such as prolonged enzyme inhibitors and receptor

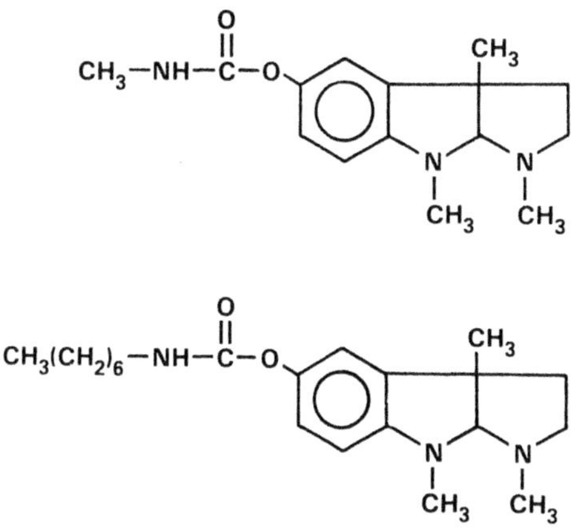

Fig. 15. Chemical structures of physostigmine (*top*) and of its long-acting derivative heptyl-physostigmine (*bottom*)

antagonists, is independent of their time-dependent brain concentration. This has important implications for drug delivery, as only a transient concentration of such drugs may be required at the target for therapeutic activity. An extreme example is that of the organophosphorus cholinesterase inhibitor metrifonate. It possesses no anticholinesterase activity following its administration, but nonenzymatically generates its active component 2,2-dichlorovinyldimethylphosphate (NORDGREN et al. 1978). Following administration of metrifonate to patients with dementia of the Alzheimer's type, neither it nor its active metabolite is detected in blood at 8 h (BECKER et al. 1990). Nevertheless, cholinesterase inhibition persists at high levels for more than 1 week. Phosphorylated enzyme intermediates are highly stable, and metrifonate-induced cholinesterase inhibition can be considered long term or irreversible, with synthesis of new enzyme occurring after 1 week. Similarly, most receptor antagonists have affinities far higher than agonists, and their duration of specific binding is long compared to their nonspecific time-dependent brain concentration, which makes them useful probes for imaging receptors by positron tomography (WAGNER et al. 1983; WILSON et al. 1991).

This brief review of attempts to develop lipophilic analogues and prodrugs to enhance drug delivery and action in brain has primarily focused on esters, and on specific agents in order to compare alternative approaches. As described, esterification represents one of several ways of modifying the octanol/water partition of a drug either irreversibly (lipophilic analogue) or reversibly (prodrug). Carbamates of alcohols as well as of esters could also prove of value for specific drug candidates, as could N-Mannich bases of NH-acidic compounds. The kinetics of hydrolysis of a large number of N-Mannich bases in aqueous solution have been studied (BUNDGAARD and JOHANSEN 1980, 1982a). *N*-Hydroxymethyl and *N*-acyloxyalkyl derivatives have been developed of assorted amides, imides, and hydantoins to vary their physicochemical properties (JOHANSEN and BUNDGAARD 1979). Additionally, oxazolidines, although often unstable in aqueous solution, represent a means of masking β-aminoalcohols, ketones, and aldehydes (BUNDGAARD and JOHANSEN 1982b; BURR et al. 1985). In the final analysis, success requires tailoring the modification to the specific drug candidate and determining whether or not effective concentrations can be achieved at the target for a desired pharmacologic endpoint without toxicity.

II. Carrier-Mediated Transport

An alternative approach to the development of lipophilic therapeutics with the required physicochemical characteristics to allow their passive diffusion into brain is to design drugs that structurally resemble or can be linked to endogenous compounds that are shuttled into brain by naturally occurring cerebrovascular transport systems. Carriers for essential water-soluble nutrients include those for L-amino acids, D-glucose, and assorted nucleosides (SMITH and TAKASATO 1986; DURICIC and MRSULJA 1988), and the previously described administration of the endogenous amino acid L-DOPA in the treatment of Parkinson's disease demonstrates the potential value of this approach.

Recent studies on anticancer alkylating agents reiterate the potential value of this goal. Using an in situ brain perfusion technique, GREIG et al. (1987a) demonstrated that the widely used therapeutic agent melphalan (Fig. 16), which is a nitrogen mustard derivative of L-phenylalanine, is carried into brain via the facilitated transport system for large neutral amino acids, the L-system. Melphalan transport was concentration dependent, saturable, and inhibited by L-phenylalanine. However, its affinity for the carrier was dramatically lower than those of the endogenous L-amino acids with which it competes under physiologic conditions (GREIG et al. 1987a; GREIG 1989a). The L-system is a high capacity carrier which is saturated under physiologic conditions, and hence transport of each L-amino acid is highly dependent on the concentration and affinity of the endogenous L-amino acids (SMITH et al. 1987).

Extensive studies by SMITH et al. (1987) using the same perfusion technique have clarified the key properties that determine the affinities of the large neutral L-amino acids for the L-system carrier. As illustrated in Fig. 17, the affinity of each ($1/K_m$) is proportional to its octanol/water partition coefficient, which is governed by its side-chain hydrophobicity. As a consequence, L-phenylalanine and L-tryptophan, which possess the largest hydrophobic side-chains, have the highest affinities of the endogenous L-amino acids for the L-system. Conversely, L-alanine and L-serine, with small hydrophilic side-chains, have the lowest affinity. Quantitative analysis of approximately 100 synthetic amino acids was further undertaken (SMITH et al. 1989), and their affinities were determined by their ability to inhibit the transport of radiolabelled L-phenylalanine at the rat blood-brain barrier. These studies confirmed that the L-system carrier accepts a wide variety of ligands, but requires that they possess a free carboxylic acid and an α-amino group. Furthermore, the calculated affinity of these agents depended, over a 10000-fold range, on side-chain hydrophobicity, with weak requirement with regard to size. These meagre structural requirements governing affinity, together with the high capacity of the large neutral L-amino acid transport system, make it ideal as a target for drug development.

i. D,L-NAM

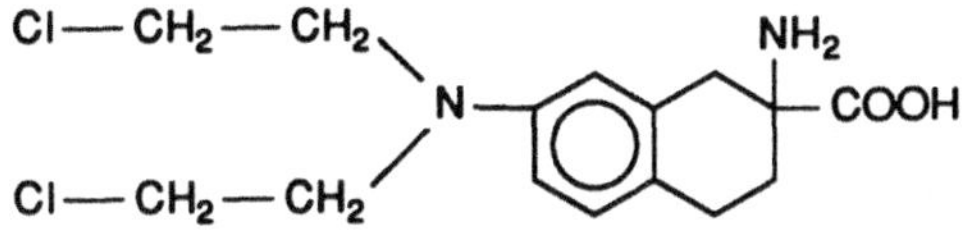

ii. Melphalan

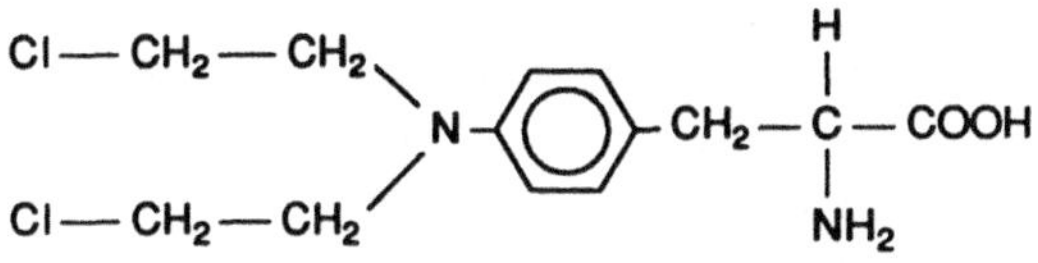

iii. Azaserine

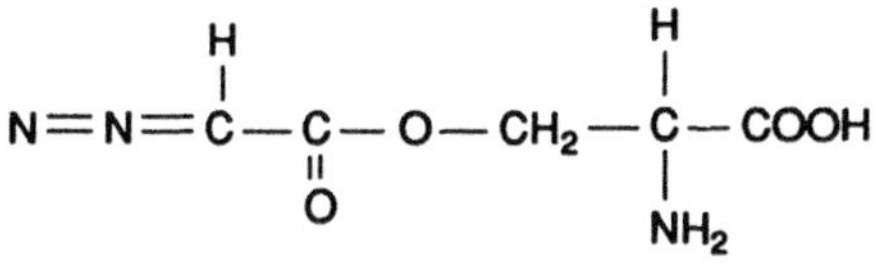

iv. L-DON

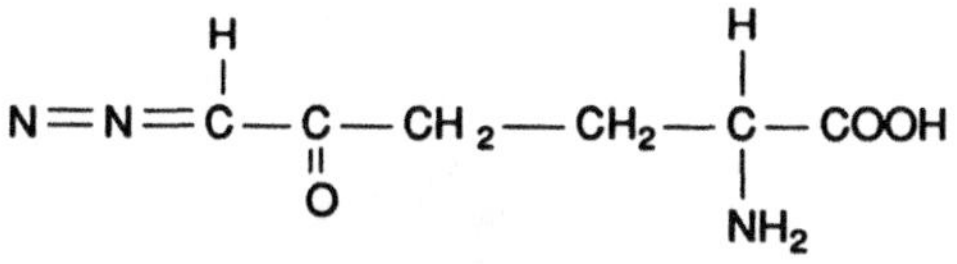

v. Acivicin

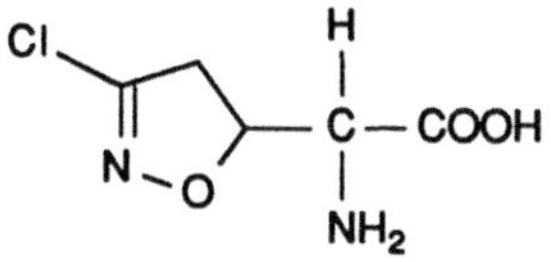

vi. Buthionine Sulfoximine (BSO)

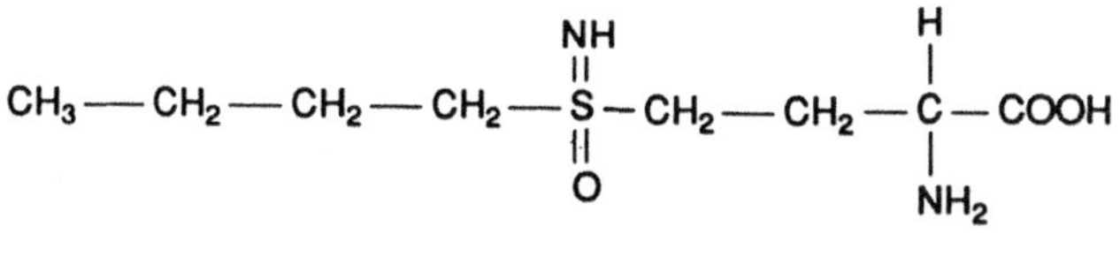

Fig. 16. Chemical structures of chemotherapeutic amino acid drugs (*L-DON*, 6-diazo-5-oxo-L-norleucine)

On the basis of the described studies, it can be predicted that L-amino acid drugs that are similar to melphalan but with a more hydrophobic side-chain, would possesses a high affinity for the L-system carrier. Recent studies by TAKADA et al. (1991) have confirmed this. The nitrogen mustard DL-2-amino-7-bis[(2-chloroethyl)amino]-1,2,3,4-tetrahydro-2-naphthoic acid (DL-NAM; Fig. 16) is structurally similar to melphalan but possesses an additional naphthoic side-chain that provides it greater lipophilicity (HAINES et al. 1987). DL-NAM possessed a remarkable affinity for the blood-brain barrier L-system carrier (TAKADA et al. 1991) which was more than 100-fold greater than that of melphalan, and approximately 20-fold greater than that of L-phenylalanine (the endogenous amino acid with the highest affinity; Fig. 17). Similar to melphalan, DL-NAM demonstrated concentration-dependent transport and saturability at the L-system carrier (Fig. 18). This agent has alkylating activity that is approximately 1.5-fold that of melphalan. It is only 20% bound to plasma proteins, rather than 80% for melphalan (GREIG et al. 1987b; TAKADA, personal communication) and therefore has a high free plasma concentration. DL-NAM also has a higher therapeutic index than melphalan (HAINES et al. 1987) and, although it possesses a lower V_{max} for the transport system than the endogenous L-amino acids (TAKADA et al. 1991), its remarkably high affinity for the transporter (which for the pure L-isomer would be yet higher) imparts to the compound a high cerebrovascular permeability and makes it a potentially valuable therapeutic agent for brain tumor therapy. It warrants further study. As tumor cells possess an L-system carrier on their surface membrane

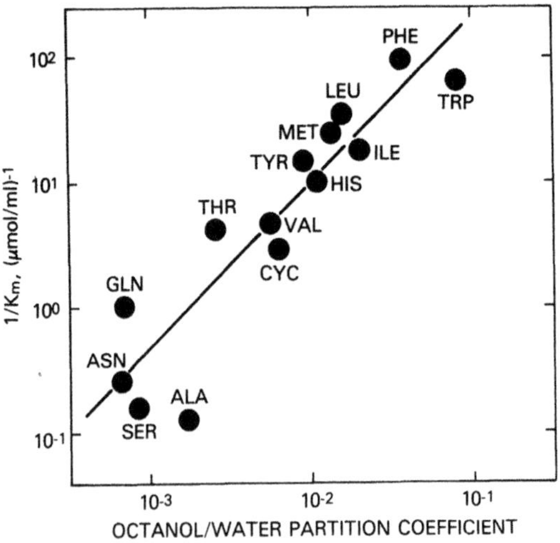

Fig. 17. Relation between the affinity ($1/k_m$, where k_m is the half-saturation concentration) of 14 large neutral L-amino acids for the blood-brain barrier L-system carrier and octanol/water partition coefficient. *CYC*, cycloleucine. (From SMITH et al. 1987)

(VISTICA et al. 1986), differences in carrier number between neoplastic and normal neuronal cells may also impart to DL-NAM a selectivity of action once inside the brain.

The affinity of four further anticancer amino acid drugs (Fig. 16) for the blood-brain barrier L-system carrier has also been determined (TAKADA et al. 1991). All were low (Fig. 18), and markedly less than those of the endogenous L-amino acids with which they would compete under physiologic conditions. Interestingly, the transport kinetics of some of these agents were examined in several tumor cell lines (HUBER et al. 1988; FEKETE et al. 1990), and affinities were similar to those determined at the blood-brain barrier (TAKADA et al. 1991). In certain circumstances, competition with endogenous L-amino acids, which too often reduces transport capability into brain, can be beneficial in protecting the brain from a neurotoxic agent. This appears to occur for acivicin (Figs. 16, 18), which possesses dose-limiting CNS toxicity. WILLIAMS et al. (1990) recently used neutral amino acid infusions to prevent acivicin neurotoxicity in cats.

These studies clearly support the idea that facilitated transport mechanisms at the blood-brain barrier may provide a valuable route to enhance the brain delivery of water-soluble drugs. Comparable approaches have been employed to utilize other blood-brain barrier carriers. However, most possess a low capacity compared to the L-system. As an example, it is probable that the antiviral nucleoside derivative azidothymidine (AZT; log *P* value 0.05) is transported into cerebrospinal fluid, but not brain, by a transport mechanism for thymidine which is present at the level of the choroid plexus (SPECTOR 1986) but not at brain capillary endothelium

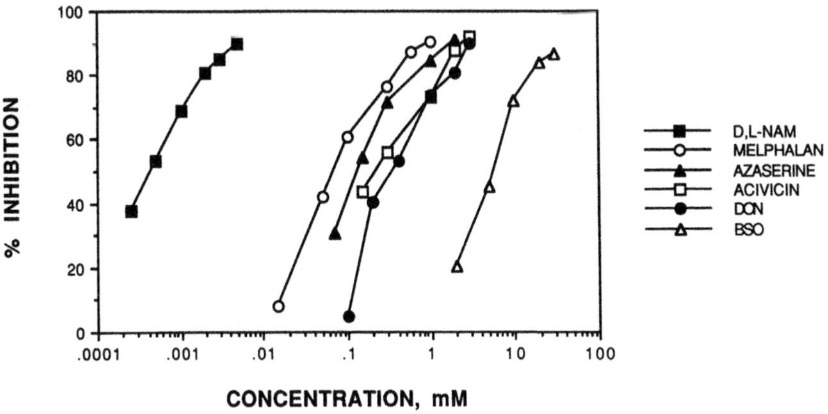

Fig. 18. Inhibition of brain L-[^{14}C]leucine uptake by chemotherapeutic amino acid drugs (from Fig. 16). Inhibition is expressed as the percent reduction in the *PA* product to L-[^{14}C]leucine, and rats were perfused with tracer levels of labelled L-leucine and chemotherapeutic amino acid drugs, with no competing endogenous amino acids. *BSO*, buthionine sulfoxime; *DON*, 6-diazo-5-oxo-L-norleucine. (From TAKADA et al. 1991)

(Terasaki and Pardridge 1988). Steady-state systemic infusions of AZT may therefore optimize entry into cerebrospinal fluid and allow slow penetration of drug into the brain parenchyma. Similarly, several studies have indicated that β-lactam antibiotics, including benzylpenicillin (Suzuki et al. 1989) and ceftriaxone (Spector 1987), are transported at the blood-brain barrier by a probenecid-sensitive mechanism for organic anions. The cerebrovascular permeabilities of these and similar compounds, however, were low and are suggestive of a low capacity system. Finally, stereoselective, carrier-mediated transport of the GABA analogue and muscle relaxing agent 4-amino-3-(*p*-chlorophenyl)butyric acid (baclofen) has been reported at the blood-brain barrier (Van Bree et al. 1988, 1991).

It is now recognized that peptides can cross the blood-brain barrier in small but significant amounts, as peripheral administration of pharmacologic concentrations can result in centrally mediated actions. Additionally, the degree of passage of small peptides across the barrier exceeds that of intravascular markers. Banks and Kastin (1990), Segal and Zloković (1990), and Begley (Chap. 6) have extensively reviewed the mechanisms by which circulating small and large peptides gain access to brain. Earlier studies were hampered by the fact that quantitative techniques commonly used to measure the brain transport kinetics of compounds are difficult to apply to peptides. Many peptides, for example enkephalin(leucine) (Enk(leu)) and thyrotrophin releasing hormone (TRH), are rapidly degraded by plasma enzymes, particularly by aminopeptidases. Nevertheless, the facilitated transport and passive diffusion of a number of small peptides, including Enk(leu), TRH, vasotocin arginine, and sleep-inducing peptide delta, have been measured (Zloković et al. 1989, 1990). Additionally, studies with human brain capillaries have identified the presence of insulin receptors (Pardridge et al. 1985), receptors for insulin-like growth factor (IGF-I and IGF-II; Frank et al. 1986), and transferrin receptors (Jefferies et al. 1984; Pardridge et al. 1987). It has been suggested that these form a large peptide receptor-mediated transport mechanism (Pardridge 1988).

The recognition that receptor-mediated transport of certain peptides occurs across the blood-brain barrier under physiologic conditions has led to the development of chimeric peptides as a strategy to deliver neuropeptides into brain (Pardridge 1988). In this approach, a therapeutically valuable peptide that is restricted from entering brain is covalently coupled to one that is transported, e.g., transferrin or insulin. Additionally, polycationic proteins such as cationic albumin and cationic immunoglobulin have been used as carriers as these also have been reported to enter brain via a receptor-mediated process involving absorptive-mediated transcytosis (Pardridge 1988; Pardridge et al. 1989). Ideally, the chimeric peptide is transported into brain and the nontransportable, therapeutically valuable peptide is then enzymatically released.

Although large numbers of receptors for transferrin and insulin are present on the luminal surface of cerebral microvessels, their transport capacity for chimeric peptides remains undetermined. Furthermore, circulating endogenous concentrations of transferrin and insulin are relatively high and would be likely to compete for receptors. As the entry of iron, the primary ligand for transferrin, into brain is extremely slow (Chap. 10) it is possible that many receptors are involved in neuromodulatory processes rather than in transport and that the transport capacity of this system is very low. The transport capacity of cationic albumin also remains undetermined. However, competition from endogenous cationic protein is probably minimal. By using a disulfide link to cationic albumin which was stable in plasma but labile intracellularly due to the presence of glutathione-dependent disulfide reductase, increased brain delivery of [D-Ala^{2}]-β-endorphin has been reported in rodents (Pardridge et al. 1990).

E. Summary

The therapeutic value of many promising drugs, peptides, and proteins is diminished by the presence of a blood-brain barrier, which possesses both structural and enzymatic components, at the level of the cerebral capillaries, choroid plexus epithelium, and arachnoid membranes. In addition to the permeability of a therapeutic agent at the blood-brain barrier depending on its octanol/water partition coefficient, four further factors codetermine the amount of an agent that eventually enters brain. These include (1) the binding of the compound to plasma constituents, and whether or not such binding is restrictive or nonrestrictive for brain uptake; (2) the ionization of the compound at physiologic pH; (3) its time-dependent plasma concentration profile, which depends on its distribution volume as well as on metabolism and other removal processes; and finally (4) cerebral blood flow, which determines the access of drug to the cerebral vasculature.

For many therapeutic agents, simple manipulations in timing (i.e., continuous infusion) or route of administration (i.e., intraarterial rather than oral or intravenous routes) will elevate brain concentrations above the threshold for therapeutic action. For other agents, however, alternative strategies are required to optimize brain delivery. Two broad approaches to enhance the uptake of a compound into brain are "chemical/pharmacologic" and "physical" techniques. The specific advantages of each approach depend on the agent to be delivered and the location of its target within brain. For water-soluble drugs that are restricted from entering brain as a consequence of one or a few ionized groups, these can be masked transiently (lipophilic prodrugs) or irreversibly (lipophilic analogues), depending on whether or not the presence of these moieties is required for neuropharmacologic activity. Few generic approaches have proved of value in chemically modifying water-soluble drugs to increase their brain delivery. However, careful drug selection and the tailoring of chemical modification to the specific compound have resulted in development of a number of useful neurologically active therapeutic agents. Additionally, the use of endogenous facilitated transport systems present at the blood/brain interface represents a further means to augment the brain delivery of water-soluble agents. The L-system carrier for large neutral L-amino acids has a high capacity and requires that its substrates possess a free carboxylic acid and an α-amino group, together with a hydrophobic side-chain, for transport. Amongst other carrier systems present are those for large peptides and polycationic proteins. Although the capacity of these systems is low, attachment of drugs, and in particular highly active neuropeptides, to substrates of these systems represents a plausible means of delivering them to brain. In the years ahead, it is likely that the application of molecular biological techniques to define therapeutic targets in brain and carrier systems at the blood/brain interface, together with a clearer understanding of differential enzyme systems in brain versus peripheral tissue, will aid in the rational design of selective neuropharmaceutical compounds.

References

Ali-Osman F, Greig NH, John V, Lieberburg IM, Rapoport SI (1991) Activity of tertiary butyl chlorambucil ester against 2-chloroethylnitrosourea-resistant human malignant glioma cell lines. Proc Am Assoc Cancer Res 32:318

Anderson WR, Simpkins JW, Woodard PA, Winwood D, Stern WC, Bodor N (1987) Anxiolytic activity of a brain delivery system for GABA. Psychopharmacology (Berlin) 92:157–163

Atack JR, Yu QS, Soncrant TT, Brossi A, Rapoport SI (1989) Comparative inhibitory effects of various physostigmine analogs against acetyl- and butyrylcholinesterases. J Pharmacol Exp Ther 249:194–202

Augustinsson KB (1961) Multiple forms of esterase in vertebrate blood plasma. Ann NY Acad Sci 94:844–860

Banks WA, Kastin AJ (1990) Peptide transport systems for opiates across the blood-brain barrier. Am J Physiol 259:E1–E10

Barachi J, Marquez VE, Driscoll JS, Ford H, Mitsuya H, Shirasaka T, Aoki S, Kelly JA (1991) Potential anti-AIDS drugs. Lipophilic, adenosine deaminase-activated prodrugs. J Med Chem 34:1647–1655

Bartholini G, Pletscher A (1969) Effects of various decarboxylase inhibitors on the cerebral metabolism of dihydroxyphenylalanine. J Pharm Pharmacol 21:323–324

Becker RE, Colliver J, Elble R, Feldman E, Giacobini E, Kumar V, Markwell S, Moriearty P, Parks R, Shillant SD, Unni L, Vicari S, Wamack C, Zec RF (1990) Effects of metrifonate, a long-acting cholinesterase inhibitor, in Alzheimer's disease. Drug Dev Res 19:424–434

Bertler A, Falck B, Owman C, Rosengren E (1966) Localization of monoaminergic blood-brain barrier mechanisms. Pharmacol Rev 18:369–385

Birkmayer W, Riederer P, Ambrozi L (1977) Implications of combined treatment with 'Madopar' and L-deprenyl in Parkinson's disease. Lancet 2:439–443

Boddy AV, Aarons L, Petrak K (1989) Efficiency of drug targeting: steady-state considerations using a three-compartmental model. Pharm Res 6:367–372

Boddy AV, Zhang K, Lepage F, Tombret F, Slatter JG, Baillie TA, Levy RH (1991) In vitro and in vivo investigations of dihydropyridine-based chemical delivery systems for anticonvulsants. Pharm Res 8:690–697

Bodor N (1987) Redox delivery system for targeting drugs to brain. Ann NY Acad Sci 507:289–306

Bodor N, Brewster M (1982) Problems of drug delivery to the brain. Pharmacol, Ther 19:337–386

Bodor N, Kaminski JJ (1987) Prodrugs and site-specific chemical delivery systems. Annu Rep Med Chem 22:303–313
Bodor N, Venkatraghavan V, Winwood D, Estes K, Brewster M (1989) Improved delivery through biological membranes. XLI. Brain enhanced delivery of chlorambucil. Int J Pharm 53:195–208
Boyer PD (1971) The enzymes, vol 5. Academic, New York
Brem H (1990) Controlled delivery to the brain. In: Gregoriadis G (ed) Targeting of drugs. Plenum, New York, pp 155–174
Brightman MW (1989) The anatomic basis of the blood-brain barrier. In: Neuwelt EA (ed) Implications of the blood-brain barrier and its manipulation, vol 1, Basic science aspects. Plenum, New York, pp 53–83
Brossi A (1990) Bioactive alkaloids. IV. Results of recent investigations with colchicine and physostigmine. J Med Chem 33:2311–2319
Bundgaard H (1987) Design of bioreductive derivatives and the utility of the double prodrug concept. In: Roche EB (ed) Bioreversible carriers in prodrug design, theory and application. Pergamon, New York, pp 13–94
Bundgaard H, Johansen M (1980) Prodrugs as delivery systems. XV. Bioreversible derivitization of phenytoin, acetazolomide, chlorzoazone and various other NH-acidic compounds by N-aminomethylation to effect enhances disolution rates. Int J Pharm 7:129–136
Bundgaard H, Johansen M (1982a) Prodrugs as delivery systems. XIX. Bioreversible derivitization of aromatic amines by formation of N-Mannich bases with succinimide. Int J Pharm 8:183–192
Bundgaard H, Johansen M (1982b) Prodrugs as drug delivery systems. XX. Oxazolidines as potential pro-drug types of β-aminoalcohols, aldehydes and ketones. Int J Pharm 10:165–175
Burr A, Bundgaard H, Falch E (1985) Prodrugs of 5-fluorouracil. IV. Hydrolysis kinetics, bioactivation and physicochemical properities of various N-acyloxymethyl derivitives of 5-fluorouracil. Int J Pharm 24:43–60
Cefalu WT, Pardridge WM (1985) Restrictive transport of a lipid-soluble peptide (cyclosporin) through the blood-brain barrier. J Neurochem 45:1954–1956
Chong CW (1970) Inhibition of human plasma esterases by 2′-dimethylaminoethyl-2,2-diphenylvalerate.HCl (SK&F 525A). Thesis, Temple University, Philadelphia
Cornford EM (1984) Blood-brain barrier permiability to anticonvulsant drugs. In: Levy RH, Pitlick WH, Echelbaum M, Meijer J (eds) Metabolism of antiepileptic drugs. Raven, New York, pp 129–142
Crevelling C, Daly J, Tokuyama T (1969) Labile lipophilic derivative of norepinephrine capable of crossing the blood-brain barrier. Experientia 25: 26–27
Croucher MJ, Meldrum BS, Krogsgaard-Larsen P (1983) Anticonvulsant activity of GABA uptake inhibitors and their prodrugs following central or systemic administration. Eur J Pharmacol 89:217–228
Daly JW, Creveling CR, Witkop B (1966) The chemorelease of norepinephrine from mouse hearts. Structure-activity relationships. I. Sympathomimetic and related amines. J Med Chem 9:273–280
Davson H (1967) Physiology of the cerebrospinal fluid. Churchill Livingstone, London
Davson H (1989) History of the blood-brain barrier concept. In: Neuwelt EA (ed) Implications of the blood-brain barrier and its manipulation, vol 1, Basic science aspects. Plenum, New York, pp 27–52
Davson H, Welch K, Segal L (1987) The blood-brain barrier. In: The physiology and pathophysiology of the cerebrospinal fluid. Livingstone, London, pp 65–103
De Sarno P, Pomponi M, Giacobini E, Tang XC, Williams E (1989) The effect of heptyl-physostigmine, a new cholinesterase inhibitor, on the central cholinergic system of the rat. Neurochem Res 14:971–977
Dietzel K, Keuth V, Estes KS, Brewster ME, Clemmons RM, Vistelle R, Bodor N, Derendorf H (1990) A redox-based system that enhances delivery of estradiol to the brain: pharmacokinetic evaluation in the dog. Pharm Res 7:879–883

Duričić BM, Mršulja BB (1988) Transport and barrier systems of the cerebral vasculature; enzymatic aspects. In: Rakić L, Begley DJ, Davson H, Zloković BV (eds) Peptides and amino acid transport mechanisms in the central nervous system. Stockton, New York, pp 269–278
Falch E, Krogsgaard-Larsen P, Christensen A (1981) Esters of isoguvacine as potential prodrugs. J Med Chem 24:285–289
Fekete I, Griffith OW, Schlageter KE, Bigner DD, Friedman HS, Groothius DR (1990) Rate of buthionine sulfoximine entry into brain and xenotransplanted human gliomas. Cancer Res 50:1251–1256
Fenstermacher JD, Cowles AL (1977) Theoretic limitations of intracarotid infusions in brain tumor chemotherapy. Cancer Treat Rep 61:519–526
Fenstermacher JD, Rapoport SI (1984) Blood-brain barrier. In: Renkin EM, Michel CC (eds) Handbook of Physiology, The cerebrovascular system IV: Microcirculation, part 2. American Physiological Society, Bethesda, pp 969–1001
Finklestein J, Shern J, Chabner B (1970) Pharmacologic studies of tritiated cytosine arabinoside (NSC 63878) in children. Cancer Chemother Rep 54:35–41
Frank HJL, Pardridge WM, Morris WM, Rosenfeld RG, Choi TB (1986) Binding and internalization of insulin and insulin-like growth factors by isolated brain microvessels. Diabetes 35:654–658
Freedman F, Johnson J (1969) Equilbrium and kinetic properties of the Evans blue albumin system. Am J Physiol 216:675–681
Freedman SB, Iversen LL, Rugarli PL, Harley EA (1991) Heptyl-physostigmine: a potent inhibitor of acetylcholinesterase with long duration of activity. 2nd International Springfield symposium on advances in Alzheimer's disease, Springfield, Ill, p 19
Genka S, Shetty HU, Stahle PL, John V, Lieberburg IM, Ali-Osman F, Rapoport SI, Greig NH (1991a) Development of lipophilic anticancer agents for the treatment of brain tumors by the esterification of water-soluble chlorambucil. Clin Exp Metastasis (in Press)
Genka S, Greig NH, Nariai T, DeGeorge J, Noronha JG, Schmall B, Rapoport SI (1991b) Brain tumor imaging with radiolabelled fatty acids in rats. Proc Am Assoc Cancer Res 32:73
Glick D (1941) Some additional observations on the specificity of cholinesterases. J Biol Chem 137:357–362
Gomori G (1941) The distribution of phosphatase in normal organs and tissues. J Cell Comp Physiol 17:71–83
Greig NH (1984) Chemotherapy of brain metastases: current status. Cancer Treat Rev 11:157–186
Greig NH (1987) Optimizing drug delivery to brain tumors. Cancer Treat Rev 14:1–28
Greig NH (1989a) Drug delivery to the brain by blood-brain barrier circumvention and drug modification. In: Neuwelt EA (ed) Implications of the blood-brain barrier and its modification, vol 1, Basic science studies. Plenum, New York, pp 311–367
Greig NH (1989b) Brain tumors and the blood-tumor barrier. In: Neuwelt EA (ed) Implications of the blood-brain barrier and its manipulation, vol. 2, Clinical studies. Plenum, New York, pp 77–106
Greig NH, Momma S, Sweeney DJ, Smith QR, Rapoport SI (1987a) Facilitated transport of melphalan at the rat blood-brain barrier by the large neutral amino acid carrier system. Cancer Res 47:1571–1576
Greig NH, Sweeney DJ, Rapoport SI (1987b) Melphalan concentration dependent plasma protein binding in healthy humans and rats. Eur J Clin Pharmacol 32:179–185
Greig NH, Sweeny DJ, Rapoport SI (1988) Comparative brain and plasma pharmacokinetics and anticancer activities of chlorambucil and melphalan in the rat. Cancer Chemother Pharmacol 21:1–8

Greig NH, Soncrant TT, Shetty HU, Momma S, Smith QR, Rapoport SI (1990a) Brain uptakes and anticancer activities of vincristine and vinblastine are restricted by their low cerebrovascular permeability and binding to plasma constituents in rat. Cancer Chemother Pharmacol 26:263–268

Greig NH, Genka S, Rapoport SI (1990b) Delivery of vital drugs to the brain for the treatment of brain tumors. J Controlled Release 11:61–78

Greig NH, Genka S, Daly EM, Sweeney DJ, Rapoport SI (1990c) Physicochemical and pharmacokinetic parameters of seven lipophilic chlorambucil esters designed for brain penetration. Cancer Chemother Pharmacol 25:311–319

Greig NH, Daly EM, Sweeney DJ, Rapoport SI (1990d) Pharmacokinetics of chlorambucil-tertiary butyl ester, a lipophilic chlorambucil derivative that achieves and maintains high concentrations in brain. Cancer Chemother Pharmacol 25:311–319

Greig NH, Stahle PL, Shetty HU, Genka S, John V, Rapoport SI (1990e) High performance liquid chromatography analysis of chlorambucil-tertiary butyl ester and its active metabolites, chlorambucil and phenylacetic mustard, in plasma and tissue samples. J Chromatogr 534:279–286

Greig NH, Wozniak KM, Tolliver T, Holloway HW, Freo U, Rapoport SI, Soncrant TT (1991a) Age-dependent pharmacokinetics of m-chlorophenylpiperazine in brain and plasma of Fischer-344 rats. Psychopharmacology (Berlin) (in Press)

Greig NH, Ali-Osman F, Genka S, Shetty HU, John V, Stahle PL, Tung J, Soncrant TRT, Lieberburg IM, Rapoport SI (1991b) Chlorambucil-tertiary butyl ester, an agent designed for brain tumor therapy: pharmacokinetics and activity in rats. Proc Am Assoc Cancer Res 32:333

Greig NH, Nariai T, Noronha JG, Schmall B, Larson DM, Soncrant TT, Rapoport SI (1991c) Brain tumor imaging in rats using positron emitting fatty acid radionuclide dl-erythro-9,10-[18F]difluoropalmitate. Clin Exp Metastasis 9: 77–84

Haines DR, Fuller RW, Ahmad S, Vistica DT, Marquez VE (1987) Selective cytotoxicity of a system L specific amino acid nitrogen mustard. J Med Chem 30:542–547

Hansch C (1972) Strategy in drug design. Cancer Chemother. Rep 56:433–441

Harbaugh RE, Saunders RL, Reeder R (1988) Use of implantable pumps for central nervous system drug infusions to treat neurological disease. Neurosurgery 23:693–700

Higuchi T (1987) Prodrug and drug delivery; an overview. In: Roche EB (ed) Bioreversible carriers in drug design, theory and application. Pergamon, Oxford, pp 1–12

Higuchi T, Stella V (1975) Prodrugs as novel drug delivery systems. American Chemical Society, Washington (American Chemical Society symposium series 14)

Ho D, Frie E (1971) Clinical pharmacology of 1-β-arabinofuransylcytosine. Clin Pharmacol Ther 12:944–954

Hochberg FH, Pruitt AA, Beck DO, DeBrun G, Davis K (1985) The rationale and methodology for intra-arterial chemotherapy with BCNU as treatment for glioblastoma. J Neurosurg 63:876–880

Hollingsworth J, Davson H (1973) Transport of sulfate in the rabbit's brain. J Neurobiol 4:389–396

Horn A, Grol C, Dijkstra D (1978a) Facile syntheses of potent dopaminergic agonists and their effects on neurotransmitter release. J Med Chem 21:825–828

Horn A, DeKaste D, Dijkstra D (1978b) A new dopaminergic prodrug. Nature 276:405–407

Horn A, Kelly P, Westerink B (1979) A prodrug of ADTN: selectivity of dopaminergic action and brain levels of ADTN. Eur J Pharmacol 60:95–99

Huber KR, Rosenfeld H, Roberts J (1988) Uptake of glutamine antimetabolites 6-diazo-5-oxo-L-norleucine (DON) and acivicin in sensitive and resistant tumor cell lines. Int J Cancer 41:752–755

Hunt CA, MacGregor RD, Siegel RA (1986) Engineering targeted in vivo drug delivery. 1. The physiological and physicochemical principles governing opportunities and limitations. Pharm Res 3:333–344
Jacob JN, Shashoua VE, Campbell A, Baldessarini RJ (1985) γ-Aminobutyric acid esters. 2. Synthesis, brain uptake, and pharmacological properties of lipid esters of γ-aminobutyric acid. J Med Chem 28:106–110
Jansen ABA, Russell TJ (1965) Some novel penicillin derivatives. J Chem Soc Ularch 1965:2127–2132
Jefferies WA, Brandon MR, Hunt SV, Williams AF, Gatter KC, Mason DY (1984) Transferrin receptor on endothelium of brain capillaries. Nature 312:162–164
Johansen M, Bundgaard H (1979) Prodrugs as delivery systems. VI. Kinetic and mechanisms of the decomposition of N-hydroxymethylated amides and imides in aqueous solution and assessment of their stability as possible pro-drugs. Arch Pharm Chem Sci Ed 7:175–192
Johanson C (1989) Ontogeny and phylogeny of the blood-brain barrier. In: Neuwelt EA (ed) Implications of the blood-brain barrier and its manipulation, vol 1, Basic science aspects. Plenum, New York, pp 157–198
Jones DR, Hall SD, Jackson EK, Branch RA, Wilkinson GR (1988) Brain uptake of benzodiazepines: effects of lipophilicity and plasma protein binding. J Pharmacol Exp Ther 245:816–822
Jusko W, Gretch M (1976) Plasma and protein binding of drugs in pharmacokinetics. Drug Metab Rev 5:43–140
Kalaria R, Mitchell M, Harik SI (1987) Correlation of 1-methyl-4-phenyl-1,2,36-tetrahydropyridine neurotoxicity with blood-brain barrier monoamine oxidase activity. Proc Natl Acad Sci USA 84:3521–3525
Kapetanović IM, Sweeney DJ, Rapoport SI (1982a) Phenobarbitol pharmacokinetics in rat as a function of age. Drug Metab Dispos 10:586–589
Kapetanović IM, Sweeney DJ, Rapoport SI (1982b) Age-effects of haloperidol pharmacokinetics in male, Fischer-344 rats. J Pharm Exp Ther 221:434–438
Karnovsky MJ (1967) The ultrastrucural basis of capillary permeability studied with peroxidase as a tracer. J Cell Biol 35:213–236
Kragh-Hansen U (1981) Molecular aspects of ligand binding to serum albumin. Phamacol Rev 33:17–53
Krogsgaard-Larsen P, Christensen A (1979) GABA agonists. Synthesis and structure-activity studies on analogues of isoguvacine and THIP. Eur J Med Chem 14:157–164
Krogsgaard-Larsen P, Falch E, Mikkelsen H, Jacobsen P (1982) Development of structural analogs and prodrugs of GABA agonists with desirable pharmacokinetic properties. In: Bundgaard H, Hansen AB, Kofod H (eds) Optimization of drug delivery. Munksgaard, Copenhagen, pp 225–234 (Alfred Benzon Symposium 17)
Leo A, Hansch C, Elkins D (1971) Partition coefficients and their uses. Chem Rev 71:525–616
Levin VA (1980) Relationship of octanol/water partition coefficient and molecular weight to rat brain capillary permeability. J Med Chem 23:682–684
Levine R, Fredericks W, Rapoport SI (1982) Entry of bilirubin into the brain due to opening of the blood-brain barrier. Pediatrics 69:255–259
Levitan H, Ziylan Z, Smith QR, Takasato Y, Rapoport SI (1984) Brain uptake of a food dye, erythrosin B, prevented by plasma protein binding. Brain Res 322:131–134
MacKichan J (1984) Pharmacokinetic consequences of drug displacement from blood and tissue proteins. J Pharmacokinet 9:32–41
Nariai T, DeGeorge JJ, Greig NH, Rapoport SI (1991a) In vivo incorporation of [9,10-3H]palmitate into a rat metastatic brain tumor model. J Neurosurg 74:643–649

Nariai T, DeGeorge J, Greig NH, Genka S, Rapoport SI (1991b) Use of intravenously injected radiolabelled fatty acids for in vivo brain tumor imaging. Clin Exp Metastasis (in Press)
Neuwelt EA, Dahlborg SA (1989) Blood-brain barrier in the treatment of brain tumors: clinical implications. In: Neuwelt EA (ed) Implications of the blood-brain barrier and its manipulation, vol 2, Clinical Studies. Plenum, New York, pp 195–262
Nordgren I, Bergstrom M, Holmstedt B, Sandoz M (1978) Transformation and action of metrifonate. Arch Toxicol 41:31–41
Noronha JG, Bell JM, Rapoport SI (1990) Quantitative brain autoradiography of [9,10-3H]palmitic acid incorporation into brain lipids. J Neurosci Res 26: 196–208
Notari E (1981) Prodrug design. Pharmacol Ther 14:25–53
Pardridge WM (1985) Strategies for delivery of drugs through the blood-brain barrier. Annu Rep Med Chem 20:305–313
Pardridge WM (1988) Recent advances in blood-brain barrier transport. Ann Neurol Pharmacol Toxicol 28:25–39
Pardridge WM, Eisenberg J, Jank J (1985) Human blood-brain barrier insulin receptor. J Neurochem 44:1771–1780
Pardridge WM, Eisenberg J, Yang J (1987) Human blood-brain barrier transferrin receptor. J Neurochem 49:1394–1401
Pardridge WM, Triguero D, Buciak JB (1989) Transport of histone through the blood-brain barrier. J Pharmacol Exp Ther 251:821–826
Pardridge WM, Triguero D, Buciak JB (1990) B-endorphin chimeric peptides: transport through the blood-brain barrier in vivo and cleavage of disulfide linkage by brain. Endocrinology 126:977–984
Pomponi M, Giacobini E, Brufani M (1990) Present state and future development of the therapy for Alzheimer's disease. Aging 2:125–153
Raffaele KC, Berardi A, Asthana S, Morris PP, Haxby JV, Soncrant TT (1991) Effects of long-term continuous infusion of the muscarinic cholinergic agonist arecoline on verbal memory in dementia of the Alzheimer's type. Psychopharm Bull 27:315–319
Raiteri R, Marietta G, Sclofaro C, Sinicco A (1989) Adenosine deaminase and HIV infection. Med Sci Res 17:187–188
Rapoport SI (1976) Blood-brain barrier in physiology and medicine. Raven, New York
Rapoport SI, Levitan H (1974) Neurotoxicity of X-ray contrast media: relation to lipid solubility and blood-brain barrier permeability. AJR 122:186–193
Reese T, Karnovsky M (1967) Fine structural localization of a blood-brain barrier to exogenous peroxidase. J Cell Biol 34:207–217
Ricardi R, Bleyer WA, Poplack DG (1983) Enhancement of delivery of antineoplastic drugs into cerebrospinal fluid. In: Wood JH (ed) Neurobiology of cerebrospinal fluid, vol 2. Plenum, New York, pp 453–466
Rinne V, Sonninen V, Siirtola T (1973) Plasma concentration of levodopa in patients with Parkinson's disease. Response to administration of levodopa alone or combined with a decarboxylase inhibitor and clinical correlations. Eur Neurol 10:301–310
Robinson PJ, Rapoport SI (1986) Kinetics of protein binding determine rates of uptake of drugs by brain. Am J Physiol 251:R1212–R1220
Robinson PJ, Rapoport SI (1990) Model for drug uptake by brain tumors: effects of osmotic treatment and of diffusion in brain. J Cer Blood Flow Metab 10: 153–161
Rowland M (1984) Protein binding and clearance. Clin Pharmacokinet 9:10–17
Scholtan W (1968) Die hydrophobe Bindung der Pharmaka an Humanalbumin und Ribonucleinsäure. Arzheimittel forschung 18:505–517

Schulman DS, Kaufman JJ, Eisenstein MM, Rapoport SI (1984) Blood pH and brain uptake of [^{14}C]morphine. Anesthesiology 61:540–543
Segal MD, Zloković BV (1990) The blood-brain barrier amino acids and peptides. Kluwer, Lancaster
Shetty HU, Daly EM, Greig NH, Rapoport SI, Soncrant TT (1991) An automatic reaction control chemical ionization technique in ion trap detector for quantitative plasma profiling of arecoline in treated Alzheimer's patients. J Am Soc Mass Spectrom 2:168–173
Sinkula A, Yalkowsky S (1975) Rationale for design of biologically reversible drug derivatives: prodrugs J Pharm Sci 64:181–210
Smith QR, Takasato Y (1986) Kinetics of amino acid transport at the blood-brain barrier studied using an in situ brain perfusion technique. Ann NY Acad Sci 481:186–201
Smith QR, Momma S, Aoyagi M, Rapoport SI (1987) Kinetics of neutral amino acid transport across the blood-brain barrier. J Neurochem 49:1651–1658
Smith QR, Aoyagi M, Rapoport SI (1989) Structural specificity of the brain capillary neutral amino acid transporter. Soc Neurosci Abstr 15:1025
Soncrant TT, Holloway HW, Greig NH, Rapoport SI (1989a) Regional brain metabolic responsivity to the muscarinic cholinergic agonist arecoline is similar in young and aged Fischer-344 rats. Brain Res 487:255–266
Soncrant TT, Morris PP, Raffaele KC, Shetty HU, Greig NH, Haxby JV, Daly EM, Rapoport SI (1989b) Rigerous evaluation of chloinergic enhancement therapy in Alzheimer's disease. Abstr Am Coll Neuropsychopharmacol 146
Soncrant TT, Raffaele KC, Asthana S, Berardi A, Morris PP, Haxby JV (1991) Memory improvement without toxicity during chronic low dose intravenous arecoline in Alheimer's disease. Neurology (in press)
Spector R (1986) Nucleoside and vitamin homeostasis in the mammalian central nervous system. Ann NY Acad Sci 481:221–230
Spector R (1987) Ceftriaxone transport through the blood-brain barrier. J Infect Dis 156:209–211
Stein WD (1986) Simple diffusion across the membrane bilayer. In: Stein WD, Lieb WR (eds) Transport and diffusion across cell membranes. Academic, London, pp 69–107
Stella VJ, Charman WNA, Naringrekar VH (1985) Prodrugs, do they have advantages in clinical practice? Drugs 29:455–473
Sturge LM, Whittaker VP (1951) The esterases of horse blood. The specificity of plasma cholinesterase and ali-esterase. Biochem J 47:518–525
Suzuki F, Hayashi H, Ito S, Hayaishi O (1987) Methyl ester of prostaglandin D2 as a delivery system of prostaglandin D2 into brain. Biochim Biophys Acta 917: 224–230
Suzuki H, Sawada Y, Sugiyama Y, Iga T, Hanano M (1989) Facilitated transport of benzylpenicillin through the blood-brain barrier in rats. J Pharmacobiodyn 12:182–185
Takada T, Greig NH, Vistica DT, Rapoport SI, Smith QR (1991) Affinity of antineoplastic amino acid drugs for the large neutral amino acid transporter of the blood-brain barrier. Cancer Chemother Pharmacol (in Press)
Takasato Y, Rapoport SI, Smith QR (1984) An in situ brain perfusion technique to study cerebrovascular transport on the rat. Am J Physiol 247:H484–H493
Terasaki T, Pardridge WM (1988) Restricted transport of 3′-azido-3′-deoxythymidine and dideoxynucleosides through the blood-brain barrier. J Infect Dis 158: 630–632
Valner J (1977) Binding of drugs by albumin and plasma proteins. J Pharm Sci 66:447–465
Van Bree JB, Audus KL, Brochardt RT (1988) Carrier-mediated transport of baclofen across monolayers of bovine brain endothelial cells in primary culture. Pharm Res 5:369–371

Van Bree JB, Heijligers-Feijen CD, DeBoer AG, Danhof M, Breimer DD (1991) Stereoselective transport of baclofen across the blood-brain barrier in rats as determined by the unit impulse response methodology. Pharm Res 8:259–262

Vistica DT, Ahmad S, Fuller R, Hill J (1986) Transport and cytotoxicity of amino acid nitrogen mustards: implications for design of more selective antitumor agents. Fed Proc 45:2447–2450

Wagner HN, Burns HD, Dannals RF, Wong DF, Langstrom B, Duelfer T, Frost JJ, Ravert HT, Links JM, Rosenbloom SB, Lukas SE, Kramer AV, Kuhar MJ (1983) Imaging dopamine receptors in the human brain by positron tomography. Science 221:1264–1266

Weinstein H, Griffin T, Feeney J (1982) Pharmacokinetics of continuous intravenous and subcutaneous infusion of cytosine arabinoside. Blood 59:1351–1353

Wermuth GC (1984) Chemical aspects of pro-drug design. In: Jolles G, Wooldridge KRH (eds) Drug design: fact or fantasy. Academic, London, pp 47–71

Williams MG, Earhart RH, Bailey H, McGovren JP (1990) Prevention of central nervous system toxicity of the antitumor antibiotic acivicin by concomitant infusion of an amino acid mixture. Cancer Res 50:5475–5480

Wilson AA, Scheffel VA, Dannal RF, Strathis M, Ravert HT, Wagner HN (1991) In vivo biodistribution of two [^{18}F]-labelled muscarinic cholinergic receptor ligands, 2-[^{18}F]- and 4-[^{18}F]-fluorodexetimide. Life Sci 48:1385–1394

Zloković BV, Sušić VT, Davson HJG, Begley DJ, Jankov RM, Mitrović DM, Lipovac MN (1989) Saturable mechanism of sleep-inducing peptide (DSIP) at the blood-brain barrier of the vascularly perfused guinea pig brain. Peptides 10:249–254

Zloković BV, Segal MB, Davson HJG, Lipovac MN, Hyman S, McComb JG (1990) Circulating neuroactive peptides and the blood-brain and blood-cerebrospinal fluid barriers. Endocrinol Exp 24:9–17

CHAPTER 21

Therapeutic Opening of the Blood-Brain Barrier in Man

M.K. Gumerlock and E.A. Neuwelt

A. Introduction

The blood-brain barrier (BBB) continues to interest neuroscientists, and is studied widely in a number of animal models and disease states. The breakdown of the BBB as a consequence of disease or injury is also studied extensively in animals and humans. However, the idea of transiently opening the BBB in a controlled fashion for the purpose of achieving increased drug permeability across the barrier is a relatively new concept. To adapt these techniques to the clinical setting has been considered desirable for improved treatment of central nervous system (CNS) disease, and recent progress has made this possible.

Having over 30 years of experience in the treatment of high grade malignant gliomas with adjuvant chemotherapy following surgery and/or radiation, neuroscientists still remain frustrated by the poor prognosis in this disease. In the face of improved survival with chemotherapy in many systemic cancers, primary malignant brain tumors continue to hold a median survival of some 56 weeks in the face of vigorous multimodality therapy (Hildebrand 1986).

Given this current situation, which differs little from that of a decade ago, the question arises whether this lack of success is in part the result of the BBB preventing drugs access to CNS parenchyma. This BBB, comprised primarily of CNS capillary endothelial cell tight junctions, limits the amount of chemotherapeutic drug reaching target tumor cells (Gumerlock and

---

Abbreviations

ACNU	3-[(4-Amino-2-methyl-5-pyrimidinyl)methyl]-1-(2-chloroethyl)-1-nitrosourea
BBB	Blood-brain barrier
BCNU	1,3-bis(2-chloroethyl)-1-nitrosourea
CNS	Central nervous system
CT	Computerized tomography
DTPA	Diethylenetriaminepentaacetic acid
FDA	Federal Drug Agency
ICP	Intracranial pressure
Ig	Immunoglobulin
MAb	Monoclonal antibody
MRI	Magnetic resonance imaging
PET	Positron emission tomography
PFI	Progression-free interval
SPECT	Single photon emission computerized tomography
VER	Visual evoked response

NEUWELT 1987). Just how much the blood-brain and blood-tumor barriers hamper cytoreductive drug delivery remains quite controversial. Drug permeability is variable within any single neoplasm and often does not correlate with tumor type, size, or anatomic location (GROOTHUIS et al. 1984). At the growing tumor edge, with neoplastic cells infiltrating normal CNS parenchyma, the BBB remains intact.

While a hallmark of malignancy in CNS tumors has been increased BBB permeability, clinical studies demonstrate that up to 30% of patients with highly anaplastic astrocytomas have nonenhancing (i.e., no breakdown of the BBB) lesions on computerized tomography (CT) scan (CHAMBERLAIN et al. 1988). Such a situation would drastically limit the CNS penetration of systemically administered drugs. Furthermore, malignant gliomas responding to treatment show improved BBB function with less contrast enhancement on imaging studies. Unfortunately, this response is often transient, suggesting that the therapeutic advantage at the initiation of drug treatment in the face of at least partial BBB permeability is lost as the tumor responds and barrier permeability decreases. Clinical evidence in support of this theory includes the observation of STEWART et al. (1985) that tumor bridging two circulations, one infused with intra-arterial drug and one not, showed regression in the treated region, with tumor progression in the noninfused area. Thus, with high BBB permeability, there is attendant good drug delivery intra-arterially, but as the tumor responds, BBB permeability decreases, with subsequent impaired delivery of chemotherapeutic agents (GUMERLOCK and NEUWELT 1987). Perhaps drug delivery is more of a problem than drug resistance in these cases.

Because normal brain and cerebrospinal fluid dynamics insure rapid equilibration and elimination of substances introduced into brain tumors and the CNS parenchyma in general, and because the CNS parenchyma has an intact BBB in the area around brain tumors, any advantage in drug delivery to the tumor itself (blood-tumor barrier open) is rapidly lost by drug diffusion along the very steep concentration gradients in surrounding brain tissue (sump or sink effect). As chemotherapeutic principles require an effective drug concentration in contact with tumor cells for a sufficient period of time, it is imperative to improve not only the amount of drug delivered, but also the length of time this drug is in contact with tumor cells before diffusing away (GUMERLOCK and NEUWELT 1987). By transiently opening the BBB through intra-arterial infusion of hyperosmolar solutions, one can improve drug delivery to tumor, not only by increasing the amount of drug entering the tumor, but also by decreasing the sump effect (eliminating steep concentration gradient by increasing drug in surrounding brain, too). Thus, by increasing drug delivery to the surrounding brain tissue as well as the tumor itself, one improves *both* variables of the "concentration × time" equation. The challenge then becomes one of finding cytoreductive agents which are effective against CNS neoplasia yet not too toxic to normal brain tissue when delivered in therapeutically sufficient quantities.

To address this factor of the BBB in the treatment of patients with malignant glioma, an initial clinical trial administering a multiagent chemotherapeutic regimen (methotrexate, cyclophosphamide, procarbazine) after osmotic BBB disruption using hyperosmolar 25% mannitol was initiated to demonstrate the safety of this procedure (NEUWELT et al. 1980). This regimen allowed drugs to be administered in conjunction with a transient opening of the BBB to insure passage of such agents into the CNS in adequate concentration over sufficient time to be therapeutic.

The protocol as initially described involves rapid infusion (via carotid or vertebral artery) of 25% mannitol at a rate to essentially replace blood flow (5–12 ml/min) for a period of 30 s. The patient is under general anesthesia (WILLIAMS et al. 1986). Methotrexate is infused intra-arterially following the mannitol. Because cyclophosphamide requires hepatic activation, it is administered intravenously 10 min prior to mannitol. The patient receives a 2-week course of oral procarbazine thereafter. The methotrexate is rescued with leucovorin, and the patients receive dexamethasone as needed.

B. Chemotherapy for Malignant Brain Tumors

I. Published Clinical Series

Chemotherapy with BBB disruption is not yet in widespread clinical use, but several centers are beginning to report their experience. While protocols at the different institutions vary in terms of mannitol infusion time, documentation of BBB disruption, and chemotherapeutic agents used (Table 1), these data emphasize the clinical feasibility of such a treatment approach and suggest that this form of adjuvant chemotherapy and hyperosmolar BBB disruption is associated with a significant advance in the median survival of patients with this dread disease.

BONSTELLE et al. (1983) reported on a series of 18 glioblastoma patients receiving intracarotid 5-fluorouracil and adriamycin after intracarotid 25% mannitol, 120 ml/2 min. Six of seven comatose patients improved to an ambulatory self-caring status for 3–12 months and of the 11 less neurologically compromised patients nine were ambulatory and self-caring. They noted two problems with treatment – necrotic mass with minimal tumor and a rare (one patient) swelling/herniation syndrome following BBB disruption (Table 2).

MIYAGAMI et al. (1985) reported on five patients with malignant glioma who underwent intracarotid 20% mannitol infusion, 200 ml at 1.3–1.6 ml/s, followed by intra-arterial 3-[(4-amino-2-methyl-5-pyrimidinyl)methyl]-1-(2-chloroethyl)-1-nitrosurea (ACNU). The patients also received 60% Conray, 100 ml/5 min intracarotid, to document barrier disruption. Changes in CT enhancement were documented, and the patients then had intraoperative tumor sampling for ACNU levels at various time intervals (5–30 min) after

Table 1. BBB disruption/chemotherapy protocols

Reference	Mannitol infusion rate (ml/s)	Chemotherapeutic agents[a]
Bonstelle et al. (1983)	1	5-Fluorouracil Adriamycin
Miyagami et al. (1985, 1990)	1.3–1.6	ACNU
Sato et al. (1985a,b)	1.0–1.5	None
Fauchon et al. (1986), Chiras et al. (1988)	4	Cisplatin Adriamycin Bleomycin Cytarabine
Neuwelt et al. (1986b)	5–12	Cyclophasphamide Methotrexate
Heimberger et al. (1987)	5.7	Cyclophosphamide Methotrexate
Li et al. (1988)	renograffin[b]	Thiotepa BCNU
Yamada et al. (1989)	0.7	5-Fluorouracil ACNU β-interferon
Markowsky et al. (1991)	5–12	Cyclophosphamide Methotrexate
Gumerlock et al. (1991)	5–12	Cyclophosphamide Methotrexate

[a] Drugs given at the time of BBB opening. All agents were given intra-arterially except cyclophosphamide (intravenous).
[b] Renograffin used as hyperosmolar opening agent; infusion rate not stated.

drug infusion. They demonstrated increased enhancement on CT scan following mannitol infusion – a mean increase in CT number 2.8 times that of the unenhanced CT scan. This compared with a mean 1.7-fold increase in CT number following contrast infusion without BBB disruption. At 10 min postdisruption mean ACNU levels in solid tumor measured 11.24 μg/g compared with 2.05 μg/g in tumor of nondisrupted patients. The ACNU tissue concentrations rapidly decreased to values of 2.41 μg/g and 1.09 μg/g respectively at 40 min.

Miyagami et al. (1990) have recently expanded their series, and fully report on a group of 16 patients undergoing 57 BBB disruption/chemotherapy procedures with intra-arterial ACNU. They have included a control group of five patients with malignant glioma who underwent 16 intra-arterial infusions of ACNU without BBB disruption. In their treatment group, there were 12 patients with malignant glioma who received 44 courses of BBB disruption/chemotherapy. Four of these patients died 12–35 months after diagnosis, while eight patients remain alive 20–48 months following diagnosis. The five patients receiving intra-arterial ACNU alone without BBB disruption have all died with median survival of 14 months.

Complications in this series of Miyagami et al. (1990) were evaluated in 32 patients undergoing 75 BBB disruption/chemotherapy procedures. They

Table 2. Complications of BBB disruption/chemotherapy

Series	No. of patients (no. of disruptions)	Stroke	Death	Seizures	Ocular toxicity	Temporary neurological deterioration
Bonstelle et al. (1983)	18 (NM)	NM	1	2	18	4
Fauchon et al. (1986, 1988)	16 (64)	2	1	2	8	5
Neuwelt et al. (1986b)	38 (227)	3	0	21	1	22
Heimberger et al. (1987)	10 (>20)	1	NM	NM	NM	7
Li et al. (1988)	10 (NM)	NM	2	NM	2	1
Yamada et al. (1989)	10 (NM)	NM	NM	NM	NM	2
Miyagami et al. (1990)	32 (75)	NM	1	2	NM	5
Markowsky et al. (1991)	7 (NM)	0	0	4	NM	NM
Gumerlock et al. (1991)	37 (246)	5	1	16	2	11

NM, not mentioned.

included transient neurologic deficits, vomiting, headache, leukopenia, thrombocytopenia, and liver dysfunction; all in a small number of patients. They describe one patient suffering a brain swelling/herniation course who died.

These same authors also elaborate on their intraoperative tumor sampling from 5 to 60 min after BBB disruption/chemotherapy. ACNU concentrations in the solid tumor were maximum at 5 min, and were higher than those in the necrotic portions of the tumor and the surrounding normal brain; they had equilibrated with the necrotic tumor tissue and normal surrounding brain by 60 min. Values were consistent with those noted in their earlier paper.

In another intraoperative study, Sato et al. (1985a) demonstrated fluorescein extravasation, particularly from venules, following infusion of either 20% mannitol or 15% glycerol 4 ml/kg at a rate of 1–1.5 ml/s. They studied five patients (two glioblastomas, two astrocytomas, one metastatic tumor) and demonstrated more marked disruption in glioblastoma multiforme than in low grade astrocytoma. They also showed that the degree of barrier disruption as determined by fluorescein extravasation was dependent on the volume and speed of mannitol infusion.

Sato et al. (1985b) further studied 21 patients undergoing 23 BBB disruption procedures with follow-up contrast-enhanced CT scanning. They again infused mannitol (15 patients) or glycerol (six patients) at their previous rate (which calculates to 280 ml/5 min in a 70-kg patient) and demonstrated a spreading increase in the contrast-enhanced area and an increased density of enhancement in the previously enhancing tumor region as compared with pre-BBB disruption studies. The increased enhancement again was most marked for patients with high grade gliomas. Low grade astrocytomas showed increased tumor enhancement without any spread to adjacent tissue. No enhancement was seen in four patients with arachnoid cyst, pituitary adenoma, or normal brain.

In a series of 38 patients with malignant glioma, NEUWELT et al. (1986a) elaborated on their initial results with BBB disruption/chemotherapy. They reported a significantly longer expected survival (17.5 months) in patients with high grade malignant glioma treated by osmotic BBB disruption/chemotherapy compared with local historic controls (surgery and radiation or surgery, radiation, and systemic chemotherapy).

FAUCHON et al.'s (1986) initial report involved their first 16 malignant glioma patients treated with intra-arterial chemotherapy after osmotic BBB disruption. Four of these patients had not received antecedent radiation therapy following biopsy. Because intracarotid mannitol induces facial pain, the patients received "neuroleptanalgesia." Mannitol 25% was infused at rates of 120 ml/30 s in the carotid or 70 ml/30 s in the vertebral arteries respectively. This was followed immediately by intra-arterial cisplatin 100 mg/10 min, adriamycin 80 mg/10 min, and bleomycin 30 mg/10 min sequentially. Five patients received this regimen; two of these plus an additional 11 patients received cisplatin 100 mg/20 min and cytarabine 300 mg/20 min. Of these 16 patients, five improved, six stabilized, and five deteriorated. Of the 11 responders, four were alive 11–14 months later.

CHIRAS et al. (1988) have updated their series to 64 cases of intra-arterial (carotid or vertebral) mannitol infusion 80 ml/30 s followed by chemotherapy. Their complications now include 2/64 cases with seizures, and a 20% incidence of transient neurologic deterioration improving within 24 h, and felt to be secondary to transient edema and increased intracranial pressure.

The BBB disruption experience of HEIMBERGER et al. (1987) includes ten patients with malignant glioma (nine cases) and CNS lymphoma (one case). These patients followed a specific protocol, and all patients had an entry Karnofsky Performance Score of at least 60%. BBB breakdown was documented using single photon emission computerized tomography (SPECT). Patients received high dose methotrexate intra-arterially following infusion of hyperosmolar 25% mannitol, 172 ml/30 s. During the disruption procedure, hemodynamic changes were documented including first a decrease in systemic vascular resistance, followed by a preload-independent rise in cardiac index (HIESMAYR et al. 1987). Patients receiving chemotherapy (methotrexate, cyclophosphamide, and procarbazine) were evaluated in terms of a "progression-free interval" (PFI), with a mean PFI of 7.75 months in the four patients in whom tumor progression resumed, and an ongoing mean PFI of 12.2 months in the five patients still without tumor progression. The patients received two to five chemotherapy courses in the reported time. Complications included one patient with a stroke-like neurologic deterioration. There was transient aggravation of neurologic deficits (<24 h duration) after seven procedures. One patient suffered hemorrhagic cystitis and one patient with secondary anemia required a blood transfusion.

Li et al. (1988) report ten patients with recurrent high grade malignant astrocytomas receiving intra-arterial thiotepa or 1,3-bis(2-chloroethyl)-1 nitrosourea (BCNU) following BBB disruption using Renograffin-76. Nine of the patients showed initial improvement on CT scan, one remained unchanged. Steroid doses were decreased in six, and unchanged in two (two were increased). Karnofsky Performance Scores improved in four, stabilized in two, and deteriorated in four. Mean survival after recurrence was 41 weeks. Two patients died during therapy. Other complications included transient neurologic deficit in one and decreased visual acuity in two, as well as patients with thrombocytopenia, deep venous thrombophlebitis, and pulmonary embolism.

Most recently, Yamada et al. (1989) have utilized 20% mannitol for osmotic BBB disruption in a series of ten patients with malignant glioma. Following BBB disruption, the patients received intra-carotid 5-fluorouracil, ACNU, and β-interferon. A response, as determined by CT scan, was seen in four of the nine evaluable patients. None of the patients showed progressive disease. Eye pain during mannitol infusion was seen in patients in whom selective catheterization of the internal carotid was below the ophthalmic artery origin. A transient increase in neurologic deficit was seen in 20%.

Markowsky et al. (1991) evaluated seven patients undergoing BBB disruption/chemotherapy for malignant glioma. While four of the patients suffered periprocedural focal seizures, there were no permanent neurologic sequelae. These authors were particularly interested in methotrexate pharmacokinetics and studied systemic drug clearance, renal clearance, central compartment volume, and half-lives of the exponential phases of the serum concentration-time profile. These data were best described by a three-compartment distribution model. In five of the seven patients, serum drug levels exceeded 0.1 μmol/l at 72 h. The authors concluded that following BBB disruption/chemotherapy, the brain acts as a reservoir with slow drug transport back to the systemic circulation. Such might be considered indirect clinical evidence of a significant increase in drug delivery to the brain after BBB disruption.

Finally, another series of 37 patients with malignant glioma receiving BBB disruption/chemotherapy demonstrated a median survivorship of 22 months (Gumerlock et al. 1992). Mean Karnofsky Performance Scores improved after six courses of treatment, and six patients had complete tumor remission by CT scan. The mean PFI of the 37 patients was 15 months, including 17 patients who continue stable with an ongoing mean PFI of 28 months, and 20 patients with tumor progression and a mean PFI of 5 months.

II. Malignant Glioma

Of the patients reported above, 176 gliomas underwent treatment. In 78% (138/176) follow-up data was available, with 53% (73/138) showing clinical and radiographic improvement. Thirty-two patients had stabilization of their disease, and 33 had tumor progression. The following is an illustrative case.

A 31-year-old white female first developed seizures in 1983. Two years later, a CT scan revealed an enhancing left parietal mass which was resected, revealing a grade II–III astrocytoma. The patient subsequently received 50 Gy whole brain radiation. Follow-up magnetic resonance imaging (MRI) scans showed no evidence of tumor recurrence or progression over the ensuing 3.5 years. Recurrent tumor then developed posterior to the initial surgical site, extending to the thalamus and across the corpus callosum. One year later, the patient was referred for BBB disruption/chemotherapy, steroid-dependent with a dense right hemiparesis. An MRI scan showed further development of multiple lesions (Fig. 1). Needle biopsy revealed high grade malignant glioma and the patient began the cyclophosphamide, methotrexate, procarbazine protocol. She received intra-arterial methotrexate following hyperosmolar BBB disruption in the right carotid and vertebrobasilar distribution during the next 10 months. The patient improved, and is now off dexamethasone with a normal neurologic examination. Follow-up MRI scan, after completing treatment, shows a single small enhancing lesion (Fig. 2).

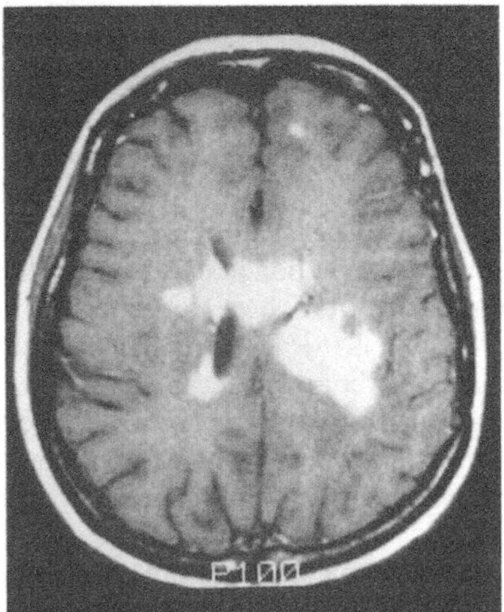

Fig. 1. T_1-weighted magnetic resonance imaging (MRI) study with gadolinium (Gd) enhancement shows evidence of multiple lesions bilaterally. Biopsy revealed high grade malignant glioma

This patient emphasizes the potential for glioblastoma to spread throughout the nervous system, thereby insuring that attempts at local treatment will be doomed to failure. The ability to administer adequate doses of antineoplastic drugs in each of the CNS vascular distributions markedly improves the likelihood of therapeutic response.

In the therapy of malignant brain tumors, it has been standard practice to treat with radiation immediately following surgical diagnosis, with any consideration for chemotherapy following later. FAUCHON et al. (1986) note that four of their 16 patients received BBB disruption/chemotherapy prior to radiation treatment, as did four of 16 patients in the series of MIYAGAMI et al. (1990). While this practice (preradiation chemotherapy) has been gaining popularity in the treatment of pediatric CNS malignancy, its use in adults with glioblastoma is unproven. Three of four patients in the Japanese series remain alive 24–44 months following diagnosis. Small unpublished series (Gumerlock and Neuwelt) of approximately 20 patients receiving BBB disruption/chemotherapy prior to radiotherapy demonstrate tumor response to BBB disruption/chemotherapy as measured by CT scan. The question of whether chemotherapy prior to radiation will offer improved survival or decreased toxicity is yet to be answered.

In summary, the treatment of patients with high grade malignant gliomas using adjuvant chemotherapy administered in conjunction with osmotic BBB disruption affords a significant chance for improved survival and a stable Karnofsky Performance Score. The intra-arterial infusion of mannitol prior to the administration of chemotherapeutic agents is not

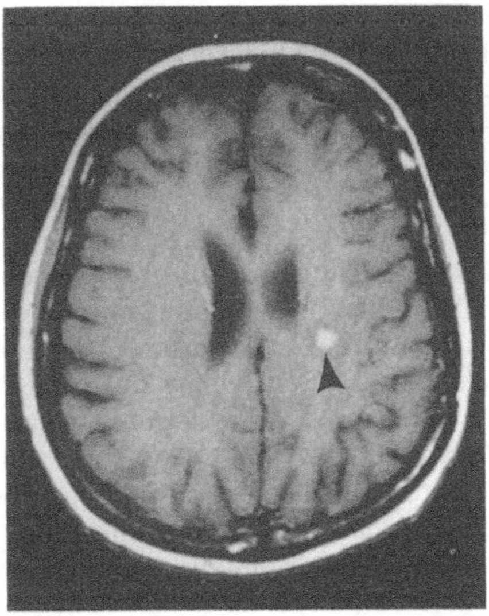

Fig. 2. T_1-weighted Gd-enhanced MRI after BBB disruption/chemotherapy (nine courses) reveals a single small lesion (*arrow*)

associated with increased or additional risks beyond those of standard intra-arterial chemotherapy. This route of drug administration offers advantage over standard intravenous, intra-arterial, intrathecal, and intratumoral methods with respect to both the length of time a therapeutic level is achieved in tumor as well as the ability to treat infiltrating tumor cells percolating beyond the defined tumor edge into normal brain tissue. Full quantitation of the advantage of BBB disruption/chemotherapy over more conventional brain tumor chemotherapy remains to be determined.

III. Cerebral Lymphoma

The results of BBB disruption/chemotherapy in the treatment of primary cerebral lymphoma are most impressive, and serve to emphasize the fact that increasing drug delivery to malignant brain tumors improves survival, particularly when the tumors are sensitive to those chemotherapeutic agents used.

Although combination chemotherapy for systemic non-Hodgkin lymphoma results in long-term remission for the majority of patients, it has had only modest efficacy in the treatment of CNS lymphoma. The best results have been with high dose methotrexate protocols, resulting in transient responses (ERVIN and CANELLOS 1980). The recent evidence that chemotherapeutic agents are effective against systemic lymphoma suggests that these drugs may also be effective against cerebral lymphoma if they could be delivered across the BBB and into the CNS to achieve therapeutic tissue levels. Administration of high dose intravenous methotrexate is an attempt to improve drug delivery across the BBB. However, methotrexate penetrates the CNS poorly with resultant subtherapeutic levels, especially in infiltrative brain tumors such as lymphoma. Intraventricular and intrathecal methotrexate infusion also fails to achieve therapeutic levels except in the superficial CNS parenchyma (BLASBERG et al. 1985). Animal studies have shown that the tissue level of methotrexate can be increased in both tumor and surrounding tumor-infiltrated brain if the BBB is transiently opened prior to infusion (NEUWELT et al. 1985).

Clinical studies have documented the efficacy of chemotherapy in conjunction with osmotic BBB disruption for patients with primary cerebral lymphoma (NEUWELT et al. 1986). The most recent report extends observations to include 30 consecutive patients with CNS lymphoma treated with BBB disruption/chemotherapy using methotrexate, cyclophosphamide, procarbazine, and dexamethasone (NEUWELT et al. 1991). Group 1 ($n = 13$) patients were initially treated with cranial radiation and subsequently received BBB disruption/chemotherapy for persistent or recurrent tumor. Group 2 ($n = 17$) patients received initial BBB disruption/chemotherapy and underwent radiation for persistent or recurrent tumor.

Survival differences between these two groups have been assessed using the Fisher's exact test and the logrank statistic applied to the Kaplan-Meier survival curves. The difference in median survivals from diagnosis, 17.8 months for group 1 and 44.5 months for group 2, is statistically significant ($p < 0.04$). One patient in group 1 and eight in group 2 remain alive and disease-free with follow-up from diagnosis ranging from 15 to 98 months. Morbidity and mortality in these 30 patients undergoing 471 BBB disruption/chemotherapy procedures included three deaths (two from sepsis and one from disease progression) within 30 days of the last procedure, two cerebral infarctions, three episodes of prolonged obtundation, and a periprocedural seizure incidence of 7%. Neuropsychological testing at 1 year showed stable or improved function in the majority of those patients tested.

IV. Metastatic Tumors to the Brain

Several of the investigators above have treated patients with metastatic tumors, and NEUWELT and DAHLBORG (1987, 1989) have documented a series of nine patients with a variety of metastatic lesions including breast, lung, and germ cell tumors as well as malignant melanoma. Of particular interest in this series of patients was one who had been receiving BBB disruption/chemotherapy for a single brainstem lesion with mannitol and methotrexate infusions in the vertebral artery. The tumor responded to treatment; however, in the face of documented tumor regression in the treated vascular distribution, new tumor nodules developed in the frontal region bilaterally. One might consider that these additional lesions had received "conventional systemic" doses of chemotherapy, but not the enhanced drug delivery afforded by BBB disruption/chemotherapy. This differential tumor response is most likely the result of differences in drug delivery.

C. Imaging BBB Disruption

Because a linear relationship exists between CT scan number (Hounsfield units) and the amount of iodinated contrast agent crossing the BBB to CNS parenchyma, *qualitative* judgment of the extent, localization, and time course of BBB disruption based on CT scan can be *quantitated* by comparing the change in CT number in the disrupted hemisphere with that of the contralateral hemisphere, or even in a specific region of interest before and after disruption (MIYAGAMI et al. 1985, 1990; NEUWELT and DAHLBORG 1989; SATO et al. 1985b). While CT scanning is an excellent method of documenting BBB disruption, patients may suffer periprocedural seizures (focal or generalized) 15%–20% of the time. Radionuclide brain scans (glucoheptonate or diethylenetriaminepentaacetic acid, DTPA) have been used more recently to image barrier opening and may be performed up to 3 h after BBB disruption (NEUWELT et al. 1983). BBB disruption is assessed on anterior, vertex, and posterior views by comparing a region of interest in the barrier-modified brain with that of non-mannitol-infused brain.

Good to excellent disruptions have been documented in 53% of 356 procedures, with only 10% showing no disruption (NEUWELT and DAHLBORG 1989). A more recent series has documented 65% good or excellent barrier opening with a 7% "no disruption" rate; additionally, in 11 of 232 procedures without parenchymal radionuclide uptake, the tumor showed increased uptake compared to control predisruption brain scans (GUMERLOCK et al. 1991). SATO et al. (1985b) have also found an increased tumor enhancement in malignant tumors following BBB disruption compared with pre-BBB disruption scans. NEUWELT et al. (1980) demonstrated the ability to detect previously nonvisualized tumor after osmotic BBB disruption by performing a contrast-enhanced CT scan immediately following BBB disruption/chemotherapy.

SPECT combines the advantages of standard radionuclide brain imaging with increased resolution of tomography and has also been used to clinically document barrier opening (HEIMBERGER et al. 1987). A small number of patients have undergone BBB disruption immediately preoperatively, with barrier opening documented visually by flourescein extravasation noted at surgery (SATO et al. 1985a).

Positron emission tomography (PET) has been used to quantitate hyperosmolar BBB opening in a small unpublished series of patients at the National Institutes of Health (B. Zunkeler, personal communication). Vascular permeability can be monitored and influx constants estimated using this technique.

While MRI has been used in animal studies of BBB disruption (RUNGE et al. 1985), published results of its clinical use in this setting are still awaited. Animal studies further suggest that contrast-enhanced (gadolinium) MRI after BBB disruption is associated with significant canine toxicity (ROMAN-GOLDSTEIN et al. 1991). Magnetic resonance spectroscopy may also eventually prove useful clinically to document changes following BBB disruption.

D. Complications of BBB Disruption/Chemotherapy

Complications of this form of chemotherapy may be divided into neurologic and nonneurologic events. The latter complications are those associated with femoral artery catheterization and chemotherapy, and include arterial damage, deep venous thrombosis, pulmonary embolism, severe granulocytopenia, anemia requiring transfusion, sepsis, hemorrhagic cystitis, and interstitial pneumonitis (NEUWELT and DAHLBORG 1989). The incidence of these ranges from 2% to 11%. Neurologic complications as noted in the various patient series include mortality, permanent neurologic worsening (stroke), visual changes, seizures, and transient neurologic deficits (Table 2).

Much has been made about the ocular complications of intracarotid chemotherapy, particularly in association with BCNU therapy. To date there have been two reports which specifically address ocular changes associated with BBB disruption/chemotherapy. MILLAY et al. (1986) described a change in the retinal pigment epithelium localized to the posterior pole in the foveal and parafoveal regions. Although these changes were seen frequently, they were not usually associated with visual change. While some patients did demonstrate visual acuity changes, these were felt to be multifactorial in origin. Suda described a single patients who received intra-arterial ACNU following hyperosmolar disruption of the BBB with mannitol (SUDA et al. 1985). This patient suffered acute onset of retro-orbital pain, conjunctival injection, palpebral edema, and total ophthalmoplegia, which all resolved in several days. Whether these changes were the result of chemotherapy, mannitol, or disruption of the blood-ocular barrier remains to be determined. Because of the various protocols used at the different centers, it is difficult to make formal comparisons, and it appears that the incidence of visual changes in these series is in keeping with or less than that of intracarotid therapy in general (JOHNSON et al. 1987; MILLER et al. 1985).

BONSTELLE et al. (1983), NEUWELT and DAHLBORG (1987), MIYAGAMI et al. (1990), and GUMERLOCK et al. (1991) all report the occasional patient suffering a swelling/herniation syndrome shortly after BBB disruption/chemotherapy. Animal data suggests an increase in brain water content and subsequent intracranial pressure (ICP) rise coincident with osmotic BBB disruption (NEUWELT et al. 1980). In an attempt to document this clinically, the initial five patients in the NEUWELT series underwent subdural pressure monitoring which documented a transient ICP rise from baseline pressures of 3–9 cmH_2O to peaks of 16–23 cmH_2O (NEUWELT and DAHLBORG 1989). A less invasive method of evaluating intracranial pressure changes involves the use of flash visual evoked response (VER) monitoring of the N_2 latency (YORK et al. 1984). In another series of patients undergoing BBB disruption/chemotherapy, ICP changes were documented and followed before, during, and after disruption. Eight patients underwent 23 BBB disruption/chemotherapy procedures (16 carotid and 7 vertebral) with a mean increase in N_2 latency from 88 to 96 (M.K. GUMERLOCK, unpublished data).

E. Tumor-Specific Monoclonal Antibody Infusion after BBB Disruption

Recent advances, particularly in tumor chemotherapy, have evolved around the use of tumor-associated antibodies as chemotherapeutic agents. Because immunoglobulin G (IgG) seems to be the most promising antibody, and because Fab fragments of IgG have less nonspecific binding, the Fab fragment of IgG (monomeric molecular weight 60000) holds considerable promise for clinical studies. To date, evidence on the administration of these antibodies to patients with brain tumors suggests that penetration into the tumor has been poor (HOUGHTON et al. 1985; PHILLIPS et al. 1983). This poor drug delivery, likely the result of the BBB and the large molecular weight of the antibody, might be enhanced by the technique of osmotic BBB disruption.

To evaluate this, a pilot study of three patients with malignant melanoma metastatic to the CNS was undertaken (NEUWELT et al. 1987). Two Federal Drug Agency (FDA) approved antibodies were used, both directed against melanoma-associated antigens (BEAUMIER et al. 1986): monoclonal antibody (MAb) 96.5, a transferrin-like cell surface glycoprotein (M_r 97000), present on approximately 50% of human melanomas with an affinity constant of $2.4 \times 10^9 M^{-1}$ and MAb 48.7, also an IgG specific antibody, a melanoma-associated cell surface proteoglycan expressed by approximately 90% of melanomas with a binding affinity of $1.7 \times 10^7 M^{-1}$. Of the three tumor samples evaluated in this pilot study, two were highly reactive with only one of the antibodies and one was highy reactive with both antibodies. Intravenous infusions of ^{131}I-labelled tumor-specific MAb were compared with a similar infusion of ^{131}I-labelled non-specific MAb. After intravenous infusion, no uptake of specific or non-specific antibody in the region of tumor was documented by brain scan in the patients. The radiolabelled tumor-associated MAb was then administered in conjunction with osmotic BBB disruption. The uptake of radiolabelled tumor-specific MAb in the tumor-bearing hemisphere was increased. Most of the radiolabelled MAb had disappeared from the brain by 72h. Persistence of antibody as measured by radioactivity in the region of tumor was not seen; however, in one patient there seemed to be increased delivery to the region of the tumor, but the washout of antibody from that region occurred at the same rate as that from surrounding tumor-free brain, suggesting minimal tumor-specific antibody binding. Malignant melanoma cells in the cerebrospinal fluid of one patient demonstrated antibody binding.

These clinical observations provide important evidence that drug delivery across the BBB is a factor in our current ability to manage patients effectively with primary and metastatic CNS tumors. In conclusion, the critical nature of the barrier proved by the above studies demands serious attention to methods of more effective drug delivery.

F. Treatment of Fungal Brain Abscess with BBB Disruption/Chemotherapy

Increasing antifungal drug delivery with intra-arterial administration and associated osmotic BBB disruption has been tried in a single patient with fungal brain abscess (NEUWELT et al. 1989).

A patient with biopsy-proven fungal sinusitis involving maxillary and ethmoid sinuses was treated with intravenous amphotericin and resolution of sinusitis. However, 2.5

years later the patient had further fungal sinusitis and a bifrontal intracranial mass. The extracranial fungal hyphae were eliminated with further amphotericin treatment, and the frontal lobe process improved by CT scan. Some 6 months later, the patient again deteriorated with biopsy-proven fungal cerebritis and no evidence of extracranial fungus disease. She received intra-arterial (carotid and vertebral arteries) amphotericin in association with osmotic opening of the BBB. After 3 weeks of intra-arterial therapy, carotid angiography suggested endothelial damage and the infusions were discontinued. Her neurologic status improved, although she remained with a left hemiparesis and marked frontal lobe signs. In spite of her gradual neurologic improvement, the patient died suddenly approximately 6 months later. An autopsy revealed fungi in the frontal lobes.

In this single patient, systemically administered amphotericin cured the extraneural fungal infection but not the CNS component. Intra-arterial drug combined with osmotic BBB disruption appeared to improve the response compared with that due to intravenous administration. Whether therapeutic advantage outweighs toxicity in such treatment remains to be seen.

G. Conclusion

A variety of protocols for intra-arterial hyperosmolar infusion (BBB disruption/chemotherapy) have been developed. As the infusion rates for mannitol (or other hyperosmolar solutions) vary considerably, one would anticipate a wide range in the degree of BBB disruption. Animal studies suggest variable drug delivery correlating with Evans blue staining and, although no drug level data is available for patients, radionuclide brain scan analysis suggests a similar situation in humans (Neuwelt et al. 1981, 1984, 1986b; Gumerlock et al. 1991). Miyagami et al. (1985, 1990) have been the first to assay drug levels in patients following BBB disruption/chemotherapy. As noted above, they document enhanced drug delivery compared to patients receiving chemotherapy without BBB disruption. Not all research groups monitor the degree of BBB opening, and it remains unclear as to how much "barrier disruption" is necessary to effect the therapeutic advantage.

This review has emphasized the use of hyperosmolar disruption of the BBB in the treatment of brain tumors; however, this means of increasing drug delivery to the brain is not limited to the treatment of tumors. As can be seen by the preliminary work with monoclonal antibody delivery and fungal brain abscess treatment, the adjunct use of BBB disruption has several other potentials. Future directions include its use in the treatment of enzyme deficiency disorders where delivery of high molecular weight proteins is necessary and possibly in genetic disorders requiring gene replacement. Viral diseases such as the acquired immunodeficiency syndrome may require the enhanced drug delivery provided by BBB disruption/chemotherapy to be effectively controlled. Finally, osmotic BBB disruption has proven to be generally safe and clinically effective; efforts continue to develop other methods of reproducibly opening the BBB in a transient, controlled fashion that is clinically applicable. The therapy of a variety of CNS diseases is likely to be improved from the additional treatment options made available by this means of increasing drug delivery to brain parenchyma.

References

Beaumier PL, Neuzil D, Yang HM et al. (1986) Immunoreactivity assay for labeled antimelanoma monclonal antibodies. J Nucl Med 27:824–828

Blasberg RG, Patlak C, Fenstermacher JD (1985) Intrathecal chemotherapy: brain tissue profiles after ventriculocisternal perfusion. J Pharmacol Exp Ther 195: 73–83

Bonstelle CT, Kori SH, Rekate H (1983) Intracarotid chemotherapy of glioblastoma after induced blood-brain barrier disruption. AJNR 4:810–812

Chamberlain MC, Murovic JA, Levin VA (1988) Absence of contrast enhancement on CT brain scans of patients with supratentorial malignant gliomas. Neurology 28:1371–1374

Chiras J, Dormont D, Fauchon F, Debussche C, Bories J (1988) Intra-arterial chemotherapy of malignant gliomas. J Neuroradiol 15:31–48

Ervin T, Canellos GP (1980) Successful treatment of recurrent primary central nervous system lymphoma with high-dose methotrexate. Cancer 45:1556–1557

Fauchon F, Chiras J, Poisson M, Rose M, Terrier L, Bories J, Guerin RA (1986) Intra-arterial chemotherapy by cisplatin and cytarabine after temporary disruption of the blood-brain barrier for the treatment of malignant gliomas in adults. J Neuroradiol 13:151–162

Groothuis DR, Molnar P, Blasberg RG (1984) Regional blood flow and blood-to-tissue transport in five brain tumor models. Prog Exp Tumor Res 27:132–153

Gumerlock MK, Neuwelt EA (1987) Principles of chemotherapy in brain neoplasia. In: Jellinger K (ed) Therapy of malignant brain tumors. Springer, Berlin Heidelberg New York, pp 277–348

Gumerlock MK, Belshe BD, Madsen R, Watts C (1992) Osmotic blood-brain barrier disruption and chemotherapy in the treatment of high grade malignant glioma: patient series and literature review. J Neurooncol 12:33–46

Heimberger K, Samec P, Podreka I, Binder H, Suess E, Reisner T, Deecke L, Steger G, Hiesmayr M, Dittrich C, Horaczek A, Zimpfer M (1987) Reversible opening of the blood-brain barrier in the chemotherapy of malignant gliomas (in German) Wien Klin Wochenschr 99:385–388

Hiesmayr M, Dirnberger H, Aloy A, Heimberger K, Horaczek A, Brandstatter B, Zimpfer M (1987) Hemodynamic effects of an osmotic bolus for the reversible opening of the blood-brain barrier (in German). Schweiz Med Wochenschr 117:450–454

Hildebrand J (1986) Radiotherapy and chemotherapy of malignant brain gliomas. Drugs Clin Res 12:167–175

Houghton AN, Mintzer D, Cordon-Cardo C et al. (1985) Mouse monoclonal IgG_3 antibody detecting GD_3 ganglioside: a phase I trial in patients with malignant melanoma. Proc Natl Acad Sci USA 82:1242–1246

Johnson DW, Parkinson D, Wolpert SM, Kasdon D, Kwan ES, Laucella M, Anderson ML (1987) Intracarotid chemotherapy with 1,3-Bis-(2-chloroethyl)-1-nitrosourea (BCNU) in 5% dextrose in water in the treatment of malignant glioma. Neurosurgery 20:577–583

Li V, Levin AB, Turski P (1988) Intra-arterial chemotherapy following blood-brain barrier disruption in patients with recurrent high grade astrocytomas (abstract). Proc Am Ass Neurol Surg 274:404

Markowsky SJ, Zimmerman CL, Tholl D, Soria I, Castillo R (1991) Methotrexate disposition following disruption of the blood-brain barrier. Ther Drug Monit 13:24–31

Millay RH, Klein ML, Shults WT, Dahlborg SA, Neuwelt EA (1986) Maculopathy associated with combination chemotherapy and osmotic opening of the blood-brain barrier. Am J Ophthmol 102:626–632

Miller DF, Bay JW, Lederman RJ, Purvis JD, Rogers LR, Tomsak RL (1985) Ocular and orbital toxicity following intracarotid injection of BCNU (carmustine) and cisplatinum for malignant gliomas. Ophthalmology 92:402–406

Miyagami M, Kagawa Y, Tusbokawa T (1985) ACNU delivery to malignant glioma tissue by osmotic blood brain barrier modification with intracarotid infusion of hyperosmolar mannitol. No Shinkei Geka 13:955–963

Miyagami M, Tsubokawa T, Tazoe M, Kagawa Y (1990) Intra-arterial ACNU chemotherapy employing 20% mannitol osmotic blood-brain barrier disruption for malignant brain tumors. Neurol Med Chir (Tokyo) 30:582–590

Neuwelt EA, Dahlborg SA (1987) Chemotherapy administered in conjunction with osmotic blood-brain barrier modification in patients with brain metastases. J Neurooncol 4:195–207

Neuwelt EA, Dahlborg SA (1989) Blood-brain barrier disruption in the treatment of brain tumors: Clinical implications. In: Neuwelt EA (ed) Implications of the blood-brain barrier and its manipulation, vol 2. Plenum, New York, pp 195–261

Neuwelt EA, Frenkel EP, Diehl JT et al (1980) Reversible osmotic blood-brain barrier disruption in humans: implications for chemotherapy of malignant brain tumors. Neurosurgery 7:44–52

Neuwelt EA, Glasberg M, Diehl J, Frenkel EP, Barnett P (1981) Osmotic blood-brain barrier disruption the posterior fossa of the dog. J Neurosurg 55:742–748

Neuwelt EA, Specht HD, Howieson J (1983) Osmotic blood-brain barrier modification: clinical documentation by enhanced CT scanning and/or radionuclide brain scanning. AJNR 4:907–913

Neuwelt EA, Barnett PA, Frenkel EP (1984) Chemotherapeutic agent permeability to normal brain and delivery to avian sarcoma virus-induced brain tumors in the rodent: observations on problems of drug delivery. Neurosurgery 14:154–159

Neuwelt EA, Frenkel E, d'Agostino AN (1985) Growth of human lung tumor in the brain of the nude rat as a model to evaluate antitumor agent delivery across the blood-brain barrier. Cancer Res 45:2827–2833

Neuwelt EA, Frenkel EP, Gumerlock MK, Braziel R, Dana B, Hill SA (1986a) Developments in the diagnosis and treatment of primary CNS lymphoma: a prospective series. Cancer 58:1609–1620

Neuwelt EA, Howieson J, Frenkel EP, Specht HD, Weigel R, Buchan CG, Hill SA (1986b) Therapeutic efficacy of multiagent chemotherapy with drug delivery enhancement by blood-brain barrier modification in glioblastoma. Neurosurgery 19:573–582

Neuwelt EA, Specht HD, Larson S et al. (1987) Increased delivery of tumor-specific monoclonal antibodies to brain after osmotic blood-brain barrier modification in patients with melanomaa metastatic to the central nervous system. Neurosurgery 20:885–895

Neuwelt EA, Enzmann DR, Pagel MA, Miller G (1989) Bacterial and fungal brain abscess and the blood-brain barrier. In: Neuwelt EA (ed) Implications of the blood-brain barrier and its manipulation, vol 2. Plenum, New York, pp 263–305

Neuwelt EA, Goldman DL, Dahlborg SA, Crossen J, Ramsey F, Roman-Goldstein S, Braziel R, Dana B (1991) Primary central nervous system lymphoma treated with osmotic blood-brain barrier disruption: prolonged survival and preservation of cognitive function. J Clin Oncol 9:1580–1590

Phillips JP, Alderson T, Sikora K et al. (1983) Localization of malignant glioma by a radiolabeled human monoclonal antibody. J Neurosurg 46:388–392

Roman-Goldstein SM, Barnett PA, McCormick CI, Ball MJ, Ramsey F, Neuwelt EA (1991) Imaging and toxicity of gadopentate dimeglumine administered after osmotic blood-brain barrier disruption. AJNR 12:885–890

Runge VM, Price AC, Wehr CJ, Atkinson JB, Tweedle MF (1985) Contrast enhanced MRI evaluation of a canine model of osmotic blood-brain barrier disruption. Invest Radiol 20:830–844

Sato S, Toya S, Otani M (1985a) Barrier opening microcirculation in human brain tumor. No To Shinkei 37:109–113

Sato S, Yoshinori A, Kodama R, Fujioka M, Otani M, Inoue H, Toya S (1985b) Blood-brain barrier opening CT. Prog Comput Tomogr 7:43–48

Schag CC, Heinrich RL, Ganz PA (1984) Karnofsky performance status revisited: reliability, validity, and guidelines. J Clin Oncol 2:187–193

Stewart DJ, Grahovak Z, Maroun J et al. (1985) Intraarterial chemotherapy for brain tumors. Proc Am Soc Clin Oncol 4:C513

Suda K, Nakasu S, Saito A (1985) Ocular complication of combined intracarotid chemotherapy and osmotic blood-brain barrier disruption. Nippon Geka Hokan 54:359–363

Williams WT, Lowry RL, Eggers GWN (1986) Anesthetic management during therapeutic disruption of the blood-brain barrier. Anesth Analg 65:188–190

Yamada K, Takahama H, Nakai O, Takanashi T, Hosoya T (1989) Intra-arterial chemotherapy for malignant glioma after osmotic blood-brain barrier disruption. Gan To Kagaku Ryoho 16:2692–2696

York D, Legan M, Benner S, Watts C (1984) Further studies with a noninvasive method of intracranial pressure estimation. Neurosurgery 14:456–461

Subject Index

Handbook of Experimental Pharmacology

Volume 57
Tissue Growth Factors

Volume 58
Cyclic Nucleotides
Part 1: **Biochemistry**
Part 2: **Physiology and Pharmacology**

Volume 59
Mediators and Drugs in Gastrointestinal Motility
Part 1: **Morphological Basis and Neurophysiological Control**
Part 2: **Endogenous and Exogenous Agents**

Volume 60
Pyretics and Antipyretics

Volume 61
Chemotherapy of Viral Infections

Volume 62
Aminoglycoside Antibiotics

Volume 63
Allergic Reactions to Drugs

Volume 64
Inhibition of Folate Metabolism in Chemotherapy

Volume 65
Teratogenesis and Reproductive Toxicology

Volume 66
Part 1: **Glucagon I**
Part 2: **Glucagon II**

Volume 67
Part 1: **Antibiotics Containing the Beta-Lactam Structure I**
Part 2: **Antibiotics Containing the Beta-Lactam Structure II**

Volume 68
Antimalarial Drugs
Part 1: **Biological Background, Experimental Methods and Drug Resistance**
Part 2: **Current Antimalarias and New Drug Developments**

Volume 69
Pharmacology of the Eye

Volume 70
Part 1: **Pharmacology of Intestinal Permeation I**
Part 2: **Pharmacology of Intestinal Permeation II**

Volume 71
Interferons and Their Applications

Volume 72
Antitumor Drug Resistance

Volume 73
Radiocontrast Agents

Volume 74
Antieliptic Drugs

Volume 75
Toxicology of Inhaled Materials

Volume 76
Clinical Pharmacology of Antiangial Drugs

Volume 77
Chemotherapy of Gastrointestinal Helminths

Volume 78
The Tetracyclines

Springer-Verlag
Berlin
Heidelberg
New York
London
Paris
Tokyo
Hong Kong
Barcelona
Budapest

Handbook of Experimental Pharmacology

Volume 79
New Neuromuscular Blocking Agents

Volume 80
Cadmium

Volume 81
Local Anesthetics

Volume 82
Radioimmunoassay in Basic and Clinical Pharmacology

Volume 83
Calcium in Drug Actions

Volume 84
Antituberculosis Drugs

Volume 85
The Pharmacology of Lymphocytes

Volume 86
The Cholinergic Synapse

Volume 87
Part 1: **Pharmacology of the Skin I**
Part 2: **Pharmacology of the Skin II**

Volume 88
Drugs for the Treatment of Parkinson's Disease

Volume 89
Antiarrhythmic Drugs

Volume 90
Part 1: **Catecholamines I**
Part 2: **Catecholamines II**

Volume 91
Microbial Resistance to Drugs

Volume 92
Insulin

Volume 93
Pharmacology of Antihypertensive Therapeutics

Volume 94
Part 1: **Chemical Carcinogenesis and Mutagenesis I**
Part 2: **Chemical Carcinogenesis and Mutagenesis II**

Volume 95
Part 1: **Peptide Growth Factors and Their Receptors I**
Part 2: **Peptide Growth Factors and Their Receptors II**

Volume 96
Chemotherapy of Fungal Diseases

Volume 97
Histamine and Histamine Antagonists

Volume 98
Pharmacology of Asthma

Volume 99
Pharmacology of Peptic Ulcer Disease

Volume 100
Targeted Drug Delivery

Volume 101
Biochemical Pharmacology of Blood and Bloodforming Organs

Volume 102
Selective Neurotoxicity

Springer-Verlag
Berlin
Heidelberg
New York
London
Paris
Tokyo
Hong Kong
Barcelona
Budapest